SMALL AREA ESTIMATION

WILEY SERIES IN SURVEY METHODOLOGY

Established in Part by WALTER A. SHEWHART AND SAMUEL S. WILKS

A complete list of the titles in this series appears at the end of this volume.

SMALL AREA ESTIMATION

Second Edition

J.N.K. RAO AND ISABEL MOLINA

Wiley Series in Survey Methodology

WILEY

Published by John Wiley & Sons, Inc., Hoboken, New Jersey
Published simultaneously in Canada

Library of Congress Cataloging-in-Publication Data:

Rao, J. N. K., 1937- author.
Small area estimation / J.N.K. Rao and Isabel Molina. – Second edition.
pages cm – (Wiley series in survey methodology)
Includes bibliographical references and index.
ISBN 978-1-118-73578-7 (cloth)
1. Small area statistics. 2. Sampling (Statistics) 3. Estimation theory. I. Molina, Isabel, 1975- author. II. Title. III. Series: Wiley series in survey methodology.
QA276.6.R344 2015
519.5′2–dc23

2015012610

10 9 8 7 6 5 4 3 2 1

2 2015

To Neela and Ángeles

CONTENTS

LIST OF FIGURES

LIST OF TABLES

FOREWORD TO THE FIRST EDITION

The history of modern sample surveys dates back to the nineteenth century, but the field did not fully emerge until the 1930s. It grew considerably during the World War II, and has been expanding at a tremendous rate ever since. Over time, the range of topics investigated using survey methods has broadened enormously as policy makers and researchers have learned to appreciate the value of quantitative data and as survey researchers—in response to policy makers' demands—have tackled topics previously considered unsuitable for study using survey methods. The range of analyses of survey data has also expanded, as users of survey data have become more sophisticated and as major developments in computing power and software have simplified the computations involved. In the early days, users were mostly satisfied with national estimates and estimates for major geographic regions and other large domains. The situation is very different today: more and more policy makers are demanding estimates for small domains for use in making policy decisions. For example, population surveys are often required to provide estimates of adequate precision for domains defined in terms of some combination of factors such as age, sex, race/ethnicity, and poverty status. A particularly widespread demand from policy makers is for estimates at a finer level of geographic detail than the broad regions that were commonly used in the past. Thus, estimates are frequently needed for such entities as states, provinces, counties, school districts, and health service areas.

The need to provide estimates for small domains has led to developments in two directions. One direction is toward the use of sample designs that can produce domain estimates of adequate precision within the standard design-based mode of inference used in survey analysis (i.e., "direct estimates"). Many sample surveys are now designed to yield sufficient sample sizes for key domains to satisfy the precision requirements for those domains. This approach is generally used for socio-economic domains and for some larger geographic domains. However, the increase in overall sample size that this approach entails may well exceed the survey's funding resources and capabilities, particularly so when estimates are required for many geographic areas. In the United States, for example, few surveys are large enough to be capable of providing reliable subpopulation estimates for all 50 states, even if the sample is optimally allocated across states for this purpose. For very small geographic areas such as school districts, either a complete census or a sample of at least the size of the census of long-form sample (on average about 1 in 6 households nationwide) is required. Even censuses, however, although valuable, cannot be the complete solution for the production of small area estimates. In most countries, censuses are conducted only once a decade. They cannot, therefore, provide satisfactory small area estimates for intermediate time points during a decade for population characteristics that change markedly over time. Furthermore, census content is inherently severely restricted, so a census cannot provide small area estimates for all the characteristics that are of interest. Hence, another approach is needed.

The other direction for producing small area estimates is to turn away from conventional direct estimates toward the use of indirect model-dependent estimates. The model-dependent approach employs a statistical model that "borrows strength" in making an estimate for one small area from sample survey data collected in other small areas or at other time periods. This approach moves away from the design-based estimation of conventional direct estimates to indirect model-dependent estimates. Naturally, concerns are raised about the reliance on models for the production of such small area estimates. However, the demand for small area estimates is strong and increasing, and models are needed to satisfy that demand in many cases. As a result, many survey statisticians have come to accept the model-dependent approach in the right circumstances, and the approach is being used in a number of important cases. Examples of major small area estimation programs in the United States include the following: the Census Bureau's Small Area Income and Poverty Estimates program, which regularly produces estimates of income and poverty measures for various population subgroups for states, counties, and school districts; the Bureau of Labor Statistics' Local Area Unemployment Statistics program, which produces monthly estimates of employment and unemployment for states, metropolitan areas, counties, and certain subcounty areas; the National Agricultural Statistics Service's County Estimates Program, which produces county estimates of crop yield; and the estimates of substance abuse in states and metropolitan areas, which are produced by the Substance Abuse and Mental Health Services Administration (see Chapter 1).

The essence of all small area methods is the use of auxiliary data available at the small area level, such as administrative data or data from the last census. These data are used to construct predictor variables for use in a statistical model that can be used to predict the estimate of interest for all small areas. The effectiveness of small area estimation depends initially on the availability of good predictor variables that are uniformly measured over the total area. It next depends on the choice of a good prediction model. Effective use of small area estimation methods further depends on a careful, thorough evaluation of the quality of the model. Finally, when small area estimates are produced, they should be accompanied by valid measures of their precision.

Early applications of small area estimation methods employed only simple methods. At that time, the choice of the method for use in particular case was relatively simple, being limited by the computable methods then in existence. However, the situation has changed enormously in recent years, and particularly in the last decade. There now exist a wide range of different, often complex, models that can be used, depending on the nature of the measurement of the small area estimate (e.g., a binary or continuous variable) and on the auxiliary data available. One key distinction in model construction is between situations where the auxiliary data are available for the individual units in the population and those where they are available only at the aggregate level for each small area. In the former case, the data can be used in unit level models, whereas in the latter they can be used only in area level models. Another feature involved in the choice of model is whether the model borrows strength cross-sectionally, over time, or both. There are also now a number of different approaches, such as empirical best linear prediction (EBLUP), empirical Bayes (EB), and hierarchical Bayes (HB), which can be used to estimate the models and the variability of the model-dependent small area estimates. Moreover, complex procedures that would have been extremely difficult to apply a few years ago can now be implemented fairly straightforwardly, taking advantage of the continuing increases in computing power and the latest developments in software.

The wide range of possible models and approaches now available for use can be confusing to those working in this area. J.N.K. Rao's book is therefore a timely contribution, coming at a point in the subject's development when an integrated, systematic treatment is needed. Rao has done a great service in producing this authoritative and comprehensive account of the subject. This book will help to advance the subject and be a valuable resource for practitioners and theorists alike.

GRAHAM KALTON

PREFACE TO THE SECOND EDITION

Small area estimation (SAE) deals with the problem of producing reliable estimates of parameters of interest and the associated measures of uncertainty for subpopulations (areas or domains) of a finite population for which samples of inadequate sizes or no samples are available. Traditional "direct estimates," based only on the area-specific sample data, are not suitable for SAE, and it is necessary to "borrow strength" across related small areas through supplementary information to produce reliable "indirect" estimates for small areas. Indirect model-based estimation methods, based on explicit linking models, are now widely used.

The first edition of Small Area Estimation (Rao 2003a) provided a comprehensive account of model-based methods for SAE up to the end of 2002. It is gratifying to see the enthusiastic reception it has received, as judged by the significant number of citations and the rapid growth in SAE literature over the past 12 years. Demand for reliable small area estimates has also greatly increased worldwide. As an example, the estimation of complex poverty measures at the municipality level is of current interest, and World Bank uses a model-based method, based on simulating multiple censuses, in more than 50 countries worldwide to produce poverty statistics for small areas.

The main aim of the present second edition is to update the first edition by providing a comprehensive account of important theoretical developments from 2003 to 2014. New SAE literature is quite extensive and often involves complex theory to handle model misspecifications and other complexities. We have retained a large portion of the material from the first edition to make the book self-contained, and supplemented it with selected new developments in theory and methods of SAE. Notations and terminology used in the first edition are largely retained. As in the first edition, applications are included throughout the chapters. An added feature of the second edition is the inclusion of sections (Sections 5.5, 6.5, 7.7, 8.11, and 9.11) describing specific R software for SAE, concretely the R package `sae` (Molina and Marhuenda 2013; Molina and Marhuenda 2015). These sections include examples of SAE applications using data sets included in the package and provide all the necessary R codes, so that the user can exactly replicate the applications. New sections and old sections with significant changes are indicated by an asterisk in the book. Chapter 3 on "Traditional Demographic Methods" from first edition is deleted partly due to page constraints and the fact that the material is somewhat unrelated to mainstream model-based methods. Also, we have not been able to keep up to date with the new developments in demographic methods.

Chapter 1 introduces basic terminology related to SAE and presents selected important applications as motivating examples. Chapter 2, as in the first edition, presents a concise account of direct estimation of totals or means for small areas and addresses survey design issues that have a bearing on SAE. New Section 2.7 deals with optimal sample allocation for planned domains and the estimation of marginal row and column strata means in the presence of two-way stratification. Chapter 3 gives a fairly detailed account of traditional indirect estimation based on implicit linking models. The well-known James–Stein method of composite estimation is also studied in the context of sample survey data. New Section 3.2.7 studies generalized structure preserving estimation (GSPREE) based on relaxing some interaction assumptions made in the traditional SPREE, which is often used in practice because it makes fuller use of reliable direct estimates at a higher level to produce synthetic estimates. Another important addition is weight sharing (or splitting) methods studied in Section 3.2.8. The weight-sharing methods produce a two-way table of weights with rows as the units in the full sample and columns as the areas such that the cell weights in each row add up to the original sample weight. Such methods are especially useful in micro-simulation modeling that can involve a large number of variables of interest.

Explicit small area models that account for between-area variability are introduced in Chapter 4 (previous Chapter 5), including linear mixed models and generalized linear mixed models such as logistic linear mixed models with random area effects. The models are classified into two broad categories: (i) area level models that relate the small area means or totals to area level covariates; and (ii) unit level models that relate the unit values of a study variable to unit-specific auxiliary variables. Extensions of the models to handle complex data structures, such as spatial dependence and time series structures, are also considered. New Section 4.6.5 introduces semi-parametric mixed models, which are studied later. Chapter 5 (previous Chapter 6) studies linear mixed models involving fixed and random effects. It gives general results on empirical best linear-unbiased prediction (EBLUP) and the estimation of mean squared error (MSE) of the EBLUP. A detailed account of model identification and checking for linear mixed models is presented in the new Section 5.4. Available SAS software and R statistical software for linear mixed models are summarized in the new Section 5.5. The R package `sae` specifically designed for SAE is also described.

Chapter 6 of the First Edition provided a detailed account of EBLUP estimation of small area means or totals for the basic area level and unit level models, using the general theory given in Chapter 5. In the past 10 years or so, researchers have done extensive work on those two models, especially addressing problems related to model misspecification and other practical issues. As a result, we decided to split the old Chapter 6 into two new chapters, with Chapter 6 focusing on area level models and Chapter 7 addressing unit level models. New topics covered in Chapter 6 include bootstrap MSE estimation (Section 6.2.4) and robust estimation in the presence of outliers (Section 6.3). Section 6.4 deals with practical issues related to the basic area level model. It includes important topics such as covariates subject to sampling errors (Section 6.4.4), misspecification of linking models (Section 6.4.7), benchmarking of model-based area estimators to ensure agreement with a reliable direct estimate when aggregated (Section 6.4.6), and the use of "big data" as possible covariates in area level models (Section 6.4.5). Functions of the R package `sae` designed for estimation under the area level model are described in Section 6.5. An example illustrating the use of these functions is provided. New topics introduced in Chapter 7 include bootstrap MSE estimation (Section 7.2.4), outlier robust EBLUP estimation (Section 7.4), and M-quantile regression (Section 7.5). Section 7.6 deals with practical issues related to the basic unit level model. It presents methods to deal with important topics, including measurement errors in covariates (Section 7.6.4), model misspecification (Section 7.6.5), and semi-parametric nested error models (Sections 7.6.6 and 7.6.7). Most of the published literature assumes that the assumed model for the population values also holds for the sample. However, in many applications, this assumption may not be true due to informative sampling leading to sample selection bias. Section 7.6.3 gives a detailed treatment of methods to make valid inferences under informative sampling. Functions of R package `sae` dealing with the basic unit level model are described in Section 7.7. The use of these functions is illustrated through an application to the County Crop Areas data of Battese, Harter, and Fuller (1988). This application includes calculation of model diagnostics and drawing residual plots. Several important applications are also presented in Chapters 6 and 7.

New chapters 8, 9, and 10 cover the same material as the corresponding chapters in the first edition. Chapter 8 contains EBLUP theory for various extensions of the basic area level and unit level models, providing updates to the sections in the first edition, in particular a more detailed account of spatial and two-level models. Section 8.4 on spatial models is updated, and functions of the R package `sae` dealing with spatial area level models are described in Section 8.11. An example illustrating the use of these functions is provided. Section 8.5 presents theory for two-fold subarea level models, which are natural extensions of the basic area level models. Chapter 9 presents empirical Bayes (EB) estimation. The EB method (also called empirical best) is more generally applicable than the EBLUP method. New Section 9.2.4 gives an account of methods for constructing confidence intervals in the case of basic area level model. EB estimation of general area parameters is the theme of Section 9.4, in particular complex poverty indicators studied by the World Bank. EB method is compared to the World Bank method in simulation experiments (Section 9.4.6). R software for EB estimation of general area parameters is described in Section 9.11, which includes an example on estimation of poverty indicators. Binary data and disease mapping from count data are studied in Sections 9.5 and 9.6, respectively. An important addition is Section 9.7 dealing with design-weighted EB estimation under exponential family models. Previous sections on constrained EB estimation and empirical linear Bayes estimation are retained.

Finally, Chapter 10 presents a self-contained account of the Hierarchical Bayes (HB) approach based on specifying prior distributions on the model parameters. Basic Markov chain Monte Carlo (MCMC) methods for HB inference, including model determination, are presented in Section 10.2. Several new developments are presented, including HB estimation of complex general area parameters, in particular poverty indicators (Section 10.7), two-part nested error models (Section 10.12), missing binary data (Section 10.14), and approximate HB inference (Section 10.17). Other sections in Chapter 10 more or less cover the material in the previous edition with some updates. Chapters 8–10 include brief descriptions of applications with real data sets.

As in the first edition, we discuss the advantages and limitations of different SAE methods throughout the book. We also emphasize the need for both internal and external evaluations. To this end, we have provided various methods for model selection from the data, and comparison of estimates derived from models to reliable values obtained from external sources, such as previous census or administrative data.

Proofs of some basic results are provided, but proofs of results that are technically involved or lengthy are omitted, as in the first edition. We have provided fairly self-contained accounts of direct estimation (Chapter 2), EBLUP and EB estimation (Chapters 5 and 9), and HB estimation (Chapter 10). However, prior exposure to a standard text in mathematical statistics, such as the 2001 Brooks/Cole book Statistical Inference (second edition) by G. Casella and R. L. Berger, is essential. Also, a basic course in regression and mixed models, such as the 2001 Wiley book Generalized, Linear and Mixed Models by C. E. McCulloch and S. E. Searle, would be helpful in understanding model-based SAE. A basic course in survey sampling techniques, such as the 1977 Wiley book Sampling Techniques (third edition) by W.G. Cochran is also useful but not essential.

This book is intended primarily as a research monograph, but it is also suitable for a graduate level course on SAE, as in the case of the first edition. Practitioners interested in learning SAE methods may also find portions of this text useful, in particular Chapters 3, 6, 7 and Sections 10.1–10.3 and 10.5 as well as the examples and applications presented throughout the book.

We are thankful to Emily Berg, Yves Berger, Ansu Chatterjee, Gauri Datta, Laura Dumitrescu, Wayne Fuller, Malay Ghosh, David Haziza, Jiming Jiang, Partha Lahiri, Bal Nandram, Jean Opsomer, Cynthia Bocci and Mikhail Sverchkov for reading portions of the book and providing helpful comments and suggestions, to Domingo Morales for providing a very helpful list of publications in SAE and to Pedro Dulce for providing us with tailor made software for making author and subject indices.

J. N. K. Rao and Isabel Molina
January, 2015

PREFACE TO THE FIRST EDITION

Sample surveys are widely used to provide estimates of totals, means, and other parameters not only for the total population of interest but also for subpopulations (or domains) such as geographic areas and socio-demographic groups. Direct estimates of a domain parameter are based only on the domain-specific sample data. In particular, direct estimates are generally "design-based" in the sense that they make use of "survey weights," and the associated inferences (standard errors, confidence intervals, etc.) are based on the probability distribution induced by the sample design, with the population values held fixed. Standard sampling texts (e.g., the 1977 Wiley book *Sampling Techniques* by W.G. Cochran) provide extensive accounts of design-based direct estimation. Models that treat the population values as random may also be used to obtain model-dependent direct estimates. Such estimates in general do not depend on survey weights, and the associated inferences are based on the probability distribution induced by the assumed model (e.g., the 2001 Wiley book *Finite Population Sampling and Inference: A Prediction Approach* by R. Valliant, A.H. Dorfman, and R.M. Royall).

We regard a domain as large as if the domain sample size is large enough to yield direct estimates of adequate precision; otherwise, the domain is regarded as small. In this text, we generally use the term "small area" to denote any subpopulation for which direct estimates of adequate precision cannot be produced. Typically, domain sample sizes tend to increase with the population size of the domains, but this is not always the case. For example, due to oversampling of certain domains in the US Third Health and Nutrition Examination Survey, sample sizes in many states were small (or even zero).

It is seldom possible to have a large enough overall sample size to support reliable direct estimates for all the domains of interest. Therefore, it is often necessary to use indirect estimates that "borrow strength" by using values of the variables of interest from related areas, thus increasing the "effective" sample size. These values are brought into the estimation process through a model (either implicit or explicit) that provides a link to related areas (domains) through the use of supplementary information related to the variables of interest, such as recent census counts and current administrative records. Availability of good auxiliary data and determination of suitable linking models are crucial to the formation of indirect estimates.

In recent years, the demand for reliable small area estimates has greatly increased worldwide. This is due, among other things, to their growing use in formulating policies and programs, the allocation of government funds, and in regional planning. Demand from the private sector has also increased because business decisions, particularly those related to small businesses, rely heavily on the local conditions. Small area estimation (SAE) is particularly important for studying the economies in transition in central and eastern European countries and the former Soviet Union countries because these countries are moving away from centralized decision making.

The main aim of this text is to provide a comprehensive account of the methods and theory of SAE, particularly indirect estimation based on explicit small area linking models. The model-based approach to SAE offers several advantages, most importantly increased precision. Other advantages include the derivation of "optimal" estimates and associated measures of variability under an assumed model, and the validation of models from the sample data.

Chapter 1 introduces some basic terminology related to SAE and presents some important applications as motivating examples. Chapter 2 contains a brief account of direct estimation, which provides a background for later chapters. It also addresses survey design issues that have a bearing on SAE. Traditional demographic methods that employ indirect estimates based on implicit linking models are studied in Chapter 3. Typically, demographic methods only use administrative and census data and sampling is not involved, whereas indirect estimation methods studied in later chapters are largely based on sample survey data in conjunction with auxiliary population information. Chapter 4 gives a detailed account of traditional indirect estimation based on implicit linking models. The well-known James–Stein method of composite estimation is also studied in the context of sample surveys.

Explicit small area models that account for between-area variation are presented in Chapter 5, including linear mixed models and generalized linear mixed models, such as logistic models with random area effects. The models are classified into two broad groups: (i) area level models that relate the small area means to area-specific auxiliary variables; (ii) unit level models that relate the unit values of study variables to unit-specific auxiliary variables. Several extensions to handle complex data structures, such as spatial dependence and time series structures, are also presented. Chapters 6–8 study in more detail linear mixed models involving fixed and random effects. General results on empirical best linear unbiased prediction (EBLUP) under the frequentist approach are presented in Chapter 6. The more difficult problem of estimating the mean squared error (MSE) of EBLUP estimators is also considered. A basic area level model and a basic unit level model are studied thoroughly in

Chapter 7 by applying the EBLUP results developed in Chapter 6. Several important applications are also presented in this chapter. Various extensions of the basic models are considered in Chapter 8.

Chapter 9 presents empirical Bayes (EB) estimation. This method is more generally applicable than the EBLUP method. Various approaches to measuring the variability of EB estimators are presented. Finally, Chapter 10 presents a self-contained account of hierarchical Bayes (HB) estimation, by assuming prior distributions on model parameters. Both chapters include actual applications with real data sets.

Throughout the text, we discuss the advantages and limitations of the different methods for SAE. We also emphasize the need for both internal and external evaluations for model selection. To this end, we provide various methods of model validation, including comparisons of estimates derived from a model with reliable values obtained from external sources, such as previous census values.

Proofs of basic results are given in Sections 2.7, 3.5, 4.4, 6.4, 9.9, and 10.14, but proofs of results that are technically involved or lengthy are omitted. The reader is referred to relevant papers for details of omitted proofs. We provide self-contained accounts of direct estimation (Chapter 2), linear mixed models (Chapter 6), EB estimation (Chapter 9), and HB estimation (Chapter 10). But prior exposure to a standard course in mathematical statistics, such as the 1990 Wadsworth & Brooks/Cole book *Statistical Inference* by G. Casella and R.L. Berger, is essential. Also, a course in linear mixed models, such as the 1992 Wiley book *Variance Components* by S.R. Searle, G. Casella, and C.E. McCulloch, would be helpful in understanding model-based SAE. A basic course in survey sampling methods, such as the 1977 Wiley book *Sampling Techniques* by W.G. Cochran, is also useful but not essential.

This book is intended primarily as a research monograph, but it is also suitable for a graduate level course on SAE. Practitioners interested in learning SAE methods may also find portions of this text useful; in particular, Chapters 4, 7, 9, and Sections 10.1–10.3 and 10.5 as well as the applications presented throughout the book.

Special thanks are due to Gauri Datta, Sharon Lohr, Danny Pfeffermann, Graham Kalton, M.P. Singh, Jack Gambino, and Fred Smith for providing many helpful comments and constructive suggestions. I am also thankful to Yong You, Ming Yu, and Wesley Yung for typing portions of this text; to Gill Murray for the final typesetting and preparation of the text; and to Roberto Guido of Statistics Canada for designing the logo on the cover page. Finally, I am grateful to my wife Neela for her long enduring patience and encouragement and to my son, Sunil, and daughter, Supriya, for their understanding and support.

J. N. K. Rao
Ottawa, Canada
January, 2003

1

*INTRODUCTION

1.1 WHAT IS A SMALL AREA?

Sample surveys have long been recognized as cost-effective means of obtaining information on wide-ranging topics of interest at frequent intervals over time. They are widely used in practice to provide estimates not only for the total population of interest but also for a variety of subpopulations (domains). Domains may be defined by geographic areas or socio-demographic groups or other subpopulations. Examples of a geographic domain (area) include a state/province, county, municipality, school district, unemployment insurance (UI) region, metropolitan area, and health service area. On the other hand, a socio-demographic domain may refer to a specific age-sex-race group within a large geographic area. An example of "other domains" is the set of business firms belonging to a census division by industry group.

In the context of sample surveys, we refer to a domain estimator as "direct" if it is based only on the domain-specific sample data. A direct estimator may also use the known auxiliary information, such as the total of an auxiliary variable, x, related to the variable of interest, y. A direct estimator is typically "design based," but it can also be motivated by and justified under models (see Section 2.1). Design-based estimators make use of survey weights, and the associated inferences are based on the probability distribution induced by the sampling design with the population values held fixed (see Chapter 2). "Model-assisted" direct estimators that make use of "working" models are also design based, aiming at making the inferences "robust" to possible model misspecification (see Chapter 2).

Small Area Estimation, Second Edition. J. N. K. Rao and Isabel Molina.

A domain (area) is regarded as large (or major) if the domain-specific sample is large enough to yield "direct estimates" of adequate precision. A domain is regarded as "small" if the domain-specific sample is not large enough to support direct estimates of adequate precision. Some other terms used to denote a domain with small sample size include "local area," "subdomain," "small subgroup," "subprovince," and "minor domain." In some applications, many domains of interest (such as counties) may have zero sample size.

In this text, we generally use the term "small area" to denote any domain for which direct estimates of adequate precision cannot be produced. Typically, domain sample size tends to increase with the population size of the domain, but this is not always the case. Sometimes, the sampling fraction is made larger than the average fraction in small domains in order to increase the domain sample sizes and thereby increase the precision of domain estimates. Such oversampling was, for example, used in the US Third Health and Nutrition Examination Survey (NHANES III) for certain domains in the cross-classification of sex, race/ethnicity, and age, in order that direct estimates of acceptable precision could be produced for those domains. This oversampling led to a greater concentration of the sample in certain states (e.g., California and Texas) than normal, and thereby exacerbated the common problem in national surveys that sample sizes in many states are small (or even zero). Thus, while direct estimates may be used to estimate characteristics of demographic domains with NHANES III, they cannot be used to estimate characteristics of many states. States may therefore be regarded as small areas in this survey. Even when a survey has large enough state sample sizes to support the production of direct estimates for the total state populations, these sample sizes may well not be large enough to support direct estimates for subgroups of the state populations, such as school-age children or persons in poverty. Due to cost considerations, it is often not possible to have a large enough overall sample size to support reliable direct estimates for all domains. Furthermore, in practice, it is not possible to anticipate all uses of the survey data, and "the client will always require more than is specified at the design stage" (Fuller 1999, p. 344).

In making estimates for small areas with adequate level of precision, it is often necessary to use "indirect" estimators that "borrow strength" by using values of the variable of interest, y, from related areas and/or time periods and thus increase the "effective" sample size. These values are brought into the estimation process through a model (either implicit or explicit) that provides a link to related areas and/or time periods through the use of supplementary information related to y, such as recent census counts and current administrative records. Three types of indirect estimators can be identified (Schaible 1996, Chapter 1): "domain indirect," "time indirect," and "domain and time indirect." A domain indirect estimator makes use of y-values from another domain but not from another time period. A time indirect estimator uses y-values from another time period for the domain of interest but not from another domain. On the other hand, a domain and time indirect estimator uses y-values from another domain as well as from another time period. Some other terms used to denote an indirect estimator include "non-traditional," "small area," "model dependent," and "synthetic."

Availability of good auxiliary data and determination of suitable linking models are crucial to the formation of indirect estimators. As noted by Schaible (1996, Chapter 10), expanded access to auxiliary information through coordination and cooperation among different agencies is needed.

1.2 DEMAND FOR SMALL AREA STATISTICS

Historically, small area statistics have long been used. For example, such statistics existed in eleventh-century England and seventeenth-century Canada based on either census or administrative records (Brackstone 1987). Demographers have long been using a variety of indirect methods for small area estimation (SAE) of population and other characteristics of interest in postcensal years. Typically, sampling is not involved in the traditional demographic methods (see Chapter 3 of Rao 2003a).

In recent years, the demand for small area statistics has greatly increased worldwide. This is due, among other things, to their growing use in formulating policies and programs, in the allocation of government funds and in regional planning. Legislative acts by national governments have increasingly created a need for small area statistics, and this trend has accelerated in recent years. Demand from the private sector has also increased significantly because business decisions, particularly those related to small businesses, rely heavily on the local socio-economic, environmental, and other conditions. Schaible (1996) provides an excellent account of the use of traditional and model-based indirect estimators in US Federal Statistical Programs.

SAE is of particular interest for the economies in transition in central and eastern European countries and the former Soviet Union countries. In the 1990s, these countries have moved away from centralized decision making. As a result, sample surveys are now used to produce estimates for large areas as well as small areas. Prompted by the demand for small area statistics, an International Scientific Conference on Small Area Statistics and Survey Designs was held in Warsaw, Poland, in 1992 and an International Satellite Conference on SAE was held in Riga, Latvia, in 1999 to disseminate knowledge on SAE (see Kalton, Kordos, and Platek (1993) and IASS Satellite Conference (1999) for the published conference proceedings).

Some other proceedings of conferences on SAE include National Institute on Drug Abuse (1979), Platek and Singh (1986), and Platek, Rao, Särndal, and Singh (1987). Rapid growth in SAE research in recent years, both theoretical and applied, led to a series of international conferences starting in 2005: Jyvaskyla (Finland, 2005), Pisa (Italy, 2007), Elche (Spain, 2009), Trier (Germany, 2011), Bangkok (Thailand, 2013), and Poznan (Poland, 2014). Three European projects dealing with SAE, namely EURAREA, SAMPLE and AMELI, have been funded by the European Commission. Many research institutions and National Statistical Offices spread across Europe have participated in these projects. Centers for SAE research have been established in the Statistical Office in Poznan (Poland) and in the Statistical Research Division of the US Census Bureau.

Review papers on SAE include Rao (1986, 1999, 2001b, 2003b, 2005, 2008), Chaudhuri (1994), Ghosh and Rao (1994), Marker (1999), Pfeffermann (2002,

2013), Jiang and Lahiri (2006), Datta (2009), and Lehtonen and Veijanen (2009). Text books on SAE have also appeared (Mukhopadhyay 1998, Rao 2003a, Longford 2005, Chaudhuri 2012). Good accounts of SAE theory are also given in the books by Fuller (2009) and Chambers and Clark (2012).

1.3 TRADITIONAL INDIRECT ESTIMATORS

Traditional indirect estimators, based on implicit linking models, include synthetic and composite estimators (Chapter 3). These estimators are generally design based, and their design variances (i.e., variances with respect to the probability distribution induced by the sampling design) are usually small relative to the design variances of direct estimators. However, the indirect estimators will be generally design biased, and the design bias will not decrease as the overall sample size increases. If the implicit linking model is approximately true, then the design bias is likely to be small, leading to significantly smaller design mean-squared error (MSE) compared to the MSE of a direct estimator. Reduction in MSE is the main reason for using indirect estimators.

1.4 SMALL AREA MODELS

Explicit linking models with random area-specific effects accounting for the between-area variation that is not explained by auxiliary variables will be called "small area models" (Chapter 4). Indirect estimators based on small area models will be called "model-based estimators." We classify small area models into two broad types. (i) Aggregate (or area) level models are the models that relate small area direct estimators to area-specific covariates. Such models are necessary if unit (or element) level data are not available. (ii) Unit level models are the models that relate the unit values of a study variable to unit-specific covariates. A basic area level model and a basic unit level model are introduced in Sections 4.2 and 4.3, respectively. Various extensions of the basic area level and unit level models are outlined in Sections 4.4 and 4.5, respectively. Sections 4.2–4.5 are relevant for continuous responses y and may be regarded as special cases of a general linear mixed model (Section 5.2). However, for binary or count variables y, generalized linear mixed models (GLMMs) are often used (Section 4.6): in particular, logistic linear mixed models for the binary case and loglinear mixed models for the count case.

A critical assumption for the unit level models is that the sample values within an area obey the assumed population model, that is, sample selection bias is absent (see Section 4.3). For area level models, we assume the absence of informative sampling of the areas in situations where only some of the areas are selected to the sample, that is, the sample area values (the direct estimates) obey the assumed population model.

Inferences from model-based estimators refer to the distribution implied by the assumed model. Model selection and validation, therefore, play a vital role in model-based estimation. If the assumed models do not provide a good fit to the data, the model-based estimators will be model biased which, in turn, can lead to erroneous inferences. Several methods of model selection and validation are presented throughout the book. It is also useful to conduct external evaluations by comparing indirect estimates (both traditional and model-based) to more reliable estimates or census values based on past data (see Examples 6.1.1 and 6.1.2 for both internal and external evaluations).

1.5 MODEL-BASED ESTIMATION

It is now generally accepted that, when indirect estimators are to be used, they should be based on explicit small area models. Such models define the way that the related data are incorporated in the estimation process. The model-based approach to SAE offers several advantages: (i) "Optimal" estimators can be derived under the assumed model. (ii) Area-specific measures of variability can be associated with each estimator unlike global measures (averaged over small areas) often used with traditional indirect estimators. (iii) Models can be validated from the sample data. (iv) A variety of models can be entertained depending on the nature of the response variables and the complexity of data structures (such as spatial dependence and time series structures).

In this text, we focus on empirical best linear unbiased prediction (EBLUP) (Chapters 5–8), parametric empirical Bayes (EB) (Chapter 9), and parametric hierarchical Bayes (HB) estimators (Chapter 10) derived from small area models. For the HB method, a further assumption on the prior distribution of model parameters is also needed. EBLUP is designed for estimating linear small area characteristics under linear mixed models, whereas EB and HB are more generally applicable.

The EBLUP method for general linear mixed models has been extensively used in animal breeding and other applications to estimate realized values of linear combinations of fixed and random effects. An EBLUP estimator is obtained in two steps: (i) The best linear unbiased predictor (BLUP), which minimizes the model MSE in the class of linear model unbiased estimators of the quantity of interest is first obtained. It depends on the variances (and covariances) of random effects in the model. (ii) An EBLUP estimator is obtained from the BLUP by substituting suitable estimators of the variance and covariance parameters. Chapter 5 presents some unified theory of the EBLUP method for the general linear mixed model, which covers many specific small area models considered in the literature (Chapters 6 and 8). Estimation of model MSE of EBLUP estimators is studied in detail in Chapters 6–8. Illustration of methods using specific R software for SAE is also provided.

Under squared error loss, the best predictor (BP) of a (random) small area quantity of interest such as mean, proportion, or more complex parameter is the conditional expectation of the quantity given the data and the model parameters. Distributional assumptions are needed for calculating the BP. The empirical BP (or EB) estimator is obtained from BP by substituting suitable estimators of model parameters (Chapter 9). On the other hand, the HB estimator under squared error loss is obtained by integrating the BP with respect to the (Bayes) posterior distribution derived from an assumed prior distribution of model parameters. The HB estimator is equal to the posterior mean of the estimand, where the expectation is with respect to the posterior distribution of the quantity of interest given the data. The HB method uses the posterior variance as a measure of uncertainty associated with the HB estimator. Posterior (or credible) intervals for the quantity of interest can also be constructed from the posterior distribution of the quantity of interest. The HB method is being extensively used for SAE because it is straightforward, inferences are "exact," and complex problems can be handled using Markov chain Monte Carlo (MCMC) methods. Software for implementing the HB method is also available (Section 10.2.4). Chapter 10 gives a self-contained account of the HB method and its applications to SAE.

"Optimal" model-based estimates of small area totals or means may not be suitable if the objective is to produce an ensemble of estimates whose distribution is in some sense close enough to the distribution of the corresponding estimands. We are also often interested in the ranks (e.g., ranks of schools, hospitals, or geographical areas) or in identifying domains (areas) with extreme values. Ideally, it is desirable to construct a set of "triple-goal" estimates that can produce good ranks, a good histogram, and good area-specific estimates. However, simultaneous optimization is not feasible, and it is necessary to seek a compromise set that can strike an effective balance between the three goals. Triple-goal EB estimation and constrained EB estimation that preserves the ensemble variance are studied in Section 9.8.

1.6 SOME EXAMPLES

We conclude the introduction by presenting some important applications of SAE as motivating examples. Details of some of these applications, including auxiliary information used, are given in Chapters 6–10.

1.6.1 Health

SAE of health-related characteristics has attracted a lot of attention in the United States because of a continuing need to assess health status, health practices, and health resources at both the national and subnational levels. Reliable estimates of health-related characteristics help in evaluating the demand for health care and the access that individuals have to it. Healthcare planning often takes place at the state and substate levels because health characteristics are known to vary geographically. Health System Agencies in the United States, mandated by the National Health Planning Resource Development Act of 1994, are required to collect and analyze data related to the health status of the residents and to the health delivery systems in their health service areas (Nandram, Sedransk, and Pickle 1999).

(i) The US National Center for Health Statistics (NCHS) pioneered the use of synthetic estimation based on implicit linking models. NCHS produced state synthetic estimates of disability and other health characteristics for different groups from the National Health Interview Survey (NHIS). Examples 3.2.2 and 10.13.3 give health applications from national surveys. Malec, Davis and Cao (1999) studied HB estimation of overweight prevalence for adults by states, using data from NHANES III. Folsom, Shah, and Vaish (1999) produced survey-weighted HB estimates of small area prevalence rates for states and age groups, for up to 20 binary variables related to drug use, using data from pooled National Household Surveys on Drug Abuse. Chattopadhyay et al. (1999) studied EB estimates of state-wide prevalences of the use of alcohol and drugs (e.g., marijuana) among civilian non-institutionalized adults and adolescents in the United States. These estimates are used for planning and resource allocation and to project the treatment needs of dependent users.

(ii) Mapping of small area mortality (or incidence) rates of diseases, such as cancer, is a widely used tool in public health research. Such maps permit the analysis of geographical variation that may be useful for formulating and assessing etiological hypotheses, resource allocation, and the identification of areas of unusually high-risk warranting intervention (see Section 9.6). Direct (or crude) estimates of rates called standardized mortality ratios (SMRs) can be very unreliable, and a map of crude rates can badly distort the geographical distribution of disease incidence or mortality because the map tends to be dominated by areas of low population. Disease mapping, using model-based estimators, has received considerable attention. We give several examples of disease mapping in this text (see Examples 9.6.1, 9.9.1, 10.11.1, and 10.11.3). Typically, sampling is not involved in disease mapping applications.

1.6.2 Agriculture

The US National Agricultural Statistics Service (NASS) publishes model-based county estimates of crop acreage using remote sensing satellite data as auxiliary information (see Example 7.3.1 for an application). County estimates assist the agricultural authorities in local agricultural decision making. Also, county crop yield estimates are used to administer federal programs involving payments to farmers if crop yields fall below certain levels. Another application, similar to Example 7.3.1, to estimate crop acreage in small areas using ground survey and remote sensing data, is reported in Ambrosio Flores and Iglesias Martínez (2000). Remote sensing satellite data and crop surveys are used in India to produce direct estimates of crop yield at the district level (Singh and Goel 2000). SAE methods are also used in India to obtain estimates of crop production at lower administrative units such as "tehsil" or block, using remote sensing satellite data as auxiliary information (Singh et al. 2002).

An application of synthetic estimation to produce county estimates of wheat production in the state of Kansas based on a non-probability sample of farms is presented in Example 3.2.4. Chapters 6 and 7 of Schaible (1996) provide details of traditional and model-based indirect estimation methods used by NASS for county crop acreage and production.

1.6.3 Income for Small Places

Example 6.1.1 gives details of an application of the EB (EBLUP) method of estimation of small area incomes, based on a basic area level linking model (see Section 6.1). The US Census Bureau adopted this method, proposed originally by Fay and Herriot (1979), to form updated per capita income (PCI) for small places. This was the largest application (prior to 1990) of model-based estimators in a US Federal Statistical Program. The PCI estimates are used to determine fund allocations to local government units (places) under the General Revenue Sharing Program.

1.6.4 Poverty Counts

The Fay–Herriot (FH) method is also used to produce model-based county estimates of poor school-age children in the United States (National Research Council 2000). Using these estimates, the US Department of Education allocates annually several billions of dollars called Title I funds to counties, and then states distribute the funds among school districts. The allocated funds support compensatory education programs to meet the needs of educationally disadvantaged children. In the past, funds were allocated on the basis of updated counts from the previous census, but this allocation system had to be changed since the poverty counts vary significantly over time. EBLUP county estimates in this application are currently obtained from the American Community Survey (ACS) using administrative data as auxiliary information. Example 6.1.2 presents details of this application.

1.6.5 Median Income of Four-Person Families

Estimates of the current median income of four-person families in each of the states of the United States are used to determine the eligibility for a program of energy assistance to low-income families administered by the US Department of Health and Human Services. Current Population Survey (CPS) data and administrative information are used to produce model-based estimates, using extensions of the FH area level model (see Examples 8.1.1 and 8.3.1).

1.6.6 Poverty Mapping

Poverty measures are typically complex non-linear parameters, for example, poverty measures used by the World Bank to produce poverty maps in many countries all over the World. EB and HB methods for estimating poverty measures in Spanish provinces are illustrated in Example 10.7.1.

2

DIRECT DOMAIN ESTIMATION

2.1 INTRODUCTION

Sample survey data are extensively used to provide reliable direct estimates of totals and means for the whole population and large areas or domains. As noted in Chapter 1, a direct estimator for a domain uses values of the variable of interest, y, only from the sample units in the domain. Sections 2.2–2.5 provide a brief account of direct estimation under the design-based or repeated sampling framework. We refer the reader to standard textbooks on sampling theory (e.g., Cochran 1977, Hedayat and Sinha 1991, Särndal, Swensson, and Wretman 1992, Thompson 1997, Lohr 2010, Fuller 2009) for a more extensive treatment of direct estimation.

Model-based methods have also been used to develop direct estimators and the associated inferences. Such methods provide valid conditional inferences referring to the particular sample that has been drawn, regardless of the sampling design (see Brewer 1963, Royall 1970, Valliant, Dorfman, and Royall 2001). But unfortunately, model-based strategies can perform poorly under model misspecification as the sample size in the domain increases. For instance, Hansen, Madow, and Tepping (1983) introduced a model misspecification that is not detectable through tests of significance from sample sizes as large as 400, and then showed that the repeated sampling coverage probabilities of model-based confidence intervals on the population mean, $\overline{Y}$, are substantially less than the desired level, and that the understatement becomes worse as the sample size increases. This poor performance is largely due to asymptotic design-inconsistency of the model-based estimator with respect to the

Small Area Estimation, Second Edition. J. N. K. Rao and Isabel Molina.

stratified random sampling design employed by Hansen et al. (1983). We consider model-based direct estimators only briefly in this book, but model-based methods will be extensively used in the context of indirect estimators and small sample sizes in the domains of interest. As noted in Chapter 1, an indirect estimator for a domain "borrows strength" by using the values of the study variable, y, from sample units outside the domain of interest.

The main intention of Chapter 2 is to provide some background material for later chapters and to indicate that direct estimation methods may sometimes suffice, particularly after addressing survey design issues that have a bearing on small area estimation (see Section 2.6). Effective use of auxiliary information through ratio and regression estimation is also useful in reducing the need for indirect small area estimators (Sections 2.3–2.5).

2.2 DESIGN-BASED APPROACH

We assume the following somewhat idealized set-up and focus on the estimation of a population total or mean in Section 2.3. Direct estimators for domains are obtained in Section 2.4 using the results for population totals. A survey population U consists of N distinct elements (or ultimate units) identified through the labels $j = 1, \ldots, N$. We assume that a characteristic of interest, y, associated with element j can be measured exactly by observing element j. Thus, measurement errors are assumed to be absent. The parameter of interest is the population total $Y = \sum_U y_j$ or the population mean $\overline{Y} = Y/N$, where $\sum_U$ denotes summation over the population elements j.

A sampling design is used to select a sample (subset) s from U with probability $p(s)$. The sample selection probability $p(s)$ can depend on known design variables such as stratum indicator variables and size measures of clusters. In practice, a sampling scheme is used to implement a sampling design. For example, a simple random sample of size n can be obtained by drawing n random numbers from 1 to N without replacement. Commonly used sampling designs include stratified simple random sampling (e.g., establishment surveys) and stratified multistage sampling (e.g., large-scale socio-economic surveys such as the Canadian Labor Force Survey and the Current Population Survey (CPS) of the United States).

To make inferences on the total Y, we observe the y-values associated with the selected sample s. For simplicity, we assume that all the elements $j \in s$ can be observed, that is, complete response. In the design-based approach, an estimator $\hat{Y}$ of Y is said to be design-unbiased (or p-unbiased) if the design expectation of $\hat{Y}$ equals Y; that is,

$$E_p(\hat{Y}) = \sum p(s)\hat{Y}_s = Y, \tag{2.2.1}$$

where the summation is overall possible samples s under the specified design and $\hat{Y}_s$ is the value of $\hat{Y}$ for the sample s. The design variance of $\hat{Y}$ is denoted as $V_p(\hat{Y}) = E_p[\hat{Y} - E_p(\hat{Y})]^2$. An estimator of $V_p(\hat{Y})$ is denoted as $v(\hat{Y}) = s^2(\hat{Y})$, and the variance estimator $v(\hat{Y})$ is p-unbiased for $V(\hat{Y})$ if $E_p[v(\hat{Y})] \equiv V_p(\hat{Y})$. An estimator $\hat{Y}$ is called design-consistent (or p-consistent) for Y if the p-bias of $\hat{Y}/N$ tends to zero as the sample size increases and $N^{-2}V_p(\hat{Y})$ tends to zero as the sample size increases. Strictly speaking, we need to consider p-consistency in the context of a sequence of populations U_ν such that both the sample size n_ν and the population size N_ν tend to ∞ as $\nu \to \infty$. p-consistency of the variance estimator $v(\hat{Y})$ is similarly defined in terms of $N^{-2}v(\hat{Y})$. If the estimator $\hat{Y}$ and variance estimator $v(\hat{Y})$ are both p-consistent, then the design-based approach provides valid inferences on Y, regardless of the population values in the sense that the pivotal $t = (\hat{Y} - Y)/s(\hat{Y})$ converges in distribution ($\to_d$) to a $N(0, 1)$ variable as the sample size increases. Thus, in repeated sampling, about $100(1 - \alpha)\%$ of the confidence intervals $[\hat{Y} - z_{\alpha/2}s(\hat{Y}), \hat{Y} + z_{\alpha/2}s(\hat{Y})]$ contain the true value Y as the sample size increases, where $z_{\alpha/2}$ is the upper $(\alpha/2)$-point of a $N(0, 1)$ variable. In practice, one often reports only the estimate (realized value of $\hat{Y}$) and the associated standard error (realized value of $s(\hat{Y})$) or coefficient of variation (realized value of $\text{cv}(\hat{Y}) = s(\hat{Y})/\hat{Y}$). Coefficient of variation or standard error is used as a measure of variability associated with the estimate.

The design-based (or probability sampling) approach has been criticized on the grounds that the associated inferences, although assumption-free, refer to repeated sampling instead of just the particular sample s that has been drawn. A conditional design-based approach that allows us to restrict the set of samples used for inference to a "relevant" subset has also been proposed. This approach leads to conditionally valid inferences. For example, in the context of poststratification (i.e., stratification after selection of the sample), it makes sense to make design-based inferences conditional on the realized poststrata sample sizes (Holt and Smith 1979, Rao 1985). Similarly, when the population total X of an auxiliary variable x is known, conditioning on the estimator $\hat{X}$ of X is justified because the distance $|\hat{X} - X|/X$ provides a measure of imbalance in the realized sample (Robinson 1987, Rao 1992, Casady and Valliant 1993).

2.3 ESTIMATION OF TOTALS

2.3.1 Design-Unbiased Estimator

Design weights $w_j(s)$ play an important role in constructing design-based estimators $\hat{Y}$ of Y. These basic weights may depend both on s and the element j ($j \in s$). An important choice is $w_j(s) = 1/\pi_j$, where $\pi_j = \sum_{\{s:j\in s\}} p(s), j = 1, 2, \ldots, N$ are the inclusion probabilities and $\{s : j \in s\}$ denotes summation over all samples s containing the element j. To simplify the notation, we write $w_j(s) = w_j$ except when the full notation $w_j(s)$ is needed. The weight w_j may be interpreted as the number of elements in the population represented by the sample element j.

In the absence of auxiliary population information, we use the expansion estimator

$$\hat{Y} = \sum_s w_j y_j, \qquad (2.3.1)$$

where $\sum_s$ denotes summation over $j \in s$. In this case, the design-unbiasedness condition (2.2.1) reduces to

$$\sum_{\{s:j\in s\}} p(s)w_j(s) = 1; \; j = 1, \ldots, N. \tag{2.3.2}$$

The choice $w_j(s) = 1/\pi_j$ satisfies the unbiasedness condition (2.3.2) and leads to the well-known Horvitz–Thompson (H–T) estimator (see Cochran 1977, p. 259).

It is convenient to denote $\hat{Y} = \sum_s w_j y_j$ in an operator notation as $\hat{Y} = \hat{Y}(y)$. Using this notation, for another variable x with values $x_j (j = 1, \ldots, N)$ we use $\hat{Y}(x) = \sum_s w_j x_j$, whereas the traditional notation is to denote $\sum_s w_j x_j$ as $\hat{X}$. Similarly, we denote a variance estimator of $\hat{Y}$ as $v(\hat{Y}) = v(y)$. Using this notation, for the variance estimator of $\hat{X}$, we have $v(\hat{X}) = v[\hat{Y}(x)] = v(x)$. Note that the formulae for $\hat{Y}(x)$ and $v(x)$ are obtained by attaching the subscript j to the character, x, in the brackets and then replacing y_j by x_j in the formulae for $\hat{Y}(y)$ and $v(y)$, respectively. Hartley (1959) introduced the operator notation. We refer the reader to Cochran (1977), Särndal et al. (1992), and Wolter (2007) for details on variance estimation. Rao (1979) has shown that a nonnegative-unbiased quadratic estimator of the variance of $\hat{Y}$ is necessarily of the form

$$v(\hat{Y}) = v(y) = -\sum\sum_{j<k,\, j,k\in s} w_{jk}(s) b_j b_k \left(\frac{y_j}{b_j} - \frac{y_k}{b_k}\right)^2, \tag{2.3.3}$$

where the weights $w_{jk}(s)$ satisfy the unbiasedness condition, and the nonzero constants b_j are such that the variance of $\hat{Y}$ becomes zero when $y_j \propto b_j$ for all j. For example, in the special case of $w_j = 1/\pi_j$ and a fixed sample size design, we have $b_j = \pi_j$ and $w_{jk}(s) = (\pi_{jk} - \pi_j\pi_k)/(\pi_{jk}\pi_j\pi_k)$ where $\pi_{jk} = \sum_{\{s:(j,k)\in s\}} p(s)$, $j < k = 1, \ldots, N$ are the joint inclusion probabilities assumed to be all positive. The variance estimator (2.3.3) in this case reduces to the well-known Sen-Yates–Grundy (S–Y–G) variance estimator (see Cochran 1977, p. 261).

Under stratified multistage sampling, the expansion estimator $\hat{Y}$ may be written as

$$\hat{Y} = \hat{Y}(y) = \sum_s w_{hlk}\, y_{hlk}, \tag{2.3.4}$$

where w_{hlk} is the design weight associated with the kth element in the lth primary sampling unit (cluster) belonging to the hth stratum, y_{hlk} is the associated y-value, and $\sum_s$ is the summation over all elements $j = (hlk) \in s (h = 1, \ldots, L; l = 1, \ldots, n_{(h)}; k = 1, \ldots, n_{(hk)})$. It is common practice to treat the sample as if the clusters are sampled with replacement, and subsampling is done independently each time a cluster is selected. This leads to overestimation of the variance, but the variance estimator is greatly simplified. We have

$$v(\hat{Y}) = v(y) = \sum_{h=1}^{L} \frac{1}{n_{(h)}(n_{(h)} - 1)} \sum_{l=1}^{n_{(h)}} (y_{hl} - \bar{y}_h)^2, \tag{2.3.5}$$

where $y_{hl} = \sum_{k=1}^{n_{(hk)}} (n_{(h)} w_{hlk}) y_{hlk}$ are weighted sample cluster totals and $\bar{y}_h = \sum_{l=1}^{n_{(h)}} y_{hl}/n_{(h)}$. Note that $v(\hat{Y})$ depends on the y_{hlk}'s only through the totals y_{hl}. The relative bias of $v(\hat{Y})$ is small if the first-stage sampling fraction is small in each stratum. Note that $\hat{Y}(x)$ and $v(x)$ are obtained by attaching the subscripts hlk to the character, x, in the brackets and then replacing y_{hlk} by x_{hlk} in the formulae (2.3.4) and (2.3.5) for $\hat{Y}(x)$ and $v(y)$, respectively.

2.3.2 Generalized Regression Estimator

Suppose now that auxiliary information in the form of known population totals $\mathbf{X} = (X_1, \ldots, X_p)^T$ is available, and that the auxiliary vector $\mathbf{x}_j$ for $j \in s$ is also observed, that is, the data $(y_j, \mathbf{x}_j)$ for each element $j \in s$ are observed. An estimator that makes efficient use of this auxiliary information is the generalized regression (GREG) estimator, which may be written as

$$\hat{Y}_{\text{GR}} = \hat{Y} + (\mathbf{X} - \hat{\mathbf{X}})^T \hat{\mathbf{B}}, \tag{2.3.6}$$

where $\hat{\mathbf{X}} = \sum_s w_j \mathbf{x}_j = \hat{Y}(\mathbf{x})$ and $\hat{\mathbf{B}} = (\hat{B}_1, \ldots, \hat{B}_p)^T =: \hat{\mathbf{B}}(y)$ is the solution of the sample-weighted least squares equations:

$$\left(\sum_s w_j \mathbf{x}_j \mathbf{x}_j^T / c_j \right) \hat{\mathbf{B}} = \sum_s w_j \mathbf{x}_j y_j / c_j, \tag{2.3.7}$$

with specified constants $c_j (> 0)$. It is also useful to write $\hat{Y}_{\text{GR}}$ in the expansion form with design weights w_j changed to "revised" weights w_j^*. We have

$$\hat{Y}_{\text{GR}} = \sum_s w_j^* y_j =: \hat{Y}_{\text{GR}}(y) \tag{2.3.8}$$

in operator notation, where the revised weight $w_j^* = w_j^*(s)$ is the product of the design weight $w_j(s)$ and the estimation weight $g_j = g_j(s)$, that is, $w_j^* = w_j g_j$, with

$$g_j = 1 + (\mathbf{X} - \hat{\mathbf{X}})^T \left(\sum_s w_j \mathbf{x}_j \mathbf{x}_j^T / c_j \right)^{-1} \mathbf{x}_j / c_j. \tag{2.3.9}$$

Note that w_j^* in (2.3.8) does not depend on the y-values. Thus, the same weight w_j^* is applied to all variables of interest treated as y, as in the case of the expansion estimator $\hat{Y}$. This ensures consistency of results when aggregated over different variables $y_1, \ldots, y_r$ attached to each unit, that is,

$$\hat{Y}_{\text{GR}}(y_1) + \cdots + \hat{Y}_{\text{GR}}(y_r) = \hat{Y}_{\text{GR}}(y_1 + \cdots + y_r).$$

An important property of the GREG estimator is that it ensures consistency with the known auxiliary totals $\mathbf{X}$ in the sense

$$\hat{Y}_{\mathrm{GR}}(\mathbf{x}) = \sum_{s} w_j^* \mathbf{x}_j = \mathbf{X}. \tag{2.3.10}$$

A proof of (2.3.10) is given in Section 2.8.1. This property does not hold for the basic expansion estimator $\hat{Y}$. Many statistical agencies regard this property as desirable from the user's viewpoint. Because of the property (2.3.10), the GREG estimator is also called a calibration estimator (Deville and Särndal 1992). In fact, among all calibration estimators of the form $\sum_s h_j y_j$ with weights h_j satisfying the calibration constraints $\sum_s h_j \mathbf{x}_j = \mathbf{X}$, the GREG weights w_j^* minimize a chi-squared distance, $\sum_s c_j (w_j - h_j)^2 / w_j$, between the basic weights w_j and the calibration weights h_j (see Section 2.8.2). Thus, the GREG weights w_j^* modify the design weights as little as possible subject to the calibration constraints. However, these weights w_j^* may take negative or very large values for some units $j \in s$, especially when the number of calibration constraints is not small. Alternative methods have been proposed to address this issue of negative or too large calibration weights (Deville and Särndal 1992, Huang and Fuller 1978, Rao and Singh 2009).

The GREG estimator takes a simpler form when the constants c_j in (2.3.7) (or in the chi-squared distance) are taken as a linear combination of the auxiliary variables $c_j = \mathbf{v}^T \mathbf{x}_j$ for all $j \in U$ and for a vector $\mathbf{v}$ of specified constants. For this choice of c_j, we have

$$\hat{Y}_{\mathrm{GR}} = \mathbf{X}^T \hat{\mathbf{B}} = \sum_{s} \tilde{w}_j y_j \tag{2.3.11}$$

because in this case $\hat{Y} - \hat{\mathbf{X}}^T \hat{\mathbf{B}} = \sum_s w_j e_j(s) = 0$ (see Section 2.8.3 for the proof), where $e_j(s) = e_j = y_j - \mathbf{x}_j^T \hat{\mathbf{B}}$ are the sample residuals and $\tilde{w}_j = \tilde{w}_j(s) = w_j(s) \tilde{g}_j(s)$ with

$$\tilde{g}_j(s) = \mathbf{X}^T \left(\sum_{s} w_j \mathbf{x}_j \mathbf{x}_j^T / c_j \right)^{-1} \mathbf{x}_j / c_j. \tag{2.3.12}$$

The simplified GREG estimator (2.3.11) is known as the "projection" estimator.

The GREG estimator (2.3.11) covers many practically useful estimators as special cases. For example, in the case of a single auxiliary variable x with values $x_j (j = 1 \ldots, N)$, setting $c_j = x_j$ in (2.3.12) and noting that $\tilde{g}_j(s) = X / \hat{X}$, we get the well-known ratio estimator

$$\hat{Y}_{\mathrm{R}} = \frac{\hat{Y}}{\hat{X}} X. \tag{2.3.13}$$

The ratio estimator $\hat{Y}_{\mathrm{R}}$ uses the weights $\tilde{w}_j = w_j (X / \hat{X})$. If only the population size N is known, we set $x_j = 1$ in (2.3.13) so that $X = N$ and $\hat{X} = \hat{N} = \sum_s w_j$. If we set $\mathbf{x}_j = (1, x_j)^T$ and $c_j = 1$, then $\mathbf{v} = (1, 0)^T$ and (2.3.11) reduces to the familiar linear regression estimator

$$\hat{Y}_{\mathrm{LR}} = \hat{Y} + \hat{B}_{\mathrm{LR}} (X - \hat{X}), \tag{2.3.14}$$

where

$$\hat{B}_{\text{LR}} = \sum_s w_j(x_j - \hat{\bar{X}})(y_j - \hat{\bar{Y}}) / \sum_s w_j(x_j - \hat{\bar{X}})^2$$

with $\hat{\bar{Y}} = \hat{Y}/\hat{N}$ and $\hat{\bar{X}} = \hat{X}/\hat{N}$. The GREG estimator (2.3.11) also covers the familiar poststratified estimator as a special case. Suppose that we partition U into G poststrata U_g (e.g., age/sex groups) with known population counts $N_g (g = 1, \ldots, G)$. Then we set $\mathbf{x}_j = (x_{1j}, \ldots, x_{Gj})^T$ with $x_{gj} = 1$ if $j \in U_g$ and $x_{gj} = 0$ otherwise so that $\mathbf{X} = (N_1, \ldots, N_G)^T$. Since the poststrata are mutually exclusive, $\mathbf{1}^T \mathbf{x}_j = 1$ for all $j \in U$. Taking $c_j = 1$ and $\boldsymbol{\nu} = \mathbf{1}$, the GREG estimator (2.3.11) reduces to the poststratified estimator

$$\hat{Y}_{\text{PS}} = \sum_{g=1}^{G} \frac{N_{\cdot g}}{\hat{N}_{\cdot g}} \hat{Y}_{\cdot g}, \tag{2.3.15}$$

where $\hat{N}_{\cdot g} = \sum_{s_{\cdot g}} w_j$ and $\hat{Y}_{\cdot g} = \sum_{s_{\cdot g}} w_j y_j$, with $s_{\cdot g}$ denoting the sample of elements belonging to poststratum g.

Turning to variance estimation, the traditional Taylor linearization method gives

$$v_{\text{L}}(\hat{Y}_{\text{GR}}) = v(e), \tag{2.3.16}$$

which is obtained by substituting the residuals e_j for y_j in $v(y)$. Simulation studies have indicated that $v_{\text{L}}(\hat{Y}_{\text{GR}})$ may lead to slight underestimation, whereas the alternative variance estimator

$$v(\hat{Y}_{\text{GR}}) = v(ge), \tag{2.3.17}$$

obtained by substituting $g_j e_j$ for y_j in $v(y)$ reduces this underestimation (Fuller 1975, Estevao, Hidiroglou, and Särndal 1995). The alternative variance estimator $v(\hat{Y}_{\text{GR}})$ also performs better for conditional inference in the sense that it is approximately unbiased for the model variance of $\hat{Y}_{\text{GR}}$ conditionally on s for several designs, under the following linear regression model (or GREG model):

$$y_j = \mathbf{x}_j^T \boldsymbol{\beta} + \epsilon_j, \ \ j \in U, \tag{2.3.18}$$

where $E_m(\epsilon_j) = 0$, $V_m(\epsilon_j) = c_j \sigma^2$, $\text{Cov}_m(e_j, e_k) = 0$ for $j \neq k$; and E_m, V_m, and Cov_m, respectively, denote the model expectation, variance, and covariance (Särndal, Swensson, and Wretman 1989, Rao 1994). In the model-based framework, y_j is a random variable and s is fixed. The GREG estimator is also model-unbiased under (2.3.18) in the sense $E_m(\hat{Y}_{\text{GR}}) = E_m(Y)$ for every s. In the design-based framework, $\hat{Y}_{\text{GR}}$, $v_{\text{L}}(\hat{Y}_{\text{GR}})$, and $v(\hat{Y}_{\text{GR}})$ are p-consistent.

The GREG estimator (2.3.6) may be expressed as

$$\hat{Y}_{\text{GR}} = \sum_{j \in U} \hat{y}_j + \sum_{j \in s} w_j(y_j - \hat{y}_j), \tag{2.3.19}$$

where $\hat{y}_j = \mathbf{x}_j^T\hat{\mathbf{B}}$ is the predictor of y_j under the "working" model (2.3.18), that is, the estimator $\hat{Y}_{\text{GR}}$ is the sum of all predicted values $\hat{y}_j, j \in U$ and the p-unbiased expansion estimator of the total prediction error $\sum_{j\in U}(y_j - \hat{y}_j)$. The above approach for deriving a p-consistent estimator of a total using a working model is called model-assisted approach. The choice of the working model affects the efficiency of the estimator although the estimator remains p-consistent, regardless of the validity of the model. It may be noted that the projection GREG estimator (2.3.11) obtained for $c_j = \mathbf{v}^T\mathbf{x}_j$ for all j may be written as the sum of all predicted values,

$$\hat{Y}_{\text{GR}} = \sum_{j\in U} \hat{y}_j.$$

We refer the reader to Särndal et al. (1992) for a detailed account of GREG estimation and to Estevao et al. (1995) for the highlights of a Generalized Estimation System at Statistics, Canada, based on GREG estimation theory.

2.4 DOMAIN ESTIMATION

2.4.1 Case of No Auxiliary Information

Suppose U_i denotes a domain (or subpopulation) of interest and that we are required to estimate the domain total $Y_i = \sum_{U_i} y_j$ or the domain mean $\overline{Y}_i = Y_i/N_i$, where N_i, the number of elements in U_i, may or may not be known. If y_j is binary (1 or 0), then $\overline{Y}_i$ reduces to the domain proportion P_i; for example, the proportion in poverty in the ith domain. Much of the theory in Section 2.3 for a total can be adapted to domain estimation by using the following relationships. Writing the total $Y = \sum_{j\in U} y_j$ in the operator notation as $Y(y)$ and defining

$$y_{ij} = \begin{cases} y_j & \text{if } j \in U_i \\ 0 & \text{otherwise,} \end{cases}$$

$$a_{ij} = \begin{cases} 1 & \text{if } j \in U_i \\ 0 & \text{otherwise,} \end{cases}$$

we can express the domain total Y_i and the domain size as

$$Y(y_i) = \sum_{j\in U} y_{ij} = \sum_{j\in U} a_{ij} y_j = \sum_{j\in U_i} y_j = Y_i \tag{2.4.1}$$

and

$$Y(a_i) = \sum_{j\in U} a_{ij} = \sum_{j\in U_i} 1 = N_i. \tag{2.4.2}$$

Therefore, the formulae of Section 2.3 for estimating a total Y can be applied by changing y_j by y_{ij}. If the domains of interest, say, $U_1, \ldots, U_m$, form a partition of U (or of a larger domain), it is desirable from a user's viewpoint to ensure that the estimates of domain totals add up to the estimate of the population total (or the larger domain).

In the absence of auxiliary population information, we use the expansion estimator

$$\hat{Y}_i = \hat{Y}(y_i) = \sum_{j\in s} w_j y_{ij} = \sum_{j\in s} w_j a_{ij} y_j = \sum_{j\in s_i} w_j y_j, \tag{2.4.3}$$

where s_i denotes the sample of elements belonging to domain U_i. It readily follows from (2.4.1) and (2.4.3) that $\hat{Y}_i$ is p-unbiased for Y_i if $\hat{Y}$ is p-unbiased for Y. It is also p-consistent if the expected domain sample size is large. Similarly, $\hat{N}_i = \hat{Y}(a_i)$ is p-unbiased for N_i, using (2.4.2). We note from (2.4.3) that the additive property is satisfied: $\hat{Y}_1 + \cdots + \hat{Y}_m = \hat{Y}$.

Noting that $v(\hat{Y}) = v(y)$, an estimator of the variance of $\hat{Y}_i$ is simply obtained from $v(y)$ by changing y_j to y_{ij}, that is,

$$v(\hat{Y}_i) = v(y_i). \tag{2.4.4}$$

It follows from (2.4.3) and (2.4.4) that no new theory is required for estimation of a domain total.

The domain mean $\overline{Y}_i = Y(y_i)/Y(a_i)$ is estimated by

$$\hat{\overline{Y}}_i = \frac{\hat{Y}(y_i)}{\hat{Y}(a_i)} = \frac{\hat{Y}_i}{\hat{N}_i}. \tag{2.4.5}$$

For the special case of a binary variable $y_j \in \{0, 1\}$, $\hat{\overline{Y}}_i$ reduces to $\hat{P}_i$, an estimator of the domain proportion P_i. If the expected domain sample size is large, the ratio estimator (2.4.5) is p-consistent in the sense that it is approximately unbiased and its variance goes to zero as the sample size increases. A Taylor linearization variance estimator is given by

$$v_{\mathrm{L}}(\hat{\overline{Y}}_i) = v(\tilde{e}_i)/\hat{N}_i^2, \tag{2.4.6}$$

where $\tilde{e}_{ij} = y_{ij} - \hat{\overline{Y}}_i a_{ij} = (y_j - \hat{\overline{Y}}_i)a_{ij}$. Note that $v(\tilde{e}_i)$ is obtained from $v(y)$ by changing y_j to $\tilde{e}_{ij}$. It follows from (2.4.5) and (2.4.6) that no new theory is required for domain means as well. Note that $\tilde{e}_{ij} = 0$ for a unit $j \in s$ not belonging to U_i. We refer the reader to Hartley (1959) for details on domain estimation.

2.4.2 GREG Domain Estimation

GREG estimation of Y is also easily adapted to estimation of a domain total Y_i. It follows from (2.3.8) that the GREG estimator of domain total Y_i is

$$\hat{Y}_{i\mathrm{GR}} = \hat{Y}_{\mathrm{GR}}(y_i) = \sum_{j\in s_i} w_j^* y_j, \tag{2.4.7}$$

when the population total of the auxiliary vector $\mathbf{x}$ is known. It follows from (2.4.7) that the GREG estimator also satisfies the additive property, namely $\hat{Y}_{1\text{GR}} + \cdots + \hat{Y}_{m\text{GR}} = \hat{Y}_{\text{GR}}$. The estimator $\hat{Y}_{i\text{GR}}$ is approximately p-unbiased if the overall sample size is large, but p-consistency requires a large expected domain sample size as well.

The GREG domain estimator (2.4.7) may be expressed as

$$\hat{Y}_{i\text{GR}} = \sum_{j \in U_i} \hat{y}_j + \sum_{j \in s_i} w_j (y_j - \hat{y}_j), \tag{2.4.8}$$

for predicted values $\hat{y}_j = \mathbf{x}_j^T \hat{\mathbf{B}}(y_i)$ under the working model (2.3.18), where $\hat{\mathbf{B}}(y_i)$ is obtained from $\hat{\mathbf{B}}(y)$ by changing y_j to y_{ij}. It may be noted that the last term in the estimator of the domain total (2.4.8) is not zero when $c_j = \mathbf{v}^T \mathbf{x}_j$, unlike the term $\sum_{j \in s} w_j e_j(s)$ in the estimator of the population total (2.3.11). As a result, the projection-GREG domain estimator $\sum_{j \in U_i} \hat{y}_j$ is not asymptotically p-unbiased unlike the bias-adjusted estimator (2.4.8).

For the special case of ratio estimation, by changing y_j to $a_{ij} y_j$ in (2.3.13) we get

$$\hat{Y}_{i\text{R}} = \frac{\hat{Y}_i}{\hat{X}} X. \tag{2.4.9}$$

Similarly, a poststratified domain estimator is obtained from (2.3.15) by changing y_j to $a_{ij} y_j$:

$$\hat{Y}_{i\text{PS}} = \sum_{g=1}^{G} \frac{N_{\cdot g}}{\hat{N}_{\cdot g}} \sum_{j \in s_{ig}} w_j y_j, \tag{2.4.10}$$

where s_{ig} is the sample falling in the (ig)th cell of the cross-classification of domains and poststrata.

A Taylor linearization variance estimator of $\hat{Y}_{i\text{GR}}$ is simply obtained from $v(y)$ by changing y_j to $e_{ij} = y_{ij} - \mathbf{x}_j^T \hat{\mathbf{B}}(y_i)$. Note that $e_{ij} = -\mathbf{x}_j^T \hat{\mathbf{B}}(y_i)$ for a unit $j \in s$ not belonging to U_i. The large negative residuals for all sampled elements not in U_i lead to inefficiency, unlike in the case of $\hat{Y}_{\text{GR}}$ where the variability of residuals e_j will be small relative to the variability of the y_j's. This inefficiency of the GREG domain estimator $\hat{Y}_{i\text{GR}}$ is due to the fact that the auxiliary population information used here is not domain specific. But $\hat{Y}_{i\text{GR}}$ has the advantage that it is approximately p-unbiased even if the expected domain sample size is small, whereas the GREG estimator based on domain-specific auxiliary population information is p-biased unless the expected domain sample size is also large.

2.4.3 Domain-Specific Auxiliary Information

We now turn to GREG estimation of a domain total Y_i under domain-specific auxiliary information. We assume that the domain totals $\mathbf{X}_i = \sum_{j \in U_i} \mathbf{x}_j = Y(\mathbf{x}_i)$ are known, where $\mathbf{x}_{ij} = \mathbf{x}_j$ if $j \in U_i$ and $\mathbf{x}_{ij} = 0$ otherwise; that is, $\mathbf{x}_{ij} = a_{ij}\mathbf{x}_j$. In this case, a GREG estimator of Y_i is given by

$$\hat{Y}^*_{i\text{GR}} = \hat{Y}_i + (\mathbf{X}_i - \hat{\mathbf{X}}_i)^T \hat{\mathbf{B}}_i, \tag{2.4.11}$$

where

$$\left(\sum_{j\in s} w_j \mathbf{x}_{ij}\mathbf{x}_{ij}^T / c_j\right)\hat{\mathbf{B}}_i = \sum_{j\in s} w_j \mathbf{x}_{ij} y_{ij}/c_j \tag{2.4.12}$$

and $\hat{\mathbf{X}}_i = \sum_{s_i} w_j \mathbf{x}_j = \hat{Y}(\mathbf{x}_i)$. We may also write (2.4.11) as

$$\hat{Y}^*_{i\text{GR}} = \sum_{j\in s} w^*_{ij} y_{ij}, \tag{2.4.13}$$

where $w^*_{ij} = w_j g^*_{ij}$ with

$$g^*_{ij} = 1 + (\mathbf{X}_i - \hat{\mathbf{X}}_i)^T \left(\sum_{j\in s} w_j \mathbf{x}_{ij}\mathbf{x}_{ij}^T / c_j\right)^{-1} \mathbf{x}_{ij}/c_j.$$

Note that the weights w^*_{ij} now depend on i unlike the weights w^*_j. Therefore, the estimators $\hat{Y}^*_{i\text{GR}}$ do not add up to $\hat{Y}_{\text{GR}}$. Also, $\hat{Y}^*_{i\text{GR}}$ is not approximately p-unbiased unless the domain sample size is large. If the condition $c_j = \mathbf{v}^T\mathbf{x}_j$ holds, then $\hat{Y}^*_{i\text{GR}}$ takes the projection form

$$\hat{Y}^*_{i\text{GR}} = \mathbf{X}_i^T \hat{\mathbf{B}}_i. \tag{2.4.14}$$

Lehtonen, Särndal, and Veijanen (2003) named the estimator (2.4.14) as GREG-D.

We now give three special cases of the projection GREG estimator given by (2.4.14). In the first case, we consider a single auxiliary variable x with known domain total X_i and set $c_j = x_j$ in (2.4.14), where $\hat{\mathbf{B}}_i$ is given in (2.4.12). This leads to the ratio estimator

$$\hat{Y}^*_{i\text{R}} = \frac{\hat{Y}_i}{\hat{X}_i} X_i. \tag{2.4.15}$$

In the second case, we consider known domain-specific poststrata counts N_{ig} and set $c_j = 1$ and $\mathbf{x}_j = (x_{1j}, \ldots, x_{Gj})^T$ with $x_{gj} = 1$ if $j \in U_g$ and $x_{gj} = 0$ otherwise. This leads to the poststratified count (PS/C) estimator

$$\hat{Y}^*_{i\text{PS/C}} = \sum_{g=1}^{G} \frac{N_{ig}}{\hat{N}_{ig}} \hat{Y}_{ig}, \tag{2.4.16}$$

where $\hat{N}_{ig} = \sum_{s_{ig}} w_j$ and $\hat{Y}_{ig} = \sum_{s_{ig}} w_j y_j$. In the third case, we consider known cell totals X_{ig} of an auxiliary variable x and set $c_j = 1$ and $\mathbf{x}_j = (x_{1j}, \ldots, x_{Gj})^T$ with $x_{gj} = x_j$, if $j \in U_g$ and $x_{gj} = 0$ otherwise. This leads to the poststratified-ratio (PS/R) estimator

$$\hat{Y}^*_{i\text{PS/R}} = \sum_{g=1}^{G} \frac{X_{ig}}{\hat{X}_{ig}} \hat{Y}_{ig}, \tag{2.4.17}$$

where $\hat{X}_{ig} = \sum_{s_{ig}} w_j x_j$.

If the expected domain sample size is large, then a Taylor linearization variance estimator of $\hat{Y}^*_{i\text{GR}}$ is obtained from $v(y)$ by changing y_j to $e^*_{ij} = y_{ij} - \mathbf{x}^T_{ij}\hat{\mathbf{B}}_i$. Note that $e^*_{ij} = 0$ if $j \in s$ and $j \notin U_i$ unlike the large negative residuals e_{ij} in the case of $\hat{Y}_{i\text{GR}}$. Thus, the domain-specific GREG estimator $\hat{Y}^*_{i\text{GR}}$ will be more efficient than $\hat{Y}_{i\text{GR}}$, provided the expected domain-specific sample size is large.

Example 2.4.1. Wages and Salaries. Särndal and Hidiroglou (1989) and Rao and Choudhry (1995) considered a population U of $N = 1,678$ unincorporated tax filers (units) from the province of Nova Scotia, Canada, divided into 18 census divisions. This population is actually a simple random sample, but it was treated as a population for simulation purposes. The population is also classified into four mutually exclusive industry groups: retail (515 units), construction (496 units), accommodation (114 units), and others (553 units). Domains (small areas) are formed by a cross-classification of the four industry types with the 18 census divisions. This leads to 70 nonempty domains out of 72 possible domains. The objective is to estimate the domain totals Y_i of the y-variable (wages and salaries), utilizing the auxiliary variable x (gross business income) assumed to be known for all the N units in the population. A simple random sample, s, of size n is drawn from U and y-values observed. The data consist of (y_j, x_j) for $j \in s$ and the auxiliary population information. The sample s_i has $n_i (\geq 0)$ units.

Under the above set-up, we have $\pi_j = n/N$, $w_j = N/n$, and the expansion estimator (2.4.3) reduces to

$$\hat{Y}_i = \begin{cases} (N/n) \sum_{s_i} y_j & \text{if } n_i \geq 1, \\ 0 & \text{if } n_i = 0. \end{cases} \tag{2.4.18}$$

The estimator $\hat{Y}_i$ is p-unbiased for Y_i because

$$E_p(\hat{Y}_i) = E_p[\hat{Y}(y_i)] = Y(y_i) = Y_i.$$

However, it is p-biased conditional on the realized domain sample size n_i. In fact, conditional on n_i, the sample s_i is a simple random sample of size n_i from U_i, and the conditional p-bias of $\hat{Y}_i$ is

$$B_2(\hat{Y}_i) = E_2(\hat{Y}_i) - Y_i = N\left(\frac{n_i}{n} - \frac{N_i}{N}\right)\overline{Y}_i, \quad n_i \geq 1,$$

where E_2 denotes conditional expectation. Thus, the conditional bias is zero only if the sample proportion n_i/n equals the population proportion N_i/N.

Suppose we form G poststrata based on the x-variable known for all the N units. Then (2.4.16) and (2.4.17) reduce to

$$\hat{Y}^*_{i\text{PS/C}} = \sum_{g=1}^{G} N_{ig}\bar{y}_{ig} \tag{2.4.19}$$

and

$$\hat{Y}^*_{i\text{PS/R}} = \sum_{g=1}^{G} X_{ig}\frac{\bar{y}_{ig}}{\bar{x}_{ig}}, \tag{2.4.20}$$

where $\bar{y}_{ig}$ and $\bar{x}_{ig}$ are the sample means for the n_{ig} units falling in the cell (ig), and the cell counts N_{ig} and the cell totals X_{ig} are assumed to be known. If $n_{ig} = 0$, we set $\bar{y}_{ig} = 0$ in (2.4.19) and $\bar{y}_{ig}/\bar{x}_{ig} = 0$ in (2.4.20). The estimator $\hat{Y}^*_{i\text{PS/C}}$ is p-unbiased conditional on the realized sample sizes $n_{ig}(\geq 1$ for all $g)$, whereas $\hat{Y}^*_{i\text{PS/R}}$ is only approximately p-unbiased conditional on the n_{ig}'s, provided all the expected sample sizes $E(n_{ig})$ are large. It is desirable to make inferences conditional on the realized sample sizes, but this may not be possible under designs more complex than simple random sampling (Rao 1985). Even under simple random sampling, we require $n_{ig} \geq 1$ for all g, which limits the use of poststratification when n_i is small.

If poststratification is not used, we can use the ratio estimator (2.4.15), which reduces to

$$\hat{Y}^*_{i\text{R}} = X_i\frac{\bar{y}_i}{\bar{x}_i}, \tag{2.4.21}$$

under simple random sampling, where $\bar{y}_i$ and $\bar{x}_i$ are the sample means for the n_i units from domain i. If an x-variable is not observed but N_i is known, we can use an alternative estimator

$$\hat{Y}^*_{i\text{C}} = N_i\bar{y}_i. \tag{2.4.22}$$

This estimator is p-unbiased conditional on the realized sample size n_i (≥ 1).

2.5 MODIFIED GREG ESTIMATOR

We now consider modified GREG estimators that use y-values from outside the domain but remain p-unbiased or approximately p-unbiased as the overall sample size increases. In particular, we replace $\hat{\mathbf{B}}_i$ in (2.4.11) by the overall regression coefficient $\hat{\mathbf{B}}$, given by (2.3.7), to get

$$\tilde{Y}_{i\text{GR}} = \hat{Y}_i + (\mathbf{X}_i - \hat{\mathbf{X}}_i)^T\hat{\mathbf{B}} = \sum_{j\in s} \tilde{w}_{ij}y_j \tag{2.5.1}$$

with

$$\tilde{w}_{ij} = w_j a_{ij} + (\mathbf{X}_i - \hat{\mathbf{X}}_i)\left(\sum_s w_j \mathbf{x}_j \mathbf{x}_j^T / c_j\right)^{-1} (w_j \mathbf{x}_j / c_j),$$

where a_{ij} is the domain indicator variable. The estimator $\tilde{Y}_{i\text{GR}}$ given by (2.5.1) is approximately p-unbiased as the overall sample size increases, even if the domain sample size is small. This estimator is also called the "survey regression" estimator (Battese, Harter, and Fuller 1988; Woodruff 1966). The modified estimator (2.5.1) may also be viewed as a calibration estimator $\sum_s \tilde{w}_{ij} y_j$ with weights $h_{ij} = \tilde{w}_{ij}$ minimizing a chi-squared distance $\sum_s c_j (w_j a_{ij} - h_{ij})^2 / w_j$ subject to the constraints $\sum_s h_{ij} \mathbf{x}_j = \mathbf{X}_i$ (Singh and Mian 1995). The survey regression estimators satisfy the additive property in the sense

$$\sum_{i=1}^{m} \tilde{Y}_{i\text{GR}} = \hat{Y} + (\mathbf{X} - \hat{\mathbf{X}})^T \hat{\mathbf{B}} = \hat{Y}_{\text{GR}}.$$

In the case of a single auxiliary variable x with known domain total X_i, a ratio form of (2.5.1) is given by

$$\tilde{Y}_{i\text{R}} = \hat{Y}_i + \frac{\hat{Y}}{\hat{X}}(X_i - \hat{X}_i). \tag{2.5.2}$$

Although the modified GREG estimator borrows strength for estimating the regression coefficient, it does not increase the "effective" sample size, unlike indirect estimators studied in Chapter 3. To illustrate this point, consider the simple random sampling of Example 2.4.1. In this case, $\tilde{Y}_{i\text{R}}$ reduces to

$$\tilde{Y}_{i\text{R}} = N_i \left[\bar{y}_i + \frac{\bar{y}}{\bar{x}}(\bar{X}_i - \bar{x}_i)\right], \tag{2.5.3}$$

where $\bar{y}$ and $\bar{x}$ are the overall sample means and $\bar{X}_i = X_i / N_i$. For large n, we can replace the sample ratio $\bar{y}/\bar{x}$ by the population ratio $R = \bar{Y}/\bar{X}$ and the conditional variance given n_i becomes

$$V_2(\tilde{Y}_{i\text{R}}) \approx N_i^2 \left(\frac{1}{n_i} - \frac{1}{N_i}\right) S_{Ei}^2, \tag{2.5.4}$$

where $S_{Ei}^2 = \sum_{j \in U_i} (E_j - \bar{E}_i)^2 / (N_i - 1)$ with $E_j = y_j - R x_j$ and $\bar{E}_i$ is the domain mean of the E_j's. It follows from (2.5.4) that $V_2(\tilde{Y}_{i\text{R}})/N_i^2$ is of order n_i^{-1} so that the effective sample is not increased, although the variability of the E_j's may be smaller than the variability of the y_j's for $j \in U_i$. Note that the variability of the E_j's will be larger than the variability of the domain-specific residuals $y_j - R_i x_j$ for $j \in U_i$ where $R_i = \bar{Y}_i / \bar{X}_i$, unless $R_i \approx R$.

The modified GREG estimator (2.5.1) may be expressed as

$$\tilde{Y}_{i\text{GR}} = \mathbf{X}_i^T \hat{\mathbf{B}} + \sum_{j \in s_i} w_j e_j. \tag{2.5.5}$$

The first term $\mathbf{X}_i^T\hat{\mathbf{B}}$ is the synthetic regression estimator (see Chapter 3) and the second term $\sum_{s_i} w_j e_j$ approximately corrects the p-bias of the synthetic regression estimator. We can improve on $\tilde{Y}_{iGR}$ by replacing the expansion estimator $\sum_{s_i} w_j e_j$ in (2.5.5) with a ratio estimator (Särndal and Hidiroglou 1989):

$$\hat{E}_{iR} = N_i \left(\sum_{s_i} w_j \right)^{-1} \sum_{s_i} w_j e_j; \tag{2.5.6}$$

note that $\hat{N}_i = \sum_{s_i} w_j$. The resulting estimator

$$\tilde{Y}_{iSH} = \mathbf{X}_i^T\hat{\mathbf{B}} + \hat{E}_{iR}, \tag{2.5.7}$$

however, suffers from the ratio bias when the domain sample size is small, unlike $\tilde{Y}_{iGR}$.

A Taylor linearization variance estimator of $\tilde{Y}_{iGR}$ is obtained from $v(y)$ by changing y_j to $a_{ij}e_j$, that is,

$$v_L(\tilde{Y}_{iGR}) = v(a_i e). \tag{2.5.8}$$

This variance estimator is valid even when the small area sample size is small, provided the overall sample size is large.

2.6 DESIGN ISSUES

"Optimal" design of samples for use with direct estimators of totals or means in large areas has received a lot of attention over the past 60 years or so. In particular, design issues, such as number of strata, construction of strata, sample allocation, and selection probabilities, have been addressed (see, e.g., Cochran 1977). The ideal goal here is to find an "optimal" design that minimizes the MSE of a direct estimator subject to a given cost. This goal is seldom achieved in practice due to operational constraints and other factors. As a result, survey practitioners often adopt a "compromise" design that is "close" to the optimal design.

In practice, it is not possible to anticipate and plan for all possible areas (or domains) and uses of survey data as "the client will always require more than is specified at the design stage" (Fuller 1999, p. 344). As a result, indirect estimators will always be needed in practice, given the growing demand for reliable small area statistics. However, it is important to consider design issues that have an impact on small area estimation, particularly in the context of planning and designing large-scale surveys. In this section, we present a brief discussion on some of the design issues. A proper resolution of these issues could lead to enhancement in the reliability of direct (and also indirect) estimates for both planned and unplanned domains. For a more detailed discussion of design issues, we refer the reader to Singh, Gambino, and Mantel (1994) and Marker (2001).

2.6.1 Minimization of Clustering

Most large-scale surveys, use clustering to a varying degree in order to reduce the survey costs. Clustering, however, results in a decrease in the "effective" sample size. It can also adversely affect the estimation for unplanned domains because it can lead to situations where some domains become sample rich while others may have no sample at all. It is therefore useful to minimize the clustering in the sample. The choice of sampling frame plays an important role in reducing clustering of units; for example, the use of a list frame, replacing clusters wherever possible, such as Business Registers for business surveys and Address Registers for household surveys, is an effective tool. Also, the choice of sampling units, their sizes, and the number of sampling stages have significant impact on the effective sample size.

2.6.2 Stratification

One method of providing better sample size distribution at the small area level is to replace large strata by many small strata from which samples are drawn. By this approach, it may be possible to make an unplanned small domain contain mostly complete strata. For example, each Canadian province is partitioned into both Economic Regions (ERs) and Unemployment Insurance Regions (UIRs), and there are 71 ERs and 61 UIRs in Canada. In this case, the number of strata may be increased by treating all the areas created by the intersections of the two partitions as strata. This strategy will lead to 133 intersections (strata). As another example of stratification, the United States National Health Interview Survey (NHIS) used stratification by region, metropolitan area status, labor force data, income, and racial composition until 1994. The resulting sample sizes for individual states did not support state-level direct estimates for several states; in fact, two of the states did not have NHIS sampled units. The NHIS stratification scheme was replaced by state and metropolitan area status in 1995, thus enabling state-level direct estimation for all states (see Marker 2001 for details).

2.6.3 Sample Allocation

By adopting compromise sample allocations, it may be possible to satisfy reliability requirements at a small area level as well as large area level, using only direct estimates. Singh et al. (1994) presented an excellent illustration of compromise sample allocation in the Canadian LFS to satisfy reliability requirements at the provincial level as well as subprovincial level. For the LFS with a monthly sample of 59,000 households, "optimizing" at the provincial level yields a coefficient of variation (CV) of the direct estimate for "unemployed" as high as 17% for some UIR's (small areas). They adopted a two-step compromise allocation, with 42,000 households allocated at the first step to get reliable provincial estimates and the remaining 17,000 households allocated at the second step to produce best possible UIR estimates. This allocation reduced the worst CV of 17% for UIR to 9.4% at the expense of a small increase at the provincial and national levels: CV for Ontario increased from 2.8% to 3.4% and for Canada from 1.36% to 1.51%. Thus, by oversampling small areas, it is possible to decrease the CV of direct estimates for these areas significantly at the expense of a small increase in CV at the national level. The U.S. National Household Survey on Drug Abuse used stratification and oversampling to produce direct estimates for every state. The 2000 Danish Health and Morbidity Survey used two national samples, each of 6,000 respondents, and distributed an additional 8,000 respondents to guarantee at least 1,000 respondents in each county (small area). Similarly, the Canadian Community Health Survey (CCHS) conducted by Statistics Canada accomplishes its sample allocation in two steps. First, it allocates 500 households to each of its 133 Health Regions, and then the remaining sample (about one half of 130,000 households) is allocated to maximize the efficiency of provincial estimates (see Béland et al. 2000 for details). Section 2.7 gives some methods of compromise sample allocation, including an "optimal" method that uses a nonlinear programming formulation.

2.6.4 Integration of Surveys

Harmonizing questions across surveys of the same population leads to increased effective sample sizes for the harmonized items. The increased sample sizes, in turn, lead to improved direct estimates for small areas. However, caution should be exercised because the data may not be comparable across surveys even if the questionnaire wording is consistent. As noted by Groves (1989), different modes of data collection and the placement of questions can cause differences.

A number of current surveys in Europe are harmonized both within countries and between countries. For example, the European Community Household Panel Survey (ECHP) collects consistent data across member countries. Statistics Netherlands uses a common procedure to collect basic information across social surveys.

2.6.5 Dual-Frame Surveys

Dual-frame surveys can be used to increase the effective sample size in a small area. In a dual-frame survey, samples are drawn independently from two overlapping frames that together cover the population of interest. For example, suppose frame A is a complete area frame and data collected by personal interviewing, while frame B is an incomplete list frame and data collected by telephone interviewing. In this case, the dual-frame design augments the expensive frame A information with inexpensive additional information from B. There are many surveys using dual-frame designs. For example, the Dutch Housing Demand Survey collects data by personal interviewing, but uses telephone supplementation in over 100 municipalities to produce dual-frame estimates for those municipalities (small areas). Statistics Canada's CCHS is another recent example, where an area sample is augmented with a telephone list sample in selected health regions.

Hartley (1974) discussed dual-frame designs and developed a unified theory for dual-frame estimation of totals by combining information from the two samples. Skinner and Rao (1996) developed dual-frame estimators that use the same survey weights for all the variables. Lohr and Rao (2000) applied the jackknife method to obtain variance estimators for dual-frame estimators of totals.

2.6.6 Repeated Surveys

Many surveys are repeated over time and effective sample size can be increased by combining data from two or more consecutive surveys. For example, the United States National Health Interview Survey (NHIS) is an annual survey that uses nonoverlapping samples across years. Combining consecutive annual NHIS samples leads to improved estimates, although the correlation between years, due to the use of the same psu's, reduces the effective sample size. Such estimates, however, can lead to significant bias if the characteristic of interest is not stable over the time period.

Marker (2001) studied the level of accuracy for state estimates by combining the 1995 NHIS sample with the previous year sample or sample of the previous 2 years. He showed that aggregation helps achieve CV's of 30% and 20% for four selected variables, but 10% CV cannot be achieved for many states even after aggregation across 3 years.

Kish (1999) recommended "rolling samples" as a method of cumulating data over time. Unlike the customary periodic surveys, such as the NHIS with the same psu's over time or the Canadian LFS and the United States CPS with the same psu's and partial overlap of sample elements, rolling samples (RS) aim at a much greater spread to facilitate maximal spatial range for cumulation over time. This, in turn, will lead to improved small area estimates when the periodic samples are cumulated. The American Community Survey (ACS), which started in 2005, is an excellent example of RS design (http://www.census.gov/acs/www/). It aims to provide monthly samples of 250,000 households and detailed annual statistics based on 3 million households spread across all counties in the United States. It will also provide quinquennial and decennial census samples.

2.7 *OPTIMAL SAMPLE ALLOCATION FOR PLANNED DOMAINS

We now study optimal sample allocation in the context of estimating domain means $\overline{Y}_i$ as well as the aggregate population mean $\overline{Y}$ under stratified simple random sampling. We assume that the domains are specified in advance of the sample allocation stage (or planned) and consider two cases: (i) Domains are identical to strata and simple random samples are drawn independently in different domains; (ii) domains cut across strata and domain frames are not available for selecting independent samples in different domains.

2.7.1 Case (i)

Suppose that the population U is partitioned into L strata $U_{(h)}(h = 1, \ldots, L)$ of sizes $N_{(h)}$, and samples of sizes $n_{(h)}$ are to be selected from the strata subject to fixed cost $C = c_0 + \sum_{h=1}^{L} c_{(h)}n_{(h)}$, where c_0 is the overhead constant and $c_{(h)}$ is the cost per unit in stratum h; for simplicity we take $c_{(h)} = 1$ for all h and let $C - c_0 = \sum_{h=1}^{L} n_{(h)} = n$. The strata means $\overline{Y}_{(h)}$ and the overall mean $\overline{Y}$ are estimated, respectively, by the sample means $\overline{y}_{(h)}$ and the weighted mean $\overline{y}_{st} = \sum_{h=1}^{L} W_{(h)}\overline{y}_{(h)}$, where $W_{(h)} = N_{(h)}/N$ is the relative size of stratum h. Sampling literature (see e.g., Cochran 1977, Chapters 5 and 5A) has largely focused on the sample allocation $\{n_{(h)}\}$ for estimating $\overline{Y}$ that minimizes $V_p(\overline{y}_{st})$ subject to $\sum_{h=1}^{L} n_{(h)} = n$, leading to the well-known Neyman allocation

$$n_{(h)}^{N} = n\frac{N_{(h)}S_{y(h)}}{\sum\limits_{h=1}^{L} N_{(h)}S_{y(h)}}, \tag{2.7.1}$$

where $S_{y(h)}^2 = \sum_{j\in U_{(h)}}(y_{hj} - \overline{y}_{(h)})^2/(N_{(h)} - 1)$ is the stratum variance. Neyman allocation depends on the unknown population variances $S_{y(h)}^2$ of the variable y, and in practice a proxy variable z with known strata variances $S_{z(h)}^2$ is used in place of y. On the other hand, proportional allocation $n_{(h)} = nW_{(h)}$, equal allocation $n_{(h)} = n/L$ and square root allocation $n_{(h)} = n\sqrt{N_{(h)}}/\sum_{h=1}^{L}\sqrt{N_{(h)}}$ (Bankier 1988) do not depend on y, but are then suboptimal for estimating $\overline{Y}$.

Neyman and proportional allocations may cause some strata to have large coefficients of variation (CV) of means $\overline{y}_{(h)}$. On the other hand, equal allocation is efficient for estimating strata means, but it may lead to a much larger CV of $\overline{y}_{st}$ compared to Neyman allocation. A compromise allocation that is good for estimating both strata means and the aggregate population mean is needed. Costa, Satorra, and Ventura (2004) proposed a compromise allocation given by

$$n_{(h)}^{C} = d(nW_{(h)}) + (1 - d)(n/L), \tag{2.7.2}$$

for a specified constant d ($0 \le d \le 1$), assuming $n/L \le N_{(h)}$ for all h. This allocation reduces to equal allocation when $d = 0$ and to proportional allocation when $d = 1$.

Longford (2006) attempted to simultaneously control the reliability of the strata means $\overline{y}_{(h)}$ and the weighted mean $\overline{y}_{st}$ by minimizing the objective function

$$\phi(n_{(1)}, \dots, n_{(L)}) = \sum_{h=1}^{L} P_{(h)}V_p(\overline{y}_{(h)}) + G\,P_{+}V_p(\overline{y}_{st}) \tag{2.7.3}$$

with respect to the strata sample sizes $n_{(h)}$ subject to $\sum_{h=1}^{L} n_{(h)} = n$, where $P_{+} = \sum_{h=1}^{L} P_{(h)}$. The first component of ϕ specifies relative strata importances $P_{(h)}$ while the second component attaches relative importance to $\overline{y}_{st}$ through the weight G; the term P_{+} offsets the effect of the strata importances $P_{(h)}$ and the number of strata, L, on the weight G. Longford (2006) proposed to consider $P_{(h)} = N_{(h)}^{q}$ for some q ($0 \le q \le 2$). Longford's allocation, obtained by minimizing (2.7.3), is given by

$$n_{(h)} = n\frac{S_{y(h)}\sqrt{P_{(h)}^{*}}}{\sum\limits_{h=1}^{L} S_{y(h)}\sqrt{P_{(h)}^{*}}}, \tag{2.7.4}$$

where $P^*_{(h)} = P_{(h)} + G\,P_+ W^2_{(h)}$. If $P_{(h)} = N^q_{(h)}$ and $q = 2$, then it reduces to the Neyman allocation $\{n^N_{(h)}\}$ given by (2.7.1).

None of the above compromise allocations are optimal for estimating both strata means and the population mean. Choudhry, Rao, and Hidiroglou (2012) proposed a nonlinear programming (NLP) method of optimal compromise allocation by minimizing $\sum_{h=1}^{L} n_{(h)}$ subject to $\text{CV}_p(\bar{y}_{(h)}) \le \text{CV}_{0(h)}, \text{CV}_p(\bar{y}_{st}) \le \text{CV}_0$, and $0 \le n_{(h)} \le N_{(h)}, h = 1, \ldots, L$, where $\text{CV}_{0(h)}$ and CV_0 are specified tolerances on the CVs of $\bar{y}_{(h)}$ and $\bar{y}_{st}$, respectively. It is easy to see that the optimization problem reduces to minimizing the separable convex function

$$\psi(k_{(1)}, \ldots, k_{(L)}) = \sum_{h=1}^{L} N_{(h)} k^{-1}_{(h)} \tag{2.7.5}$$

with respect to the variables $k_{(h)} = N_{(h)}/n_{(h)}$ subject to the following linear constraints in $k_{(h)}$,

$$\text{RV}(\bar{y}_{(h)}) = \frac{k_{(h)} - 1}{N_{(h)}} C^2_{y(h)} \le \text{RV}_{0(h)}, \quad h = 1, \ldots, L, \tag{2.7.6}$$

$$\text{RV}(\bar{y}_{st}) = \overline{Y} \sum_{h=1}^{L} W^2_{(h)} \frac{k_{(h)} - 1}{N_{(h)}} S^2_{y(h)} \le \text{RV}_0, \tag{2.7.7}$$

$$k_{(h)} \ge 1, \quad h = 1, \ldots, L, \tag{2.7.8}$$

where RV denotes relative variance (or squared CV) and $C_{y(h)} = S_{y(h)}/\overline{Y}_{(h)}$ is the within-stratum CV. The constraint (2.7.8) can be modified to $1 \le k_{(h)} \le N_{(h)}/2$ to ensure that the optimal sample size satisfies $n^0_{(h)} \ge 2$ for all h, which permits unbiased variance estimation. Choudhry et al. (2012) used the SAS procedure NLP with the Newton–Raphson option to determine the optimal $k^0_{(h)}$ (or equivalently $n^0_{(h)} = N_{(h)}/k^0_{(h)}$). The NLP method can be readily extended to handle reliability requirements for multiple variables of interest $y_1, \ldots, y_p (p \ge 2)$.

Example 2.7.1. Monthly Retail Trade Survey (MRTS). Choudhry et al. (2012) studied the relative performance of the three compromise allocation methods (Costa et al., NLP and Longford) using population data from Statistics Canada's Monthly Retail Trade Survey (MRTS) of single establishments. The ten provinces in Canada were treated as strata, and the tolerances for the NLP method were taken as $\text{CV}_{0(h)} =$ 15% for all the provinces h and $\text{CV}_0 = 6\%$ for Canada.

Table 1 in Choudhry et al. (2012) reported the population values $N_{(h)}, \overline{Y}_{(h)}, S_{y(h)}$, and $C_{y(h)}$. Using those values, the optimal NLP allocations $\{n^0_{(h)}\}$ and the associated $\text{CV}(\bar{y}_{(h)})$ and $\text{CV}(\bar{y}_{st})$ were computed and reported in Table 2. This table shows that the NLP allocation respects the specified tolerance $\text{CV}_0 = 6\%$, gives CVs smaller than the specified 15% for two of the larger provinces (Quebec 11.4% and Ontario

11.0%), and attains a 15% CV for the remaining provinces. The resulting optimal overall sample size is $n^0 = \sum_{h=1}^{L} n_{(h)}^0 = 3,446$.

Using the optimal sample size n^0 for n, Choudhry et al. (2012) computed $\text{CV}(\overline{y}_{(h)})$ and $\text{CV}(\overline{y}_{st})$ under the square root allocation, Costa et al. allocation $n_{(h)}^C$, and the Longford allocation $n_{(h)}^L$ with selected q. The study indicated that no suitable combination of q and G in Longford's method can be found that ensures all the specified reliability requirements are satisfied even approximately. On the other hand, the Costa et al. allocation with $d = 0.5$ performed quite well, leading to $\text{CV}(\overline{y}_{st}) = 6.4\%$ and $\text{CV}(\overline{y}_{(h)})$ around 15% except for the provinces of Nova Scotia and Alberta with CVs of 17.0% and 16.5%. Square root allocation performed somewhat similarly.

2.7.2 Case (ii)

We now turn to the case of domains i cutting across design strata h. Suppose that the population is partitioned into domains $U_i (i = 1, \ldots, m)$, and that the estimates of domain means need to satisfy relative variance tolerances $\text{RV}_{0i}, i = 1, \ldots, m$. If $\{n_{(h)}^0\}$ is the current optimal sample allocation that satisfies the specified requirements for the strata, we need to find the optimal additional strata sample sizes $\{\tilde{n}_{(h)} - n_{(h)}^0\}$ that are needed to satisfy also the domain tolerances, where $\tilde{n}_{(h)}$ denotes the revised total sample size in stratum h. For this purpose, we use again the NLP method.

To estimate the domain mean $\overline{Y}_i = N_i^{-1} \sum_{j \in U_i} y_j$, we use the ratio estimator

$$\hat{\overline{Y}}_i = \frac{\sum_{h=1}^{L} N_{(h)} \tilde{n}_{(h)}^{-1} \sum_{j \in s_{(h)}} a_{ij} y_j}{\sum_{h=1}^{L} N_{(h)} \tilde{n}_{(h)}^{-1} \sum_{j \in s_{(h)}} a_{ij}}, \tag{2.7.9}$$

where $a_{ij} = 1$ if $j \in U_i$ and $a_{ij} = 0$ otherwise, and $s_{(h)}$ is the sample from stratum h. The variance of $\hat{\overline{Y}}_i$ is obtained by the Taylor linearization formula for a ratio, and $\text{RV}(\hat{\overline{Y}}_i) = V_p(\hat{\overline{Y}}_i)/\hat{\overline{Y}}_i^2$.

We then obtain the optimal allocation $\{\tilde{n}_{(h)}\}$ by minimizing the total sample size increase

$$\tilde{\phi}(\tilde{f}_{(1)}, \ldots, \tilde{f}_{(L)}) - \sum_{h=1}^{L} n_{(h)}^0 = \sum_{h=1}^{L} (\tilde{f}_{(h)} - f_{(h)}^0) N_{(h)}, \tag{2.7.10}$$

with respect to the variables $\tilde{f}_{(h)} = \tilde{n}_{(h)}/N_{(h)}$, where $f_{(h)}^0 = n_{(h)}^0/N_{(h)}$, subject to

$$\text{RV}(\hat{\overline{Y}}_i) \leq \text{RV}_{0i}, \quad i = 1, \ldots, m, \tag{2.7.11}$$

$$f_{(h)}^0 \leq \tilde{f}_{(h)} \leq 1, \quad h = 1, \ldots, L, \tag{2.7.12}$$

where RV_{0i} is the specified tolerance on the RV of the domain mean $\hat{\overline{Y}}_i$. The constraint (2.7.12) ensures that $n^0_{(h)} \le \tilde{n}_{(h)} \le N_{(h)}$ for all h.

It is easy to see that the above optimization problem reduces to minimizing a separable convex function

$$\psi(\tilde{k}_{(1)}, \ldots, \tilde{k}_{(L)}) = \sum_{h=1}^{L} N_{(h)} \tilde{k}^{-1}_{(h)} \tag{2.7.13}$$

with respect to the variables $\tilde{k}_{(h)} = N_{(h)}/\tilde{n}_{(h)}$ subject to the following linear constraints in $\tilde{k}_{(h)}$:

$$\text{RV}(\hat{\overline{Y}}_i) \approx \overline{Y}_i^{-2} \sum_{h=1}^{L} \left(\frac{N_{(h)}}{N_i}\right)^2 \frac{\tilde{k}_{(h)} - 1}{N_{(h)}} S^2_{e(h),i} \le \text{RV}_{0i}, \quad i = 1, \ldots, m, \tag{2.7.14}$$

$$1 \le \tilde{k}_{(h)} \le k^0_{(h)}, \quad h = 1, \ldots, L. \tag{2.7.15}$$

Here, $S^2_{e(h),i}$ is the stratum variance of the residuals $e_{ij} = a_{ij}(y_j - \overline{Y}_i)$ for $j \in U_i$. Note that the formula for $\text{RV}(\hat{\overline{Y}}_i)$ used in (2.7.14) is based on Taylor linearization. Denoting the resulting optimal $\tilde{k}_{(h)}$ and $\tilde{n}_{(h)}$ by $\tilde{k}^0_{(h)}$ and $\tilde{n}^0_{(h)}$, respectively, the optimal sample size increase in stratum h is $\tilde{n}^0_{(h)} - n^0_{(h)}$.

In practice, domain reliability requirements are often specified after determining $\tilde{n}^0_{(h)}$ and in that case the proposed two-step method is suitable. However, if domain reliability requirements are simultaneously specified, then we can minimize the total sample size $\sum_{h=1}^{L} n_{(h)}$ subject to all the reliability constraints:

$$\text{RV}(\bar{y}_{(h)}) \le \text{RV}_{0(h)}, \quad h = 1, \ldots, L,$$

$$\text{RV}(\hat{\overline{Y}}_i) \le \text{RV}_{0i}, \quad i = 1, \ldots, m,$$

$$\text{RV}(\bar{y}_{st}) \le \text{RV}_0,$$

$$0 < f_{(h)} = n_{(h)}/N_{(h)} \le 1, \quad h = 1, \ldots, L.$$

The resulting optimal solution $\tilde{n}^0_{(h)}$ is identical to the previous two-step solution.

Example 2.7.2. Choudhry et al. (2012) applied the domain allocation method described above to the MRTS population data on single establishments. They took trade groups that cut across provinces as domains of interest. Applying NLP based on (2.7.13)–(2.7.15), they obtained the following results: (i) if domain tolerances are set to $\text{CV}_{0i} = 30\%$ for all i, then the optimal revised sample size is $\tilde{n}^0 = \sum_{h=1}^{L} \tilde{n}^0_{(h)} = n^0$, that is, no increase in n^0 is needed; (ii) if domain tolerances CV_{0i} are reduced to 25%, then the optimal increase in the total sample size equals 622 or the total optimal sample size is $\tilde{n}^0 = 4,068$, noting that $n^0 = 3,446$; (iii) if tolerances CV_{0i} are further reduced to 20%, then $\tilde{n}^0 = 5,546$ satisfies reliability requirements. It should be noted that as the total sample size increases, the CVs of $\overline{y}_{(h)}$ and $\overline{y}_{st}$ decrease.

2.7.3 Two-Way Stratification: Balanced Sampling

In some applications, it may be of interest to consider stratification according to two different partitions of the population U. We denote the two partitions as $U_{(1h)}, h = 1, \ldots, L_1$, and $U_{(2h)}, h = 1, \ldots, L_2$. For example, Falorsi and Righi (2008) considered a population of enterprises in Italy partitioned according to geographical region with $L_1 = 20$ marginal strata and economic activity group by size class with $L_2 = 24$ marginal strata. The parameters of interest are the L $(= L_1 + L_2)$ marginal strata totals or means. The overall sample size, n, is allocated to the two partitions separately, leading to fixed marginal strata sample sizes $n_{(th)}, h = 1, \ldots, L_t$, such that $\sum_{h=1}^{L_t} n_{(th)} = n$ for partition $t = 1, 2$. For example, the Costa et al. (2004) compromise allocation method, mentioned in Section 2.7.1, leads to marginal allocations given by

$$n_{(th)}^C = d_t(nW_{(th)}) + (1 - d_t)(n/L_t), \tag{2.7.16}$$

for specified constants $d_t (0 \leq d_t \leq 1)$, assuming $n/L_t \leq N_{(th)}$, where $N_{(th)}$ is the known size of stratum h in partition t.

Given the allocations, $n_{(th)}$, the next step is to obtain calibrated inclusion probabilities $\pi_j^*, j \in U$, by minimizing a distance function $\sum_{j \in U} G(\pi_j^*, \pi_j)$ subject to $\sum_{j \in U} \pi_j^* = n$ and $\sum_{j \in U_{(th)}} \pi_j^* = n_{(th)}, h = 1, \ldots, L_t - 1, t = 1, 2$, where $\pi_j = n/N$ for all $j \in U$ are preliminary inclusion probabilities. The choice $G(\pi_j^*, \pi_j) = \pi_j^* \log(\pi_j^*/\pi_j) - (\pi_j^* + \pi_j)$ avoids negative values for π_j^*.

The final step uses balanced sampling (Tillé 2006, Chapter 8) to select a sample that leads to realized strata sample sizes $\tilde{n}_{(th)}$ exactly equal to the desired fixed sizes $n_{(th)}$. In general, balanced sampling ensures that the H–T estimator $\sum_{j \in s} \mathbf{z}_j^T / \pi_j^*$ is exactly equal to the total $\sum_{j \in U} \mathbf{z}_j^T$ for specified vectors $\mathbf{z}_j^T$. The "cube method" allows the selection of balanced samples for a large set of auxiliary variables in the vector $\mathbf{z}_j^T$. For the problem at hand, we let

$$\mathbf{z}_j^T = \pi_j^*(a_{(11)j}, \ldots, a_{(1L_1)j}; a_{(21)j}, \ldots, a_{(2L_2)j}), \tag{2.7.17}$$

where $a_{(th)j} = 1$ if unit j belongs to $U_{(th)}$ and $a_{(th)j} = 0$ otherwise. Substituting (2.7.17) into the H–T estimator, we obtain $\sum_{j \in s} \mathbf{z}_j^T / \pi_j^* = (\tilde{n}_{(11)}, \ldots, \tilde{n}_{(1L_1)}; \tilde{n}_{(21)}, \ldots, \tilde{n}_{(2L_2)})$. On the other hand, the corresponding true total for $\mathbf{z}_j^T$ given in (2.7.17) is $\sum_{j \in U} \mathbf{z}_j^T = (n_{(11)}, \ldots, n_{(1L_1)}; n_{(21)}, \ldots, n_{(2L_2)})$ noting that $\sum_{j \in U} \pi_j^* a_{(th)j} = \sum_{j \in U_{(th)}} \pi_j^* = n_{(th)}$. Thus, balanced sampling leads to the specified strata sizes $n_{(th)}$.

A direct expansion estimation of the domain (stratum) total $Y_{(th)}$ is given by $\hat{Y}_{(th)} = \sum_{j \in s} a_{(th)j} y_j / \pi_j^*$. A GREG estimator of $Y_{(th)}$ may be used if an auxiliary vector $\mathbf{x}_j$ with known total $\mathbf{X}$ is available. A variance estimator for $\hat{Y}_{(th)}$ is obtained by viewing balanced sampling as a "conditional Poisson" sampling. Falorsi and Righi (2008) used a conditional Poisson sampling approximation to obtain "optimal" π_j by minimizing $\sum_{j\in U}\pi_j$ subject to specified tolerances on the variances of the GREG estimators, using mathematical programming, similar to the method used in Section 2.7.2.

2.8 PROOFS

2.8.1 Proof of $\hat{Y}_{\text{GR}}(\mathbf{x}) = \mathbf{X}$

Writing $\hat{Y}_{\text{GR}}(\mathbf{x}^T)$ in the expansion form with weights $w_j^* = w_j g_j$ and g_j given by (2.3.9), we have

$$\begin{aligned}\hat{Y}_{\text{GR}}(\mathbf{x}^T) &= \sum_s w_j g_j \mathbf{x}_j^T \\ &= \sum_s w_j \left[\mathbf{x}_j^T + (\mathbf{X}-\hat{\mathbf{X}})^T \left(\sum_s w_j \mathbf{x}_j \mathbf{x}_j^T/c_j\right)^{-1} \mathbf{x}_j \mathbf{x}_j^T/c_j\right] \\ &= \hat{\mathbf{X}}^T + (\mathbf{X}-\hat{\mathbf{X}})^T = \mathbf{X}^T.\end{aligned}$$

2.8.2 Derivation of Calibration Weights w_j^*

We minimize the chi-squared distance $\sum_s c_j(w_j - h_j)^2/w_j$ with respect to the h_j's subject to the calibration constraints $\sum_s h_j \mathbf{x}_j = \mathbf{X}$, that is, we minimize the Lagrangian function $\phi = \sum_s c_j(w_j - h_j)^2/w_j - 2\boldsymbol{\lambda}^T(\sum_s h_j \mathbf{x}_j - \mathbf{X})$, where $\boldsymbol{\lambda}$ is the vector of Lagrange multipliers. Taking derivatives of ϕ with respect to h_j and $\boldsymbol{\lambda}$ and equating to zero, we get $h_j = w_j(1 + \mathbf{x}_j^T\boldsymbol{\lambda}/c_j)$, where

$$\boldsymbol{\lambda} = \sum_s (w_j \mathbf{x}_j \mathbf{x}_j^T/c_j)^{-1}(\mathbf{X}-\hat{\mathbf{X}}).$$

Thus, $h_j = w_j^*$, where $w_j^* = w_j g_j$ and g_j is given by (2.3.9).

2.8.3 Proof of $\hat{Y} = \hat{\mathbf{X}}^T\hat{\mathbf{B}}$ when $c_j = \mathbf{v}^T\mathbf{x}_j$

Since $\hat{\mathbf{X}}^T = \sum_s w_j \mathbf{x}_j^T$, multiplying and dividing by $c_j = \mathbf{v}^T\mathbf{x}_j$ within the previous sum and using the definition of $\hat{\mathbf{B}}$ in (2.3.7), we get

$$\begin{aligned}\hat{\mathbf{X}}^T\hat{\mathbf{B}} &= \left(\sum_s w_j \mathbf{x}_j^T\right)\hat{\mathbf{B}} \\ &= \mathbf{v}^T\left(\sum_s w_j \mathbf{x}_j \mathbf{x}_j^T/c_j\right)\hat{\mathbf{B}}\end{aligned}$$

$$
\begin{aligned}
&= \mathbf{v}^T \left(\sum_s w_j \mathbf{x}_j y_j / c_j \right) \\
&= \sum_s w_j (\mathbf{v}^T \mathbf{x}_j) y_j / c_j \\
&= \sum_s w_j y_j = \hat{Y}.
\end{aligned}
$$

This establishes the result $\sum_s w_j e_j(s) = 0$ noted below (2.3.11).

3

INDIRECT DOMAIN ESTIMATION

3.1 INTRODUCTION

In Chapter 2, we studied direct estimators for domains with sufficiently large sample sizes. We also introduced a "survey regression" estimator of a domain total that borrows strength for estimating the regression coefficient, but it is essentially a direct estimator. In the context of small area estimation, direct estimators lead to unacceptably large standard errors for areas with unduly small sample sizes; in fact, no sample units may be selected from some small domains. This makes it necessary to find indirect estimators that increase the "effective" sample size and thus decrease the standard error. In this chapter, we study indirect domain estimators based on implicit models that provide a link to related small areas. Such estimators include synthetic estimators (Section 3.2), composite estimators (Section 3.3), and James–Stein (JS) (or shrinkage) estimators (Section 3.4); JS estimators have attracted much attention in mainstream statistics. We study their statistical properties in the design-based framework outlined in Chapter 2. Explicit small area models that account for local variation will be presented in Chapter 4, and model-based indirect estimators and associated inferences will be studied in subsequent chapters. Indirect estimators studied in Chapter 3 and later chapters are largely based on sample survey data in conjunction with auxiliary population data, such as a census or an administrative register.

Small Area Estimation, Second Edition. J. N. K. Rao and Isabel Molina.

3.2 SYNTHETIC ESTIMATION

An estimator is called a synthetic estimator if a reliable direct estimator for a large area, covering several small areas, is used to derive an indirect estimator for a small area under the assumption that the small areas have the same characteristics as the large area (Gonzalez 1973). Hansen, Hurwitz, and Madow (1953, pp. 483–486) described the first application of a synthetic regression estimator in the context of a radio listening survey (see Section 3.2.2). The National Center for Health Statistics (1968) in the United States pioneered the use of synthetic estimation for developing state estimates of disability and other health characteristics from the National Health Interview Survey (NHIS). Sample sizes in many states were too small to provide reliable direct state estimates.

3.2.1 No Auxiliary Information

Suppose that auxiliary population information is not available and that we are interested in estimating the small area mean $\overline{Y}_i$, for example, the proportion P_i of persons in poverty in small area i ($\overline{Y}_i = P_i$). In this case, a synthetic estimator of $\overline{Y}_i$ is given by

$$\hat{\overline{Y}}_{iS} = \hat{\overline{Y}} = \frac{\hat{Y}}{\hat{N}}, \tag{3.2.1}$$

where $\hat{\overline{Y}}$ is the direct estimator of the overall population mean $\overline{Y}$, $\hat{Y} = \sum_s w_j y_j$, and $\hat{N} = \sum_s w_j$. The p-bias of $\hat{\overline{Y}}_{iS}$ is approximately equal to $\overline{Y} - \overline{Y}_i$, which is small relative to $\overline{Y}_i$ if $\overline{Y}_i \approx \overline{Y}$. If the latter implicit model that the small area mean is approximately equal to the overall mean is satisfied, then the synthetic estimator will be very efficient because its mean squared error (MSE) will be small. On the other hand, it can be heavily biased for areas exhibiting strong individual effects which, in turn, can lead to large MSE. The condition $\overline{Y}_i \approx \overline{Y}$ may be relaxed to $\overline{Y}_i \approx \overline{Y}(r)$ where $\overline{Y}(r)$ is the mean of a larger area (region) covering the small area. In this case, we use $\hat{\overline{Y}}_{iS} = \hat{\overline{Y}}(r)$ where $\hat{\overline{Y}}(r)$ is the direct estimator of the regional mean $\overline{Y}(r)$. The p-bias of $\hat{\overline{Y}}(r)$ is approximately equal to $\overline{Y}(r) - \overline{Y}_i$, which is negligible relative to $\overline{Y}_i$ under the weaker condition $\overline{Y}_i \approx \overline{Y}(r)$, and the MSE of $\hat{\overline{Y}}(r)$ will be small provided that the regional sample size is large.

3.2.2 *Area Level Auxiliary Information

Suppose survey estimates $\hat{Y}_i$ of area totals Y_i and related area level auxiliary variables $x_{i1}, \ldots, x_{ip}$ are available for m out of M local areas ($i = 1, \ldots, m$). We can then fit by least squares a linear regression to the data $(\hat{Y}_i, x_{i1}, \ldots, x_{ip})$ from the m sampled areas. Resulting estimators $\hat{\beta}_0, \hat{\beta}_1, \ldots, \hat{\beta}_p$ of the regression coefficients lead to the regression-synthetic predictors (estimators) for all the M areas given by

$$\tilde{Y}_i = \hat{\beta}_0 + \hat{\beta}_1 x_{i1} + \cdots + \hat{\beta}_p x_{ip}, \quad i = 1, \ldots, M. \tag{3.2.2}$$

The estimators (3.2.2) can be heavily biased if the underlying model assumptions are not valid.

Example 3.2.1. Radio Listening. Hansen, Hurwitz, and Madow (1953, pp. 483–486) applied (3.2.2) to estimate the median number of stations heard during the day in each of the more than 500 county areas in the United States. In this application, the direct estimate, y_i, of the true median y_{0i}, obtained from a radio listening survey based on personal interviews, plays the role of $\hat{Y}_i$. The estimate x_i of the true median y_{0i}, obtained from a mail survey, was used as a single covariate in the linear regression of y_i on x_i. The mail survey was first conducted by sampling 1,000 families from each county area and mailing questionnaires. Incomplete list frames were used for this purpose and the response rate was about 20%. The estimates, x_i, are biased due to nonresponse and incomplete coverage but are expected to have high correlation with the true median values, y_{0i}. Direct estimates y_i for a sample of 85 county areas were obtained through an intensive interview survey. The sample county areas were selected by first grouping the population county areas into 85 primary strata on the basis of geographical region and type of radio service available, and then selecting one county area from each stratum with probability proportional to the estimated number of families in the county. A subsample of area segments was selected from each of the sampled county areas, and the families within the selected area segments were interviewed. The correlation between y_i and x_i was 0.70.

Regression-synthetic estimates were calculated for the nonsampled counties as $\tilde{y}_i = 0.52 + 0.74x_i$, using the estimates x_i obtained from the mail survey. Ericksen (1974) used the fitted regression equation (3.2.2) for both nonsampled and sampled counties.

3.2.3 *Unit Level Auxiliary Information

Suppose that unit level sample data $\{y_j, \mathbf{x}_j; j \in s\}$ and domain-specific auxiliary information are available in the form of known totals $\mathbf{X}_i$. Then, we can estimate the domain total Y_i using the GREG-synthetic estimator

$$\hat{Y}_{i\text{GRS}} = \mathbf{X}_i^T \hat{\mathbf{B}}, \tag{3.2.3}$$

where $\hat{\mathbf{B}}$ is given by (2.3.7). We can express (3.2.3) as

$$\hat{Y}_{i\text{GRS}} = \sum_{j \in s} \tilde{w}_{ij} y_j, \tag{3.2.4}$$

where

$$\tilde{w}_{ij} = \mathbf{X}_i^T \left(\sum_{j \in s} w_j \mathbf{x}_j \mathbf{x}_j^T / c_j \right)^{-1} w_j \mathbf{x}_j / c_j \tag{3.2.5}$$

and $c_j = \mathbf{v}^T\mathbf{x}_j$ for a vector of constants $\mathbf{v}$. The above form (3.2.4) shows that the same weight $\tilde{w}_{ij}$ is attached to all variables of interest treated as y. Furthermore, we have the weight-sharing (WS) property

$$\sum_{i=1}^{m} \tilde{w}_{ij} = \mathbf{X}^T \left(\sum_{j\in s} w_j \mathbf{x}_j \mathbf{x}_j^T / c_j \right)^{-1} w_j \mathbf{x}_j / c_j = \tilde{w}_j$$

and the calibration property $\sum_{j\in s} \tilde{w}_{ij}\mathbf{x}_j^T = \mathbf{X}_i^T$. The WS property implies that the GREG-synthetic estimators (3.2.3) add up to the projection-GREG estimator of the population total $\hat{Y}_{\text{GR}} = \sum_{j\in s} \tilde{w}_j y_j$, obtained as a special case of the GREG estimator with $c_j = \mathbf{v}^T\mathbf{x}_j$. Note that the projection-GREG estimator at a large area level is considered to be reliable.

The p-bias of $\hat{Y}_{i\text{GRS}}$ is approximately equal to $\mathbf{X}_i^T\mathbf{B} - Y_i$, where $\mathbf{B} = (\sum_U \mathbf{x}_j\mathbf{x}_j^T/c_j)^{-1}\sum_U \mathbf{x}_j y_j/c_j$ is the population regression coefficient. This p-bias will be small relative to Y_i if the domain-specific regression coefficient $\mathbf{B}_i = (\sum_{U_i} \mathbf{x}_j\mathbf{x}_j^T/c_j)^{-1}\sum_{U_i} \mathbf{x}_j y_j/c_j$ is close to $\mathbf{B}$ and $Y_i = \mathbf{X}_i^T\mathbf{B}_i$. The last condition $Y_i = \mathbf{X}_i^T\mathbf{B}_i$ is satisfied if $c_j = \mathbf{v}^T\mathbf{x}_j$. Thus, the GREG-synthetic estimator will be very efficient when the small area i does not exhibit strong individual effect with respect to the regression coefficient.

A special case of (3.2.3) is the ratio-synthetic estimator in the case of a single auxiliary variable x. It is obtained by letting $c_j = x_j$ in (3.2.3) and it is given by

$$\hat{Y}_{i\text{RS}} = X_i \frac{\hat{Y}}{\hat{X}}. \tag{3.2.6}$$

The p-bias of $\hat{Y}_{i\text{RS}}$ relative to Y_i is approximately equal to $X_i(R - R_i)/Y_i$, which will be small if the area-specific ratio $R_i = Y_i/X_i$ is close to the overall ratio $R = Y/X$. The ratio-synthetic estimators (3.2.6) add up to the direct ratio estimator $\hat{Y}_{\text{R}} = (\hat{Y}/\hat{X})X$.

We now consider poststratified synthetic estimators when the cell counts N_{ig} are known for poststrata $g = 1, \ldots, G$. In this case, a count-synthetic poststratified estimator is obtained as a special case of the GREG-synthetic estimator (3.2.3) by letting $\mathbf{x}_j = (x_{1j}, \ldots, x_{Gj})^T$ with $x_{gj} = 1$ if $j \in U_g$ and $x_{gj} = 0$ otherwise. It is given by

$$\hat{Y}_{i\text{S/C}} = \sum_{g=1}^{G} N_{ig} \frac{\hat{Y}_{\cdot g}}{\hat{N}_{\cdot g}}, \tag{3.2.7}$$

where $\hat{Y}_{\cdot g}$ and $\hat{N}_{\cdot g}$ are estimators of the poststratum total $Y_{\cdot g}$ and size $N_{\cdot g}$ (National Center for Health Statistics 1968). In the special case of a binary variable $y_j \in \{0, 1\}$, a count-synthetic estimator of the domain proportion P_i is obtained from (3.2.7) as

$$\hat{P}_{i\text{S/C}} = \left(\sum_{g=1}^{G} N_{ig} \right)^{-1} \left(\sum_{g=1}^{G} N_{ig} \hat{P}_{\cdot g} \right), \tag{3.2.8}$$

where $\hat{P}_{\cdot g}$ is the direct estimator of the gth poststratum proportion, $P_{\cdot g}$.

More generally, a ratio-synthetic poststratified estimator is obtained if the cell totals X_{ig} of an auxiliary variable x are known. By setting $x_{gj} = x_j$ if $j \in U_g$ and $x_{gj} = 0$ otherwise in (3.2.3), we get the ratio-synthetic poststratified estimator

$$\hat{Y}_{iS/R} = \sum_{g=1}^{G} X_{ig} \frac{\hat{Y}_{\cdot g}}{\hat{X}_{\cdot g}}, \tag{3.2.9}$$

where $\hat{X}_{\cdot g} = \sum_{s_{\cdot g}} w_j x_j$ (Ghangurde and Singh 1977). The p-bias of $\hat{Y}_{iS/R}$ is approximately equal to $\sum_{g=1}^{G} X_{ig}(Y_{\cdot g}/X_{\cdot g} - Y_{ig}/X_{ig}) = \sum_{g=1}^{G} X_{ig}(R_{\cdot g} - R_{ig})$. Thus, the p-bias relative to Y_i will be small if the area-specific ratio R_{ig} is close to the poststratum ratio $R_{\cdot g}$ for each g. In the special case of counts, the latter implicit model is equivalent to saying that the small area mean $\overline{Y}_{ig}$ is close to the poststratum mean $\overline{Y}_{\cdot g}$ for each g. If poststrata can be formed to satisfy this model, then the count-synthetic estimator will be very efficient, provided the direct poststrata estimators $\hat{Y}_{\cdot g}/\hat{N}_{\cdot g}$ are reliable. Furthermore, note that the estimators $\hat{Y}_{iS/C}$ and $\hat{Y}_{iS/R}$ add up to the direct poststratified estimators $\sum_{g=1}^{G}(N_{\cdot g}/\hat{N}_{\cdot g})\hat{Y}_{\cdot g}$ and $\sum_{g=1}^{G}(X_{\cdot g}/\hat{X}_{\cdot g})\hat{Y}_{\cdot g}$ of Y, respectively.

In the poststratification context, changing $\hat{N}_{\cdot g}$ to $N_{\cdot g}$ in (3.2.7) and $\hat{X}_{\cdot g}$ to $X_{\cdot g}$ in (3.2.9), we obtain the following alternative synthetic estimators:

$$\tilde{Y}_{iS/C} = \sum_{g=1}^{G} N_{ig} \frac{\hat{Y}_{\cdot g}}{N_{\cdot g}} \tag{3.2.10}$$

and

$$\tilde{Y}_{iS/R} = \sum_{g=1}^{G} X_{ig} \frac{\hat{Y}_{\cdot g}}{X_{\cdot g}}; \tag{3.2.11}$$

see Purcell and Linacre (1976) and Singh and Tessier (1976). In large samples, the alternative synthetic estimators (3.2.10) and (3.2.11) are less efficient than the estimators (3.2.7) and (3.2.9), respectively, when $\hat{Y}_{\cdot g}$ and $\hat{X}_{\cdot g}$ (or $\hat{N}_{\cdot g}$) are positively correlated, as in the case of the ratio estimator $\hat{Y}/\hat{X}$ compared to the expansion estimator $\hat{Y}$. The p-bias of the alternative estimator $\tilde{Y}_{iS/R}$ ($\tilde{Y}_{iS/C}$) in large samples remains the same as the p-bias of $\hat{Y}_{iS/R}$ ($\hat{Y}_{iS/C}$), but in moderate samples the p-bias will be smaller because the ratio bias is not present. Moreover, the alternative synthetic estimators add up to the direct estimator $\hat{Y} = \sum_{g=1}^{G} \hat{Y}_{\cdot g} = \sum_s w_j y_j$.

It should be noted that all synthetic estimators given earlier can be used also for nonsampled areas. This is an attractive property of synthetic estimation. The synthetic method can also be used even when sampling is not involved. For example, suppose that Y, but not Y_i, is known from some administrative source, and that X_i and X are also known. Then a synthetic estimator of Y_i may be taken as $(X_i/X)Y$, whose bias relative to Y_i will be small when $R_i \approx R$. Note that $(X_i/X)Y$ is not an estimator in the usual sense of a random quantity.

Example 3.2.2. Health Variables. The count-synthetic estimator (3.2.7) has been used to produce state estimates of proportions for certain health variables from the 1980 U.S. National Natality Survey (NNS). This survey was based on a probability sample of 9,941 live births with a fourfold oversampling of low-birth-weight infants. Information was collected from birth certificates and questionnaires sent to married mothers and hospitals. $G = 25$ poststrata were formed according to mother's race (white, all others), mother's age group (6 groups), and live birth order (1, 1–2, 1–3, 2, 2+, 3, 3+, 4+).

In this application (Gonzalez, Placek, and Scott 1996), a state is a small area. To illustrate the calculation of the count-synthetic estimate (3.2.7), suppose i denotes Pennsylvania and y_j is a binary variable taking the value 1 if the jth live birth is jaundiced and 0 otherwise. Direct estimates of percent jaundiced, $\hat{P}_{\cdot g}$, were obtained from the NNS for each of the 25 poststrata. The number of hospital births in each cell, N_{ig}, were obtained from State Vital Registration data. Multiplying N_{ig} by $\hat{P}_{\cdot g}$ and summing over the poststrata g, we get the numerator of (3.2.8) as 33,806. Now dividing 33,806 by the total hospital births $\sum_{g=1}^{G} N_{ig} = 156{,}799$ in Pennsylvania, the count-synthetic estimate of percent jaundiced live births in Pennsylvania is given by $(33{,}806/156{,}799) \times 100 = 21.6\%$.

External Evaluation: Gonzalez, Placek, and Scott (1996) also evaluated the accuracy of NNS synthetic estimates by comparing the estimates with "true" state values, P_i, of selected health variables: percent low birth weight, percent late or no prenatal care, and percent low 1-minute "Apgar" scores. Five states covering a wide range of annual number of births (15,000–160,000) were selected for this purpose. True values P_i were ascertained from the State Vital Registration System. Direct state estimates were also calculated from the NNS data. Standard errors (SE) of the direct estimates were estimated using balanced repeated replication method with 20 replicate half-samples; see Rust and Rao (1996) for an overview of replication methods for estimating SE. The MSE of the synthetic estimator $\hat{P}_{iS/C}$ was estimated as $(\hat{P}_{iS/C} - P_i)^2$. This MSE estimator is unbiased but very unstable.

Table 3.1 reports true values, direct estimates, and synthetic estimates of state proportions, as well as the estimated values of relative root mean squared error (RRMSE), where RRMSE= $\sqrt{\text{MSE}}/(\text{true value})$. It is clear from Table 3.1 that the synthetic estimates performed better than the direct estimates in terms of estimated RRMSE, especially for states with small numbers of sample cases (e.g., Montana). The values of estimated RRMSE ranged from 0.14 (Pennsylvania) to 0.62 (Montana) for the direct estimates, whereas those for the synthetic estimates ranged from 0.000 (Pennsylvania) to 0.32 (Kansas). The National Center for Health Statistics (NCHS) used a maximum estimated RRMSE of 25% as the standard for reliability of estimates, and most of the synthetic estimates met this criterion for reliability, unlike the direct estimates. But this conclusion should be interpreted with caution due to the instability of the MSE estimator of $\hat{P}_{iS/C}$ that was used.

Example 3.2.3. *Labor Force Counts in Lombardy, Italy. Bartoloni (2008) studied the performance of the direct poststratified count estimator (2.4.16) and the alternative count-synthetic poststratified estimator (3.2.10) for the industrial districts in

TABLE 3.1 True State Proportions, Direct and Synthetic Estimates, and Associated Estimates of RRMSE

	True	Direct Estimate		Synthetic Estimate	
Variable/State	(%)	Estimates(%)	RRMSE(%)	Estimates(%)	RRMSE(%)
Low birth:					
Pennsylvania	6.5	6.6	15	6.5	0
Indiana	6.3	6.8	22	6.5	3
Tennessee	8.0	8.5	23	7.2	10
Kansas	5.8	6.8	36	6.4	10
Montana	5.6	9.2	71	6.3	13
Prenatal care:					
Pennsylvania	3.9	4.3	21	4.3	10
Indiana	3.8	2.0	21	4.7	24
Tennessee	5.4	4.7	26	5.0	7
Kansas	3.4	2.1	35	4.5	32
Montana	3.7	3.0	62	4.3	16
Apgar score:					
Pennsylvania	7.9	7.7	14	9.4	19
Indiana	10.9	9.5	16	9.4	14
Tennessee	9.6	7.3	18	9.7	1
Kansas	11.1	12.3	25	9.4	15
Montana	11.6	12.9	40	9.4	19

Source: Adapted from Tables 4 and 5 in Gonzalez et al. (1996).

Lombardy (Italy). Those districts are underrepresented in the Italian Labor Force Survey (LFS), thus resulting in small sample sizes. Bartoloni (2008) conducted a simulation study using Lombardy's 1991 census data. Repeated samples were selected from this census applying the LFS design used in Lombardy. Poststratification was done based on age–sex grouping. Results of the simulation study indicated superior performance of the alternative count-synthetic poststratified estimator in terms of RRMSE. This is largely due to the low values of absolute relative bias (ARB) averaged over the areas.

Example 3.2.4. County Crop Production. Stasny, Goel, and Rumsey (1991) used a regression-synthetic estimator to produce county estimates of wheat production in the state of Kansas. County estimates of farm production are often used in local decision making and by companies selling fertilizers, pesticides, crop insurance, and farm equipment. Stasny et al. (1991) used a nonprobability sample of farms, assuming a linear regression model relating wheat production of the jth farm in the ith county, y_{ij}, to a vector of predictors $\mathbf{x}_{ij} = (1, x_{ij1}, \ldots, x_{ijp})^T$. The predictor variables $x_{ijk}(k = 1, \ldots, p)$ selected in the model have known county totals X_{ik} and include a measure of farm size, which might account for the fact that the sample is not a probability sample.

The regression-synthetic estimator of the ith county total Y_i is obtained as $\tilde{Y}_{iS} = \sum_{j=1}^{N_i} \hat{y}_{ij}$, where $\hat{y}_{ij} = \hat{\beta}_0 + \hat{\beta}_1 x_{ij1} + \cdots + \hat{\beta}_p x_{ijp}$ is the least squares predictor of y_{ij}, $j = 1, \ldots, N_i$, and N_i is the total number of farms in the ith county. The least squares estimators, $\hat{\beta}_k$, are obtained from the linear regression model $y_{ij} = \beta_0 + \beta_1 x_{ij1} + \cdots + \beta_p x_{ijp} + \epsilon_{ij}$ with independent and identically distributed (iid) errors ϵ_{ij}, using the sample data $\{(y_{ij}, \mathbf{x}_{ij}); j = 1, \ldots, n_i; i = 1, \ldots, m\}$, where n_i is the number of sample farms from the ith county. The estimator $\tilde{Y}_{iS}$ reduces to

$$\tilde{Y}_{iS} = N_i \hat{\beta}_0 + X_{i1} \hat{\beta}_1 + \cdots + X_{ip} \hat{\beta}_p,$$

which requires only the known county totals $X_{ik}, k = 1, \ldots, p$. It is not necessary to know the individual values $\mathbf{x}_{ij}$ for the nonsampled farms in the ith county.

The synthetic estimates $\tilde{Y}_{iS}$ do not add up to the direct state estimate $\hat{Y}$ of wheat production obtained from a large probability sample. The state estimate $\hat{Y}$ is regarded as more accurate than the total $\sum_{i=1}^{m} \tilde{Y}_{iS}$. A simple ratio benchmarking of the form

$$\tilde{Y}_{iS}(a) = \frac{\tilde{Y}_{iS}}{\sum_{i=1}^{m} \tilde{Y}_{iS}} \hat{Y}$$

was therefore used to ensure that the adjusted estimates $\tilde{Y}_{iS}(a)$ add up to the reliable direct estimate $\hat{Y}$.

The predictor variables x_{ijk} chosen for this application consist of acres planted in wheat and district indicators. A more complex model involving the interaction between acres planted and district indicators was also studied, but the two models gave similar fits.

If the sampling fractions $f_i = n_i / N_i$ are not negligible, a more efficient estimator of Y_i is given by

$$\hat{Y}_{iS}^{*} = \sum_{j \in s_i} y_{ij} + \sum_{j \in r_i} \hat{y}_{ij},$$

where r_i is the set of nonsampled units from area i (Holt, Smith, and Tomberlin 1979). The estimator $\hat{Y}_{iS}^{*}$ reduces to

$$\hat{Y}_{iS}^{*} = \sum_{j \in s_i} y_{ij} + (N_i - n_i)\hat{\beta}_0 + X_{i1}^{*} \hat{\beta}_1 + \cdots + X_{ip}^{*} \hat{\beta}_p,$$

where $X_{ik}^{*} = X_{ik} - \sum_{j \in s_i} x_{ijk}$ is the total of x_{ijk} for the nonsampled units r_i. This estimator also requires only the county totals X_{ik}.

3.2.4 Regression-Adjusted Synthetic Estimator

Levy (1971) proposed a regression-adjusted synthetic estimator that attempts to account for local variation by combining area-specific covariates with the synthetic estimator. Covariates $\mathbf{z}_i$ are used to model the relative bias (RB) $B_i = (\overline{Y}_i - \hat{\overline{Y}}_{iS})/\hat{\overline{Y}}_{iS}$ associated with the synthetic estimator $\hat{\overline{Y}}_{iS}$ of the mean $\overline{Y}_i$:

$$B_i = \gamma_0 + \boldsymbol{\gamma}^T \mathbf{z}_i + \varepsilon_i,$$

where γ_0 and $\boldsymbol{\gamma}$ are the regression parameters and ε_i is a random error. Since B_i is not observable, the regression model is fitted by least squares to estimated bias values $\hat{B}_a = (\hat{\overline{Y}}_a - \hat{\overline{Y}}_{aS})/\hat{\overline{Y}}_{aS}$ for large areas a using reliable direct estimators $\hat{\overline{Y}}_a$ and synthetic estimators $\hat{\overline{Y}}_{aS}, a = 1, \ldots, A$. Denoting the resulting least squares estimates as $\hat{\gamma}_0$ and $\hat{\boldsymbol{\gamma}}, B_i$ is estimated as $\hat{\gamma}_0 + \hat{\boldsymbol{\gamma}}^T \mathbf{z}_i$, assuming that the above area level model holds for the large areas. This in turn leads to a regression-adjusted synthetic estimator of $\overline{Y}_i$ given by

$$\hat{\overline{Y}}_{iS}(a) = \hat{\overline{Y}}_{iS}(1 + \hat{\gamma}_0 + \hat{\boldsymbol{\gamma}}^T \mathbf{z}_i). \tag{3.2.12}$$

Levy (1971) obtained state level regression-adjusted synthetic estimates by fitting the model to the estimated bias values $\hat{B}_i$ at the regional level a.

3.2.5 Estimation of MSE

The p-variance of a synthetic estimator $\hat{Y}_{iS}$ will be small relative to the p-variance of a direct estimator $\hat{Y}_i$ because it depends only on the precision of direct estimators at a large area level. The p-variance is readily estimated using standard design-based methods, but it is more difficult to estimate the MSE of $\hat{Y}_{iS}$. For example, the p-variance of the ratio-synthetic estimator (3.2.9) or of the count-synthetic estimator (3.2.7) can be estimated using Taylor linearization. Similarly, the p-variance of the GREG-synthetic estimator (3.2.3) can be estimated using the results of Fuller (1975) on the large sample covariance matrix of $\hat{\mathbf{B}}$ or by using a resampling method such as the jackknife. We refer the readers to Wolter (2007), Shao and Tu (1995, Chapter 6), and Rust and Rao (1996) for a detailed account of resampling methods for sample surveys.

An approximately p-unbiased estimator of MSE of $\hat{Y}_{iS}$ can be obtained using a p-unbiased direct estimator $\hat{Y}_i$ for sampled area i. We have

$$\begin{aligned}
\text{MSE}_p(\hat{Y}_{iS}) &= E_p(\hat{Y}_{iS} - Y_i)^2 \\
&= E_p(\hat{Y}_{iS} - \hat{Y}_i + \hat{Y}_i - Y_i)^2 \\
&= E_p(\hat{Y}_{iS} - \hat{Y}_i)^2 - V_p(\hat{Y}_i) + 2\,\text{Cov}_p(\hat{Y}_{iS}, \hat{Y}_i) \\
&= E_p(\hat{Y}_{iS} - \hat{Y}_i)^2 - V_p(\hat{Y}_{iS} - \hat{Y}_i) + V_p(\hat{Y}_{iS}).
\end{aligned} \tag{3.2.13}$$

It now follows from (3.2.13) that an approximately p-unbiased estimator of $\text{MSE}_p(\hat{Y}_{iS})$ is

$$\text{mse}(\hat{Y}_{iS}) = (\hat{Y}_{iS} - \hat{Y}_i)^2 - v(\hat{Y}_{iS} - \hat{Y}_i) + v(\hat{Y}_{iS}), \tag{3.2.14}$$

where $v(\cdot)$ is a p-based estimator of $V_p(\cdot)$; for example, a jackknife estimator. The estimator (3.2.14), however, can be very unstable and it can take negative values. Consequently, it is customary to average the MSE estimator over small areas $i(= 1,\ldots,m)$ belonging to a large area to get a stable estimator (Gonzalez and Waksberg 1973). Let $\hat{\bar{Y}}_{iS} = \hat{Y}_{iS}/N_i$ be the estimated area mean, so that $\text{mse}(\hat{\bar{Y}}_{iS}) = \text{mse}(\hat{Y}_{iS})/N_i^2$. We take $m^{-1}\sum_{\ell=1}^{m}\text{mse}(\hat{\bar{Y}}_{\ell S})$ as an estimator of $\text{MSE}(\hat{\bar{Y}}_{iS})$ so that we get $\text{mse}_a(\hat{Y}_{iS}) = N_i^2\text{mse}_a(\hat{\bar{Y}}_{iS})$ as an estimator of $\text{MSE}(\hat{Y}_{iS})$, where

$$\text{mse}_a(\hat{\bar{Y}}_{iS}) = \frac{1}{m}\sum_{\ell=1}^{m}\frac{1}{N_\ell^2}(\hat{Y}_{\ell S} - \hat{Y}_\ell)^2 - \frac{1}{m}\sum_{\ell=1}^{m}\frac{1}{N_\ell^2}\,v(\hat{Y}_{\ell S} - \hat{Y}_\ell) + \frac{1}{m}\sum_{\ell=1}^{m}\frac{1}{N_\ell^2}\,v(\hat{Y}_{\ell S}). \tag{3.2.15}$$

But such a global measure of uncertainty can be misleading since it refers to the average MSE rather than to the area-specific MSE.

A good approximation to (3.2.14) is given by

$$\text{mse}(\hat{Y}_{iS}) \approx (\hat{Y}_{iS} - \hat{Y}_i)^2 - v(\hat{Y}_i), \tag{3.2.16}$$

noting that the variance of the synthetic estimator $\hat{Y}_{iS}$ is small relative to the variance of the direct estimator $\hat{Y}_i$. Using the approximation (3.2.16) and applying the above averaging idea, we obtain

$$\text{mse}_a(\hat{\bar{Y}}_{iS}) \approx \frac{1}{m}\sum_{\ell=1}^{m}\frac{1}{N_\ell^2}(\hat{Y}_{\ell S} - \hat{Y}_\ell)^2 - \frac{1}{m}\sum_{\ell=1}^{m}\frac{1}{N_\ell^2}\,v(\hat{Y}_\ell). \tag{3.2.17}$$

Marker (1995) proposed a simple method of getting a more stable area-specific estimator of MSE of $\hat{Y}_{iS}$ than (3.2.14) and (3.2.16). It uses the assumption that the squared p-bias $B_p^2(\hat{\bar{Y}}_{iS})$ is approximately equal to the average squared bias:

$$B_p^2(\hat{\bar{Y}}_{iS}) \approx \frac{1}{m}\sum_{\ell=1}^{m}B_p^2(\hat{\bar{Y}}_{\ell S}) =: B_a^2(\hat{\bar{Y}}_{iS}). \tag{3.2.18}$$

The average squared bias (3.2.18) can be estimated as

$$b_a^2(\hat{\bar{Y}}_{iS}) = \text{mse}_a(\hat{\bar{Y}}_{iS}) - \frac{1}{m}\sum_{\ell=1}^{m}v(\hat{\bar{Y}}_{\ell S}), \tag{3.2.19}$$

noting that average MSE = average variance + average (bias)^2. The variance estimator $v(\hat{\bar{Y}}_{\ell S}) = v(\hat{Y}_{\ell S})/N_\ell^2$ is readily obtained using traditional methods, as noted earlier. It now follows under the assumption (3.2.18) that $\text{MSE}_p(\hat{Y}_{iS})$ can be estimated as

$$\text{mse}_{\text{M}}(\hat{Y}_{iS}) = v(\hat{Y}_{iS}) + N_i^2 b_a^2(\hat{\bar{Y}}_{iS}), \tag{3.2.20}$$

which is area-specific if $v(\hat{Y}_{iS})$ depends on the area. However, the assumption (3.2.18) may not be satisfied for areas exhibiting strong individual effects. Nevertheless, (3.2.20) is an improvement over the global measure (3.2.17), provided the variance term dominates the bias term in (3.2.20). Note that both $\text{mse}_{\text{M}}(\hat{Y}_{iS})$ and $\text{mse}_a(\hat{Y}_{iS})$ require the knowledge of the domain sizes, N_i. Furthermore, averaging the area-specific estimator $\text{mse}_{\text{M}}(\hat{\bar{Y}}_{iS}) = N_i^{-2}\text{mse}_{\text{M}}(\hat{Y}_{iS})$ over $i = 1, \ldots, m$ leads exactly to the average MSE estimator $\text{mse}_a(\hat{\bar{Y}}_{iS})$ given by (3.2.17), as noted by Lahiri and Pramanik (2013). It is important to note that the average MSE estimator (3.2.17), as well as (3.2.15), may take negative values in practice. Lahiri and Pramanik (2013) proposed modifications to $\text{mse}_a(\hat{\bar{Y}}_{iS})$ that always lead to positive average MSE estimates.

To illustrate the calculation of (3.2.20), suppose $\hat{Y}_{iS}$ is the ratio-synthetic estimator $(\hat{Y}/\hat{X})X_i = \hat{R}X_i$ and $\hat{Y}_i$ is the expansion estimator $\sum_{s_i} w_j y_j$. Then $v(\hat{Y}_i) = v(y_i)$ and $v(\hat{Y}_{iS}) \approx (X_i/\hat{X})^2 v(e)$ in the operator notation introduced in Section 2.4, where $e_j = y_j - \hat{R}x_j, j \in s$. Using these variance estimators we can compute $\text{mse}_{\text{M}}(\hat{Y}_{iS})$ from (3.2.17), (3.2.19), and (3.2.20). Note that $v(y_i)$ is obtained from $v(y)$ by changing y_j to y_{ij}, and $v(e)$ is obtained by changing y_j to e_j.

3.2.6 Structure Preserving Estimation

Structure preserving estimation (SPREE) is a generalization of synthetic estimation in the sense that it makes a fuller use of reliable direct estimates. The parameter of interest is a count such as the number of employed in a small area. SPREE uses the well-known method of iterative proportional fitting (IPF) (Deming and Stephan 1940) to adjust the cell counts of a multiway table such that the adjusted counts satisfy specified margins. The cell counts are obtained from the last census while the specified margins represent reliable direct survey estimates of current margins. Thus, SPREE provides intercensal estimates of small area totals of characteristics that are also measured in the census.

We illustrate SPREE in the context of a three-way table of census counts $\{N_{iab}\}$, where $i(= 1, \ldots, m)$ denotes the small areas, $a(= 1, \ldots, A)$ the categories of the variable of interest y (e.g., employed/unemployed) and $b(= 1, \ldots, B)$ the categories of some variable closely related to y (e.g., white/nonwhite). The unknown current counts are denoted by $\{M_{iab}\}$, and the parameters of interest are the marginal counts $M_{ia\cdot} = \sum_{b=1}^{B} M_{iab}$. We assume that reliable survey estimates of some of the margins are available. In particular, we consider two cases: 1) Survey estimates $\{\hat{M}_{\cdot ab}\}$ of the margins $\{M_{\cdot ab}\}$ are available, where $M_{\cdot ab} = \sum_{i=1}^{m} M_{iab}$. Note that the margins correspond to a larger area covering the small areas. 2) In addition to $\{\hat{M}_{\cdot ab}\}$, estimates $\{\hat{M}_{i\cdot\cdot}\}$ of the margins $\{M_{i\cdot\cdot}\}$ are also available, where $M_{i\cdot\cdot} = \sum_{a=1}^{A} M_{ia\cdot}$. Such estimates of current small area population counts $M_{i\cdot\cdot}$ may be obtained using demographic methods such as the sample regression method considered in Section 3.3 of Rao (2003a).

SPREE is similar to calibration estimation studied in Section 2.3. We seek values $\{x_{iab}\}$ of M_{iab}, which minimize a distance measure to $\{N_{iab}\}$ subject to the constraints $\sum_{i=1}^{m} x_{iab} = \hat{M}_{\cdot ab}$ in case 1, and $\sum_{i=1}^{m} x_{iab} = \hat{M}_{\cdot ab}$ and $\sum_{a=1}^{A}\sum_{b=1}^{B} x_{iab} = \hat{M}_{i\cdot\cdot}$ in case 2.

Denoting $\tilde{M}_{iab}$ as the resulting values, the estimate of $M_{ia\cdot}$ is then obtained as $\tilde{M}_{ia\cdot} = \sum_{b=1}^{B} \tilde{M}_{iab}$.

Case 1 *One-Step SPREE*

Using a chi-squared distance, we minimize

$$D_{\mathcal{X}} = \sum_{i=1}^{m}\sum_{a=1}^{A}\sum_{b=1}^{B}(N_{iab} - x_{iab})^2/c_{iab} - \sum_{a=1}^{A}\sum_{b=1}^{B}\lambda_{ab}\left(\sum_{i=1}^{m}x_{iab} - \hat{M}_{\cdot ab}\right)$$

with respect to $\{x_{iab}\}$, where c_{iab} are some prespecified weights and λ_{ab} are the Lagrange multipliers. If we choose $c_{iab} = N_{iab}$, then the "optimal" value of x_{iab} is given by

$$\tilde{M}_{iab} = \frac{N_{iab}}{N_{\cdot ab}}\hat{M}_{\cdot ab}.$$

The resulting estimate of $M_{ia\cdot}$ is

$$\tilde{M}_{ia\cdot} = \sum_{b=1}^{B}\frac{N_{iab}}{N_{\cdot ab}}\hat{M}_{\cdot ab}, \qquad (3.2.21)$$

which has the same form as the alternative count-synthetic estimator (3.2.10). Note that the structure preserving estimators $\tilde{M}_{ia\cdot}$ satisfy the additive property when summed over i, that is, $\sum_{i=1}^{m}\tilde{M}_{ia\cdot} = \sum_{b=1}^{B}\hat{M}_{\cdot ab} = \hat{M}_{\cdot a\cdot}$.

The same optimal estimates $\tilde{M}_{iab}$ are obtained minimizing the discrimination information measure

$$D_{\mathcal{I}}(\{N_{iab}\},\{x_{iab}\}) = \sum_{i=1}^{m}\sum_{a=1}^{A}\sum_{b=1}^{B}N_{iab}\log\frac{N_{iab}}{x_{iab}} \qquad (3.2.22)$$

instead of the chi-squared distance.

The estimates $\tilde{M}_{iab}$ preserve the *association structure* in the three-way table of counts $\{N_{iab}\}$ without modifying the marginal *allocation structure* $\{\hat{M}_{\cdot ab}\}$; that is, the interactions as defined by cross-product ratios of the cell counts are preserved. It is easy to verify that the values $\{\tilde{M}_{iab}\}$ preserve census area effects in the sense

$$\frac{\tilde{M}_{iab}}{\tilde{M}_{i'ab}} = \frac{N_{iab}}{N_{i'ab}}, \qquad (3.2.23)$$

for all pairs of areas (i, i'). The values $\tilde{M}_{iab}$ also preserve the two-way interactions of area i and variable of interest a from the census, in the sense that the cross-product ratios remain the same:

$$\frac{\tilde{M}_{iab}\tilde{M}_{i'a'b}}{\tilde{M}_{i'ab}\tilde{M}_{ia'b}} = \frac{N_{iab}N_{i'a'b}}{N_{i'ab}N_{ia'b}} \quad \text{for all } i, i', a, a'.$$

The two-way interactions of area i and associated variable b, and the three-way interactions of area i, variable of interest a and associated variable b, that is, the ratios of cross-product ratios, also remain the same. Because of this structure preserving property, the method is called SPREE. This property is desirable because one would expect the association structure to remain fairly stable over the intercensal period. Under the implicit assumption that the association structures of $\{N_{iab}\}$ and $\{M_{iab}\}$ are identical in the above sense, SPREE gives an exactly p-unbiased estimator of $M_{ia\cdot}$, provided $\hat{M}_{\cdot ab}$ is p-unbiased for $M_{\cdot ab}$. To prove the p-unbiasedness of $\tilde{M}_{ia\cdot}$, first note that

$$E_p(\tilde{M}_{ia\cdot}) = \sum_{b=1}^{B} \frac{N_{iab}}{N_{\cdot ab}} M_{\cdot ab}.$$

Second, applying the condition (3.2.23) to the true counts, M_{iab} gives

$$M_{iab} N_{i'ab} = N_{iab} M_{i'ab},$$

which, when summed over i', leads to $N_{iab}/N_{\cdot ab} = M_{iab}/M_{\cdot ab}$.

Case 2 *Two-Step SPREE*

In case 2 with additional estimated margins $\{M_{i\cdot\cdot}\}$ available, the values that minimize the chi-squared distance do not preserve the association structure, unlike those based on the discrimination information measure (3.2.22). We therefore consider only the latter measure and minimize $D_I(\{N_{iab}\}, \{x_{iab}\})$ subject to the constraints $\sum_{i=1}^{m} x_{iab} = \hat{M}_{\cdot ab}$ for all a, b and $\sum_{a=1}^{A} \sum_{b=1}^{B} x_{iab} = \hat{M}_{i\cdot\cdot}$ for all i. The "optimal" solution $\tilde{M}_{iab}$ cannot be obtained in a closed form, but the well-known method of IPF can be used to obtain $\tilde{M}_{iab}$ iteratively. The IPF procedure involves a sequence of iteration cycles each consisting of two steps. At the kth iteration cycle, the values $\{\tilde{M}_{iab}^{(k-1)}\}$ at the end of $(k-1)$th iteration cycle are first adjusted to the set of constraints, $\sum_{i=1}^{m} x_{iab} = \hat{M}_{\cdot ab}$, as follows:

$${}_1\tilde{M}_{iab}^{(k)} = \frac{\tilde{M}_{iab}^{(k-1)}}{\tilde{M}_{\cdot ab}^{(k-1)}} \hat{M}_{\cdot ab}.$$

The solution to the previous adjustment, ${}_1\tilde{M}_{iab}^{(k)}$, is then adjusted to the second set of constraints $\sum_{a=1}^{A} \sum_{b=1}^{B} x_{iab} = \hat{M}_{i\cdot\cdot}$ as

$$\tilde{M}_{iab}^{(k)} = \frac{{}_1\tilde{M}_{iab}^{(k)}}{{}_1\tilde{M}_{i\cdot\cdot}^{(k)}} \hat{M}_{i\cdot\cdot}.$$

The starting values for the iteration, $\tilde{M}_{iab}^{(0)}$, are set equal to the census counts N_{iab} specifying the initial association structure. If all the counts N_{iab} are strictly positive, then the sequence $\{\tilde{M}_{iab}^{(k)}\}$ converges to the optimal solution $\{\tilde{M}_{iab}\}$ as $k \to \infty$ (Ireland and Kullback 1968). The resulting estimator of $M_{ia\cdot}$ is given by the marginal value

$\tilde{M}_{ia\cdot} = \sum_{b=1}^{B} \tilde{M}_{iab}$. The estimator $\tilde{M}_{ia\cdot}$ will be approximately p-unbiased for $M_{ia\cdot}$ if the population association structure remains stable over the intercensal period. Purcell and Kish (1980) studied SPREE under different types of marginal constraints, including the above-mentioned Cases 1 and 2.

The two-step estimator obtained by doing just one iteration, $\tilde{M}_{ia\cdot}^{(1)}$, is simpler than the SPREE $\tilde{M}_{ia\cdot}$. It also makes effective use of the current marginal information $\{\hat{M}_{\cdot ab}\}$ and $\{\hat{M}_{i\cdot\cdot}\}$, although both marginal constraints may not be exactly satisfied. Rao (1986) derived a Taylor linearization variance estimator of the two-step estimator $\tilde{M}_{ia\cdot}^{(1)}$. Chambers and Feeney (1977) gave an estimator of the asymptotic covariance matrix of the SPREE estimators $\tilde{M}_{ia\cdot}$ in a general form without details needed for computation.

Example 3.2.5. Vital Statistics. Purcell and Kish (1980) made an evaluation of the one-step and the two-step SPREE estimators by comparing the estimates to true counts obtained from the Vital Statistics registration system. In this study, SPREE estimates of mortality due to each of four different causes (a) and for each state (i) in the United States were calculated for five individual years ranging over the postcensal period 1960–1970. Here, the categories b denote 36 age-sex-race groups, $\{N_{iab}\}$ the 1960 census counts and $\{\hat{M}_{\cdot ab} = M_{\cdot ab}\}$, $\{\hat{M}_{i\cdot\cdot} = M_{i\cdot\cdot}\}$ the known current counts.

Table 3.2 reports the medians of the state percent absolute relative errors ARE = |estimate − true value|/true value of the one-step and the two-step SPREE estimates. It is clear from Table 3.2 that the two-step estimator performs significantly better

TABLE 3.2 Medians of Percent ARE of SPREE Estimates

Cause of Death	Year	One-Step	Two-Step
Malignant neoplasms	1961	1.97	1.85
	1964	3.50	2.21
	1967	5.58	3.22
	1970	8.18	2.75
Major CVR diseases	1961	1.47	0.73
	1964	1.98	1.03
	1967	3.47	1.20
	1970	4.72	2.22
Suicides	1961	5.56	6.49
	1964	8.98	8.64
	1967	7.76	6.32
	1970	13.41	8.52
Total others	1961	1.92	1.39
	1964	3.28	2.20
	1967	4.89	3.36
	1970	6.65	3.85

Source: Adapted from Table 3 in Purcell and Kish (1980).

than the one-step estimator (3.2.21) in terms of median ARE. Thus, it is important to incorporate, through the allocation structure, the maximum available current data into SPREE.

3.2.7 *Generalized SPREE

GLSM Approach Zhang and Chambers (2004) proposed a generalized linear structural model (GLSM) for estimating cell counts in two-way or three-way tables with given margins. The GLSM model assumes that the interactions in the table of true counts, measured in terms of ratios of counts, are proportional (but not necessarily equal) to the corresponding interactions in the census table. SPREE is a special case of this model, in which the proportionality constant is equal to one.

For simplicity, we introduce GLSM only for a two-way table. Let $\{M_{ia}\}$ be the true counts and $\{N_{ia}\}$ be the known census counts, for $a = 1, \ldots, A$ and $i = 1, \ldots, m$. Also, let $\theta_{ia}^M = M_{ia}/M_{i\cdot}$ and $\theta_{ia}^N = N_{ia}/N_{i\cdot}$ be, respectively, the true and census proportions of individuals in the category a of the variable of interest within domain i, where $N_{i\cdot} = \sum_{a=1}^A N_{ia}$ and the true margins $M_{i\cdot} = \sum_{a=1}^A M_{ia}$ are assumed to be known. Consider the deviations of these proportions in the log scale from their means,

$$\mu_{ia}^M = \log(\theta_{ia}^M) - \frac{1}{A}\sum_{a'=1}^{A} \log(\theta_{ia'}^M) =: g_a(\theta_i^M),$$

$$\mu_{ia}^N = \log(\theta_{ia}^N) - \frac{1}{A}\sum_{a'=1}^{A} \log(\theta_{ia'}^N) =: g_a(\theta_i^N),$$

where $\boldsymbol{\theta}_i^M = (\theta_{i1}^M, \ldots, \theta_{iA}^M)^T$ and $\boldsymbol{\theta}_i^N = (\theta_{i1}^N, \ldots, \theta_{iA}^N)^T$. The model proposed by Zhang and Chambers (2004) assumes that

$$\mu_{ia}^M = \lambda_a + \beta\mu_{ia}^N, \quad i = 1, \ldots, m, \tag{3.2.24}$$

where the condition $\sum_{a=1}^A \lambda_a = 0$ ensures that the deviations add up to zero, that is, $\sum_{a=1}^A \mu_{ia}^M = 0$ for each i.

Model (3.2.24) has an interpretation in terms of log-linear models. Consider the saturated log-linear model representations for the census and the true counts:

$$\log(M_{ia}) = \alpha_0^M + \alpha_i^M + \alpha_a^M + \alpha_{ia}^M, \tag{3.2.25}$$

$$\log(N_{ia}) = \alpha_0^N + \alpha_i^N + \alpha_a^N + \alpha_{ia}^N, \tag{3.2.26}$$

where α_0^M and α_0^N are the overall means, α_i^M and α_i^N are the marginal effects of the areas, α_a^M and α_a^N are the marginal effects of the categories, and α_{ia}^M and α_{ia}^N are the interaction effects of each crossing area-category. Note that the log-linear models (3.2.25) and (3.2.26) lead to

$$\mu_{ia}^M = \alpha_a^M + \alpha_{ia}^M, \quad \mu_{ia}^N = \alpha_a^N + \alpha_{ia}^N. \tag{3.2.27}$$

Using (3.2.27), model (3.2.24) implies the following:

(a) $\alpha_a^M = \lambda_a + \beta\alpha_a^N$, that is, the marginal effects of the categories a in the above log-linear model for the true counts $\{M_{ia}\}$ are a linear function of the corresponding effects in the model for the census counts $\{N_{ia}\}$.

(b) $\alpha_{ia}^M = \beta\alpha_{ia}^N$, that is, the interaction effects (category by area) in the log-linear model for the true counts $\{M_{ia}\}$ are proportional to those in the model for the census counts $\{N_{ia}\}$.

From (b), it is easy to see that SPREE is obtained by setting $\beta = 1$, that is, when the interaction effects in the log-linear models for true and census counts are exactly equal.

Let us now express model (3.2.24) in matrix notation. For this, let us define the vectors

$$\boldsymbol{\mu}_i^M = (\mu_{i1}^M, \ldots, \mu_{iA}^M)^T, \quad \boldsymbol{\mu}_i^N = (\mu_{i1}^N, \ldots, \mu_{iA}^N)^T, \quad \xi = (\lambda_2, \ldots, \lambda_A, \beta)^T$$

and the matrix

$$\mathbf{Z}_i = \left(\begin{array}{c|c} -1'_{A-1} & \\ I_{(A-1)\times(A-1)} & \boldsymbol{\mu}_i^N \end{array} \right),$$

where $\mathbf{1}_r$ denotes a column vector of ones of size r. Then the model (3.2.24) may be represented as a multivariate generalized linear model in the form

$$\mathbf{g}(\boldsymbol{\theta}_i^M) = \mathbf{Z}_i\xi, \quad i = 1, \ldots, m,$$

where $\mathbf{g}(\boldsymbol{\theta}_i^M)$ is a column vector with elements $g_a(\boldsymbol{\theta}_i^M), a = 1, \ldots, A$.

Let us denote $\eta_i = \mathbf{Z}_i\xi = (\eta_{i1}, \ldots, \eta_{iA})^T$. The link function $\mathbf{g}(\cdot)$ is one-to-one and its inverse is given by $\boldsymbol{\theta}_i^M = \mathbf{h}(\eta_i) = (h_1(\eta_i), \ldots, h_A(\eta_i))^T$, where

$$h_a(\eta_i) = \left[\sum_{a'=1}^{A} \exp(\eta_{ia'}) \right]^{-1} \exp(\eta_{ia}).$$

The model is fitted using design-unbiased direct estimators $\hat{\boldsymbol{\theta}}_i^M$ of $\boldsymbol{\theta}_i^M$ obtained from independent samples of size n_i from each domain i. The design-based covariance matrix of $\hat{\boldsymbol{\theta}}_i^M$, denoted $\mathbf{G}_i$, is assumed to be known for each area i. Then the model parameters ξ are estimated by using iterative weighted least squares (IWLS). This method relies on linearizing $\mathbf{g}(\hat{\boldsymbol{\theta}}_i^M)$ as

$$\mathbf{g}(\hat{\boldsymbol{\theta}}_i^M) \approx \mathbf{g}(\boldsymbol{\theta}_i^M) + \mathbf{g}'(\boldsymbol{\theta}_i^M)(\hat{\boldsymbol{\theta}}_i^M - \boldsymbol{\theta}_i^M) =: \mathbf{u}_i, \quad i = 1, \ldots, m,$$

where $\mathbf{g}'(\boldsymbol{\theta}_i^M)$ is the matrix of partial derivatives of $\mathbf{g}(\boldsymbol{\theta}_i^M)$. Now noting that $E_p(\mathbf{u}_i) = \mathbf{g}(\boldsymbol{\theta}_i^M) = \mathbf{Z}_i\xi$ and $\text{Cov}_p(\mathbf{u}_i) = \mathbf{g}'(\boldsymbol{\theta}_i^M)\mathbf{G}_i[\mathbf{g}'(\boldsymbol{\theta}_i^M)]^T =: \mathbf{V}_i$, we can represent

$\mathbf{u}_i = (u_{i1}, \ldots, u_{iA})^T$ in terms of a linear model $\mathbf{u}_i = \mathbf{Z}_i\xi + \mathbf{e}_i, i = 1, \ldots, m$, where the errors $\mathbf{e}_i$ are independent with zero mean vector and covariance matrix $\mathbf{V}_i$. However, $\mathbf{V}_i$ is singular because $\sum_{a=1}^{A} u_{ia} = 0$. Therefore, one of the elements of $\mathbf{u}_i$ is redundant. Removing the first element $a = 1$ from $\mathbf{u}_i$ leads to the model

$$\mathbf{u}_{i(1)} = \mathbf{Z}_{i(1)}\xi + \mathbf{e}_{i(1)}, \tag{3.2.28}$$

where $\mathbf{u}_{i(1)} = (u_{i2}, \ldots, u_{iA})^T, \mathbf{Z}_{i(1)} = (I_{(A-1)\times(A-1)}|\boldsymbol{\mu}_{i(1)})$ with $\boldsymbol{\mu}_{i(1)} = (\mu_{i2}^N, \ldots, \mu_{iA}^N)^T$, and $\mathbf{e}_{i(1)} = (e_{i2}, \ldots, e_{iA})^T$. Let $\mathbf{V}_{i(1)} = V_p(\mathbf{e}_{i(1)})$ be the resulting covariance matrix, which is equal to $\mathbf{V}_i$ with the first row and column deleted. Model (3.2.28) can be fit by IWLS. The updating equation for ξ is given by

$$\xi^{(k)} = \left[\sum_{i=1}^{m} \mathbf{Z}_{i(1)}^T(\mathbf{V}_{i(1)}^{(k-1)})^{-1}\mathbf{Z}_{i(1)}\right]^{-1} \sum_{i=1}^{m} \mathbf{Z}_{i(1)}^T(\mathbf{V}_{i(1)}^{(k-1)})^{-1}\mathbf{u}_{i(1)}^{(k-1)}. \tag{3.2.29}$$

In this equation, $\mathbf{V}_{i(1)}^{(k-1)}$ and $\mathbf{u}_{i(1)}^{(k-1)}$ are equal to $\mathbf{V}_{i(1)}$ and $\mathbf{u}_{i(1)}$, respectively, evaluated at the current vector of proportions $\boldsymbol{\theta}_i^{M,(k-1)}$, which is obtained from the estimator of ξ in the previous iteration of the algorithm, $\xi^{(k-1)}$, through the inverse link function $\boldsymbol{\theta}_i^{M,(k-1)} = \mathbf{h}(\mathbf{Z}_i\xi^{(k-1)})$. If $\hat{\xi} = \xi^{(K)}$ is the estimate of ξ obtained in the last iteration K, then $\tilde{\boldsymbol{\theta}}_i^M = \mathbf{h}(\mathbf{Z}_i\hat{\xi}) = (\tilde{\theta}_{i1}^M, \ldots, \tilde{\theta}_{iA}^M)$ is the vector of estimated proportions. The estimated counts are in turn given by $\tilde{M}_{ia} = M_{i\cdot}\tilde{\theta}_{ia}, a = 1, \ldots, A$.

Composite SPREE Another generalization of the one-way SPREE is the composite SPREE (Molina, Rao, and Hidiroglou 2008). The composite SPREE finds estimates of the cell counts that tend to be close (in terms of a chi-squared distance) to the census counts for cells in which direct estimates are not reliable due to small sample sizes, and close to direct estimates for cells in which the sample sizes are large enough.

In the composite SPREE, the counts for the categories in the ith domain $\{M_{ia}; a = 1, \ldots, A\}$ are estimated by minimizing a sum of composite distances to the two available sets of counts, namely census counts $\{N_{ia}; a = 1, \ldots, A\}$ and direct estimates $\{\hat{M}_{ia}; a = 1, \ldots, A\}$, subject to the restriction that they add up to the given domain margin $M_{i\cdot}$. More concretely, the counts $\{M_{ia}; a = 1, \ldots, A\}$ are estimated by solving the problem in x_{ia}

$$\begin{aligned} &\min_{(x_{i1},\ldots,x_{iA})} \quad \sum_{a=1}^{A} \left[\alpha_{ia}\frac{(N_{ia}-x_{ia})^2}{N_{ia}} + (1-\alpha_{ia})\frac{(\hat{M}_{ia}-x_{ia})^2}{\hat{M}_{ia}}\right] \\ &\text{s.t.} \quad \sum_{a=1}^{A} x_{ia} = M_{i\cdot}, \end{aligned} \tag{3.2.30}$$

where $\alpha_{ia} \in [0, 1], a = 1, \ldots, A$, are specified constants. Note that the problem (3.2.30) is solved separately for each domain i. Using the Lagrange multiplier method, the optimal solution x_{ia} to problem (3.2.30) for domain i is given by

$$\tilde{M}_{ia} = (\delta_{ia}/\delta_{i\cdot})\, M_{i\cdot}, \quad a = 1, \ldots, A, \tag{3.2.31}$$

where $\delta_{i\cdot} = \sum_{a=1}^{A} \delta_{ia}$ and

$$\delta_{ia} = \left(\frac{\alpha_{ia}}{N_{ia}} + \frac{1-\alpha_{ia}}{\hat{M}_{ia}} \right)^{-1}, \quad a = 1, \ldots, A.$$

We now turn to the choice of the constants α_{ia} by considering the special case of a simple random sample of size n_i drawn independently from each domain i, with sample sizes $n_{i1}, \ldots, n_{iA}$ in the A categories, where $n_i = \sum_{a=1}^{A} n_{ia}$. Then, following ideas similar to those in Section 3.3.2, a possible choice of α_{ia} is given by

$$\alpha_{ia} = \frac{1}{1 + n_{ia}/k^*}, \tag{3.2.32}$$

where $k^* > 0$ is a constant that can be interpreted as a critical sample size, satisfying $\alpha_{ia} < 0.5$ when $n_{ia} > k^*$ and more weight is given to the direct estimates $\hat{M}_{ia}$ than to the census counts N_{ia} in (3.2.30). If $n_{ia} < k^*$, then $\alpha_{ia} > 0.5$ and the composite distance in (3.2.30) gives more weight to the census counts. The constant k^* may be taken as the minimum sample size under which the user considers a direct estimator as a minimally reliable estimator.

Example 3.2.6. Count Estimation: Canadian LFS. Molina, Rao, and Hidiroglou (2008) applied the GLSM and composite SPREE methods to data from the 2001 Canadian census and LFS. Employed people are classified into different occupational classes labeled by two digits. Three-digit codes are used for a subclassification on different sectors of industry/services within each two-digit occupational class. The goal is to estimate the counts $\{M_{ia}\}$ in the cross-classification by province ($i = 1, \ldots, 10$) and three-digit category ($a = 1, \ldots, A$) separately for each two-digit category. The number of three-digit occupation categories A is different for each two-digit category. Direct estimates of the two-digit totals of employed people in each province were available from the LFS with a large enough sample size. In this application, they were treated as true margins, that is, $M_{i\cdot} = \sum_{a=1}^{A} \hat{M}_{ia}, i = 1, \ldots, m$. The composite SPREE weights α_{ia} were taken as in (3.2.32) with $k^* = 20$, a value close to the median sample size, which is also approximately half of the average sample size over the three-digit categories.

GLSM was fitted using, as starting values of λ_a and β, those corresponding to a model that preserves the marginal and interaction effects of the census counts, that is, $\lambda_a^{(0)} = 0, a = 1, \ldots, A$, and $\beta^{(0)} = 1$. The starting value for ξ is then $\xi^{(0)} = (0, \ldots, 0, 1)^T$. This method requires also the design-based covariance matrix $\mathbf{G}_i$ of the vector of direct estimators $\hat{\theta}_i^M$ for each province i. For province i, considering the total $M_{i\cdot}$ as a fixed quantity, the covariance matrix of the vector $\hat{\theta}_i^M$ is taken as

$$\mathbf{G}_i = \left[\hat{\Theta}_i - \hat{\theta}_i^M \left(\hat{\theta}_i^M \right)^T \right] / n_i, \quad \hat{\Theta}_i = \text{diag}(\hat{\theta}_{i1}^M, \ldots, \hat{\theta}_{iA}^M),$$

assuming simple random sampling within the provinces. In the calculation of $\mathbf{V}_i$, the Jacobian matrix $\mathbf{g}'(\boldsymbol{\theta}_i^M)$ is given by

$$\mathbf{g}'(\boldsymbol{\theta}_i^M) = \text{diag}(\theta_{i1}^M, \dots, \theta_{iA}^M) - \frac{1}{A}[(\theta_{i1}^M)^{-1}\mathbf{1}_A, \dots, (\theta_{iA}^M)^{-1}\mathbf{1}_A].$$

We now show results for the two-digit occupation classes A1 and B5, which have $A \geq 4$ three-digit categories and for which census and direct estimators θ_{ia}^N and $\hat{\theta}_{ia}^M$ show clear differences for at least some categories a. Figure 3.1 plots direct (labeled LFS), census, composite SPREE (CSPREE), and GLSM estimates of the row profiles $\boldsymbol{\theta}_i^M = (\theta_{i1}^M, \dots, \theta_{iA}^M)^T$ for two Canadian provinces and the two-digit class A1 with $A = 4$ categories. Figure 3.2 shows results for the two-digit occupation class B5 with $A = 7$ categories in two other Canadian provinces.

Figures 3.1 and 3.2 clearly show that the composite SPREE estimates are always approximately in between the direct and the census estimates, in contrast to the GLSM estimates, which appear, for several three-digit categories, either below both direct and census estimates, or above. Note also that the GLSM estimates are not respecting the profiles displayed by the census and LFS counts for the different occupation classes, see for example the plot for Newfoundland and Labrador in Figure 3.2a, where the GLSM estimated proportion for the three-digit category `B54` is larger than for `B55`, whereas for all the other estimates it is exactly the opposite.

3.2.8 *Weight-Sharing Methods

In this section, we study methods of producing weights w_{ij}^* for each area i by sharing the weight w_j^* attached to unit $j \in s$ among the areas $i = 1, \dots, m$, such that the

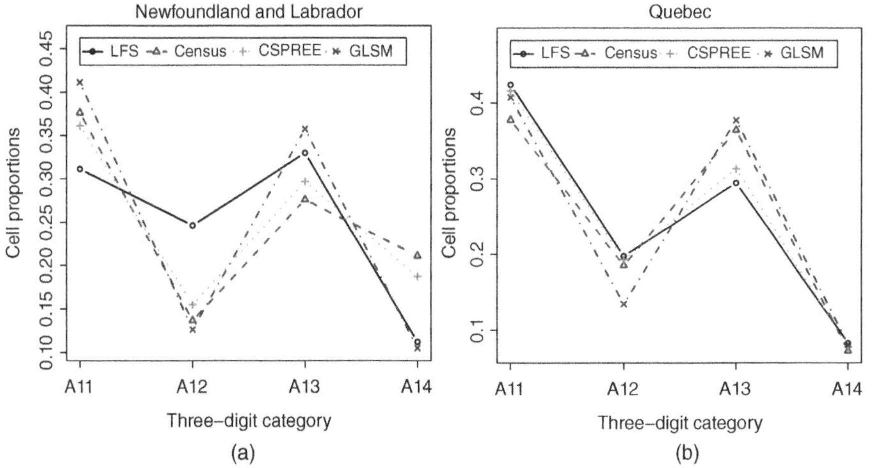

Figure 3.1 Direct, Census, composite SPREE, and GLSM estimates of Row Profiles $\theta_i^M = (\theta_{i1}^M, \dots, \theta_{iA}^M)^T$ for Canadian provinces Newfoundland and Labrador (a) and Quebec (b), for Two-Digit Occupation class A1.

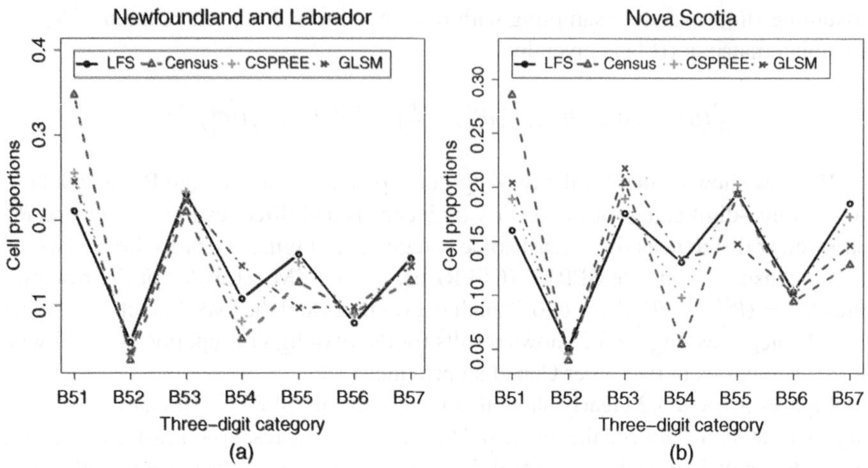

Figure 3.2 Direct, Census, composite SPREE, and GLSM estimates of row profiles $\theta_i^M = (\theta_{i1}^M, \ldots, \theta_{iA}^M)^T$ for Canadian provinces Newfoundland and Labrador (a) and Nova Scotia (b), for two-digit occupation class B5.

new weights satisfy calibration constraints. The WS property is desirable in practice because it ensures that the associated area synthetic estimators add up to the GREG estimator of the population total. The WS synthetic estimator of the area total Y_i is simply obtained as

$$\hat{Y}_{iWS} = \sum_{j\in s} w_{ij}^* y_j, \tag{3.2.33}$$

where the weights w_{ij}^* satisfy the calibration property

$$\sum_{j\in s} w_{ij}^* \mathbf{x}_j = \mathbf{X}_i, \quad i = 1, \ldots, m, \tag{3.2.34}$$

and the WS property

$$\sum_{i=1}^{m} w_{ij}^* = w_j^*, \quad j \in s. \tag{3.2.35}$$

The use of the same weight, w_{ij}^*, for all variables of interest used as y to produce small area estimates is of practical interest, particularly in microsimulation modeling that can involve a large number of y-variables.

Method 1 A simple method of calibration to known totals $\mathbf{X}_i$ for area i is to find weights h_{ij} of unit j in area i that minimize a chi-squared distance to the original weights, $D_{\mathcal{X},i} = \sum_{j\in s} c_j(w_j - h_{ij})^2/w_j$, subject to $\sum_{j\in s} h_{ij}\mathbf{x}_j = \mathbf{X}_i$ for each i. The resulting calibration weights agree with the weights $\tilde{w}_{ij}$ associated with the GREG-synthetic estimator (3.2.3), provided $c_j = \boldsymbol{\nu}^T\mathbf{x}_j$ is used. However, the weights can take negative values and do not satisfy range restrictions on the ratios

$\tilde{w}_{ij}/w_j$, although satisfy the calibration property (3.2.34) and the WS property (3.2.35). Following Deville and Särndal (1992), National Center for Social and Economic Modeling (NATSEM) in Australia uses an alternative distance function. to obtain weights satisfying also range restrictions. This distance function is given by $D_i = \sum_{j\in s} G(h_{ij}, w_j)$, where

$$G(h_{ij}, w_j) = (g_{ij} - L)\log\frac{g_{ij} - L}{1 - L} + (U - g_{ij})\log\frac{U - g_{ij}}{U - 1}. \tag{3.2.36}$$

Here, $g_{ij} = h_{ij}/w_j$ and $L(< 1)$ and $U(> 1)$ are specified lower and upper limits on the ratios g_{ij}. Minimizing D_i with respect to the variables h_{ij}, subject to calibration constraints $\sum_{j\in s} h_{ij}\mathbf{x}_j = \mathbf{X}_i$ for each i, the range restrictions are satisfied if a solution, h_{ij}, exists. NATSEM method adjusts L and U iteratively to find a solution such that $L \le g_{ij} \le U$ for all i and $j \in s$. The resulting weights w_{ij}^*, however, may not satisfy the WS property $\sum_{i=1}^m w_{ij}^* = w_j^*, j \in s$.

Using Lagrange multipliers, $\boldsymbol{\lambda}_i$, to minimize D_i subject to (3.2.34), it can be shown that $\boldsymbol{\lambda}_i$ is the solution of the equation

$$(\mathbf{X}_i - \hat{\mathbf{X}}_i) - \sum_{j\in s} w_j[g^{-1}(\mathbf{x}_j^T\boldsymbol{\lambda}_i) - 1]\mathbf{x}_j = \mathbf{0}, \tag{3.2.37}$$

where

$$g^{-1}(u) = \frac{L(U-1) + U(1-L)\exp(\alpha u)}{(U-1) + (1-L)\exp(\alpha u)},$$

with $\alpha = (U - L)/[(1 - L)(U - 1)]$. If a solution $\boldsymbol{\lambda}_i^*$ to (3.2.37) exists, then the optimal ratios are $g_{ij}^* = g^{-1}(\mathbf{x}_j^T\boldsymbol{\lambda}_i^*)$ and the optimal weights are $w_{ij}^* = w_j g_{ij}^*$; Newton's iterative method may be used to find the solution $\boldsymbol{\lambda}_i^*$ to the nonlinear equation (3.2.37).

Method 2 An alternative WS method that requires modeling the weights w_{ij}^* was proposed by Schirm and Zaslavsky (1997). They assumed a "Poisson" model on the weights,

$$w_{ij}^* = \gamma_{ij}\exp(\mathbf{x}_j^T\boldsymbol{\beta}_i + \delta_j), \quad j \in s, \tag{3.2.38}$$

where $\boldsymbol{\beta}_i$ and δ_j are unknown coefficients and γ_{ij} is an indicator equal to one if area i is allowed to borrow from the area to which unit j belongs, and zero otherwise. This is called "restricted borrowing" and is typically specified by the user. We impose the WS and calibration constraints (3.2.35) and (3.2.34), respectively.

Schirm and Zaslavsky (1997) proposed an iterative two-step method to estimate the parameters $\boldsymbol{\beta}_i$ and δ_j in the Poisson model (3.2.38) subject to the specified constraints. Letting $\boldsymbol{\beta}_i^{(t-1)}$ and $\delta_j^{(t-1)}$ denote the values of the parameters in iteration $t - 1$, the two steps are given by

$$\text{Step 1:} \quad \delta_j^{(t)} = \log\left\{w_j^* \Big/ \sum_{i=1}^m \gamma_{ij}\exp\left(\mathbf{x}_j^T\boldsymbol{\beta}_i^{(t-1)}\right)\right\}$$

and

$$\text{Step 2:}\quad \boldsymbol{\beta}_i^{(t)} = \boldsymbol{\beta}_i^{(t-1)} + \left(\sum_{j\in s} w_{ij}^{*(t-1)}\mathbf{x}_j\mathbf{x}_j^T\right)^{-1}\left(\mathbf{X}_i - \sum_{j\in s} w_{ij}^{*(t-1)}\mathbf{x}_j\right),$$

where $w_{ij}^{*(t-1)}$ is obtained by substituting $\boldsymbol{\beta}_i^{(t-1)}$ and $\delta_j^{(t)}$ in the formula (3.2.38) for w_{ij}^*. Iteration is continued until convergence, to get $\hat{\boldsymbol{\beta}}_i$ and $\hat{\delta}_j$, which in turn lead to the desired weights w_{ij}^* satisfying both constraints.

Method 3 Randrianasolo and Tillé (2013) proposed an alternative method of weight sharing that avoids modeling the area-specific weights w_{ij}^*. Letting $w_{ij}^* = w_j^* q_{ij}$, the WS condition is equivalent to $\sum_{i=1}^m q_{ij} = 1$ for each $j \in s$. The fractions q_{ij} that satisfy the calibration conditions $\sum_{j\in s}(w_j^* q_{ij})\mathbf{x}_j = \mathbf{X}_i$ for $i = 1, \ldots, m$ and the WS condition are obtained by a two-step iterative procedure. First, the calibration weights w_j^* are obtained by minimizing the Kullback–Leibler distance measure $D_K = \sum_{j\in s} K(h_j, w_j)$ with respect to the variables h_j subject to the calibration constraint $\sum_{j\in s} h_j\mathbf{x}_j = \mathbf{X}$, where

$$K(h_j, w_j) = h_j \log(h_j/w_j) + w_j - h_j. \tag{3.2.39}$$

Using the Lagrange multiplier method, the desired weights $h_j = w_j^*$ are given by

$$w_j^* = w_j \exp(\mathbf{x}_j^T \boldsymbol{\lambda}), \quad j \in s, \tag{3.2.40}$$

where the Lagrange multiplier $\boldsymbol{\lambda}$ is obtained by solving the calibration equations

$$\sum_{j\in s} w_j \exp(\mathbf{x}_j^T \boldsymbol{\lambda}) = \mathbf{X}. \tag{3.2.41}$$

Equations (3.2.41) can be solved iteratively by the Newton–Raphson method. It follows from (3.2.40) that the weights w_j^* are always positive. However, some of these calibration weights may take large values.

The calibration weights w_j^* given by (3.2.40) are used to determine the fractions q_{ij}. Taking as starting values $q_{ij}^{(0)} = N_i/N$, which satisfy $\sum_{i=1}^m q_{ij}^{(0)} = 1$, where N_i is the known size of area i, new values $q_{ij}^{(1)}$ are obtained by minimizing separately for each i

$$D_{K,i} = \sum_{j\in s} K(q_{ij}^{(1)}, q_{ij}^{(0)}) \tag{3.2.42}$$

with respect to the variables $q_{ij}^{(1)}$ subject to the area-specific calibration condition

$$\sum_{j\in s} (w_j^* q_{ij})\mathbf{x}_j = \mathbf{X}_i. \tag{3.2.43}$$

Using again the Lagrange multiplier method, the solution $q_{ij}^{(1)}$ is given by

$$q_{ij}^{(1)} = q_{ij}^{(0)} \exp(\mathbf{x}_j^T \boldsymbol{\lambda}_i), \tag{3.2.44}$$

where the Lagrange multiplier λ_i is obtained by solving iteratively the calibration equations

$$\sum_{j\in s}(w_j^* q_{ij}^{(0)}) \exp(\mathbf{x}_j^T \lambda_i) = \mathbf{X}_i. \tag{3.2.45}$$

Note that m sets of calibration equations need to be solved to obtain $\lambda_1, \ldots, \lambda_m$. In the second step of the method, $q_{ij}^{(1)}$ values are revised to satisfy the WS property. Revised fractions are given by

$$q_{ij}^{(2)} = q_{ij}^{(1)} / \sum_{i=1}^{m} q_{ij}^{(1)}. \tag{3.2.46}$$

The two-step iteration is repeated until convergence to obtain the desired fractions q_{ij}, which are strictly positive. The resulting weights $w_{ij}^* = w_j^* q_{ij}$ satisfy the area-specific calibration conditions and the WS property. Note that this method ensures strictly positive weights w_{ij}^* because $q_{ij} > 0$ and $w_j^* > 0$.

3.3 COMPOSITE ESTIMATION

A natural way to balance the potential bias of a synthetic estimator, say $\hat{Y}_{i2}$, against the instability of a direct estimator, say $\hat{Y}_{i1}$, is to take a weighted average of $\hat{Y}_{i1}$ and $\hat{Y}_{i2}$. Such composite estimators of the small area total Y_i may be written as

$$\hat{Y}_{iC} = \phi_i \hat{Y}_{i1} + (1 - \phi_i)\hat{Y}_{i2} \tag{3.3.1}$$

for a suitably chosen weight $\phi_i (0 \le \phi_i \le 1)$. Many of the estimators proposed in the literature, both design-based and model-based, have the composite form (3.3.1). In latter chapters we study model-based composite estimators derived from more realistic small area models that account for local variation.

3.3.1 Optimal Estimator

The design MSE of the composite estimator is given by

$$\begin{aligned} \mathrm{MSE}_p(\hat{Y}_{iC}) = {} & \phi_i^2 \mathrm{MSE}_p(\hat{Y}_{i1}) + (1 - \phi_i)^2 \mathrm{MSE}_p(\hat{Y}_{i2}) \\ & + 2\,\phi_i(1 - \phi_i) E_p(\hat{Y}_{i1} - Y_i)(\hat{Y}_{i2} - Y_i). \end{aligned} \tag{3.3.2}$$

By minimizing (3.3.2) with respect to ϕ_i, we get the optimal weight ϕ_i as

$$\begin{aligned} \phi_i^* &= \frac{\mathrm{MSE}_p(\hat{Y}_{i2}) - E_p(\hat{Y}_{i1} - Y_i)(\hat{Y}_{i2} - Y_i)}{\mathrm{MSE}_p(\hat{Y}_{i1}) + \mathrm{MSE}_p(\hat{Y}_{i2}) - 2E_p(\hat{Y}_{i1} - Y_i)(\hat{Y}_{i2} - Y_i)} \\ &\approx \mathrm{MSE}_p(\hat{Y}_{i2}) / [\mathrm{MSE}_p(\hat{Y}_{i1}) + \mathrm{MSE}_p(\hat{Y}_{i2})], \end{aligned} \tag{3.3.3}$$

assuming that the covariance term $E_p(\hat{Y}_{i1} - Y_i)(\hat{Y}_{i2} - Y_i)$ is small relative to $\mathrm{MSE}_p(\hat{Y}_{i2})$.

Note that the approximate optimal ϕ_i^*, given by (3.3.3), lies in the interval $[0, 1]$ and depends only on the ratio of the MSEs $F_i = \text{MSE}_p(\hat{Y}_{i1})/\text{MSE}_p(\hat{Y}_{i2})$ as

$$\phi_i^* = 1/(1 + F_i). \tag{3.3.4}$$

Furthermore, the MSE of the resulting composite estimator $\hat{Y}_{iC}$ obtained using the optimal weight ϕ_i^* reduces to

$$\text{MSE}_p^*(\hat{Y}_{iC}) = \phi_i^* \text{MSE}_p(\hat{Y}_{i1}) = (1 - \phi_i^*)\text{MSE}_p(\hat{Y}_{i2}). \tag{3.3.5}$$

It now follows from (3.3.5) that the reduction in MSE achieved by the optimal estimator relative to the smaller of the MSEs of the component estimators is given by ϕ_i^* when $0 \le \phi_i^* \le 1/2$, and it equals $1 - \phi_i^*$ when $1/2 \le \phi_i^* \le 1$. Thus, the maximum reduction of 50 percent is achieved when $\phi_i^* = 1/2$ (or equivalently $F_i = 1$).

The ratio of $\text{MSE}_p(\hat{Y}_{iC})$ with a fixed weight ϕ_i and $\text{MSE}_p(\hat{Y}_{i2})$ may be expressed as

$$\frac{\text{MSE}_p(\hat{Y}_{iC})}{\text{MSE}_p(\hat{Y}_{i2})} = (F_i + 1)\phi_i^2 - 2\phi_i + 1. \tag{3.3.6}$$

Schaible (1978) studied the behavior of the MSE ratio (3.3.6) as a function of ϕ_i for selected values of $F_i (= 1, 2, 6)$. His results suggest that sizable deviations from the optimal weight ϕ_i^* do not produce a significant increase in the MSE of the composite estimator, that is, the curve (3.3.6) in ϕ_i is fairly flat in the neighborhood of the optimal weight. Moreover, both the reduction in MSE and the range of ϕ_i for which the composite estimator has a smaller MSE than either component estimators depend on the size of F_i. When F_i is close to one, we get the most advantage in terms of both situations.

It is easy to show that $\hat{Y}_{iC}$ is better than either component estimator in terms of MSE when $\max(0, 2\phi_i^* - 1) \le \phi_i \le \min(2\phi_i^*, 1)$. The latter interval reduces to the whole range $0 \le \phi_i \le 1$ when $F_i = 1$, and it becomes narrower as F_i deviates from one. The optimal weight ϕ_i^* will be close to zero or one when one of the component estimators has a much larger MSE than the other, that is, when F_i is either large or small. In this case, the estimator with larger MSE adds little information, and therefore it is better to use the component estimator with smaller MSE in preference to the composite estimator.

In practice, we use either a prior guess of the optimal value ϕ_i^* or estimate ϕ_i^* from the sample data. Assuming that the direct estimator $\hat{Y}_{i1}$ is either p-unbiased or approximately p-unbiased as the overall sample size increases, we can estimate the approximate optimal weight (3.3.3) using (3.2.16). We substitute the estimator $\text{mse}(\hat{Y}_{i2})$ given by (3.2.16) for the numerator $\text{MSE}_p(\hat{Y}_{i2})$ and $(\hat{Y}_{i2} - \hat{Y}_{i1})^2$ for the denominator $\text{MSE}_p(\hat{Y}_{i1}) + \text{MSE}_p(\hat{Y}_{i2})$, leading to the estimated optimal weight

$$\hat{\phi}_i^* = \frac{\text{mse}(\hat{Y}_{i2})}{(\hat{Y}_{i2} - \hat{Y}_{i1})^2}. \tag{3.3.7}$$

Nevertheless, the estimator (3.3.7) of ϕ_i^* can be very unstable. One way to overcome this difficulty is to average the estimated weights $\hat{\phi}_i^*$ over several variables or "similar" areas or both. The resulting composite estimator should perform well in view of the insensitivity to deviations from the optimal weight.

Estimation of MSE of the composite estimator, even with a fixed weight, runs into difficulties similar to those for the synthetic estimators. It is possible to extend the methods in Section 3.2.5 to composite estimators (see Example 3.3.1).

Example 3.3.1. Labor Force Characteristics. Griffiths (1996) used a SPREE-based composite estimator to provide indirect estimates of labor force characteristics for congressional districts (CDs) i in the United States. This estimator is of the form (3.3.1) with the one-step SPREE estimator $\tilde{M}_{ia\cdot}$ (3.2.21) as the synthetic component, $\hat{Y}_{i2}$, and a sample-based estimator $\hat{M}_{ia\cdot}$ as the direct component, $\hat{Y}_{i1}$. To calculate the SPREE estimate, the population counts $\{N_{iab}\}$ were obtained from the 1990 Decennial Census, while the estimated marginals $\{\hat{M}_{\cdot ab}\}$ were obtained from the 1994 March Current Population Survey (CPS). The CPS sample is a stratified two-stage probability sample drawn independently from each of the 50 states and the District of Columbia. It was not designed to provide reliable sample-based estimates $\hat{M}_{ia\cdot}$ at the CD level; the CD sample sizes tend to be too small for direct estimation with desired reliability.

Estimates of the optimal weights ϕ_i^* were obtained from (3.3.3), using (3.2.14) to estimate $\text{MSE}_p(\tilde{M}_{ia\cdot})$ and $v(\hat{M}_{ia\cdot})$ to estimate $\text{MSE}_p(\hat{M}_{ia\cdot})$. Note that we replace $\hat{Y}_{iS}$ by $\tilde{M}_{ia\cdot}$ and $\hat{Y}_i$ by $\hat{M}_{ia\cdot}$ in (3.2.14). One could also use the simpler estimate, $\hat{\phi}_i^*$, given by (3.3.7), but both estimates of ϕ_i^* are highly unstable. Griffiths (1996) used again (3.2.14) to estimate the MSE of the composite estimator, by replacing $\hat{Y}_{iS}$ and $\hat{Y}_i$ in (3.2.14) by the composite estimator and $\hat{M}_{ia\cdot}$, respectively. This MSE estimate is also highly unstable.

Griffiths (1996) evaluated the efficiencies of the composite estimator and the SPREE estimator, $\tilde{M}_{ia\cdot}$, relative to the sample-based estimator $\hat{M}_{ia\cdot}$ for the five CDs in the state of Iowa. The MSE estimates based on (3.2.14) were again used for this purpose. The composite estimator provided an improvement over the sample-based estimator in terms of estimated MSE. The reduction in estimated MSE, when averaged over the five CDs, ranged from 17% to 78%. The one-step SPREE estimator did not perform better than the direct estimator, $\hat{M}_{ia\cdot}$, for three of the categories: employed, others, and household income in the range \$10,000–\$25,000.

3.3.2 Sample-Size-Dependent Estimators

Sample-size-dependent (SSD) estimators are composite estimators with simple weights ϕ_i that depend only on the domain counts $\hat{N}_i$ and N_i or the domain totals $\hat{X}_i$ and X_i of an auxiliary variable x. These estimators were originally designed to handle domains for which the expected sample sizes are large enough, such that direct estimators for domains with realized sample sizes exceeding the expected sample sizes satisfy reliability requirements (Drew, Singh, and Choudhry 1982).

The SSD estimator proposed by Drew et al. (1982) is a composite estimator of the form (3.3.1) with weight

$$\phi_i(\mathrm{S1}) = \begin{cases} 1 & \text{if } \hat{N}_i/N_i \geq \delta; \\ \hat{N}_i/(\delta N_i) & \text{if } \hat{N}_i/N_i < \delta, \end{cases} \quad (3.3.8)$$

where $\hat{N}_i = \sum_{s_i} w_j$ is the direct expansion estimator of N_i and $\delta > 0$ is subjectively chosen to control the contribution of the synthetic estimator. The form of $\hat{N}_i$ suggests that $\phi_i(\mathrm{S1})$ increases with the domain sample size. Another choice of weight is obtained by substituting $\hat{X}_i/X_i$ for $\hat{N}_i/N_i$ in (3.3.8). Under this choice, Drew et al. (1982) used the poststratified-ratio estimator (2.4.17) as the direct estimator $\hat{Y}_{i1}$ and the ratio-synthetic estimator (3.2.9) as the synthetic estimator $\hat{Y}_{i2}$. The direct estimator (2.4.17), however, suffers from the ratio bias unless the domain sample size is not small. To avoid the ratio bias, we could use the modified direct estimator $\hat{Y}_i + \sum_{g=1}^{G}(X_{ig} - \hat{X}_{ig})(\hat{Y}_g/\hat{X}_g)$, whose p-bias goes to zero as the overall sample size increases, even if the domain sample size is small. Generally, we can use the modified GREG estimator $\hat{Y}_i + (\mathbf{X}_i - \hat{\mathbf{X}}_i)^T\hat{\mathbf{B}}$ as the direct estimator and the regression-synthetic estimator $\mathbf{X}_i^T\hat{\mathbf{B}}$ as the synthetic estimator in conjunction with the weight $\phi_i(\mathrm{S1})$. A general-purpose choice of δ in (3.3.8) is $\delta = 1$. The Canadian Labour Force Survey (LFS) used the SSD estimator with $\delta = 2/3$ to produce Census Division level estimates.

Särndal and Hidiroglou (1989) proposed the "dampened regression estimator," which is obtained from the alternative modified GREG estimator (2.5.7) by dampening the effect of the direct component $\sum_{s_i} w_j e_j$ whenever $\hat{N}_i < N_i$. This estimator is given by

$$\hat{Y}_{i\mathrm{DR}} = \mathbf{X}_i^T\hat{\mathbf{B}} + (\hat{N}_i/N_i)^{H-1}\sum_{s_i} w_j e_j \quad (3.3.9)$$

with $H = 0$ if $\hat{N}_i \geq N_i$ and $H = h$ if $\hat{N}_i < N_i$, where $h > 0$ is a suitably chosen constant. This estimator can be written as a composite estimator with the alternative modified GREG estimator as the direct estimator and $\mathbf{X}_i^T\hat{\mathbf{B}}$ as the synthetic estimator in conjunction with the weight

$$\phi_i(\mathrm{S2}) = \begin{cases} 1 & \text{if } \hat{N}_i/N_i \geq 1, \\ (\hat{N}_i/N_i)^h & \text{if } \hat{N}_i/N_i < 1. \end{cases} \quad (3.3.10)$$

A general-purpose choice of h is $h = 2$.

To study the nature of the SSD weight $\phi_i(\mathrm{S1})$, we consider the special case of simple random sampling from the population U. In this case, $\hat{N}_i = N(n_i/n)$. Taking $\delta = 1$ in (3.3.8), it now follows that $\phi_i(\mathrm{S1}) = 1$ if $n_i \geq E(n_i) = n(N_i/N)$. Therefore, the SSD estimator can fail to borrow strength from other domains when $E(n_i)$ is not large enough. On the other hand, when $\hat{N}_i < N_i$, the weight $\phi_i(\mathrm{S1}) = \hat{N}_i/N_i = Nn_i/(nN_i)$ decreases as n_i decreases. As a result, more weight is given to the synthetic component when n_i is small. Thus, in the case $\hat{N}_i < N_i$, the weight $\phi_i(\mathrm{S1})$ behaves well, unlike in the case $\hat{N}_i \geq N_i$. Similar comments apply to the SSD estimator based on the weight $\phi_i(\text{S2})$. Another disadvantage of SSD estimators is that the weights do not take account of the size of between-area variation relative to within-area variation for the characteristic of interest. Then, all the characteristics get the same weight regardless of their differences with respect to between-area heterogeneity. In Example 7.3.2 of Chapter 7, we demonstrate that large efficiency gains over SSD estimators can be achieved by using model-based estimators when the between-area heterogeneity is relatively small.

General SSD estimators provide consistency when aggregated over different characteristics because the same weight is used for all of them. However, they do not add up to a direct estimator at a large area level. A simple ratio adjustment gives

$$\hat{Y}_{i\text{C}}(a) = \frac{\hat{Y}_{i\text{C}}}{\sum_{i=1}^{m} \hat{Y}_{i\text{C}}} \hat{Y}_{\text{GR}}. \tag{3.3.11}$$

The adjusted estimators $\hat{Y}_{i\text{C}}(a)$ add up to the direct estimator $\hat{Y}_{\text{GR}}$ at the large area level.

The SSD estimators with weights $\phi_i(\text{S1})$ may also be viewed as calibration estimators $\sum_s w_{ij}^* y_j$ with weights w_{ij}^* minimizing the chi-squared distance

$$\sum_{i=1}^{m} \sum_{j \in s} c_j [w_j a_{ij} \phi_i(\text{S1}) - h_{ij}^*]^2 / w_j \tag{3.3.12}$$

with respect to h_{ij}^* subject to the constraints $\sum_{j \in s} h_{ij}^* \mathbf{x}_j = \mathbf{X}_i, i = 1, \ldots, m$. Here, a_{ij} is the domain indicator variable and c_j is a specified constant $j \in s$; that is, the "optimal" h_{ij}^* equals w_{ij}^*. Using this distance measure, we are calibrating the dampened weights $w_j a_{ij} \phi_i(\text{S1})$ rather than the original weights $w_j a_{ij}$. Singh and Mian (1995) used the calibration approach to take account of different sets of constraints simultaneously. For example, the additivity constraint $\sum_{i=1}^{m} \sum_{j \in s} h_{ij}^* y_j = \hat{Y}_{\text{GR}}$ can be introduced along with the calibration constraints $\sum_{j \in s} h_{ij}^* \mathbf{x}_j = \mathbf{X}_i, i = 1, \ldots, m$. Note that, using this approach, the calibration weights w_{ij}^* are obtained simultaneously for all the areas.

Estimation of the MSE of SSD estimators runs into difficulties similar to those for the synthetic estimator and the optimal composite estimator. One ad hoc approach to variance estimation is to use the variance estimator of the modified direct estimator $\hat{Y}_i + (\mathbf{X}_i - \hat{\mathbf{X}}_i)^T \hat{\mathbf{B}}$, namely, $v(a_i e)$ in operator notation, as an overestimator of the true variance of the SSD estimator using either $\phi_i(\text{S1})$ or $\phi_i(\text{S2})$ as the weight attached to the direct estimator (Särndal and Hidiroglou 1989). Another approach is to treat the weight $\phi_i(\text{S1})$ as fixed and estimate the variance as $\{\phi_i(\text{S1})\}^2 v(a_i e)$, noting that the variance of the synthetic component $\mathbf{X}_i^T \hat{\mathbf{B}}$ is small relative to the variance of the direct component. This variance estimator will underestimate the true variance, unlike $v(a_i e)$. Resampling methods, such as the jackknife or the bootstrap (Rao, Wu, and Yue 1992), can also be readily used to get a variance estimator. Properties of resampling variance estimators in the context of SSD estimators have not been studied.

Example 3.3.2. Unemployed Counts. Falorsi, Falorsi, and Russo (1994) compared the performances of the direct estimator, the synthetic estimator, the SSD estimator with $\delta = 1$, and the optimal composite estimator. They conducted a simulation study in which the Italian LFS design (stratified two-stage sampling) was simulated using data from the 1981 Italian census. The "optimal" weight ϕ_i^* was obtained from the census data. In this study, health service areas (HSAs) are the small areas (unplanned domains) that cut across design strata. The study was confined to the $m = 14$ HSAs of the Friuli region, and the sample design was based on the selection of 39 primary sampling units (PSUs) and 2,290 second-stage units (SSUs); PSU is a municipality and SSU is a household. The variable of interest, y, is the number of unemployed in a household.

In this simulation study, the performance of the estimators was evaluated in terms of ARB and relative root mean square error (RRMSE). The RB and MSE of a generic estimator of the total Y_i, say $\tilde{Y}_i$, are given by

$$\text{RB} = \frac{1}{R}\sum_{r=1}^{R}\left(\frac{\tilde{Y}_i^{(r)}}{Y_i} - 1\right); \quad \text{MSE} = \frac{1}{R}\sum_{r=1}^{R}(\tilde{Y}_i^{(r)} - Y_i)^2,$$

where $\tilde{Y}_i^{(r)}$ is the value of the estimator of Y_i for rth simulated sample ($r = 1, \ldots, R$). Note that $\text{RRMSE} = \sqrt{\text{MSE}}/Y_i$ and $\text{ARB} = |\text{RB}|$. Falorsi et al. (1994) used $R = 400$ simulated samples to calculate ARB and RRMSE for each HSA and each estimator.

Table 3.3 reports, for each estimator, average values over the $m = 14$ HSAs of ARB, denoted $\overline{\text{ARB}}$, and of RRMSE, denoted $\overline{\text{RRMSE}}$. It is clear from Table 3.3 that $\overline{\text{ARB}}$ values of the direct estimator and SSD estimator are negligible (< 2.5%), whereas those of the composite and synthetic estimators are somewhat large. The synthetic estimator has the largest $\overline{\text{ARB}}$% (about 9%). However, in terms of $\overline{\text{RRMSE}}$%, synthetic and composite estimators have the smallest values (almost one half of the value for the direct estimator) followed by the SSD estimator.

Falorsi et al. (1994) also examined area-specific values of ARB and RRMSE. Synthetic and composite estimators were found to be badly biased in small areas with low values of the ratio (population of HSA)/(population of the set of strata including the HSA), but exhibited low RRMSE compared to the other alternatives. Considering

TABLE 3.3 Percent Average Absolute Relative Bias ($\overline{\text{ARB}}$%) and Percent Average RRMSE ($\overline{\text{RRMSE}}$%) of Estimators

Estimator	$\overline{\text{ARB}}$%	$\overline{\text{RRMSE}}$%
Direct	1.75	42.08
Synthetic	8.97	23.80
Composite	6.00	23.57
SSD	2.39	31.08

Source: Adapted from Table 1 in Falorsi et al. (1994).

both bias and efficiency, they concluded that SSD estimator is preferable over the other estimators. It may be noted that the sampling fractions were large enough, leading to large enough expected domain sample sizes, a case favorable to the SSD estimator.

3.4 JAMES–STEIN METHOD

3.4.1 Common Weight

Another approach to composite estimation is to use a common weight for all areas, $\phi_i = \phi$, and then minimize the total MSE, $\sum_{i=1}^{m} \text{MSE}_p(\hat{Y}_{iC})$, with respect to ϕ (Purcell and Kish 1979). This ensures good overall efficiency for the group of small areas but not necessarily for each of the small areas in the group. We have

$$\sum_{i=1}^{m} \text{MSE}_p(\hat{Y}_{iC}) \approx \phi^2 \sum_{i=1}^{m} \text{MSE}_p(\hat{Y}_{i1}) + (1-\phi)^2 \sum_{i=1}^{m} \text{MSE}_p(\hat{Y}_{i2}). \tag{3.4.1}$$

Minimizing (3.4.1) with respect to ϕ gives the optimal weight

$$\phi^* = \frac{\sum_{i=1}^{m} \text{MSE}_p(\hat{Y}_{i2})}{\sum_{i=1}^{m} [\text{MSE}_p(\hat{Y}_{i1}) + \text{MSE}_p(\hat{Y}_{i2})]}. \tag{3.4.2}$$

Suppose we take $\hat{Y}_{i1}$ as the direct expansion estimator $\hat{Y}_i$ and $\hat{Y}_{i2}$ as the synthetic estimator $\hat{Y}_{iS}$. Then from (3.2.16), ϕ^* may be estimated as

$$\hat{\phi}^* = \frac{\sum_{i=1}^{m} [(\hat{Y}_{iS} - \hat{Y}_i)^2 - v(\hat{Y}_i)]}{\sum_{i=1}^{m} (\hat{Y}_{iS} - \hat{Y}_i)^2} = 1 - \frac{\sum_{i=1}^{m} v(\hat{Y}_i)}{\sum_{i=1}^{m} (\hat{Y}_{iS} - \hat{Y}_i)^2}. \tag{3.4.3}$$

The estimator $\hat{\phi}^*$ is quite reliable, unlike $\hat{\phi}_i^*$ given by (3.3.7), because we are pooling over several small areas. However, the use of a common weight may not be reasonable if the individual variances, $V_p(\hat{Y}_i)$, vary considerably.

Rivest (1995) used weights of the form $\hat{\phi}^*$ in the context of adjustment for population undercount in the 1991 Canadian Census. He also obtained an approximation to the total MSE of the resulting composite estimators as well as an estimator of this total MSE. Overall performance of the composite estimators relative to the direct estimators was studied by comparing their estimated total MSE.

The composite estimator based on $\hat{\phi}^*$ is similar to the well-known JS estimator, which has attracted a lot of attention in the mainstream statistical literature. We give a brief account of the JS methods in this section and refer the reader to Efron and Morris (1972a, 1973, 1975) and Brandwein and Strawderman (1990) for a more extensive treatment. Efron (1975) gave an excellent expository account of the JS methods as well as examples of their practical application, including the popular example of predicting batting averages of baseball players.

Suppose the small area means $\overline{Y}_i$ are the parameters of interest. Let $\theta_i = g(\overline{Y}_i)$ be a specified transformation of $\overline{Y}_i$, which induces normality of the corresponding estimators $\hat{\theta}_i = g(\hat{\overline{Y}}_i)$ and stabilizes the variances of $\hat{\theta}_i$. For example, if $\overline{Y}_i$ is a proportion, we can use an arc-sine transformation. Some other choices of $g(\overline{Y}_i)$ include $\theta_i = \overline{Y}_i$ and $\theta_i = \log \overline{Y}_i$. Let $\boldsymbol{\theta} = (\theta_1, \dots, \theta_m)^T$ and $\hat{\boldsymbol{\theta}} = (\hat{\theta}_1, \dots, \hat{\theta}_m)^T$. We assume $\hat{\theta}_i \overset{\text{ind}}{\sim} N(\theta_i, \psi_i)$ with known variances $\psi_i (i = 1, \dots, m)$ where $\overset{\text{ind}}{\sim}$ denotes "independently distributed as" and $N(a, b)$ denotes a normal variable with mean a and variance b. We further assume that a prior guess of $\boldsymbol{\theta} = (\theta_1, \dots, \theta_m)^T$, say $\boldsymbol{\theta}^0 = (\theta_1^0, \dots, \theta_m^0)^T$, is available or can be evaluated from the data. For example, if θ_i is linearly related to a p-vector of auxiliary variables, $\mathbf{z}_i$, then we can take the least squares predictor $\mathbf{z}_i^T(\mathbf{Z}^T\mathbf{Z})^{-1}\mathbf{Z}^T\hat{\boldsymbol{\theta}} = \mathbf{z}_i^T\hat{\boldsymbol{\beta}}_{\text{LS}}$ as θ_i^0, where $\mathbf{Z}^T = (\mathbf{z}_1, \dots, \mathbf{z}_m)$. In the absence of such auxiliary information, we set $\mathbf{z}_i = 1$ so that $\theta_i^0 = \sum_{i=1}^m \hat{\theta}_i/m = \hat{\theta}_.$ for all i. The performance of an estimator of $\boldsymbol{\theta}$, say $\tilde{\boldsymbol{\theta}}$, will be measured in terms of its total MSE (total squared error risk) given by

$$R(\boldsymbol{\theta}, \tilde{\boldsymbol{\theta}}) = \sum_{i=1}^{m} E_p(\tilde{\theta}_i - \theta_i)^2. \tag{3.4.4}$$

3.4.2 Equal Variances $\psi_i = \psi$

In the special case of equal sampling variances, $\psi_i = \psi$, the JS estimator of θ_i is given by

$$\hat{\theta}_{i,\text{JS}} = \theta_i^0 + \left[1 - \frac{(m-2)\psi}{S}\right](\hat{\theta}_i - \theta_i^0), \quad m \geq 3 \tag{3.4.5}$$

assuming $\boldsymbol{\theta}^0 = (\theta_1^0, \dots, \theta_m^0)^T$ is fixed, where $S = \sum_{i=1}^m (\hat{\theta}_i - \theta_i^0)^2$. If θ_i^0 is the least squares predictor, then we replace $m - 2$ by $m - p - 2$ in (3.4.5), where p is the number of estimated parameters in the regression equation. Note that (3.4.5) may also be expressed as a composite estimator with weight $\hat{\phi}_{\text{JS}} = 1 - [(m-2)\psi]/S$ attached to $\hat{\theta}_i$ and $1 - \hat{\phi}_{\text{JS}}$ to the prior guess θ_i^0. The JS estimator is also called a shrinkage estimator because it shrinks the direct estimator $\hat{\theta}_i$ toward the guess θ_i^0.

James and Stein (1961) established the following remarkable results on the superiority of $\hat{\boldsymbol{\theta}}_{\text{JS}} = (\hat{\theta}_{1,\text{JS}}, \dots, \hat{\theta}_{m,\text{JS}})^T$ over $\hat{\boldsymbol{\theta}} = (\hat{\theta}_1, \dots, \hat{\theta}_m)^T$ in terms of total MSE:

Theorem 3.4.1. Suppose the direct estimators $\hat{\theta}_i$ are independent $N(\theta_i, \psi)$ with known sampling variance ψ, and the guess θ_i^0 is fixed. Then,

(a) $R(\boldsymbol{\theta}, \hat{\boldsymbol{\theta}}_{\text{JS}}) < R(\boldsymbol{\theta}, \hat{\boldsymbol{\theta}}) = m\psi$ for all $\boldsymbol{\theta}$, that is, $\hat{\boldsymbol{\theta}}_{\text{JS}}$ dominates $\hat{\boldsymbol{\theta}}$ with respect to total MSE.

(b) $R(\boldsymbol{\theta}, \hat{\boldsymbol{\theta}}_{\text{JS}}) = 2\psi$ at $\boldsymbol{\theta} = \boldsymbol{\theta}^0$.

(c) $R(\boldsymbol{\theta}, \hat{\boldsymbol{\theta}}_{\text{JS}}) \leq m\psi - \frac{(m-2)^2\psi^2}{(m-2)\psi + \sum_{i=1}^m(\theta_i - \theta_i^0)^2}$, so that $R(\boldsymbol{\theta}, \hat{\boldsymbol{\theta}}_{\text{JS}})$ is minimized when $\boldsymbol{\theta} = \boldsymbol{\theta}^0$.

A proof of Theorem 3.4.1 is given in Section 3.5. It follows from Theorem 3.4.1 that $\hat{\boldsymbol{\theta}}_{\text{JS}}$ leads to large reduction in total MSE when the true value $\boldsymbol{\theta}$ is close to the guess $\boldsymbol{\theta}^0$ and m is not very small. For example, the relative total MSE $R(\boldsymbol{\theta}^0, \hat{\boldsymbol{\theta}}_{\text{JS}})/R(\boldsymbol{\theta}^0, \hat{\boldsymbol{\theta}}) = 2/10 = 0.2$ when $m = 10$ so that the total MSE of $\hat{\boldsymbol{\theta}}_{\text{JS}}$ is only one-fifth of the total MSE of $\hat{\boldsymbol{\theta}}$ when $\boldsymbol{\theta} = \boldsymbol{\theta}^0$. On the other hand, the reduction in total MSE will be small if the variability of the errors in guessing $\boldsymbol{\theta}$ is large, that is, as $\sum_{i=1}^m (\theta_i - \theta_i^0)^2$ increases, $R(\boldsymbol{\theta}, \hat{\boldsymbol{\theta}}_{\text{JS}})$ tends to $R(\boldsymbol{\theta}, \hat{\boldsymbol{\theta}}) = m\psi$.

We now discuss several salient features of the JS method:

(1) The JS method is attractive to users wanting good overall efficiency for the group of small areas because large gains in efficiency can be achieved in the traditional design-based framework without assuming a model on the small area parameters θ_i.

(2) The JS estimator arises quite naturally in the empirical best linear unbiased prediction (EBLUP) approach or the empirical Bayes (EB) approach, assuming a random-effects model with $\theta_i \overset{\text{ind}}{\sim} N(\mathbf{z}_i^T\boldsymbol{\beta}, \sigma_v^2)$; see Section 6.1.

(3) A "plus-rule" estimator $\hat{\theta}_{i,\text{JS}}^+$ is obtained from (3.4.5) by changing the factor $1 - (m-2)\psi/S$ to 0 whenever $S < (m-2)\psi$, that is,

$$\hat{\theta}_{i,\text{JS}}^+ = \begin{cases} \theta_i^0 & \text{if } S < (m-2)\psi, \\ \hat{\theta}_{i,\text{JS}} & \text{if } S \geq (m-2)\psi. \end{cases} \tag{3.4.6}$$

The estimator $\hat{\boldsymbol{\theta}}_{\text{JS}}^+ = (\hat{\theta}_{1,\text{JS}}^+, \ldots, \hat{\theta}_{m,\text{JS}}^+)^T$ dominates $\hat{\boldsymbol{\theta}}_{\text{JS}}$ in terms of total MSE, that is,

$$R(\boldsymbol{\theta}, \hat{\boldsymbol{\theta}}_{\text{JS}}^+) < R(\boldsymbol{\theta}, \hat{\boldsymbol{\theta}}_{\text{JS}}), \quad \text{for all } \boldsymbol{\theta},$$

see Efron and Morris (1973).

(4) The dominance property is not necessarily true for the retransform of $\hat{\theta}_{i,\text{JS}}$, that is, the estimators $g^{-1}(\hat{\theta}_{i,\text{JS}}) = \hat{\overline{Y}}_{i,\text{JS}}$ may not dominate the direct estimators $g^{-1}(\hat{\theta}_i) = \hat{\overline{Y}}_i$ in terms of total MSE. Moreover, the estimator of domain totals resulting from $\hat{\overline{Y}}_{i,\text{JS}}$, namely, $N_i\hat{\overline{Y}}_{i,\text{JS}}$, may not dominate the direct estimators $N_i\hat{\overline{Y}}_i$ even when $g(\overline{Y}_i) = \overline{Y}_i$. Users may find the above lack of invariance undesirable.

(5) The assumption $E_p(\hat{\theta}_i) = \theta_i$ for nonlinear $g(\cdot)$ may not hold if the area sample size is very small. Moreover, the common sampling variance ψ is not known in practice. If an independent estimator $\hat{\psi}$ is available such that $\nu\,\hat{\psi}/\psi$ is a χ^2 variable with ν degrees of freedom (df), then we modify the estimator (3.4.5) by changing ψ to $\nu\,\hat{\psi}/(\nu + 2)$. The modified estimators retain the dominance property. For example, suppose that $\theta_i = \overline{Y}_i$ and we draw simple random samples $\{y_{ij}; j = 1, \ldots, \overline{n}\}$ independently from each small area i. The estimator of θ_i is $\hat{\theta}_i = \overline{y}_i$, the sample mean in the ith area. Under the assumption $y_{ij} \overset{\text{ind}}{\sim} N(\overline{Y}_i, \sigma^2)$, we have $\nu = m(\overline{n} - 1)$ and $\nu\hat{\psi} = \sum_{i=1}^m \sum_{j=1}^{\overline{n}} (y_{ij} - \overline{y}_i)^2$. If no

guess of $\overline{Y}_i$ is available, we take θ_i^0 as the overall mean $\overline{y}$ and change $m-2$ to $m-3$ $(m>3)$:

$$\hat{\overline{Y}}_{i,\mathrm{JS}} = \overline{y} + \left[1 - \frac{(m-3)\nu\,\hat{\psi}}{(\nu+2)S}\right](\overline{y}_i - \overline{y}), \tag{3.4.7}$$

where $S = \sum_{i=1}^{m}(\overline{y}_i - \overline{y})^2$. Estimators (3.4.7) dominate direct estimators $\overline{y}_i$ in terms of total MSE (Rao 1976). The normality assumption on $\{y_{ij}\}$ is used in establishing the dominance result. The assumptions of normality, equal sample sizes, and common variance are quite restrictive in the sample survey context.

(6) We have assumed that the direct estimators $\hat{\theta}_i$ are independent, but this may not hold when the small areas cut across the design strata and clusters. In this case, we assume that $\hat{\boldsymbol{\theta}}$ is m-variate normal, $N_m(\boldsymbol{\theta}, \boldsymbol{\Psi})$, and that an independent estimator of $\boldsymbol{\Psi}$ is based on a statistic $\mathbf{S}$ that is distributed as $W_m(\boldsymbol{\Psi}, \nu)$, a Wishart distribution with ν df. In this case, the JS estimator of θ_i is given by

$$\hat{\theta}_{i,\mathrm{JS}} = \theta_i^0 + \left(1 - \frac{(m-2)}{(\nu-m+3)Q}\right)(\hat{\theta}_i - \theta_i^0), \tag{3.4.8}$$

where $Q = (\hat{\boldsymbol{\theta}} - \boldsymbol{\theta}^0)^T \mathbf{S}^{-1}(\hat{\boldsymbol{\theta}} - \boldsymbol{\theta}^0)$ (see James and Stein 1961 and Bilodeau and Srivastava 1988).

(7) The JS method assumes no specific relationship between the θ_i's, such as $\theta_i \overset{\text{iid}}{\sim} N(\theta_i^0, \sigma_v^2)$, and yet uses all data $\{\hat{\theta}_\ell; \ell = 1, \ldots, m\}$ to estimate θ_i. Thus, the total MSE is reduced even when the different θ_i refer to obviously disjoint problems (e.g., some θ_i refer to batting averages of baseball players and the rest to per capita income of small areas in the United States), although the total MSE has no practical relevance in such situations.

(8) The JS method may perform poorly in estimating those components θ_i with unusually large or small deviations $\theta_i - \theta_i^0$. In fact, for large m, the maximum MSE for an individual area parameter θ_i over all areas can be as large as $(m/4)\psi$, using $\hat{\theta}_{i,\mathrm{JS}}$; for example, for $m = 16$ this maximum MSE is exactly equal to 4.41 times the MSE of $\hat{\theta}_i$ (Efron and Morris 1972a). To reduce this undesirable effect, Efron and Morris (1972a) proposed "limited translation estimators," which offer a compromise between the JS estimator and the direct estimator. These estimators have both good ensemble and good individual properties, unlike $\hat{\theta}_{i,\mathrm{JS}}$. A straightforward compromise estimator is obtained by restricting the amount by which $\hat{\theta}_{i,\mathrm{JS}}$ differs from $\hat{\theta}_i$ to a multiple of the standard error of $\hat{\theta}_i$:

$$\hat{\theta}^*_{i,\mathrm{JS}} = \begin{cases} \hat{\theta}_{i,\mathrm{JS}} & \text{if } \hat{\theta}_i - c\psi_i^{1/2} \le \hat{\theta}_{i,\mathrm{JS}} \le \hat{\theta}_i + c\psi_i^{1/2}, \\ \hat{\theta}_i - c\psi_i^{1/2} & \text{if } \hat{\theta}_{i,\mathrm{JS}} < \hat{\theta}_i - c\psi_i^{1/2}, \\ \hat{\theta}_i + c\psi_i^{1/2} & \text{if } \hat{\theta}_{i,\mathrm{JS}} > \hat{\theta}_i + c\psi_i^{1/2}, \end{cases} \tag{3.4.9}$$

where $c > 0$ is a suitably chosen constant. The choice $c = 1$, for example, ensures that $\text{MSE}_p(\hat{\theta}^*_{i,\text{JS}}) < 2\psi = 2\ \text{MSE}_p(\hat{\theta}_i)$, while retaining more than 80% of the gain of $\hat{\boldsymbol{\theta}}_{\text{JS}}$ over $\hat{\boldsymbol{\theta}}$ in terms of total MSE. We refer the reader to Efron and Morris (1972a) for further details on limited translation estimators.

Example 3.4.1. Batting Averages. Efron (1975) gave an amusing example of batting averages of major league baseball players in the United States to illustrate the superiority of JS estimators over direct estimators. Table 3.4 gives the batting averages, $\hat{P}_i$, of $m = 18$ players after their first 45 times at bat during the 1970 season. These estimates are taken as the direct estimates $\hat{\theta}_i = \hat{P}_i$. The JS estimates were calculated from (3.4.5) using as prior guess $\theta_i^0 = \sum_{i=1}^m \hat{P}_i/18 = 0.265 =: \hat{P}_.$ and ψ is taken as $\hat{P}_.(1 - \hat{P}_.)/45 = 0.0043$, the binomial variance. Note that $\hat{P}_i$ is treated as $N(P_i, \psi)$. The compromise JS estimates were obtained from (3.4.9) using $c = 1$. To compare the accuracies of the estimates, the batting average for each player i during the remainder of the season (about 370 more times at bat on average) was taken as the true value $\theta_i = P_i$. Table 3.4 also reports the values of the JS estimator $\hat{P}_{i,\text{JS}}$ and the compromise JS estimator $\hat{P}^*_{i,\text{JS}}$.

Because the true values P_i are assumed to be known, we can compare the relative overall accuracies using the ratios

$$R_1 = \sum_{i=1}^m (\hat{P}_i - P_i)^2 / \sum_{i=1}^m (\hat{P}_{i,\text{JS}} - P_i)^2$$

and

$$R_2 = \sum_{i=1}^m (\hat{P}_i - P_i)^2 / \sum_{i=1}^m (\hat{P}^*_{i,\text{JS}} - P_i)^2.$$

We get $R_1 = 3.50$, which means that the JS estimates outperform the direct estimates by a factor of 3.50. Also, $R_2 = 4.09$, so that the compromise JS estimates perform even better than the JS estimates in this example. It may be noted that the compromise JS estimator protects the proportion $\hat{P}_1 = 0.400$ of player 1 (Roberto Clemente) from overshrinking toward the common proportion $\hat{P}_. = 0.265$.

We treated $\hat{P}_i \overset{\text{ind}}{\sim} N(P_i, \psi)$ in calculating the JS estimates in Table 3.4. But it may be more reasonable to assume that $n\hat{P}_i \overset{\text{ind}}{\sim} \text{Bin}(n, P_i)$ with $n = 45$. Under this assumption, the variance of $n\hat{P}_i$ depends on P_i. Efron and Morris (1975) used the well-known arc-sine transformation to stabilize the variance of a binomial distribution. This transformation leads to $\hat{\theta}_i = \sqrt{n}\arcsin(2\hat{P}_i - 1) = g(\hat{P}_i)$ and $\theta_i = \sqrt{n}\arcsin(2P_i - 1) = g(P_i)$, and we have approximately $\hat{\theta}_i \overset{\text{ind}}{\sim} N(\theta_i, 1)$. The JS estimate of θ_i was calculated from (3.4.5) using the guess $\theta_i^0 = \sum_{i=1}^m \hat{\theta}_i/18 = \hat{\theta}_{..}$. The resulting estimates $\hat{\theta}_{i,\text{JS}}$ were retransformed to provide estimates $g^{-1}(\hat{\theta}_{i,\text{JS}})$ of the true proportions P_i. Efron and Morris (1975) calculated the overall accuracy of the estimators $\hat{\theta}_{i,\text{JS}}$ relative to the direct estimators $\hat{\theta}_i$ as $R = 3.50$. They also noted that $\hat{\theta}_{i,\text{JS}}$ is closer to θ_i than $\hat{\theta}_i$ for 15 out of $m = 18$ batters, and is worse only for batters 1, 10, and 15.

TABLE 3.4 Batting Averages for 18 Baseball Players

Player	Direct Estimate	True Value	JS Estimate	Compromise JS Estimate
1	0.400	0.346	0.293	0.334
2	0.378	0.298	0.289	0.312
3	0.356	0.276	0.284	0.290
4	0.333	0.221	0.279	0.279
5	0.311	0.273	0.275	0.275
6	0.311	0.270	0.275	0.275
7	0.289	0.263	0.270	0.270
8	0.267	0.210	0.265	0.265
9	0.244	0.269	0.261	0.261
10	0.244	0.230	0.261	0.261
11	0.222	0.264	0.256	0.256
12	0.222	0.256	0.256	0.256
13	0.222	0.304	0.256	0.256
14	0.222	0.264	0.256	0.256
15	0.222	0.226	0.256	0.256
16	0.200	0.285	0.251	0.251
17	0.178	0.319	0.247	0.243
18	0.156	0.200	0.242	0.221

Source: Adapted from Table 1 in Efron (1975).

3.4.3 Estimation of Component MSE

We now turn to the estimation of MSE of $\hat{\theta}_{i,\text{JS}}$ for each area i, assuming $\hat{\theta}_i \overset{\text{ind}}{\sim} N(\theta_i, \psi), i = 1, \ldots, m$ with known sampling variance ψ. For this purpose, we express $\hat{\theta}_{i,\text{JS}}$ as

$$\hat{\theta}_{i,\text{JS}} = \hat{\theta}_i + h_i(\hat{\boldsymbol{\theta}}) \tag{3.4.10}$$

with

$$h_i(\hat{\boldsymbol{\theta}}) = \frac{(m-2)\psi}{S}(\theta_i^0 - \hat{\theta}_i). \tag{3.4.11}$$

Using the representation (3.4.10), we have

$$\begin{aligned} \text{MSE}_p(\hat{\theta}_{i,\text{JS}}) &= E_p[\hat{\theta}_i + h_i(\hat{\boldsymbol{\theta}}) - \theta_i]^2 \\ &= \psi + 2E_p[(\hat{\theta}_i - \theta_i)h_i(\hat{\boldsymbol{\theta}})] + E_p[h_i^2(\hat{\boldsymbol{\theta}})]. \end{aligned}$$

By Corollary 3.5.1 to Stein's lemma given in Section 3.5, we can write

$$E_p[(\hat{\theta}_i - \theta_i)h_i(\hat{\boldsymbol{\theta}})] = \psi E_p[\partial h_i(\hat{\boldsymbol{\theta}})/\partial\hat{\theta}_i].$$

Thus,

$$\text{MSE}_p(\hat{\theta}_{i,\text{JS}}) = E_p[\psi + 2\psi\ \partial h_i(\hat{\boldsymbol{\theta}})/\partial\hat{\theta}_i + h_i^2(\hat{\boldsymbol{\theta}})].$$

Hence, an unbiased estimator of the MSE of $\hat{\theta}_{i,\text{JS}}$ is given by

$$\text{mse}_p(\hat{\theta}_{i,\text{JS}}) = \psi + 2\psi\ \partial h_i(\hat{\boldsymbol{\theta}})/\partial\hat{\theta}_i + h_i^2(\hat{\boldsymbol{\theta}}). \tag{3.4.12}$$

In fact, (3.4.12) is the minimum variance-unbiased estimator due to complete sufficiency of the statistic $\hat{\boldsymbol{\theta}}$ under normality. This estimator, however, can take negative values. Then, a better estimator is given by $\text{mse}_p^+(\hat{\theta}_{i,\text{JS}}) = \max[0, \text{mse}_p(\hat{\theta}_{i,\text{JS}})]$.

If the guess θ_i^0 is fixed, then using (3.4.11) we have

$$\frac{\partial h_i(\hat{\boldsymbol{\theta}})}{\partial\hat{\theta}_i} = -\frac{(m-2)\psi}{S}\left[1 - \frac{2(\hat{\theta}_i - \theta_i^0)^2}{S}\right]. \tag{3.4.13}$$

If θ_i^0 is taken as the mean $\hat{\theta}_{\cdot} = m^{-1}\sum_{i=1}^m \hat{\theta}_i$, then

$$h_i(\hat{\boldsymbol{\theta}}) = \frac{(m-3)\psi}{S}(\hat{\theta}_{\cdot} - \hat{\theta}_i) \tag{3.4.14}$$

and

$$\frac{\partial h_i(\hat{\boldsymbol{\theta}})}{\partial\hat{\theta}_i} = -\left(1 - \frac{1}{m}\right)\frac{(m-3)\psi}{S}\left[1 - \frac{2(\hat{\theta}_i - \hat{\theta}_{\cdot})^2}{S}\right], \tag{3.4.15}$$

The derivations for the case of a least squares predictor, $\theta_i^0 = \mathbf{z}_i^T\hat{\boldsymbol{\beta}}_{\text{LS}}$, can be obtained in a similar manner. Note that (3.4.12) is valid for a general differentiable function $h_i(\hat{\boldsymbol{\theta}})$, assuming $\hat{\theta}_i \overset{\text{ind}}{\sim} N(\theta_i, \psi)$.

Although the MSE estimator (3.4.12) is the minimum variance unbiased estimator, its coefficient of variation (CV) can be quite large, that is, it can be very unstable. Model-based estimators of MSE, considered in Chapter 6, are more stable but may be somewhat biased in the design-based framework.

Bilodeau and Srivastava (1988) considered the general case of correlated estimators $\hat{\theta}_i, i = 1, \ldots, m$, and the resulting estimator $\hat{\boldsymbol{\theta}}_{\text{JS}} = (\hat{\theta}_{1,\text{JS}}, \ldots, \hat{\theta}_{m,\text{JS}})^T$ given by (3.4.8). They obtained the minimum variance-unbiased estimator of the MSE matrix $\mathbf{M}(\hat{\boldsymbol{\theta}}_{\text{JS}}) = E_p(\hat{\boldsymbol{\theta}}_{\text{JS}} - \boldsymbol{\theta})(\hat{\boldsymbol{\theta}}_{\text{JS}} - \boldsymbol{\theta})^T$ as

$$\hat{\mathbf{M}}(\hat{\boldsymbol{\theta}}_{\text{JS}}) = \left(1 - \frac{2\kappa}{Q}\right)\frac{\mathbf{S}}{\nu} + \kappa\left[\kappa + \frac{4}{\nu}\frac{(\nu+1)}{(\nu-m+3)}\right]\frac{(\hat{\boldsymbol{\theta}}_{\text{JS}} - \boldsymbol{\theta}^0)(\hat{\boldsymbol{\theta}}_{\text{JS}} - \boldsymbol{\theta}^0)^T}{Q^2}, \tag{3.4.16}$$

where $\kappa = (m-2)/(\nu - m + 3)$. The ith diagonal element of (3.4.16) is the estimator of $\text{MSE}_p(\hat{\theta}_{i,\text{JS}})$.

Rivest and Belmonte (2000) generalized the MSE estimator (3.4.12) to the case of $\hat{\boldsymbol{\theta}} \sim N_m(\boldsymbol{\theta}, \boldsymbol{\Psi})$ with known covariance matrix $\boldsymbol{\Psi} = (\psi_{i\ell})$. In particular, if the estimators $\hat{\theta}_i$ are independent with $\psi_{i\ell} = 0$ when $i \neq \ell$, then

$$\text{mse}_p(\hat{\theta}_{i,\text{JS}}) = \psi_{ii} + 2\psi_{ii}\ \partial h_i(\hat{\boldsymbol{\theta}})/\partial\hat{\theta}_i + h_i^2(\hat{\boldsymbol{\theta}}), \tag{3.4.17}$$

where ψ_{ii} is the ith diagonal element of $\boldsymbol{\Psi}$.

Rivest and Belmonte (2000) noted that the derivative $\partial h_i(\hat{\theta})/\partial\hat{\theta}_i$ may be evaluated numerically when $h_i(\cdot)$ has no explicit form. For example, suppose that $\theta_i = \overline{Y}_i$ and a reliable direct estimator $\hat{Y} = \sum_{i=1}^m \hat{Y}_i$ of the population total is available, where $\hat{Y}_i = N_i\hat{\overline{Y}}_i$ and the domain size N_i is known. Then, it is desirable to adjust the JS estimator $\hat{Y}_{i,\text{JS}} = N_i\hat{\overline{Y}}_{i,\text{JS}}$ to add up to the direct estimator $\hat{Y}$. For example, a simple ratio adjustment of the form (3.3.11) may be used to ensure that the adjusted estimators add up to $\hat{Y}$. The ratio-adjusted JS estimator of the ith area total Y_i is given by

$$\hat{Y}_{i,\text{JS}}(a) = \frac{\hat{Y}_{i,\text{JS}}}{\sum_{i=1}^m \hat{Y}_{i,\text{JS}}}\hat{Y}, \tag{3.4.18}$$

where $\hat{Y}_{i,\text{JS}} = N_i[\hat{\overline{Y}}_i + h_i(\hat{\overline{Y}})] = \hat{Y}_i + N_i h_i(\hat{\overline{Y}})$. The estimator (3.4.18) may be written as $\hat{Y}_{i,\text{JS}}(a) = \hat{Y}_i + h_i^*(\hat{Y})$, where

$$h_i^*(\hat{Y}) = \frac{\sum_{i=1}^m \hat{Y}_i}{\sum_{i=1}^m \hat{Y}_{i,\text{JS}}} N_i h_i(\hat{\overline{Y}}) + \left(\frac{\sum_{i=1}^m \hat{Y}_i}{\sum_{i=1}^m \hat{Y}_{i,\text{JS}}} - 1\right)\hat{Y}_i. \tag{3.4.19}$$

Note that $\sum_{i=1}^m h_i^*(\hat{Y}) = 0$, which ensures that the adjusted JS estimators add up to $\hat{Y}$. Numerical differentiation may be used to evaluate $\partial h_i^*(\hat{Y})/\partial\hat{Y}_i$. An estimator of MSE of $\hat{Y}_{i,\text{JS}}(a)$ may then be obtained from (3.4.17) with $h_i(\cdot)$ changed to $h_i^*(\cdot)$.

3.4.4 Unequal Variances ψ_i

We now turn to the case of unequal but known sampling variances ψ_i. A straightforward way to generalize the JS method is to consider the rescaled parameters $\delta_i = \theta_i/\sqrt{\psi_i}$ and estimators $\hat{\delta}_i = \hat{\theta}_i/\sqrt{\psi_i}$. Then, it holds that $\hat{\delta}_i \overset{\text{ind}}{\sim} N(\delta_i, 1)$ and $\delta_i^0 = \theta_i^0/\sqrt{\psi_i}$ is taken as the guess of δ_i. We can now apply the JS estimator (3.4.5) to the transformed data and then transform back to the original coordinates. This leads to

$$\hat{\theta}_{i,\text{JS}} = \theta_i^0 + \left(1 - \frac{m-2}{\tilde{S}}\right)(\hat{\theta}_i - \theta_i^0), \quad m \geq 3, \tag{3.4.20}$$

where $\tilde{S} = \sum_{i=1}^m (\hat{\theta}_i - \theta_i^0)^2/\psi_i$. The estimator (3.4.20) dominates the direct estimator $\hat{\theta}_i$ in terms of a weighted MSE with weights $1/\psi_i$, but not in terms of the total MSE, that is,

$$\sum_{i=1}^m \frac{1}{\psi_i}\text{MSE}_p(\hat{\theta}_{i,\text{JS}}) < \sum_{i=1}^m \frac{1}{\psi_i}\text{MSE}_p(\hat{\theta}_i). \tag{3.4.21}$$

Moreover, it gives the common weight $\tilde{\phi}_{\text{JS}} = 1 - (m-2)/\tilde{S}$ to $\hat{\theta}_i$ and $1 - \tilde{\phi}_{\text{JS}}$ to the guess θ_i^0, that is, each $\hat{\theta}_i$ is shrunk toward the guess θ_i^0 by the same factor $\tilde{\phi}_{\text{JS}}$ regardless of its sampling variance ψ_i. This is not appealing to the user as in the case of the composite estimator with common weight $\hat{\phi}^*$ given by (3.4.3). One would like to have more shrinkage of $\hat{\theta}_i$, the larger the ψ_i is. The model-based methods of Chapters 5–8 provide such unequal shrinkage.

3.4.5 Extensions

Various extensions of the JS method have been studied in the literature. In particular, the dominance result holds under spherically symmetric or more generally, elliptical family of distributions for $\hat{\boldsymbol{\theta}}$ (Brandwein and Strawderman 1990, Srivastava and Bilodeau 1989) and for the exponential family of distributions (Ghosh and Auer 1983). The elliptical family of distributions include the normal, Student-t, and double exponential distributions.

Efron and Morris (1972b) extended the JS method to the case of a vector $\boldsymbol{\theta}_i$ of q different area characteristics. In particular, assume that $\hat{\boldsymbol{\theta}}_i \overset{\text{ind}}{\sim} N(\boldsymbol{\theta}_i, \boldsymbol{\Sigma})$ with known $\boldsymbol{\Sigma}$, and define the composite risk of an estimator $\tilde{\boldsymbol{\theta}} = (\tilde{\boldsymbol{\theta}}_1, \ldots, \tilde{\boldsymbol{\theta}}_m)$ of the $q \times m$ matrix of parameters $\boldsymbol{\theta} = (\boldsymbol{\theta}_1, \ldots, \boldsymbol{\theta}_m)$ as

$$R(\boldsymbol{\theta}, \tilde{\boldsymbol{\theta}}) = E_p\left[\text{tr}(\boldsymbol{\theta} - \tilde{\boldsymbol{\theta}})^T \boldsymbol{\Sigma}^{-1} (\boldsymbol{\theta} - \tilde{\boldsymbol{\theta}})\right],$$

where tr denotes the trace operator. Then, the estimators

$$\hat{\boldsymbol{\theta}}_{i,\text{JS}} = \boldsymbol{\theta}_i^0 + \left[\mathbf{I} - (m - q - 1)\boldsymbol{\Sigma}\mathbf{S}^{-1}\right]\hat{\boldsymbol{\theta}}_i, \quad i = 1, \ldots, m \tag{3.4.22}$$

dominate the direct estimators $\hat{\boldsymbol{\theta}}_i, i = 1, \ldots, m$, in terms of the above composite risk. In (3.4.22), $\boldsymbol{\theta}_i^0$ is a guess of $\boldsymbol{\theta}_i$ and

$$\mathbf{S} = (\hat{\boldsymbol{\theta}} - \boldsymbol{\theta}^0)(\hat{\boldsymbol{\theta}} - \boldsymbol{\theta}^0)^T$$

with $\hat{\boldsymbol{\theta}} = (\hat{\boldsymbol{\theta}}_1, \ldots, \hat{\boldsymbol{\theta}}_m)$ and $\boldsymbol{\theta}^0 = (\boldsymbol{\theta}_1^0, \ldots, \boldsymbol{\theta}_m^0)$; note that $\mathbf{S}$ is a $q \times q$ matrix. In the univariate case with $q = 1$, (3.4.22) reduces to the JS estimator (3.4.5).

3.5 PROOFS

To prove Theorem 3.4.1 in section 3.4.2, we need the following lemma attributed to Stein (1981) (see also Brandwein and Strawderman 1990).

Lemma 3.5.1. Let $Z \sim N(\mu, 1)$. Then, $E[h(Z)(Z - \mu)] = E[h'(Z)]$, provided the expectations exist and

$$\lim_{z \to \pm\infty} h(z) \exp\left[-\frac{1}{2}(z - \mu)^2\right] = 0, \tag{3.5.1}$$

where $h'(Z) = \partial h(Z)/\partial Z$.

Proof:

$$\begin{aligned} E[h(Z)(Z - \mu)] &= \frac{1}{\sqrt{2\pi}} \int_{-\infty}^{\infty} h(z)(z - \mu) \exp\left\{-\frac{1}{2}(z - \mu)^2\right\} dz \\ &= -\frac{1}{\sqrt{2\pi}} \int_{-\infty}^{\infty} h(z) \frac{d}{dz}\left[\exp\left\{-\frac{1}{2}(z - \mu)^2\right\}\right] dz. \end{aligned}$$

Integration by parts and (3.5.1) yield

$$E[h(Z)(Z-\mu)] = -\frac{1}{\sqrt{2\pi}}h(z)\exp\left[-\frac{1}{2}(z-\mu)^2\right]\Bigg|_{-\infty}^{\infty}$$
$$+\frac{1}{\sqrt{2\pi}}\int_{-\infty}^{\infty} h'(z)\exp\left[-\frac{1}{2}(z-\mu)\right]^2 dz$$
$$= E[h'(Z)].$$

■

Corollary 3.5.1. Let $\mathbf{Z} = (Z_1, \ldots, Z_m)^T$ with $Z_i \overset{\text{ind}}{\sim} N(\mu_i, 1), i = 1, \ldots, m$. If condition (3.5.1) holds, then $E[h(\mathbf{Z})(Z_i - \mu_i)] = E[\partial h(\mathbf{Z})/\partial Z_i]$, for a real-valued function $h(\cdot)$.

To establish Theorem 3.4.1, it is sufficient to prove Theorem 3.5.1, dealing with the canonical case $Z_i \overset{\text{ind}}{\sim} N(\mu_i, 1)$

Theorem 3.5.1. Let $\mathbf{Z} = (Z_1, \ldots, Z_m)^T$ with $Z_i \overset{\text{ind}}{\sim} N(\mu_i, 1), i = 1, \ldots, m$, for $m \geq 3$, and

$$\hat{\boldsymbol{\mu}}_{\text{JS}}(a) = \left(1 - \frac{a}{\|\mathbf{Z}\|^2}\right)\mathbf{Z}, \quad 0 < a < 2(m-2),$$

where $\|\mathbf{Z}\|^2 = \mathbf{Z}^T\mathbf{Z}$. Then,

(a) $\hat{\boldsymbol{\mu}}_{\text{JS}}(a)$ dominates $\mathbf{Z}$ for $0 < a < 2(m-2)$ in terms of total MSE, and the optimal choice of a is $a = m-2$, which ensures

$$R(\boldsymbol{\mu}, \hat{\boldsymbol{\mu}}_{\text{JS}}) < R(\boldsymbol{\mu}, \mathbf{Z}) = m \quad \text{for all } \boldsymbol{\mu} = (\mu_1, \ldots, \mu_m)^T,$$

where $\hat{\boldsymbol{\mu}}_{\text{JS}}$ is equal to $\hat{\boldsymbol{\mu}}_{\text{JS}}(a)$ evaluated at $a = m-2$.

(b) $R(\boldsymbol{\mu}, \hat{\boldsymbol{\mu}}_{\text{JS}}) = 2$ at $\boldsymbol{\mu} = \mathbf{0}$.

(c) $R(\boldsymbol{\mu}, \hat{\boldsymbol{\mu}}_{\text{JS}}) \leq m - \frac{(m-2)^2}{(m-2)+\|\boldsymbol{\mu}\|^2}$, so that $R(\boldsymbol{\mu}, \hat{\boldsymbol{\mu}}_{\text{JS}})$ is minimized when $\boldsymbol{\mu} = \mathbf{0}$.

Proof:

(a) We follow Brandwein and Strawderman (1990). We have

$$R[\boldsymbol{\mu}, \hat{\boldsymbol{\mu}}_{\text{JS}}(a)] = E\|\hat{\boldsymbol{\mu}}_{\text{JS}}(a) - \boldsymbol{\mu}\|^2$$
$$= E\|\mathbf{Z} - \boldsymbol{\mu}\|^2 + a^2 E\left(\frac{1}{\mathbf{Z}^T\mathbf{Z}}\right) - 2aE\left[\frac{\mathbf{Z}^T(\mathbf{Z}-\boldsymbol{\mu})}{\mathbf{Z}^T\mathbf{Z}}\right]$$
$$= m + a^2 E\left(\frac{1}{\mathbf{Z}^T\mathbf{Z}}\right) - 2a\sum_{i=1}^{m} E\left[\frac{Z_i(Z_i-\mu_i)}{\mathbf{Z}^T\mathbf{Z}}\right].$$

Using Corollary 3.5.1 with $h(\mathbf{Z}) = Z_i/(\mathbf{Z}^T\mathbf{Z})$, we get

$$R[\boldsymbol{\mu}, \hat{\boldsymbol{\mu}}_{\rm JS}(a)] = m + a^2 E\left(\frac{1}{\mathbf{Z}^T\mathbf{Z}}\right) - 2a \sum_{i=1}^{m} E\left[\frac{\partial}{\partial Z_i}\left(\frac{Z_i}{\mathbf{Z}^T\mathbf{Z}}\right)\right]. \qquad (3.5.2)$$

Next, evaluating the derivative in the last term of (3.5.2), we get

$$\begin{aligned} R(\boldsymbol{\mu}, \hat{\boldsymbol{\mu}}_{\rm JS}(a)) &= m + a^2 E\left(\frac{1}{\mathbf{Z}^T\mathbf{Z}}\right) - 2a \sum_{i=1}^{m} E\left[\frac{\mathbf{Z}^T\mathbf{Z} - 2Z_i^2}{(\mathbf{Z}^T\mathbf{Z})^2}\right] \\ &= m + a^2 E\left(\frac{1}{\mathbf{Z}^T\mathbf{Z}}\right) - 2a(m-2)E\left(\frac{1}{\mathbf{Z}^T\mathbf{Z}}\right) \\ &= m + [a^2 - 2a(m-2)]E\left(\frac{1}{\mathbf{Z}^T\mathbf{Z}}\right). \qquad (3.5.3) \end{aligned}$$

The quadratic function $a^2 - 2a(m-2)$ in a is negative in the range $0 < a < 2(m-2)$ so that $\hat{\boldsymbol{\mu}}_{\rm JS}(a)$ dominates $\mathbf{Z}$ in that range, noting that $R(\boldsymbol{\mu}, \mathbf{Z}) = m$. Furthermore, it attains its minimum at $a = m-2$ so that the optimum choice of a is $m-2$.

(b) At $\boldsymbol{\mu} = 0$, $\mathbf{Z}^T\mathbf{Z}$ is a chi-squared random variable with m df so that $E(1/\mathbf{Z}^T\mathbf{Z}) = 1/(m-2)$ and (3.5.3) for $a = m-2$ reduces to

$$R(\boldsymbol{\mu}, \hat{\boldsymbol{\mu}}_{\rm JS}) = m - (m-2) = 2.$$

(c) We follow Casella and Hwang (1982). We note that $\mathbf{Z}^T\mathbf{Z}$ is a noncentral chi-squared variable with m df and noncentrality parameter $\tau = \|\boldsymbol{\mu}\|^2/2$, denoted by $\chi^2_m(\tau)$. Furthermore, we use the well-known fact that a noncentral chi-squared distribution is the infinite sum of central chi-squared distributions with Poisson weights. It now follows that

$$E\left(\frac{1}{\mathbf{Z}^T\mathbf{Z}}\right) = E\left[\frac{1}{\chi^2_m(\tau)}\right] = E\left[E\left(\frac{1}{\chi^2_{m+2K}}\,\middle|\, K\right)\right],$$

where χ^2_{m+2K} is a (central) chi-squared variable with $m + 2K$ df and K is a Poisson random variable with mean τ. Therefore, by Jensen's inequality $E(1/X) \geq 1/E(X)$ for a positive random variable X, it follows that

$$E\left(\frac{1}{\mathbf{Z}^T\mathbf{Z}}\right) = E\left(\frac{1}{m+2K-2}\right) \geq \frac{1}{m-2+2\tau} = \frac{1}{m-2+\|\boldsymbol{\mu}\|^2}.$$

It now follows from (3.5.3) that

$$R(\boldsymbol{\mu}, \hat{\boldsymbol{\mu}}_{\rm JS}) = m - (m-2)^2 E\left(\frac{1}{\mathbf{Z}^T\mathbf{Z}}\right) \leq m - \frac{(m-2)^2}{(m-2) + \|\boldsymbol{\mu}\|^2}.$$

■

Proof of Theorem 3.4.1. Let $Z_i = (\hat{\theta}_i - \theta_i^0)/\sqrt{\psi}$ in Theorem 3.5.1, so that $\mu_i = (\theta_i - \theta_i^0)/\sqrt{\psi}$, and

$$R(\boldsymbol{\mu}, \mathbf{Z}) = \frac{1}{\psi} R(\boldsymbol{\theta}, \hat{\boldsymbol{\theta}}), \quad R(\boldsymbol{\mu}, \hat{\boldsymbol{\mu}}_{\text{JS}}) = \frac{1}{\psi} R(\boldsymbol{\theta}, \hat{\boldsymbol{\theta}}_{\text{JS}}).$$

Hence, Theorem 3.4.1 follows from Theorem 3.5.1.

4

SMALL AREA MODELS

4.1 INTRODUCTION

Traditional methods of indirect estimation, studied in Chapter 3, are based on implicit models that provide a link to related small areas through supplementary data. We now turn to explicit small area models that make specific allowance for between-area variation. In particular, we introduce mixed models involving random area-specific effects that account for between-area variation beyond that explained by auxiliary variables included in the model. The use of explicit models offers several advantages: (i) Model diagnostics can be used to find suitable model(s) that fit the data well. Such model diagnostics include residual analysis to detect departures from the assumed model, selection of auxiliary variables for the model, and case-deletion diagnostics to detect influential observations. (ii) Area-specific measures of precision can be associated with each small area estimate, unlike the global measures (averaged over small areas) often used with synthetic estimates (Chapter 3, Section 3.2.5). (iii) Linear mixed models as well as nonlinear models, such as logistic regression models and generalized linear models with random area effects, can be entertained. Complex data structures, such as spatial dependence and time series, can also be handled. (iv) Recent methodological developments for random effects models can be utilized to achieve accurate small area inferences.

Although we present a variety of models for small area estimation in this chapter, it is important to note that the subject matter specialists or end users should have influence on the choice of models, particularly on the choice of auxiliary variables. Also,

Small Area Estimation, Second Edition. J. N. K. Rao and Isabel Molina.

the success of any model-based method depends on the availability of good auxiliary data. More attention should therefore be given to the compilation of auxiliary variables that are good predictors of the study variables.

We present models that permit empirical best linear unbiased prediction (EBLUP) and empirical Bayes (EB) inferences (Chapters 5–9). Additional assumptions on the model parameters, in the form of prior distributions, are needed to implement the hierarchical Bayes (HB) approach (Chapter 10). The models we study may be classified into two broad types: (i) Aggregate level (or area level) models that relate the small area means to area-specific auxiliary variables. Such models are essential if unit (element) level data are not available. (ii) Unit level models that relate the unit values of the study variable to unit-specific auxiliary variables. Models involving both unit level and area level auxiliary variables will also be studied.

4.2 BASIC AREA LEVEL MODEL

We assume that $\theta_i = g(\overline{Y}_i)$, for some specified $g(\cdot)$, is related to area-specific auxiliary data $\mathbf{z}_i = (z_{1i}, \ldots, z_{pi})^T$ through a linear model

$$\theta_i = \mathbf{z}_i^T \boldsymbol{\beta} + b_i v_i, \quad i = 1, \ldots, m, \tag{4.2.1}$$

where the b_i's are known as positive constants and $\boldsymbol{\beta} = (\beta_1, \ldots, \beta_p)^T$ is the $p \times 1$ vector of regression coefficients ($m > p$). Furthermore, the v_i's are area-specific random effects assumed to be independent and identically distributed (iid) with

$$E_m(v_i) = 0, \quad V_m(v_i) = \sigma_v^2 \ (\geq 0), \tag{4.2.2}$$

where E_m denotes the model expectation and V_m the model variance. We denote this assumption as $v_i \stackrel{\text{iid}}{\sim} (0, \sigma_v^2)$. Normality of the random effects v_i is also often used, but it is possible to make "robust" inferences by relaxing the normality assumption (Chapter 6, Section 6.3). The parameter σ_v^2 is a measure of homogeneity of the areas after accounting for the covariates $\mathbf{z}_i$.

In some applications, not all areas are selected in the sample (see Example 4.2.2). Suppose that we have M areas in the population and only m areas are selected in the sample. We assume a model of the form (4.2.1) for the population, that is, $\theta_i = \mathbf{z}_i^T \boldsymbol{\beta} + b_i v_i$, $i = 1, \ldots, M$. We further assume that the sample areas obey the population model, that is, the bias in the sample selection of areas is absent so that (4.2.1) holds for the sampled areas.

For making inferences about the small area means $\overline{Y}_i$ under model (4.2.1), we assume that direct estimators $\hat{\overline{Y}}_i$ are available. As in the James–Stein method (Chapter 3, Section 3.4), we assume that

$$\hat{\theta}_i = g(\hat{\overline{Y}}_i) = \theta_i + e_i, \quad i = 1, \ldots, m, \tag{4.2.3}$$

where the sampling errors e_i are independent with

$$E_p(e_i|\theta_i) = 0, \quad V_p(e_i|\theta_i) = \psi_i. \tag{4.2.4}$$

It is also customary to assume that the sampling variances, ψ_i, are known. The above assumptions may be quite restrictive in some applications. For example, the direct estimator $\hat{\theta}_i$ may be design-biased for θ_i if $g(\cdot)$ is a nonlinear function and the area sample size n_i is small. The sampling errors may not be independent if the small areas cut across the strata or clusters of the sampling design. The assumption of known sampling variances, ψ_i, can be relaxed by estimating ψ_i from the unit level sample data and then smoothing the estimated variances $\hat{\psi}_i$ to get a more stable estimate of ψ_i. This smoothing is called the generalized variance function (GVF) approach (Wolter 2007). Normality of the estimator $\hat{\theta}_i$ is also often assumed, but this may not be as restrictive as the normality of the random effects, due to the central limit theorem's effect on $\hat{\theta}_i$.

Deterministic models on θ_i are obtained by setting $\sigma_v^2 = 0$, that is, $\theta_i = \mathbf{z}_i^T \boldsymbol{\beta}$. Such models lead to synthetic estimators that do not account for local variation other than the variation reflected in the auxiliary variables $\mathbf{z}_i$.

Combining (4.2.1) with (4.2.3), we obtain the model

$$\hat{\theta}_i = \mathbf{z}_i^T \boldsymbol{\beta} + b_i v_i + e_i, \quad i = 1, \ldots, m. \tag{4.2.5}$$

Note that (4.2.5) involves design-induced errors e_i as well as model errors v_i. We assume that v_i and e_i are independent. Model (4.2.5) is a special case of a linear mixed model (Chapter 5).

The assumption $E_p(e_i|\theta_i) = 0$ in the sampling model (4.2.3) may not be valid if the sample size n_i in the ith area is small and θ_i is a nonlinear function of the total Y_i, even if the direct estimator $\hat{Y}_i$ is design-unbiased. Then, a more realistic sampling model is given by

$$\hat{Y}_i = Y_i + \varepsilon_i, \quad i = 1, \ldots, m, \tag{4.2.6}$$

with $E_p(\varepsilon_i|Y_i) = 0$, that is, $\hat{Y}_i$ is design-unbiased for the total Y_i. In this case, the sampling and linking models are not matched. As a result, we cannot combine (4.2.6) with the linking model (4.2.1) to produce a single linear mixed model of the form (4.2.5). Therefore, standard results in linear mixed model theory, presented in Chapter 5, do not apply. In Section 10.4, we use a HB approach to handle unmatched sampling and linking models.

Example 4.2.1. Income for Small Places. In the context of estimating per capita income (PCI) for small places in the United States with population less than 1,000, Fay and Herriot (1979) used model (4.2.5) with $b_i = 1$. In their application, $\theta_i = \log(\overline{Y}_i)$, where $\overline{Y}_i$ is the PCI in the ith area. Model (4.2.5) is called the Fay–Herriot (FH) model in the small area literature because they were the first to use such models for small area estimation. Some details of this application are given in Example 6.1.1.

Example 4.2.2. US Poverty Counts. The basic area level model (4.2.5) has been used in the small area income and poverty estimation (SAIPE) program in the

United States to produce model-based county estimates of poor school-age children (National Research Council 2000). Using these estimates, the US Department of Education allocates annually several billion dollars of general funds to counties, and then states distribute these funds among school districts. In the past, funds were allocated on the basis of estimated counts from the previous census, but the poverty counts have changed significantly over time.

In this application, $\theta_i = \log(Y_i)$, where Y_i is the true poverty count in the ith county (small area) and a direct estimator $\hat{Y}_i$ of Y_i was obtained from the current population survey (CPS). Area level predictor variables, $\mathbf{z}_i$, were obtained from census and administrative records. Current details of this application are given in Example 6.1.2.

Example 4.2.3. Census Undercount. In the context of estimating the undercount in the decennial census of the United States, Ericksen and Kadane (1985) used model (4.2.5) with $b_i = 1$ and treating σ_v^2 as known. In their application, $\theta_i = (T_i - C_i)/T_i$ is the census undercount for the ith state (area), where T_i is the true (unknown) count and C_i is the census count in the ith area, and $\hat{\theta}_i$ is the direct estimate of θ_i from a post-enumeration survey (PES). Cressie (1989) also used model (4.2.5) with $b_i = 1/\sqrt{C_i}$ to estimate the census adjustment factors $\theta_i = T_i/C_i$, using the PES estimates, $\hat{\theta}_i$, of θ_i. In these applications, the PES estimates $\hat{\theta}_i$ could be seriously biased, as noted by Freedman and Navidi (1986).

In the context of the Canadian Census, Dick (1995) used model (4.2.5) with $\theta_i = T_i/C_i$ and $b_i = 1$, where i denotes province × age × sex combination. He used smoothed estimates of the sampling variances ψ_i, by assuming ψ_i to be proportional to some power of the true census count C_i. Some details of this application are given in Example 6.1.3.

4.3 BASIC UNIT LEVEL MODEL

We assume that unit-specific auxiliary data $\mathbf{x}_{ij} = (x_{ij1}, \ldots, x_{ijp})^T$ are available for each population element j in each small area i. It is often sufficient to assume that only the population means $\overline{\mathbf{X}}_i$ are known. Furthermore, the variable of interest, y_{ij}, is assumed to be related to $\mathbf{x}_{ij}$ through the basic nested error linear regression model given by

$$y_{ij} = \mathbf{x}_{ij}^T \boldsymbol{\beta} + v_i + e_{ij}, \quad j = 1, \ldots, N_i, \quad i = 1, \ \ldots \ m. \tag{4.3.1}$$

Here the area-specific effects v_i are assumed to be iid random variables satisfying (4.2.2), $e_{ij} = k_{ij}\tilde{e}_{ij}$ for known constants k_{ij} and the $\tilde{e}_{ij}$'s are iid random variables independent of the v_i's, with

$$E_m(\tilde{e}_{ij}) = 0, \quad V_m(\tilde{e}_{ij}) = \sigma_e^2. \tag{4.3.2}$$

In addition, normality of the v_i's and e_{ij}'s is often assumed. The parameters of interest are the small area means $\overline{Y}_i$ or the totals Y_i. Standard regression models are obtained by setting $\sigma_v^2 = 0$ or equivalently $v_i = 0$ in (4.3.1). Such models lead to synthetic-type estimators (Chapter 3, Section 3.2.3).

We assume that a sample s_i of size n_i is taken from the N_i units in the ith area ($i = 1, \ldots, m$), and that the sample values also obey the assumed model (4.3.1). The latter assumption is satisfied under simple random sampling from each area or more generally for sampling designs that use the auxiliary information $\mathbf{x}_{ij}$ in the selection of the samples s_i. To see this, we write (4.3.1) in matrix form as

$$\mathbf{y}_i^P = \mathbf{X}_i^P \boldsymbol{\beta} + v_i \mathbf{1}_i^P + \mathbf{e}_i^P, \quad i = 1, \ldots, m, \tag{4.3.3}$$

where $\mathbf{X}_i^P$ is a $N_i \times p$ matrix, $\mathbf{y}_i^P$ and $\mathbf{e}_i^P$ are $N_i \times 1$ vectors, and $\mathbf{1}_i^P$ is the $N_i \times 1$ vector of ones. We next partition (4.3.3) into sampled and nonsampled parts:

$$\mathbf{y}_i^P = \begin{bmatrix} \mathbf{y}_i \\ \mathbf{y}_{ir} \end{bmatrix} = \begin{bmatrix} \mathbf{X}_i \\ \mathbf{X}_{ir} \end{bmatrix} \boldsymbol{\beta} + v_i \begin{bmatrix} \mathbf{1}_i \\ \mathbf{1}_{ir} \end{bmatrix} + \begin{bmatrix} \mathbf{e}_i \\ \mathbf{e}_{ir} \end{bmatrix}, \tag{4.3.4}$$

where the subscript r denotes the nonsampled units. If the model holds for the sample, that is, if selection bias is absent, then inferences on $\boldsymbol{\lambda} = (\boldsymbol{\beta}^T, \sigma_v^2, \sigma_e^2)^T$ are based on

$$f(\mathbf{y}_i | \mathbf{X}_i^P, \boldsymbol{\lambda}) = \int f(\mathbf{y}_i, \mathbf{y}_{ir} | \mathbf{X}_i^P, \boldsymbol{\lambda}) d\mathbf{y}_{ir}, \quad i = 1, \ldots, m, \tag{4.3.5}$$

where $f(\mathbf{y}_i, \mathbf{y}_{ir} | \mathbf{X}_i^P, \boldsymbol{\lambda})$ is the assumed joint distribution of $\mathbf{y}_i$ and $\mathbf{y}_{ir}$. On the other hand, defining the vector of indicators $\mathbf{a}_i = (a_{i1}, \ldots, a_{iN_i})^T$, where $a_{ij} = 1$ if $j \in s_i$ and $a_{ij} = 0$ otherwise, the joint distribution of sample data $(\mathbf{y}_i, \mathbf{a}_i)$ is given by

$$\begin{aligned} f(\mathbf{y}_i, \mathbf{a}_i | \mathbf{X}_i^P, \boldsymbol{\lambda}) &= \int f(\mathbf{y}_i, \mathbf{y}_{ir} | \mathbf{X}_i^P, \boldsymbol{\lambda}) f(\mathbf{a}_i | \mathbf{y}_i, \mathbf{y}_{ir}, \mathbf{X}_i^P) d\mathbf{y}_{ir} \\ &= \left[\int f\left(\mathbf{y}_i, \mathbf{y}_{ir} | \mathbf{X}_i^P, \boldsymbol{\lambda}\right) d\mathbf{y}_{ir} \right] f(\mathbf{a}_i | \mathbf{X}_i^P), \end{aligned}$$

provided

$$f(\mathbf{a}_i | \mathbf{y}_i, \mathbf{y}_{ir}, \mathbf{X}_i^P) = f(\mathbf{a}_i | \mathbf{X}_i^P),$$

that is, the sample selection probabilities do not depend on $\mathbf{y}_i^P$ but may depend on $\mathbf{X}_i^P$. In this case, selection bias is absent and we may assume that the sample values also obey the assumed model, that is, use $f(\mathbf{y}_i | \mathbf{X}_i^P, \boldsymbol{\lambda})$ for inferences on $\boldsymbol{\lambda}$ (Smith 1983).

If the sample selection probabilities depend on an auxiliary variable $\mathbf{z}_i^P$ that is not included in $\mathbf{X}_i^P$, then the distribution of sample data $(\mathbf{y}_i, \mathbf{a}_i)$ is

$$f(\mathbf{y}_i, \mathbf{a}_i | \mathbf{X}_i^P, \mathbf{z}_i^P, \boldsymbol{\lambda}) = \left[\int f\left(\mathbf{y}_i, \mathbf{y}_{ir} | \mathbf{X}_i^P, \mathbf{z}_i^P, \boldsymbol{\lambda}\right) d\mathbf{y}_{ir} \right] f(\mathbf{a}_i | \mathbf{z}_i^P, \mathbf{X}_i^P).$$

In this case, inference on $\boldsymbol{\lambda}$ is based on $f(\mathbf{y}_i | \mathbf{X}_i^P, \mathbf{z}_i^P, \boldsymbol{\lambda})$, which is different from (4.3.5) unless $\mathbf{z}_i^P$ is unrelated to $\mathbf{y}_i^P$ given $\mathbf{X}_i^P$. Therefore, we have sample selection bias and we cannot assume that the model (4.3.3) holds for the sample values. We could extend model (4.3.3) by including $\mathbf{z}_i^P$ and then test for the significance of the associated regression coefficient using the sample data. If the null hypothesis is not rejected, then we could assume that the original model (4.3.3) also holds for the sample values (Skinner 1994).

The model (4.3.3) is also not appropriate under two-stage cluster sampling within small areas because random cluster effects are not incorporated. But we can extend the model to account for such features (Section 4.5.2).

We write the small area mean $\overline{Y}_i$ as

$$\overline{Y}_i = f_i\overline{y}_i + (1-f_i)\overline{y}_{ir} \tag{4.3.6}$$

with $f_i = n_i/N_i$, where $\overline{y}_i$ and $\overline{y}_{ir}$ denote, respectively, the means of the sampled and nonsampled elements. It follows from (4.3.6) that estimating the small area mean $\overline{Y}_i$ is equivalent to estimating the realization of the random variable $\overline{y}_{ir}$ given the sample data $\{\mathbf{y}_i\}$ and the auxiliary data $\{\mathbf{X}_i^P\}$.

If the population size N_i is large, then the small area mean $\overline{Y}_i$ is approximately equal to

$$\mu_i = \overline{\mathbf{X}}_i^T\boldsymbol{\beta} + v_i \tag{4.3.7}$$

noting that $\overline{Y}_i = \overline{\mathbf{X}}_i^T\boldsymbol{\beta} + v_i + \overline{E}_i$ and $\overline{E}_i \approx 0$, where $\overline{E}_i$ is the mean of the N_i errors e_{ij} and $\overline{\mathbf{X}}_i$ is the known mean of $\mathbf{X}_i^P$. It follows that the estimation of $\overline{Y}_i$ is approximately equivalent to the estimation of a linear combination of $\boldsymbol{\beta}$ and the realization of the random variable v_i.

We now give some examples of model (4.3.1). Some details of these applications are given in Chapter 6, Section 7.3.

Example 4.3.1. County Crop Areas. Battese, Harter, and Fuller (1988) used the nested error regression model (4.3.1) to estimate county crop areas using sample survey data in conjunction with satellite information. In particular, they were interested in estimating the area under corn and soybeans for each of $m = 12$ counties in North-Central Iowa using farm-interview data as $\{\mathbf{y}_i\}$ and LANDSAT satellite data as $\{\mathbf{X}_i^P\}$. Each county was divided into area segments, and the areas under corn and soybeans were ascertained for a sample of segments by interviewing farm operators. The number of sampled segments in a county, n_i, ranged from 1 to 6. Auxiliary data in the form of number of pixels (a term used for "picture elements" of about 0.45 ha) classified as corn and soybeans were also obtained for all the area segments, including the sampled segments, in each county using the LANDSAT satellite readings. Battese et al. (1988) proposed the model

$$y_{ij} = \beta_0 + \beta_1 x_{ij1} + \beta_2 x_{ij2} + v_i + \tilde{e}_{ij}, \tag{4.3.8}$$

which is a special case of model (4.3.1) with $k_{ij} = 1$, $\mathbf{x}_{ij} = (1, x_{ij1}, x_{ij2})^{\mathrm{T}}$, and $\boldsymbol{\beta} = (\beta_0, \beta_1, \beta_2)^{\mathrm{T}}$. Here, y_{ij} = number of hectares of corn (or soybeans), x_{ij1} = number of pixels classified as corn, and x_{ij2} = number of pixels classified as soybeans in the jth area segment of the ith county. Some details of this application are given in Example 7.3.1.

Example 4.3.2. Wages and Salaries. Rao and Choudhry (1995) studied the population of unincorporated tax filers from the province of Nova Scotia, Canada (Example 2.4.1). They proposed the model

$$y_{ij} = \beta_0 + \beta_1 x_{ij} + v_i + x_{ij}^{1/2}\tilde{e}_{ij}, \tag{4.3.9}$$

which is a special case of (4.3.1) with $k_{ij} = x_{ij}^{1/2}$. Here, y_{ij} and x_{ij} denote, respectively, the total wages and salaries and gross business income for the jth firm in the ith area. Simple random sampling from the overall population was used to estimate the area totals Y_i or the means $\overline{Y}_i$. Some details of a simulation study based on this data are given in Example 7.3.2.

4.4 EXTENSIONS: AREA LEVEL MODELS

We now consider various extensions of the basic area level model (4.2.5).

4.4.1 Multivariate Fay–Herriot Model

Suppose we want to estimate an $r \times 1$ vector of area characteristics $\boldsymbol{\theta}_i = (\theta_{i1}, \ldots, \theta_{ir})^T$, where $\theta_{ij} = g_j(\overline{Y}_{ij})$ and $\overline{Y}_{ij}$ is the ith small area mean for the jth characteristic, $j = 1, \ldots, r$, and $\hat{\boldsymbol{\theta}}_i = (\hat{\theta}_{i1}, \ldots, \hat{\theta}_{ir})^T$ is the vector of survey estimators. We consider the multivariate sampling model

$$\hat{\boldsymbol{\theta}}_i = \boldsymbol{\theta}_i + \mathbf{e}_i, \quad i = 1, \ldots, m, \tag{4.4.1}$$

where the sampling errors $\mathbf{e}_i = (e_{i1}, \ldots, e_{ir})^T$ are independent r-variate normal, $N_r(\mathbf{0}, \boldsymbol{\Psi}_i)$, with mean $\mathbf{0}$ and known covariance matrices $\boldsymbol{\Psi}_i$ conditional on $\boldsymbol{\theta}_i$. Here $\mathbf{0}$ is the $r \times 1$ null vector. We further assume that $\boldsymbol{\theta}_i$ is related to area-specific auxiliary data $\{\mathbf{z}_{ij}\}$ through the linear model

$$\boldsymbol{\theta}_i = \mathbf{Z}_i\boldsymbol{\beta} + \mathbf{v}_i, \quad i = 1, \ldots, m, \tag{4.4.2}$$

where the area-specific random effects $\mathbf{v}_i$ are independent $N_r(\mathbf{0}, \boldsymbol{\Sigma}_v)$, $\mathbf{Z}_i$ is an $r \times rp$ matrix with jth row given by $(\mathbf{0}^T, \ldots, \mathbf{0}^T, \mathbf{z}_{ij}^T, \mathbf{0}^T, \ldots, \mathbf{0}^T)$, and $\boldsymbol{\beta}$ is the rp-vector of regression coefficients. Here, $\mathbf{0}$ is the $p \times 1$ null vector and $\mathbf{z}_{ij}^T$ occurs in the jth position of the row vector (j th row).

Combining (4.4.1) with (4.4.2), we obtain a multivariate mixed linear model

$$\hat{\boldsymbol{\theta}}_i = \mathbf{Z}_i\boldsymbol{\beta} + \mathbf{v}_i + \mathbf{e}_i. \tag{4.4.3}$$

The model (4.4.3) is a natural extension of the FH model (4.2.5) with $b_i = 1$. Fay (1987) and Datta, Fay, and Ghosh (1991) proposed the multivariate extension (4.4.3) and demonstrated that it can lead to more efficient estimators of the small area means $\overline{Y}_{ij}$ because it takes advantage of the correlations between the components of $\widehat{\boldsymbol{\theta}}_i$ unlike the univariate model (4.2.5).

Example 4.4.1. Median Income. Datta, Fay, and Ghosh (1991) applied the multivariate model (4.4.3) to estimate the current median income for four-person families in each of the American states (small areas). These estimates are used in a formula to determine the eligibility for a program of energy assistance to low-income families administered by the U.S. Department of Health and Human Services. In this application, $\boldsymbol{\theta}_i = (\theta_{i1}, \theta_{i2})^T$ with θ_{i1} = population median income of four-person families in state i and $\theta_{i2} = \frac{3}{4}$ (population median income of five-person families in state i) $+\frac{1}{4}$ (population median income of three-person families in state i). Direct estimates $\widehat{\boldsymbol{\theta}}_i$ and the associated covariance matrix $\widehat{\boldsymbol{\Psi}}_i$ were obtained from the CPS, and $\boldsymbol{\Psi}_i$ was treated as known by letting $\boldsymbol{\Psi}_i = \widehat{\boldsymbol{\Psi}}_i$ and ignoring the variability of the estimate $\widehat{\boldsymbol{\Psi}}_i$. Here θ_{i2} is not a parameter of interest, but the associated direct estimator $\widehat{\theta}_{i2}$ is strongly related to $\widehat{\theta}_{i1}$. By taking advantage of this association, an improved estimator of the parameter of interest θ_{i1} can be obtained through the multivariate model (4.4.3).

The auxiliary data $\{\mathbf{z}_{ij}\}$ was based on census data: $\mathbf{z}_{ij} = (1, z_{ij1}, z_{ij2})^T$, $j = 1, 2$, where z_{i11} and z_{i12} denote, respectively, the adjusted census median income for the current year and the base-year census median income for four-person families in the ith state, and z_{i21} and z_{i22} denote, respectively, the weighted average (with weights $\frac{3}{4}$ and $\frac{1}{4}$) of adjusted census median incomes for three-person and five-person families and the corresponding weighted average (with the same weights) of the base-year census medians in the ith state. The adjusted census median incomes for the current year were obtained by multiplying the census median incomes by adjustment factors produced by the Bureau of Economic Analysis of the U.S. Department of Commerce. Some details of this application are given in Examples 4.4.3 and 10.10.1.

4.4.2 Model with Correlated Sampling Errors

A natural extension of the basic FH model with independent sampling errors (4.2.5) is to consider the correlated sampling errors e_i. Define the vectors $\widehat{\boldsymbol{\theta}} = (\widehat{\theta}_1, \ldots, \widehat{\theta}_m)^T$, $\boldsymbol{\theta} = (\theta_1, \ldots, \theta_m)^T$, and $\mathbf{e} = (e_1, \ldots, e_m)^T$, and assume that

$$\widehat{\boldsymbol{\theta}} = \boldsymbol{\theta} + \mathbf{e}, \tag{4.4.4}$$

with $\mathbf{e}|\boldsymbol{\theta} \sim N_m(\mathbf{0}, \boldsymbol{\Psi})$, where the sampling error covariance matrix $\boldsymbol{\Psi} = (\psi_{i\ell})$ is known. Combining (4.4.4) with the model (4.2.1) for the θ_i's, we obtain a generalization of the basic FH model (4.2.5). If $b_i = 1$, $i = 1, \ldots, m$, in (4.2.1), then the combined model may be written as

$$\widehat{\boldsymbol{\theta}} = \mathbf{Z}\boldsymbol{\beta} + \mathbf{v} + \mathbf{e}, \tag{4.4.5}$$

where $\mathbf{v} = (v_1, \ldots, v_m)^T$ and $\mathbf{Z}$ is an $m \times p$ matrix with ith row equal to $\mathbf{z}_i^T$. In practice, $\boldsymbol{\Psi}$ is replaced by a survey estimator $\widehat{\boldsymbol{\Psi}}$ or a smoothed estimator, but the variability associated with the estimator is often ignored.

Example 4.4.2. US Census Undercoverage. In the context of estimating the undercount in the 1990 Census of the United States, the population was divided into $m = 357$ poststrata composed of 51 poststratum groups, each of which was subdivided into 7 age–sex categories. The 51 poststratum groups were defined on the basis of race/ethnicity, tenure (owner, renter), type of area, and region. Dual-system estimates $\hat{\theta}_i$ for each poststratum $i = 1, \ldots, 357$ were obtained using the data from the 1990 PES. We refer the reader to Section 3.4, Chapter 3 of Rao (2003a), for details of dual system estimation. Here, $\hat{\theta}_i$ is the estimated census adjustment factor for the ith poststratum. A model of the form (4.4.5) was employed to obtain smoothed estimates of the adjustment factors θ_i (Isaki, Tsay, and Fuller 2000). In a previous study (Isaki, Hwang, and Tsay 1991), 1392 poststrata were employed, and smoothed estimates of adjustment factors were obtained. Some details of this application are given in Example 8.2.1.

4.4.3 Time Series and Cross-Sectional Models

Many sample surveys are repeated in time with partial replacement of the sample elements. For example, in the monthly US CPS, an individual household remains in the sample for four consecutive months, then drops out of the sample for the eight succeeding months, and then comes back for another four consecutive months. In the monthly Canadian labour force survey (LFS), an individual household remains in the sample for six consecutive months and then drops out of the sample. For such repeated surveys, considerable gain in efficiency can be achieved by borrowing strength across both areas and time.

Rao and Yu (1992, 1994) proposed an extension of the basic FH model (4.2.5) to handle time series and cross-sectional data. Their model consists of a sampling error model

$$\hat{\theta}_{it} = \theta_{it} + e_{it}, \quad t = 1, \ldots, T, \ \ i = 1, \ldots, m \tag{4.4.6}$$

and a linking model

$$\theta_{it} = \mathbf{z}_{it}^T \boldsymbol{\beta} + v_i + u_{it}. \tag{4.4.7}$$

Here $\theta_{it} = g(\overline{Y}_{it})$ is a function of the small area mean $\overline{Y}_{it}$, $\hat{\theta}_{it}$ is the direct survey estimator for small area i at time t, the sampling errors e_{it} are normally distributed given the θ_{it}'s with zero means and a known block diagonal covariance matrix $\boldsymbol{\Psi}$ with blocks $\boldsymbol{\Psi}_i$, and $\mathbf{z}_{it}$ is a vector of area-specific covariates, some of which may change with t, for example, administrative data. Furthermore, $v_i \overset{\text{iid}}{\sim} N(0, \sigma_v^2)$ and the u_{it}'s are assumed to follow a common first-order autoregressive process for each area i, that is,

$$u_{it} = \rho u_{i,t-1} + \varepsilon_{it}, \quad |\rho| < 1, \tag{4.4.8}$$

with $\varepsilon_{it} \overset{\text{iid}}{\sim} N(0, \sigma^2)$. The errors $\{e_{it}\}$, $\{v_i\}$, and $\{\varepsilon_{it}\}$ are also assumed to be independent of each other. Models of the form (4.4.7) and (4.4.8) have been extensively used in the econometrics literature (Anderson and Hsiao 1981), but ignoring sampling errors e_{it}.

The model (4.4.7) on the θ_{it}'s depends on both area effects v_i and the area-by-time effects u_{it} that are correlated across time for each area i. We can also express (4.4.7) as a distributed-lag model

$$\theta_{it} = \rho\theta_{i,t-1} + (\mathbf{z}_{it} - \rho\mathbf{z}_{i,t-1})^T\boldsymbol{\beta} + (1-\rho)v_i + \varepsilon_{it}. \tag{4.4.9}$$

The alternative form (4.4.9) relates θ_{it} to the previous period area parameter $\theta_{i,t-1}$, the values of the auxiliary variables for the time points t and $t-1$, the random area effects v_i, and the area-by-time effects ε_{it}. More complex models on the u_{it}'s than (4.4.8) can be formulated by assuming an autoregressive moving average (ARMA) process, but the resulting efficiency gains relative to (4.4.8) are unlikely to be significant in the small area estimation context.

Ghosh, Nangia, and Kim (1996) proposed a different time series cross-sectional model for small area estimation given by

$$\hat{\theta}_{it}|\theta_{it} \overset{\text{ind}}{\sim} N(\theta_{it}, \psi_{it}), \tag{4.4.10}$$

$$\theta_{it}|\boldsymbol{\alpha}_t \overset{\text{ind}}{\sim} N(\mathbf{z}_{it}^T\boldsymbol{\beta} + \mathbf{w}_{it}^T\boldsymbol{\alpha}_t, \sigma_t^2), \tag{4.4.11}$$

and

$$\boldsymbol{\alpha}_t|\boldsymbol{\alpha}_{t-1} \overset{\text{ind}}{\sim} N_r(\mathbf{H}_t\boldsymbol{\alpha}_{t-1}, \boldsymbol{\Delta}). \tag{4.4.12}$$

Here $\mathbf{z}_{it}$ and $\mathbf{w}_{it}$ are vectors of area-specific covariates, the sampling variances ψ_{it} are assumed to be known, $\boldsymbol{\alpha}_t$ is a $r \times 1$ vector of time-specific random effects, and $\mathbf{H}_t$ is a known $r \times r$ matrix. The dynamic (or state-space) model (4.4.12) in the univariate case ($r = 1$) with $H_t = 1$ reduces to the well-known random walk model. The above model suffers from two major limitations: (i) The direct estimators $\hat{\theta}_{it}$ are assumed to be independent over time for each i. This assumption is not realistic in the context of repeated surveys with overlapping samples, such as the CPS and the Canadian LFS. (ii) Area-specific random effects are not included in the model, which leads to excessive shrinkage of small area estimators similar to synthetic estimators.

Datta, Lahiri, and Maiti (2002) and You (1999) used the Rao–Yu sampling and linking models (4.4.6) and (4.4.7), but replaced the AR(1) model (4.4.8) on the u_{it}'s by a random walk model given by (4.4.8) with $\rho = 1$, that is, $u_{it} = u_{it-1} + \varepsilon_{it}$. Datta et al. (1999) considered a similar model, but added extra terms to the linking model to reflect seasonal variation in their application.

Pfeffermann and Burck (1990) proposed a general model involving area-by-time random effects. Their model is of the form

$$\hat{\theta}_{it} = \theta_{it} + e_{it}, \tag{4.4.13}$$

$$\theta_{it} = \mathbf{z}_{it}^T\boldsymbol{\beta}_{it}, \tag{4.4.14}$$

where the coefficients $\boldsymbol{\beta}_{it} = (\beta_{it0}, \ldots, \beta_{itp})^T$ are allowed to vary cross-sectionally (over areas) and over time, and the sampling errors e_{it} for each area i are assumed to be serially uncorrelated with mean 0 and variance ψ_{it}. The variation of β_{it} over time is specified by the following state-space model:

$$\begin{bmatrix} \beta_{itj} \\ \beta_{ij} \end{bmatrix} = \mathbf{T}_j \begin{bmatrix} \beta_{i,t-1,j} \\ \beta_{ij} \end{bmatrix} + \begin{bmatrix} 1 \\ 0 \end{bmatrix} v_{itj}, \quad j = 0, 1, \ldots, p. \tag{4.4.15}$$

Here the β_{ij}'s are fixed coefficients, $\mathbf{T}_j$ is a known 2×2 matrix with $(0, 1)$ as the second row, and the model errors $\{v_{itj}\}$ for each area i are uncorrelated over time with mean 0 and covariances $E_m(v_{itj}v_{it\ell}) = \sigma_{vj\ell}; j, \ell = 0, 1, \ldots, p$.

The formulation (4.4.15) covers several useful models. First, the choice $\mathbf{T}_j = \begin{bmatrix} 0 & 1 \\ 0 & 1 \end{bmatrix}$ gives the well-known random-coefficient regression model $\beta_{itj} = \beta_{ij} + v_{itj}$ (Swamy 1971). The familiar random walk model $\beta_{itj} = \beta_{i,t-1,j} + v_{itj}$ is obtained by choosing $\mathbf{T}_j = \begin{bmatrix} 1 & 0 \\ 0 & 1 \end{bmatrix}$. In this case, the coefficient β_{ij} in (4.4.15) is redundant and should be omitted so that $\mathbf{T}_j = 1$. The choice $\mathbf{T}_j = \begin{bmatrix} \rho & 1-\rho \\ 0 & 1 \end{bmatrix}$ gives the AR(1) model $\beta_{itj} - \beta_{ij} = \rho(\beta_{i,t-1,j} - \beta_{ij}) + v_{itj}$. The state space model (4.4.15) is quite general, but the assumption of serially uncorrelated sampling errors e_{it} in (4.4.13) is restrictive in the context of repeated surveys with overlapping samples.

Example 4.4.3. Median Income. Ghosh, Nangia, and Kim (1996) applied the model (4.4.10)–(4.4.12) to estimate the median income of four-person families for the 50 American states and the District of Columbia ($m = 51$). The US Department of Health and Human Services uses these estimates to formulate its energy assistance program for low-income families. They used CPS data for $T = 9$ years (1981–1989) to estimate θ_{iT}, the median incomes for year 1989 ($i = 1, \ldots, 51$). (Here $\mathbf{z}_{it} = (1, z_{it1})^T$ with z_{it1} denoting the "adjusted" census median income for year t and area i, which is obtained by adjusting the base year (1979) census median income by the proportional growth in PCI. Using the same data, Datta, Lahiri, and Maiti (2002) estimated median income of four-person families applying model (4.4.6) and (4.4.7) with a random walk for u_{it}. Some details of this application are given in Example 8.3.1.

Example 4.4.4. Canadian Unemployment Rates. You, Rao, and Gambino (2003) applied the Rao–Yu models (4.4.6) and (4.4.7) to estimate monthly unemployment rates for cities with population over 100,000, called Census Metropolitan Areas (CMAs), and other urban centers, called Census Agglomerations (CAs), in Canada. Reliable estimates at the CMA and CA levels are used by the employment insurance (EI) program to determine the rules used to administer the program. Direct estimates of unemployment rates from the Canadian LFS are reliable for the nation and the provinces, but many CMAs and CAs do not have a large enough sample to produce reliable direct estimates from the LFS.

EI beneficiary rates were used as auxiliary data, z_{it}, in the linking model (4.4.7). Both AR(1) and random walk models on the area-by-time random effects, u_{it}, were considered. Some details of this application are given in Example 10.9.1.

Example 4.4.5. US Unemployment Rates. Datta et al. (1999) applied models (4.4.6) and (4.4.7) with a random walk for u_{it} and seasonal variation to estimate monthly unemployment rates for 49 American states (excluding the state of New York) and the District of Columbia ($m = 50$). These estimates are used by various federal agencies for the allocation of funds and policy formulation. They considered the period January 1985–December 1988 and used the CPS estimates as $\hat{\theta}_{it}$ and the unemployment insurance (UI) claims rate (percentage of unemployed workers claiming UI benefits among the total nonagricultural employment) as auxiliary data, $\mathbf{z}_{it}$. Seasonal variation in monthly unemployment rates was accounted for by introducing random month and year effects into the model. Some details of this application are given in Example 10.9.2.

4.4.4 *Spatial Models

The basic FH model (4.2.5) assumes iid area effects v_i, but in some applications it may be more realistic to entertain models that allow correlations among the v_i's. Spatial models on the area effects v_i are used when "neighboring" areas can be defined for each area i. Such models induce correlations among the v_i's depending, for example, on geographical proximity in the context of estimating local disease and mortality rates. Cressie (1991) used a spatial model for small area estimation in the context of US census undercount.

If A_i denotes a set of "neighboring" areas of area i, then a conditional autoregression (CAR) spatial model assumes that the conditional distribution of $b_i v_i$, given the area effects for the other areas $\{v_\ell : \ell \neq i\}$, is given by

$$b_i v_i | \{v_\ell : \ell \neq i\} \sim N\left(\rho \sum_{\ell \in A_i} q_{i\ell} b_\ell v_\ell, b_i^2 \sigma_v^2\right). \tag{4.4.16}$$

Here $\{q_{i\ell}\}$ are known constants satisfying $q_{i\ell} b_\ell^2 = q_{\ell i} b_i^2$ $(i < \ell)$, and $\boldsymbol{\delta} = (\rho, \sigma_v^2)^T$ is the unknown parameter vector. The model (4.4.16) implies that

$$\mathbf{B}^{1/2}\mathbf{v} \sim N_m(\mathbf{0}, \boldsymbol{\Gamma}(\boldsymbol{\delta}) = \sigma_v^2(\mathbf{I} - \rho\mathbf{Q})^{-1}\mathbf{B}), \tag{4.4.17}$$

where $\mathbf{B} = \text{diag}(b_1^2, \ldots, b_m^2)$ and $\mathbf{Q} = (q_{i\ell})$ is an $m \times m$ matrix with $q_{i\ell} = 0$ whenever $\ell \notin A_i$ (including $q_{ii} = 0$) and $\mathbf{v} = (v_1, \ldots, v_m)^T$ (see Besag 1974). Using (4.4.17) in (4.2.5), we obtain a spatial small area model. Note that $\boldsymbol{\delta} = (\rho, \sigma_v^2)^T$ appears nonlinearly in $\boldsymbol{\Gamma}(\boldsymbol{\delta})$.

In the geostatistics literature, covariance structures of the form (i) $\boldsymbol{\Gamma}(\delta) = \sigma_v^2(\delta_1\mathbf{I} + \delta_2\mathbf{D})$ and (ii) $\boldsymbol{\Gamma}(\delta) = \sigma_v^2[\delta_1\mathbf{I} + \delta_2\mathbf{D}(\delta_3)]$ have been used, where $\mathbf{D} = (e^{-d_{i\ell}})$ and $\mathbf{D}(\delta_3) = (\delta_3^{d_{i\ell}})$ are $m \times m$ matrices with $d_{i\ell}$ denoting a "distance" (not necessarily

Euclidean) between small areas i and ℓ. Note that in case (i), the parameters δ_1 and δ_2 appear linearly in $\boldsymbol{\Gamma}(\boldsymbol{\delta})$, whereas in case (ii), δ_2 and δ_3 appear nonlinearly in $\boldsymbol{\Gamma}(\boldsymbol{\delta})$.

Consider now the basic FH model (4.2.5) with $b_i = 1$, $i = 1, \ldots, m$. An alternative approach to assuming a model for $v_i|\{v_\ell : \ell \neq i\}$, referred to as conditional approach (Besag 1974), is to assume a model for the joint distribution of the vector $\mathbf{v} = (v_1, \ldots, v_m)^T$ based on the simultaneous equations

$$\mathbf{v} = \phi\mathbf{W}\mathbf{v} + \mathbf{u}, \quad \mathbf{u} \sim N(\mathbf{0}, \sigma_u^2\mathbf{I}). \tag{4.4.18}$$

Model (4.4.18) is referred to as simultaneously autoregressive process (SAR). Similarly to $\mathbf{Q}$ in (4.4.17), the matrix $\mathbf{W}$ in (4.4.18) describes the neighborhood structure of the small areas and ϕ represents the strength of the spatial relationship among the random effects associated with neighboring areas and the only requirement is that $\mathbf{I} - \phi\mathbf{W}$ is nonsingular. Then, (4.4.18) is equivalent to

$$\mathbf{v} = (\mathbf{I} - \phi\mathbf{W})^{-1}\mathbf{u} \sim N(\mathbf{0}, \mathbf{G}(\boldsymbol{\delta})), \tag{4.4.19}$$

with $\mathbf{G}(\boldsymbol{\delta}) = \sigma_u^2[(\mathbf{I} - \phi\mathbf{W})(\mathbf{I} - \phi\mathbf{W})^T]^{-1}$, for $\boldsymbol{\delta} = (\phi, \sigma_u^2)^T$.

A simple choice of $\mathbf{W}$ is the symmetric binary contiguity matrix $\tilde{\mathbf{W}} = (\tilde{w}_{i\ell})$, in which $\tilde{w}_{i\ell} = 1$ if $\ell \neq i$ and $\ell \in A_i$, and $\tilde{w}_{i\ell} = 0$ otherwise, where $A_i \subseteq \{1, \ldots, m\}$ is again the set of neighbor areas of area i. Instead, if one considers the row-standardized matrix $\mathbf{W} = (w_{i\ell})$ with $w_{i\ell} = \tilde{w}_{i\ell}/\sum_{k=1}^m \tilde{w}_{ik}$, then $\phi \in (-1, 1)$. In this case, ϕ can be interpreted as a correlation coefficient and is called spatial autocorrelation parameter (Banerjee, Carlin and Gelfand 2004). Petrucci and Salvati (2006) used the spatial model defined by (4.2.5) and (4.4.18) to estimate the amount of erosion delivered to streams in the Rathbun Lake Watershed in Iowa by $m = 61$ sub-watersheds. Pratesi and Salvati (2008) used the same model to estimate the mean PCI in $m = 43$ sub-regions of Tuscany, named Local Economic Systems, using data from the 2001 Life Conditions Survey from Tuscany.

A drawback of the spatial models (4.4.17) or (4.4.18) is that they depend on how the neighborhoods A_i are defined. Therefore, these models introduce some subjectivity (Marshall 1991).

Example 4.4.6. US Census Undercount. Cressie (1991) extended his model with $b_i = 1/\sqrt{C_i}$ for estimating the US census undercount (Example 4.2.3), by allowing spatial dependence through the CAR model (4.4.17). By exploratory spatial data analysis, he defined

$$q_{i\ell} = \begin{cases} \sqrt{C_\ell/C_i} & \text{if } d_{i\ell} \leq 700 \text{ miles}, \quad i \neq \ell, \\ 0 & \text{otherwise,} \end{cases}$$

where C_i is the census count in the ith state and $d_{i\ell}$ is the distance between the centers of gravity of the ith and ℓth states (small areas). Other choices of $q_{i\ell}$, not necessarily distance-based, may be chosen. Cressie (1991) noted that the sociologist's or ethnographer's map of the small areas may be quite different from the geographer's map.

For example, New York City and the rest of New York State may not be "neighbors" for the purpose of undercount estimation and it is more reasonable to regard other big cities such as Detroit, Chicago, and Los Angeles as "neighbors" of New York City. Some details of this application are given in Example 8.4.1.

4.4.5 Two-Fold Subarea Level Models

We now consider that each area i is subdivided into N_i subareas, and the parameters of interest are the area means θ_i together with the subarea means θ_{ij} ($j = 1, \ldots, N_i$, $i = 1, \ldots, m$). A linking model in this case is given by $\theta_{ij} = \mathbf{z}_{ij}^T \boldsymbol{\beta} + v_i + u_{ij}$, where $\mathbf{z}_{ij}$ is a $p \times 1$ vector of subarea level auxiliary variables ($m > p$), $\boldsymbol{\beta}$ is the $p \times 1$ vector of regression parameters and area effects $v_i \overset{\text{iid}}{\sim} N(0, \sigma_v^2)$ are independent of subarea effects $u_{ij} \overset{\text{iid}}{\sim} N(0, \sigma_u^2)$. We assume that n_i subareas are sampled from the N_i subareas in area i. Furthermore, $\hat{\theta}_{ij}$ is a direct estimator of the subarea mean θ_{ij}, based on a sample of n_{ij} units selected from N_{ij} units within subarea ij. The sampling model is given by $\hat{\theta}_{ij} = \theta_{ij} + e_{ij}$, where $e_{ij}|\theta_{ij} \overset{\text{ind}}{\sim} N(0, \psi_{ij})$, with known ψ_{ij}. Now combining the sampling model with the linking model leads to the two-fold subarea level model:

$$\hat{\theta}_{ij} = \mathbf{z}_{ij}^T \boldsymbol{\beta} + v_i + u_{ij} + e_{ij}, \quad j = 1, \ldots, n_i, \quad i = 1, \ldots, m, \tag{4.4.20}$$

see Torabi and Rao (2014). Note that the mean of area i is given by $\theta_i = \sum_{j=1}^{N_i} N_{ij}\theta_{ij}/N_{i\cdot}$, where $N_{i\cdot} = \sum_{j=1}^{N_i} N_{ij}$ is the population count in area i. Model (4.4.20) enables us to estimate both area means θ_i and subarea means θ_{ij} ($j = 1, \ldots, N_i$, $i = 1, \ldots, m$), by borrowing strength from related areas and subareas.

Fuller and Goyeneche (1998) proposed a subarea model similar to (4.4.20) in the context of SAIPE (see Example 4.2.2). In this application, county is the subarea, which is nested within a state (area) and direct county estimates were obtained from the CPS data. County level auxiliary variables were obtained from census and administrative records, as noted in Example 4.2.2.

4.5 EXTENSIONS: UNIT LEVEL MODELS

We now consider various extensions of the basic unit level model (4.3.1).

4.5.1 Multivariate Nested Error Regression Model

As in Section 4.3, we assume that unit-specific auxiliary data $\mathbf{x}_{ij}$ are available for all the population elements j in each small area i. We further assume that an $r \times 1$ vector of variables of interest, $\mathbf{y}_{ij}$, is related to $\mathbf{x}_{ij}$ through a multivariate nested error regression model (Fuller and Harter 1987):

$$\mathbf{y}_{ij} = \mathbf{B}\mathbf{x}_{ij} + \mathbf{v}_i + \mathbf{e}_{ij}, \quad j = 1, \ldots, N_i, \quad i = 1, \ldots, m. \tag{4.5.1}$$

Here $\mathbf{B}$ is $r \times p$ matrix of regression coefficients, the $r \times 1$ vectors of area effects $\mathbf{v}_i$ are assumed to be iid with mean $\mathbf{0}$ and covariance matrix $\mathbf{\Sigma}_v$, and the $r \times 1$ vectors of errors $\mathbf{e}_{ij}$ are iid with mean $\mathbf{0}$ and covariance matrix $\mathbf{\Sigma}_e$ and independent of $\mathbf{v}_i$. In addition, normality of the $\mathbf{v}_i$'s and $\mathbf{e}_{ij}$'s is often assumed. The model (4.5.1) is a natural extension of the (univariate) nested error regression model (4.3.1) with $k_{ij} = 1$.

The target parameters are the vectors of area means $\overline{\mathbf{Y}}_i = N_i^{-1} \sum_{j=1}^{N_i} \mathbf{y}_{ij}$, which can be approximated by $\boldsymbol{\mu}_i = \mathbf{B}\overline{\mathbf{X}}_i + \mathbf{v}_i$ if the population size N_i is large. In the latter case, it follows that the estimation of $\overline{\mathbf{Y}}_i$ is equivalent to the estimation of a linear combination of $\mathbf{B}$ and the realization of the random vector $\mathbf{v}_i$. As in the multivariate FH model, the unit level model (4.5.1) can lead to more efficient estimators by taking advantage of the correlations between the components of $\mathbf{y}_{ij}$, unlike the univariate model (4.3.1).

Example 4.5.1. County Crop Areas. In Example 4.3.1, we could take $\mathbf{y}_{ij} = (y_{ij1}, y_{ij2})^T$ with y_{ij1} = number of hectares of corn and y_{ij2} = number of hectares of soybeans, and retain the same $\mathbf{x}_{ij} = (1, x_{ij1}, x_{ij2})^T$ with x_{ij1} = number of pixels classified as corn and x_{ij2} = number of pixels classified as soybeans. By taking advantage of the correlation between y_{ij1} and y_{ij2}, an improved estimator of $\boldsymbol{\mu}_i = (\mu_{i1}, \mu_{i2})^T$ can be obtained through the multivariate nested error model (4.5.1). Some details of a simulation study based on this data are given in Example 8.6.1.

4.5.2 Two-Fold Nested Error Regression Model

Suppose that the ith small area contains M_i primary units (or clusters), and that the jth primary unit (cluster) in the ith area contains N_{ij} subunits (elements). Let $(y_{ij\ell}, \mathbf{x}_{ij\ell})$ be the y and $\mathbf{x}$-values for the ℓth element in the jth primary unit from the ith area ($\ell = 1, \ldots, N_{ij}, j = 1, \ldots, M_i, i = 1, \ldots, m$). Under this population structure, it is a common practice to employ two-stage cluster sampling in each small area. A sample, s_i, of m_i clusters is selected from the ith area. From the jth sampled cluster, a subsample, s_{ij}, of n_{ij} elements is selected and the associated y and x-values are observed.

The foregoing population structure is reflected by the two-fold nested error regression model (Stukel and Rao 1999) given by

$$y_{ij\ell} = \mathbf{x}_{ij\ell}^T \beta + v_i + u_{ij} + e_{ij\ell} \quad \ell = 1, \ldots, N_{ij}, \; j = 1, \ldots, M_i, . \quad i = 1, \ldots, m. \tag{4.5.2}$$

Here the area effects $\{v_i\}$, the cluster effects $\{u_{ij}\}$, and the residual errors $\{e_{ij\ell}\}$ with $e_{ij\ell} = k_{ij\ell}\tilde{e}_{ij\ell}$, for known constants $k_{ij\ell}$, are assumed to be mutually independent. Furthermore, $v_i \overset{\text{iid}}{\sim} (0, \sigma_v^2)$, $u_{ij} \overset{\text{iid}}{\sim} (0, \sigma_u^2)$, and $\tilde{e}_{ij\ell} \overset{\text{iid}}{\sim} (0, \sigma_e^2)$; normality of the random components v_i, u_{ij}, and $\tilde{e}_{ij\ell}$ is also often assumed. We consider that model (4.5.2) holds also for the sample values, which is true under simple random sampling of clusters and subunits within sampled clusters or, more generally, for sampling designs that use the auxiliary information $\mathbf{x}_{ij\ell}$ in the selection of the sample. Datta and Ghosh (1991) used the model (4.5.2) for the special case of cluster-specific covariates, that is, with $\mathbf{x}_{ij\ell} = \mathbf{x}_{ij}$ for $\ell = 1, \ldots, N_{ij}$. Ghosh and Lahiri (1998) studied the case of no auxiliary information, in which $\mathbf{x}_{ij\ell}^T \boldsymbol{\beta} = \beta$ for all i, j, and ℓ.

The parameters of interest are the small area means, which can be expressed as

$$\overline{Y}_i = \frac{1}{N_i}\left[\sum_{j\in s_i}\sum_{\ell\in s_{ij}} y_{ij\ell} + \sum_{j\in s_i}\sum_{\ell\in r_{ij}} y_{ij\ell} + \sum_{j\in r_i}\sum_{\ell=1}^{N_{ij}} y_{ij\ell}\right], \tag{4.5.3}$$

where r_i and r_{ij} denote, respectively, the set of nonsampled clusters in area i and the set of nonsampled subunits in cluster ij. If the population size of area i, $N_i = \sum_{j=1}^{N_i} N_{ij}$, is large, then $\overline{Y}_i$ may be approximated as

$$\overline{Y}_i \approx \overline{\mathbf{X}}_i^T \boldsymbol{\beta} + v_i, \tag{4.5.4}$$

noting that $\overline{Y}_i = \overline{\mathbf{X}}_i^T \boldsymbol{\beta} + v_i + \overline{U}_i + \overline{E}_i$ with $\overline{U}_i \approx 0$ and $\overline{E}_i \approx 0$, where $\overline{U}_i$ and $\overline{E}_i$ are the area means of u_{ij} and $e_{ij\ell}$ and $\overline{\mathbf{X}}_i$ is the known mean of the $\mathbf{x}_{ij\ell}$'s. It follows from (4.5.4) that the estimation of $\overline{Y}_i$ is equivalent to the estimation of a linear combination of $\boldsymbol{\beta}$ and the realization of the random variable v_i.

4.5.3 Two-Level Model

The basic unit level model (4.3.1) with intercept β_1 may be expressed as a model with random area-specific intercepts $\beta_{1i} = \beta_1 + v_i$ and common slopes $\beta_2, \ldots, \beta_p$, that is, $y_{ij} = \beta_{1i} + \beta_2 x_{ij2} + \cdots + \beta_p x_{ijp} + e_{ij}$. This model form suggests a more general model that allows between-area variation in the slopes beyond that in the intercepts. Thus, we consider random-coefficients $\boldsymbol{\beta}_i = (\beta_{i1}, \ldots, \beta_{ip})^T$ and then model $\boldsymbol{\beta}_i$ in terms of area level covariates $\tilde{\mathbf{Z}}_i$, which leads to the two-level small area model (Moura and Holt 1999) given by

$$y_{ij} = \mathbf{x}_{ij}^T \boldsymbol{\beta}_i + e_{ij}, \quad j = 1, \ldots, N_i, \quad i = 1, \ldots, m \tag{4.5.5}$$

with

$$\boldsymbol{\beta}_i = \tilde{\mathbf{Z}}_i \boldsymbol{\alpha} + \mathbf{v}_i, \tag{4.5.6}$$

where $\tilde{\mathbf{Z}}_i$ is a $p \times q$ matrix, $\boldsymbol{\alpha}$ is a $q \times 1$ vector of regression parameters, $\mathbf{v}_i \stackrel{\text{iid}}{\sim} (\mathbf{0}, \Sigma_v)$ and $e_{ij} = k_{ij}\tilde{e}_{ij}$ with $\tilde{e}_{ij} \stackrel{\text{iid}}{\sim} (0, \sigma_e^2)$. We may express (4.5.5) in a matrix form as

$$\mathbf{y}_i^P = \mathbf{X}_i^P \boldsymbol{\beta}_i + \mathbf{e}_i^P. \tag{4.5.7}$$

The two-level model given by (4.5.6)–(4.5.7) effectively integrates the use of unit level and area level covariates into a single model:

$$\mathbf{y}_i^P = \mathbf{X}_i^P \tilde{\mathbf{Z}}_i \boldsymbol{\alpha} + \mathbf{X}_i^P \mathbf{v}_i + \mathbf{e}_i^P. \tag{4.5.8}$$

Furthermore, the use of random slopes, $\boldsymbol{\beta}_i$, permits greater flexibility in modeling. The sample values $\{(y_{ij}, \mathbf{x}_{ij}); j = 1, \ldots, n_i, i = 1, \ldots, m\}$ are also assumed to obey the model (4.5.8), that is, there is no sample selection bias. If N_i is large, we can express the mean $\overline{Y}_i$ under (4.5.8) as

$$\mu_i = \overline{\mathbf{X}}_i^T \tilde{\mathbf{Z}}_i \boldsymbol{\alpha} + \overline{\mathbf{X}}_i^T \mathbf{v}_i. \tag{4.5.9}$$

It follows from (4.5.9) that the estimation of $\overline{Y}_i$ is approximately equivalent to the estimation of a linear combination of $\boldsymbol{\alpha}$ and the realization of the random vector $\mathbf{v}_i$ with unknown covariance matrix $\boldsymbol{\Sigma}_v$.

The model (4.5.8) is a special case of a general linear mixed model used extensively for longitudinal data (Laird and Ware 1982). This model allows arbitrary matrices $\mathbf{X}_{1i}^P$ and $\mathbf{X}_{2i}^P$ to be associated with $\boldsymbol{\alpha}$ and $\mathbf{v}_i$ in the form

$$\mathbf{y}_i^P = \mathbf{X}_{1i}^P \boldsymbol{\alpha} + \mathbf{X}_{2i}^P \mathbf{v}_i + \mathbf{e}_i^P, \quad i = 1, \ldots, m. \tag{4.5.10}$$

The choice $\mathbf{X}_{1i}^P = \mathbf{X}_i^P \tilde{\mathbf{Z}}_i$ and $\mathbf{X}_{2i}^P = \mathbf{X}_i^P$ gives the two-level model (4.5.8). The general model (4.5.10) covers many of the small area models considered in the literature.

4.5.4 General Linear Mixed Model

Datta and Ghosh (1991) considered a general linear mixed model for the population that covers the univariate unit level models as special cases:

$$\mathbf{y}^P = \mathbf{X}^P \boldsymbol{\beta} + \mathbf{Z}^P \mathbf{v} + \mathbf{e}^P. \tag{4.5.11}$$

Here $\mathbf{e}^P$ and $\mathbf{v}$ are independent, with $\mathbf{e}^P \sim N(\mathbf{0}, \sigma^2 \boldsymbol{\Psi}^P)$ and $\mathbf{v} \sim N(\mathbf{0}, \sigma^2 \mathbf{D}(\boldsymbol{\lambda}))$, where $\boldsymbol{\Psi}^P$ is a known positive definite (p.d.) matrix, and $\mathbf{D}(\boldsymbol{\lambda})$ is a p.d. matrix, which is structurally known except for some parameters $\boldsymbol{\lambda}$ typically involving ratios of variance components of the form σ_k^2/σ^2. Furthermore, $\mathbf{X}^P$ and $\mathbf{Z}^P$ are known design matrices and $\mathbf{y}^P$ is the $N \times 1$ vector of population y-values.

Similar to (4.3.4), we can partition (4.5.11) as

$$\mathbf{y}^P = \begin{bmatrix} \mathbf{y} \\ \mathbf{y}_r \end{bmatrix} = \begin{bmatrix} \mathbf{X} \\ \mathbf{X}_r \end{bmatrix} \boldsymbol{\beta} + \begin{bmatrix} \mathbf{Z} \\ \mathbf{Z}_r \end{bmatrix} \mathbf{v} + \begin{bmatrix} \mathbf{e} \\ \mathbf{e}_r \end{bmatrix}, \tag{4.5.12}$$

where the subscript r denotes nonsampled units. Let $\oplus$ denote the direct sum of matrices, that is, for matrices $\mathbf{B}_1, \ldots, \mathbf{B}_m$, we have $\oplus_{i=1}^m \mathbf{B}_i = \text{blockdiag}(\mathbf{B}_1, \ldots, \mathbf{B}_m)$. Then, the vector of area totals $\mathbf{Y} = (Y_1, \ldots, Y_m)^T$ has the form $\mathbf{Ay} + \mathbf{Cy}_r$ with $\mathbf{A} = \oplus_{i=1}^m \mathbf{1}_{n_i}^T$ and $\mathbf{C} = \oplus_{i=1}^m \mathbf{1}_{N_i - n_i}^T$.

Datta and Ghosh (1991) considered a cross-classification model that is covered by the general model (4.5.11) but not by the "longitudinal" model (4.5.10). Suppose the units in a small area are classified into C subgroups (e.g., age, socio-economic class) labeled by $j = 1, \ldots, C$ and the area-by-subgroup cell sizes N_{ij} are known. The cross-classification model is given by

$$y_{ij\ell} = \mathbf{x}_{ij\ell}^T \boldsymbol{\beta} + v_i + a_j + u_{ij} + e_{ij\ell},$$
$$\ell = 1, \ldots, N_{ij}, \;\; j = 1, \ldots, C, \;\; i = 1, \ldots, m, \tag{4.5.13}$$

where all random terms $\{v_i\}$, $\{a_j\}$, $\{u_{ij}\}$, and $\{e_{ij\ell}\}$ are mutually independent, satisfying $e_{ij\ell} \overset{\text{iid}}{\sim} N(0, \sigma^2)$, $v_i \overset{\text{iid}}{\sim} N(0, \lambda_1 \sigma^2)$, $a_j \overset{\text{iid}}{\sim} N(0, \lambda_2 \sigma^2)$, and $u_{ij} \overset{\text{iid}}{\sim} N(0, \lambda_3 \sigma^2)$ for $\lambda_k = \sigma_k^2 / \sigma^2$, $k = 1, 2, 3$.

4.6 GENERALIZED LINEAR MIXED MODELS

We now consider generalized linear mixed models that are especially suited for binary and count y-values, often encountered in practice.

4.6.1 Logistic Mixed Models

Suppose y_{ij} is binary, that is, $y_{ij} = 0$ or 1, and the parameters of interest are the small area proportions $\overline{Y}_i = P_i = \sum_{j=1}^{N_i} y_{ij} / N_i$, $i = 1, \ldots, m$. MacGibbon and Tomberlin (1989) used a logistic regression model with random area-specific effects to estimate P_i. The y_{ij}'s are assumed to be independent Bernoulli(p_{ij}) conditional on the p_{ij}'s. Then, the p_{ij}'s are assumed to obey the following linking model with random area effects v_i:

$$\text{logit}(p_{ij}) = \log\left(\frac{p_{ij}}{1 - p_{ij}}\right) = \mathbf{x}_{ij}^T \boldsymbol{\beta} + v_i, \tag{4.6.1}$$

where $v_i \overset{\text{iid}}{\sim} N(0, \sigma_v^2)$ and the $\mathbf{x}_{ij}$ are unit-specific covariates. The model-based estimator of P_i is of the form $\left(\sum_{j \in s_i} y_{ij} + \sum_{j \in r_i} \hat{p}_{ij}\right) / N_i$, where $\hat{p}_{ij}$ is obtained from (4.6.1) by estimating $\boldsymbol{\beta}$ and the realization of v_i, using the empirical Bayes or empirical Best (EB) or HB methods.

Malec et al. (1997) considered a different logistic regression model with random regression coefficients. Suppose the units in area i are grouped into classes indexed by h. The counts y_{ihj} ($j = 1, \ldots, N_{ih}$) in the cell (i, h) are independent Bernoulli variables with constant probability p_{ih}, conditional on the p_{ih}'s. Furthermore, the cell probabilities p_{ih} are assumed to obey the following random-slopes logistic regression model:

$$\theta_{ih} = \text{logit}(p_{ih}) = \mathbf{x}_h^T \boldsymbol{\beta}_i, \tag{4.6.2}$$

where

$$\boldsymbol{\beta}_i = \mathbf{Z}_i \boldsymbol{\alpha} + \mathbf{v}_i, \tag{4.6.3}$$

with $\mathbf{v}_i \stackrel{\text{iid}}{\sim} N(\mathbf{0}, \mathbf{\Sigma}_v)$, where $\mathbf{Z}_i$ is a $p \times q$ matrix of area level covariates and $\mathbf{x}_h$ is a class-specific covariate vector.

Example 4.6.1. Visits to Physicians. Malec et al. (1997) applied the model given by (4.6.2) and (4.6.3) to estimate the proportion of persons that visited a physician in the past year in the 50 states and the District of Columbia and in counties by demographic classes, using data from the US National Health Interview Survey (NHIS). Here i denotes a county and h a demographic class; $\mathbf{x}_h$ is a vector of covariates that characterizes the demographic class h. Some details of this application are given in Example 10.13.3.

4.6.2 *Models for Multinomial Counts

We now consider the case of counts in K categories, $(y_{ij1}, \ldots, y_{ijK})$, which follow a multinomial distribution with size $m_{ij} \geq 1$ and probabilities $(p_{ij1}, \ldots, p_{ijK})$ with $\sum_{k=1}^{K} p_{ijk} = 1$. The probabilities p_{ijk} follow the logistic mixed model

$$\log(p_{ijk}/p_{ijK}) = \mathbf{x}_{ij}^T \boldsymbol{\beta}_k + v_{ik}, \quad k = 1, \ldots, K-1, \;\; j = 1, \ldots, n_i, \;\; i = 1, \ldots, m, \tag{4.6.4}$$

where the vectors of random effects for area i, $(v_{i1}, \ldots, v_{i,K-1})^T$, are iid $N_{K-1}(\mathbf{0}, \mathbf{\Sigma}_v)$. In the special case of $m_{ij} = 1$, $y_{ijk} \in \{0, 1\}$, $k = 1, \ldots, K$, such that $\sum_{k=1}^{K} y_{ijk} = 1$. Furthermore, letting $K = 2$, we obtain the logistic mixed model (4.6.1).

Example 4.6.2. Labor Force Estimates. Molina, Saei, and Lombardía (2007) estimated labor force characteristics in UK small areas under model (4.6.4) with $v_{ik} = v_i$ for the sample counts of unemployed, employed, and not in labor force ($K = 3$) in gender–age class j within area i. The covariates $\mathbf{x}_{ij}$ are the log-proportion of registered unemployed and 22 dummy indicators including gender–age categories and regions covering the small areas. López-Vizcaíno, Lombardía, and Morales (2013) considered the model (4.6.4) with $n_i = 1$ for all i and with independent random effects v_{ik}, $k = 1, \ldots, K-1$, that is, with diagonal $\mathbf{\Sigma}_v$. López-Vizcaíno, Lombardía, and Morales (2014) extended this model by including, independently for each category k and area i, time-correlated random effects following an AR(1) process.

4.6.3 Models for Mortality and Disease Rates

Mortality and disease rates for small areas in a region are often used to construct disease maps such as cancer atlases. Such maps are used to display geographical variability of a disease and then identify high-rate areas warranting intervention. A simple area model is obtained by assuming that the observed area counts, y_i, are independent Poisson variables with conditional mean $E(y_i|\lambda_i) = n_i\lambda_i$ and that $\lambda_i \stackrel{\text{iid}}{\sim} \text{gamma}(\alpha, \nu)$. Here λ_i and n_i are the true rate and number exposed in the ith area, and (α, ν) are the scale and shape parameters of the gamma distribution. Under this model, smoothed estimates of λ_i are obtained using EB or HB methods (Clayton and Kaldor 1987; Datta, Ghosh, and Waller 2000). CAR spatial models of the form (4.4.16) on log rates $\theta_i = \log(\lambda_i)$ have also been proposed (Clayton and Kaldor 1987). The model on λ_i can be extended to incorporate area level covariates $\mathbf{z}_i$, for example, $\theta_i = \mathbf{z}_i^T\boldsymbol{\beta} + v_i$ with $v_i \overset{\text{iid}}{\sim} N(0, \sigma_v^2)$. Nandram, Sedransk, and Pickle (1999) studied regression models with random slopes for the age-specific log rates $\theta_{ij} = \log(\lambda_{ij})$, where j denotes age.

Joint mortality rates (y_{1i}, y_{2i}) can also be modeled by assuming that (y_{1i}, y_{2i}) are conditionally independent Poisson variables with $E(y_{1i}|\lambda_{1i}) = n_{1i}\lambda_{1i}$ and $E(y_{2i}|\lambda_{2i}) = n_{2i}\lambda_{2i}$. In addition, $\boldsymbol{\theta}_i = (\log(\lambda_{1i}), \log(\lambda_{2i}))^T$ are assumed to be iid $N_2(\boldsymbol{\mu}, \boldsymbol{\Sigma})$. This mixture model induces dependence between y_{1i} and y_{2i} marginally. As an example of this bivariate model, y_{1i} and y_{2i} denote the number of deaths due to cancer at sites 1 and 2 and (n_{1i}, n_{2i}) the population at risk at sites 1 and 2 in area i. DeSouza (1992) showed that the bivariate model leads to improved estimates of the rates $(\lambda_{1i}, \lambda_{2i})$ compared to estimates based on separate univariate models.

Example 4.6.3. Lip Cancer. Maiti (1998) modeled $\theta_i = \log(\lambda_i)$ as iid $N(\mu, \sigma^2)$. He also considered a CAR spatial model on the θ_i's that relates each θ_i to a set of neighborhood areas of area i. He developed model-based estimates of lip cancer incidence in Scotland for each of $m = 56$ counties. Some details of this application are given in Examples 9.6.1 and 10.11.1.

4.6.4 Natural Exponential Family Models

Ghosh et al. (1998) proposed generalized linear models with random area effects. Conditional on the θ_{ij}'s, the sample values y_{ij} $(j = 1, \ldots, n_i,\ i = 1, \ldots, m)$ are assumed to be independently distributed with probability density function belonging to the natural exponential family with canonical parameters θ_{ij}:

$$f(y_{ij}|\theta_{ij}) = \exp\left\{\frac{1}{\phi_{ij}}\left[\theta_{ij}y_{ij} - a\left(\theta_{ij}\right)\right] + b(y_{ij}, \phi_{ij})\right\} \tag{4.6.5}$$

for known scale parameters ϕ_{ij} (> 0) and specified functions $a(\cdot)$ and $b(\cdot)$. The exponential family (4.6.5) covers well-known distributions including the normal, binomial, and Poisson distributions. For example, when y_{ij} is binomial(n_{ij}, p_{ij}), we have $\theta_{ij} = \text{logit}(p_{ij})$ and $\phi_{ij} = 1$. Similarly, when y_{ij} is Poisson(λ_{ij}), we have $\theta_{ij} = \log(\lambda_{ij})$ and $\phi_{ij} = 1$. Furthermore, the θ_{ij}'s are modeled as

$$\theta_{ij} = x_{ij}^T\boldsymbol{\beta} + v_i + u_{ij}, \tag{4.6.6}$$

where v_i and u_{ij} are mutually independent of $v_i \overset{\text{iid}}{\sim} N(0, \sigma_v^2)$ and $u_{ij} \overset{\text{iid}}{\sim} N(0, \sigma_u^2)$. Ghosh et al. (1999) extended the linking model (5.6.5) to handle spatial data and applied the model to disease mapping.

4.6.5 *Semi-parametric Mixed Models

Only Moments Specified Semi-parametric models based only on the specification of the first two moments of the responses y_{ij} conditional on the area means μ_i and of the μ_i's have also been proposed. In the absence of covariates, Ghosh and Lahiri (1987) assumed the following unit level model: (i) For each area i, conditional on the θ_i's, the y_{ij}'s are iid with mean θ_i and variance $\mu_2(\theta_i)$, denoted $y_{ij}|\theta_i \overset{\text{iid}}{\sim} (\theta_i, \mu_2(\theta_i))$, $j = 1, \ldots, N_i, i = 1, \ldots, m$; (ii) $\theta_i \overset{\text{iid}}{\sim} (\mu, \sigma_v^2)$; (iii) $0 < \sigma_e^2 = E\mu_2(\theta_i) < \infty$.

Raghunathan (1993) incorporated area level covariate information $\mathbf{z}_i$ as follows: (i) $y_{ij}|\theta_i \overset{\text{iid}}{\sim} (\theta_i, b_1(\phi, \theta_i, a_{ij}))$ where $b_1(\cdot)$ is a known positive function of a "dispersion" parameter ϕ, small area mean θ_i, and known constant a_{ij}; (ii) $\theta_i \overset{\text{ind}}{\sim} (\tau_i = h(\mathbf{z}_i^T\boldsymbol{\beta}), b_2(\psi, \tau_i, a_i))$ where $h(\cdot)$ is a known function and $b_2(\cdot)$ is a known positive function of a "dispersion" parameter ψ, the mean τ_i, and a known constant a_i.

The "longitudinal" model (4.5.10) with unit level covariates can be generalized by letting

$$E(y_{ij}|\mathbf{v}_i) = \mu_{ij}, \quad V(y_{ij}|\mathbf{v}_i) = \phi b(\mu_{ij}) \tag{4.6.7}$$

and

$$h(\mu_{ij}) = \mathbf{x}_{ij1}^T\boldsymbol{\beta} + \mathbf{x}_{ij2}^T\mathbf{v}_i, \quad \mathbf{v}_i \overset{\text{iid}}{\sim} (\mathbf{0}, \boldsymbol{\Sigma}_v); \tag{4.6.8}$$

that is, $\mathbf{v}_i$ are independent and identically distributed with mean $\mathbf{0}$ and covariance matrix $\boldsymbol{\Sigma}_v$ (Breslow and Clayton 1993).

Example 4.6.4. Hospital Admissions. Raghunathan (1993) obtained model-based estimates of the mean number of hospital admissions for cancer chemotherapy per 1,000 individuals in each county from the state of Washington, using 1987 hospital discharge data. The model considered for those data has the form

$$E(y_{ij}|\theta_i) = \theta_i, \quad V(y_{ij}|\theta_i) = \theta_i + \phi\theta_i^2$$

and

$$E(\theta_i) = \beta, \quad V(\theta_i) = \psi,$$

where y_{ij} is the number of cancer admissions for individual j in county i, restricting to individuals aged 18 years or older.

Spline Models The assumption of a specified parametric linear regression model for the mean function in the basic unit level model (4.3.1) may be replaced by the weaker assumption of a semi-parametric regression. Approximating the semi-parametric regression mean model by a penalized spline model results in a P-spline unit level model. Opsomer et al. (2008) studied the estimation of small area means under P-spline unit level models.

5

EMPIRICAL BEST LINEAR UNBIASED PREDICTION (EBLUP): THEORY

5.1 INTRODUCTION

In Chapter 4, we presented several small area models that may be regarded as special cases of a general linear mixed model involving fixed and random effects. Moreover, when the population sizes of the areas are large, small area means can be expressed as linear combinations of fixed and random effects of the model. Best linear unbiased prediction (BLUP) estimators of such parameters can be obtained in the classical frequentist framework by appealing to general results on BLUP estimation. BLUP estimators minimize the MSE among the class of linear unbiased estimators and do not depend on normality of the random effects. But they depend on the variances (and possibly covariances) of random effects, called variance components. These parameters can be estimated by a method of moments such as the method of fitting constants (Henderson 1953), or alternatively by maximum likelihood (ML) or restricted maximum likelihood (REML) based on the normal likelihood (Hartley and Rao 1967, Patterson and Thompson 1971). Substituting the estimated variance components into the BLUP estimator, we obtain a two-stage estimator referred to as the empirical BLUP or EBLUP estimator (Harville 1991), in analogy with the empirical Bayes (EB) estimator (Chapter 9).

In this chapter, we present general results on EBLUP estimation. We also consider the more difficult problem of estimating the MSE of EBLUP estimators, taking account of the variability in the estimated variance and covariance components.

Small Area Estimation, Second Edition. J. N. K. Rao and Isabel Molina.

Results are spelled out for the special case of a linear mixed model with block diagonal covariance structure, which covers many commonly used small area models.

5.2 GENERAL LINEAR MIXED MODEL

Suppose that the sample data obey the general linear mixed model

$$\mathbf{y} = \mathbf{X}\boldsymbol{\beta} + \mathbf{Z}\mathbf{v} + \mathbf{e}. \tag{5.2.1}$$

Here $\mathbf{y}$ is the $n \times 1$ vector of sample observations, $\mathbf{X}$ and $\mathbf{Z}$ are known $n \times p$ and $n \times h$ matrices of full rank, and $\mathbf{v}$ and $\mathbf{e}$ are independently distributed with means $\mathbf{0}$ and covariance matrices $\mathbf{G}$ and $\mathbf{R}$ depending on some variance parameters $\boldsymbol{\delta} = (\delta_1, \ldots, \delta_q)^T$. We assume that $\boldsymbol{\delta}$ belongs to a specified subset of Euclidean q-space such that the variance–covariance matrix of $\mathbf{y}$, given by $\mathbf{V} = \mathbf{V}(\boldsymbol{\delta}) := \mathbf{R} + \mathbf{Z}\mathbf{G}\mathbf{Z}^T$, is nonsingular for all $\boldsymbol{\delta}$ belonging to the subset.

We are interested in estimating a linear combination, $\mu = \mathbf{l}^T\boldsymbol{\beta} + \mathbf{m}^T\mathbf{v}$, of the regression parameters $\boldsymbol{\beta}$ and the realization of $\mathbf{v}$, for specified vectors of constants $\mathbf{l}$ and $\mathbf{m}$. A linear estimator of μ is of the form $\hat{\mu} = \mathbf{a}^T\mathbf{y} + b$ for $\mathbf{a}$ and b known. It is model-unbiased for μ if

$$E(\hat{\mu}) = E(\mu), \tag{5.2.2}$$

where E denotes the expectation with respect to the model (5.2.1). The MSE of $\hat{\mu}$ is given by

$$\text{MSE}(\hat{\mu}) = E(\hat{\mu} - \mu)^2, \tag{5.2.3}$$

which reduces to the variance of the estimation error, $\text{MSE}(\hat{\mu}) = V(\hat{\mu} - \mu)$, if $\hat{\mu}$ is unbiased for μ. MSE of $\hat{\mu}$ is also called mean squared prediction error (MSPE) or prediction mean squared error (PMSE); see Pfeffermann (2013). We are interested in finding the BLUP estimator, which minimizes the MSE in the class of linear unbiased estimators $\hat{\mu}$.

5.2.1 BLUP Estimator

For known $\boldsymbol{\delta}$, the BLUP estimator of μ is given by

$$\tilde{\mu}^H = t(\boldsymbol{\delta}, \mathbf{y}) = \mathbf{l}^T\tilde{\boldsymbol{\beta}} + \mathbf{m}^T\tilde{\mathbf{v}} = \mathbf{l}^T\tilde{\boldsymbol{\beta}} + \mathbf{m}^T\mathbf{G}\mathbf{Z}^T\mathbf{V}^{-1}(\mathbf{y} - \mathbf{X}\tilde{\boldsymbol{\beta}}), \tag{5.2.4}$$

where

$$\tilde{\boldsymbol{\beta}} = \tilde{\boldsymbol{\beta}}(\boldsymbol{\delta}) = (\mathbf{X}^T\mathbf{V}^{-1}\mathbf{X})^{-1}\mathbf{X}^T\mathbf{V}^{-1}\mathbf{y} \tag{5.2.5}$$

is the best linear unbiased estimator (BLUE) of $\boldsymbol{\beta}$,

$$\tilde{\mathbf{v}} = \tilde{\mathbf{v}}(\boldsymbol{\delta}) = \mathbf{G}\mathbf{Z}^T\mathbf{V}^{-1}(\mathbf{y} - \mathbf{X}\tilde{\boldsymbol{\beta}}), \tag{5.2.6}$$

and the superscript H on $\tilde{\mu}$ stands for Henderson, who proposed (5.2.4); see Henderson (1950). A direct proof that (5.2.4) is the BLUP estimator is given in Section 5.6.1, following Henderson (1963). Robinson (1991) gave an alternative proof by writing $\hat{\mu}$ as $\hat{\mu} = t(\boldsymbol{\delta}, \mathbf{y}) + \mathbf{c}^T\mathbf{y}$ with $E(\mathbf{c}^T\mathbf{y}) = 0$, that is, $\mathbf{X}^T\mathbf{c} = \mathbf{0}$, and then showing that $E[\mathbf{l}^T(\tilde{\boldsymbol{\beta}} - \boldsymbol{\beta})\mathbf{y}^T\mathbf{c}] = 0$ and $E[\mathbf{m}^T(\tilde{\mathbf{v}} - \mathbf{v})\mathbf{y}^T\mathbf{c}] = 0$. This leads to

$$\text{MSE}(\hat{\mu}) = \text{MSE}[t(\boldsymbol{\delta}, \mathbf{y})] + E(\mathbf{c}^T\mathbf{y})^2 \geq \text{MSE}[t(\boldsymbol{\delta}, \mathbf{y})].$$

Robinson's proof assumes the knowledge of the BLUP $t(\boldsymbol{\delta}, \mathbf{y})$.

Henderson et al. (1959) assumed normality of $\mathbf{v}$ and $\mathbf{e}$ and maximized the joint density of $\mathbf{y}$ and $\mathbf{v}$ with respect to $\boldsymbol{\beta}$ and $\mathbf{v}$. This is equivalent to maximizing the function

$$\phi = -\frac{1}{2}(\mathbf{y} - \mathbf{X}\boldsymbol{\beta} - \mathbf{Z}\mathbf{v})^T\mathbf{R}^{-1}(\mathbf{y} - \mathbf{X}\boldsymbol{\beta} - \mathbf{Z}\mathbf{v}) - \frac{1}{2}\mathbf{v}^T\mathbf{G}^{-1}\mathbf{v}, \tag{5.2.7}$$

which leads to the following "mixed model" equations:

$$\begin{bmatrix} \mathbf{X}^T\mathbf{R}^{-1}\mathbf{X} & \mathbf{X}^T\mathbf{R}^{-1}\mathbf{Z} \\ \mathbf{Z}^T\mathbf{R}^{-1}\mathbf{X} & \mathbf{Z}^T\mathbf{R}^{-1}\mathbf{Z} + \mathbf{G}^{-1} \end{bmatrix} \begin{bmatrix} \boldsymbol{\beta}^* \\ \mathbf{v}^* \end{bmatrix} = \begin{bmatrix} \mathbf{X}^T\mathbf{R}^{-1}\mathbf{y} \\ \mathbf{Z}^T\mathbf{R}^{-1}\mathbf{y} \end{bmatrix}. \tag{5.2.8}$$

The solution of (5.2.8) is identical to the BLUP estimators of $\boldsymbol{\beta}$ and $\mathbf{v}$, that is, $\boldsymbol{\beta}^* = \tilde{\boldsymbol{\beta}}$ and $\mathbf{v}^* = \tilde{\mathbf{v}}$. This follows by noting that

$$\mathbf{R}^{-1} - \mathbf{R}^{-1}\mathbf{Z}(\mathbf{Z}^T\mathbf{R}^{-1}\mathbf{Z} + \mathbf{G}^{-1})^{-1}\mathbf{Z}^T\mathbf{R}^{-1} = \mathbf{V}^{-1}$$

and

$$(\mathbf{Z}^T\mathbf{R}^{-1}\mathbf{Z} + \mathbf{G}^{-1})^{-1}\mathbf{Z}^T\mathbf{R}^{-1} = \mathbf{G}\mathbf{Z}^T\mathbf{V}^{-1}.$$

When $\mathbf{V}$ has no simple inverse but $\mathbf{G}$ and $\mathbf{R}$ are easily invertible (e.g., diagonal), the mixed model equations (5.2.8) are often computationally simpler than (5.2.5) and (5.2.6).

In view of the equivalence of $(\boldsymbol{\beta}^*, \mathbf{v}^*)$ and $(\tilde{\boldsymbol{\beta}}, \tilde{\mathbf{v}})$, the BLUP estimates are often called "joint maximum-likelihood estimates," but the function being maximized, ϕ, is not a log likelihood in the usual sense because $\mathbf{v}$ is nonobservable (Robinson 1991). It is called a "penalized likelihood" because a "penalty" $-\mathbf{v}^T\mathbf{G}^{-1}\mathbf{v}/2$ is added to the log likelihood obtained regarding $\mathbf{v}$ as fixed.

The best prediction (BP) estimator of μ is given by its conditional expectation $E(\mu|\mathbf{y})$, because for any estimator $d(\mathbf{y})$ of μ, not necessarily linear or unbiased, it holds

$$E[d(\mathbf{y}) - \mu]^2 \geq E[E(\mu|\mathbf{y}) - \mu]^2.$$

This result follows by noting that

$$\begin{aligned} E[(d(\mathbf{y}) - \mu)^2|\mathbf{y}] &= E[(d(\mathbf{y}) - E(\mu|\mathbf{y}) + E(\mu|\mathbf{y}) - \mu)^2|\mathbf{y}] \\ &= [d(\mathbf{y}) - E(\mu|\mathbf{y})]^2 + E[(E(\mu|\mathbf{y}) - \mu)^2|\mathbf{y}] \\ &\geq E[(E(\mu|\mathbf{y}) - \mu)^2|\mathbf{y}] \end{aligned}$$

with equality if and only if $d(\mathbf{y}) = E(\mu|\mathbf{y})$. Under normality, the BP estimator $E(\mu|\mathbf{y})$ reduces to the BLUP estimator (5.2.5) with $\tilde{\boldsymbol{\beta}}$ replaced by $\boldsymbol{\beta}$, that is, it depends on the unknown $\boldsymbol{\beta}$. In particular, the BP estimator of $\mathbf{m}^T\mathbf{v}$ is given by

$$E(\mathbf{m}^T\mathbf{v}|\mathbf{y}) = \mathbf{m}^T E(\mathbf{v}|\mathbf{y}) = \mathbf{m}^T\mathbf{G}\mathbf{Z}^T\mathbf{V}^{-1}(\mathbf{y} - \mathbf{X}\boldsymbol{\beta}).$$

This estimator is also the best linear prediction (BLP) estimator of $\mathbf{m}^T\mathbf{v}$ without assuming normality.

Since we do not want the estimator to depend on $\boldsymbol{\beta}$, we transform $\mathbf{y}$ to all error contrasts $\mathbf{A}^T\mathbf{y}$ with mean $\mathbf{0}$, that is, satisfying $\mathbf{A}^T\mathbf{X} = \mathbf{0}$, where $\mathbf{A}$ is any $n \times (n-p)$ full-rank matrix that is orthogonal to the $n \times p$ model matrix $\mathbf{X}$. For the transformed data $\mathbf{A}^T\mathbf{y}$, the BP estimator $E(\mathbf{m}^T\mathbf{v}|\mathbf{A}^T\mathbf{y})$ of $\mathbf{m}^T\mathbf{v}$ in fact reduces to the BLUP estimator $\mathbf{m}^T\tilde{\mathbf{v}}$ (see Section 5.6.2). This result provides an alternative justification of BLUP without linearity and unbiasedness, but assuming normality. It is interesting to note that the transformed data $\mathbf{A}^T\mathbf{y}$ are also used to obtain the restricted (or residual) maximum-likelihood (REML) estimators of the variance parameters $\boldsymbol{\delta}$ (see Section 5.2.3).

We have considered the BLUP estimation of a single linear combination $\mu = \mathbf{l}^T\boldsymbol{\beta} + \mathbf{m}^T\mathbf{v}$, but the method readily extends to simultaneous estimation of $r(\geq 2)$ linear combinations, $\boldsymbol{\mu} = \mathbf{L}\boldsymbol{\beta} + \mathbf{M}\mathbf{v}$, where $\boldsymbol{\mu} = (\mu_1, \ldots, \mu_r)^T$. The BLUP estimator of $\boldsymbol{\mu}$ is given by

$$\mathbf{t}(\boldsymbol{\delta}, \mathbf{y}) = \mathbf{L}\tilde{\boldsymbol{\beta}} + \mathbf{M}\tilde{\mathbf{v}} = \mathbf{L}\tilde{\boldsymbol{\beta}} + \mathbf{M}\mathbf{G}\mathbf{Z}^T\mathbf{V}^{-1}(\mathbf{y} - \mathbf{X}\tilde{\boldsymbol{\beta}}). \tag{5.2.9}$$

The estimator $\mathbf{t}(\boldsymbol{\delta}, \mathbf{y})$ is optimal in the sense that for any other linear unbiased estimator $\mathbf{t}^*(\mathbf{y})$ of $\boldsymbol{\mu}$, the matrix $E(\mathbf{t}^* - \boldsymbol{\mu})(\mathbf{t}^* - \boldsymbol{\mu})^T - E(\mathbf{t} - \boldsymbol{\mu})(\mathbf{t} - \boldsymbol{\mu})^T$ is positive semi-definite (psd). Note that $E(\mathbf{t} - \boldsymbol{\mu})(\mathbf{t} - \boldsymbol{\mu})^T$ is the dispersion matrix of $\mathbf{t} - \boldsymbol{\mu}$.

5.2.2 MSE of BLUP

The BLUP estimator $t(\boldsymbol{\delta}, \mathbf{y})$ may be also expressed as

$$t(\boldsymbol{\delta}, \mathbf{y}) = t^*(\boldsymbol{\delta}, \boldsymbol{\beta}, \mathbf{y}) + \mathbf{d}^T(\tilde{\boldsymbol{\beta}} - \boldsymbol{\beta}),$$

where $t^*(\boldsymbol{\delta}, \boldsymbol{\beta}, \mathbf{y})$ is the BLUP estimator when $\boldsymbol{\beta}$ is known, given by

$$t^*(\boldsymbol{\delta}, \boldsymbol{\beta}, \mathbf{y}) = \mathbf{l}^T\boldsymbol{\beta} + \mathbf{b}^T(\mathbf{y} - \mathbf{X}\boldsymbol{\beta}), \tag{5.2.10}$$

with $\mathbf{b}^T = \mathbf{m}^T\mathbf{G}\mathbf{Z}^T\mathbf{V}^{-1}$ and $\mathbf{d}^T = \mathbf{l}^T - \mathbf{b}^T\mathbf{X}$. It now follows that $t^*(\boldsymbol{\delta}, \boldsymbol{\beta}, \mathbf{y}) - \mu$ and $\mathbf{d}^T(\tilde{\boldsymbol{\beta}} - \boldsymbol{\beta})$ are uncorrelated, noting that

$$E[(\mathbf{b}^T(\mathbf{Z}\mathbf{v} + \mathbf{e}) - \mathbf{m}^T\mathbf{v})(\mathbf{v}^T\mathbf{Z}^T + \mathbf{e}^T)\mathbf{V}^{-1}] = \mathbf{0}.$$

Therefore, taking expectation of $[t(\boldsymbol{\delta}, \mathbf{y}) - \mu]^2$, we obtain

$$\text{MSE}[t(\boldsymbol{\delta}, \mathbf{y})] = \text{MSE}[t^*(\boldsymbol{\delta}, \boldsymbol{\beta}, \mathbf{y})] + V[\mathbf{d}^T(\tilde{\boldsymbol{\beta}} - \boldsymbol{\beta})] = g_1(\boldsymbol{\delta}) + g_2(\boldsymbol{\delta}), \tag{5.2.11}$$

where

$$g_1(\boldsymbol{\delta}) = V[t^*(\boldsymbol{\delta}, \boldsymbol{\beta}, \mathbf{y}) - \mu] = \mathbf{m}^T(\mathbf{G} - \mathbf{G}\mathbf{Z}^T\mathbf{V}^{-1}\mathbf{Z}\mathbf{G})\mathbf{m} \tag{5.2.12}$$

and

$$g_2(\boldsymbol{\delta}) = \mathbf{d}^T(\mathbf{X}^T\mathbf{V}^{-1}\mathbf{X})^{-1}\mathbf{d}. \tag{5.2.13}$$

The second term in (5.2.11), $g_2(\boldsymbol{\delta})$, accounts for the variability in the estimator $\tilde{\boldsymbol{\beta}}$ of $\boldsymbol{\beta}$.

Henderson (1975) used the mixed model equations (5.2.8) to obtain an alternative formula for MSE$[t(\boldsymbol{\delta}, \mathbf{y})]$, given by

$$\text{MSE}[t(\boldsymbol{\delta}, \mathbf{y})] = (\mathbf{l}^T, \mathbf{m}^T)\begin{bmatrix} \mathbf{C}_{11} & \mathbf{C}_{12} \\ \mathbf{C}_{21} & \mathbf{C}_{22} \end{bmatrix}\begin{pmatrix} \mathbf{l} \\ \mathbf{m} \end{pmatrix}, \tag{5.2.14}$$

where the matrix $\mathbf{C}$ with blocks $\mathbf{C}_{ij}(i, j = 1, 2)$ is the inverse of the coefficient matrix of the mixed model equations (5.2.8). This form is computationally simpler than (5.2.11) when $\mathbf{G}$ and $\mathbf{R}$ are more easily invertible than $\mathbf{V}$.

5.2.3 EBLUP Estimator

The BLUP estimator $t(\boldsymbol{\delta}, \mathbf{y})$ given by (5.2.4) depends on the variance parameters $\boldsymbol{\delta}$, which are unknown in practical applications. Replacing $\boldsymbol{\delta}$ by an estimator $\hat{\boldsymbol{\delta}} = \hat{\boldsymbol{\delta}}(\mathbf{y})$, we obtain a two-stage estimator $\hat{\mu}^H = t(\hat{\boldsymbol{\delta}}, \mathbf{y})$, which is referred to as the empirical BLUP (or EBLUP) estimator. For convenience, we also write $t(\hat{\boldsymbol{\delta}}, \mathbf{y})$ and $t(\boldsymbol{\delta}, \mathbf{y})$ as $t(\hat{\boldsymbol{\delta}})$ and $t(\boldsymbol{\delta})$.

The two-stage estimator $t(\hat{\boldsymbol{\delta}})$ remains unbiased for μ, that is, $E[t(\hat{\boldsymbol{\delta}}) - \mu] = 0$, provided (i) $E[t(\hat{\boldsymbol{\delta}})]$ is finite; (ii) $\hat{\boldsymbol{\delta}}$ is any even translation-invariant estimator of $\boldsymbol{\delta}$, that is, $\hat{\boldsymbol{\delta}}(-\mathbf{y}) = \hat{\boldsymbol{\delta}}(\mathbf{y})$ and $\hat{\boldsymbol{\delta}}(\mathbf{y} - \mathbf{X}\mathbf{b}) = \hat{\boldsymbol{\delta}}(\mathbf{y})$ for all $\mathbf{y}$ and $\mathbf{b}$; (iii) The distributions of $\mathbf{v}$ and $\mathbf{e}$ are both symmetric around $\mathbf{0}$ (not necessarily normal). A proof of unbiasedness of $t(\hat{\boldsymbol{\delta}})$, due to Kackar and Harville (1981), uses the following results: (a) $\hat{\boldsymbol{\delta}}(\mathbf{y}) = \hat{\boldsymbol{\delta}}(\mathbf{Z}\mathbf{v} + \mathbf{e}) = \hat{\boldsymbol{\delta}}(-\mathbf{Z}\mathbf{v} - \mathbf{e})$; (b) $t(\hat{\boldsymbol{\delta}}) - \mu = \phi(\mathbf{v}, \mathbf{e}) - \mathbf{m}^T\mathbf{v}$, where $\phi(\mathbf{v}, \mathbf{e})$ is an odd function of $\mathbf{v}$ and $\mathbf{e}$, that is, $\phi(-\mathbf{v}, -\mathbf{e}) = -\phi(\mathbf{v}, \mathbf{e})$. Result (b) implies that $E\phi(\mathbf{v}, \mathbf{e}) = E\phi(-\mathbf{v}, -\mathbf{e}) = -E\phi(\mathbf{v}, \mathbf{e})$, or equivalently $E\phi(\mathbf{v}, \mathbf{e}) = 0$, which implies $E[t(\hat{\boldsymbol{\delta}}) - \mu] = 0$.

Kackar and Harville (1981) have also shown that standard procedures for estimating $\boldsymbol{\delta}$ yield even translation-invariant estimators, in particular, ML, REML, and the method of fitting constants (also called Henderson's method 3). We refer the reader to Searle, Casella, and McCulloch (1992) and Rao (1997) for details of these methods for the analysis of variance (ANOVA) model, which is a special case of the general linear mixed model (5.2.1). The ANOVA model is given by

$$\mathbf{y} = \mathbf{X}\boldsymbol{\beta} + \mathbf{Z}_1\mathbf{v}_1 + \cdots + \mathbf{Z}_r\mathbf{v}_r + \mathbf{e}, \tag{5.2.15}$$

where $\mathbf{v}_1, \ldots, \mathbf{v}_r$ and $\mathbf{e}$ are independently distributed with means $\mathbf{0}$ and covariance matrices $\sigma_1^2\mathbf{I}_{h_1}, \ldots, \sigma_r^2\mathbf{I}_{h_r}$ and $\sigma_e^2\mathbf{I}_n$. The parameters $\boldsymbol{\delta} = (\sigma_0^2, \ldots, \sigma_r^2)^T$

with $\sigma_k^2 \geq 0 (k = 1, \ldots, r)$ and $\sigma_0^2 = \sigma_e^2 > 0$ are the variance components. Note that $\mathbf{G}$ is now block diagonal with blocks $\sigma_k^2 \mathbf{I}_{h_k}, k = 1, \ldots, r, \mathbf{R} = \sigma_e^2 \mathbf{I}_n$, and $\mathbf{V} = \sigma_e^2 \mathbf{I}_n + \sum_{k=1}^r \sigma_k^2 \mathbf{Z}_k \mathbf{Z}_k^T$, which is a special case of a covariance matrix with linear structure, $\mathbf{V} = \sum_{k=1}^r \delta_k \mathbf{H}_k$, for known symmetric matrices $\mathbf{H}_k$.

5.2.4 ML and REML Estimators

We now provide formulas for the ML and REML estimators of $\boldsymbol{\beta}$ and $\boldsymbol{\delta}$ under the general linear mixed model (5.2.1) and the associated asymptotic covariance matrices, assuming normality (Cressie 1992). Under normality, the log-likelihood function is given by

$$l(\boldsymbol{\beta}, \boldsymbol{\delta}) = c - \frac{1}{2} \log |\mathbf{V}| - \frac{1}{2} (\mathbf{y} - \mathbf{X}\boldsymbol{\beta})^T \mathbf{V}^{-1} (\mathbf{y} - \mathbf{X}\boldsymbol{\beta}), \tag{5.2.16}$$

where c denotes a generic constant. The partial derivative of $l(\boldsymbol{\beta}, \boldsymbol{\delta})$ with respect to $\boldsymbol{\delta}$ is given by $\mathbf{s}(\boldsymbol{\beta}, \boldsymbol{\delta})$, with the jth element

$$s_j(\boldsymbol{\beta}, \boldsymbol{\delta}) = \partial l(\boldsymbol{\beta}, \boldsymbol{\delta}) / \partial \delta_j = -\frac{1}{2} \text{tr}(\mathbf{V}^{-1} \mathbf{V}_{(j)}) - \frac{1}{2} (\mathbf{y} - \mathbf{X}\boldsymbol{\beta})^T \mathbf{V}^{(j)} (\mathbf{y} - \mathbf{X}\boldsymbol{\beta}),$$

where $\mathbf{V}_{(j)} = \partial \mathbf{V} / \partial \delta_j$ and $\mathbf{V}^{(j)} = \partial \mathbf{V}^{-1} / \partial \delta_j = -\mathbf{V}^{-1} \mathbf{V}_{(j)} \mathbf{V}^{-1}$, noting that $\mathbf{V} = \mathbf{V}(\boldsymbol{\delta})$. Also, the matrix of expected second-order derivatives of $-l(\boldsymbol{\beta}, \boldsymbol{\delta})$ with respect to $\boldsymbol{\delta}$ is given by $\mathcal{I}(\boldsymbol{\delta})$ with (j, k) th element

$$\mathcal{I}_{jk}(\boldsymbol{\delta}) = \frac{1}{2} \text{tr}(\mathbf{V}^{-1} \mathbf{V}_{(j)} \mathbf{V}^{-1} \mathbf{V}_{(k)}). \tag{5.2.17}$$

The ML estimator of $\boldsymbol{\delta}$ is obtained iteratively using the Fisher-scoring algorithm, with updating equation

$$\boldsymbol{\delta}^{(a+1)} = \boldsymbol{\delta}^{(a)} + [\mathcal{I}(\boldsymbol{\delta}^{(a)})]^{-1} \mathbf{s}[\tilde{\boldsymbol{\beta}}(\boldsymbol{\delta}^{(a)}), \boldsymbol{\delta}^{(a)}], \tag{5.2.18}$$

where the superscript (a) indicates that the specified terms are evaluated at the values of $\boldsymbol{\delta}$ and $\tilde{\boldsymbol{\beta}} = \tilde{\boldsymbol{\beta}}(\boldsymbol{\delta})$ at the ath iteration $\boldsymbol{\delta} = \boldsymbol{\delta}^{(a)}$ and $\tilde{\boldsymbol{\beta}} = \tilde{\boldsymbol{\beta}}(\boldsymbol{\delta}^{(a)}), (a = 0, 1, 2, \ldots)$. At convergence of the iterations (5.2.18), we get the ML estimators $\hat{\boldsymbol{\delta}}_{\text{ML}}$ of $\boldsymbol{\delta}$ and $\hat{\boldsymbol{\beta}}_{\text{ML}} = \tilde{\boldsymbol{\beta}}(\hat{\boldsymbol{\delta}}_{\text{ML}})$ of $\boldsymbol{\beta}$. The asymptotic covariance matrix of $\hat{\boldsymbol{\beta}}_{\text{ML}}$ and $\hat{\boldsymbol{\delta}}_{\text{ML}}$ has a block diagonal structure, $\text{diag}[\overline{\mathbf{V}}(\hat{\boldsymbol{\beta}}_{\text{ML}}), \overline{\mathbf{V}}(\hat{\boldsymbol{\delta}}_{\text{ML}})]$, with

$$\overline{\mathbf{V}}(\hat{\boldsymbol{\beta}}_{\text{ML}}) = (\mathbf{X}^T \mathbf{V}^{-1} \mathbf{X})^{-1}, \quad \overline{\mathbf{V}}(\hat{\boldsymbol{\delta}}_{\text{ML}}) = \mathcal{I}^{-1}(\boldsymbol{\delta}). \tag{5.2.19}$$

A drawback of the ML estimator of $\boldsymbol{\delta}$ is that it does not take account of the loss in degrees of freedom (df) due to estimating $\boldsymbol{\beta}$. For example, when $y_1, \ldots, y_n$ are iid $N(\mu, \sigma^2)$, the ML estimator $\hat{\sigma}^2 = [(n-1)/n] s^2$ is not equal to the customary unbiased estimator of $\sigma^2, s^2 = \sum_{i=1}^n (y_i - \bar{y})^2 / (n-1)$. The REML method takes account of the loss in df by using the transformed data $\mathbf{y}^* = \mathbf{A}^T \mathbf{y}$, where $\mathbf{A}$ is any $n \times (n-p)$ full-rank matrix that is orthogonal to the $n \times p$ matrix $\mathbf{X}$. It follows that $\mathbf{y}^* = \mathbf{A}^T \mathbf{y}$

is distributed as a $(n-p)$-variate normal with mean $\mathbf{0}$ and covariance matrix $\mathbf{A}^T\mathbf{VA}$. The logarithm of the joint density of $\mathbf{y}^*$, expressed as function of $\boldsymbol{\delta}$, is called restricted log-likelihood function and is given by

$$l_R(\boldsymbol{\delta}) = c - \frac{1}{2}\log|\mathbf{V}| - \frac{1}{2}\log|\mathbf{X}^T\mathbf{V}^{-1}\mathbf{X}| - \frac{1}{2}\mathbf{y}^T\mathbf{P}\mathbf{y}, \tag{5.2.20}$$

where

$$\mathbf{P} = \mathbf{V}^{-1} - \mathbf{V}^{-1}\mathbf{X}(\mathbf{X}^T\mathbf{V}^{-1}\mathbf{X})^{-1}\mathbf{X}^T\mathbf{V}^{-1}. \tag{5.2.21}$$

Note that $\mathbf{Py} = \mathbf{V}^{-1}(\mathbf{y} - \mathbf{X}\tilde{\boldsymbol{\beta}})$. The partial derivative of $l_R(\boldsymbol{\delta})$ with respect to $\boldsymbol{\delta}$ is given by $\mathbf{s}_R(\boldsymbol{\delta})$, with the jth element

$$s_{Rj}(\boldsymbol{\delta}) = \partial l_R(\boldsymbol{\delta})/\partial\delta_j = -\frac{1}{2}\text{tr}(\mathbf{PV}_{(j)}) + \frac{1}{2}\mathbf{y}^T\mathbf{PV}_{(j)}\mathbf{Py}.$$

Also, the matrix of expected second-order derivatives of $-l_R(\boldsymbol{\delta})$ with respect to $\boldsymbol{\delta}$ is given by $\mathcal{I}_R(\boldsymbol{\delta})$ with (j,k)th element

$$\mathcal{I}_{R,jk}(\boldsymbol{\delta}) = \frac{1}{2}\text{tr}(\mathbf{PV}_{(j)}\mathbf{PV}_{(k)}). \tag{5.2.22}$$

Note that both $\mathbf{s}_R(\boldsymbol{\delta})$ and $\mathcal{I}_R(\boldsymbol{\delta})$ are invariant to the choice of $\mathbf{A}$.

The REML estimator of $\boldsymbol{\delta}$ is obtained iteratively from (5.2.18) by replacing $\mathcal{I}(\boldsymbol{\delta}^{(a)})$ and $\mathbf{s}[\tilde{\boldsymbol{\beta}}(\boldsymbol{\delta}^{(a)}), \boldsymbol{\delta}^{(a)}]$ by $\mathcal{I}_R(\boldsymbol{\delta}^{(a)})$ and $\mathbf{s}_R(\boldsymbol{\delta}^{(a)})$, respectively. At convergence of the iterations, we get REML estimators $\hat{\boldsymbol{\delta}}_{\text{RE}}$ and $\hat{\boldsymbol{\beta}}_{\text{RE}} = \tilde{\boldsymbol{\beta}}(\hat{\boldsymbol{\delta}}_{\text{RE}})$. Asymptotically, $\overline{\mathbf{V}}(\hat{\boldsymbol{\delta}}_{\text{RE}}) \approx \overline{\mathbf{V}}(\hat{\boldsymbol{\delta}}_{\text{ML}}) = \mathcal{I}^{-1}(\boldsymbol{\delta})$ and $\overline{\mathbf{V}}(\hat{\boldsymbol{\beta}}_{\text{RE}}) \approx \overline{\mathbf{V}}(\hat{\boldsymbol{\beta}}_{\text{ML}}) = (\mathbf{X}^T\mathbf{V}^{-1}\mathbf{X})^{-1}$, provided p is fixed.

For the ANOVA model (5.2.15), the ML estimator of $\boldsymbol{\delta}$ can be obtained iteratively using the BLUP estimators $\tilde{\boldsymbol{\beta}}$ and $\tilde{\mathbf{v}}$ (Hartley and Rao 1967, Henderson 1973). The updating equations are given by

$$\sigma_i^{2(a+1)} = \frac{1}{h_i}[\tilde{\mathbf{v}}_i^{(a)T}\tilde{\mathbf{v}}_i^{(a)} + \sigma_i^{2(a)}\text{tr}(\mathbf{T}_{ii}^{*(a)})] \tag{5.2.23}$$

and

$$\sigma_e^{2(a+1)} = \frac{1}{n}\mathbf{y}^T(\mathbf{y} - \mathbf{X}\tilde{\boldsymbol{\beta}}^{(a)} - \mathbf{Z}\tilde{\mathbf{v}}^{(a)}), \tag{5.2.24}$$

where

$$\mathbf{T}_{ii}^* = (\mathbf{I} + \mathbf{Z}^T\mathbf{R}^{-1}\mathbf{ZG})^{-1}\mathbf{F}_{ii},$$

with $\text{tr}(\mathbf{T}_{ii}^*) > 0$. Here, $\mathbf{F}_{ii}$ is given by $\mathbf{G}$ with unity in place of σ_i^2 and zero in place of σ_j^2 $(j \neq i)$. The values of $\tilde{\boldsymbol{\beta}}$ and $\tilde{\mathbf{v}}$ for a specified $\boldsymbol{\delta}$ are readily obtained from the mixed model equations (5.2.8), without evaluating $\mathbf{V}^{-1}$. The algorithm given by (5.2.23) and (5.2.24) is similar to the EM algorithm (Dempster, Laird, and Rubin 1977).

The asymptotic covariance matrices of $\hat{\boldsymbol{\beta}}_{\text{ML}}$ and $\hat{\boldsymbol{\delta}}_{\text{ML}}$ are given by (5.2.19) using $\mathbf{V}_{(j)} = \mathbf{Z}_j\mathbf{Z}_j^T (j = 0, 1, \ldots, r)$ and $\mathbf{Z}_0 = \mathbf{I}_n$ in (5.2.17).

For the ANOVA model (5.2.15), the REML estimator of $\boldsymbol{\delta}$ can be obtained iteratively from (5.2.23) and (5.2.24) by changing n to $n - p$ in (5.2.24) and $\mathbf{T}_{ii}^*$ to $\mathbf{T}_{ii}$ in (5.2.23), where

$$\mathbf{T}_{ii} = (\mathbf{I} + \mathbf{Z}^T\mathbf{QZG})^{-1}\mathbf{F}_{ii}$$

with

$$\mathbf{Q} = \mathbf{R}^{-1} - \mathbf{R}^{-1}\mathbf{X}(\mathbf{X}^T\mathbf{R}^{-1}\mathbf{X})^{-1}\mathbf{X}^T\mathbf{R}^{-1}$$

(Harville 1977). The elements of the information matrix, $\mathcal{I}_R(\boldsymbol{\delta})$, are given by (5.2.22) using $\mathbf{V}_{(j)} = \mathbf{Z}_j\mathbf{Z}_j^T$.

Anderson (1973) suggested an alternative iterative algorithm to obtain $\hat{\boldsymbol{\delta}}_{\text{ML}}$ with updating equation

$$\boldsymbol{\delta}^{(a+1)} = [\mathcal{I}(\boldsymbol{\delta}^{(a)})]^{-1}\mathbf{b}(\boldsymbol{\delta}^{(a)}), \tag{5.2.25}$$

where the ith element of $\mathbf{b}(\boldsymbol{\delta})$ is given by

$$b_i(\boldsymbol{\delta}) = \frac{1}{2}\mathbf{y}^T\mathbf{PZ}_i\mathbf{Z}_i^T\mathbf{Py}. \tag{5.2.26}$$

This algorithm is equivalent to the Fisher-scoring algorithm for solving ML equations (Rao 1974). The algorithm is also applicable to any covariance matrix with a linear structure, $\mathbf{V} = \sum_{i=1}^r \delta_i\mathbf{H}_i$. We simply replace $\mathbf{Z}_i\mathbf{Z}_i^T$ by $\mathbf{H}_i$ to define $\mathcal{I}(\boldsymbol{\delta})$ and $\mathbf{b}(\boldsymbol{\delta})$ using (5.2.17) and (5.2.26).

For REML estimation of $\boldsymbol{\delta}$, an iterative algorithm similar to (5.2.25) is given by

$$\boldsymbol{\delta}^{(a+1)} = [\mathcal{I}_R(\boldsymbol{\delta}^{(a)})]^{-1}\mathbf{b}(\boldsymbol{\delta}^{(a)}). \tag{5.2.27}$$

The algorithm (5.2.27) is also equivalent to the Fisher-scoring algorithm for solving REML equations (Hocking and Kutner 1975).

Rao (1971) proposed the method of minimum norm quadratic unbiased (MINQU) estimation, which does not require the normality assumption, unlike ML and REML. The MINQU estimators depend on a preassigned value $\boldsymbol{\delta}_0$ for $\boldsymbol{\delta}$ and are identical to the first iterative solution, $\boldsymbol{\delta}^{(1)}$, of the REML updating equation (5.2.27) using $\boldsymbol{\delta}_0$ as the starting value, that is, taking $\boldsymbol{\delta}^{(0)} = \boldsymbol{\delta}_0$. This result suggests that REML (or ML) estimators of $\boldsymbol{\delta}$, derived under normality, may perform well even under nonnormal distributions. In fact, Jiang (1996) established asymptotic consistency of the REML estimator of $\boldsymbol{\delta}$ for the ANOVA model (5.2.15) when normality may not hold.

Moment methods such as fitting-of-constants are also used to estimate $\boldsymbol{\delta}$. We study those methods for particular cases of the general linear mixed model or the ANOVA model in Chapters 6–8.

Henderson's (1973) iterative algorithms defined by (5.2.23) and (5.2.24), for computing ML and REML estimates of variance components σ_e^2 and $\sigma_i^2 (i = 1, \ldots, r)$ in the ANOVA model (5.2.15), are not affected by the constraints on the parameter space: $\sigma_e^2 > 0; \sigma_i^2 \geq 0, i = 1, \ldots, r$ (Harville 1977). If the starting values $\sigma_e^{2(0)}$ and $\sigma_i^{2(0)}$ are strictly positive, then, at every iteration a, the values $\sigma_e^{2(a+1)}$ and $\sigma_i^{2(a+1)}$ remain positive, although it is possible for some of them to be arbitrarily close to 0. The Fisher-scoring method for general linear mixed models (5.2.1) and the equivalent Anderson's method for models with linear covariance structures do not enjoy this desirable property. Modifications are needed to accommodate constraints on $\boldsymbol{\delta} = (\delta_1, \ldots, \delta_q)^T$. We refer the reader to Harville (1977) for details. MINQUE and moment methods also require modifications to account for constraints on $\boldsymbol{\delta}$.

5.2.5 MSE of EBLUP

The error in the EBLUP estimator $t(\hat{\boldsymbol{\delta}})$ may be decomposed as

$$t(\hat{\boldsymbol{\delta}}) - \mu = [t(\boldsymbol{\delta}) - \mu] + [t(\hat{\boldsymbol{\delta}}) - t(\boldsymbol{\delta})].$$

Taking expectation of the squared error, we obtain

$$\text{MSE}[t(\hat{\boldsymbol{\delta}})] = \text{MSE}[t(\boldsymbol{\delta})] + E[t(\hat{\boldsymbol{\delta}}) - t(\boldsymbol{\delta})]^2 + 2\, E[t(\boldsymbol{\delta}) - \mu][t(\hat{\boldsymbol{\delta}}) - t(\boldsymbol{\delta})]. \quad (5.2.28)$$

Under normality of the random effects $\mathbf{v}$ and errors $\mathbf{e}$, the cross-product term in (5.2.28) is zero provided $\hat{\boldsymbol{\delta}}$ is translation invariant (see Section 5.6.3) so that

$$\text{MSE}[t(\hat{\boldsymbol{\delta}})] = \text{MSE}[t(\boldsymbol{\delta})] + E[t(\hat{\boldsymbol{\delta}}) - t(\boldsymbol{\delta})]^2. \quad (5.2.29)$$

It is clear from (5.2.29) that the MSE of the EBLUP estimator is always larger than that of the BLUP estimator $t(\boldsymbol{\delta})$ under normality. The common practice of approximating MSE$[t(\hat{\boldsymbol{\delta}})]$ by MSE$[t(\boldsymbol{\delta})]$ could therefore lead to significant understatement of MSE$[t(\hat{\boldsymbol{\delta}})]$, especially in cases where $t(\boldsymbol{\delta})$ varies with $\boldsymbol{\delta}$ to a significant extent and where the variability of $\hat{\boldsymbol{\delta}}$ is not small.

The last term of (5.2.29) is generally intractable except in special cases, such as the balanced one-way model ANOVA $y_{ij} = \mu + v_i + e_{ij}, i = 1, \ldots, m,$ and $j = 1, \ldots, \overline{n}$ (Peixoto and Harville 1986). It is therefore necessary to obtain an approximation to this term. Das, Jiang, and Rao (2004) obtained a second-order approximation to $E[t(\hat{\boldsymbol{\delta}}) - t(\boldsymbol{\delta})]^2$ for general linear mixed models. Here we only give a sketch of the proof along the lines of Kackar and Harville (1984). By a first-order Taylor expansion, denoting $\mathbf{d}(\boldsymbol{\delta}) = \partial t(\boldsymbol{\delta})/\partial \boldsymbol{\delta}$, we obtain

$$t(\hat{\boldsymbol{\delta}}) - t(\boldsymbol{\delta}) \approx \mathbf{d}(\boldsymbol{\delta})^T(\hat{\boldsymbol{\delta}} - \boldsymbol{\delta}), \quad (5.2.30)$$

where $\approx$ is used to indicate that the remainder terms in the approximation are of lower order (second-order approximation). Result (5.2.30) is obtained assuming that the terms involving higher powers of $\hat{\boldsymbol{\delta}} - \boldsymbol{\delta}$ are of lower order than $\mathbf{d}(\boldsymbol{\delta})^T(\hat{\boldsymbol{\delta}} - \boldsymbol{\delta})$. Furthermore, under normality, noting that the terms involving the derivatives of $\tilde{\boldsymbol{\beta}} - \boldsymbol{\beta}$ with respect to $\boldsymbol{\delta}$ are of lower order, we have

$$\mathbf{d}(\boldsymbol{\delta}) \approx \partial t^*(\boldsymbol{\delta}, \boldsymbol{\beta})/\partial \boldsymbol{\delta} = (\partial \mathbf{b}^T/\partial \boldsymbol{\delta})(\mathbf{y} - \mathbf{X}\boldsymbol{\beta}) =: \mathbf{d}^*(\boldsymbol{\delta}),$$

where $t^*(\boldsymbol{\delta}, \boldsymbol{\beta})$ is given by (5.2.10). Thus, we have

$$E[\mathbf{d}(\boldsymbol{\delta})^T(\hat{\boldsymbol{\delta}} - \boldsymbol{\delta})]^2 \approx E[\mathbf{d}^*(\boldsymbol{\delta})^T(\hat{\boldsymbol{\delta}} - \boldsymbol{\delta})]^2. \tag{5.2.31}$$

Furthermore,

$$\begin{aligned} E[\mathbf{d}^*(\boldsymbol{\delta})^T(\hat{\boldsymbol{\delta}} - \boldsymbol{\delta})]^2 &\approx \text{tr}[E(\mathbf{d}^*(\boldsymbol{\delta})\mathbf{d}^*(\boldsymbol{\delta})^T)\overline{V}(\hat{\boldsymbol{\delta}})] \\ &= \text{tr}[(\partial\mathbf{b}^T/\partial\boldsymbol{\delta})\mathbf{V}(\partial\mathbf{b}^T/\partial\boldsymbol{\delta})^T\overline{V}(\hat{\boldsymbol{\delta}})] =: g_3(\boldsymbol{\delta}), \end{aligned} \tag{5.2.32}$$

where $\overline{V}(\hat{\boldsymbol{\delta}})$ is the asymptotic covariance matrix of $\hat{\boldsymbol{\delta}}$. It now follows from (5.2.30), (5.2.31), and (5.2.32) that

$$E[t(\hat{\boldsymbol{\delta}}) - t(\boldsymbol{\delta})]^2 \approx g_3(\boldsymbol{\delta}). \tag{5.2.33}$$

Replacing (5.2.11) and (5.2.33) in (5.2.29), we get a second-order approximation to the MSE of $t(\hat{\boldsymbol{\delta}})$ as

$$\text{MSE}[t(\hat{\boldsymbol{\delta}})] \approx g_1(\boldsymbol{\delta}) + g_2(\boldsymbol{\delta}) + g_3(\boldsymbol{\delta}). \tag{5.2.34}$$

The terms $g_2(\boldsymbol{\delta})$ and $g_3(\boldsymbol{\delta})$, due to estimating $\boldsymbol{\beta}$ and $\boldsymbol{\delta}$, are of lower order than the leading term $g_1(\boldsymbol{\delta})$.

5.2.6 Estimation of MSE of EBLUP

In practical applications, we need an estimator of $\text{MSE}[t(\hat{\boldsymbol{\delta}})]$ as a measure of variability (or uncertainty) associated with the estimator $t(\hat{\boldsymbol{\delta}})$. A naive approach approximates $\text{MSE}[t(\hat{\boldsymbol{\delta}})]$ by $\text{MSE}[t(\boldsymbol{\delta})]$ in (5.2.11) and then substitutes $\hat{\boldsymbol{\delta}}$ for $\boldsymbol{\delta}$ in $\text{MSE}[t(\boldsymbol{\delta})]$. The resulting naive estimator of MSE is given by

$$\text{mse}_N[t(\hat{\boldsymbol{\delta}})] = g_1(\hat{\boldsymbol{\delta}}) + g_2(\hat{\boldsymbol{\delta}}), \tag{5.2.35}$$

where $g_1(\boldsymbol{\delta})$ and $g_2(\boldsymbol{\delta})$ are given in (5.2.12) and (5.2.13), respectively. Another MSE estimator is obtained by substituting $\hat{\boldsymbol{\delta}}$ for $\boldsymbol{\delta}$ in the MSE approximation (5.2.34), leading to

$$\text{mse}_1[t(\hat{\boldsymbol{\delta}})] = g_1(\hat{\boldsymbol{\delta}}) + g_2(\hat{\boldsymbol{\delta}}) + g_3(\hat{\boldsymbol{\delta}}). \tag{5.2.36}$$

It holds that $Eg_2(\hat{\boldsymbol{\delta}}) \approx g_2(\boldsymbol{\delta})$ and $Eg_3(\hat{\boldsymbol{\delta}}) \approx g_3(\boldsymbol{\delta})$ with neglected terms of lower order, but $g_1(\hat{\boldsymbol{\delta}})$ is not a second-order unbiased estimator of $g_1(\boldsymbol{\delta})$ because its bias is generally of the same order as $g_2(\boldsymbol{\delta})$ and $g_3(\boldsymbol{\delta})$.

To evaluate the bias of $g_1(\hat{\boldsymbol{\delta}})$, take a second-order Taylor expansion of $g_1(\hat{\boldsymbol{\delta}})$ around $\boldsymbol{\delta}$:

$$\begin{aligned} g_1(\hat{\boldsymbol{\delta}}) &\approx g_1(\boldsymbol{\delta}) + (\hat{\boldsymbol{\delta}} - \boldsymbol{\delta})^T\nabla g_1(\boldsymbol{\delta}) + \frac{1}{2}(\hat{\boldsymbol{\delta}} - \boldsymbol{\delta})^T\nabla^2 g_1(\boldsymbol{\delta})(\hat{\boldsymbol{\delta}} - \boldsymbol{\delta}) \\ &=: g_1(\boldsymbol{\delta}) + \Delta_1 + \Delta_2, \end{aligned}$$

where $\nabla g_1(\boldsymbol{\delta})$ is the vector of first-order derivatives of $g_1(\boldsymbol{\delta})$ with respect to $\boldsymbol{\delta}$ and $\nabla^2 g_1(\boldsymbol{\delta})$ is the matrix of second-order derivatives of $g_1(\boldsymbol{\delta})$ with respect to $\boldsymbol{\delta}$. If $\hat{\boldsymbol{\delta}}$ is unbiased for $\boldsymbol{\delta}$, then $E(\Delta_1) = 0$. In general, if $E(\Delta_1) = E(\hat{\boldsymbol{\delta}} - \boldsymbol{\delta})^T \nabla g_1(\boldsymbol{\delta})$ is of lower order than $E(\Delta_2)$, then

$$Eg_1(\hat{\boldsymbol{\delta}}) \approx g_1(\boldsymbol{\delta}) + \frac{1}{2}\text{tr}[\nabla^2 g_1(\boldsymbol{\delta})\overline{\mathbf{V}}(\hat{\boldsymbol{\delta}})]. \tag{5.2.37}$$

Furthermore, when the covariance matrix $\mathbf{V}$ has a linear structure $\mathbf{V} = \sum_{i=1}^{r} \delta_i \mathbf{H}_i$, noting that the second derivatives of $\mathbf{G}$ and $\mathbf{V}$ with respect to the parameters are zero when $\mathbf{G}$ and $\mathbf{V}$ are linear in the parameters $\boldsymbol{\delta}$, the second term on the right-hand side of (5.2.37) reduces to $-g_3(\boldsymbol{\delta})$ and then (5.2.37) reduces to

$$Eg_1(\hat{\boldsymbol{\delta}}) \approx g_1(\boldsymbol{\delta}) - g_3(\boldsymbol{\delta}). \tag{5.2.38}$$

It now follows from (5.2.35), (5.2.36), and (5.2.38) that the biases of $\text{mse}_N[t(\hat{\boldsymbol{\delta}})]$ and $\text{mse}_1[t(\hat{\boldsymbol{\delta}})]$ are, respectively, given by

$$B_N \approx -2g_3(\boldsymbol{\delta}), \quad B_1 \approx -g_3(\boldsymbol{\delta}).$$

Noting that $E[g_1(\hat{\boldsymbol{\delta}}) + g_3(\hat{\boldsymbol{\delta}})] \approx g_1(\boldsymbol{\delta})$ from (5.2.38), a second-order unbiased estimator of $\text{MSE}[t(\hat{\boldsymbol{\delta}})]$ is given by

$$\text{mse}[t(\hat{\boldsymbol{\delta}})] \approx g_1(\hat{\boldsymbol{\delta}}) + g_2(\hat{\boldsymbol{\delta}}) + 2g_3(\hat{\boldsymbol{\delta}}). \tag{5.2.39}$$

Consequently, we have

$$E\{\text{mse}[t(\hat{\boldsymbol{\delta}})]\} \approx \text{MSE}[t(\hat{\boldsymbol{\delta}})].$$

Formula (5.2.39) holds for the REML estimator, $\hat{\boldsymbol{\delta}}_{\text{RE}}$, and some moment estimators.

If $E(\Delta_1)$ is of the same order as $E(\Delta_2)$, as in the case of the ML estimator $\hat{\boldsymbol{\delta}}_{\text{ML}}$, then an extra term needs to be subtracted from (5.2.39). Using the additional approximation $E(\Delta_1) \approx \mathbf{b}^T(\hat{\boldsymbol{\delta}};\boldsymbol{\delta})\nabla g_1(\boldsymbol{\delta})$, where $\mathbf{b}(\hat{\boldsymbol{\delta}};\boldsymbol{\delta})$ is the bias of $\hat{\boldsymbol{\delta}}$ up to terms of lower order, a second-order unbiased estimator of $\text{MSE}[t(\hat{\boldsymbol{\delta}})]$ is given by

$$\text{mse}_*[t(\hat{\boldsymbol{\delta}})] \approx g_1(\hat{\boldsymbol{\delta}}) - \mathbf{b}^T(\hat{\boldsymbol{\delta}};\hat{\boldsymbol{\delta}})\nabla g_1(\hat{\boldsymbol{\delta}}) + g_2(\hat{\boldsymbol{\delta}}) + 2g_3(\hat{\boldsymbol{\delta}}). \tag{5.2.40}$$

The term $\mathbf{b}(\hat{\boldsymbol{\delta}};\boldsymbol{\delta})$ is spelled out in Section 5.3.2 for the special case of block diagonal covariance matrix $\mathbf{V} = \mathbf{V}(\boldsymbol{\delta})$ and $\hat{\boldsymbol{\delta}} = \hat{\boldsymbol{\delta}}_{\text{ML}}$.

Prasad and Rao (1990) derived the MSE estimator (5.2.39) for special cases covered by the general linear mixed model with a block diagonal covariance structure (see Section 5.3 and Chapter 6). Following Prasad and Rao (1990), Harville and Jeske (1992) proposed (5.2.39) for the general linear mixed model (5.2.1), assuming $E(\hat{\boldsymbol{\delta}}) = \boldsymbol{\delta}$, and referred to (5.2.39) as the Prasad–Rao estimator. Das, Jiang, and Rao (2004) provide rigorous proofs of the approximations (5.2.39) and (5.2.40) for REML and ML methods, respectively.

5.3 BLOCK DIAGONAL COVARIANCE STRUCTURE

5.3.1 EBLUP Estimator

A special case of the general linear mixed model is obtained when the vectors and matrices involved in (5.2.1) are partitioned into m components (typically the small areas) as follows:

$$\mathbf{y} = \text{col}_{1\le i\le m}(\mathbf{y}_i) = (\mathbf{y}_1^T, \ldots, \mathbf{y}_m^T)^T, \quad \mathbf{X} = \text{col}_{1\le i\le m}(\mathbf{X}_i),$$
$$\mathbf{Z} = \text{diag}_{1\le i\le m}(\mathbf{Z}_i), \quad \mathbf{v} = \text{col}_{1\le i\le m}(\mathbf{v}_i), \quad \mathbf{e} = \text{col}_{1\le i\le m}(\mathbf{e}_i),$$

where $\mathbf{X}_i$ is $n_i \times p$, $\mathbf{Z}_i$ is $n_i \times h_i$, and $\mathbf{y}_i$ is an $n_i \times 1$ vector, with $\sum_{i=1}^m n_i = n$ and $\sum_{i=1}^m h_i = h$. Furthermore,

$$\mathbf{R} = \text{diag}_{1\le i\le m}(\mathbf{R}_i), \quad \mathbf{G} = \text{diag}_{1\le i\le m}(\mathbf{G}_i),$$

so that $\mathbf{V}$ has a block diagonal structure, that is, $\mathbf{V} = \text{diag}_{1\le i\le m}(\mathbf{V}_i)$, with $\mathbf{V}_i = \mathbf{R}_i + \mathbf{Z}_i\mathbf{G}_i\mathbf{Z}_i^T, i = 1, \ldots, m$. The model, therefore, may be expressed as follows:

$$\mathbf{y}_i = \mathbf{X}_i\boldsymbol{\beta} + \mathbf{Z}_i\mathbf{v}_i + \mathbf{e}_i, \quad i = 1, \ldots, m. \tag{5.3.1}$$

Model (5.3.1) covers many of the small area models considered in the literature. We are interested in estimating linear combinations of the form $\mu_i = \mathbf{l}_i^T\boldsymbol{\beta} + \mathbf{m}_i^T\mathbf{v}_i$, $i = 1, \ldots, m$.

It follows from (5.2.4) that the BLUP estimator of μ_i reduces to

$$\tilde{\mu}_i^H = t_i(\boldsymbol{\delta}, \mathbf{y}) = \mathbf{l}_i^T\tilde{\boldsymbol{\beta}} + \mathbf{m}_i^T\tilde{\mathbf{v}}_i, \tag{5.3.2}$$

where

$$\tilde{\mathbf{v}}_i = \tilde{\mathbf{v}}_i(\boldsymbol{\delta}) = \mathbf{G}_i\mathbf{Z}_i^T\mathbf{V}_i^{-1}(\mathbf{y}_i - \mathbf{X}_i\tilde{\boldsymbol{\beta}}), \tag{5.3.3}$$

and

$$\tilde{\boldsymbol{\beta}} = \tilde{\boldsymbol{\beta}}(\boldsymbol{\delta}) = \left(\sum_{i=1}^m \mathbf{X}_i^T\mathbf{V}_i^{-1}\mathbf{X}_i\right)^{-1} \sum_{i=1}^m \mathbf{X}_i^T\mathbf{V}_i^{-1}\mathbf{y}_i \tag{5.3.4}$$

From (5.2.11), the MSE of the BLUP estimator $\tilde{\mu}_i^H$ reduces to

$$\text{MSE}(\tilde{\mu}_i^H) = g_{1i}(\boldsymbol{\delta}) + g_{2i}(\boldsymbol{\delta}) \tag{5.3.5}$$

with

$$g_{1i}(\boldsymbol{\delta}) = \mathbf{m}_i^T(\mathbf{G}_i - \mathbf{G}_i\mathbf{Z}_i^T\mathbf{V}_i^{-1}\mathbf{Z}_i\mathbf{G}_i)\mathbf{m}_i \tag{5.3.6}$$

and

$$g_{2i}(\boldsymbol{\delta}) = \mathbf{d}_i^T\left(\sum_{i=1}^m \mathbf{X}_i^T\mathbf{V}_i^{-1}\mathbf{X}_i\right)^{-1}\mathbf{d}_i, \tag{5.3.7}$$

where $\mathbf{d}_i^T = \mathbf{l}_i^T - \mathbf{b}_i^T\mathbf{X}_i$, with $\mathbf{b}_i^T = \mathbf{m}_i^T\mathbf{G}_i\mathbf{Z}_i^T\mathbf{V}_i^{-1}$. Replacing $\boldsymbol{\delta}$ by an estimator $\hat{\boldsymbol{\delta}}$ in (5.3.2), we get the EBLUP estimator

$$\hat{\mu}_i^H = t_i(\hat{\boldsymbol{\delta}}, \mathbf{y}) = \mathbf{l}_i^T\hat{\boldsymbol{\beta}} + \mathbf{m}_i^T\hat{\mathbf{v}}_i, \tag{5.3.8}$$

where $\hat{\boldsymbol{\beta}} = \tilde{\boldsymbol{\beta}}(\hat{\boldsymbol{\delta}})$ and $\hat{\mathbf{v}}_i = \tilde{\mathbf{v}}_i(\hat{\boldsymbol{\delta}})$.

5.3.2 Estimation of MSE

The second-order approximation (5.2.34) of the MSE of $\hat{\mu}_i^H$ reduces to

$$\text{MSE}(\hat{\mu}_i^H) \approx g_{1i}(\boldsymbol{\delta}) + g_{2i}(\boldsymbol{\delta}) + g_{3i}(\boldsymbol{\delta}) \tag{5.3.9}$$

with

$$g_{3i}(\boldsymbol{\delta}) = \text{tr}[(\partial\mathbf{b}_i^T/\partial\boldsymbol{\delta})\mathbf{V}_i(\partial\mathbf{b}_i^T/\partial\boldsymbol{\delta})^T\overline{V}(\hat{\boldsymbol{\delta}})]. \tag{5.3.10}$$

Neglected terms in the approximation (5.3.9) are of order $o(m^{-1})$ for large m. For REML and some moment estimators of $\boldsymbol{\delta}$, the estimator of MSE of $\hat{\mu}_i^H = t_i(\hat{\boldsymbol{\delta}}, \mathbf{y})$ given by (5.2.39) reduces to

$$\text{mse}(\hat{\mu}_i^H) \approx g_{1i}(\hat{\boldsymbol{\delta}}) + g_{2i}(\hat{\boldsymbol{\delta}}) + 2g_{3i}(\hat{\boldsymbol{\delta}}) \tag{5.3.11}$$

for large m. For the ML estimator $\hat{\boldsymbol{\delta}}_{\text{ML}}$, the MSE estimator given by (5.2.40) reduces to

$$\text{mse}_*(\hat{\mu}_i^H) = g_{1i}(\hat{\boldsymbol{\delta}}) - \mathbf{b}^T(\hat{\boldsymbol{\delta}};\hat{\boldsymbol{\delta}})\nabla g_{1i}(\hat{\boldsymbol{\delta}}) + g_{2i}(\hat{\boldsymbol{\delta}}) + 2g_{3i}(\hat{\boldsymbol{\delta}}) \tag{5.3.12}$$

for large m, where in this case

$$\mathbf{b}^T(\hat{\boldsymbol{\delta}}_{\text{ML}};\boldsymbol{\delta}) = \frac{1}{2m}\left\{\mathcal{I}^{-1}(\boldsymbol{\delta}) \underset{1\le j\le m}{\text{col}} \text{tr}\left[\sum_{i=1}^m (\mathbf{X}_i^T\mathbf{V}_i^{-1}\mathbf{X}_i)^{-1}\sum_{i=1}^m \mathbf{X}_i^T\mathbf{V}_i^{(j)}\mathbf{X}_i\right]\right\} \tag{5.3.13}$$

with

$$\mathbf{V}_i^{(j)} = \partial\mathbf{V}_i^{-1}/\partial\delta_j = -\mathbf{V}_i^{-1}(\partial\mathbf{V}_i/\partial\delta_j)\mathbf{V}_i^{-1}$$

and

$$\mathcal{I}_{jk}(\boldsymbol{\delta}) = \frac{1}{2}\sum_{i=1}^m \text{tr}(\mathbf{V}_i^{-1}\partial\mathbf{V}_i/\partial\delta_j)(\mathbf{V}_i^{-1}\partial\mathbf{V}_i/\partial\delta_k).$$

The neglected terms in the second-order approximation (5.3.9) to $\text{MSE}(\hat{\mu}_i^H)$ are of order $o(m^{-1})$ and the MSE estimators (5.3.11) for REML and (5.3.12) for ML are second-order unbiased in the sense $E[\text{mse}(\hat{\mu}_i^H)] - \text{MSE}(\hat{\mu}_i^H) = o(m^{-1})$ for large m, under the following regularity assumptions (Datta and Lahiri 2000):

(i) The elements of $\mathbf{X}_i$ and $\mathbf{Z}_i$ are uniformly bounded such that $\sum_{i=1}^m \mathbf{X}_i^T\mathbf{V}_i^{-1}\mathbf{X}_i = [O(m)]_{p\times p}$.

(ii) $\sup_{i\geq 1} n_i \ll \infty$ and $\sup_{i\geq 1} h_i \ll \infty$, where h_i is the number of columns in $\mathbf{Z}_i$.

(iii) Covariance matrices $\mathbf{G}_i$ and $\mathbf{R}_i$ have linear structures of the form $\mathbf{G}_i = \sum_{j=0}^{q} \delta_j \mathbf{A}_{ij} \mathbf{A}_{ij}^T$ and $\mathbf{R}_i = \sum_{j=0}^{q} \delta_j \mathbf{B}_{ij} \mathbf{B}_{ij}^T$, where $\delta_0 = 1, \mathbf{A}_{ij}$ and $\mathbf{B}_{ij} (i = 1, \ldots, m, j = 0, \ldots, q)$ are known matrices of order $n_i \times h_i$ and $h_i \times h_i$ respectively, and the elements of $\mathbf{A}_{ij}$ and $\mathbf{B}_{ij}$ are uniformly bounded such that $\mathbf{G}_i$ and $\mathbf{R}_i$ are positive definite matrices for $i = 1, \ldots, m$.

The MSE estimators (5.3.11) and (5.3.12) are not area-specific in the sense that they do not depend directly on the area-specific data $\mathbf{y}_i$, but using the form (5.3.10) for $g_{3i}(\boldsymbol{\delta})$, it is easy to define other MSE estimators that are area-specific. For example, $\tilde{\mathbf{V}}_i(\boldsymbol{\delta}, \mathbf{y}_i) = (\mathbf{y}_i - \mathbf{X}_i^T \tilde{\boldsymbol{\beta}})(\mathbf{y}_i - \mathbf{X}_i^T \tilde{\boldsymbol{\beta}})^T$ is area-specific and approximately unbiased for $\mathbf{V}_i$. Using this estimator of $\mathbf{V}_i$, we get the following alternative area-specific estimator of $g_{3i}(\boldsymbol{\delta})$:

$$g_{3i}^*(\boldsymbol{\delta}, \mathbf{y}_i) = \mathrm{tr}[(\partial \mathbf{b}_i^T / \partial \boldsymbol{\delta}) \tilde{\mathbf{V}}_i(\boldsymbol{\delta}, \mathbf{y}_i) (\partial \mathbf{b}_i^T / \partial \boldsymbol{\delta})^T \overline{\mathbf{V}}(\hat{\boldsymbol{\delta}})]. \tag{5.3.14}$$

This choice leads to two alternative area-specific versions of (5.3.11) in the case of REML estimation:

$$\mathrm{mse}_1(\hat{\mu}_i^H) \approx g_{1i}(\hat{\boldsymbol{\delta}}) + g_{2i}(\hat{\boldsymbol{\delta}}) + 2 g_{3i}^*(\hat{\boldsymbol{\delta}}, \mathbf{y}_i) \tag{5.3.15}$$

and

$$\mathrm{mse}_2(\hat{\mu}_i^H) \approx g_{1i}(\hat{\boldsymbol{\delta}}) + g_{2i}(\hat{\boldsymbol{\delta}}) + g_{3i}(\hat{\boldsymbol{\delta}}) + g_{3i}^*(\hat{\boldsymbol{\delta}}, \mathbf{y}_i). \tag{5.3.16}$$

Similarly, area-specific versions of the MSE estimator (5.3.12) for ML estimation are given by

$$\mathrm{mse}_{*1}(\hat{\mu}_i^H) \approx g_{1i}(\hat{\boldsymbol{\delta}}) - \mathbf{b}^T(\hat{\boldsymbol{\delta}}; \hat{\boldsymbol{\delta}}) \nabla g_{1i}(\hat{\boldsymbol{\delta}}) + g_{2i}(\hat{\boldsymbol{\delta}}) + 2 g_{3i}^*(\hat{\boldsymbol{\delta}}, \mathbf{y}_i) \tag{5.3.17}$$

and

$$\mathrm{mse}_{*2}(\hat{\mu}_i^H) \approx g_{1i}(\hat{\boldsymbol{\delta}}) - \mathbf{b}^T(\hat{\boldsymbol{\delta}}; \hat{\boldsymbol{\delta}}) \nabla g_{1i}(\hat{\boldsymbol{\delta}}) + g_{2i}(\hat{\boldsymbol{\delta}}) + g_{3i}(\hat{\boldsymbol{\delta}}) + g_{3i}^*(\hat{\boldsymbol{\delta}}, \mathbf{y}_i). \tag{5.3.18}$$

The use of the area-specific matrix $\tilde{\mathbf{V}}_i(\hat{\boldsymbol{\delta}}, \mathbf{y}_i)$, based on the residuals $\mathbf{y}_i - \mathbf{X}_i \hat{\boldsymbol{\beta}}$, induces instability in the MSE estimators, but its effect should be small for large m because the term $g_{3i}^*(\hat{\boldsymbol{\delta}}, \mathbf{y}_i)$ is of order $O(m^{-1})$.

5.3.3 Extension to Multidimensional Area Parameters

The BLUP estimator (5.3.2) readily extends to the multidimensional case of $\boldsymbol{\mu}_i = \mathbf{L}_i \boldsymbol{\beta} + \mathbf{M}_i \mathbf{v}_i$ for specified matrices $\mathbf{L}_i$ and $\mathbf{M}_i$, and is thus given by

$$\begin{aligned} \tilde{\boldsymbol{\mu}}_i^H &= \mathbf{t}_i(\boldsymbol{\delta}, \mathbf{y}) = \mathbf{L}_i \tilde{\boldsymbol{\beta}} + \mathbf{M}_i \tilde{\mathbf{v}}_i \\ &= \mathbf{L}_i \tilde{\boldsymbol{\beta}} + \mathbf{M}_i \mathbf{G}_i \mathbf{Z}_i^T \mathbf{V}_i^{-1} (\mathbf{y}_i - \mathbf{X}_i \tilde{\boldsymbol{\beta}}). \end{aligned} \tag{5.3.19}$$

The covariance matrix of $\tilde{\boldsymbol{\mu}}_i^H - \boldsymbol{\mu}_i$ follows from (5.3.5) by changing $\mathbf{l}_i^T$ and $\mathbf{m}_i^T$ to $\mathbf{L}_i$ and $\mathbf{M}_i$, respectively, that is,

$$\begin{aligned}\text{MSE}(\tilde{\boldsymbol{\mu}}_i^H) &= E(\tilde{\boldsymbol{\mu}}_i^H - \boldsymbol{\mu}_i)(\tilde{\boldsymbol{\mu}}_i^H - \boldsymbol{\mu}_i)^T \\ &= \mathbf{M}_i(\mathbf{G}_i - \mathbf{G}_i\mathbf{Z}_i^T\mathbf{V}_i^{-1}\mathbf{Z}_i\mathbf{G}_i)\mathbf{M}_i^T \\ &\quad + \mathbf{D}_i\left(\sum_{i=1}^m \mathbf{X}_i^T\mathbf{V}_i^{-1}\mathbf{X}_i\right)^{-1}\mathbf{D}_i^T, \end{aligned} \qquad (5.3.20)$$

where $\mathbf{D}_i = \mathbf{L}_i - \mathbf{M}_i\mathbf{G}_i\mathbf{Z}_i^T\mathbf{V}_i^{-1}\mathbf{X}_i$.

The EBLUP estimator is given by $\hat{\boldsymbol{\mu}}_i^H = \mathbf{t}_i(\hat{\boldsymbol{\delta}}, \mathbf{y})$. An estimator of $\text{MSE}(\hat{\boldsymbol{\mu}}_i^H)$ that accounts for the uncertainty due to estimating $\boldsymbol{\delta}$ may be obtained along the lines of Section 5.3.2, but details are omitted here for simplicity.

5.4 *MODEL IDENTIFICATION AND CHECKING

5.4.1 Variable Selection

AIC-Type Methods For the standard linear regression model, $\mathbf{y} = \mathbf{X}\boldsymbol{\beta} + \mathbf{e}$ with $\mathbf{e} \sim N(\mathbf{0}, \sigma^2\mathbf{I})$, several methods have been proposed for the selection of covariates (or fixed effects) for inclusion in the model. Methods studied include stepwise regression, Mallows' C_p statistic, Akaike Information Criterion (AIC), and Bayesian Information Criterion (BIC). Extensions of those methods to the linear mixed model (5.2.1) have been developed in recent years and we provide a brief account. Müller, Scealy, and Welsh (2013) provide an excellent review of methods for model selection in linear mixed models.

The AIC for the general linear mixed model (5.2.1) uses either the marginal log likelihood $l(\boldsymbol{\beta}, \boldsymbol{\delta})$ or the marginal restricted log likelihood, $l_R(\boldsymbol{\delta})$, based on normality of random effects $\mathbf{v}$ and errors $\mathbf{e}$. In the case of $l(\boldsymbol{\beta}, \boldsymbol{\delta})$, it is given by

$$\text{mAIC} = -2l(\hat{\boldsymbol{\beta}}_{\text{ML}}, \hat{\boldsymbol{\delta}}_{\text{ML}}) + 2(q^* + p), \qquad (5.4.1)$$

where p is the dimension of $\boldsymbol{\beta}$ and q^* is the effective number of variance parameters $\boldsymbol{\delta}$, that is, those not estimated to be on the boundary of the parameter space. In the case of $l_R(\boldsymbol{\delta})$, the mAIC is given by (5.4.1) with $l(\hat{\boldsymbol{\beta}}_{\text{ML}}, \hat{\boldsymbol{\delta}}_{\text{ML}})$ changed to $l_R(\hat{\boldsymbol{\delta}})$ and $q^* + p$ to q^*. Several refinements to mAIC have also been proposed (see Müller et al. 2013). The BIC is obtained from (5.4.1) by replacing the penalty term $2(q^* + p)$ by $\log(n)(q^* + p)$ in the case of $l(\boldsymbol{\beta}, \boldsymbol{\delta})$ and by $\log(n^*)q^*$ in the case of $l_R(\boldsymbol{\delta})$, where $n^* = n - p$. Models with smaller mAIC or BIC values are preferred.

Conditional AIC is more relevant than AIC when the focus is on estimation of the realized random effects $\mathbf{v}$ and the regression parameters $\boldsymbol{\beta}$. As noted in Chapter 4, small area estimation falls into this category. In this case, the marginal log likelihood $l(\boldsymbol{\beta}, \boldsymbol{\delta})$ is replaced by the conditional log likelihood $l(\boldsymbol{\beta}, \boldsymbol{\delta}|\mathbf{v})$ obtained from the conditional density $f(\mathbf{y}|\mathbf{v}, \boldsymbol{\beta}, \boldsymbol{\delta})$:

$$l(\boldsymbol{\beta}, \boldsymbol{\delta}|\mathbf{v}) = \text{const} - \frac{1}{2}[\log |\mathbf{R}| + (\mathbf{y} - \mathbf{X}\boldsymbol{\beta} - \mathbf{Z}\mathbf{v})^T \mathbf{R}^{-1}(\mathbf{y} - \mathbf{X}\boldsymbol{\beta} - \mathbf{Z}\mathbf{v})]. \tag{5.4.2}$$

For given $\boldsymbol{\delta}$, we can write $\mathbf{X}\boldsymbol{\beta}^* + \mathbf{Z}\mathbf{v}^* = \mathbf{H}_1\mathbf{y}$, where $\boldsymbol{\beta}^*$ and $\mathbf{v}^*$ are the solutions of the mixed model equations (5.2.8) and $\mathbf{H}_1 = \mathbf{H}_1(\boldsymbol{\delta})$ is treated as a "hat" matrix. Effective degrees of freedom used in estimating $\boldsymbol{\beta}$ and $\mathbf{v}$ are defined as

$$\varrho(\boldsymbol{\delta}) = \text{tr}[\mathbf{H}_1(\boldsymbol{\delta})] = \text{tr}[(\mathbf{X}'\mathbf{V}^{-1}\mathbf{X})^{-1}\mathbf{X}'\mathbf{V}^{-1}\mathbf{R}\mathbf{V}^{-1}\mathbf{X}] + n - \text{tr}(\mathbf{R}\mathbf{V}^{-1}), \tag{5.4.3}$$

(see Müller et al. 2013). A naive estimator of $\varrho(\boldsymbol{\delta})$ is taken as $\varrho(\hat{\boldsymbol{\delta}}_{\text{ML}})$. A simple conditional AIC is then given by

$$\text{cAIC} = -2l(\hat{\boldsymbol{\beta}}_{\text{ML}}, \hat{\boldsymbol{\delta}}_{\text{ML}}|\hat{\mathbf{v}}^H) + 2[\varrho(\hat{\boldsymbol{\delta}}_{\text{ML}}) + q^*], \tag{5.4.4}$$

where $\hat{\mathbf{v}}^H = \tilde{\mathbf{v}}(\hat{\boldsymbol{\delta}}_{\text{ML}})$ is the EBLUP estimator of $\mathbf{v}$. Refinements to the penalty term in (5.4.4) have been proposed (see Müller et al. 2013). In particular, for the special case of $\mathbf{R} = \sigma^2\mathbf{I}_n$, Vaida and Blanchard (2005) proposed the penalty term

$$\alpha_{n,\text{VB}}(\hat{\boldsymbol{\beta}}_{\text{ML}}, \hat{\boldsymbol{\delta}}_{\text{ML}}) = \frac{2n}{n-p-2}\left[\varrho(\hat{\boldsymbol{\delta}}_{\text{ML}}) + 1 - \frac{\varrho(\hat{\boldsymbol{\delta}}_{\text{ML}}) - p}{n-p}\right], \tag{5.4.5}$$

which tends to $2[\varrho(\hat{\boldsymbol{\delta}}_{\text{ML}}) + 1]$ as $n \to \infty$ with p fixed. Note that $\varrho(\boldsymbol{\delta}) + 1$ is the effective degrees of freedom for estimating $\boldsymbol{\beta}$, $\mathbf{v}$, and the scalar σ^2, for given $\boldsymbol{\delta}^* := \boldsymbol{\delta}/\sigma^2$. Han (2013) studied cAIC for the Fay–Herriot area level model (4.2.5). The proposed cAIC depends on the method of estimating σ_v^2, the variance of the random effect v_i in the model. It performs better than the simple cAIC, given by (5.4.4), which ignores the error in estimating σ_v^2.

Jiang and Rao (2003) studied the selection of fixed and random effects for the general linear mixed model (5.2.1), avoiding the estimation of $\boldsymbol{\delta}$ and using only the ordinary least squares (OLS) estimator of $\boldsymbol{\beta}$. Two cases are considered: (i) selection of fixed covariates, $\mathbf{X}$, from a set of candidate covariates when the random effects are not subject to selection; (ii) selection of both covariates and random effects. Normality of $\mathbf{v}$ and $\mathbf{e}$ is not assumed, unlike in the case of AIC and BIC. For case (i), suppose $\mathbf{x}_1, \ldots, \mathbf{x}_q$ denote the $n \times 1$ candidate vectors of covariates from which the columns of the $n \times p$ matrix are to be selected ($p \leq q$). Let $\mathbf{X}(a)$ denote the matrix for a subset, a, of the q column vectors, and let $\hat{\boldsymbol{\beta}}(a) = [\mathbf{X}(a)^T\mathbf{X}(a)]^{-1}\mathbf{X}(a)^T\mathbf{y}$ be the OLS estimator of $\boldsymbol{\beta}(a)$ for the model $\mathbf{y} = \mathbf{X}(a)\boldsymbol{\beta}(a) + \boldsymbol{\epsilon}$, with $\boldsymbol{\epsilon} = \mathbf{Z}\mathbf{v} + \mathbf{e}$. A generalized information criterion (GIC) for the selection of covariates only is given by

$$C_n(a) = [\mathbf{y} - \mathbf{X}(a)\hat{\boldsymbol{\beta}}(a)]^T[\mathbf{y} - \mathbf{X}(a)\hat{\boldsymbol{\beta}}(a)] + \lambda_n|a|, \tag{5.4.6}$$

where $|a|$ is the dimension of the subset a and λ_n is a positive number satisfying certain conditions.

GIC selects the subset a^* that minimizes $C_n(a)$ over all (or specified) subsets of the q column vectors. GIC is consistent in the sense that, if a_0 is the subset associated with the true model, assuming that the true model is among the candidate models, then $P(a^* \neq a_0) \to 0$ as $n \to \infty$. The statistic $C_n(a)$ is computationally attractive because the number of candidate models can be very large and $C_n(a)$ does not require estimates of variance parameters, unlike AIC of BIC.

GIC, given by (5.4.6), does not work for the case (ii) involving the selection of both fixed and random effects. Jiang and Rao (2003) proposed an alternative method for the general ANOVA model (5.2.15). This method divides the random effect factors into several groups and then applies different model selection procedures for the different groups simultaneously.

If the true underlying model is based on a single random factor, as in the case of the basic unit level model (5.3.1), the BIC choice $\lambda_n = \log n$ in (5.4.6) satisfies conditions needed for consistency but not the mAIC choice $\lambda_n = 2$.

Fence Method Information-based criteria have several limitations when applied to mixed models, as noted by Jiang, Rao, Gu, and Nguyen (2008): (i) n should be replaced by the effective sample size n^* because observations are correlated, but the choice of n^* is not always clear. (ii) A log-likelihood function is used, but this is not available if a parametric family is not assumed. (iii) Finite sample performance of the criteria may be sensitive to the choice of penalty term. Fence methods, proposed by Jiang et al. (2008), are free of the above limitations. The basic idea behind fence methods consists of two steps: (1) Isolate a subgroup of "correct" models by constructing a statistical fence that eliminates incorrect models. (2) Select the "optimal" model from the subgroup according to a suitable criterion.

The fence is constructed by first choosing a lack-of-fit measure $Q_M = Q_M(\mathbf{y}, \boldsymbol{\beta}_M, \boldsymbol{\delta}_M)$ of a candidate model M with associated parameters $\boldsymbol{\beta}_M$ and $\boldsymbol{\delta}_M$ such that $E(Q_M)$ is minimized when M is a true model (a correct model but not necessarily optimal) and $\boldsymbol{\beta}_M$ and $\boldsymbol{\delta}_M$ are the true parameter vectors. A simple choice of Q_M is $Q_M = |\mathbf{y} - \mathbf{X}_M\boldsymbol{\beta}_M|^2$, where $\mathbf{X}_M$ is the matrix of covariates associated with M. Now let $\hat{Q}_M = Q(\mathbf{y}, \hat{\boldsymbol{\beta}}_M, \hat{\boldsymbol{\delta}}_M)$, where $(\hat{\boldsymbol{\beta}}_M, \hat{\boldsymbol{\delta}}_M)$ minimize Q_M. Using $\hat{Q}_M$ for each candidate model M, find the model $\tilde{M} \in \mathcal{M}$ such that $Q_{\tilde{M}} = \min_{M \in \mathcal{M}} \hat{Q}_M$, where $\mathcal{M}$ denotes the set of candidate models. Note that $\tilde{M} = M_f$ if $\mathcal{M}$ contains the full model M_f. In general, $\mathcal{M}$ may not contain M_f but, under certain regularity conditions, $\tilde{M}$ is a true model with probability tending to one.

The identified model $\tilde{M}$ does not rule out the possibility of other correct models with smaller dimension than $\tilde{M}$. Therefore, a fence is constructed around $\tilde{M}$ to eliminate incorrect models, and then select the optimal model from the set of models within the fence using an optimality criterion such as minimal dimension of the model. The fence is defined by the subset of models M belonging to $\mathcal{M}$ that satisfy the following inequality:

$$\hat{Q}_M \leq \hat{Q}_{\tilde{M}} + a_n[v(\hat{Q}_M - \hat{Q}_{\tilde{M}})]^{1/2}, \qquad (5.4.7)$$

where $v(\hat{Q}_M - \hat{Q}_{\tilde{M}})$ denotes an estimate of the variance of $\hat{Q}_M - \hat{Q}_{\tilde{M}}$ and a_n is a tuning constant increasing slowly with n.

If the minimal dimension criterion is used as optimality criterion, then the implementation of the fence method can be simplified by checking each candidate model, from the simplest to most complex, and stopping the search once a model that falls within the fence is identified, and other models of the same dimension are checked to see if they belong to the fence. However, it is often not easy to construct the fence because of the difficulty in estimating the variance of $\hat{Q}_M - \hat{Q}_{\tilde{M}}$. A simplified adaptive fence (see below) may be constructed by replacing the last term of (5.4.7) by a tuning constant c_n and then select c_n optimally, then avoiding the computation of $v(\hat{Q}_M - \hat{Q}_{\tilde{M}})$. Jiang, Nguyen, and Rao (2009) developed a bootstrap method (named adaptive fence) to determine c_n. The method consists of the following steps: (i) Generate parametric bootstrap samples from either the full model or any large model known to be correct but not optimal. (ii) For each $M \in \mathcal{M}$, calculate the proportion of bootstrap samples $p^*(M; c_n)$ in which M is selected by the simplified fence method with a specified tuning constant c_n. (iii) Compute $p^*(c_n) = \max_{M \in \mathcal{M}} p^*(M; c_n)$ and choose c_n that maximizes $p^*(c_n)$.

5.4.2 Model Diagnostics

In this section, we give a brief account of model diagnostics based on residuals and influence measures for particular cases of the general linear mixed model (5.2.1).

Residual Analysis Calvin and Sedransk (1991) describe two basic approaches to define residuals under model (5.3.1) with block diagonal covariance structure. The first approach is based on transforming the model into a standard linear regression model and then obtaining the usual residuals for these models based on the transformed data. Let $\mathbf{V}_i = \sigma_e^2 \mathbf{A}_i$ where σ_e^2 denotes the error variance. Using $\mathbf{A}_i$, we transform the model (5.3.1) to $\tilde{\mathbf{y}}_i = \tilde{\mathbf{X}}_i \boldsymbol{\beta} + \tilde{\boldsymbol{\epsilon}}_i, i = 1, \ldots, m$, where $\tilde{\mathbf{y}}_i = \mathbf{A}_i^{-1/2} \mathbf{y}_i, \tilde{\mathbf{X}}_i = \mathbf{A}_i^{-1/2} \mathbf{X}_i$, and $\tilde{\boldsymbol{\epsilon}}_i = \mathbf{A}_i^{-1/2}(\mathbf{Z}_i \mathbf{v}_i + \mathbf{e}_i)$. We may express the transformed model as a standard linear regression model, $\tilde{\mathbf{y}} = \tilde{\mathbf{X}} \boldsymbol{\beta} + \tilde{\boldsymbol{\epsilon}}$ with iid errors $\tilde{\boldsymbol{\epsilon}}$, that is, $\tilde{\boldsymbol{\epsilon}}$ has mean $\mathbf{0}$ and covariance matrix $\sigma_e^2 \mathbf{I}_n$. Therefore, we can apply standard linear regression methods for model selection and validation such as the selection of fixed effects, residual analysis, and influence analysis. In practice, model parameters are not known and residuals are defined as $\hat{\boldsymbol{\epsilon}}_i = \hat{\mathbf{A}}_i^{-1/2}(\mathbf{y}_i - \mathbf{X}_i \hat{\boldsymbol{\beta}})$, where $\hat{\mathbf{A}}_i^{-1/2}$ is a square root of $\hat{\mathbf{V}}_i^{-1}$ and $\hat{\mathbf{V}}_i^{-1}$ is obtained by replacing the unknown variance components in $\mathbf{V}_i^{-1}$ by suitable estimates. Jacqmin-Gadda et al. (2007) take $\hat{\mathbf{A}}_i^{-1/2}$ as the triangular matrix derived from the Cholesky decomposition of $\hat{\mathbf{V}}_i^{-1}$. This transformation method looks appealing because residuals are approximately uncorrelated, and therefore residual plots can be interpreted in the same way as for linear regression models. However, transformation leads to residuals defined as weighted averages of many data points, and therefore if a single data point is an outlier, its effect will be masked by the smoothing effect of averaging. Moreover, these residuals only focus on the fixed effects while random effects are also of interest in many applications. In fact, if a model departure is detected based on these residuals, it is difficult to say if this departure is on the distribution of the random effects or the model errors. Moreover, the sensitivity of this transformation method to the estimates of variance components has not been well studied.

The second approach is based on the BLUP residuals $\tilde{\mathbf{e}}_i = \mathbf{y}_i - \mathbf{X}_i\tilde{\boldsymbol{\beta}} - \mathbf{Z}_i\tilde{\mathbf{v}}_i$ defined by Calvin and Sedransk (1991), also called conditional residuals by Jacqmin-Gadda et al. (2007). These residuals are correlated and therefore interpretation must be careful, but they make use of the estimated random effects and are used as substitutes for the model errors; thus, they can be used more specifically to check departures on the distribution of the model errors.

In the ANOVA model (5.2.15) with r independent random factors and with normality, Zewotir and Galpin (2007) studied the detection of outliers and high leverage points. Assuming that the variance ratios $\lambda_k = \sigma_k^2/\sigma_e^2$ are known for $k = 1, \ldots, r$, they noted that BLUP residuals $\tilde{\mathbf{e}} = \mathbf{y} - \tilde{\mathbf{y}}$, for $\tilde{\mathbf{y}} = \mathbf{X}\tilde{\boldsymbol{\beta}} - \mathbf{Z}\tilde{\mathbf{v}}$ satisfy $\tilde{\mathbf{e}} = \mathbf{S}\mathbf{y} \sim N(\mathbf{0}, \sigma_e^2\mathbf{S})$, for $\mathbf{S} = (s_{ij}) = \sigma_e^2\mathbf{P}$. They also noted that $\tilde{\mathbf{y}} = (\mathbf{I} - \mathbf{S})\mathbf{y}$, or equivalently, for the ith observation, we have $\tilde{y}_i = (1 - s_{ii})y_i + \sum_{j\neq i}^{n} s_{ij}y_j$, where $0 \leq s_{ii} \leq 1$ and $s_{ij} \to 0$ for $i \neq j$ when $s_{ii} \to 0$. This means that data points y_i with a small value s_{ii} attract the predicted value toward themselves, that is, they have a high leverage effect in the regression. Based on this, they propose to use the diagonal elements s_{ii} of the matrix $\mathbf{S} = \sigma_e^2\mathbf{P}$ to detect high leverage points. To detect both outliers and high leverage points, they propose to look at a plot of s_{ii} versus $\tilde{e}_i^2/\tilde{\mathbf{e}}^T\tilde{\mathbf{e}}$. Points are expected to concentrate around the upper-left corner of the plot. Points separated from the main cloud of points that fall in the lower-left corner (small s_{ii}) are regarded as high leverage points, and points that appear separated on the right side (large relative squared residual $\tilde{e}_i^2/\tilde{\mathbf{e}}^T\tilde{\mathbf{e}}$) are regarded as outliers. They provide a rough cutoff value for s_{ii} below which an observation is considered as a high leverage point, given by $(1 - 2p/n)\left\{1 - \sum_{k=1}^{r}[\sigma_e^2/\sigma_k^2 + n/(2h_k)]^{-1}\right\}$.

Concerning residuals, Zewotir and Galpin (2007) define internally Studentized residuals $t_i = \tilde{e}_i/\hat{\sigma}_e\sqrt{s_{ii}}$ and externally Studentized residuals $t_i^* = \tilde{e}_i/\hat{\sigma}_{e(i)}\sqrt{s_{ii}}$, where $\hat{\sigma}_e^2 = n^{-1}(\mathbf{y} - \mathbf{X}\tilde{\boldsymbol{\beta}})^T(\mathbf{I} + \sum_{k=1}^{r}\lambda_k\mathbf{Z}_k\mathbf{Z}_k^T)^{-1}(\mathbf{y} - \mathbf{X}\tilde{\boldsymbol{\beta}})$ and $\hat{\sigma}_{e(i)}^2$ is the same estimator with observation i removed from the data. The two types of squared Studentized residuals are related monotonically as $t_i^2 = n[1 + (n-1)/(t_i^*)^2]^{-1}$. When variance ratios $\lambda_k = \sigma_k^2/\sigma_e^2$ are known, $(t_i^*)^2 \sim \frac{n-1}{n-p-1}\mathcal{F}(1, n-p-1)$ and when variance components are estimated using ML, $(t_i^*)^2 \xrightarrow{d} \mathcal{X}^2(1)$ as $n \to \infty$. Based on the critical value 2 for the Student t distribution, they give the cutoff point $\sqrt{4n/(n-p-3)}$ for t_i, above which an observation is considered as an outlier. They also provide a formal test based on $\max_i|t_i|$ to detect a single outlier. For the random factors $\mathbf{v}_k, k = 1, \ldots, r$, Zewotir and Galpin (2007) argued that the BLUP estimators $\tilde{\mathbf{v}}_k = \lambda_k\mathbf{Z}_k^T\mathbf{S}\mathbf{y} = \lambda_k\mathbf{Z}_k^T\tilde{\mathbf{e}}$ can be interpreted as residuals and can then be used for diagnostic purposes to detect subjects (blocks) that evolve differently from the other subjects in the data set. They propose to standardize the elements of $\tilde{\mathbf{v}}_k$ by the square root of the diagonal elements of the covariance matrix $V(\tilde{\mathbf{v}}_k) = \sigma_e^2\lambda_k^2\mathbf{Z}_k^T\mathbf{S}\mathbf{Z}_k$. When using ML estimates of the variance components, they suggest to pinpoint the jth level of the factor $\mathbf{v}_k$ as an outlier if the standardized BLUP estimator exceeds

$t(1-\alpha/2; n-\text{rank}[\mathbf{X}\ \mathbf{Z}])$, where $t(\beta; m)$ denotes the lower β level of a Student t distribution with m degrees of freedom. In practice, we calculate the empirical versions of residuals (EBLUP residuals), $\hat{\mathbf{e}} = \mathbf{y} - \mathbf{X}\hat{\boldsymbol{\beta}} - \mathbf{Z}\hat{\mathbf{v}}$, and of the BLUP estimator of $\mathbf{v}_k$, given by $\hat{\mathbf{v}}_k = \hat{\lambda}_k \mathbf{Z}_k^T \hat{\mathbf{S}}\mathbf{y}$ for $\hat{\mathbf{S}} = \hat{\sigma}_e^2 \hat{\mathbf{P}}$, but the above results for known λ_k serve as approximations.

To check the normality assumption for the random effects distribution in the special case of model (5.3.1) with $\mathbf{R}_i = \sigma^2 \mathbf{I}_i$ and $\mathbf{G}_i = \boldsymbol{\Delta}$ for all i, Lange and Ryan (1989) developed weighted normal probability plots. Their approach, followed also by Calvin and Sedransk (1991), uses the BLUP estimates of linear functions of the random effects $\mathbf{v}_i$ suitably standardized by their estimated standard deviations. Discarding the effect of estimating $\boldsymbol{\beta}$, the vector of estimated random effects is $\tilde{\mathbf{v}}_i^B = \mathbf{G}_i \mathbf{Z}_i^T \mathbf{V}_i^{-1}(\mathbf{y}_i - \mathbf{X}_i \boldsymbol{\beta})$, with covariance matrix given by $V(\tilde{\mathbf{v}}_i^B) = \boldsymbol{\Delta}\mathbf{Z}_i^T \mathbf{V}_i^{-1} \mathbf{Z}_i \boldsymbol{\Delta}$. They define a normal Q–Q plot of $z_i = \mathbf{c}^T \tilde{\mathbf{v}}_i^B$, for a suitably chosen vector of constants $\mathbf{c}$, versus $\Phi^{-1}[F_m^*(z_i)]$, where Φ is the standard normal cumulative distribution function (c.d.f.) and $F_m^*(z)$ is a weighted empirical c.d.f. of z_i defined as

$$F_m^*(z) = \frac{\sum_{i=1}^m w_i I(z_i \le z)}{\sum_{i=1}^m w_i},$$

with weights $w_i = \mathbf{c}^T V(\tilde{\mathbf{v}}_i^B)\mathbf{c}$, where $I(z_i \le z) = 1$ if $z_i \le z$ and $I(z_i \le z) = 0$ otherwise. They prove large sample (for $m \to \infty$) normality of $\hat{F}_m^*(z)$ obtained from $F_m^*(z)$ by estimating the unknown model parameters, and propose confidence bands for the normal Q–Q plot based on the estimated large sample standard errors of $\hat{F}_m^*(z)$. Formal normality tests for the random effects and model errors are studied by Claeskens and Hart (2009).

Influence Diagnostics Cook's (1977) distance is widely used in standard linear regression to study the effect of deleting a "case" on the estimate of $\boldsymbol{\beta}$. Banerjee and Frees (1997) extended Cook's distance to the model (5.3.1) with block diagonal covariance structure to study the effect of deleting a block (small area) on the estimate of $\boldsymbol{\beta}$.

Let $\hat{\boldsymbol{\beta}}_{(i)}$ be the estimator of $\boldsymbol{\beta}$ after dropping block i. The influence of block i on the estimate $\hat{\boldsymbol{\beta}} = (\sum_{i=1}^m \mathbf{X}_i^T \hat{\mathbf{V}}_i^{-1} \mathbf{X}_i)^{-1} \sum_{i=1}^m \mathbf{X}_i^T \hat{\mathbf{V}}_i^{-1} \mathbf{y}_i$ may be measured by a Mahalanobis distance given by

$$B_i(\hat{\boldsymbol{\beta}}) = \frac{1}{p}(\hat{\boldsymbol{\beta}} - \hat{\boldsymbol{\beta}}_{(i)})^T \left(\sum_{i=1}^m \mathbf{X}_i^T \hat{\mathbf{V}}_i^{-1} \mathbf{X}_i \right) (\hat{\boldsymbol{\beta}} - \hat{\boldsymbol{\beta}}_{(i)}). \tag{5.4.8}$$

Note that $B_i(\hat{\boldsymbol{\beta}})$ is the squared distance from $\hat{\boldsymbol{\beta}}$ to $\hat{\boldsymbol{\beta}}_{(i)}$ relative to the estimated covariance matrix $(\sum_{i=1}^m \mathbf{X}_i^T \hat{\mathbf{V}}_i^{-1} \mathbf{X}_i)^{-1}$ of $\hat{\boldsymbol{\beta}}$. The measure $B_i(\hat{\boldsymbol{\beta}})$ may be simplified to

$$B_i(\hat{\boldsymbol{\beta}}) = \frac{1}{p}(\mathbf{y}_i - \mathbf{X}_i\hat{\boldsymbol{\beta}})^T (\hat{\mathbf{V}}_i - \mathbf{H}_i)^{-1} \mathbf{H}_i (\hat{\mathbf{V}}_i - \mathbf{H}_i)^{-1} (\mathbf{y}_i - \mathbf{X}_i \hat{\boldsymbol{\beta}}), \tag{5.4.9}$$

where

$$\mathbf{H}_i = \mathbf{X}_i \left(\sum_{i=1}^{m} \mathbf{X}_i^T \hat{\mathbf{V}}_i^{-1} \mathbf{X}_i \right)^{-1} \mathbf{X}_i^T. \tag{5.4.10}$$

An index plot of the measure $B_i(\hat{\boldsymbol{\beta}})$ for $i = 1, \ldots, m$ is useful in identifying influential blocks (or small areas). Note that we are assuming that $\mathbf{V}_i$ is correctly specified.

Christiansen, Pearson, and Johnson (1992) studied case-deletion diagnostics for mixed ANOVA models. They extended Cook's distance to measure influence on the fixed effects as well as on the variance components. Computational procedures are also provided. These measures are useful for identifying influential observations (or cases).

Cook (1986) developed a "local" influence approach to diagnostics for assessing to what extent slight perturbations to the model can influence ML inferences. Beckman, Nachtsheim, and Cook (1987) applied this approach to linear mixed ANOVA models. Let $\boldsymbol{\omega}$ be a $q \times 1$ vector of perturbations and $l_{\boldsymbol{\omega}}(\boldsymbol{\beta}, \boldsymbol{\delta})$ be the corresponding log-likelihood function. For example, to check the assumption of homogeneity of error variances in the ANOVA model, that is, $\mathbf{R} = \sigma_e^2 \mathbf{I}_n$, we introduce perturbation of the form $\mathbf{R}_{\boldsymbol{\omega}} = \sigma_e^2 \mathbf{D}(\boldsymbol{\omega})$, where $\mathbf{D}(\boldsymbol{\omega}) = \text{diag}(\omega_1, \ldots, \omega_n)$. We denote the "null" perturbations that yield the original model by $\boldsymbol{\omega}_0$ so that $l_{\boldsymbol{\omega}_0}(\boldsymbol{\beta}, \boldsymbol{\delta}) = l(\boldsymbol{\beta}, \boldsymbol{\delta})$, the original log- likelihood function. Note that $\boldsymbol{\omega}_0 = \mathbf{1}$ in the above example.

The influence of the perturbation $\boldsymbol{\omega}$ can be assessed by using the "likelihood displacement" given by

$$\text{LD}(\boldsymbol{\omega}) = 2[l(\hat{\boldsymbol{\beta}}, \hat{\boldsymbol{\delta}}) - l(\hat{\boldsymbol{\beta}}_{\boldsymbol{\omega}}, \hat{\boldsymbol{\delta}}_{\boldsymbol{\omega}})], \tag{5.4.11}$$

where $(\hat{\boldsymbol{\beta}}_{\boldsymbol{\omega}}, \hat{\boldsymbol{\delta}}_{\boldsymbol{\omega}})$ are the ML estimators of $\boldsymbol{\beta}$ and $\boldsymbol{\delta}$ under the perturbed log-likelihood $l_{\boldsymbol{\omega}}(\boldsymbol{\beta}, \boldsymbol{\delta})$. Note that $\text{LD}(\boldsymbol{\omega})$ is nonnegative and achieves its minimum value of 0 when $\boldsymbol{\omega} = \boldsymbol{\omega}_0$. Large values of $\text{LD}(\boldsymbol{\omega})$ indicate that $(\hat{\boldsymbol{\beta}}_{\boldsymbol{\omega}}, \hat{\boldsymbol{\delta}}_{\boldsymbol{\omega}})$ differ considerably from $(\hat{\boldsymbol{\beta}}, \hat{\boldsymbol{\delta}})$ relative to the contours of the unperturbed log-likelihood $l(\boldsymbol{\beta}, \boldsymbol{\delta})$. Rather than calculating $\text{LD}(\boldsymbol{\omega})$, we examine the behavior of $\text{LD}(\boldsymbol{\omega})$ as a function of $\boldsymbol{\omega}$ for values that are "local" to $\boldsymbol{\omega}_0$; in particular, we examine curvatures of $\text{LD}(\boldsymbol{\omega})$. Let $\boldsymbol{\omega}_0 + a\, \boldsymbol{v}$ be a vector through $\boldsymbol{\omega}_0$ in the direction $\boldsymbol{v}$. Then, the "normal curvature" of the surface $(\boldsymbol{\omega}^T, \text{LD}(\boldsymbol{\omega}))$ in the direction of $\boldsymbol{v}$ is given by

$$C_{\boldsymbol{v}} = \partial^2 \text{LD}(\boldsymbol{\omega}_0 + a\boldsymbol{v}) / \partial a^2 |_{a=0}. \tag{5.4.12}$$

Large values of $C_{\boldsymbol{v}}$ indicate sensitivity to the induced perturbations in the direction $\boldsymbol{v}$. We find the largest curvature $C_{\max}$ and its corresponding direction $\boldsymbol{v}_{\max}$. An index plot of the elements of the normalized vector $\boldsymbol{v}_{\max}/\|\boldsymbol{v}_{\max}\|$ is useful for identifying influential perturbations; large elements indicate influential perturbations. We refer to Beckman et al. (1987) for computational details. Hartless, Booth, and Littell (2000) applied the above method to the data of county crop areas of Example 4.3.1. By perturbing the error variances, that is, using $\mathbf{R}_{\boldsymbol{\omega}} = \sigma_e^2 \mathbf{D}(\boldsymbol{\omega})$, they identified an erroneous observation in the original data (see Example 7.3.1 of Chapter 7).

5.5 *SOFTWARE

PROC MIXED (SAS/STAT 13.1 Users Guide 2013, Chapter 63) implements ML and REML estimation of model parameters $\boldsymbol{\beta}$ and $\boldsymbol{\delta}$ for the linear mixed model (5.2.1), using the Newton–Raphson algorithm. Inverse of the observed information matrix is used to estimate the covariance matrix of parameter estimates $\hat{\boldsymbol{\beta}}$ and $\hat{\boldsymbol{\delta}}$. SCORING option in PROC MIXED uses instead Fisher-scoring algorithm to estimate $\boldsymbol{\beta}$ and $\boldsymbol{\delta}$, and the inverse of the expected information matrix to estimate the covariance matrix of parameter estimates. PROC MIXED also gives the EBLUP $\hat{\mu} = t(\hat{\boldsymbol{\delta}}, \mathbf{y})$ of a specified parameter $\mu = \mathbf{l}^T\boldsymbol{\beta} + \mathbf{m}^T\mathbf{v}$ along with the naive MSE estimator, $\text{mse}_N(\hat{\mu})$, given by (5.2.35). Option DDFM=KENWARDROGER computes the second-order unbiased MSE estimator (5.2.39), using the estimated covariance matrix of $\hat{\boldsymbol{\beta}}_{\text{RE}}$ and $\hat{\boldsymbol{\delta}}_{\text{RE}}$ based on the observed information matrix; note that (5.2.39) is only first-order unbiased if the ML estimators $\hat{\boldsymbol{\beta}}_{\text{ML}}$ and $\hat{\delta}_{\text{ML}}$ are used.

Model selection criteria mAIC and BIC for ML and REML methods (see Section 5.4.1) are also implemented in PROC MIXED; mAIC is denoted as AIC. PROC MIXED warns against mechanical use of model selection criteria, by noting that "subject matter considerations and objectives are of great importance when selecting a model."

Several model diagnostics such as plots of within-area residuals and normal probability plots of estimated residuals can also be obtained with PROC MIXED (see PLOTS option). INFLUENCE option in the MODEL statement can be used to compute case-deletion diagnostics (see Section 5.4.2 for some model checking methods in the linear mixed model with a block diagonal covariance structure).

Mukhopadhyay and McDowell (2011) illustrate small area estimation for the basic area level model (4.2.5) and the unit level model (4.3.1) using SAS software. They also considered unmatched area level models, studied in Section 10.4, using the MCMC procedure in SAS.

Specific SAS software for small area estimation has been developed by Estevao, Hidiroglou, and You (2014). This software covers EBLUP and pseudo-EBLUP estimation and linearization-based MSE estimation for the basic area level model (4.2.5) and the basic unit level model (4.3.1) including some model diagnostics.

In R statistical software (R Core Team 2013), the function `lme` from the package `nlme` (Pinheiro et al. 2014) fits the linear mixed model (5.3.1) with block diagonal covariance matrices composed of m blocks and with iid random effects $\mathbf{v}_i$, that is, with $\mathbf{G}_i = \sigma_v^2\mathbf{I}, i = 1, \ldots, m$, for $\sigma_v^2 \geq 0$ unknown. The function is designed for nested random effects and it accepts also random slopes and general correlation structures for $\mathbf{R}_i$, as long as the structure of $\mathbf{R}_i$ only depends on i through its dimension n_i. The function uses the E-M algorithm to obtain simultaneously the estimates of the variance parameters $\boldsymbol{\delta}$, fixed effects $\boldsymbol{\beta}$, and random effects $\mathbf{v}_i$. Specifying `method=ML`, the E-M algorithm obtains the ML estimate of $\boldsymbol{\delta}$, $\hat{\boldsymbol{\delta}}_{\text{ML}}$, whereas setting `method=REML` or omitting this option, the E-M algorithm implements the REML approach that yields $\hat{\boldsymbol{\delta}}_{\text{RE}}$. The estimates of $\boldsymbol{\beta}$ and $\mathbf{v}$ are then obtained using equations (5.2.5) and (5.2.6) with $\boldsymbol{\delta}$ replaced by $\hat{\boldsymbol{\delta}}_{\text{ML}}$ or $\hat{\boldsymbol{\delta}}_{\text{RE}}$, depending on the choice of `method`. For more details on the fitting algorithm, see Laird and Ware (1982).

Function `lme` also returns the usual goodness-of-fit measures mAIC, called AIC in the function output, BIC, and the value of the log-likelihood at the estimated parameters. The output of the linear mixed model fit is an object of class `lme`. Fitted values $\hat{\mathbf{y}}_i$ and EBLUP residuals $\hat{\mathbf{e}}_i = \mathbf{y}_i - \mathbf{X}_i\hat{\boldsymbol{\beta}} - \mathbf{Z}_i\hat{\mathbf{v}}_i, i = 1, \ldots, m$, can be obtained by applying, respectively, the functions `fitted()` and `resid()` to that object. They can then automatically be used to obtain usual diagnostic plots by means of standard R plotting functions.

Linear, generalized linear, and nonlinear mixed models can also be fitted and analyzed using functions `lmer`, `glmer`, and `nlmer`, respectively, from R package `lme4` (Bates et al. 2014). This package is designed as a memory-efficient extension to the mentioned `nlme` package.

The R package `sae` (Molina and Marhuenda 2013; Molina and Marhuenda 2015) is specific for small area estimation. It estimates area means under two special cases of model (5.3.1) that are widely used in small area estimation, namely the basic area level model (4.2.5) and the unit level model (4.3.1). It also includes functions for more complex area level models with spatial and spatiotemporal correlation, together with estimation of general nonlinear small area parameters under the basic unit level model (4.3.1) based on the theory given in Section 9.4. The package incorporates functions for MSE estimation of small area estimators. In the case of the basic area level model, MSE estimates are obtained analytically, whereas for the rest of models, bootstrap procedures are implemented. Finally, it includes also functions for basic small area estimators such as the direct Horvitz–Thompson estimators (2.3.1), the count-synthetic poststratified estimators (3.2.7), and sample-size-dependent estimators defined by (3.3.1) with weights ϕ_i given in (3.3.8). Examples of use of the functions in the `sae` package are included in the remaining Software sections of the book.

5.6 PROOFS

5.6.1 Derivation of BLUP

A linear estimator $\hat{\mu} = \mathbf{a}^T\mathbf{y} + b$ is unbiased for $\mu = \mathbf{l}^T\boldsymbol{\beta} + \mathbf{m}^T\mathbf{v}$ under the linear mixed model (5.2.1), that is, $E(\hat{\mu}) = E(\mu)$, if and only if $\mathbf{a}^T\mathbf{X} = \mathbf{l}^T$ and $b = 0$. The MSE of a linear unbiased estimator $\hat{\mu}$ is given by

$$\text{MSE}(\hat{\mu}) = V(\hat{\mu} - \mu) = \mathbf{a}^T\mathbf{V}\mathbf{a} - 2\mathbf{a}^T\mathbf{Z}\mathbf{G}\mathbf{m} + \mathbf{m}^T\mathbf{G}\mathbf{m}.$$

Minimizing $V(\hat{\mu} - \mu)$ subject to the unbiasedness condition $\mathbf{a}^T\mathbf{X} = \mathbf{l}^T$ using Lagrange multiplier $2\boldsymbol{\lambda}$, we get

$$\mathbf{V}\mathbf{a} + \mathbf{X}\boldsymbol{\lambda} = \mathbf{Z}\mathbf{G}\mathbf{m}$$

and solving for $\mathbf{a}$ we obtain

$$\mathbf{a} = -\mathbf{V}^{-1}\mathbf{X}\boldsymbol{\lambda} + \mathbf{V}^{-1}\mathbf{Z}\mathbf{G}\mathbf{m}. \tag{5.6.1}$$

Substituting (5.6.1) for $\mathbf{a}$ into the constraint $\mathbf{a}^T\mathbf{X} = \mathbf{l}^T$, we now solve for $\boldsymbol{\lambda}$:

$$\boldsymbol{\lambda} = -(\mathbf{X}^T\mathbf{V}^{-1}\mathbf{X})^{-1}\mathbf{l} + (\mathbf{X}^T\mathbf{V}^{-1}\mathbf{X})^{-1}\mathbf{X}^T\mathbf{V}^{-1}\mathbf{Z}\mathbf{G}\mathbf{m}. \qquad (5.6.2)$$

Again, substituting (5.6.2) for $\boldsymbol{\lambda}$ in (5.6.1) and using $\tilde{\boldsymbol{\beta}} = (\mathbf{X}^T\mathbf{V}^{-1}\mathbf{X})^{-1}\mathbf{X}^T\mathbf{V}^{-1}\mathbf{y}$, we obtain

$$\mathbf{a}^T\mathbf{y} = \mathbf{l}^T\tilde{\boldsymbol{\beta}} + \mathbf{m}^T\mathbf{G}\mathbf{Z}^T\mathbf{V}^{-1}(\mathbf{y} - \mathbf{X}\tilde{\boldsymbol{\beta}}),$$

which is identical to the BLUP given by (5.2.4).

5.6.2 Equivalence of BLUP and Best Predictor $E(\mathbf{m}^T\mathbf{v}|\mathbf{A}^T\mathbf{y})$

Under the transformed data $\mathbf{A}^T\mathbf{y}$ with $\mathbf{A}^T\mathbf{X} = \mathbf{0}$, we show here that the BP estimator, $E(\mathbf{m}^T\mathbf{v}|\mathbf{A}^T\mathbf{y})$, of $\mathbf{m}^T\mathbf{v}$ reduces to the BLUP estimator $\mathbf{m}^T\tilde{\mathbf{v}}$, where $\tilde{\mathbf{v}}$ is given by (5.2.6). Under normality, it is easy to verify that the conditional distribution of $\mathbf{m}^T\mathbf{v}$ given $\mathbf{A}^T\mathbf{y}$ is normal with mean

$$E(\mathbf{m}^T\mathbf{v}|\mathbf{A}^T\mathbf{y}) = \mathbf{C}\mathbf{A}(\mathbf{A}^T\mathbf{V}\mathbf{A})^{-1}\mathbf{A}^T\mathbf{y}, \qquad (5.6.3)$$

where $\mathbf{C} = \mathbf{m}^T\mathbf{G}\mathbf{Z}^T$. Since $\mathbf{V}^{1/2}\mathbf{A}$ and $\mathbf{V}^{-1/2}\mathbf{X}$ are orthogonal to each other, that is, $\mathbf{V}^{1/2}\mathbf{A}\mathbf{X}^T\mathbf{V}^{-1/2} = \mathbf{0}$, and $\text{rank}(\mathbf{V}^{1/2}\mathbf{A}) + \text{rank}(\mathbf{V}^{-1/2}\mathbf{X}) = n$, the following decomposition of projections holds:

$$\mathbf{I} = \mathbf{P}_{\mathbf{V}^{1/2}\mathbf{A}} + \mathbf{P}_{\mathbf{V}^{-1/2}\mathbf{X}},$$

where $\mathbf{P}_\mathbf{B} = \mathbf{B}(\mathbf{B}^T\mathbf{B})^{-1}\mathbf{B}^T$ for a matrix $\mathbf{B}$. Hence,

$$\mathbf{I} = \mathbf{V}^{1/2}\mathbf{A}(\mathbf{A}^T\mathbf{V}\mathbf{A})^{-1}\mathbf{A}^T\mathbf{V}^{1/2} + \mathbf{V}^{-1/2}\mathbf{X}(\mathbf{X}^T\mathbf{V}^{-1}\mathbf{X})^{-1}\mathbf{X}^T\mathbf{V}^{-1/2},$$

or equivalently,

$$\mathbf{A}(\mathbf{A}^T\mathbf{V}\mathbf{A})^{-1}\mathbf{A}^T = \mathbf{V}^{-1} - \mathbf{V}^{-1}\mathbf{X}(\mathbf{X}^T\mathbf{V}^{-1}\mathbf{X})^{-1}\mathbf{X}^T\mathbf{V}^{-1}. \qquad (5.6.4)$$

Replacing (5.6.4) in (5.6.3), we obtain

$$E(\mathbf{m}^T\mathbf{v}|\mathbf{A}^T\mathbf{y}) = \mathbf{C}[\mathbf{V}^{-1} - \mathbf{V}^{-1}\mathbf{X}(\mathbf{X}^T\mathbf{V}^{-1}\mathbf{X})^{-1}\mathbf{X}^T\mathbf{V}^{-1}]\mathbf{y} = \mathbf{C}\mathbf{V}^{-1}(\mathbf{y} - \mathbf{X}\tilde{\boldsymbol{\beta}}),$$

which does not depend on the choice of matrix $\mathbf{A}$ and it is equal to the BLUP estimator $\mathbf{m}^T\tilde{\mathbf{v}}$. This proof is due to Jiang (1997).

5.6.3 Derivation of MSE Decomposition (5.2.29)

The result of Section 5.6.2 may be used to provide a simple proof of the MSE decomposition $\text{MSE}[t(\hat{\boldsymbol{\delta}})] = \text{MSE}[t(\boldsymbol{\delta})] + E[t(\hat{\boldsymbol{\delta}}) - t(\boldsymbol{\delta})]^2$ given in (5.2.29), where $\hat{\boldsymbol{\delta}}$ is a function of $\mathbf{A}^T\mathbf{y}$. Define $\hat{\mathbf{V}} := \mathbf{V}(\hat{\boldsymbol{\delta}})$, $\hat{\boldsymbol{\beta}} := \tilde{\boldsymbol{\beta}}(\hat{\boldsymbol{\delta}})$ and $\hat{\mathbf{v}} := \tilde{\mathbf{v}}(\hat{\boldsymbol{\delta}})$. Also, write $t(\hat{\boldsymbol{\delta}}) - \mu = t(\hat{\boldsymbol{\delta}}) - t(\boldsymbol{\delta}) + t(\boldsymbol{\delta}) - \mu$. First, note that $t(\hat{\boldsymbol{\delta}}) - t(\boldsymbol{\delta}) = \mathbf{l}^T(\hat{\boldsymbol{\beta}} - \tilde{\boldsymbol{\beta}}) + \mathbf{m}^T(\hat{\mathbf{v}} - \tilde{\mathbf{v}})$ is a function of $\mathbf{A}^T\mathbf{y}$. This follows by writing

$$\hat{\boldsymbol{\beta}} - \tilde{\boldsymbol{\beta}} = [(\mathbf{X}^T\hat{\mathbf{V}}^{-1}\mathbf{X})^{-1}\mathbf{X}\hat{\mathbf{V}}^{-1} - (\mathbf{X}^T\mathbf{V}^{-1}\mathbf{X})^{-1}\mathbf{X}^T\mathbf{V}^{-1}](\mathbf{y} - \mathbf{X}\tilde{\boldsymbol{\beta}})$$

and noting that both $\hat{\mathbf{V}}$ and $\mathbf{y} - \mathbf{X}\tilde{\boldsymbol{\beta}}$ are functions of $\mathbf{A}^T\mathbf{y}$. Similarly, $\hat{\mathbf{v}} - \tilde{\mathbf{v}}$ also depends only on $\mathbf{A}^T\mathbf{y}$. Furthermore, the first term on the right-hand side of $t(\boldsymbol{\delta}) - \mu = \mathbf{l}^T(\tilde{\boldsymbol{\beta}} - \boldsymbol{\beta}) + \mathbf{m}^T(\tilde{\mathbf{v}} - \mathbf{v})$ is independent of $\mathbf{A}^T\mathbf{y}$ because of the condition $\mathbf{A}^T\mathbf{X} = \mathbf{0}$, and the last term equals $\mathbf{m}^T[E(\mathbf{v}|\mathbf{A}^T\mathbf{y}) - \mathbf{v}]$, using the result of Section 5.6.2. Hence, we have

$$\begin{aligned}
\text{MSE}[t(\hat{\boldsymbol{\delta}})] &= \text{MSE}[t(\boldsymbol{\delta})] + E[t(\hat{\boldsymbol{\delta}}) - t(\boldsymbol{\delta})]^2 \\
&\quad + 2\,E\{[t(\hat{\boldsymbol{\delta}}) - t(\boldsymbol{\delta})]\mathbf{l}^T E(\tilde{\boldsymbol{\beta}} - \boldsymbol{\beta}|\mathbf{A}^T\mathbf{y})\} \\
&\quad + 2\,E\{[t(\hat{\boldsymbol{\delta}}) - t(\boldsymbol{\delta})]\mathbf{m}^T E[E(\mathbf{v}|\mathbf{A}^T\mathbf{y}) - \mathbf{v}|\mathbf{A}^T\mathbf{y}]\} \\
&= \text{MSE}[t(\boldsymbol{\delta})] + E[t(\hat{\boldsymbol{\delta}}) - t(\boldsymbol{\delta})]^2,
\end{aligned} \tag{5.6.5}$$

noting that $E(\tilde{\boldsymbol{\beta}} - \boldsymbol{\beta}|\mathbf{A}^T\mathbf{y}) = E(\tilde{\boldsymbol{\beta}} - \boldsymbol{\beta}) = \mathbf{0}$ and $E[E(\mathbf{v}|\mathbf{A}^T\mathbf{y}) - \mathbf{v}|\mathbf{A}^T\mathbf{y}] = \mathbf{0}$.

The above proof of (5.6.5) is due to Jiang (2001). Kackar and Harville (1984) gave a somewhat different proof of (5.6.5).

6

EMPIRICAL BEST LINEAR UNBIASED PREDICTION (EBLUP): BASIC AREA LEVEL MODEL

We presented the empirical best linear unbiased prediction (EBLUP) theory in Chapter 5 under a general linear mixed model (MM) given by (5.2.1). We also studied the special case of a linear MM with a block diagonal covariance structure given by (5.3.1). Model (5.3.1) covers many small area models used in practice. In this chapter we apply the EBLUP results in Section 5.3 to the basic area level (4.2.5), also called the Fay–Herriot (FH) model because it was first proposed by Fay and Herriot (1979) (see Example 4.2.1). Section 6.1 spells out EBLUP estimation and also gives details of the major applications mentioned in Examples 4.2.1 and 4.2.2, Chapter 4. Section 6.2 covers second-order unbiased mean squared error (MSE) estimation, using general results given in Section 5.3.2. Several practical issues associated with the FH model are studied in Section 6.4 and methods that address those issues are presented.

6.1 EBLUP ESTIMATION

In this section we consider the basic area level model (4.2.5) and spell out EBLUP estimation, using the results in Section 5.3, Chapter 5, for the general linear MM with block diagonal covariance structure.

Small Area Estimation, Second Edition. J. N. K. Rao and Isabel Molina.

6.1.1 BLUP Estimator

The basic area level model is given by

$$\hat{\theta}_i = \mathbf{z}_i^T \boldsymbol{\beta} + b_i v_i + e_i, \quad i = 1, \ldots, m, \tag{6.1.1}$$

where $\mathbf{z}_i$ is a $p \times 1$ vector of area level covariates, area effects $v_i \overset{\text{iid}}{\sim} (0, \sigma_v^2)$ are independent of the sampling errors $e_i \overset{\text{ind}}{\sim} (0, \psi_i)$ with known variance ψ_i, $\hat{\theta}_i$ is a direct estimator of ith area parameter $\theta_i = g(\overline{Y}_i)$, and b_i is a known positive constant. Model (6.1.1) is obtained as a special case of the general linear MM with block diagonal covariance structure, given by (5.3.1), by setting

$$\mathbf{y}_i = \hat{\theta}_i, \quad \mathbf{X}_i = \mathbf{z}_i^T, \quad \mathbf{Z}_i = b_i$$

and

$$\mathbf{v}_i = v_i, \quad \mathbf{e}_i = e_i, \quad \boldsymbol{\beta} = (\beta_1, \ldots, \beta_p)^T.$$

Furthermore, in this special case, the covariance matrices of $\mathbf{v}_i$ and $\mathbf{e}_i$ become scalars, given by

$$\mathbf{G}_i = \sigma_v^2, \quad \mathbf{R}_i = \psi_i$$

and the variance–covariance matrix of $\mathbf{y}_i = \hat{\theta}_i$ becomes

$$\mathbf{V}_i = \psi_i + \sigma_v^2 b_i^2.$$

Also, the target parameter is in this case $\mu_i = \theta_i = \mathbf{z}_i^T \boldsymbol{\beta} + b_i v_i$, which is a special case of the general parameter $\mathbf{l}_i^T \boldsymbol{\beta} + \mathbf{m}_i^T \mathbf{v}_i$ with $\mathbf{l}_i = \mathbf{z}_i$ and $\mathbf{m}_i = b_i$.

Making the above substitutions in the general formula (5.3.2) for the BLUP estimator of μ_i, we get the BLUP estimator of θ_i as

$$\tilde{\theta}_i^H = \mathbf{z}_i^T \tilde{\boldsymbol{\beta}} + \gamma_i (\hat{\theta}_i - \mathbf{z}_i^T \tilde{\boldsymbol{\beta}}) \tag{6.1.2}$$

$$= \gamma_i \hat{\theta}_i + (1 - \gamma_i) \mathbf{z}_i^T \tilde{\boldsymbol{\beta}}, \tag{6.1.3}$$

where

$$\gamma_i = \sigma_v^2 b_i^2 / (\psi_i + \sigma_v^2 b_i^2) \tag{6.1.4}$$

and $\tilde{\boldsymbol{\beta}}$ is the best linear unbiased estimator (BLUE) of $\boldsymbol{\beta}$, given in this case by

$$\tilde{\boldsymbol{\beta}} = \tilde{\boldsymbol{\beta}}(\sigma_v^2) = \left[\sum_{i=1}^m \mathbf{z}_i \mathbf{z}_i^T / \left(\psi_i + \sigma_v^2 b_i^2\right) \right]^{-1} \left[\sum_{i=1}^m \mathbf{z}_i \hat{\theta}_i / \left(\psi_i + \sigma_v^2 b_i^2\right) \right]. \tag{6.1.5}$$

It is clear from (6.1.3) that the BLUP estimator, $\tilde{\theta}_i^H$, can be expressed as a weighted average of the direct estimator $\hat{\theta}_i$ and the regression-synthetic estimator $\mathbf{z}_i^T \tilde{\boldsymbol{\beta}}$, where the weight γ_i $(0 \le \gamma_i \le 1)$, given by (6.1.4), measures the uncertainty in modeling the θ_i's, namely, the model variance $\sigma_v^2 b_i^2$ relative to the total variance $\psi_i + \sigma_v^2 b_i^2$. Thus, $\tilde{\theta}_i^H$ takes proper account of the between-area variation relative to the precision of the direct estimator. If the model variance $\sigma_v^2 b_i^2$ is relatively small, then γ_i will be small and more weight is attached to the synthetic estimator. Similarly, more weight is attached to the direct estimator if the design variance ψ_i is relatively small, or equivalently γ_i is large. The form (6.1.2) for $\tilde{\theta}_i^H$ suggests that it adjusts the regression-synthetic estimator $\mathbf{z}_i^T \tilde{\boldsymbol{\beta}}$ to account for potential model deviation.

It is important to note that $\tilde{\theta}_i^H$ is valid for general sampling designs because we are modeling only the $\hat{\theta}_i$'s and not the individual elements in the population, unlike unit level models, and the direct estimator $\hat{\theta}_i$ uses the design weights. Furthermore, $\tilde{\theta}_i^H$ is design-consistent because $\gamma_i \to 1$ as the sampling variance $\psi_i \to 0$. The design-bias of $\tilde{\theta}_i^H$ is given by

$$B_p(\tilde{\theta}_i^H) \approx (1 - \gamma_i)(\mathbf{z}_i^T \boldsymbol{\beta}^* - \theta_i), \tag{6.1.6}$$

where $\boldsymbol{\beta}^* = E_2(\tilde{\boldsymbol{\beta}})$ is the conditional expectation of $\tilde{\boldsymbol{\beta}}$ given $\boldsymbol{\theta} = (\theta_1, \dots, \theta_m)^T$. It follows from (6.1.6) that the design bias relative to θ_i tends to zero as $\psi_i \to 0$ or $\gamma_i \to 1$. Note that $E_m(\mathbf{z}_i^T \boldsymbol{\beta}^*) = E_m(\theta_i)$, where E_m denotes expectation under the linking model $\theta_i = \mathbf{z}_i^T \boldsymbol{\beta} + b_i v_i$, so that the average bias is zero when the linking model holds.

The MSE of the BLUP estimator $\tilde{\theta}_i^H$ is easily obtained either from the general result (5.3.5) or by direct calculation. It is given by

$$\text{MSE}(\tilde{\theta}_i^H) = E(\tilde{\theta}_i^H - \theta_i)^2 = g_{1i}(\sigma_v^2) + g_{2i}(\sigma_v^2), \tag{6.1.7}$$

where

$$g_{1i}(\sigma_v^2) = \sigma_v^2 b_i^2 \psi_i / (\psi_i + \sigma_v^2 b_i^2) = \gamma_i \psi_i \tag{6.1.8}$$

and

$$g_{2i}(\sigma_v^2) = (1 - \gamma_i)^2 \mathbf{z}_i^T \left[\sum_{i=1}^m \mathbf{z}_i \mathbf{z}_i^T / \left(\psi_i + \sigma_v^2 b_i^2\right) \right]^{-1} \mathbf{z}_i. \tag{6.1.9}$$

Let us define $\tilde{\mathbf{z}}_i = \mathbf{z}_i / b_i$ and $\tilde{h}_{ii} = \tilde{\mathbf{z}}_i^T \left(\sum_{i=1}^m \tilde{\mathbf{z}}_i \tilde{\mathbf{z}}_i^T \right)^{-1} \tilde{\mathbf{z}}_i$, $i = 1, \dots, m$. Under the regularity conditions (i) and (ii) below, the first term in (6.1.7), $g_{1i}(\sigma_v^2)$, is $O(1)$, whereas the second term, $g_{2i}(\sigma_v^2)$, due to estimating $\boldsymbol{\beta}$, is $O(m^{-1})$ for large m:

(i) ψ_i and b_i are uniformly bounded; (6.1.10)

(ii) $\sup_{1 \le i \le m} \tilde{h}_{ii} = O(m^{-1})$. (6.1.11)

Condition (ii) is a standard condition in linear regression analysis (Wu 1986).

Comparing the leading term $g_{1i}(\sigma_v^2) = \gamma_i \psi_i$ with $\text{MSE}(\hat{\theta}_i) = \psi_i$, the MSE of the direct estimator $\hat{\theta}_i$, it is clear that $\tilde{\theta}_i^H$ leads to large gains in efficiency when γ_i is small, that is, when the variability of the model error $b_i v_i$ is small relative to the total variability. Note that ψ_i is also the design variance of $\hat{\theta}_i$.

The BLUP estimator (6.1.3) depends on the variance component σ_v^2, which is unknown in practical applications. Replacing σ_v^2 by an estimator $\hat{\sigma}_v^2$, we obtain an EBLUP estimator $\hat{\theta}_i^H$:

$$\hat{\theta}_i^H = \hat{\gamma}_i \hat{\theta}_i + (1 - \hat{\gamma}_i)\mathbf{z}_i^T \hat{\boldsymbol{\beta}}, \tag{6.1.12}$$

where $\hat{\gamma}_i$ and $\hat{\boldsymbol{\beta}}$ are the values of γ_i and $\hat{\boldsymbol{\beta}}$ when σ_v^2 is replaced by $\hat{\sigma}_v^2$. If $\theta_i = g(\overline{Y}_i)$ and $\overline{Y}_i$ is the parameter of interest, then $\hat{\theta}_i^H$ is transformed back to the original scale to obtain an estimator of the area mean $\overline{Y}_i$ as $\hat{\overline{Y}}_i^H = h(\hat{\theta}_i^H) := g^{-1}(\hat{\theta}_i^H)$. Note that $\hat{\overline{Y}}_i^H$ does not retain the EBLUP property of $\hat{\theta}_i^H$. The EB and HB approaches (Chapters 9 and 10) are better suited to handle nonlinear cases, $h(\theta_i)$.

Fay and Herriot (1979) recommended the use of a compromise EBLUP estimator, $\hat{\theta}_{ic}^H$, similar to the compromise James–Stein (J–S) estimator (3.4.9) in Chapter 3 and obtained as follows: (i) take $\hat{\theta}_{ic}^H = \hat{\theta}_i^H$ if $\hat{\theta}_i^H$ lies in the interval $[\hat{\theta}_i - c\sqrt{\psi_i}, \hat{\theta}_i + c\sqrt{\psi_i}]$ for a specified constant c (typically $c = 1$); and (ii) take $\hat{\theta}_{ic}^H = \hat{\theta}_i^H - c\sqrt{\psi_i}$ if $\hat{\theta}_i^H < \hat{\theta}_i - c\sqrt{\psi_i}$; (iii) take $\hat{\theta}_{ic}^H = \hat{\theta}_i^H + c\sqrt{\psi_i}$ if $\hat{\theta}_i^H > \hat{\theta}_i + c\sqrt{\psi_i}$. The compromise EBLUP estimator $\hat{\theta}_{ic}^H$ is transformed back to the original scale to obtain an estimator of the ith area mean $\overline{Y}_i$ as $\hat{\overline{Y}}_{ic}^H = h(\hat{\theta}_{ic}^H) = g^{-1}(\hat{\theta}_{ic}^H)$.

We now turn to the case where not all areas are sampled. We assume that the model (6.1.1) holds for both the sampled areas $i = 1, \ldots, m$ and the nonsampled areas $\ell = m + 1, \ldots, M$. For the nonsampled areas, direct estimates $\hat{\theta}_i$ are not available and as a result we use the regression-synthetic estimator of θ_i based on the covariates, $\mathbf{z}_\ell$, observed from the nonsampled areas:

$$\hat{\theta}_\ell^{\text{RS}} = \mathbf{z}_\ell^T \hat{\boldsymbol{\beta}}, \quad \ell = m + 1, \ldots, M, \tag{6.1.13}$$

where $\hat{\boldsymbol{\beta}} = \tilde{\boldsymbol{\beta}}(\hat{\sigma}_v^2)$, obtained from (6.1.5), is computed from the sample data $\{(\hat{\theta}_i, \mathbf{z}_i); i = 1, \ldots, m\}$. If $\theta_\ell = g(\overline{Y}_\ell)$, then the estimator of $\overline{Y}_\ell$ is taken as $\hat{\overline{Y}}_\ell^{\text{RS}} = h(\hat{\theta}_\ell^{\text{RS}}) = g^{-1}(\hat{\theta}_\ell^{\text{RS}})$.

6.1.2 Estimation of σ_v^2

A method of moment estimator $\hat{\sigma}_{vm}^2$ can be obtained by noting that

$$E\left[\sum_{i=1}^m (\hat{\theta}_i - \mathbf{z}_i^T \tilde{\boldsymbol{\beta}})^2 / (\psi_i + \sigma_v^2 b_i^2)\right] = E[a(\sigma_v^2)] = m - p,$$

where $\tilde{\boldsymbol{\beta}} = \tilde{\boldsymbol{\beta}}(\sigma_v^2)$. It follows that $\hat{\sigma}_{vm}^2$ is obtained by solving

$$a(\sigma_v^2) = m - p$$

iteratively and letting $\hat{\sigma}^2_{vm} = 0$ when no positive solution $\tilde{\sigma}^2_{vm}$ exists. Fay and Herriot (1979) suggested the following iterative solution: using a starting value $\sigma_v^{2(0)}$, define

$$\sigma_v^{2(k+1)} = \sigma_v^{2(k)} + \frac{1}{a'_*(\sigma_v^{2(k)})}[m - p - a(\sigma_v^{2(k)})], \tag{6.1.14}$$

where

$$a'_*(\sigma_v^2) = -\sum_{i=1}^m b_i^2(\hat{\theta}_i - \mathbf{z}_i^T\tilde{\boldsymbol{\beta}})^2/(\psi_i + \sigma_v^2 b_i^2)^2$$

is an approximation to the derivative $a'(\sigma_v^2)$; FH used $\sigma_v^{2(0)} = 0$. Convergence of the iterative procedure (6.1.14) is rapid, generally requiring less than 10 iterations.

Alternatively, a simple moment estimator is given by $\hat{\sigma}^2_{vs} = \max(\tilde{\sigma}^2_{vs}, 0)$, where

$$\tilde{\sigma}^2_{vs} = \frac{1}{m-p}\left[\sum_{i=1}^m (b_i^{-1}\hat{\theta}_i - \tilde{\mathbf{z}}_i^T\hat{\boldsymbol{\beta}}_{\text{WLS}})^2 - \sum_{i=1}^m \frac{\psi_i}{b_i^2}(1 - \tilde{h}_{ii})\right] \tag{6.1.15}$$

and

$$\hat{\boldsymbol{\beta}}_{\text{WLS}} = \left(\sum_i \tilde{\mathbf{z}}_i\tilde{\mathbf{z}}_i^T\right)^{-1}\left(\sum_{i=1}^m \tilde{\mathbf{z}}_i\hat{\theta}_i/b_i\right)$$

is a weighted least squares estimator of $\boldsymbol{\beta}$. If $b_i = 1$, then (6.1.15) reduces to the formula of Prasad and Rao (1990). Neither of these moment estimators of σ_v^2 require normality, and both lead to consistent estimators as $m \to \infty$.

The Fisher-scoring algorithm (5.2.18) for ML estimation of σ_v^2 reduces to

$$\sigma_v^{2(a+1)} = \sigma_v^{2(a)} + [\mathcal{I}(\sigma_v^{2(a)})]^{-1}s(\tilde{\boldsymbol{\beta}}^{(a)}, \sigma_v^{2(a)}), \tag{6.1.16}$$

where

$$\mathcal{I}(\sigma_v^2) = \frac{1}{2}\sum_{i=1}^m \frac{b_i^4}{(\sigma_v^2 b_i^2 + \psi_i)^2} \tag{6.1.17}$$

and

$$s(\tilde{\boldsymbol{\beta}}, \sigma_v^2) = -\frac{1}{2}\sum_{i=1}^m \frac{b_i^2}{\sigma_v^2 b_i^2 + \psi_i} + \frac{1}{2}\sum_{i=1}^m b_i^2 \frac{(\hat{\theta}_i - \mathbf{z}_i^T\tilde{\boldsymbol{\beta}})^2}{(\sigma_v^2 b_i^2 + \psi_i)^2}.$$

The final ML estimator $\hat{\sigma}^2_{v\text{ML}}$ is taken as $\hat{\sigma}^2_{v\text{ML}} = \max(\tilde{\sigma}^2_{v\text{ML}}, 0)$, where $\tilde{\sigma}^2_{v\text{ML}}$ is the solution obtained from (6.1.16). Similarly, the Fisher-scoring algorithm for REML estimation of σ_v^2 reduces to

$$\sigma_v^{2(a+1)} = \sigma_v^{2(a)} + [\mathcal{I}_R(\sigma_v^{2(a)})]^{-1}s_R(\sigma_v^{2(a)}), \tag{6.1.18}$$

where

$$\mathcal{I}_R(\sigma_v^2) = \frac{1}{2}\text{tr}(\mathbf{PBPB}) \tag{6.1.19}$$

and

$$s_R(\sigma_v^2) = -\frac{1}{2}\text{tr}(\mathbf{PB}) + \frac{1}{2}\mathbf{y}^T\mathbf{PBPy},$$

where $\mathbf{B} = \text{diag}(b_1^2, \ldots, b_m^2)$ and $\mathbf{P}$ is defined in (5.2.21) (see Cressie 1992). Asymptotically, $\mathcal{I}(\sigma_v^2)/\mathcal{I}_R(\sigma_v^2) \to 1$ as $m \to \infty$. The final REML estimator $\hat{\sigma}_{v\text{RE}}^2$ is taken as $\hat{\sigma}_{v\text{RE}}^2 = \max(\tilde{\sigma}_{v\text{RE}}^2, 0)$, where $\tilde{\sigma}_{v\text{RE}}^2$ is the solution obtained from (6.1.18).

The EBLUP estimator $\hat{\theta}_i^H$, based on a moment, ML or REML estimator of σ_v^2, remains model-unbiased if the errors v_i and e_i are symmetrically distributed around 0. In particular, $\hat{\theta}_i^H$ is model-unbiased for θ_i if v_i and e_i are normally distributed.

For the special case of $b_i = 1$ and equal sampling variances $\psi_i = \psi$, the BLUP estimator (6.1.3) reduces to

$$\tilde{\theta}_i^H = \gamma\hat{\theta}_i + (1-\gamma)\mathbf{z}_i^T\hat{\boldsymbol{\beta}}_{\text{LS}}$$

with $1 - \gamma = \psi/(\psi + \sigma_v^2)$, where $\hat{\boldsymbol{\beta}}_{\text{LS}}$ is the least squares estimator of $\boldsymbol{\beta}$. Let $S = \sum_{i=1}^m (\hat{\theta}_i - \mathbf{z}_i^T\hat{\boldsymbol{\beta}}_{\text{LS}})^2$ be the residual sum of squares. Under normality, it holds that $S/(\psi + \sigma_v^2) \sim \chi_{m-p}^2$. Using this result, an unbiased estimator of $1 - \gamma$ is given by

$$1 - \gamma^* = \psi(m - p - 2)/S,$$

and an alternative EBLUP estimator of θ_i is therefore given by

$$\hat{\theta}_i^H = \gamma^*\hat{\theta}_i + (1-\gamma^*)\mathbf{z}_i^T\hat{\boldsymbol{\beta}}_{\text{LS}}.$$

This estimator is identical to the J–S estimator, studied in Section 3.4.2, with guess $\theta_i^0 = \mathbf{z}_i^T\hat{\boldsymbol{\beta}}_{\text{LS}}$. Note that the plug-in estimator $\hat{\gamma}$ of γ obtained using either $\hat{\sigma}_{vs}^2$ or $\hat{\sigma}_{vm}^2$ leads to

$$1 - \hat{\gamma} = \psi(m - p)/S,$$

which is approximately equal to $1 - \gamma^*$ for large m and fixed p.

6.1.3 Relative Efficiency of Estimators of σ_v^2

Asymptotic variances (as $m \to \infty$) of ML and REML estimators are equal:

$$\overline{V}(\hat{\sigma}_{v\text{ML}}^2) = \overline{V}(\hat{\sigma}_{v\text{RE}}^2) = [\mathcal{I}(\sigma_v^2)]^{-1} = 2\left[\sum_{i=1}^m b_i^4/(\sigma_v^2 b_i^2 + \psi_i)^2\right]^{-1}, \quad (6.1.20)$$

where the Fisher information, $\mathcal{I}(\sigma_v^2)$, is given by (6.1.17). The asymptotic variance of the simple moment estimator $\hat{\sigma}_{vs}^2$ is given by

$$\overline{V}(\hat{\sigma}_{vs}^2) = 2m^{-2}\sum_{i=1}^m (\sigma_v^2 b_i^2 + \psi_i)^2/b_i^4 \quad (6.1.21)$$

(Prasad and Rao 1990). Datta, Rao, and Smith (2005) derived the asymptotic variance of the FH moment estimator, $\hat{\sigma}^2_{vm}$, as

$$\overline{V}(\hat{\sigma}^2_{vm}) = 2m\left[\sum_{i=1}^{m} b_i^2 / \left(\sigma_v^2 b_i^2 + \psi_i\right)\right]^{-2}. \qquad (6.1.22)$$

Using the Cauchy–Schwarz inequality and the fact that the arithmetic mean is greater than or equal to the harmonic mean, it follows from (6.1.20) to (6.1.22) that

$$\overline{V}(\hat{\sigma}^2_{v\text{RE}}) = \overline{V}(\hat{\sigma}^2_{v\text{ML}}) \leq \overline{V}(\hat{\sigma}^2_{vm}) \leq \overline{V}(\hat{\sigma}^2_{vs}). \qquad (6.1.23)$$

Equality in (6.1.23) holds if $\psi_i = \psi$ and $b_i = 1$. The FH moment estimator, σ^2_{vm}, becomes significantly more efficient relative to $\hat{\sigma}^2_{vs}$ as the variability of the terms $\sigma_v^2 + \psi_i/b_i^2$ increases. On the other hand, the loss in efficiency of σ^2_{vm} relative to the REML (ML) estimator is relatively small as it depends on the variability of the terms $(\sigma_v^2 + \psi_i/b_i^2)^{-1}$.

6.1.4 *Applications

We now provide some details of the applications in Examples 4.2.1 and 4.2.2, Chapter 4.

Example 6.1.1. Income for Small Places. The U.S. Bureau of the Census is required to provide the Treasury Department with the estimates of per capita income (PCI) and other statistics for state and local governments receiving funds under the General Revenue Sharing Program. Those statistics are then used by the Treasury Department to determine allocations to the local government units (places) within the different states by dividing the corresponding state allocations.

Initially, the Census Bureau determined the current PCI estimate for a place by multiplying the 1970 census estimate of PCI in 1969 (based on a 20% sample) by the ratio of an administrative estimate of PCI in the current year and a similarly derived estimate for 1969. But the sampling error of the PCI estimates turned out to be quite large for places having fewer than 500 persons in 1970, with coefficient of variation (CV) of about 13% for a place of 500 persons and 30% for a place of 100 persons. As a result, the Census Bureau initially decided to set aside the census estimates for these small places and to substitute the corresponding county estimates in their place. But this solution turned out to be unsatisfactory because the census estimates for many small places differed significantly from the corresponding county estimates after accounting for sampling errors.

Using the EBLUP estimator (6.1.12) with $b_i = 1$ and $\hat{\sigma}_v^2 = \hat{\sigma}^2_{vm}$, Fay and Herriot (1979) presented empirical evidence that the EBLUP estimates of log(PCI) for small places have average error smaller than either the census estimates or the county estimates. The EBLUP estimator used by them is a weighted average of the direct estimator $\hat{\theta}_i$ and a regression-synthetic estimator $\mathbf{z}_i^T\hat{\boldsymbol{\beta}}$, obtained by fitting a linear regression equation to $(\hat{\theta}_i, \mathbf{z}_i)$, where $\mathbf{z}_i = (z_{1i}, z_{2i}, \ldots, z_{pi})^T$ with $z_{1i} = 1$, $i = 1, \ldots, m$, and the independent variables $z_{2i}, \ldots, z_{pi}$ are based on the associated county PCI, tax return data for 1969, and data on housing from the 1970 census. The method used by Fay and Herriot (1979) was adopted by the Census Bureau in 1974 to form updated PCI estimates for small places. This was the largest application (prior to 1990) of EBLUP methods in a U.S. Federal Statistical Program.

We now present some details of the FH application and the results of an external evaluation. First, based on past studies, the CV of the direct estimate, $\hat{\bar{Y}}_i$, of PCI was taken as $3/\hat{N}_i^{1/2}$ for the ith small place, where $\hat{N}_i$ is the weighted sample count; $\hat{\bar{Y}}_i$ and $\hat{N}_i$ were available for almost all places. This suggested the use of logarithmic transformation, $\hat{\theta}_i = \log(\hat{\bar{Y}}_i)$, with $V(\hat{\theta}_i) \approx [\mathrm{CV}(\hat{\bar{Y}}_i)]^2 = 9/\hat{N}_i = \psi_i$. Second, four separate regression models were evaluated to determine a suitable combined model, treating the sampling variances, ψ_i, as known. The independent variables, $z_1, z_2, \ldots, z_p$, for the four models are, respectively, given by (1) $p = 2$ with $z_1 = 1$ and $z_2 = \log$(county PCI); (2) $p = 4$ with z_1, z_2, $z_3 = \log$(value of owner-occupied housing for the place) and $z_4 = \log$(value of owner-occupied housing for the county); (3) $p = 4$ with z_1, z_2, $z_5 = \log$(adjusted gross income per exemption from the 1969 tax returns for the place) and $z_6 = \log$(adjusted gross income per exemption from the 1969 tax returns for the county); (4) $p = 6$ with $z_1, z_2, \ldots, z_6$.

Fay and Herriot (1979) calculated the values of $\hat{\sigma}_{vm}^2$ for each of the four models, using the iterative algorithm (6.1.14). A small value of $\hat{\sigma}_{vm}^2$ indicates a better average fit of the regression models to the sample data, after allowing for the sampling errors in the direct estimators $\hat{\theta}_i$. For a place of estimated size $\hat{N}_i = 200$, we have $\psi_i = 9/200 = 0.045$. If one desires to attach equal weight $\hat{\gamma}_{im} = \hat{\sigma}_{vm}^2/(\hat{\sigma}_{vm}^2 + \psi_i) = 1/2$ to the direct estimate $\hat{\theta}_i$ and the regression-synthetic estimate $\mathbf{z}_i^T\hat{\boldsymbol{\beta}}$, we then need $\hat{\sigma}_{vm}^2 = 0.045$ as well. In this case, the resulting MSE, based on the leading term $g_{1i}(\hat{\sigma}_{vm}^2) = \hat{\gamma}_{im}\psi_i$, is one-half of the sampling variance ψ_i; that is, the EBLUP estimate for a place of 200 persons has roughly the same precision as the direct estimate for a place of 400 persons. Fay and Herriot used the values of $\hat{\sigma}_v^2$ relative to ψ_i as the criterion for model selection.

Table 6.1 reports the values of $\hat{\sigma}_{vm}^2$ for the states with more than 500 small places of size less than 500. It is clear from Table 6.1 that regressions involving either tax (models 3 and 4) or housing data (models 2 and 4), but especially those involving both types of covariates are significantly better than the regression on the county values alone; that is, model 4 and to a lesser extent models 2 and 3 provide better fits to the data in terms of $\hat{\sigma}_{vm}^2$-values than model 1. Note that the values of $\hat{\sigma}_{vm}^2$ for model 4 are much smaller than 0.045, especially for North Dakota, Nebraska, Wisconsin, and Iowa, which suggests that large gains in MSE can be achieved for those states.

Fay and Herriot obtained compromise EBLUP estimates, $\hat{\theta}_{i*}^H$, from $\hat{\theta}_i^H$ and then transformed $\hat{\theta}_{i*}^H$ back to the original scale to get the estimate of $\bar{Y}_i$ given by $\hat{\bar{Y}}_{i*}^H = \exp(\hat{\theta}_{i*}^H)$. The latter estimates were then subjected to a two-step raking (benchmarking) to ensure consistency with the following aggregate sample estimates: (i) For each of the population size classes ($< 500, 500 - 1000$, and > 1000),

TABLE 6.1 Values of $\hat{\sigma}^2_{vm}$ for States with More Than 500 Small Places

State	Model (1)	(2)	(3)	(4)
Illinois	0.036	0.032	0.019	0.017
Iowa	0.029	0.011	0.017	0.000
Kansas	0.064	0.048	0.016	0.020
Minnesota	0.063	0.055	0.014	0.019
Missouri	0.061	0.033	0.034	0.017
Nebraska	0.065	0.041	0.019	0.000
North Dakota	0.072	0.081	0.020	0.004
South Dakota	0.138	0.138	0.014	*
Wisconsin	0.042	0.025	0.025	0.004

*Not fitted because some z-values for model 4 were not available for several places in South Dakota.

Source: Adapted from Table 1 in Fay and Herriot (1979).

the total estimated income for all places equals the direct estimate at the state level. (ii) The total estimated income for all places in a county equals the direct county estimate of total income.

External Evaluation Fay and Herriot (1979) also conducted an external evaluation by comparing the 1972 estimates to "true" values obtained from a special complete census of a random sample of places in 1973. The 1972 estimates for each place were obtained by multiplying the 1970 estimates by an updating factor derived from administrative sources. Table 6.2 reports the values of percentage absolute relative error ARE = (|estimate − true value|/true value) × 100, for the special complete census areas using direct, county and EBLUP estimates. Values of average ARE are also reported. Table 6.2 shows that the EBLUP estimates exhibit smaller average ARE and a lower incidence of extreme errors than either the direct estimates or the county estimates: the average ARE for places with population less than 500 is 22% compared to 28.6% for the direct estimates and 31.6% for the county estimates. The EBLUP estimates were consistently higher than the special census values. But missing income was not imputed in the special census, unlike in the 1970 census. As a result, the special census values, which are based on only completed cases, may be subject to a downward bias.

Example 6.1.2. U.S. Poverty Counts. Current county estimates of school-age children in poverty in the United States are used by the U.S. Department of Education to allocate government funds, called Title I funds, annually to counties, and then states distribute the funds among school districts. The allocated funds support compensatory education programs to meet the needs of educationally disadvantaged children. Title I funds of about 14.5 billion U.S. dollars were allocated in 2009.

The U.S. Census Bureau, under their small area income and poverty estimation (SAIPE) program, uses the basic area level model (6.1.1) with $b_i = 1$ to produce

TABLE 6.2 Values of Percentage Absolute Relative Error of Estimates from True Values: Places with Population Less Than 500

Special Census Area	Direct Estimate	EBLUP Estimate	County Estimate
1	10.2	14.0	12.9
2	4.4	10.3	30.9
3	34.1	26.2	9.1
4	1.3	8.3	24.6
5	34.7	21.8	6.6
6	22.1	19.8	14.6
7	14.1	4.1	18.7
8	18.1	4.7	25.9
9	60.7	78.7	99.7
10	47.7	54.7	95.3
11	89.1	65.8	86.5
12	1.7	9.1	12.7
13	11.4	1.4	6.6
14	8.6	5.7	23.5
15	23.6	25.3	34.3
16	53.6	10.5	11.7
17	51.4	14.4	23.7
Average	28.6	22.0	31.6

Source: Adapted from Table 3 in Fay and Herriot (1979).

EBLUP estimates of annual poverty rates and counts of poor school-age children (between ages 5 and 17) for counties and states. In this application, $\hat{\theta}_i = \log(\hat{Y}_i)$, where $\hat{Y}_i$ is the direct estimator of the poverty count, Y_i, for county i.

Prior to 2005, direct estimates $\hat{Y}_i$ were obtained from the Annual Social and Economic Supplement (ASEC) of the Current Population Survey (CPS), see National Research Council (2000) and Rao (2003a, Example 7.1.2) for details of past methodology. In 2005, SAIPE made a major switch by replacing CPS estimates by more reliable direct estimates obtained from the American Community Survey (ACS). ACS is based on a much larger sample size than CPS, with monthly rolling samples of 250,000 households spread across all counties. ACS direct estimates are reliable for counties with population size 65,000 or larger, but reliable estimates are needed for all the roughly 3,140 counties. Hence, ACS direct estimates, $\hat{Y}_i$, and associated covariates, $\mathbf{z}_i$, are used to produce EBLUP estimates, $\hat{\theta}_i^H$, of $\theta_i = \log(Y_i)$ and then transformed back to the original scale to obtain $\hat{Y}_i^H = \exp(\hat{\theta}_i^H)$. The estimates $\hat{Y}_i^H$ are then adjusted for bias, assuming normality of the errors v_i and e_i in the model.

For the SAIPE program, very good covariates are available from the decennial census and other administrative sources. Covariates included in the county model, $\hat{\theta}_i = \mathbf{z}_i^T\boldsymbol{\beta} + v_i + e_i$, are $z_{1i} = 1$, $z_{2i} = $ log(number of child exemptions claimed by families in poverty on tax returns), $z_{3i} = $ log(number of people receiving food stamps), and $z_{4i} = $ log(estimated population under age 18), $z_{5i} = $ log(number of poor school-age children estimated from the previous census). For a small number of counties with small sample sizes, $\hat{Y}_i$ may be zero because of no poor children in the sample, and those counties were excluded in fitting the model because $\hat{\theta}_i = \log(\hat{Y}_i)$ is not defined if $\hat{Y}_i = 0$. Moreover, direct estimates, $\hat{\psi}_i$, of the sampling variances ψ_i, obtained from replication methods, are used as proxies for the unknown ψ_i.

Model parameters are estimated by the ML method, and the estimator of θ_i is the EBLUP estimator $\hat{\theta}_i^H$. For the few counties excluded from the model fitting, the regression-synthetic estimator, $\hat{\theta}_i^{\text{RS}} = \mathbf{z}_i^T\hat{\boldsymbol{\beta}}$, is used. To estimate the poverty count Y_i, a bias-adjusted estimator, obtained from $\hat{Y}_i^H = \exp(\hat{\theta}_i^H)$, is used:

$$\hat{Y}_{ia}^H = \hat{Y}_i^H \exp[\hat{\sigma}_v^2(1-\hat{\gamma}_i)/2]. \tag{6.1.24}$$

The bias-adjustment factor $\hat{F}_i$ for estimating a general function of θ_i, $h(\theta_i)$, is given by

$$F_i = E[h(\mathbf{z}_i^T\boldsymbol{\beta} + v_i)]/E\{h[\mathbf{z}_i^T\boldsymbol{\beta} + \gamma_i(v_i + e_i)]\} \tag{6.1.25}$$

evaluated at $(\boldsymbol{\beta}, \sigma_v^2) = (\hat{\boldsymbol{\beta}}, \hat{\sigma}_v^2)$ (Slud and Maiti 2006). For the special case of $h(a) = \exp(a)$, $\hat{F}_i$ reduces to $\hat{F}_i = \exp[\hat{\sigma}_v^2(1-\hat{\gamma}_i)/2]$, using the fact that

$$E(e^X) = \exp(\mu + \sigma^2/2). \tag{6.1.26}$$

if $X \sim N(\mu, \sigma^2)$. In the SAIPE methodology, the term $\hat{\sigma}_v^2(1-\hat{\gamma}_i) = g_{1i}(\hat{\sigma}_v^2)$ in the exponent of (6.1.24) is replaced by the naive MSE estimates, $\text{mse}_N(\hat{\theta}_i^H) = g_{1i}(\hat{\sigma}_v^2) + g_{2i}(\hat{\sigma}_v^2)$. However, the term $g_{2i}(\hat{\sigma}_v^2)$, due to estimating $\boldsymbol{\beta}$ is negligible relative to $g_{1i}(\hat{\sigma}_v^2)$ in the context of SAIPE since $m = M$, is large and g_{2i} is $O(m^{-1})$ for large m, whereas g_{1i} is $O(1)$. Note that $\hat{Y}_{ia}^H$ is not exactly unbiased for the total Y_i.

Luery (2011) compared the CVs of the 2006 SAIPE county estimates $\hat{Y}_{ia}^H$ with the corresponding CVs of the ACS estimates $\hat{Y}_i$. His results showed that the SAIPE estimates can lead to significant reduction in CV relative to ACS estimates, especially for small counties.

State estimates of poverty counts are obtained from a model similar to the county model, except that the model is applied to state poverty rates, instead of logarithms of poverty counts. The covariates used in the model are state proportions corresponding to the same census and administrative sources used in the county model. The EBLUP estimates of state poverty rates are multiplied by the corresponding counts of school-age children obtained from the U.S. Census Bureau's program of population estimates to obtain the EBLUP estimates of state poverty counts. Reduction in CV due to using SAIPE state estimates is marginal for most states, unlike the case of county estimates, because ACS direct estimates are based on much larger sample sizes at the state level.

The SAIPE state estimates of poverty counts are first ratio adjusted (or raked) to agree with the ACS national estimate. In the second step, the SAIPE county estimates are ratio adjusted to agree with the raked SAIPE state estimates. The two-step raking method ensures that the SAIPE estimates at the county level or state level add up to the ACS national estimate.

SAIPE computes estimates of poor school-age children also for school districts, but in this case using simple synthetic estimation, due to lack of suitable data at the school district level. Luery (2011) describes the methodology that is currently applied for school districts.

Prior to the adaptation of the current SAIPE county and state models, both internal and external evaluations were conducted for model selection and for checking the validity of the chosen models. In an internal evaluation, the validity of the underlying assumptions and features of the model are examined. Additionally, an external evaluation compares the estimates derived from a model to "true" values that were not used in the development of the model. For internal evaluation, standard methods for linear regression models were used without taking account of random county or state effects. Model features examined included linearity of the regression model, choice of predictor variables, normality of the standardized residuals through quantile–quantile (q–q) plots and residual analysis to detect outliers. For external evaluation, EBLUP estimates based on the 1989 CPS direct estimates and several candidate models were compared to the 1990 census estimates by treating the latter as true values. We refer the reader to National Research Council (2000) and Rao (2003a, Example 7.1.2) for further details.

Example 6.1.3. Canadian Census Undercoverage. We have already noted in Example 4.2.3, Chapter 4, that the basic area level model (6.1.1) has been used to estimate the undercount in the decennial census of the United States and in the Canadian census. We now provide a brief account of the application to the 1991 Canadian census (Dick 1995).

Let T_i be the true (unknown) count and C_i is the census count in the ith domain. The objective here is to estimate census adjustment factors $\theta_i = T_i/C_i$ for the $m = 96 = 2 \times 4 \times 12$ domains obtained by crossing the categories of the variables such as sex (2 genders), age (4 classes), and province (12). The net undercoverage rate in the ith domain is given by $U_i = 1 - \theta_i^{-1}$. Direct estimates $\hat{\theta}_i$ were obtained from a post-enumeration survey. The associated sampling variances, ψ_i, were derived through smoothing of the estimated variances. In particular, the variance of the estimated number of missing persons, $\hat{M}_i$, was assumed to be proportional to a power of the census count C_i, and a linear regression was fitted to $\{\log(v(\hat{M}_i)), \log(C_i); i = 1, \ldots, m\}$, where $v(\hat{M}_i)$ is the estimated variance of $\hat{M}_i$. The sampling variances were then predicted through the fitted line $\log(\psi_i) = -6.13 - 0.28\log(C_i)$, and treating the predicted values as the true ψ_i.

Auxiliary (predictor) variables $\mathbf{z}$ for building the model were selected from a set of 42 variables by backward stepwise regression (Draper and Smith 1981, Chapter 6). Note that the random area effects v_i are not taken into account in stepwise regression. Internal evaluation of the resulting area level model (6.1.1) with $b_i = 1$ was then performed, by treating the standardized BLUP residuals $r_i = (\hat{\theta}_i^H - \mathbf{z}_i^T\hat{\boldsymbol{\beta}})/(\hat{\sigma}_v^2 + \psi_i)^{1/2}$ as iid $N(0, 1)$ variables, where $\hat{\boldsymbol{\beta}}$ and $\hat{\sigma}_v^2$ are the REML estimates of $\boldsymbol{\beta}$ and σ_v^2. No significant departures from the assumed model were observed. In particular, to check the normality assumption and to detect outliers, a normal q–q plot of the r_i's versus $\Phi^{-1}[F_m(r_i)]$, $i = 1, \ldots, m$ was examined, where $\Phi(x)$ is the cumulative distribution function (CDF) of a $N(0, 1)$ variable and $F_m(x)$ is the empirical CDF of the r_i's, that is, $F_m(x) = m^{-1}\sum_{i=1}^m I(r_i \le x)$, where $I(r_i \le x) = 1$ if $r_i \le x$ and 0 otherwise. Dempster and Ryan (1985) proposed a weighted normal q–q plot based on the r_i's that is more sensitive to departures from normality than the unweighted normal q–q plot. This plot is similar to the weighted normal probability plot of Lange and Ryan (1989) for checking normality of random effects (see Section 5.4.2, Chapter 5). This weighted normal q–q plot uses the weighted empirical CDF

$$\begin{aligned} F_m^*(x) &= \left[\sum_{i=1}^m (\hat{\sigma}_v^2 + \psi_i)^{-1} I(x - r_i)\right] \Bigg/ \left[\sum_{i=1}^m (\hat{\sigma}_v^2 + \psi_i)^{-1}\right] \\ &= \sum_{i=1}^m \hat{\gamma}_i I(x - r_i) \Bigg/ \sum_{i=1}^m \hat{\gamma}_i \end{aligned}$$

for the special case $b_i = 1$, instead of the usual empirical CDF $F_m(x)$. Note that $F_m^*(x)$ assigns greater weight to those areas for which $\hat{\sigma}_v^2$ accounts for a larger part of the total estimated variance $\hat{\sigma}_v^2 + \psi_i$.

The EBLUP adjustment factors, $\hat{\theta}_i^H$, were converted to estimates of missing persons, $\hat{M}_i^H$, where $M_i = T_i - C_i$. These estimates were then subjected to two-step raking (see Section 3.2.6, Chapter 3) to ensure consistency with the reliable direct estimates of marginal totals, $\hat{M}_{p\cdot}$ and $\hat{M}_{\cdot a}$, where "p" denotes a province, "a" denotes an age–sex group, and $M_i = M_{pa}$. The raked EBLUP estimates were further divided into single year of age estimates by using simple synthetic estimation:

$$\hat{M}_{pa}^S(q) = \hat{M}_{pa}^{RH}[C_{pa}(q)/C_{pa}],$$

where C_{pa} is the census count in province p and age–sex group a, q denotes a sub-age group, and $C_{pa}(q)$ is the associated census count.

Example 6.1.4. Poverty Estimates in Spain. Molina and Morales (2009) applied the FH model to data from the 2006 Spanish Survey on Income and Living Conditions (EU-SILC) to estimate poverty rates for the 52 Spanish provinces by gender ($m = 52 \times 2 = 104$). In the application, the estimated sampling variances of the direct estimators were treated as the true variances. Domain proportions of individuals with Spanish nationality, in different age groups and in several employment categories, were used as covariates. Results of Molina and Morales (2009) indicated gains in efficiency of the EBLUPs based on the FH model with respect to the direct estimators for most of the domains. However, in general, the gains in efficiency when using the FH model with those covariates are modest (see Table 10.5 in Section 10.7.3).

6.2 MSE ESTIMATION

6.2.1 Unconditional MSE of EBLUP

In this section, we apply the general results of Section 5.3 to obtain the MSE of the EBLUP, $\hat{\theta}_i^H$, and MSE estimators that are second-order unbiased. The MSE and the MSE estimators are unconditional in the sense that they are valid under the model (4.2.5) obtained by combining the sampling model (4.2.3) and the linking model (4.2.1). Conditional MSE is studied in Section 6.2.7.

The second-order MSE approximation (5.3.9) is valid for the FH model, $\hat{\theta}_i = \mathbf{z}_i^T\boldsymbol{\beta} + b_i v_i + e_i$, under regularity conditions (6.1.10) and (6.1.11) and normality of the errors v_i and e_i. It reduces to

$$\text{MSE}(\hat{\theta}_i^H) \approx g_{1i}(\sigma_v^2) + g_{2i}(\sigma_v^2) + g_{3i}(\sigma_v^2), \qquad (6.2.1)$$

where $g_{1i}(\sigma_v^2)$ and $g_{2i}(\sigma_v^2)$ are given by (6.1.8) and (6.1.9), and

$$g_{3i}(\sigma_v^2) = \psi_i^2 b_i^4(\psi_i + \sigma_v^2 b_i^2)^{-3}\overline{V}(\hat{\sigma}_v^2), \qquad (6.2.2)$$

where $\overline{V}(\hat{\sigma}_v^2)$ is the asymptotic variance of an estimator, $\hat{\sigma}_v^2$, of σ_v^2. We have $\overline{V}(\hat{\sigma}_v^2) = \overline{V}(\hat{\sigma}_{v\text{ML}}^2) = \overline{V}(\hat{\sigma}_{v\text{RE}}^2)$, given by (6.1.20), if we use $\hat{\sigma}_{v\text{ML}}^2$ or $\hat{\sigma}_{v\text{RE}}^2$ to estimate σ_v^2. If we use the simple moment estimator $\hat{\sigma}_{vs}^2$, then $\overline{V}(\hat{\sigma}_v^2) = \overline{V}(\hat{\sigma}_{vs}^2)$ given by (6.1.21). For the FH moment estimator, $\hat{\sigma}_{vm}^2$, we have $\overline{V}(\hat{\sigma}_v^2) = \overline{V}(\hat{\sigma}_{vm}^2)$ given by (6.1.22). It follows from (6.1.23) that the term $g_{3i}(\sigma_v^2)$ is the smallest for ML and REML estimators of σ_v^2, followed by $\hat{\sigma}_{vm}^2$. The g_{3i} term for $\hat{\sigma}_{vm}^2$ is significantly smaller than that for $\hat{\sigma}_{vs}^2$ if the variability of $\sigma_v^2 + \psi_i/b_i^2$, $i = 1, \ldots, m$, is substantial.

We now turn to the estimation of $\text{MSE}(\hat{\theta}_i^H)$. The estimator of MSE given by (5.3.11) is valid for REML and simple moment estimators of σ_v^2 under regularity conditions (6.1.10) and (6.1.11), and normality of the errors v_i and e_i. In this case, it reduces to

$$\text{mse}(\hat{\theta}_i^H) = g_{1i}(\hat{\sigma}_v^2) + g_{2i}(\hat{\sigma}_v^2) + 2g_{3i}(\hat{\sigma}_v^2) \qquad (6.2.3)$$

if $\hat{\sigma}_v^2$ is chosen as $\hat{\sigma}_{v\text{RE}}^2$ or $\hat{\sigma}_{vs}^2$. The corresponding area-specific versions, $\text{mse}_1(\hat{\theta}_i^H)$ and $\text{mse}_2(\hat{\theta}_i^H)$, are obtained from (5.3.15) and (5.3.16) by changing $g_{3i}(\sigma_v^2, \hat{\theta}_i)$ to $g_{3i}^*(\sigma_v^2, \hat{\theta}_i)$, where $g_{3i}^*(\sigma_v^2, \hat{\theta}_i)$ is obtained from (5.3.14), which reduces to

$$g_{3i}^*(\sigma_v^2, \hat{\theta}_i) = [b_i^4\psi_i^2/(\psi_i + \sigma_v^2 b_i^2)^4](\hat{\theta}_i - \mathbf{z}_i^T\tilde{\boldsymbol{\beta}})^2\overline{V}(\hat{\sigma}_v^2). \qquad (6.2.4)$$

Then, the two area-specific MSE estimators are given by

$$\text{mse}_1(\hat{\theta}_i^H) = g_{1i}(\hat{\sigma}_v^2) + g_{2i}(\hat{\sigma}_v^2) + 2g_{3i}^*(\hat{\sigma}_v^2, \hat{\theta}_i) \qquad (6.2.5)$$

and

$$\text{mse}_2(\hat{\theta}_i^H) = g_{1i}(\hat{\sigma}_v^2) + g_{2i}(\hat{\sigma}_v^2) + g_{3i}(\hat{\sigma}_v^2) + g_{3i}^*(\hat{\sigma}_v^2, \hat{\theta}_i). \qquad (6.2.6)$$

Rao (2001a) obtained the area-specific MSE estimators (6.2.5) and (6.2.6) for the special case of $b_i = 1$.

For the ML estimator $\hat{\sigma}^2_{v\text{ML}}$ and the FH moment estimator $\hat{\sigma}^2_{vm}$, we apply the MSE estimator (5.3.12), which reduces to

$$\text{mse}_*(\hat{\theta}_i^H) = g_{1i}(\hat{\sigma}_v^2) - b_{\hat{\sigma}_v^2}(\hat{\sigma}_v^2)\nabla g_{1i}(\hat{\sigma}_v^2) + g_{2i}(\hat{\sigma}_v^2) + 2g_{3i}(\hat{\sigma}_v^2), \tag{6.2.7}$$

where in this case

$$\nabla g_{1i}(\hat{\sigma}_v^2) = b_i^2(1-\gamma_i)^2. \tag{6.2.8}$$

The bias term $b_{\hat{\sigma}_v^2}(\sigma_v^2)$ for the ML estimator is obtained from (5.3.13), which reduces to

$$b_{\hat{\sigma}^2_{v\text{ML}}}(\sigma_v^2) = -[2\mathcal{I}(\sigma_v^2)]^{-1}\text{tr}\left\{\left[\sum_{i=1}^m \left(\psi_i + \sigma_v^2 b_i^2\right)^{-1}\mathbf{z}_i\mathbf{z}_i^T\right]^{-1} \times \left[\sum_{i=1}^m b_i^2\left(\psi_i + \sigma_v^2 b_i^2\right)^{-2}\mathbf{z}_i\mathbf{z}_i^T\right]\right\}, \tag{6.2.9}$$

where $\mathcal{I}(\sigma_v^2)$ is given by (6.1.17). It follows from (6.2.8) and (6.2.9) that the term $-b_{\hat{\sigma}^2_{v\text{ML}}}(\hat{\sigma}^2_{v\text{ML}})\nabla g_{1i}(\hat{\sigma}^2_{v\text{ML}})$ in (6.2.7) is positive. Therefore, ignoring this term and using (6.2.3) with $\hat{\sigma}_v^2 = \hat{\sigma}^2_{v\text{ML}}$ would lead to underestimation of MSE approximation given by (6.2.1).

The bias term $b_{\hat{\sigma}^2_{vm}}(\sigma_v^2)$ for the FH moment estimator $\hat{\sigma}^2_{vm}$ is given by

$$b_{\hat{\sigma}^2_{vm}}(\sigma_v^2) = \frac{2\left\{m\sum_{i=1}^m\left(\psi_i + \sigma_v^2 b_i^2\right)^{-2} - \left[\sum_{i=1}^m\left(\psi_i + \sigma_v^2 b_i^2\right)^{-1}\right]^2\right\}}{\left[\sum_{i=1}^m\left(\psi_i + \sigma_v^2 b_i^2\right)^{-1}\right]^3} \tag{6.2.10}$$

(Datta, Rao, and Smith 2005). It follows from (6.2.10) that the bias of $\hat{\sigma}^2_{vm}$ is positive, unlike the bias of $\hat{\sigma}^2_{v\text{ML}}$, and it reduces to 0 if $b_i = 1$ and $\psi_i = \psi$ for all i. As a result, the term $-b_{\hat{\sigma}^2_{vm}}(\hat{\sigma}^2_{vm})\nabla g_{1i}(\hat{\sigma}^2_{vm})$ in (6.2.7) is negative. Therefore, ignoring this term and using (6.2.3) with $\hat{\sigma}_v^2 = \hat{\sigma}^2_{vm}$ would lead to overestimation of the MSE approximation given by (6.2.1).

Area-specific versions of (6.2.7) for the ML estimator, $\hat{\sigma}^2_{v\text{ML}}$, and the moment estimator, $\hat{\sigma}^2_{vm}$, are obtained from (5.3.17) and (5.3.18):

$$\text{mse}_{*1}(\hat{\theta}_i^H) = g_{1i}(\hat{\sigma}_v^2) - b_{\hat{\sigma}_v^2}(\hat{\sigma}_v^2)\nabla g_{1i}(\hat{\sigma}_v^2) + g_{2i}(\hat{\sigma}_v^2) + 2g^*_{3i}(\hat{\sigma}_v^2, \hat{\theta}_i), \tag{6.2.11}$$

and

$$\mathrm{mse}_{*2}(\hat{\theta}_i^H) = g_{1i}(\hat{\sigma}_v^2) - b_{\hat{\sigma}_v^2}(\hat{\sigma}_v^2)\nabla g_{1i}(\hat{\sigma}_v^2) + g_{2i}(\hat{\sigma}_v^2) + g_{3i}(\hat{\sigma}_v^2) + g_{3i}^*(\hat{\sigma}_v^2, \hat{\theta}_i), \quad (6.2.12)$$

where $\hat{\sigma}_v^2$ is chosen as either $\hat{\sigma}_{v\mathrm{ML}}^2$ or $\hat{\sigma}_{vm}^2$.

All the above MSE estimators are approximately unbiased in the sense of having bias of lower order than m^{-1} for large m. We assumed normality of the random effects, v_i, in deriving the MSE estimators of the EBLUP estimator $\hat{\theta}_i^H$. Lahiri and Rao (1995), however, showed that the MSE estimator (6.2.3) is also valid under nonnormal area effects v_i provided that $E|v_i|^{8+\delta} < \infty$ for $0 < \delta < 1$. This robustness result was established using the simple moment estimator $\hat{\sigma}_{vs}^2$, assuming normality of the sampling errors e_i. The latter assumption, however, is not restrictive, unlike the normality of v_i, due to the central limit theorem (CLT) effect on the direct estimator $\hat{\theta}_i$. The moment condition $E|v_i|^{8+\delta} < \infty$ is satisfied by many continuous distributions including the double exponential, "shifted" exponential and lognormal distributions with $E(v_i) = 0$. The proof of the validity of (6.2.3) under nonnormal v_i is highly technical. We refer the reader to the Appendix in Lahiri and Rao (1995) for details of the proof. It may be noted that $\mathrm{MSE}(\hat{\theta}_i^H)$ is affected by the nonnormality of the v_i's. In fact, it depends on the fourth moment of v_i, and the cross-product term $E(\hat{\theta}_i^H - \theta_i)(\hat{\theta}_i^H - \tilde{\theta}_i^H)$ in the decomposition of the MSE given in (5.2.28) is nonzero, unlike in the case of normal v_i. However, the MSE estimator (6.2.3) remains valid in the sense that $E[\mathrm{mse}(\hat{\theta}_i^H)] = \mathrm{MSE}(\hat{\theta}_i^H) + o(m^{-1})$. This robustness property of the MSE estimator obtained using $\hat{\sigma}_{vs}^2$ may not hold in the case of ML, REML, and the FH estimators of σ_v^2 (Chen, Lahiri and Rao 2008). However, simulation results indicate robustness for those estimators as well (see Example 6.2.1).

Example 6.2.1. Simulation Study. Datta, Rao, and Smith (2005) studied the relative bias (RB) of MSE estimators through simulation. They used the basic area level model (6.1.1) with $b_i = 1$ and no covariates, that is, $\hat{\theta}_i = \mu + v_i + e_i$, $i = 1, \ldots, m$. Since the MSE of an EBLUP is translation invariant, they took $\mu = 0$ without loss of generality. However, to account for the uncertainty due to estimation of unknown regression parameters, this zero mean was estimated from each simulation run. The simulation runs consisted of $R = 100{,}000$ replicates of $\hat{\theta}_i = v_i + e_i$, $i = 1, \ldots, m$, for $m = 15$ generated from $v_i \overset{\text{iid}}{\sim} N(0, \sigma_v^2 = 1)$ and $e_i \overset{\text{ind}}{\sim} N(0, \psi_i)$ for specified sampling variances ψ_i. In particular, three different ψ_i-patterns, with five groups $G_1, \ldots, G_5$ in each pattern and equal number of areas and equal ψ_i's within each group G_t, were chosen, namely patterns (a) 0.7, 0.6, 0.5, 0.4, 0.3; (b) 2.0, 0.6, 0.5, 0.4, 0.2 and (c) 4.0, 0.6, 0.5, 0.4, 0.1. Note that pattern (a) is nearly balanced, while pattern (c) has the largest variability ($\max \psi_i / \min \psi_i = 40$) and pattern (b) has intermediate variability ($\max \psi_i / \min \psi_i = 10$). Patterns similar to (c) can occur when the sample sizes for a group of areas are significantly larger than those for the remaining areas.

The true MSEs of the EBLUP estimators $\hat{\theta}_i^H$ were approximated from the simulated data sets, using $\mathrm{MSE} = R^{-1} \sum_{r=1}^R [\hat{\theta}_i^H(r) - \theta_i(r)]^2$, where $\hat{\theta}_i^H(r)$ and $\theta_i(r)$ denote, respectively, the values of $\hat{\theta}_i^H$ and $\theta_i = v_i$ for the rth simulation run. The

MSE values were approximated using the estimators of σ_v^2 discussed in Section 6.1.2, namely the Prasad–Rao (PR) simple moment estimator $\hat{\sigma}_{vs}^2$, the FH moment estimator $\hat{\sigma}_{vm}^2$, and the ML and REML estimators $\hat{\sigma}_{v\text{ML}}^2$ and $\hat{\sigma}_{v\text{RE}}^2$. The RB of a MSE estimator was computed as $\text{RB} = [R^{-1}\sum_{r=1}^{R}(\text{mse}_r - \text{MSE})]/\text{MSE}$, where mse_r is the value of an MSE estimator for the rth replicate. The MSE estimator (6.2.3) associated with PR and REML and the MSE estimator (6.2.7) for ML and FH were compared with respect to RB.

For the nearly balanced pattern (a), all four methods performed well, with an average RB (ARB) lower than 2%. However, for the extreme pattern (c), PR led to large ARB when ψ_i/σ_v^2 was small, with about 80% ARB for group G_4 and 700% for G_5. When increasing the number of areas to $m = 30$, this overestimation of MSE for G_5 decreased by 140% (Datta and Lahiri 2000). The remaining three estimation methods for σ_v^2 performed well for patterns (b) and (c) with ARB less than 13% for REML and less than 10% for FH and ML. Datta, Rao, and Smith (2005) also calculated ARB values for two nonnormal distributions for the v_i's with mean 0 and variance 1, namely double exponential and location exponential. FH also performed well for these distributions, with ARB smaller than 10% (ARB for ML smaller than 16%).

6.2.2 MSE for Nonsampled Areas

The MSE of the regression-synthetic estimator, $\hat{\theta}_\ell^{\text{RS}}$, for nonsampled areas $\ell = m+1, \ldots, M$, is given by

$$\begin{aligned}\text{MSE}(\hat{\theta}_\ell^{\text{RS}}) &= E(\mathbf{z}_\ell^T\hat{\boldsymbol{\beta}} - \theta_\ell)^2 \\ &= E[\mathbf{z}_\ell^T(\hat{\boldsymbol{\beta}} - \boldsymbol{\beta}) - v_\ell]^2. \end{aligned} \tag{6.2.13}$$

Now, noting that $\hat{\boldsymbol{\beta}}$ is independent of v_ℓ because $\hat{\boldsymbol{\beta}}$ is calculated from the sample data $\{(\hat{\theta}_i, \mathbf{z}_i); i = 1, \ldots, m\}$, it follows from (6.2.13) that

$$\begin{aligned}\text{MSE}(\hat{\theta}_\ell^{RS}) &= \sigma_v^2 + \mathbf{z}_\ell^T\left[\sum_{i=1}^{m}\mathbf{z}_i\mathbf{z}_i^T/\left(\psi_i + \sigma_v^2 b_i^2\right)\right]^{-1}\mathbf{z}_\ell + o(m^{-1}) \\ &=: \sigma_v^2 + h_\ell(\sigma_v^2) + o(m^{-1}), \end{aligned} \tag{6.2.14}$$

noting that $\|V(\hat{\boldsymbol{\beta}}) - V(\tilde{\boldsymbol{\beta}})\| = o(m^{-1})$. Using (6.2.14), a second-order unbiased MSE estimator is given by

$$\text{mse}(\hat{\theta}_\ell^{\text{RS}}) = \hat{\sigma}_v^2 + b_{\hat{\sigma}_v^2}(\hat{\sigma}_v^2) + h_\ell(\hat{\sigma}_v^2), \tag{6.2.15}$$

where $b_{\hat{\sigma}_v^2}(\sigma_v^2)$ is the bias of $\hat{\sigma}_v^2$. The bias term $b_{\hat{\sigma}_v^2}(\sigma_v^2)$ is zero up to terms of order m^{-1} for the simple moment estimator $\hat{\sigma}_{vs}^2$ and the REML estimator $\hat{\sigma}_{v\text{RE}}^2$, and the bias terms for the ML estimator $\hat{\sigma}_{v\text{ML}}^2$ and the FH moment estimator $\hat{\sigma}_{vm}^2$ are given by (6.2.9) and (6.2.10), respectively.

6.2.3 *MSE Estimation for Small Area Means

If $\theta_i = g(\overline{Y}_i) = \overline{Y}_i$, then the EBLUP estimator of the small area mean $\overline{Y}_i$ is $\hat{\overline{Y}}_i^H = \hat{\theta}_i^H$ and $\text{mse}(\hat{\overline{Y}}_i^H) = \text{mse}(\hat{\theta}_i^H)$, where $\text{mse}(\hat{\theta}_i^H)$ is given in Section 6.2.1. On the other hand, for nonlinear $g(\cdot)$ as in Examples 6.1.1 and 6.1.2, a naive estimator of $\overline{Y}_i$ is given by $\hat{\overline{Y}}_i^H = g^{-1}(\hat{\theta}_i^H) = h(\hat{\theta}_i^H)$, which is subject to bias. A bias-adjusted estimator of $\overline{Y}_i$ is taken as

$$\hat{\overline{Y}}_{ia}^H = \hat{F}_i \hat{\overline{Y}}_i^H, \tag{6.2.16}$$

where $\hat{F}_i$ is obtained by evaluating F_i in (6.1.25) at $(\boldsymbol{\beta}, \sigma_v^2) = (\hat{\boldsymbol{\beta}}, \hat{\sigma}_v^2)$. As noted in Example 6.1.2, F_i reduces to $\exp[\sigma_v^2(1-\gamma_i)/2]$ if $h(\theta_i) = \exp(\theta_i)$. It may be noted that $\hat{\overline{Y}}_{ia}^H$ is not exactly unbiased for $\overline{Y}_i$ because of the estimation of $\boldsymbol{\beta}$ and σ_v^2.

Slud and Maiti (2006) derived a second-order unbiased MSE estimator $\text{mse}(\hat{\overline{Y}}_{ia}^H)$, for the special case of $h(\theta_i) = \exp(\theta_i)$, using the ML estimator of σ_v^2. The formula for $\text{mse}(\hat{\overline{Y}}_{ia}^H)$ is much more complicated than $\text{mse}(\hat{\theta}_i^H)$ given in Section 6.2.1.

A crude MSE estimator of $\hat{\overline{Y}}_i^H$ for the special case $\overline{Y}_i = \exp(\theta_i)$ may be obtained by first considering the case of known $\boldsymbol{\beta}$ and σ_v^2. In this case, the optimal (or best) estimator of θ_i is $\hat{\theta}_i^B = \mathbf{z}_i^T\boldsymbol{\beta} + \gamma_i(\hat{\theta}_i - \mathbf{z}_i^T\boldsymbol{\beta})$ under normality of v_i and e_i. Letting $\hat{\overline{Y}}_i^B = \exp(\hat{\theta}_i^B)$, an exactly unbiased estimator of $\overline{Y}_i$ is given by

$$\begin{aligned} \hat{\overline{Y}}_{ia}^B &= \hat{\overline{Y}}_i^B E[\exp(\theta_i)]/E[\exp(\hat{\theta}_i^B)] \\ &= F_i \hat{\overline{Y}}_i^B, \end{aligned} \tag{6.2.17}$$

where again $F_i = \exp[\sigma_v^2(1-\gamma_i)/2]$. Note that $E(\hat{\overline{Y}}_{ia}^B) = E(\overline{Y}_i) = \exp(\mathbf{z}_i^T\boldsymbol{\beta} + \sigma_v^2/2)$. Using (6.1.26), the MSE of $\hat{\overline{Y}}_{ia}^B$ may be expressed as

$$\begin{aligned} \text{MSE}(\hat{\overline{Y}}_{ia}^B) &= E\left[F_i \exp\left(\hat{\theta}_i^B\right) - \exp(\theta_i)\right]^2 \\ &= F_i^2 E\left[\exp\left(2\hat{\theta}_i^B\right)\right] - 2F_i E\left[\exp\left(\hat{\theta}_i^B + \theta_i\right)\right] + E\left[\exp\left(2\theta_i\right)\right] \\ &= [E(\overline{Y}_i)]^2 \left\{\exp\left[V\left(\hat{\theta}_i^B\right)\right] - 2\exp[\text{Cov}(\hat{\theta}_i^B, \theta_i)] + \exp[V(\theta_i)]\right\} \\ &\approx [E(\overline{Y}_i)]^2 V(\hat{\theta}_i^B - \theta_i) \qquad (6.2.18) \\ &= [E(\overline{Y}_i)]^2 \text{MSE}(\hat{\theta}_i^B), \qquad (6.2.19) \end{aligned}$$

using the crude approximation $\exp(\delta) \approx 1 + \delta$. The approximation (6.2.19) suggests that a crude MSE estimator of $\hat{\overline{Y}}_{ia}^B$ is given by

$$\text{mse}(\hat{\overline{Y}}_{ia}^B) = (\hat{\overline{Y}}_{ia}^B)^2 \text{mse}(\hat{\theta}_i^B), \tag{6.2.20}$$

where $\text{MSE}(\hat{\theta}_i^B) = \gamma_i \psi_i$. We now imitate (6.2.20) to get a crude approximation to $\text{mse}(\hat{\overline{Y}}_{ia}^H)$ as

$$\text{mse}(\hat{\overline{Y}}_{ia}^H) = (\hat{\overline{Y}}_{ia}^H)^2 \text{mse}(\hat{\theta}_i^H). \tag{6.2.21}$$

Luery (2011) proposed an MSE estimator similar to (6.2.21) in the context of SAIPE (see Example 6.1.2).

6.2.4 *Bootstrap MSE Estimation

Bootstrap resampling from the sample data $\{(\hat{\theta}_i, \mathbf{z}_i);\ i = 1, \ldots, m\}$ may be used to estimate the MSE of the EBLUP, $\hat{\theta}_i^H$, and more generally the MSE of a complex estimator, $\hat{\phi}_i^H$, of $\phi_i = h(\theta_i)$, such as the bias-adjusted estimator $\hat{\overline{Y}}_{ia}^H$ of the mean $\overline{Y}_i$ or the estimator of $\overline{Y}_i$ based on the compromise estimator $\hat{\theta}_{ic}^H$, of θ_i. The generality of the bootstrap method for estimating the MSE of complex estimators is attractive because the analytical method based on linearization is not readily applicable to parameters other than θ_i.

Assuming normality of v_i and e_i and $\hat{\sigma}_v^2 > 0$, bootstrap data $(\hat{\theta}_{i*}, \mathbf{z}_i)$ are generated independently for each $i = 1, \ldots, m$ as follows: (i) generate θ_{i*} from $N(\mathbf{z}_i^T \hat{\boldsymbol{\beta}}, \hat{\sigma}_v^2)$, $i = 1, \ldots, m$ and let $\phi_{i*} = h(\theta_{i*})$; (ii) generate $\hat{\theta}_{i*}$ from $N(\theta_{i*}, \psi_i)$, $i = 1, \ldots, m$. The method used for calculating $\hat{\phi}_i^H$ is then applied to the bootstrap data $\{(\hat{\theta}_{i*}, \mathbf{z}_i);\ i = 1, \ldots, m\}$ to obtain $\hat{\phi}_{i*}^H$. Repeat the above steps a large number, B, of times to get B bootstrap estimates $\hat{\phi}_{i*}^H(1), \ldots, \hat{\phi}_{i*}^H(B)$ and the bootstrap values of ϕ_i, denoted $\phi_{i*}(1), \ldots, \phi_{i*}(B)$.

The theoretical bootstrap estimator of $\text{MSE}(\hat{\phi}_i^H)$ is given by $\text{mse}_B(\hat{\phi}_i^H) = E_*(\hat{\phi}_{i*}^H - \phi_{i*})^2$, where E_* denotes the bootstrap expectation. We approximate $\text{mse}_B(\hat{\phi}_i^H)$ by Monte Carlo, using the B bootstrap replicates

$$\text{mse}_{B1}(\hat{\phi}_i^H) = B^{-1} \sum_{b=1}^{B} [\hat{\phi}_{i*}^H(b) - \phi_{i*}(b)]^2. \tag{6.2.22}$$

Theoretical properties of (6.2.22) are not known for general parameters ϕ_i, but the motivation behind $\text{mse}_B(\hat{\phi}_i^H)$ is to mimic the true $\text{MSE}(\hat{\phi}_i^H) = E(\hat{\phi}_i^H - \phi_i)^2$, by changing $\hat{\phi}_i^H$ to $\hat{\phi}_{i*}^H$, ϕ_i to ϕ_{i*} and the expectation E to the bootstrap expectation E_*.

In the special case of $\phi_i = \theta_i$, by imitating the second-order approximation (6.2.1) to $\text{MSE}(\hat{\theta}_i^H)$, noting that the bootstrap FH model is a replica of the FH model (6.1.1) with $(\boldsymbol{\beta}, \sigma_v^2)$ changed to $(\hat{\boldsymbol{\beta}}, \hat{\sigma}_v^2)$, we get the approximation

$$\text{mse}_B(\hat{\theta}_i^H) \approx g_{1i}(\hat{\sigma}_v^2) + g_{2i}(\hat{\sigma}_v^2) + g_{3i}(\hat{\sigma}_v^2). \tag{6.2.23}$$

Now, comparing (6.2.23) to the second-order unbiased MSE estimator (6.2.3), valid for $\hat{\sigma}_{v\text{RE}}^2$ and $\hat{\sigma}_{vs}^2$, it follows that $\text{mse}_B(\hat{\theta}_i^H)$ is not second-order unbiased. A double-bootstrap MSE estimator has been proposed for the basic unit level model (see Chapter 7) to rectify this problem. A similar method may be used for the FH model.

Alternatively, for the special case of $\phi_i = \theta_i$, hybrid bootstrap MSE estimators that are second-order unbiased may be used (Butar and Lahiri 2003). The hybrid method uses the representation

$$\text{MSE}(\hat{\theta}_i^H) = [g_{1i}(\sigma_v^2) + g_{2i}(\sigma_v^2)] + E(\hat{\theta}_i^H - \tilde{\theta}_i^H)^2 \tag{6.2.24}$$

to obtain a bias-corrected bootstrap estimator of $g_{1i}(\sigma_v^2) + g_{2i}(\sigma_v^2)$ and a bootstrap estimator of the last term in (6.2.24). The mentioned bias-corrected estimator is given by $2[g_{1i}(\hat{\sigma}_v^2) + g_{2i}(\hat{\sigma}_v^2)] - E_*[g_{1i}(\hat{\sigma}_{v*}^2) + g_{2i}(\hat{\sigma}_{v*}^2)]$, where $\hat{\sigma}_{v*}^2$ is the estimator of σ_v^2 obtained from the bootstrap data. The bootstrap estimator of the last term, $E(\hat{\theta}_i^H - \tilde{\theta}_i^H)^2$, is given by $E_*[\tilde{\theta}_i^H(\hat{\sigma}_{v*}^2) - \tilde{\theta}_i^H(\hat{\sigma}_v^2)]^2$, noting that $\tilde{\theta}_i^H = \tilde{\theta}_i^H(\sigma_v^2)$ and $\hat{\theta}_i^H = \tilde{\theta}_i^H(\hat{\sigma}_v^2)$. The sum of the two terms leads to the hybrid bootstrap MSE estimator

$$\begin{aligned} \text{mse}_{\text{BL}}(\hat{\theta}_i^H) = {} & 2[g_{1i}(\hat{\sigma}_v^2) + g_{2i}(\hat{\sigma}_v^2)] - E_*[g_{1i}(\hat{\sigma}_{v*}^2) + g_{2i}(\hat{\sigma}_{v*}^2)] \\ & + E_*[\tilde{\theta}_i^H(\hat{\sigma}_{v*}^2) - \tilde{\theta}_i^H(\hat{\sigma}_v^2)]^2. \end{aligned} \tag{6.2.25}$$

Formula (6.2.25) shows that the bootstrap data are used here only to calculate $\hat{\sigma}_{v*}^2$; the bootstrap EBLUP estimator $\hat{\theta}_{i*}^H$ of θ_i is not used unlike in the direct bootstrap MSE estimator, $\text{mse}_{B1}(\hat{\theta}_i^H)$, given by (6.2.22). Butar and Lahiri (2003) obtained an analytical approximation to (6.2.25), and it is interesting to note that this approximation is identical to the area-specific MSE estimator, $\text{mse}_{*2}(\hat{\theta}_i^H)$, given by (6.2.12).

We now turn to the case of obtaining $\hat{\sigma}_v^2 = 0$. In this case, random effects v_i are absent in the model. Consequently, the true bootstrap parameter θ_{i*} is taken as the regression-synthetic estimator obtained by evaluating $\tilde{\boldsymbol{\beta}}(\sigma_v^2)$ at $\sigma_v^2 = 0$, that is, $\theta_{i*} = \mathbf{z}_i^T \tilde{\boldsymbol{\beta}}(0)$, which is fixed over bootstrap replicates. Then, $\hat{\theta}_{i*}$ is generated from $N(\theta_{i*}, \psi_i)$. Bootstrap data $\{(\hat{\theta}_{i*}, \mathbf{z}_i);\ i = 1, \ldots, m\}$, are used to calculate the regression-synthetic estimator $\hat{\theta}_{i*}^{\text{RS}} = \mathbf{z}_i^T \hat{\boldsymbol{\beta}}_{\text{WLS}}^*$, where $\hat{\boldsymbol{\beta}}_{\text{WLS}}^* = \left(\sum_{i=1}^m \mathbf{z}_i \mathbf{z}_i^T/\psi_i\right)^{-1} \sum_{i=1}^m \mathbf{z}_i \hat{\theta}_{i*}/\psi_i$. Repeating the above steps a large number, B, of times, we get B bootstrap estimates $\hat{\theta}_{i*}^{\text{RS}}(1), \ldots, \hat{\theta}_{i*}^{\text{RS}}(B)$. Theoretical bootstrap estimator of MSE is given by $\text{mse}_B(\hat{\theta}_i^H) = E_*(\hat{\theta}_{i*}^{\text{RS}} - \theta_{i*})^2$, which is approximated by

$$\text{mse}_{B1}(\hat{\theta}_i^H) = B^{-1} \sum_{b=1}^{B} [\hat{\theta}_{i*}^{RS}(b) - \theta_{i*}]^2. \tag{6.2.26}$$

Note that in the case of obtaining $\hat{\sigma}_v^2 = 0$, the EBLUP estimate of θ_i reduces to $\hat{\theta}_i^H = \mathbf{z}_i^T \tilde{\boldsymbol{\beta}}(0) =: \hat{\theta}_i^{\text{RS}}$. Moreover, by the method of imitation, we have

$$\text{mse}_B(\hat{\theta}_i^H) \approx g_{2i}(0), \tag{6.2.27}$$

that is, the bootstrap estimator (6.2.26) is tracking the correct MSE.

6.2.5 *MSE of a Weighted Estimator

The EBLUP estimator $\hat{\theta}_i^H$ runs into difficulties in the case of $\hat{\sigma}_v^2 = 0$. In this case, it reduces to the regression-synthetic estimator $\mathbf{z}_i^T\hat{\boldsymbol{\beta}}_{\text{WLS}}$ regardless of the area sample sizes. For example, in the state model of Example 6.1.2 dealing with poverty counts of school-age children, in year 1992 it turned out that $\hat{\sigma}_{v\text{ML}}^2 = \hat{\sigma}_{v\text{RE}}^2 = 0$. As a result, for that year the EBLUP attached zero weight to all the direct estimates $\hat{\theta}_i$ regardless of the CPS sample sizes n_i (number of households). Moreover, the leading term of the MSE estimate, $g_{1i}(\hat{\sigma}_v^2) = \hat{\gamma}_i\psi_i$, becomes zero when $\hat{\sigma}_v^2 = 0$. One way to get around the above problems is to use a weighted combination of $\hat{\theta}_i$ and $\mathbf{z}_i^T\hat{\boldsymbol{\beta}}$ with fixed weights a_i and $1 - a_i$:

$$\hat{\theta}_i(a_i) = a_i\hat{\theta}_i + (1 - a_i)\mathbf{z}_i^T\hat{\boldsymbol{\beta}}, \qquad 0 < a_i < 1. \tag{6.2.28}$$

A prior guess σ_{v0}^2 of σ_v^2, available from past studies, may be used to construct the weight a_i as $a_i = \sigma_{v0}^2 b_i^2/(\psi_i + \sigma_{v0}^2 b_i^2)$. Note that $\hat{\theta}_i(a_i)$ remains model-unbiased for θ_i for any fixed weight $a_i \in (0, 1)$, provided the linking model holds, that is, $E(\theta_i) = \mathbf{z}_i^T\boldsymbol{\beta}$.

Datta, Kubokawa, Molina, and Rao (2011) derived a second-order unbiased estimator of $\text{MSE}[\hat{\theta}_i(a_i)]$, assuming $b_i = 1$ in the FH model (6.1.1), given by

$$\text{mse}[\hat{\theta}_i(a_i)] = g_{1i}(\hat{\sigma}_v^2) + g_{2i}(\hat{\sigma}_v^2) + g_{ai}^*(\hat{\sigma}_v^2) - b_{\hat{\sigma}_v^2}(\hat{\sigma}_v^2), \tag{6.2.29}$$

where $b_{\hat{\sigma}_v^2}(\sigma_v^2)$ is the bias of $\hat{\sigma}_v^2$ and

$$g_{ai}^*(\sigma_v^2) = (a_i - \gamma_i)^2[\sigma_v^2 + \psi_i - g_{2i}(\sigma_v^2)/(1 - \gamma_i)^2]. \tag{6.2.30}$$

The bias term $b_{\hat{\sigma}_v^2}(\sigma_v^2)$ is zero up to $o(m^{-1})$ terms for the simple moment estimator $\hat{\sigma}_{vs}^2$ and the REML estimator $\hat{\sigma}_{v\text{RE}}^2$. The bias terms for the ML estimator $\hat{\sigma}_{v\text{ML}}^2$ and the FH estimator $\hat{\sigma}_{vm}^2$ are given by (6.2.9) and (6.2.10), respectively. Note that $E[\text{mse}(\hat{\theta}_i(a_i)] - \text{MSE}[\hat{\theta}_i(a_i)] = o(m^{-1})$ for any fixed weight a_i $(0 < a_i < 1)$.

In the case that $\hat{\sigma}_v^2 = 0$, the leading term of order $O(1)$ in (6.2.29) reduces to $a_i^2\psi_i$ noting that $\hat{\gamma}_i = 0$, unlike the leading term of $\text{mse}(\hat{\theta}_i^H)$, $g_{1i}(\hat{\sigma}_v^2) = \hat{\gamma}_i\psi_i$, which becomes zero. Note also that $a_i^2\psi_i$ decreases as the sampling variance decreases, which is a desirable property of (6.2.29).

Datta et al. (2011) conducted a limited simulation study on the performance of the weighted estimator $\hat{\theta}_i(a_i)$, using weight a_i based on a prior guess $\sigma_{v0}^2 = 0.75$ and letting true σ_v^2 equal to 1, for selected ψ_i-patterns and $m = 30$. Their results indicated that the MSE of the weighted estimator (6.2.28) is very close to that of the EBLUP estimator $\hat{\theta}_i^H$, and even slightly smaller in many cases because the weighted estimator avoids the uncertainty due to estimating σ_v^2. On the other hand, the weighted estimator that uses a constant weight for all the areas, say $a_i = 1/2$, $i = 1, \ldots, m$, does not perform well relative to $\hat{\theta}_i^H$ in terms of MSE, especially for areas with smaller ψ_i; note that the prior guess σ_{v0}^2 of σ_v^2 leads to varying weights a_i. The MSE estimator (6.2.29) performed well in terms of RB.

6.2.6 Mean Cross Product Error of Two Estimators

The small area estimators $\hat{\theta}_i^H$ are often aggregated to obtain an estimator for a larger area. To obtain the MSE of the larger area estimator, we also need the mean cross product error (MCPE) of the estimators $\hat{\theta}_i^H$ and $\hat{\theta}_t^H$ for two different areas $i \neq t$. The MCPE of $\hat{\theta}_i^H$ and $\hat{\theta}_t^H$ may be expressed as

$$\text{MCPE}(\hat{\theta}_i^H, \hat{\theta}_t^H) = E(\hat{\theta}_i^H - \theta_i)(\hat{\theta}_t^H - \theta_t) = \text{MCPE}(\tilde{\theta}_i^H, \tilde{\theta}_t^H) + \text{lower order terms}, \tag{6.2.31}$$

where the leading term in (6.2.31) is given by

$$\text{MCPE}(\tilde{\theta}_i^H, \tilde{\theta}_t^H) = (1-\gamma_i)(1-\gamma_t)\mathbf{z}_i^T \left[\sum_{i=1}^m \mathbf{z}_i\mathbf{z}_i^T / \left(\psi_i + \sigma_v^2 b_i^2\right) \right]^{-1} \mathbf{z}_t$$
$$=: g_{2it}(\sigma_v^2), \tag{6.2.32}$$

which is $O(m^{-1})$, unlike $\text{MSE}(\tilde{\theta}_i^H)$ that is $O(1)$. It follows from (6.2.31) and (6.2.32) that

$$\text{MCPE}(\hat{\theta}_i^H, \hat{\theta}_t^H) \approx \text{MCPE}(\tilde{\theta}_i^H, \tilde{\theta}_t^H) = g_{2it}(\sigma_v^2) \tag{6.2.33}$$

is correct up to $o(m^{-1})$ terms. Furthermore, an estimator of $\text{MCPE}(\hat{\theta}_i^H, \hat{\theta}_t^H)$ is given by

$$\text{mcpe}(\hat{\theta}_i^H, \hat{\theta}_t^H) = g_{2it}(\hat{\sigma}_v^2), \tag{6.2.34}$$

which is unbiased up to $o(m^{-1})$ terms, that is,

$$E[\text{mcpe}(\hat{\theta}_i^H, \hat{\theta}_t^H)] = \text{MCPE}(\hat{\theta}_i^H, \hat{\theta}_t^H) + o(m^{-1}).$$

6.2.7 *Conditional MSE

Conditional on $\boldsymbol{\theta} = (\theta_1, \ldots, \theta_m)^T$ In Section 6.2.1, we studied the estimation of unconditional MSE of the EBLUP estimator $\hat{\theta}_i^H$. It is, however, more appealing to survey practitioners to consider the estimation of conditional MSE, $\text{MSE}_p(\hat{\theta}_i^H) = E_p(\hat{\theta}_i^H - \theta_i)^2$, where the expectation E_p is with respect to the sampling model $\hat{\theta}_i = \theta_i + e_i$ only, treating the small area means, θ_i, as fixed unknown parameters.

We first consider the simple case where all the parameters, $(\boldsymbol{\beta}, \sigma_v^2)$, of the linking model are assumed to be known and $b_i = 1$ for all areas $i = 1, \ldots, m$. In this case, the minimum MSE estimator of θ_i (best prediction estimator) is given by $\hat{\theta}_i^B = \gamma_i\hat{\theta}_i + (1-\gamma_i)\mathbf{z}_i^T\boldsymbol{\beta}$, under normality of v_i and e_i. Following Section 3.4.3, Chapter 3, $\hat{\theta}_i^B$ may be expressed as $\hat{\theta}_i + h_i(\hat{\boldsymbol{\theta}})$, where $\hat{\boldsymbol{\theta}} = (\hat{\theta}_1, \ldots, \hat{\theta}_m)^T$ and $h_i(\hat{\theta}) = -(1-\gamma_i)(\hat{\theta}_i - \mathbf{z}_i^T\boldsymbol{\beta})$. Now noting that $\hat{\theta}_i|\theta_i \overset{\text{ind}}{\sim} N(\theta_i, \psi_i)$, $i = 1, \ldots, m$, and appealing to the general formula (3.4.15) for the derivative of $h_i(\hat{\boldsymbol{\theta}})$, a p-unbiased estimator, $\text{mse}_p(\hat{\theta}_i^B)$, of the conditional MSE, $\text{MSE}_p(\hat{\theta}_i^B)$, may be written as

$$\begin{aligned}\text{mse}_p(\hat{\theta}_i^B) &= \psi_i + 2\psi_i \partial h_i(\hat{\boldsymbol{\theta}})/\partial \hat{\theta}_i + h_i^2(\hat{\boldsymbol{\theta}}) \\ &= \gamma_i \psi_i + (1-\gamma_i)^2[(\hat{\theta}_i - \mathbf{z}_i^T \boldsymbol{\beta})^2 - (\psi_i + \sigma_v^2)] \end{aligned} \qquad (6.2.35)$$

(Rivest and Belmonte 2000). Note that (6.2.35) can take negative values and, in fact, when γ_i is close to zero, the probability of getting a negative value is close to 0.5.

Rivest and Belmonte (2000) studied the relative performance of $\text{mse}_p(\hat{\theta}_i^B)$ and the unconditional MSE estimator $\text{mse}_p(\hat{\theta}_i^B) = \gamma_i \psi_i$, as estimators of the conditional MSE, $\text{MSE}_p(\hat{\theta}_i^B)$, for the special case of $\psi_i = \psi$ for all i. They calculated the ratio, R, of $E_m\{\overline{\text{MSE}}_p[\text{mse}_p(\hat{\theta}_i^B)]\}$ to $E_m\{\overline{\text{MSE}}_p[\text{mse}(\hat{\theta}_i^B)]\}$, where $\overline{\text{MSE}}_p$ denotes the average of design MSE over the m areas and E_m denotes the expectation with respect to the linking model $\theta_i = \mathbf{z}_i^T \boldsymbol{\beta} + v_i$. The ratio R measures the average efficiency of $\text{mse}(\hat{\theta}_i^B)$ relative to $\text{mse}_p(\hat{\theta}_i^B)$. Note that in this case $R = (\psi^2 + 2\psi\sigma_v^2)/\sigma_v^4 > 1$ if $\sigma_v^2/\psi < 2.4$. When shrinking is appreciable, that is, when $\gamma_i = \gamma$ is small, $\text{mse}(\hat{\theta}_i^B)$ is a much more efficient estimator of $\text{MSE}_p(\hat{\theta}_i^B)$ than $\text{mse}_p(\hat{\theta}_i^B)$. For example, if $\sigma_v^2/\psi = 1$ or equivalently $\gamma = 1/2$, we get $R = 3$. It may be noted that the estimator $\hat{\theta}_i^B$ leads to significant gain in efficiency relative to the direct estimator, $\hat{\theta}_i$, only when the shrinking factor, γ, is small.

Datta et al. (2011) derived explicit formulae for $\text{mse}_p(\hat{\theta}_i^H)$ in the case of REML and FH estimators of σ_v^2, by first expressing $\hat{\theta}_i^H$ as $\hat{\theta}_i + h_i(\hat{\boldsymbol{\theta}})$ and then using the formula $\text{mse}_p(\hat{\theta}_i^H) = \psi_i + 2\psi_i \partial h_i(\hat{\boldsymbol{\theta}})/\partial \hat{\theta}_i + h_i^2(\hat{\boldsymbol{\theta}})$. They also conducted a simulation study under the conditional setup for $m = 30$. The study showed that the CV of $\text{mse}_p(\hat{\theta}_i^H)$ can be very large (ranged from 13% to 393%), especially for areas with large sampling variances ψ_i. Therefore, $\text{mse}_p(\hat{\theta}_i^{EB})$ is not reliable as an estimator of $\text{MSE}_p(\hat{\theta}_i^H)$ although it is p-unbiased.

Conditional on $\hat{\theta}_i$ Fuller (1989) proposed a compromise measure of conditional MSE, given by $\text{MSE}_c(\hat{\theta}_i^H) = E[(\hat{\theta}_i^H - \theta_i)^2 | \hat{\theta}_i]$, where the expectation is conditional on the observed $\hat{\theta}_i$ for area i only. Datta et al. (2011) made a systematic study of the Fuller measure and obtained a second-order unbiased estimator of $\text{MSE}_c(\hat{\theta}_i^H)$, using REML and FH estimators of σ_v^2. We now summarize their main results. A second-order approximation to $\text{MSE}_c(\hat{\theta}_i^H)$ is given by

$$\text{MSE}_c(\hat{\theta}_i^H) = g_{1i}(\sigma_v^2) + g_{2i}(\sigma_v^2) + [(\hat{\theta}_i - \mathbf{z}_i^T \boldsymbol{\beta})^2/(\sigma_v^2 + \psi_i)]g_{3i}(\sigma_v^2) + o_p(m^{-1}), \qquad (6.2.36)$$

where g_{1i}, g_{2i}, and g_{3i} are as defined in Section 6.2.1 and $o_p(m^{-1})$ denotes terms of lower order in probability than m^{-1}. Note that $\text{MSE}_c(\hat{\theta}_i^H)$ depends on the direct estimator $\hat{\theta}_i$. Expression (6.2.36) is valid for both REML and FH estimators of σ_v^2. Turning to conditional MSE estimation, a second-order unbiased MSE estimator is given by

$$\text{mse}_c(\hat{\theta}_i^H) = g_{1i}(\hat{\sigma}_v^2) + g_{2i}(\hat{\sigma}_v^2) - \left[g_{1i}^{(1)}(\hat{\sigma}_v^2)\right] h(\hat{\sigma}_v^2, \hat{\theta}_i - \mathbf{z}_i^T\hat{\boldsymbol{\beta}})$$
$$+\left[\frac{(\hat{\theta}_i - \mathbf{z}_i^T\hat{\boldsymbol{\beta}})^2}{\hat{\sigma}_v^2 + \psi_i} + 1\right] g_{3i}(\hat{\sigma}_v^2) + o_p(m^{-1}),$$

where $g_{1i}^{(1)}(\hat{\sigma}_v^2)$ is the first derivative of $g_{1i}(\hat{\sigma}_v^2)$ evaluated at $\sigma_v^2 = \hat{\sigma}_v^2$ and $h(\hat{\sigma}_v^2, \hat{\theta}_i - \mathbf{z}_i^T\hat{\boldsymbol{\beta}}) = E(\hat{\sigma}_v^2 - \sigma_v^2|\hat{\theta}_i)$ is the conditional bias of $\hat{\sigma}_v^2$ given $\hat{\theta}_i$. The conditional bias term depends on the choice of the estimation method for σ_v^2 (see Datta et al. 2011 for the conditional bias under REML and FH moment methods). It is interesting to note that the conditional bias of the REML estimator, $\hat{\sigma}_{v\text{RE}}^2$, is not zero up to $O_p(m^{-1})$ terms unlike the unconditional bias of $\hat{\sigma}_{v\text{RE}}^2$, which is zero up to $O(m^{-1})$ terms.

Simulation results for $m = 30$ reported by Datta et al. (2011) showed that the RB of b under the conditional setup is comparable to the corresponding RB of $\text{mse}(\hat{\theta}_i^H)$ under the unconditional setup, and both are small (median absolute RB lower than 2%). Furthermore, the CV values of the estimator $\text{mse}_c(\hat{\theta}_i^H)$ under the conditional setup and those of $\text{mse}(\hat{\theta}_i^H)$ under the unconditional setup are also similar, with CV increasing for areas with larger ψ_i.

6.3 *ROBUST ESTIMATION IN THE PRESENCE OF OUTLIERS

This section describes methods for robust estimation of small area parameters, θ_i, under the FH model (6.1.1) with normality, in the presence of outliers in the random area effects v_i, or the sampling errors e_i, or both. However, due to the Central Limit Theorem CLT effect, e_i is less likely prone to outliers.

Datta and Lahiri (1995) used a hierarchical Bayes (HB) framework (see Chapter 10) to study the effect of outliers in v_i. The distribution of v_i is assumed to be a scale mixture of a normal distribution with a general mixing distribution. This mixture family induces long-tailed distributions, in particular, t and Cauchy. HB estimators based on the assumed family are more robust to outlying v_i than the estimators based on normal distribution. If the areas with outlying v_i are a priori known, then Datta and Lahiri (1995) recommend the use of a Cauchy distribution on the v_i for those areas and a mixture distribution with tails lighter than the Cauchy on the v_i for the remaining areas.

Bell and Huang (2006) studied empirically how the HB estimator of θ_i changes when a t-distribution with small degrees of freedom, k, on the area effects v_i, is used in place of the normal distribution. Their results indicate that the use of a t-distribution with small k can diminish the effect of outliers in the sense that more weight is given to the direct estimator $\hat{\theta}_i$ than in the case of normal v_i; note that we are dealing with outliers in v_i and $\hat{\theta}_i$ is assumed to be well-behaved. On the other hand, if a $\hat{\theta}_i$ is regarded as outlier, then assuming a t-distribution on the sampling error e_i leads to an estimator that gives less weight to $\hat{\theta}_i$ than in the case of normal e_i. Bell and Huang (2006) note that outlying $\hat{\theta}_i$ might occur due to nonsampling errors.

Ghosh, Maiti, and Roy (2008) studied robust estimation of θ_i under the area level model (6.1.1) with $b_i = 1$, by robustifying the EBLUP estimator, using Huber's (1972) ψ-function. Assuming σ_v^2 known, the BLUP estimator (6.1.2) of θ_i may be written as

$$\tilde{\theta}_i^H = \hat{\theta}_i - (1-\gamma_i)(\hat{\theta}_i - \mathbf{z}_i^T\tilde{\boldsymbol{\beta}}), \tag{6.3.1}$$

where $\tilde{\boldsymbol{\beta}}$ is the BLUE of $\boldsymbol{\beta}$ given by (6.1.5). The estimator $\tilde{\theta}_i^H$ is robustified by applying the Huber function to the residuals $\hat{\theta}_i - \mathbf{z}_i^T\tilde{\boldsymbol{\beta}}$. This leads to the Ghosh–Maiti–Roy (GMR) estimator

$$\tilde{\theta}_i^{\text{GMR}} = \hat{\theta}_i - (1-\gamma_i)\psi_i^{1/2}\psi_b[\psi_i^{-1/2}(\hat{\theta}_i - \mathbf{z}_i^T\tilde{\boldsymbol{\beta}})], \tag{6.3.2}$$

where $\psi_b(u) = u\min(1, b/|u|)$ and $b > 0$ denotes a tuning constant. Note that as $b \to \infty$, $\tilde{\theta}_i^{\text{GMR}}$ tends to $\tilde{\theta}_i^H$. The robust EBLUP estimator, $\hat{\theta}_i^{\text{GMR}}$, is obtained from (6.3.1) by substituting the non robust estimators $\hat{\boldsymbol{\beta}}$ and $\hat{\sigma}_v^2$ for the unknown $\hat{\boldsymbol{\beta}}$ and σ_v^2.

GMR evaluated MSE($\tilde{\theta}_i^{\text{GMR}}$) under the assumed FH normal model and showed that it is strictly decreasing in b and that it exceeds MSE($\tilde{\theta}_i^H$), as expected. Regarding the choice of b, GMR suggest that the tuning constant b may be chosen to ensure that the increase in MSE does not exceed a pre-specified percentage value. This approach leads to an adaptive choice of b, possibly varying across areas. However, note that the adaptive b depends on the unknown σ_v^2.

GMR also derived a second-order unbiased estimator of MSE($\hat{\theta}_i^{\text{GMR}}$), denoted mse($\hat{\theta}_i^{\text{GMR}}$), under the assumed FH normal model. Using mse($\hat{\theta}_i^{\text{GMR}}$) and the MSE estimator of the EBLUP $\hat{\theta}_i^H$, the tuning constant, b, may be estimated for each area i for desired tolerance.

A limitation of $\hat{\theta}_i^{\text{GMR}}$ is that it Huberizes the composite error $v_i + e_i = \hat{\theta}_i - \mathbf{z}_i^T\boldsymbol{\beta}$. An alternative robust estimator that permits Huberizing v_i and e_i separately, or v_i only, may be obtained by robustifying the Henderson et al. (1959) MM equations (5.2.8). For the FH model, the robustified MM equation for v_i is given by

$$\psi_i^{-1/2}\psi_{b_1}[\psi_i^{-1/2}(\hat{\theta}_i - \mathbf{z}_i^T\boldsymbol{\beta} - v_i)] - \sigma_v^{-1}\psi_{b_2}(\sigma_v^{-1}v_i) = 0, \quad i = 1,\ldots,m, \tag{6.3.3}$$

where the tuning constants b_1 and b_2 may be different. If the sampling error $e_i = \hat{\theta}_i - \mathbf{z}_i^T\boldsymbol{\beta} - v_i$ is not deemed to be a possible outlier, then b_1 may be chosen as $b_1 = \infty$. Equation (6.3.3) thus offers flexibility. For given $\boldsymbol{\beta}$ and σ_v^2, equation (6.3.3) might be solved for v_i by using the Newton–Raphson (NR) method. We denote the solution as $\tilde{v}_{iR}(\boldsymbol{\beta}, \sigma_v^2)$. Robust estimators $\hat{\boldsymbol{\beta}}_R$ and $\hat{\sigma}_{vR}^2$ of $\boldsymbol{\beta}$ and σ_v^2 are obtained by Huberizing the ML equations for $\boldsymbol{\beta}$ and σ_v^2. This leads to robustified ML equations

$$\beta: \ \sum_{i=1}^{m}\mathbf{z}_i(\sigma_v^2 + \psi_i)^{-1/2}\psi_b(r_i) = \mathbf{0}, \tag{6.3.4}$$

$$\sigma_v^2 : \sum_{i=1}^{m}(\sigma_v^2 + \psi_i)[\psi_b^2(r_i) - c] = 0, \tag{6.3.5}$$

where $r_i = (\sigma_v^2 + \psi_i)^{-1/2}(\hat{\theta}_i - \mathbf{z}_i^T \boldsymbol{\beta})$ and $c = E[\psi_b^2(u)]$ with $u \sim N(0, 1)$. Methods described in Section 7.4 for unit level models can be used to solve (6.1.4) and (6.1.5), leading to robust estimators $\hat{\boldsymbol{\beta}}_R$ and $\hat{\sigma}_{vR}^2$, and in turn to the robust estimator $\hat{v}_{iR} = \tilde{v}_{iR}(\hat{\boldsymbol{\beta}}_R, \hat{\sigma}_{vR}^2)$ of v_i. Finally, a robust EBLUP (REBLUP) estimator of θ_i is given by

$$\hat{\theta}_i^{\text{RS}} = \mathbf{z}_i^T \hat{\boldsymbol{\beta}}_R + \hat{v}_{iR}. \tag{6.3.6}$$

The above method is designed to handle symmetric outliers in v_i or e_i. In the case of nonsymmetric outliers, bias-correction methods, similar to those given in Section 7.4, can be used, but we omit details here. Also, a bootstrap method for estimating MSE($\hat{\theta}_i^{\text{RS}}$) can be developed along the lines of the bootstrap method in Section 7.4. Unlike GMR, normality of v_i and e_i is not required for this bootstrap method.

GMR conducted a simulation study where outliers in the random area effects v_i were generated from a contaminated normal distribution. Their results, based on adaptive choice of b, indicate that, in terms of efficiency, both $\hat{\theta}_i^{\text{GMR}}$ and $\hat{\theta}_i^H$ perform similarly, indicating robustness of $\hat{\theta}_i^H$ to outliers in v_i. In an unpublished simulation study, Rao and Sinha found a similar efficiency result for $\hat{\theta}_i^{\text{RS}}$ relative to $\hat{\theta}_i^H$ under contaminated normal and t-distribution with small degrees of freedom and a common choice $b = 1.345$. Moreover, under the normal model, $\hat{\theta}_i^H$ and $\hat{\theta}_i^{\text{RS}}$ performed similarly in terms of efficiency. On the other hand, $\hat{\theta}_i^{\text{GMR}}$ with $b = 1.345$ led to significant loss in efficiency over $\hat{\theta}_i^H$ under the normal model.

6.4 *PRACTICAL ISSUES

6.4.1 Unknown Sampling Error Variances

The basic area level model (6.1.1) assumes that the sampling variances, ψ_i, are known. In practice, ψ_i is seldom known and must be replaced by an estimator ψ_{i0}. If a direct estimator, $\hat{\psi}_i$, based on unit level data is used, then $\psi_{i0} = \hat{\psi}_i$. Alternatively, the estimators $\hat{\psi}_i$ may be smoothed by using a generalized variance function (GVF) approach to obtain the smoothed estimators $\hat{\psi}_{iS}$ and then take $\psi_{i0} = \hat{\psi}_{iS}$.

It would be useful to study the sensitivity of small area inferences based on ψ_{i0}. Assuming normality of v_i and e_i and known model parameters $\boldsymbol{\beta}$ and σ_v^2, the best estimator of θ_i is given by its conditional expectation

$$E(\theta_i|\hat{\theta}_i) = \gamma_i\hat{\theta}_i + (1 - \gamma_i)\mathbf{z}_i^T \boldsymbol{\beta} =: \tilde{\theta}_i^B. \tag{6.4.1}$$

If γ_i is replaced by $\gamma_{i0} = \sigma_v^2 b_i^2/(\psi_{i0} + \sigma_v^2 b_i^2)$ in (6.4.1), then the resulting estimator $\tilde{\theta}_{i0}^B = \gamma_{i0}\hat{\theta}_i + (1 - \gamma_{i0})\mathbf{z}_i^T \boldsymbol{\beta}$ leads to an increase in MSE, conditional on ψ_{i0}.

In particular, noting that $\text{MSE}(\tilde{\theta}_{i0}^B) = \text{MSE}(\tilde{\theta}_i^B) + E(\tilde{\theta}_{i0}^B - \tilde{\theta}_i^B)^2$, we have

$$\text{MSE}(\tilde{\theta}_{i0}^B) - \text{MSE}(\tilde{\theta}_i^B) = (\gamma_i - \gamma_{i0})^2(\sigma_v^2 b_i^2 + \psi_i). \tag{6.4.2}$$

It follows from (6.4.2) that the relative increase in MSE is given by

$$\text{MSE}(\tilde{\theta}_{i0}^B)/\text{MSE}(\tilde{\theta}_i^B) - 1 = (\gamma_i - \gamma_{i0})^2/[\gamma_i(1 - \gamma_i)], \tag{6.4.3}$$

noting that $\text{MSE}(\tilde{\theta}_i^B) = \sigma_v^2(1 - \gamma_i)$ and $\gamma_i = \sigma_v^2 b_i^2/(\psi_i + \sigma_v^2 b_i^2)$.

It is also of interest to examine the RB of the reported MSE estimator

$$\text{mse}_R(\tilde{\theta}_{i0}^B) = \gamma_i \psi_{i0}, \tag{6.4.4}$$

which assumes that ψ_{i0} is the true sampling variance. Note that $\text{mse}_R(\tilde{\theta}_{i0}^B)$ is not random conditional on ψ_{i0}. RB of the reported MSE estimator is given by

$$\text{RB}[\text{mse}_R(\tilde{\theta}_{i0}^B)] = \frac{\text{mse}_R(\tilde{\theta}_{i0}^B)}{\text{MSE}(\tilde{\theta}_{i0}^B)} - 1 = \frac{\gamma_i(1 - \gamma_i)}{\gamma_i(1 - \gamma_i) + (\gamma_i - \gamma_{i0})^2} - 1. \tag{6.4.5}$$

Bell (2008) reported the results of an empirical study on the relative increase in MSE, given by (6.4.3), and the RB of the reported MSE estimator, given by (6.4.5), for the special case $b_i = 1$. Main conclusions are as follows: (i) Under- or overestimation of true ψ_i impacts the RB of the reported MSE estimator more than the increase in MSE. (ii) Underestimation of ψ_i is a more serious problem when σ_v^2/ψ_i is small. (iii) Overestimation of ψ_i is a more serious problem when σ_v^2/ψ_i is large, a case unlikely to occur in practice because small ψ_i typically results from a large area sample size. Bell (2008) also examined unconditional properties by assuming $\psi_{i0} = \hat{\psi}_i$ and a χ_d^2 distribution for $d\hat{\psi}_i/\psi_i$ for several values of d. His main conclusions are as follows: (i) Increase in the unconditional MSE, when d is moderate ($d = 16$) or large ($d = 80$), is quite small, and is less than 10% even for d as small as 6. (ii) Estimation error in $\hat{\psi}_i$ leads to downward bias in the reported MSE estimator. (iii) Effect of estimation error in $\hat{\psi}_i$ is relatively mild in the unconditional set up in contrast to the conditional perspective.

Rivest and Vandal (2003) also studied the case of $\psi_{i0} = \hat{\psi}_i$, assuming known $\boldsymbol{\beta}$ and σ_v^2 and $b_i = 1$ in the model (6.1.1). In particular, suppose that the direct estimator $\hat{\theta}_i$ is the mean $\bar{y}_i$ of n_i observations $y_{ij} \overset{\text{iid}}{\sim} N(\theta_i, \sigma_i^2), j = 1, \ldots, n_i$. Then $\hat{\psi}_i = s_i^2/n_i$, where $s_i^2 = \sum_{j=1}^{n_i}(y_{ij} - \bar{y}_i)^2/(n_i - 1)$ is the sample variance, and $(n_i - 1)s_i^2/\sigma_i^2$ is distributed as a $\chi_{n_i-1}^2$ variable. In this setting, $n_i\hat{\psi}_i$ may be approximated as $N(\sigma_i^2, \delta_i)$, for $\delta_i = 2\sigma_i^4/(n_i - 1)$. An MSE estimator of $\tilde{\theta}_{i0}^B$ may be then obtained as

$$\text{mse}(\tilde{\theta}_{i0}^B) = \text{mse}_R(\tilde{\theta}_{i0}^B) + 2\hat{\delta}_i\sigma_v^4/(\hat{\psi}_i + \sigma_v^2)^2, \tag{6.4.6}$$

where $\hat{\delta}_i = 2s_i^4/(n_i - 1)$ is the plug-in estimator of δ_i and $\text{mse}_R(\tilde{\theta}_{i0}^B)$ is given by (6.4.4) with $\psi_{i0} = \hat{\psi}_i$ (Rivest and Vandal 2003). It follows from (6.4.6) that the reported

MSE estimator (6.4.6) should be inflated by adding the last term that depends on the variability of the $\hat{\psi}_i$'s. Proof of (6.4.6) is based on Lemma 3.5.1 (Stein's lemma), Chapter 3.

Wang and Fuller (2003) presented comprehensive results for the case $\psi_{i0} = \hat{\psi}_i$ and $b_i = 1$ by relaxing the assumption of known $\boldsymbol{\beta}$ and σ_v^2. The sampling variance ψ_i in the EBLUP $\hat{\theta}_i^H$, given by (6.1.12), is replaced by a design-unbiased estimator $\hat{\psi}_i$. In one version of their procedure, $\boldsymbol{\beta}$ is estimated by the ordinary least squares (OLS) estimator $\hat{\boldsymbol{\beta}}_{\text{OLS}} = (\sum_{i=1}^m \mathbf{z}_i\mathbf{z}_i^T)^{-1}\sum_{i=1}^m \mathbf{z}_i\hat{\theta}_i$ and σ_v^2 by the moment estimator $\hat{\sigma}_{v\text{WF}}^2 = \max(0, \tilde{\sigma}_{v\text{WF}}^2)$, where

$$\tilde{\sigma}_{v\text{WF}}^2 = \sum_{i=1}^m \tilde{d}_i[(\hat{\theta}_i - \mathbf{z}_i^T\hat{\boldsymbol{\beta}}_{\text{OLS}})^2 - \hat{\psi}_i], \tag{6.4.7}$$

where $\tilde{d}_i = d_i/\sum_{i=1}^m d_i$ and d_i is the degrees of freedom associated with $\hat{\psi}_i$. Denote the resulting estimator as

$$\hat{\theta}_i^{\text{WF}} = \hat{\gamma}_i^{\text{WF}}\hat{\theta}_i + (1 - \hat{\gamma}_i^{\text{WF}})\mathbf{z}_i^T\hat{\boldsymbol{\beta}}_{\text{OLS}}, \tag{6.4.8}$$

where $\hat{\gamma}_i^{\text{WF}} = \hat{\sigma}_{v\text{WF}}^2/(\hat{\sigma}_{v\text{WF}}^2 + \hat{\psi}_i)$. Wang and Fuller (2003) examined alternative estimators of σ_v^2 in a simulation study. Under the regularity conditions on $\hat{\psi}_i$, ψ_i, and d_i $(i = 1, \ldots, m)$ stated in Theorem 1 of Wang and Fuller (2003), a second-order approximation of $\text{MSE}(\hat{\theta}_i^{\text{WF}})$ is given by

$$\begin{aligned}\text{MSE}_A(\hat{\theta}_i^{\text{WF}}) &= \gamma_i\psi_i + (1-\gamma_i)^2\mathbf{z}_i^T V(\hat{\boldsymbol{\beta}}_{\text{OLS}})\mathbf{z}_i \\ &\quad + (\sigma_v^2 + \psi_i)^{-3}[\psi_i^2\overline{V}(\hat{\sigma}_{v\text{WF}}^2) + \sigma_v^4 V(\hat{\psi}_i)],\end{aligned} \tag{6.4.9}$$

where $\overline{V}(\hat{\sigma}_{v\text{WF}}^2)$ is the asymptotic variance of $\hat{\sigma}_{v\text{WF}}^2$,

$$V(\hat{\boldsymbol{\beta}}_{\text{OLS}}) = \left(\sum_{i=1}^m \mathbf{z}_i\mathbf{z}_i^T\right)^{-1}\left[\sum_{i=1}^m (\sigma_v^2 + \psi_i)\,\mathbf{z}_i\mathbf{z}_i^T\right]^{-1}\left(\sum_{i=1}^m \mathbf{z}_i\mathbf{z}_i^T\right)^{-1}$$

and $V(\hat{\psi}_i)$ is the variance of $\hat{\psi}_i$. The approximation (6.4.9) to $\text{MSE}(\hat{\theta}_i^{\text{WF}})$ assumes that $d = \min(d_i)$ also increases with the number of areas m, and the error in the approximation is of order $\max(m^{-1.5}, m^{-1}d^{-1}, d^{-1.5})$. It is clear from (6.4.9) that the approximation is similar to the second-order approximation to $\text{MSE}(\hat{\theta}_i^H)$, given by (6.2.1), except that it involves the additional term $(\sigma_v^2 + \psi_i)^{-3}\sigma_v^4 V(\hat{\psi}_i)$ arising from the variance of $\hat{\psi}_i$. Note that Rivest and Vandal (2003) used $\hat{\psi}_i = s_i^2/n_i$ in which case $d_i = n_i - 1$ and $V(\hat{\psi}_i) = \delta_i = 2\sigma_i^4/(n_i - 1)$.

Rivest and Vandal (2003) suggested a generalized PR MSE estimator, using the PR simple moment estimator $\hat{\sigma}_{vs}^2$ with $b_i = 1$ and $\psi_i = \hat{\psi}_i = s_i^2/n_i$ in the EBLUP estimator $\hat{\theta}_i^H$. Denoting the latter estimator by $\hat{\theta}_i^{\text{RV}}$, the proposed MSE estimator is given by

$$\text{mse}(\hat{\theta}_i^{\text{RV}}) = \text{mse}_{\text{PR}}(\hat{\theta}_i^{\text{RV}}) + 2(\hat{\sigma}_{vs}^2 + \hat{\psi}_i)^{-3}\hat{\sigma}_{vs}^4\hat{\delta}_i, \tag{6.4.10}$$

where $\hat{\delta}_i = 2s_i^4/(n_i - 1)$ and mse_{PR} denotes the customary MSE estimator (6.2.3) with ψ_i replaced by $\hat{\psi}_i$. It is clear from (6.4.10) that $\text{mse}(\hat{\theta}_i^{\text{RV}})$ accounts for the extra variability due to estimating ψ_i. As expected, simulation results indicated that the customary MSE estimator can lead to significant underestimation, unlike the adjusted MSE estimator (6.4.10), especially when n_i is small. However, the simulations covered a narrow range $(1/3, 5/2)$ of values of ψ_i/σ_v^2. Wang and Fuller (2003) proposed two estimators of $\text{MSE}(\hat{\theta}_i^{\text{WF}})$ that performed quite well in simulations, except when σ_v^2/ψ_i is very small; range of ψ_i/σ_v^2 was (1/4, 160). In the case of very small σ_v^2/ψ_i, the MSE estimators lead to severe overestimation.

In the case of using smoothed estimators $\psi_{i0} = \hat{\psi}_{iS}$, the estimator (6.4.10) is valid for the MSE of the resulting small area estimator $\hat{\theta}_i^{\text{RV}}$, but with $\hat{\psi}_i$ replaced by $\hat{\psi}_{iS}$ and $\hat{\delta}_i$ by an estimator of $V(\hat{\psi}_{iS})$. Use of smoothed estimators should make the additional term small because $V(\hat{\psi}_{iS})$ will be significantly smaller than $V(\hat{\psi}_i)$, specially for large m. As a result, the customary estimator, mse_{PR}, should perform well. Rivest and Vandal (2003) applied (6.4.10) with smoothed estimators $\hat{\psi}_{iS}$ to Canadian census undercoverage (Example 6.1.3) and found that (6.4.10) increases mse_{PR} by only 1%. This suggests that the use of customary EBLUP estimator $\hat{\theta}_i^H$ and the associated mse_{PR} with ψ_i replaced by $\hat{\psi}_{iS}$ should perform well in practice, especially when m is not small; in the Canadian census example, $m = 96$.

González-Manteiga et al. (2010) studied the case where an estimator ψ_{i0} of ψ_i is not available, and the user has access only to the area level data $\{(\hat{\theta}_i, \mathbf{z}_i); i = 1, \ldots, m\}$. Sampling variance ψ_i is modeled as $\psi_i = \sigma_e^2 h(\mathbf{z}_i^T \boldsymbol{\beta})$, where $h(\mathbf{z}_i^T \boldsymbol{\beta})$ is a smooth function of $\mathbf{z}_i^T \boldsymbol{\beta}$. A kernel-based method is used to estimate $h(\mathbf{z}_i^T \boldsymbol{\beta})$ and in turn the model parameters $\boldsymbol{\beta}$, σ_v^2, and σ_e^2, the area means θ_i and the associated MSE estimators. Practical implementations of this method requires bandwidth selection for the kernel function (see González-Manteiga et al. 2010 for details).

6.4.2 Strictly Positive Estimators of σ_v^2

Section 6.1.2 considered the estimation of σ_v^2, the variance of the random effects v_i, and presented ML, REML, and two methods of moment estimators. All these methods can lead to negative estimates $\tilde{\sigma}_v^2$, especially for small m, which are then truncated to zero: $\hat{\sigma}_v^2 = \max(\tilde{\sigma}_v^2, 0)$. A drawback of truncating to zero is that the resulting EBLUP estimates, $\hat{\theta}_i^H$, will attach zero weight to all the direct area estimates $\hat{\theta}_i$ regardless of the area sample sizes when $\hat{\sigma}_v^2 = 0$. For example, Bell (1999) used the state model for poverty rates (Example 6.1.2) to calculate ML and REML estimates of σ_v^2 for 5 years (1989–1993), and obtained $\hat{\sigma}_v^2 = 0$ for the first 4 years. Giving a zero weight to direct estimates for states with large sample sizes, such as California and New York, is not appealing to the user, and may lead to substantial difference between the EBLUP estimate and the corresponding direct estimate due to overshrinkage induced by the zero estimate of σ_v^2.

Several methods have been proposed to avoid a zero value for $\hat{\sigma}_v^2$. We present a brief account of methods that lead to strictly positive estimators of σ_v^2. Wang and Fuller (2003) proposed a data-based truncation leading to

$$\hat{\sigma}^2_{v\text{WF}} = \max\left[\frac{1}{2}\hat{V}^{1/2}\left(\tilde{\sigma}^2_v\right), \tilde{\sigma}^2_v\right], \tag{6.4.11}$$

where $\hat{V}(\tilde{\sigma}^2_v)$ is an estimator of $V(\tilde{\sigma}^2_v)$. Wang and Fuller (2003) studied several choices of $\tilde{\sigma}^2_v$ and $\hat{V}(\tilde{\sigma}^2_v)$, but we focus on the simple moment estimator $\tilde{\sigma}^2_{vs}$ given by (6.1.15) and let $b_i = 1$ in the basic model (6.1.1). Following Wang and Fuller (2003), a simple choice for $\hat{V}(\tilde{\sigma}^2_v)$ is

$$\hat{V}(\tilde{\sigma}^2_v) = m^{-2}\sum_{i=1}^{m}\left\{\frac{m}{m-p}\left[\left(\hat{\theta}_i - \mathbf{z}_i^T\hat{\boldsymbol{\beta}}_{\text{OLS}}\right)^2 - (1-h_{ii})\psi_i\right] - \hat{\sigma}^2_{vs}\right\}^2, \tag{6.4.12}$$

where $\hat{\boldsymbol{\beta}}_{\text{OLS}}$ is the OLS estimator of $\boldsymbol{\beta}$ given by $\hat{\boldsymbol{\beta}}_{\text{OLS}} = (\sum_{i=1}^m \mathbf{z}_i\mathbf{z}_i^T)^{-1}\sum_{i=1}^m \mathbf{z}_i\hat{\theta}_i$ and $h_{ii} = \mathbf{z}_i^T(\sum_{i=1}^m \mathbf{z}_i\mathbf{z}_i^T)^{-1}\mathbf{z}_i$ (Yoshimori and Lahiri 2014a).

Methods based on adjusting the likelihood function to avoid zero estimates of σ^2_v have also been studied. Letting $b_i = 1$ in the model (6.1.1) and writing it in matrix form as $\hat{\boldsymbol{\theta}} = \mathbf{Z}\boldsymbol{\beta} + \mathbf{v} + \mathbf{e}$, the profile likelihood and the residual likelihood under normality are given by

$$L_P(\sigma^2_v) \propto |\mathbf{V}|^{-1/2}\exp\left(-\frac{1}{2}\hat{\boldsymbol{\theta}}^T\mathbf{P}\hat{\boldsymbol{\theta}}\right) \tag{6.4.13}$$

and

$$L_R(\sigma^2_v) \propto |\mathbf{Z}^T\mathbf{V}^{-1}\mathbf{Z}|^{-1/2}L_P(\sigma^2_v) \tag{6.4.14}$$

respectively, where $\mathbf{V} = \text{diag}(\sigma^2_v + \psi_1, \ldots, \sigma^2_v + \psi_m)$ and $\mathbf{P} = \mathbf{V}^{-1} - \mathbf{V}^{-1}\mathbf{Z}(\mathbf{Z}^T\mathbf{V}^{-1}\mathbf{Z})^{-1}\mathbf{Z}^T\mathbf{V}^{-1}$. An adjustment factor $h(\sigma^2_v)$ is now introduced to define an adjusted likelihood as

$$L_A(\sigma^2_v) \propto h(\sigma^2_v)L(\sigma^2_v), \tag{6.4.15}$$

where $L(\sigma^2_v)$ is either $L_P(\sigma^2_v)$ or $L_R(\sigma^2_v)$. The factor $h(\sigma^2_v)$ is chosen to ensure that the estimate maximizing $L_A(\sigma^2_v)$ with respect to σ^2_v over $[0, \infty)$ is strictly positive. This feature prevents overshrinkage of the EBLUP even for small m.

A simple choice of $h(\sigma^2_v)$ is $h(\sigma^2_v) = \sigma^2_v$ (Morris and Tang 2011, Li and Lahiri 2010). This choice gives a strictly positive estimate $\hat{\sigma}^2_{v\text{LL}}$, noting that $L_A(0) = 0$ and $L_A(\sigma^2_v) \to 0$ as $\sigma^2_v \to \infty$. Thus, the maximum cannot be achieved at $\sigma^2_v = 0$. Focusing on $L_R(\sigma^2_v)$ and the case $\psi_i = \psi$, the bias of the corresponding $\hat{\sigma}^2_{v\text{LL}}$ relative to σ^2_v is equal to $(2/m)\gamma^{-2}$, where $\gamma = \sigma^2_v/(\sigma^2_v + \psi) = \sigma^2_v\psi^{-1}/(\sigma^2_v\psi^{-1} + 1)$. It now follows that the RB of $\hat{\sigma}^2_{v\text{LL}}$ can be large when σ^2_v/ψ is small, even though it is $O(m^{-1})$ when $\sigma^2_v > 0$. Small σ^2_v/ψ induces zero REML estimates $\hat{\sigma}^2_{v\text{RE}}$ of σ^2_v more frequently, but it also causes difficulty with $\hat{\sigma}^2_{v\text{LL}}$ in terms of RB. This dual problem prompted Yoshimori and Lahiri (2014a) to explore alternative choices of $h(\sigma^2_v)$ such that $h(0) = 0$ still holds, but lead to $L_A(\sigma^2_v)$ "closer" to $L_R(\sigma^2_v)$ or $L_P(\sigma^2_v)$. The choice

$$h(\sigma^2_v) = \left\{\tan^{-1}\left(\sum_{i=1}^m \gamma_i\right)\right\}^{1/m} \tag{6.4.16}$$

satisfies these conditions, and we denote the resulting estimator of σ_v^2 as $\hat{\sigma}^2_{v\text{YL}}$.

An alternative strategy is to use $\hat{\sigma}^2_{v\text{LL}}$ only when $\hat{\sigma}^2_{v\text{RE}} = 0$ and retain $\hat{\sigma}^2_{v\text{RE}}$ otherwise (Rubin-Bleuer and You 2013, Molina, Rao, and Datta 2015). We denote this MIX estimator as $\hat{\sigma}^2_{v\text{LLM}}$. This will ensure that the weight attached to the direct estimator is always positive, and that the RB of $\hat{\sigma}^2_{v\text{LLM}}$ is closer to the corresponding RB of $\hat{\sigma}^2_{v\text{RE}}$. A similar MIX strategy can also be used with $\hat{\sigma}^2_{v\text{YL}}$ to obtain the estimator $\hat{\sigma}^2_{v\text{YLM}}$, but the reduction on RB will be smaller because the choice (6.4.16) makes $L_A(\sigma_v^2)$ closer to $L_R(\sigma_v^2)$ or $L_P(\sigma_v^2)$.

Yoshimori and Lahiri (2014b) conducted a limited simulation study on the MSE of EBLUP estimators $\hat{\theta}_i^{\text{RE}}$, $\hat{\theta}_i^{\text{LL}}$, $\hat{\theta}_i^{\text{LLM}}$, $\hat{\theta}_i^{\text{YL}}$, and $\hat{\theta}_i^{\text{YLM}}$ based on $\hat{\sigma}^2_{v\text{RE}}$, $\hat{\sigma}^2_{v\text{LL}}$, $\hat{\sigma}^2_{v\text{LLM}}$, $\hat{\sigma}^2_{v\text{YL}}$, and $\hat{\sigma}^2_{v\text{YLM}}$. The study used $m = 15$, $\mathbf{z}_i^T\boldsymbol{\beta} = \mu$, and $\psi_i = \psi$ for all i. Table 6.3 reports the average MSE values based on $R = 10^4$ simulation runs, for selected values of σ_v^2/ψ reflecting σ_v^2 small relative to ψ. As expected, Table 6.3 shows that the average MSE of all EBLUP estimators increases as σ_v^2/ψ decreases from 0.33 to 0.05. In terms of MSE, the estimator LL performs poorly as σ_v^2/ψ decreases, but the MIX version of it, LLM, rectifies this problem to some extent. Table 6.3 also shows that RE, YL, and YLM perform similarly in terms of MSE, and that it is not necessary to modify YL.

In view of the results in Table 6.3 supporting $\hat{\theta}_i^{\text{YL}}$ in terms of MSE, we focus on the estimation of MSE($\hat{\theta}_i^{\text{YL}}$). Note that $\hat{\theta}_i^{\text{RE}}$ also performs well in terms of MSE, but $\hat{\sigma}^2_{v\text{RE}}$ is not strictly positive unlike $\hat{\sigma}^2_{v\text{YL}}$.

A second-order unbiased MSE estimator of $\hat{\theta}_i^{\text{YL}}$, denoted mse($\hat{\theta}_i^{\text{YL}}$), has the same form as

$$\text{mse}(\hat{\theta}_i^{\text{RE}}) = g_{1i}(\hat{\sigma}^2_{v\text{RE}}) + g_{2i}(\hat{\sigma}^2_{v\text{RE}}) + 2g_{3i}(\hat{\sigma}^2_{v\text{RE}}), \tag{6.4.17}$$

but with $\hat{\sigma}^2_{v\text{RE}}$ replaced by $\hat{\sigma}^2_{v\text{YL}}$. This result follows from the fact that the asymptotic variances of $\hat{\sigma}^2_{v\text{RE}}$ and $\hat{\sigma}^2_{v\text{YL}}$ are identical and biases of both estimators are of lower order than m^{-1} (Yoshimori and Lahiri 2014a). A MIX estimator of MSE($\hat{\theta}_i^{\text{YL}}$) is taken as mse($\hat{\theta}_i^{\text{RE}}$) when $\hat{\sigma}^2_{v\text{RE}} > 0$ and as $g_{2i}(\hat{\sigma}^2_{v\text{YL}})$ when $\hat{\sigma}^2_{v\text{RE}} = 0$ (Yoshimori and Lahiri 2014b). We denote this option as $\text{mse}_{\text{YL}}(\hat{\theta}_i^{\text{YL}})$. Alternatively, we can use $g_{2i}(0)$ when $\hat{\sigma}^2_{v\text{RE}} = 0$ and retain mse($\hat{\theta}_i^{\text{RE}}$) when $\hat{\sigma}^2_{v\text{RE}} > 0$ (Molina, Rao, and Datta 2015). We denote this option as $\text{mse}_{\text{MRD}}(\hat{\theta}_i^{\text{YL}})$. Chen and Lahiri (2003) also suggested replacing

TABLE 6.3 Average MSE of EBLUP Estimators Based on REML, LL, LLM, YL, and YLM Methods of Estimating σ_v^2

ψ	σ_v^2/ψ	REML	LL	LLM	YL	YLM
3.0	0.33	1.06	1.19	1.03	1.04	1.05
5.7	0.18	1.44	1.86	1.45	1.42	1.42
9.0	0.11	1.85	2.66	1.95	1.83	1.83
19.0	0.05	3.02	5.04	3.46	3.02	3.00

TABLE 6.4 % Relative Bias (RB) of Estimators of MSE($\hat{\theta}_i^{\text{YL}}$)

ψ	σ_v^2/ψ	YL	YL1	YL2
3.0	0.33	53.7	19.8	20.0
5.7	0.18	104.1	40.4	40.7
9.0	0.11	149.0	59.4	59.8
19.0	0.05	218.2	88.1	88.7

second-order unbiased MSE estimators by $g_{2i}(\hat{\sigma}_v^2)$ whenever $\hat{\sigma}_v^2 = 0$. This suggestion was made in the context of the simple moment estimator $\hat{\sigma}_v^2 = \hat{\sigma}_{vs}^2$ given by (6.1.15).

Table 6.4 reports the percent RB, averaged over areas, of $\text{mse}(\hat{\theta}_i^{\text{YL}})$, $\text{mse}_{\text{YL}}(\hat{\theta}_i^{\text{YLM}})$, and $\text{mse}_{\text{MRD}}(\hat{\theta}_i^{\text{YLM}})$, denoted as YL, YL1, and YL2, respectively. As expected, Table 6.4 shows that the RB values of YL1 and YL2 are very close. Furthermore, YL has RB twice as large as the RB of YL1 and YL2.

Results from Tables 6.3 and 6.4 suggest the use of the estimator $\hat{\theta}_i^{\text{YL}}$ and the associated MIX MSE estimator $\text{mse}_{\text{YL}}(\hat{\theta}_i^{\text{YL}})$ or $\text{mse}_{\text{MRD}}(\hat{\theta}_i^{\text{YL}})$ whenever it is deemed necessary to attach a strictly positive weight to the direct estimator in the formula for the EBLUP estimator. If $\hat{\sigma}_v^2 = 0$ is allowed for an estimator of σ_v^2, then one could use $\hat{\theta}_i^{\text{RE}}$ together with a MIX MSE estimator: Use $g_{2i}(0)$ when $\hat{\sigma}_{v\text{RE}}^2 = 0$ and the formula (6.4.17) when $\hat{\sigma}_{v\text{RE}}^2 > 0$. This MIX estimator of MSE will have significantly smaller RB than (6.4.17), similar to MIX estimators of MSE for $\hat{\theta}_i^{\text{YL}}$ (Molina, Rao, and Datta 2015).

6.4.3 Preliminary Test Estimation

In some applications, it may be possible to find very good predictors, $\mathbf{z}_i$, that can explain the variability of the θ_i's without the need for the inclusion of the model errors v_i. This may be interpreted as having a small σ_v^2 in the linking model $\theta_i = \mathbf{z}_i^T\boldsymbol{\beta} + v_i$ with $v_i \sim (0, \sigma_v^2)$. A preliminary test (PT) of the hypothesis $H_0 : \sigma_v^2 = 0$ at a suitable level α may then be conducted to select between the linking models $\theta_i = \mathbf{z}_i^T\boldsymbol{\beta} + v_i$ or $\theta_i = \mathbf{z}_i^T\boldsymbol{\beta}$. In particular, if H_0 is not rejected, then we take as linking model $\theta_i = \mathbf{z}_i^T\boldsymbol{\beta}$ and estimate θ_i with the regression-synthetic estimator $\mathbf{z}_i^T\hat{\boldsymbol{\beta}}_{\text{PT}}$, where $\hat{\boldsymbol{\beta}}_{\text{PT}} = (\sum_{i=1}^m \psi_i^{-1}\mathbf{z}_i\mathbf{z}_i^T)^{-1}\sum_{i=1}^m \psi_i^{-1}\mathbf{z}_i\hat{\theta}_i$ is the WLS estimator of $\boldsymbol{\beta}$ under H_0. On the other hand, the EBLUP $\hat{\theta}_i^H$ is used when H_0 is rejected (Datta, Hall, and Mandal 2011). A large value of α (typically $\alpha = 0.2$) is recommended in the PT estimation literature (Han and Bancroft 1968, Ehsanes Saleh 2006).

In the SAE context, when H_0 is not rejected, then the regression-synthetic estimator $\mathbf{z}_i^T\hat{\boldsymbol{\beta}}_{\text{PT}}$ is used for all areas $i = 1, \ldots, m$, that is, zero weight is attached to all the direct estimates $\hat{\theta}_i$ regardless of the area sample sizes or the quality of direct estimates. But attaching a nonzero weight to direct estimates is recommendable when they are reliable, because they protect against failure of model assumptions. Since the power of the test increases with α, then to avoid overshrinking to the synthetic estimators, it is also advisable to consider a not so small α.

Datta, Hall, and Mandal (2011) proposed a simple test of H_0 assuming normality. Their test statistic in this setup is given by

$$T = \sum_{i=1}^{m} \psi_i^{-1}(\hat{\theta}_i - \mathbf{z}_i^T \hat{\beta}_{\text{PT}})^2, \tag{6.4.18}$$

and T is distributed as $\mathcal{X}^2_{m-p}$ with $m-p$ degrees of freedom under H_0. The PT estimator of θ_i, based on T, is given by

$$\hat{\theta}_i^{\text{PT}} = \begin{cases} \mathbf{z}_i^T \hat{\beta}_{\text{PT}} & \text{if } T \le \mathcal{X}^2_{m-p,\alpha}; \\ \hat{\theta}_i^H & \text{if } T > \mathcal{X}^2_{m-p,\alpha}, \end{cases} \quad i = 1, \ldots, m, \tag{6.4.19}$$

where $\mathcal{X}^2_{m-p,\alpha}$ is the upper α-point of $\mathcal{X}^2_{m-p}$. The estimator $\hat{\theta}_i^{\text{PT}}$ is recommended when m is moderate (say 15 to 20). However, simulation results with $m = 15$ and $\alpha = 0.2$ indicate that $\hat{\theta}_i^{\text{PT}}$ is practically the same as $\hat{\theta}_i^H$ in terms of bias and MSE (Molina, Rao, and Datta 2015), and therefore it may be better to always use the EBLUP $\hat{\theta}_i^H$ because it automatically gives a small weight $\hat{\gamma}_i$ to the direct estimator when H_0 is not rejected.

On the other hand, PT is useful for estimating the MSE of $\hat{\theta}_i^H$ through a MIX estimator of MSE given by

$$\text{mse}_{\text{PT}}(\hat{\theta}_i^H) = \begin{cases} g_{2i} & \text{if } T \le \mathcal{X}^2_{m-p,\alpha} \text{ or } \hat{\sigma}^2_{v\text{RE}} = 0; \\ \text{mse}\left(\hat{\theta}_i^{\text{RE}}\right) & \text{if } T > \mathcal{X}^2_{m-p,\alpha} \text{ and } \hat{\sigma}^2_{v\text{RE}} > 0. \end{cases} \tag{6.4.20}$$

Here, $g_{2i} := g_{2i}(0) = \mathbf{z}_i^T(\sum_{i=1}^m \psi_i^{-1}\mathbf{z}_i\mathbf{z}_i^T)^{-1}\mathbf{z}_i$ and $\hat{\theta}_i^{\text{RE}}$ denotes $\hat{\theta}_i^H$ obtained with $\hat{\sigma}_v^2 = \hat{\sigma}^2_{v\text{RE}}$. Note that, even when H_0 is rejected, it may happen that $\hat{\sigma}^2_{v\text{RE}} = 0$ because the test statistic T does not depend on $\hat{\sigma}^2_{v\text{RE}}$. This is the reason for including the alternative condition $\hat{\sigma}^2_{v\text{RE}} = 0$ in (6.4.20). Simulation results by Molina et al. (2015) indicate that $\text{mse}_{\text{PT}}(\hat{\theta}_i^H)$ performs better in terms of RB than the alternative MIX estimator of MSE that uses g_{2i} when $\hat{\sigma}^2_{v\text{RE}} = 0$ and the usual formula (6.4.17) when $\hat{\sigma}^2_{v\text{RE}} > 0$. Hence, it is better to use g_{2i} when H_0 is not rejected, even if the realized value of $\hat{\sigma}^2_{v\text{RE}}$ is positive.

One could also apply the PT method to estimate the MSE of the small area estimator $\hat{\theta}_i^{\text{YL}}$, which attaches a positive weight to the direct estimator $\hat{\theta}_i$. The PT-based MIX estimator of $\text{MSE}(\hat{\theta}_i^{\text{YL}})$ is given by

$$\text{mse}_{\text{PT}}(\hat{\theta}_i^{\text{YL}}) = \begin{cases} g_{2i}\left(\hat{\sigma}^2_{v\text{YL}}\right) & \text{if } T \le \mathcal{X}^2_{m-p,\alpha}; \\ \text{mse}(\hat{\theta}_i^{\text{RE}}) & \text{if } T > \mathcal{X}^2_{m-p,\alpha}. \end{cases} \tag{6.4.21}$$

Instead of excluding all random effects, v_i, $i = 1, \ldots, m$, in the linking model and then use the regression-synthetic estimator for all areas when PT does not reject H_0 at a specified α, Datta and Mandal (2014) proposed a modified approach that takes a middle ground to the PT approach and the FH model. Under their formulation, v_i is changed to $\delta_i v_i$, where δ_i equals 1 or 0 with probabilities p and $1-p$, respectively. Conditional on $\delta_i = 1$, $v_i \overset{\text{iid}}{\sim} N(0, \sigma_v^2)$, $i = 1, \ldots, m$. This modified model permits more flexible data-dependent shrinkage of $\hat{\theta}_i$ toward the regression-synthetic estimator $\mathbf{z}_i^T \hat{\boldsymbol{\beta}}_{\text{PT}}$ than the PT and FH approaches. A HB approach was used to derive the HB estimator of θ_i. An application to the estimation of poverty rates of school-age children in the states of the United States shows advantages of the proposed approach.

6.4.4 Covariates Subject to Sampling Errors

The basic area level model (6.1.1) assumes that the covariates $\mathbf{z}_i$ are population values not subject to sampling errors. However, in practice some elements of $\mathbf{z}_i$ may be subject to sampling errors. Ybarra and Lohr (2008) studied the estimation of mean body mass index for 50 small areas (demographic subgroups) using direct estimates, $\hat{\theta}_i$, obtained from the 2003–2004 U.S. National Health and Nutrition Examination Survey (NHANES). They also used the 2003 U.S. National Health Interview Survey (NHIS) direct estimates $\hat{\mathbf{z}}_i$ of mean self-reported body mass index $\mathbf{z}_i$ as the covariate. In the NHANES, body mass index for each respondent was ascertained through medical examination in contrast to the NHIS values based on self-reported responses. However, the NHIS sample size (29,652 persons) is much larger than the sample size (4,424 persons) for the NHANES, and the reliable direct estimates $\hat{\mathbf{z}}_i$ of mean self-reported body mass index are highly correlated with the direct area estimates $\hat{\theta}_i$. A plot of $\hat{\theta}_i$ against $\hat{\mathbf{z}}_i$ showed strong linear relationship.

To study the effect of sampling errors in $\mathbf{z}_i$, assume $b_i = 1$ and known linking model parameters $\boldsymbol{\beta}$ and σ_v^2. A "naive" best estimator of θ_i is obtained simply by substituting the available estimator $\hat{\mathbf{z}}_i$ for the unknown $\mathbf{z}_i$ in the best estimator $\tilde{\theta}_i^B$:

$$\tilde{\theta}_i^{\text{NB}} = \gamma_i \hat{\theta}_i + (1-\gamma_i)\hat{\mathbf{z}}_i^T \boldsymbol{\beta}, \tag{6.4.22}$$

where $\gamma_i = \sigma_v^2/(\sigma_v^2 + \psi_i)$. Now, assuming that $\hat{\mathbf{z}}_i \overset{\text{ind}}{\sim} N(\mathbf{z}_i, \mathbf{C}_i)$ with known variance–covariance matrix $\mathbf{C}_i$ and with $\hat{\mathbf{z}}_i$ independent of v_i and e_i, it follows that $\text{MSE}(\tilde{\theta}_i^{\text{NB}}) = \gamma_i \psi_i + (1-\gamma_i)^2 \boldsymbol{\beta}^T \mathbf{C}_i \boldsymbol{\beta}$, which is larger than the MSE of the direct estimator, $\text{MSE}(\hat{\theta}_i) = \psi_i$, if $\boldsymbol{\beta}^T \mathbf{C}_i \boldsymbol{\beta} > \sigma_v^2 + \psi_i$. Therefore, ignoring the sampling error in $\hat{\mathbf{z}}_i$ and using $\tilde{\theta}_i^{\text{NB}}$ is less efficient than using the direct estimator $\hat{\theta}_i$ if the previous condition on $\boldsymbol{\beta}^T \mathbf{C}_i \boldsymbol{\beta}$ holds. Thus, when covariates are estimated with error, using the naive estimator might not lead to gains in efficiency with respect to direct estimation. Moreover, the MSE that is typically reported, obtained by treating the estimate $\hat{\mathbf{z}}_i$ as if it were the true $\mathbf{z}_i$ and given by $\gamma_i \psi_i$, leads to serious understatement of the true MSE of $\tilde{\theta}_i^{\text{NB}}$.

Ybarra and Lohr (2008) considered the class of all linear combinations of $\hat{\theta}_i$ and $\hat{\mathbf{z}}_i^T \boldsymbol{\beta}$, that is, $a_i \hat{\theta}_i + (1-a_i)\hat{\mathbf{z}}_i^T \boldsymbol{\beta}$, for $0 \le a_i \le 1$. The optimal a_i that minimizes the MSE is given by

$$a_i^* = \frac{\sigma_v^2 + \boldsymbol{\beta}^T \mathbf{C}_i \boldsymbol{\beta}}{\sigma_v^2 + \boldsymbol{\beta}^T \mathbf{C}_i \boldsymbol{\beta} + \psi_i} =: \gamma_i^L, \tag{6.4.23}$$

and the resulting best linear estimator in the mentioned class is then

$$\tilde{\theta}_i^L = \gamma_i^L \hat{\theta}_i + (1 - \gamma_i^L)\hat{\mathbf{z}}_i^T \boldsymbol{\beta}. \tag{6.4.24}$$

It follows from (6.4.23) that $\gamma_i^L > \gamma_i$, which implies that $\tilde{\theta}_i^L$ gives a larger weight to the direct estimator $\hat{\theta}_i$ than the naive best estimator $\tilde{\theta}_i^{\text{NB}}$. Normality assumption is not needed to derive $\tilde{\theta}_i^L$ because it is obtained from the class of linear combinations of $\hat{\theta}_i$ and $\hat{\mathbf{z}}_i^T \boldsymbol{\beta}$.

Furthermore, it is easy to verify that $\text{MSE}(\tilde{\theta}_i^L) = \gamma_i^L \psi_i$, which is larger than $\gamma_i \psi_i$, the MSE of the best estimator $\tilde{\theta}_i^B = \gamma_i \hat{\theta}_i + (1 - \gamma_i)\mathbf{z}_i^T \boldsymbol{\beta}$ when $\mathbf{z}_i$ is known. Also, if $E(\hat{\mathbf{z}}_i) = \mathbf{z}_i$, then $E(\tilde{\theta}_i^L - \theta_i) = 0$ holds, that is, $\tilde{\theta}_i^L$ is unbiased for the area mean θ_i. The estimator $\tilde{\theta}_i^L$ is never less efficient than the direct estimator $\hat{\theta}_i$ because $0 \le \gamma_i^L \le 1$.

Restriction to the class of linear combinations of $\hat{\theta}_i$ and $\hat{\mathbf{z}}_i^T \boldsymbol{\beta}$ can be relaxed if normality is assumed. Under normality, an efficient estimator of $\mathbf{z}_i$ is obtained by considering the likelihood $L(\mathbf{z}_i) = f(\hat{\theta}_i, \hat{\mathbf{z}}_i | \mathbf{z}_i) = f(\hat{\theta}_i | \mathbf{z}_i) f(\hat{\mathbf{z}}_i | \mathbf{z}_i)$ and then maximizing $L(\mathbf{z}_i)$ with respect to $\mathbf{z}_i$. This leads to the "best supporting" estimator of $\mathbf{z}_i^T \boldsymbol{\beta}$ given by $\tilde{\mathbf{z}}_i^T \boldsymbol{\beta} = \delta_i \hat{\mathbf{z}}_i^T \boldsymbol{\beta} + (1 - \delta_i)\hat{\theta}_i$, where $\delta_i = (\sigma_v^2 + \psi_i)/(\sigma_v^2 + \psi_i + \boldsymbol{\beta}^T \mathbf{C}_i \boldsymbol{\beta})$. Interestingly, the Ybarra–Lohr estimator $\tilde{\theta}_i^L$ is obtained by substituting $\tilde{\mathbf{z}}_i^T \boldsymbol{\beta}$ for $\mathbf{z}_i^T \boldsymbol{\beta}$ in $\tilde{\theta}_i^B$ (Datta, Rao, and Torabi 2010; Arima, Datta, and Liseo 2015). It may be noted that both $\hat{\mathbf{z}}_i$ and $\hat{\theta}_i$ provide information on $\mathbf{z}_i$, noting that $\hat{\theta}_i | \mathbf{z}_i \sim N(\mathbf{z}_i^T \boldsymbol{\beta}, \sigma_v^2 + \psi_i)$ and $\hat{\mathbf{z}}_i | \mathbf{z}_i \sim N(\mathbf{z}_i, \mathbf{C}_i)$, and that is the reason for considering the likelihood $L(\mathbf{z}_i)$ based on the joint density $f(\hat{\theta}_i, \hat{\mathbf{z}}_i | \mathbf{z}_i)$. On the other hand, using only $f(\hat{\mathbf{z}}_i | \mathbf{z}_i)$ to estimate $\hat{\mathbf{z}}_i$ leads to $\hat{\mathbf{z}}_i^T \boldsymbol{\beta}$, and substituting $\hat{\mathbf{z}}_i^T \boldsymbol{\beta}$ for $\mathbf{z}_i^T \boldsymbol{\beta}$ in $\tilde{\theta}_i^B$ gives the naive best estimator $\tilde{\theta}_i^{\text{NB}}$. Carroll et al. (2006, p. 38) used $f(\hat{\mathbf{z}}_i | \mathbf{z}_i)$ to argue in favor of the naive best predictor in the context of linear regression models: "It rarely makes any sense to worry about measurement error". Ybarra and Lohr (2008) derived $\tilde{\theta}_i^L$, under normality, as the best estimator of θ_i by conditioning on the residuals $\hat{\theta}_i - \hat{\mathbf{z}}_i^T \boldsymbol{\beta}$.

The optimal estimator $\tilde{\theta}_i^L$ depends on the parameters $\boldsymbol{\beta}$ and σ_v^2, which are unknown in practical applications. It is necessary to replace $\boldsymbol{\beta}$ and σ_v^2 by suitable estimators $\hat{\boldsymbol{\beta}}$ and $\hat{\sigma}_v^2$ to obtain the empirical estimator

$$\hat{\theta}_i^L = \hat{\gamma}_i^L \hat{\theta}_i + (1 - \hat{\gamma}_i^L)\hat{\mathbf{z}}_i^T \hat{\boldsymbol{\beta}}, \tag{6.4.25}$$

where $\hat{\gamma}_i^L$ is given by (6.4.23) with σ_v^2 and $\boldsymbol{\beta}$ replaced by $\hat{\sigma}_v^2$ and $\hat{\boldsymbol{\beta}}$, respectively. Ybarra and Lohr (2008) proposed a simple moment estimator of σ_v^2 given by

$$\hat{\sigma}_{va}^2 = (m - p)^{-1} \sum_{i=1}^{m} [(\hat{\theta}_i - \hat{\mathbf{z}}_i^T \hat{\boldsymbol{\beta}}_a)^2 - \psi_i - \hat{\boldsymbol{\beta}}_a^T \mathbf{C}_i \hat{\boldsymbol{\beta}}_a], \tag{6.4.26}$$

where $\hat{\boldsymbol{\beta}}_a$ is based on modified least squares. The estimator $\hat{\boldsymbol{\beta}}_a$ satisfies the equation

$$\left[\sum_{i=1}^{m} a_i \left(\hat{\mathbf{z}}_i \hat{\mathbf{z}}_i^T - \mathbf{C}_i \right) \right] \boldsymbol{\beta} = \sum_{i=1}^{m} a_i \hat{\mathbf{z}}_i \hat{\theta}_i \tag{6.4.27}$$

for a given set of weights $a_i > 0, i = 1, \ldots, m$. In analogy with the weight $(\sigma_v^2 + \psi_i)^{-1}$ used to estimate $\boldsymbol{\beta}$ under the FH model with known $\mathbf{z}_i$, we would like to use the weights $a_i = (\sigma_v^2 + \psi_i + \hat{\boldsymbol{\beta}}^T \mathbf{C}_i \hat{\boldsymbol{\beta}})^{-1}$ in (6.4.27) to account for the measurement error in $\hat{\mathbf{z}}_i$. In practice, we can use a two-step procedure to estimate $\boldsymbol{\beta}$. In step 1, we set $a_i = 1$ and estimate $\boldsymbol{\beta}$ and σ_v^2 using (6.4.27) and (6.4.26). In step 2, we substitute these estimates into the expression for a_i to get $\hat{a}_i$, which in turn leads to the desired estimators $\hat{\boldsymbol{\beta}}_a$ and $\hat{\sigma}_{va}^2$. Both $\hat{\boldsymbol{\beta}}_a$ and $\hat{\sigma}_{va}^2$ are consistent estimators as the number of areas, m, tends to infinity.

A second-order approximation to $\text{MSE}(\hat{\theta}_i^L)$ is similar to $\text{MSE}(\hat{\theta}_i^H)$ given by (6.2.1), but involves an extra term that accounts for the sampling error in $\hat{\mathbf{z}}_i$, and it reduces to (6.2.1) when $\mathbf{C}_i = \mathbf{0}$. Ybarra and Lohr (2008) used a jackknife method to estimate $M_i := \text{MSE}(\hat{\theta}_i^L)$. The jackknife method is outlined in Section 9.2.2 for the estimator $\hat{\theta}_i^H$, but it is applicable to other estimators. In the case of $\hat{\theta}_i^L$, however, we need to ignore the cross-product term $M_{12i} = 2E(\tilde{\theta}_i^L - \theta_i)(\hat{\theta}_i^L - \tilde{\theta}_i^L)$ in the decomposition $M_i = M_{1i} + M_{2i} + M_{12i}$, where $M_{1i} = E(\tilde{\theta}_i^L - \theta_i)^2$ and $M_{2i} = E(\hat{\theta}_i^L - \tilde{\theta}_i^L)^2$ in order to apply the jackknife method, as done by Ybarra and Lohr (2008). In practice, the jackknife method performs well if M_{12i} is negligible relative to M_{2i}.

The method of Simulation Extrapolation (SIMEX) is widely used in the measurement error literature to estimate regression parameters in linear or nonlinear regression models (Carroll et al. 2006, Chapter 6). SIMEX can be adapted to estimate the area means under the basic model (6.1.1) in the presence of sampling errors in $\mathbf{z}_i$ (Singh, Wang, and Carroll 2015a). Basic idea of SIMEX is to create additional vectors $\hat{\mathbf{z}}_{b,i}(\zeta)$ of increasingly larger variances $(1+\zeta)\mathbf{C}_i$ for $b = 1, \ldots, B$, where the number of replications B is large and $\zeta \geq 0$. This is accomplished by generating vectors $\mathbf{r}_{b,i}$, $b = 1, \ldots, B$, independently from $N(\mathbf{0}, \mathbf{C}_i)$ and then letting $\hat{\mathbf{z}}_{b,i}(\zeta) = \hat{\mathbf{z}}_i + \zeta^{1/2}\mathbf{r}_{b,i}$. Carroll et al. (2006) call $\hat{\mathbf{z}}_{b,i}(\zeta)$ a "remeasurement" of $\hat{\mathbf{z}}_i$. Noting that $\hat{\mathbf{z}}_i|\mathbf{z}_i \sim N(\mathbf{z}_i, \mathbf{C}_i)$, it follows that $E[\hat{\mathbf{z}}_{b,i}(\zeta)|\mathbf{z}_i] = \mathbf{z}_i$ and $V[\hat{\mathbf{z}}_{b,i}(\zeta)|\mathbf{z}_i] = (1+\zeta)\mathbf{C}_i = (1+\zeta)V(\hat{\mathbf{z}}_i|\mathbf{z}_i)$. Therefore, $\text{MSE}[\hat{\mathbf{z}}_{b,i}(\zeta)|\mathbf{z}_i]$ converges to zero as $\zeta \to -1$.

SIMEX steps for getting the estimates of small area means are as follows: (i) Select L different values of ζ: $0 < \zeta_1 < \cdots < \zeta_L$. (ii) For each ζ_ℓ ($\ell = 1, \ldots, L$) generate $\{\hat{\mathbf{z}}_{b,i}(\zeta_\ell); b = 1, \ldots, B\}$ and then replacing $\mathbf{z}_i$ by $\hat{\mathbf{z}}_{b,i}(\zeta_\ell)$ in the model (6.1.1), calculate the naive EBLUP estimator from the data $\{\hat{\theta}_i, \hat{\mathbf{z}}_{b,i}(\zeta_\ell); i = 1, \ldots, m\}$ and denote it as $\hat{\theta}_{b,i}^H(\zeta_\ell)$, for $b = 1, \ldots, B$. (iii) For each ζ_ℓ ($\ell = 1, \ldots, L$), calculate the average over the B replicates of the naive estimators: $\hat{\theta}_{+,i}^H(\zeta_\ell) = B^{-1}\sum_{b=1}^B \hat{\theta}_{b,i}^H(\zeta_\ell)$. (iv) Plot the average $\hat{\theta}_{+,i}^H(\zeta_\ell)$ against ζ_ℓ for $\ell = 1, \ldots, L$ and fit an extrapolation function, say $\hat{\theta}_{f,i}^H(\zeta)$, to the pairs $(\zeta_\ell, \hat{\theta}_{+,i}(\zeta_\ell))$, $\ell = 1, \ldots, L$. (v) Extrapolate $\hat{\theta}_{f,i}^H(\zeta)$ to $\zeta = -1$ to get the SIMEX estimator $\hat{\theta}_{f,i}^H(-1)$ of θ_i.

Singh, Wang, and Carroll (2015b) also proposed a corrected score equations method assuming normality. Limited simulation results indicated that SIMEX and the estimator based on the corrected score equations perform better, in terms of MSE, than the Ybarra–Lohr estimator $\hat{\theta}_i^L$.

6.4.5 Big Data Covariates

"Big Data" is currently a hot topic in Statistics, and it is going to receive lot more attention in the future. These kinds of data are also called "organic" data (Groves 2011). Three characteristics of organic data are volume, velocity, and variability. Such data include transaction data (e.g., traffic flow data) and social media Internet data (e.g., Google trends or Facebook postings). We briefly mention two recent applications to small area estimation that use big data covariates as additional predictors in the area level model.

Marchetti et al. (2015) studied the estimation of poverty indicators for local areas in Tuscany region of Italy, in particular poverty rate, which is defined as the proportion of households with a measure of welfare below a specified poverty line. Poverty rate is a special case of the class of FGT poverty measures introduced in Section 9.4.

Big data on mobility comprised of different car journeys automatically tracked with a GPS device during May 2011. Mobility for a given vehicle v is defined as a measure of entropy given by

$$M_v = -\sum_{\ell_1=1}^{L}\sum_{\ell_2=1}^{L} p_v(\ell_1,\ell_2)\log[p_v(\ell_1,\ell_2)],$$

where (ℓ_1,ℓ_2) denotes a pair of locations, L is the number of locations, and $p_v(\ell_1,\ell_2)$ = (number of trips vehicle v made between locations ℓ_1 and ℓ_2)/(total number of trips of vehicle v). The area level big data covariate is taken as $\overline{M}_i = V_i^{-1}\sum_{v\in A_i} M_v$, where A_i is the set of vehicles in area i with the registered GPS device and V_i is the number of vehicles belonging to A_i. The covariate $\overline{M}_i$ is subject to error in estimating the true mobility in area i. Marchetti et al. (2015) treated the vehicles with GPS as a simple random sample and estimated the variance of the mean $\overline{M}_i$. Under this assumption, they used $\overline{M}_i$ as a covariate in the FH model and applied the Ybarra–Lohr modification of the customary EBLUP estimator, treating the estimated variance of $\overline{M}_i$ as the true variance. Direct estimators of area poverty rates were obtained from survey data.

The second application (Porter et al. 2014) analyses relative change of percent household Spanish-speaking in the eastern half of the United States, using direct estimators for the states (small areas) from the ACS and a big data covariate extracted from Google Trends of commonly used Spanish words available at the state level. Google Trends data may be regarded as functional covariates defined over time domain.

6.4.6 Benchmarking Methods

Suppose that θ_i is the ith area mean, $\theta_+ = \sum_{\ell=1}^{m} W_\ell\theta_\ell$ is the aggregate mean, where $W_\ell = N_\ell/N$ is the known proportion of units in area ℓ, and all areas are sampled ($\sum_{\ell=1}^{m} W_\ell = 1$). For example, in the U.S. poverty counts application (Example 6.1.2), θ_i is the poverty rate in state i and θ_+ is the national poverty rate. Assuming that a direct estimator $\hat{\theta}_+ = \sum_{\ell=1}^{m} W_\ell\hat{\theta}_\ell$ of θ_+ is reliable, it is desirable to ensure that the estimators of the small area means θ_i, when aggregated, agree with the reliable direct estimator $\hat{\theta}_+$, especially when $\hat{\theta}_+$ is regarded as the "gold standard". The EBLUP estimators, $\hat{\theta}_i^H$, do not satisfy this "benchmarking" property. In fact,

$$\hat{\theta}_+ - \sum_{\ell=1}^{m} W_\ell \hat{\theta}_\ell^H = \sum_{\ell=1}^{m} W_\ell (1-\hat{\gamma}_\ell)(\hat{\theta}_\ell - \mathbf{z}_\ell^T \hat{\boldsymbol{\beta}}) \neq 0. \tag{6.4.28}$$

It is therefore necessary to modify the EBLUP estimators $\hat{\theta}_i^H$, $i = 1, \ldots, m$ to ensure benchmarking.

Simple Adjustments to EBLUP A very simple but widely used modification to the EBLUPs is called ratio benchmarking. It consists of multiplying each $\hat{\theta}_i^H$ by the common adjustment factor $\sum_{\ell=1}^m W_\ell \hat{\theta}_\ell / \sum_{\ell=1}^m W_\ell \hat{\theta}_\ell^H$, leading to the ratio benchmarked estimator

$$\hat{\theta}_i^{\text{RB}} = \hat{\theta}_i^H \left(\sum_{\ell=1}^{m} W_\ell \hat{\theta}_\ell / \sum_{\ell=1}^{m} W_\ell \hat{\theta}_\ell^H \right). \tag{6.4.29}$$

It readily follows from (6.4.29) that $\sum_{i=1}^m W_i \hat{\theta}_i^{\text{RB}} = \sum_{\ell=1}^m W_\ell \hat{\theta}_\ell = \hat{\theta}_+$. Ratio benchmarking, however, has the following limitations: (i) A common adjustment factor is applied to all the estimators $\hat{\theta}_i^H$, regardless of their precision. (ii) Unlike $\hat{\theta}_i^H$, the benchmarked estimator $\hat{\theta}_i^{\text{RB}}$ is not design-consistent as the sample size in area i increases but holding the sample sizes in the remaining areas fixed. (iii) Second-order unbiased MSE estimators are not readily available, unlike in the case of $\hat{\theta}_i^H$ (Pfeffermann, Sikov, and Tiller 2014). Another simple adjustment is the difference benchmarking given by

$$\hat{\theta}_i^{\text{DB}} = \hat{\theta}_i^H + \left(\sum_{\ell=1}^{m} W_\ell \hat{\theta}_\ell - \sum_{\ell=1}^{m} W_\ell \hat{\theta}_\ell^H \right). \tag{6.4.30}$$

It readily follows from (6.4.30) that $\sum_{i=1}^m W_i \hat{\theta}_i^{\text{DB}} = \hat{\theta}_+$. Similar to $\hat{\theta}_i^{\text{RB}}$, the estimator $\hat{\theta}_i^{\text{DB}}$ also suffers from the limitations (i) and (ii), but it permits second-order unbiased MSE estimation. Steorts and Ghosh (2013) have shown that the effect of benchmarking is to increase the MSE. Assuming $b_i = 1$ in the model (6.1.1) and using the simple moment estimator σ_{vs}^2 of σ_v^2 given by (6.1.15), they derived a second-order approximation to $\text{MSE}(\hat{\theta}_i^{\text{DB}})$ obtained by adding a common term $g_4(\sigma_v^2)$ to the usual $\text{MSE}(\hat{\theta}_i^{\text{H}})$ given by (6.2.1), that is,

$$\text{MSE}(\hat{\theta}_i^{\text{DB}}) \approx \text{MSE}(\hat{\theta}_i^{\text{H}}) + g_4(\sigma_v^2). \tag{6.4.31}$$

This common term $g_4(\sigma_v^2)$ is given by

$$g_4(\sigma_v^2) = \sum_{i=1}^{m} W_i^2 (1-\gamma_i)^2 (\sigma_v^2 + \psi_i) - \sum_{i=1}^{m} \sum_{j=1}^{m} W_i W_j (1-\gamma_i)(1-\gamma_j) h_{ij}, \tag{6.4.32}$$

where $h_{ij} = \mathbf{z}_i^T[\sum_{\ell=1}^m (\sigma_v^2 + \psi_\ell)^{-1}\mathbf{z}_\ell \mathbf{z}_\ell^T]^{-1}\mathbf{z}_j$. Furthermore, a second-order unbiased estimator of MSE($\hat{\theta}_i^{\text{DB}}$) is given by

$$\text{mse}(\hat{\theta}_i^{\text{DB}}) \approx \text{mse}(\hat{\theta}_i^H) + g_4(\hat{\sigma}_v^2), \tag{6.4.33}$$

where mse($\hat{\theta}_i^H$) is given by (6.2.3). Regularity conditions (6.1.10) and (6.1.11) and the additional condition $\max_{1\le\ell\le m} W_\ell = O(m^{-1})$ are used to derive (6.4.31) and (6.4.33). Steorts and Ghosh (2013) compared mse($\hat{\theta}_i^{\text{DB}}$) to mse($\hat{\theta}_i^H$) under the state model for poverty rates (Example 6.1.2), using data sets for the years 1997 and 2000. Their results indicated that the term $g_4(\sigma_v^2)$ is very small relative to mse($\hat{\theta}_i^H$) for all states i. Hence, in this application, difference benchmarking led to negligible inflation of MSE.

"Optimal" Benchmarking Pfeffermann and Barnard (1991) showed that it is impossible to find a best estimator of θ_i for each area i in the class of linear estimators $t_i = \sum_{\ell=1}^m a_{\ell i}\hat{\theta}_\ell$ that are model unbiased and satisfy the benchmarking property. One way to get around this difficulty is to minimize a reasonable overall criterion. Among all estimators $\mathbf{t} = (t_1, \ldots, t_m)^T$ that are linear and model unbiased and satisfy the benchmark constraint $\sum_{\ell=1}^m W_\ell t_\ell = \hat{\theta}_+$, Wang, Fuller, and Qu (2008) obtained $t_1, \ldots, t_m$ that minimize

$$Q(t) = \sum_{i=1}^m \phi_i E(t_i - \theta_i)^2 \tag{6.4.34}$$

for specified positive weights ϕ_i, $i = 1, \ldots, m$. The resulting "optimal" estimator of θ_i is given by

$$\tilde{\theta}_i^{\text{WFQ}} = \tilde{\theta}_i^H + \lambda_i\left(\sum_{\ell=1}^m W_\ell\hat{\theta}_\ell - \sum_{\ell=1}^m W_\ell\tilde{\theta}_\ell^H\right), \tag{6.4.35}$$

where

$$\lambda_i = \left(\sum_{\ell=1}^m \phi_\ell^{-1} W_\ell^2\right)^{-1} \phi_i^{-1} W_i. \tag{6.4.36}$$

Noting that $\sum_{i=1}^m W_i\lambda_i = 1$, it follows from (6.4.35) that $\sum_{i=1}^m W_i\tilde{\theta}_i^{\text{WFQ}} = \hat{\theta}_+$. Replacing σ_v^2 in (6.4.35) by a suitable estimator $\hat{\sigma}_v^2$ leads to the empirical optimal estimator

$$\hat{\theta}_i^{\text{WFQ}} = \hat{\theta}_i^H + \lambda_i\left(\sum_{\ell=1}^m W_\ell\hat{\theta}_\ell - \sum_{\ell=1}^m W_\ell\hat{\theta}_\ell^H\right). \tag{6.4.37}$$

It follows from (6.4.35) and (6.4.36) that (6.4.35) does not suffer from the limitations (i) and (ii), provided $\lambda_i \to 0$ as the area sample size increases. Result (6.4.35) is a special case of a general result for multiple benchmark constraints (Bell, Datta, and Ghosh 2013). It follows from (6.4.35) that $\tilde{\theta}_i^{\text{WFQ}}$ is obtained by allocating the difference $\sum_{\ell=1}^m W_\ell\hat{\theta}_\ell - \sum_{\ell=1}^m W_\ell\tilde{\theta}_\ell^H$ using λ_i given by (6.4.36). The choice $\phi_i = W_i$ gives

$\hat{\theta}_i^{\text{DB}}$, noting that $\lambda_i = 1$ in this case. Popular choices of ϕ_i^{-1} include (i) $\phi_i^{-1} = \psi_i$, (ii) $\phi_i^{-1} = \text{MSE}(\tilde{\theta}_i^H)$, and (iii) $\phi_i^{-1} = W_i^{-1}\text{Cov}(\tilde{\theta}_i^H, \sum_{\ell=1}^m W_\ell \tilde{\theta}_\ell^H)$. These choices were originally proposed in the context of the basic unit level model (see Example 7.3.1).

Datta et al. (2011) provided an alternative, simpler approach to deriving the benchmarked estimators $\tilde{\theta}_i^{\text{WFQ}}$. They minimized the posterior expectation of the weighted squared error loss, $\sum_{i=1}^m \phi_i E[(t_i - \theta_i)^2 | \hat{\theta}]$, with respect to the t_i's subject to $\sum_{i=1}^m W_i t_i = \hat{\theta}_+$, where $\hat{\theta} = (\hat{\theta}_1, \ldots, \hat{\theta}_m)^T$ is the data vector; the t_i's are not restricted to the linear and model-unbiased class. Now, let $\tilde{\theta}_i^B = E(\theta_i | \hat{\theta})$ the Bayes (best) estimator of θ_i depending on the model parameters and $V(\theta_i | \hat{\theta})$ be the posterior variance of θ_i. Noting that

$$E[(t_i - \theta_i)^2 | \hat{\theta}] = V(\theta_i | \hat{\theta}) + (t_i - \tilde{\theta}_i^B)^2, \tag{6.4.38}$$

the problem reduces to minimizing $\sum_{i=1}^m \phi_i (t_i - \tilde{\theta}_i^B)^2$ with respect to the t_i's subject to $\sum_{i=1}^m W_i t_i = \hat{\theta}_+$. Using the Lagrange multiplier method, it readily follows that the "optimal" benchmarked estimator is

$$\tilde{\theta}_i^{\text{OB}} = \tilde{\theta}_i^B + \lambda_i \left(\sum_{\ell=1}^m W_\ell \hat{\theta}_\ell - \sum_{\ell=1}^m W_\ell \tilde{\theta}_\ell^B \right), \tag{6.4.39}$$

where λ_i is given by (6.4.36) (see Theorem 1 of Datta et al. 2011). Replacing the model parameters in (6.4.39) by suitable estimators, we get the empirical optimal benchmarked estimator denoted by $\hat{\theta}_i^{\text{OB}}$. For the special case of model (6.1.1) with normal v_i and e_i, $\hat{\theta}_i^{\text{OB}}$ reduces to $\hat{\theta}_i^{\text{WFQ}}$ noting that $\tilde{\theta}_i^B = \gamma_i \hat{\theta}_i + (1 - \gamma_i)\mathbf{z}_i^T \boldsymbol{\beta}$ and $\hat{\theta}_i^H = \hat{\gamma}_i \hat{\theta}_i + (1 - \hat{\gamma}_i)\mathbf{z}_i^T \hat{\boldsymbol{\beta}}$. Note that the result (6.4.39) is quite general, but requires parametric assumptions. The estimator is also applicable under the HB approach (Chapter 10), specifying priors on the model parameters.

An advantage of the above method is that it readily extends to multiple benchmark constraints (Datta et al. 2011). Suppose there are $q\ (< m)$ benchmark constraints $\mathbf{W}^T \mathbf{t} = \hat{\boldsymbol{\theta}}_+$, where $\mathbf{W}$ is a known $m \times q$ matrix of full rank q and $\hat{\boldsymbol{\theta}}_+ = \mathbf{W}^T \hat{\boldsymbol{\theta}}$. For example, in the U.S. poverty count application (Example 6.1.2), benchmarking to reliable direct estimates at the level of the four U.S. regions may be desired (Bell, Datta, and Ghosh 2013), in which case $q = 4$. Considering a more general loss function $L(\boldsymbol{\theta}, \mathbf{t}) = (\mathbf{t} - \boldsymbol{\theta})^T \Omega (\mathbf{t} - \boldsymbol{\theta})$ for specified positive definite matrix Ω, the desired solution is obtained by minimizing the posterior expectation $E[L(\boldsymbol{\theta}, \mathbf{t}) | \hat{\boldsymbol{\theta}}]$ with respect to $\mathbf{t}$ subject to the constraints $\mathbf{W}^T \mathbf{t} = \hat{\boldsymbol{\theta}}_+$. Now writing

$$E[L(\boldsymbol{\theta}, \mathbf{t}) | \hat{\boldsymbol{\theta}}] = E[(\boldsymbol{\theta} - \tilde{\boldsymbol{\theta}}^B)^T \Omega (\boldsymbol{\theta} - \tilde{\boldsymbol{\theta}}^B) | \hat{\boldsymbol{\theta}}] + (\mathbf{t} - \tilde{\boldsymbol{\theta}}^B)^T \Omega (\mathbf{t} - \tilde{\boldsymbol{\theta}}^B), \tag{6.4.40}$$

where $\tilde{\boldsymbol{\theta}}^B = E(\boldsymbol{\theta} | \hat{\boldsymbol{\theta}})$ is the Bayes (best) estimator of $\boldsymbol{\theta}$ depending on model parameters, it follows from (6.4.40) that the problem is reduced to minimizing $(\mathbf{t} - \tilde{\boldsymbol{\theta}}^B)^T \Omega (\mathbf{t} - \tilde{\boldsymbol{\theta}}^B)$ subject to $\mathbf{W}^T \mathbf{t} = \hat{\boldsymbol{\theta}}_+$. The desired optimal solution is given by

$$\tilde{\boldsymbol{\theta}}^{\text{OMB}} = \tilde{\boldsymbol{\theta}}^B + \Omega^{-1} \mathbf{W} (\mathbf{W}^T \Omega^{-1} \mathbf{W})^{-1} (\hat{\boldsymbol{\theta}}_+ - \mathbf{W}^T \tilde{\boldsymbol{\theta}}^B). \tag{6.4.41}$$

Replacing the model parameters in (6.4.41) by suitable estimators, we get the empirical optimal multiple benchmarked estimator, denoted by $\hat{\boldsymbol{\theta}}^{\text{OMB}}$. This estimator depends on the choice of Ω; a simple choice is $\Omega = \text{diag}(\psi_1^{-1}, \ldots, \psi_m^{-1})$.

Bell, Datta, and Ghosh (2013) also extended the result (6.4.35) of Wang, Fuller, and Qu (2008) to multiple benchmarking constraints. In the class of linear unbiased estimators satisfying the benchmarking constraints, the optimal estimator of $\boldsymbol{\theta}$ turns out to be equal to (6.4.41) with $\tilde{\boldsymbol{\theta}}^B$ replaced by $\tilde{\boldsymbol{\theta}}^H$, the vector of BLUP estimators. Replacing σ_v^2 by $\hat{\sigma}_v^2$ leads to the optimal multiple benchmarked estimator of $\boldsymbol{\theta}$, which under normality is identical to $\hat{\boldsymbol{\theta}}^{\text{OMB}}$.

Two-Stage Benchmarking In some applications, we may have a hierarchical structure consisting of areas and subareas within areas, and it is of interest to estimate both area means and subarea means. For example, in the U.S. poverty counts application (Example 6.1.2), states are areas and counties within states are subareas. Let θ_i denote the mean of area i and $\theta_+ = \sum_{\ell=1}^m W_\ell \theta_\ell$ be the aggregated mean as before. The mean of subarea j within area i is denoted by θ_{ij} ($j = 1, \ldots, N_i$) and $\theta_i = \sum_{j=1}^{N_i} W_{ij}\theta_{ij}$, where W_{ij} is the proportion of units in subarea j belonging to area i ($\sum_{j=1}^{N_i} W_{ij} = 1$). For simplicity, assume that all subareas within an area are sampled and denote the vector with all direct estimates $\hat{\theta}_{ij}$ by $\hat{\boldsymbol{\theta}}$. Furthermore, let $\tilde{\theta}_{ij}^B = E(\theta_{ij}|\hat{\boldsymbol{\theta}})$ and $\tilde{\theta}_i^B = \sum_{j=1}^{N_i} W_{ij}\tilde{\theta}_{ij}^B$ be the best (Bayes) estimators of θ_{ij} and θ_i under an assumed linking model. A simple two-stage ratio adjustment ensures benchmarking at both levels. In stage 1, the estimator $\tilde{\theta}_i^B$ is ratio adjusted to $\hat{\theta}_+ = \sum_{\ell=1}^m W_\ell \hat{\theta}_\ell$ to get $\tilde{\theta}_i^{\text{RB}} = \tilde{\theta}_i^B \left(\sum_{\ell=1}^m W_\ell \hat{\theta}_\ell / \sum_{\ell=1}^m W_\ell \tilde{\theta}_\ell^B\right)$, where $\hat{\theta}_\ell = \sum_{\ell=1}^m W_{\ell j}\hat{\theta}_{\ell j}$ is the direct estimator of θ_ℓ. It now follows that $\sum_{i=1}^m W_i \tilde{\theta}_i^{\text{RB}} = \sum_{\ell=1}^m W_\ell \hat{\theta}_\ell = \hat{\theta}_+$. In stage 2, the subarea estimator $\tilde{\theta}_{ij}^B$ is adjusted to the benchmarked estimate of the area mean $\tilde{\theta}_i^{\text{RB}}$ to get

$$\tilde{\theta}_{ij}^{\text{RB}} = \tilde{\theta}_{ij}^B \frac{\tilde{\theta}_i^{\text{RB}}}{\tilde{\theta}_i^B} = \tilde{\theta}_{ij}^B \frac{\hat{\theta}_+}{\sum_{\ell=1}^m W_\ell \tilde{\theta}_\ell^B}. \tag{6.4.42}$$

Replacing the model parameters in (6.4.42) by suitable estimators leads to the empirical versions $\hat{\theta}_{ij}^{\text{RB}}$ and $\hat{\theta}_i^{\text{RB}}$ of the benchmarked estimators $\tilde{\theta}_{ij}^{\text{RB}}$ and $\tilde{\theta}_i^{\text{RB}}$, respectively.

Ghosh and Šteorts (2013) extended the method of Datta et al. (2011) to two-stage benchmarking by using a composite loss function of the form

$$L(\boldsymbol{\theta}) = \sum_{i=1}^m \sum_{j=1}^{N_i} \xi_{ij}(t_{ij} - \theta_{ij})^2 + \sum_{i=1}^m \phi_i (t_i - \theta_i)^2 \tag{6.4.43}$$

for specified constants ϕ_i and ξ_{ij} and minimized its posterior expectation $E[L(\boldsymbol{\theta})|\hat{\boldsymbol{\theta}}]$ with respect to the t_i's and the t_{ij}'s, subject to the constraints $\sum_{i=1}^m W_i t_i = \hat{\theta}_+$ and $\sum_{j=1}^{N_i} W_{ij} t_{ij} = t_i$. The resulting "optimal" benchmarked estimators $\tilde{\theta}_{ij}^{\text{OB}}$ and $\tilde{\theta}_i^{\text{OB}}$ depend on the constants ϕ_i and ξ_{ij}. Replacing the model parameters in $\tilde{\theta}_{ij}^{\text{OB}}$ and $\tilde{\theta}_i^{\text{OB}}$ by suitable estimators, we obtain the empirical optimal benchmarked estimators $\hat{\theta}_{ij}^{\text{OB}}$ and $\hat{\theta}_i^{\text{OB}}$.

Self-Benchmarking Two alternative methods that do not suffer from limitations (i)–(iii) mentioned before have been proposed. The two methods are self-benchmarking (or self-calibrating) in the sense that adjustments are not made to EBLUP estimators; instead, estimators that automatically satisfy the benchmarking constraints are obtained. Method 1 simply replaces the optimal $\hat{\boldsymbol{\beta}}$ in the EBLUP estimator $\hat{\theta}_i^H$ by an estimator that depends on the benchmarking weights W_ℓ. Method 2, on the other hand, augments the vector $\mathbf{z}_i^T$ of covariates to $(\mathbf{z}_i^T, W_i\psi_i)$ and uses the EBLUP estimator of θ_i based on the augmented model. Method 1 was originally proposed in the context of the basic unit level model (You and Rao 2002a). For the area level model (6.1.1) with $b_i = 1$, Wang, Fuller, and Qu (2008) proposed a similar method. Both methods permit second-order unbiased MSE estimators without requiring additional theory, unlike the adjustment methods. Method 1 uses

$$\tilde{\boldsymbol{\beta}}^{\text{YR}} = \tilde{\boldsymbol{\beta}}^{\text{YR}}(\sigma_v^2) := \left[\sum_{i=1}^m W_i\,(1-\gamma_i)\,\mathbf{z}_i\mathbf{z}_i^T\right]^{-1} \sum_{i=1}^m W_i(1-\gamma_i)\mathbf{z}_i\hat{\theta}_i \tag{6.4.44}$$

and the resulting EBLUP of θ_i is given by

$$\hat{\theta}_i^{YR} = \hat{\gamma}_i\hat{\theta}_i + (1-\hat{\gamma}_i)\mathbf{z}_i^T\hat{\boldsymbol{\beta}}^{\text{YR}}, \tag{6.4.45}$$

where $\hat{\boldsymbol{\beta}}^{\text{YR}} = \tilde{\boldsymbol{\beta}}^{\text{YR}}(\hat{\sigma}_v^2)$. It readily follows from (6.4.44) that $\sum_{i=1}^m W_i(1-\hat{\gamma}_i)(\hat{\theta}_i - \mathbf{z}_i^T\hat{\boldsymbol{\beta}}^{\text{YR}}) = 0$, assuming that the intercept term is included in the model. This in turn implies that $\sum_{i=1}^m W_i\hat{\theta}_i^{YR} = \sum_{i=1}^m W_i\hat{\theta}_i = \hat{\theta}_+$.

Under REML estimation of σ_v^2, a second-order approximation to $\text{MSE}(\hat{\theta}_i^{\text{YR}})$ has the same form as (6.2.1) for $\hat{\theta}_i^H$, excepting that $g_{2i}(\sigma_v^2)$ is changed to

$$g_{2i}^{\text{YR}}(\sigma_v^2) = (1-\gamma_i)^2\mathbf{z}_i^T V(\tilde{\boldsymbol{\beta}}^{\text{YR}})\mathbf{z}_i, \tag{6.4.46}$$

where

$$V(\tilde{\boldsymbol{\beta}}^{\text{YR}}) = \sigma_v^2\left[\sum_{i=1}^m W_i\,(1-\gamma_i)\,\mathbf{z}_i\mathbf{z}_i^T\right]^{-1}\left[\sum_{i=1}^m W_i^2\,(1-\gamma_i)^2\gamma_i^{-1}\mathbf{z}_i\mathbf{z}_i^T\right]$$
$$\times\left[\sum_{i=1}^m W_i\,(1-\gamma_i)\,\mathbf{z}_i\mathbf{z}_i^T\right]^{-1} \tag{6.4.47}$$

(You, Rao, and Hidiroglou 2013). It follows from (6.4.46) that the effect of using the nonoptimal estimator $\hat{\boldsymbol{\beta}}_i^{\text{YR}}$ in place of the optimal $\hat{\boldsymbol{\beta}}$ is to increase the g_{2i} term, but the increase is likely to be small for large m because g_{2i} is $O(m^{-1})$. A second-order unbiased estimator of $\text{MSE}(\hat{\theta}_i^{\text{YR}})$ has the same form as (6.2.3) with $g_{2i}(\hat{\sigma}_v^2)$ changed to $g_{2i}^{\text{YR}}(\hat{\sigma}_v^2)$:

$$\text{mse}(\hat{\theta}_i^{\text{YR}}) \approx g_{1i}(\hat{\sigma}_v^2) + g_{2i}^{\text{YR}}(\hat{\sigma}_v^2) + 2g_{3i}(\hat{\sigma}_v^2). \tag{6.4.48}$$

Turning now to Method 2, the augmented model is obtained by including the covariate $h_i = W_i\psi_i$ in the FH model, that is,

$$\hat{\theta}_i = \mathbf{z}_i^T\boldsymbol{\delta} + h_i\lambda + u_i + e_i =: \eta_i + e_i, \tag{6.4.49}$$

where the random area effects u_i are independent of $E(u_i) = 0$, $V(u_i) = \sigma_u^2$, and u_i is independent of e_i (Wang, Fuller, and Qu 2008). Now the EBLUP of η_i under the augmented model (6.4.49), denoted $\hat{\eta}_i^H = \hat{\theta}_i^{\text{WFQ}}$, is taken as the estimator of θ_i. The estimators $\hat{\theta}_i^{\text{WFQ}}$ satisfy the benchmark constraint: $\sum_{i=1}^m W_i\hat{\theta}_i^{\text{WFQ}} = \sum_{i=1}^m W_i\hat{\theta}_i = \hat{\theta}_+$. Proof of this result follows as a special case of a general result for multiple benchmark constraints (Bell, Datta, and Ghosh 2013).

Assuming normality and using REML estimation, $\text{MSE}(\hat{\eta}_i^H) = E(\hat{\eta}_i^H - \eta_i)^2$ has the same form as (6.2.1) for $\hat{\theta}_i^H$ under model (6.4.49). It is obtained from (6.2.1) by changing $\mathbf{z}_i^T$ to $(\mathbf{z}_i^T, h_i)$, $\hat{\sigma}_v^2$ to $\hat{\sigma}_u^2$ and $\hat{\boldsymbol{\beta}}^T$ to $(\hat{\boldsymbol{\delta}}^T, \hat{\lambda})$. The estimator of $\text{MSE}(\hat{\eta}_i^H)$ is similarly obtained from (6.2.3). Under the true model (6.1.1), $\text{MSE}(\hat{\theta}_i^{\text{WFQ}}) = E(\hat{\theta}_i^{\text{WFQ}} - \theta_i)^2$ will be larger than $\text{MSE}(\hat{\theta}_i^H)$, similarly as $\text{MSE}(\hat{\theta}_i^{\text{YR}})$. However, the MSE estimator $\text{mse}(\hat{\eta}_i^H) = \text{mse}(\hat{\theta}_i^{\text{WFQ}})$ will be second-order unbiased for $\text{MSE}(\hat{\theta}_i^{\text{WFQ}})$ because the true model is a special case of (6.4.49) with $\lambda = 0$.

Simulation results have shown that both $\hat{\theta}_i^{\text{YR}}$ and $\hat{\theta}_i^{\text{WFQ}}$ perform equally well in terms of MSE under the true model and slightly less efficient than $\hat{\theta}_i^H$. However, under gross misspecification of the true model due to omitted variable, MSE is significantly increased for all three estimators, except when the omitted variable is sufficiently correlated with $h_i = W_i\psi_i$, in which case $\hat{\theta}_i^{\text{WFQ}}$ is superior to $\hat{\theta}_i^{\text{YR}}$ and $\hat{\theta}_i^H$ in terms of MSE because the associated augmented model virtually eliminates model misspecification. However, in practice it is difficult to account for model misspecification, because the omitted variables are typically unknown.

As noted above, self-benchmarking does not necessarily protect against misspecification of the linking model. Also, reporting only self-benchmarked estimators of area means will mask possible model failures that may become transparent if unbenchmarked and benchmarked estimators are compared. Adjusted benchmarked estimators have an advantage in this respect, provided that the unadjusted and the adjusted estimators are reported together. It is often claimed that the adjusted estimators provide robustness to model failure, but no formal theory to support this claim exists. However, Pfeffermann et al. (2013) provide some support for this claim in the time series context.

6.4.7 Misspecified Linking Model

The EBLUP estimator $\hat{\theta}_i^H$ with $b_i = 1$ depends on the assumed linking model $\theta_i = \mathbf{z}_i^T\boldsymbol{\beta} + v_i$ with $v_i \stackrel{\text{iid}}{\sim} (0, \sigma_v^2)$. Hence, its efficiency is affected under misspecification of the linking model. Note that the sampling model, $\hat{\theta}_i = \theta_i + e_i$, depends only on the sampling errors e_i with $e_i|\theta_i \stackrel{\text{ind}}{\sim} (0, \psi_i)$. The best estimator of θ_i under normality of v_i and e_i is given by $\tilde{\theta}_i^B = \gamma_i\hat{\theta}_i + (1 - \gamma_i)\mathbf{z}_i^T\boldsymbol{\beta}$ and it depends on the unknown model parameters $\boldsymbol{\beta}$ and σ_v^2 of the assumed linking model. Model parameters are then estimated efficiently under the assumed linking model to arrive at the EBLUP $\hat{\theta}_i^H$. Jiang, Nguyen, and Rao (2011) estimated model parameters $\boldsymbol{\beta}$ and σ_v^2 in $\tilde{\theta}_i^B$ from a predictive point of view, by minimizing the total sampling MSE of $\tilde{\boldsymbol{\theta}}^B = (\tilde{\theta}_1^B, \ldots, \tilde{\theta}_m^B)^T$ with respect to the parameters conditional on $\boldsymbol{\theta} = (\theta_1, \ldots, \theta_m)^T$, in contrast to estimating the parameters under the assumed linking model. The resulting estimators of $\boldsymbol{\beta}$ and σ_v^2 called best predictive estimators (BPEs) are then substituted into $\tilde{\theta}_i^B$ to get observed best predictor (OBP) $\hat{\theta}_i^{\text{OBP}}$ of θ_i. Since the BPEs of $\boldsymbol{\beta}$ and σ_v^2 do not appeal to the assumed linking model, the associated OBPs may be more robust to misspecification of the linking model than the customary EBLUPs $\hat{\theta}_i^H$. Jiang, Nguyen, and Rao (2011) presented examples to demonstrate the superior performance of OBP estimators, in terms of total MSE, under misspecification of the linking model. Moreover, under correct specification of the linking model, OBP and EBLUP estimators perform similarly in terms of total MSE.

The total sampling MSE of $\tilde{\boldsymbol{\theta}}^B$ is given by

$$\text{MSE}_p(\tilde{\boldsymbol{\theta}}^B) = E_p(|\tilde{\boldsymbol{\theta}}^B - \boldsymbol{\theta}|^2) = \sum_{i=1}^{m} E_p(\tilde{\theta}_i^B - \theta_i)^2, \tag{6.4.50}$$

where E_p denotes expectation with respect to the sampling model given the vector of true values $\boldsymbol{\theta} = (\theta_1, \ldots, \theta_m)^T$. Now, noting that $\tilde{\theta}_i^B - \theta_i = e_i - (1-\gamma_i)(\hat{\theta}_i - \mathbf{z}_i^T\beta)$ and $E_p(e_i|\theta_i) = 0$, it follows that

$$\begin{aligned} E_p(\tilde{\theta}_i^B - \theta_i)^2 &= (1-\gamma_i)^2 E_p(\hat{\theta}_i - \mathbf{z}_i^T\boldsymbol{\beta})^2 - 2(1-\gamma_i)\psi_i + \psi_i \\ &= E_p[(1-\gamma_i)^2(\hat{\theta}_i - \mathbf{z}_i^T\boldsymbol{\beta})^2 - 2(1-\gamma_i)\psi_i + \psi_i]. \end{aligned} \tag{6.4.51}$$

Using (6.4.50) and (6.4.51), we obtain

$$\text{MSE}_p(\tilde{\boldsymbol{\theta}}^B) = E_p\left\{\sum_{i=1}^{m}\left[(1-\gamma_i)^2(\hat{\theta}_i - \mathbf{z}_i^T\boldsymbol{\beta})^2 - 2(1-\gamma_i)\psi_i + \psi_i\right]\right\}. \tag{6.4.52}$$

Since $E_p(\hat{\theta}_i - \mathbf{z}_i^T\boldsymbol{\beta})^2 = \psi_i + (\theta_i - \mathbf{z}_i^T\boldsymbol{\beta})^2$. depends on the unknown θ_i, instead of minimizing (6.4.52) which is unknown, the BPE of $(\boldsymbol{\beta}, \sigma_v^2)$ is obtained by minimizing the expression inside the expectation of (6.4.52) or, equivalently, by minimizing with respect to $\boldsymbol{\beta}$ and σ_v^2 the "objective function"

$$Q(\boldsymbol{\beta}, \sigma_v^2) = \sum_{i=1}^{m}[(1-\gamma_i)^2(\hat{\theta}_i - \mathbf{z}_i^T\boldsymbol{\beta})^2 + 2\gamma_i\psi_i]. \tag{6.4.53}$$

Minimizing first (6.4.53) with respect to $\boldsymbol{\beta}$ for a fixed σ_v^2 leads to

$$\tilde{\boldsymbol{\beta}}_{\text{BPE}} = \tilde{\boldsymbol{\beta}}_{\text{BPE}}(\sigma_v^2) = \left[\sum_{i=1}^{m}(1-\gamma_i)^2\mathbf{z}_i\mathbf{z}_i^T\right]^{-1}\sum_{i=1}^{m}(1-\gamma_i)^2\mathbf{z}_i\hat{\theta}_i. \tag{6.4.54}$$

In the second step, $\tilde{\boldsymbol{\beta}}_{\text{BPE}}(\sigma_v^2)$ is substituted into $Q(\boldsymbol{\beta},\sigma_v^2)$ to get a profile objective function $\tilde{Q}(\sigma_v^2) = Q(\tilde{\boldsymbol{\beta}}_{\text{BPE}}(\sigma_v^2),\sigma_v^2)$. Then $\tilde{Q}(\sigma_v^2)$ is minimized with respect to σ_v^2 to get $\tilde{\sigma}^2_{vBPE}$ and in turn $\hat{\boldsymbol{\beta}}_{\text{BPE}} = \tilde{\boldsymbol{\beta}}_{\text{BPE}}(\hat{\sigma}^2_{v\text{BPE}})$, where $\hat{\sigma}^2_{v\text{BPE}} = \max(\tilde{\sigma}^2_{v\text{BPE}}, 0)$. Minimization of $\tilde{Q}(\sigma_v^2)$ does not admit a closed form solution in general, but an iterative method, such as the Fay and Herriot (1979) moment method described in Section 6.1.2, may be used to obtain the desired solution $\tilde{\sigma}^2_{v\text{BPE}}$. The OBP estimator of θ_i is now defined as

$$\hat{\theta}_i^{\text{OBP}} = \hat{\gamma}_i^{\text{BPE}}\hat{\theta}_i + (1-\hat{\gamma}_i^{\text{BPE}})\mathbf{z}_i^T\hat{\boldsymbol{\beta}}_{\text{BPE}}, \tag{6.4.55}$$

where $\hat{\gamma}_i^{\text{BPE}} = \hat{\sigma}^2_{v\text{BPE}}/(\hat{\sigma}^2_{v\text{BPE}} + \psi_i)$. In the special case of equal sampling variances, $\psi_i = \psi$, $\tilde{\boldsymbol{\beta}}_{\text{BPE}}$ is identical to the BLUE $\tilde{\boldsymbol{\beta}}(\sigma_v^2)$ given by (6.1.5) and both reduce to the OLS estimator $\hat{\boldsymbol{\beta}}_{\text{OLS}} = \left(\sum_{i=1}^m \mathbf{z}_i\mathbf{z}_i^T\right)^{-1}\sum_{i=1}^m \mathbf{z}_i\hat{\theta}_i$, noting that $\gamma_i = \gamma$. Furthermore, $\tilde{Q}(\sigma_v^2)$ reduces to

$$\tilde{Q}(\sigma_v^2) = (1-\gamma)^2\sum_{i=1}^{m}(\hat{\theta}_i - \mathbf{z}_i^T\hat{\boldsymbol{\beta}}_{\text{OLS}})^2 + 2m\gamma\psi, \tag{6.4.56}$$

where $\gamma = \sigma_v^2/(\sigma_v^2+\psi)$. Setting $\partial\tilde{Q}(\sigma_v^2)/\partial\sigma_v^2 = 0$ leads to the explicit solution $\tilde{\sigma}^2_{v\text{BPE}}$, which is identical to the ML estimator of σ_v^2 under the assumed linking model

$$\tilde{\sigma}^2_{v\text{BPE}} = \tilde{\sigma}^2_{v\text{ML}} = \frac{1}{m}\sum_{i=1}^{m}(\hat{\theta}_i - \mathbf{z}_i^T\hat{\boldsymbol{\beta}}_{\text{OLS}})^2 - \psi. \tag{6.4.57}$$

The above results show that in the balanced case $\psi_i = \psi$, the OBP estimator of θ_i is identical to the EBLUP estimator of θ_i based on the ML estimator of σ_v^2 under the assumed linking model.

Estimation of area-specific $\text{MSE}(\hat{\theta}_i^{\text{OBP}})$ encounters difficulties because the assumed linking model is misspecified. A way around this difficulty is to estimate the conditional MSE given by $\text{MSE}_p(\hat{\theta}_i^{\text{OBP}}) = E_p(\hat{\theta}_i^{\text{OBP}} - \theta_i)^2$ similar to the estimation of $\text{MSE}_p(\hat{\theta}_i^H)$ studied in Section 6.2.7. Jiang, Nguyen, and Rao (2011) derived a second-order unbiased estimator of $\text{MSE}_p(\hat{\theta}_i^{\text{OBP}})$, denoted $\text{mse}_p(\hat{\theta}_i^{\text{OBP}})$, but it involves the term $(\hat{\theta}_i^{\text{OBP}} - \hat{\theta}_i)^2$, similar to the term $(\hat{\theta}_i - \mathbf{z}_i^T\hat{\boldsymbol{\beta}})^2$ in the MSE estimator $\text{mse}_p(\hat{\theta}_i^H)$ studied in Section 6.2.7. As a result, the proposed estimator of $\text{MSE}_p(\hat{\theta}_i^{\text{OBP}})$ can be highly unstable, as in the case of $\text{mse}_p(\hat{\theta}_i^H)$, although it performs well in terms of RB. Moreover, it can take negative values as in the case of $\text{mse}_p(\hat{\theta}_i^H)$.

Jiang, Nguyen, and Rao (2011) also proposed a bootstrap MSE estimator that is always nonnegative. The bootstrap method involves repeating the following steps (i) and (ii) for $r = 1, \ldots, R$, for R large: (i) Generate $\hat{\theta}_i^{(r)}$ independently from $N(\hat{\theta}_i^{\text{OBP}}, \psi_i)$, $i = 1, \ldots, m$; (ii) Using the bootstrap data $\{(\hat{\theta}_i^{(r)}, \mathbf{z}_i); i = 1, \ldots, m\}$, calculate the corresponding OBP estimator $\{\hat{\theta}_i^{\text{OBP}(r)}; i = 1, \ldots, m\}$. Finally, The bootstrap MSE estimator of $\hat{\theta}_i^{\text{OBP}}$ is given by

$$\text{mse}_B(\hat{\theta}_i^{\text{OBP}}) = \frac{1}{R}\sum_{r=1}^{R}(\hat{\theta}_i^{\text{OBP}(r)} - \hat{\theta}_i^{\text{OBP}})^2. \tag{6.4.58}$$

A limited simulation study indicated that $\text{mse}_B(\hat{\theta}_i^{\text{OBP}})$ does not perform as well as the second-order unbiased MSE estimator $\text{mse}_p(\hat{\theta}_i^{\text{OBP}})$ in terms of RB, but it might remedy the instability problem associated with $\text{mse}_p(\hat{\theta}_i^{\text{OBP}})$ to some extent.

Simulation Study Jiang et al. (2011) reported the results of a small simulation study on the relative performance of EBLUP and OBP estimators in terms of MSE. The true model is a FH model with different means and different sampling variances for each half of the areas, that is, the sampling model is $\hat{\theta}_i = \theta_i + e_i$ with $e_i \stackrel{\text{iid}}{\sim} N(0, \psi_i)$, where

$$\psi_i = \begin{cases} a, & i = 1, \ldots, m_1, \\ b, & i = m_1 + 1, \ldots, m, \end{cases}$$

for $m = 2m_1$, and the true linking model is

$$\theta_i = \begin{cases} \mu_1 + \tilde{v}_i, & i = 1, \ldots, m_1, \\ \mu_2 + \tilde{v}_i, & i = m_1 + 1, \ldots, m, \end{cases}$$

with $\tilde{v}_i \stackrel{\text{iid}}{\sim} N(0, \tilde{\sigma}_v^2)$. The (incorrectly) assumed linking model is a model with constant mean for all areas, that is, $\theta_i = \mu + v_i$, $i = 1, \ldots, m$, for $v_i \stackrel{\text{iid}}{\sim} N(0, \sigma_v^2)$. For each simulation run r $(= 1, \ldots, R)$, for $R = 500$, data $\{\hat{\theta}_i^{(r)}; i = 1, \ldots, m\}$ were generated from the true model and both EBLUP and OBP estimates were computed from the generated data, by taking $\mu_1 = 0$; $\mu_2 = 1$ or 5; $a = 4$; $b = 1$; and $m = 50$, 100, and 200. Total empirical MSE of EBLUP and OBP were computed as $\sum_{i=1}^m R^{-1}\sum_{r=1}^R[\hat{\theta}_i^H(r) - \theta_i(r)]^2$ and $\sum_{i=1}^m R^{-1}\sum_{r=1}^R[\hat{\theta}_i^{\text{OBP}}(r) - \theta_i(r)]^2$, respectively, where $\theta_i(r)$ is the true θ_i for the rth simulation run. Results showed that the total MSE of EBLUP is 11% to 44% higher than the total MSE of OBP, with substantial increase in $\mu_2 = 5$ as compared with $\mu_2 = 1$. The latter feature makes sense because $\mu_2 = 5$ indicates a more serious misspecification of the true model than $\mu_2 = 1$. Jiang et al. (2011) also compared the relative performance of EBLUP and OBP in terms of area-specific MSE. OBP exhibited median MSE smaller than EBLUP in all cases. Furthermore, MSEs of EBLUP for half of the areas (for given number of areas m) are slightly to moderately smaller than the corresponding MSEs of OBP and for the other half of the areas, MSEs of EBLUPs are much larger than those of OBPs. Overall, the performance of OBP is much more robust than the EBLUP in terms of area-specific MSE.

To avoid misspecification problems in the linking model $\theta_i = \mathbf{z}_i^T \boldsymbol{\beta} + v_i$ for the special case of $\mathbf{z}_i = (1, z_i)^T$, Rueda, Menéndez, and Gómez (2010) relaxed the above linking model by assuming that $\theta_i = m_i(\mathbf{z}) + v_i$, where $v_i \overset{\text{iid}}{\sim} N(0, \sigma_v^2)$ and $m_i(\mathbf{z})$ is an unspecified function of $\mathbf{z} = (z_1, \ldots, z_m)^T$ such that $m_i(\mathbf{z})$ increases with z_i. Letting the areas ordered according to z_i (i.e., $z_i = z_{(i)}$), the covariates z_i are used to derive an ordering of the $m_i(\mathbf{z})$: $m_1(\mathbf{z}) \leq \cdots \leq m_m(\mathbf{z})$. An iterative procedure was used to obtain estimators of $m_1(\mathbf{z}), \ldots, m_m(\mathbf{z})$ and σ_v^2 subject to order restriction on $\{m_1(\mathbf{z}), \ldots, m_m(\mathbf{z})\}$. Denoting the resulting estimators as $\hat{m}_{1R}(\mathbf{z}), \ldots, \hat{m}_{mR}(\mathbf{z})$ and $\hat{\sigma}_{vR}^2$, the estimator of the target area parameter θ_i is taken as

$$\hat{\theta}_i^{\text{RMG}} = \hat{\gamma}_{iR}\hat{\theta}_i + (1 - \hat{\gamma}_{iR})\hat{m}_{iR}(\mathbf{z}), \tag{6.4.59}$$

where $\hat{\gamma}_{iR} = \hat{\sigma}_{vR}^2 / (\hat{\sigma}_{vR}^2 + \psi_i)$.

6.5 *SOFTWARE

The R package sae contains two specific functions for the FH area level model. Function `eblupFH` gives the EBLUP estimates (6.1.12) of θ_i, $i = 1, \ldots, m$, based on the FH model (6.1.1), and function mseFH returns the same EBLUP estimates together with MSE estimates, depending on the model fitting method specified in argument method. Both functions admit ML, REML (default), and moments fitting methods. The calls to these functions are as follows:

```
eblupFH(formula, vardir, method = "REML", MAXITER = 100,
   PRECISION = 0.0001, data)
mseFH(formula, vardir, method = "REML", MAXITER = 100,
   PRECISION = 0.0001, data)
```

These two functions require to specify, in the argument `formula`, the regression equation for the fixed (nonrandom) part of the model as a usual R regression formula. The left-hand side of the regression formula is the vector of direct estimates $\hat{\boldsymbol{\theta}} = (\hat{\theta}_1, \ldots, \hat{\theta}_m)^T$, and the right-hand side contains the auxiliary variables included in $\mathbf{z}_i$, separated by + and with an intercept included by default. The vector of sampling variances of direct estimators, ψ_i, must also be specified in the argument `vardir`. The optional arguments `MAXITER` and `PRECISION` might be used, respectively, to specify the maximum number of iterations of the Fisher-scoring fitting algorithm and the relative difference between the model parameter estimates in two consecutive iterations below which the Fisher-scoring algorithm stops.

The function `eblupFH` returns a list of two objects: the vector `eblup` with the EBLUP estimates $\hat{\theta}_i^H$ for the domains, and the list `fit`, containing the results of the fitting method. The function `mseFH` returns a list with two objects: the first object is the list `est` containing the results of the model fit (in another list called `fit`) and the vector of EBLUP estimates (`eblup`). The second object is the vector `mse`, which contains the second-order unbiased MSE estimates. The list `fit` contains the results of the fitting procedure, namely the fitting method (`method`), a logical value indicating if convergence is achieved in the specified maximum number of iterations (`convergence`), the number of iterations (`iterations`), the estimated regression coefficients (`estcoef`), the estimate of the random effects variance σ_v^2 (`refvar`), and the usual goodness-of-fit criteria mAIC, BIC, and loglikelihood (`AIC`, `BIC`, and `loglikelihood`).

Example 6.5.1. Milk Expenditure. We illustrate the use of the `mseFH` function (the function `eblupFH` is used similarly) with the predefined data set milk on fresh milk expenditure. This data set was used originally by Arora and Lahiri (1997) and later by You and Chapman (2006). It contains $m = 43$ observations on the following six variables: `SmallArea` containing the areas of inferential interest, `ni` with the area sample sizes, `yi` with the average expenditure on fresh milk for the year 1989 (direct estimates), `SD` with the estimated standard deviations of direct estimators, `CV` with the estimated coefficients of variation of direct estimators and, finally, `MajorArea` containing major areas created by You and Chapman (2006). We will obtain EBLUP estimates of average area expenditure on fresh milk for 1989 based on the FH model with fixed effects for `MajorArea` categories. We will also calculate second-order unbiased MSE estimates and coefficients of variation (CVs). The gain in efficiency of the EBLUP estimators in comparison with direct estimators will be analyzed.

The following R code loads the `sae` package and the `milk` data set:

```
R> library(sae)
R> data("milk")
R> attach(milk)
```

Next, we call the function `mseFH`. Direct estimates (left-hand side of `formula`) are given by `yi` and sampling variances of direct estimators (argument `vardir`) are given in this case by the square of `SD`. As auxiliary variables (right-hand side of `formula`), we consider the categories of the factor defined by `MajorArea`. We use the default `REML` fitting method. The code for calculating EBLUP estimates as in (6.1.12) and their corresponding second-order unbiased MSE estimates according to (6.2.3) is as follows:

```
R> FH <- mseFH(yi ~ as.factor(MajorArea), SD^2)
```

With the previous sentence, the output of the function `mseFH` has been placed in the object called `FH`. Then, `FH$est$eblup` gives the vector of EBLUP estimates and `FH$mse` gives the MSE estimates. We calculate coefficients of variation (CVs) as square root of MSE estimates divided by the point estimates, in percentage.

```
R> cv.FH <- 100 * sqrt(FH$mse) / FH$est$eblup R> detach(milk)
```

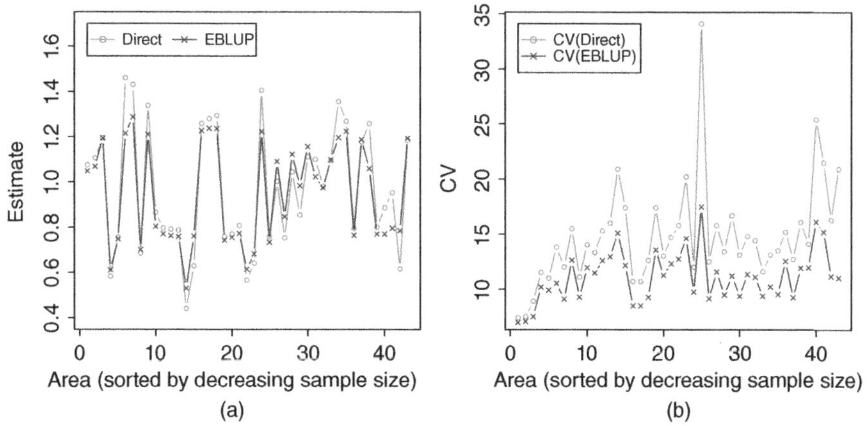

Figure 6.1 EBLUP and Direct Area Estimates of Average Expenditure on Fresh Milk for Each Small Area (a). CVs of EBLUP and Direct Estimators for Each Small Area (b). Areas are Sorted by Decreasing Sample Size.

EBLUP and direct area estimates of average expenditure are plotted for each small area in Figure 6.1a. CVs of these estimates are plotted in Figure 6.1b. In both plots, small areas have been sorted by decreasing sample size. Observe in Figure 6.1a that EBLUP estimates track direct estimates but seem to be more stable. See also in Figure 6.1b that CVs of EBLUP estimates are smaller than those of direct estimates for all the small areas. In fact, direct estimates have CVs over 20% for several areas, whereas the CVs of the EBLUP estimates do not exceed this limit for any of the areas. Moreover, the gains in efficiency of the EBLUP estimators tend to be larger for areas with smaller sample sizes (those in Figure 6.1b). Thus, in this example, EBLUP estimates based on FH model seem to be more reliable than direct estimates.

7

BASIC UNIT LEVEL MODEL

In this chapter, we study the basic unit level model (4.3.1) and spell out EBLUP inference for small area means. Section 7.1 gives EBLUP estimators, and MSE estimation is covered in Section 7.2. Several practical issues associated with the unit level model are studied in Section 7.6, and methods that address those issues are presented.

7.1 EBLUP ESTIMATION

In this section, we consider the basic unit level model (4.3.1) and spell out EBLUP estimation, using the results in Section 5.3, Chapter 5, for the general linear mixed model with a block-diagonal covariance structure. We assume that all the areas are sampled ($m = M$), and that the sampling within areas is noninformative. In this case, we can use the sample part of the population model (4.3.1) to make inference on the population mean $\overline{Y}_i = N_i^{-1} \sum_{j=1}^{N_i} y_{ij}$. Considering without loss of generality that the sample units from area i are the first n_i units, for $i = 1, \ldots, m$, the sample part of the population model (4.3.1) is given by

$$y_{ij} = \mathbf{x}_{ij}^T \boldsymbol{\beta} + v_i + e_{ij}, \quad j = 1, \ldots, n_i, \quad i = 1, \ldots, m, \tag{7.1.1}$$

where $v_i \overset{\text{iid}}{\sim} (0, \sigma_v^2)$ and $e_{ij} \overset{\text{ind}}{\sim} (0, k_{ij}^2 \sigma_e^2)$, random effects v_i are independent of sampling errors e_{ij} and $\sum_{i=1}^m n_i > p$. This model may be written in matrix notation as

$$\mathbf{y}_i = \mathbf{X}_i \boldsymbol{\beta} + v_i \mathbf{1}_{n_i} + \mathbf{e}_i, \quad i = 1, \ldots, m, \tag{7.1.2}$$

where $\mathbf{y}_i = (y_{i1}, \ldots, y_{in_i})^T, \mathbf{X}_i = (\mathbf{x}_{i1}, \ldots, \mathbf{x}_{in_i})^T$, and $\mathbf{e}_i = (\mathbf{e}_{i1}, \ldots, \mathbf{e}_{in_i})^T$.

Small Area Estimation, Second Edition. J. N. K. Rao and Isabel Molina.

7.1.1 BLUP Estimator

Model (7.1.2) is a special case of the general linear model with block-diagonal structure, given by (5.3.1), by setting

$$\mathbf{y}_i = \mathbf{y}_i, \quad \mathbf{X}_i = \mathbf{X}_i, \quad \mathbf{Z}_i = \mathbf{1}_{n_i},$$
$$\mathbf{v}_i = v_i, \quad \mathbf{e}_i = \mathbf{e}_i, \quad \boldsymbol{\beta} = (\beta_1, \ldots, \beta_p)^T,$$

where $\mathbf{y}_i$ is the $n_i \times 1$ vector of sample observations y_{ij} from the ith area. Furthermore,

$$\mathbf{G}_i = \sigma_v^2, \quad \mathbf{R}_i = \sigma_e^2 \text{diag}_{1\le j \le n_i}(k_{ij}^2)$$

so that

$$\mathbf{V}_i = \mathbf{R}_i + \sigma_v^2 \mathbf{1}_{n_i} \mathbf{1}_{n_i}^T.$$

The matrix $\mathbf{V}_i$ can be inverted explicitly. Using the following standard result on matrix inversion:

$$(\mathbf{A} + \mathbf{u}\mathbf{v}^T)^{-1} = \mathbf{A}^{-1} - \mathbf{A}^{-1}\mathbf{u}\mathbf{v}^T\mathbf{A}^{-1}/(1 + \mathbf{v}^T\mathbf{A}^{-1}\mathbf{u}) \tag{7.1.3}$$

and denoting

$$a_{ij} = k_{ij}^{-2}, \quad a_{i\cdot} = \sum_{j=1}^{n_i} a_{ij}, \quad \mathbf{a}_i = (a_{i1}, \ldots, a_{in_i})^T,$$

the inverse is given by

$$\mathbf{V}_i^{-1} = \frac{1}{\sigma_e^2}\left[\text{diag}_{1\le j\le n_i}(a_{ij}) - \frac{\gamma_i}{a_{i\cdot}}\mathbf{a}_i\mathbf{a}_i^T\right], \tag{7.1.4}$$

where

$$\gamma_i = \sigma_v^2/(\sigma_v^2 + \sigma_e^2/a_{i\cdot}). \tag{7.1.5}$$

The target parameters are the small area means $\overline{Y}_i, i = 1, \ldots, m$, which by the model assumptions are given by $\overline{Y}_i = \overline{\mathbf{X}}_i^T\boldsymbol{\beta} + v_i + \overline{E}_i, i = 1, \ldots, m$. For N_i large, by the Strong Law of Large Numbers, $\overline{E}_i$ tends almost surely to the expected value $E(e_{ij}) = 0$. According to this result, instead of focusing on $\overline{Y}_i$, Battese et al. (1988) considered as target parameter the ith area mean $\mu_i = \overline{\mathbf{X}}_i^T\boldsymbol{\beta} + v_i$, which is the mean of the conditional expectations $E(y_{ij}|v_i, \mathbf{x}_{ij}) = \mathbf{x}_{ij}^T\boldsymbol{\beta} + v_i$, where $\overline{\mathbf{X}}_i$ is the vector of known means $\overline{X}_{i1}, \ldots, \overline{X}_{ip}$. Moreover, the EBLUP estimator of $\overline{Y}_i$ approaches the EBLUP estimator of μ_i if the sampling fraction, $f_i = n_i/N_i$, is negligible.

In this section and in Section 7.1.2, we focus on the estimation of μ_i. Section 7.1.3 studies the case of $\overline{Y}_i$ when f_i is not negligible. The issue of informative sampling within areas and informative sampling of areas (when $m < M$) will be addressed in Section 7.6.3.

Note that the target parameter $\mu_i = \overline{\mathbf{X}}_i^T \boldsymbol{\beta} + v_i$ is a special case of the general area parameter $\mathbf{l}_i^T \boldsymbol{\beta} + \mathbf{m}_i^T \mathbf{v}$ taking $\mathbf{l}_i = \overline{\mathbf{X}}_i$ and $\mathbf{m}_i = 1$. Making these substitutions in the general formula (5.3.2) and noting that $(\sigma_v^2/\sigma_e^2)(1-\gamma_i) = \gamma_i/a_{i\cdot}$, we get the BLUP estimator of μ_i as

$$\tilde{\mu}_i^H = \overline{\mathbf{X}}_i^T \tilde{\boldsymbol{\beta}} + \tilde{v}_i, \quad \tilde{v}_i = \gamma_i(\overline{y}_{ia} - \overline{\mathbf{x}}_{ia}^T \tilde{\boldsymbol{\beta}}). \tag{7.1.6}$$

Here $\overline{y}_{ia}$ and $\overline{\mathbf{x}}_{ia}$ are weighted means given by

$$\overline{y}_{ia} = \sum_{j=1}^{n_i} a_{ij} y_{ij}/a_{i\cdot}, \quad \overline{\mathbf{x}}_{ia} = \sum_{j=1}^{n_i} a_{ij} \mathbf{x}_{ij}/a_{i\cdot}, \tag{7.1.7}$$

and

$$\tilde{\boldsymbol{\beta}} = \left(\sum_{i=1}^{m} \mathbf{X}_i^T \mathbf{V}_i^{-1} \mathbf{X}_i \right)^{-1} \left(\sum_{i=1}^{m} \mathbf{X}_i^T \mathbf{V}_i^{-1} \mathbf{y}_i \right) \tag{7.1.8}$$

is the BLUE of $\boldsymbol{\beta}$, where

$$\mathbf{A}_i := \mathbf{X}_i^T \mathbf{V}_i^{-1} \mathbf{X}_i = \sigma_e^{-2} \left(\sum_{j=1}^{n_i} a_{ij} \mathbf{x}_{ij} \mathbf{x}_{ij}^T - \gamma_i a_{i\cdot} \overline{\mathbf{x}}_{ia} \overline{\mathbf{x}}_{ia}^T \right) \tag{7.1.9}$$

and

$$\mathbf{X}_i^T \mathbf{V}_i^{-1} \mathbf{y}_i = \sigma_e^{-2} \left(\sum_{j=1}^{n_i} a_{ij} \mathbf{x}_{ij} y_{ij} - \gamma_i a_{i\cdot} \overline{\mathbf{x}}_{ia} \overline{y}_{ia} \right). \tag{7.1.10}$$

Note that the computation of $\tilde{\boldsymbol{\beta}}$ is reduced to the inversion of the $p \times p$ matrix $\sum_{i=1}^m \mathbf{A}_i$, where $\mathbf{A}_i$ is given in (7.1.9).

The BLUP estimator (7.1.6) can also be expressed as a weighted average of the "survey regression" estimator $\overline{y}_{ia} + (\overline{\mathbf{X}}_i - \overline{\mathbf{x}}_{ia})^T \tilde{\boldsymbol{\beta}}$ and the regression-synthetic estimator $\overline{\mathbf{X}}_i^T \tilde{\boldsymbol{\beta}}$:

$$\tilde{\mu}_i^H = \gamma_i [\overline{y}_{ia} + (\overline{\mathbf{X}}_i - \overline{\mathbf{x}}_{ia})^T \tilde{\boldsymbol{\beta}}] + (1-\gamma_i) \overline{\mathbf{X}}_i^T \tilde{\boldsymbol{\beta}}. \tag{7.1.11}$$

The weight $\gamma_i = \sigma_v^2/(\sigma_v^2 + \sigma_e^2/a_{i\cdot}) \in (0,1)$ measures the unexplained between-area variability, σ_v^2, relative to the total variability $\sigma_v^2 + \sigma_e^2/a_{i\cdot}$. If the unexplained between-area variability σ_v^2 is relatively small, then γ_i will be small and more weight is attached to the synthetic component. On the contrary, more weight is attached to the survey regression estimator as $a_{i\cdot}$ increases. Note that $a_{i\cdot}$ is $O(n_i)$ and it reduces to n_i if $k_{ij} = 1$ for all (i,j). Thus, in that case, more weight is attached to the survey regression estimator when the area sample size n_i is large.

Note also that when $k_{ij} = 1$ for all (i,j), the survey regression estimator is approximately design-unbiased for μ_i under simple random sampling (SRS), provided the total sample size $n := \sum_{i=1}^m n_i$ is large. In the case of general k_{ij}'s, it is model-unbiased for μ_i conditional on the realized area effect v_i, provided $\tilde{\boldsymbol{\beta}}$ is conditionally unbiased for $\boldsymbol{\beta}$. On the other hand, the BLUP estimator (7.1.11) is conditionally biased due to the presence of the synthetic component $\overline{\mathbf{X}}_i^T \tilde{\boldsymbol{\beta}}$. The survey regression estimator is therefore valid under weaker assumptions, but it does not increase the "effective" sample size as illustrated in Section 2.5, Chapter 2.

Under SRS and $k_{ij} = 1$ for all (i,j), the BLUP estimator is design-consistent for $\overline{Y}_i$ as n_i increases because $\gamma_i \to 1$. For general designs with unequal survey weights w_{ij}, we consider survey-weighted pseudo-BLUP estimators that are design-consistent (see Section 7.6.2).

The MSE of the BLUP estimator is easily obtained either directly or from the general result (5.3.5) by letting $\boldsymbol{\delta} = (\sigma_v^2, \sigma_e^2)^T$. It is given by

$$\text{MSE}(\tilde{\mu}_i^H) = E(\tilde{\mu}_i^H - \mu_i)^2 = g_{1i}(\sigma_v^2, \sigma_e^2) + g_{2i}(\sigma_v^2, \sigma_e^2), \tag{7.1.12}$$

where

$$g_{1i}(\sigma_v^2, \sigma_e^2) = \gamma_i(\sigma_e^2 / a_{i\cdot}) \tag{7.1.13}$$

and

$$g_{2i}(\sigma_v^2, \sigma_e^2) = (\overline{\mathbf{X}}_i - \gamma_i \overline{\mathbf{x}}_{ia})^T \left(\sum_{i=1}^m \mathbf{A}_i \right)^{-1} (\overline{\mathbf{X}}_i - \gamma_i \overline{\mathbf{x}}_{ia}) \tag{7.1.14}$$

with $\mathbf{A}_i$ given by (7.1.9). The first term, $g_{1i}(\sigma_v^2, \sigma_e^2)$, is due to prediction of the area effect v_i and is $O(1)$, whereas the second term, $g_{2i}(\sigma_v^2, \sigma_e^2)$, is due to estimating $\boldsymbol{\beta}$ and is $O(m^{-1})$ for large m, under the following regularity conditions:

(i) k_{ij} and n_i are uniformly bounded.
(ii) Elements of $\mathbf{X}_i$ are uniformly bounded such that $\mathbf{A}_i$ is $[O(1)]_{p \times p}$.

We have already mentioned that the leading term of the MSE of the BLUP estimator is $g_{1i}(\sigma_v^2, \sigma_e^2) = \gamma_i(\sigma_e^2 / a_{i\cdot})$. Comparing this term with the leading term of the MSE of the sample regression estimator, $\sigma_e^2 / a_{i\cdot}$, it is clear that the BLUP estimator provides considerable gain in efficiency over the sample regression estimator if γ_i is small. Therefore, models that achieve smaller values of γ_i should be preferred, provided they give an adequate fit in terms of residual analysis and other model diagnostics. This is similar to the model choice in Example 6.1.1 for the basic area level model.

The BLUE $\tilde{\boldsymbol{\beta}}$ and its covariance matrix $(\sum_{i=1}^m \mathbf{X}_i^T \mathbf{V}_i^{-1} \mathbf{X}_i)^{-1}$ can be calculated using only ordinary least squares (OLS) by first transforming the model (7.1.2) with correlated errors $u_{ij} = v_i + e_{ij}$ to a model with uncorrelated errors u_{ij}^*. The transformed model is given by

$$k_{ij}^{-1}(y_{ij} - \tau_i \overline{y}_{ia}) = k_{ij}^{-1}(\mathbf{x}_{ij} - \tau_i \overline{\mathbf{x}}_{ia})^T \boldsymbol{\beta} + u_{ij}^*, \tag{7.1.15}$$

where $\tau_i = 1 - (1 - \gamma_i)^{1/2}$ and the new errors u_{ij}^* have mean zero and constant variance σ_e^2 (Stukel and Rao 1997). If $k_{ij} = 1$ for all (i,j), then (7.1.15) reduces to the transformed model of Fuller and Battese (1973). Another advantage of the transformation method is that standard OLS model diagnostics may be applied to the transformed data $\{k_{ij}^{-1}(y_{ij} - \tau_i \bar{y}_{ia}), k_{ij}^{-1}(\mathbf{x}_{ij} - \tau_i \bar{\mathbf{x}}_{ia})\}$ to check the validity of the nested error regression model (7.1.2) (see Example 7.3.1). In practice, τ_i is estimated from the data (see Section 7.1.2). Meza and Lahiri (2005) demonstrated that the standard C_p statistic based on the original data $(y_{ij}, \mathbf{x}_{ij})$ is inefficient for variable selection when σ_v^2 is large because it ignores the within-area correlation structure. On the other hand, the transformed data lead to uncorrelated errors, and therefore the C_p statistic based on the transformed data leads to efficient model selection, as in the case of standard regression models with independent errors.

The BLUP estimator (7.1.11) depends on the variance ratio σ_v^2/σ_e^2, which is unknown in practice. Replacing σ_v^2 and σ_e^2 by estimators $\hat{\sigma}_v^2$ and $\hat{\sigma}_e^2$, we obtain an EBLUP estimator

$$\hat{\mu}_i^H = \hat{\gamma}_i[\bar{y}_{ia} + (\overline{\mathbf{X}}_i - \bar{\mathbf{x}}_{ia})^T \hat{\boldsymbol{\beta}}] + (1 - \hat{\gamma}_i)\overline{\mathbf{X}}_i^T \hat{\boldsymbol{\beta}}, \tag{7.1.16}$$

where $\hat{\gamma}_i$ and $\hat{\boldsymbol{\beta}}$ are the values of γ_i and $\tilde{\boldsymbol{\beta}}$ when (σ_v^2, σ_e^2) is replaced by $(\hat{\sigma}_v^2, \hat{\sigma}_e^2)$.

7.1.2 Estimation of σ_v^2 and σ_e^2

We present a simple method of estimating the variance components σ_v^2 and σ_e^2. It involves performing two ordinary least squares regressions and then using the method of moments to get unbiased estimators of σ_e^2 and σ_v^2 (Stukel and Rao 1997). Fuller and Battese (1973) proposed this method for the special case $k_{ij} = 1$ for all (i, j).

We first calculate the residual sum of squares SSE(1) with ν_1 degrees of freedom by regressing through the origin the y-deviations $k_{ij}^{-1}(y_{ij} - \bar{y}_{ia})$ on the nonzero $\mathbf{x}$-deviations $k_{ij}^{-1}(\mathbf{x}_{ij} - \bar{\mathbf{x}}_{ia})$ for those areas with sample size $n_i > 1$. This leads to the unbiased estimator of σ_e^2 given by

$$\hat{\sigma}_{em}^2 = \nu_1^{-1}\text{SSE}(1), \tag{7.1.17}$$

where $\nu_1 = n - m - p_1$ and p_1 is the number of nonzero $\mathbf{x}$-deviations. We next calculate the residual sum of squares SSE(2) by regressing y_{ij}/k_{ij} on $\mathbf{x}_{ij}/k_{ij}$. Noting that

$$E[\text{SSE}(2)] = \eta_1 \sigma_v^2 + (n - p)\sigma_e^2,$$

where

$$\eta_1 = \sum_{i=1}^{m} a_{i\cdot}\left[1 - a_{i\cdot}\bar{\mathbf{x}}_{ia}^T \left(\sum_{i=1}^{m}\sum_{j=1}^{n_i} a_{ij}\mathbf{x}_{ij}\mathbf{x}_{ij}^T\right)^{-1} \bar{\mathbf{x}}_{ia}\right], \tag{7.1.18}$$

an unbiased estimator of σ_v^2 is given by

$$\tilde{\sigma}_{vm}^2 = \eta_1^{-1}[\text{SSE}(2) - (n - p)\hat{\sigma}_{em}^2]. \tag{7.1.19}$$

The estimators $\tilde{\sigma}_{vm}^2$ and $\hat{\sigma}_{em}^2$ are equivalent to those found by using the well-known "fitting-of-constants" method attributed to Henderson (1953). However, the latter method requires OLS regression on $p_1 + m$ variables, in contrast to p_1 variables for the transformation method, and thus gets computationally slower as the number of small areas, m, increases.

Since $\tilde{\sigma}_{vm}^2$ can take negative values, we truncate it to zero when it is negative. The truncated estimator $\hat{\sigma}_{vm}^2 = \max(\tilde{\sigma}_{vm}^2, 0)$ is no longer unbiased, but it is consistent as m increases. For the special case of $k_{ij} = 1$ for all (i,j), Battese, Harter, and Fuller (1988) proposed an alternative estimator of γ_i that is approximately unbiased for γ_i.

Assuming normality of the errors v_i and e_{ij}, ML or REML can also be employed. For example, PROC MIXED in SAS can be used to calculate ML or REML estimates of σ_v^2 and σ_e^2 and the associated EBLUP estimate $\hat{\mu}_i^H$. The naive MSE estimator,

$$\text{mse}_N(\hat{\mu}_i^H) = g_{1i}(\hat{\sigma}_v^2, \hat{\sigma}_e^2) + g_{2i}(\hat{\sigma}_v^2, \hat{\sigma}_e^2), \tag{7.1.20}$$

can also be computed using PROC MIXED in SAS, where $(\hat{\sigma}_v^2, \hat{\sigma}_e^2)$ are the ML or REML estimators of (σ_v^2, σ_e^2).

7.1.3 *Nonnegligible Sampling Fractions

We now turn to EBLUP estimation of the finite population area mean $\overline{Y}_i$ when the sampling fraction $f_i = n_i/N_i$ is not negligible. We write $\overline{Y}_i$ as

$$\overline{Y}_i = f_i\overline{y}_i + (1 - f_i)\overline{y}_{ir},$$

where $\overline{y}_i$ is the sample mean and $\overline{y}_{ir}$ is the mean of the nonsampled values $y_{i\ell}, \ell = n_i + 1, \ldots, N_i$, of the ith area. Under the population model (4.3.4), we replace the unobserved $y_{i\ell}$ by its estimator $\mathbf{x}_{i\ell}^T\tilde{\boldsymbol{\beta}} + \tilde{v}_i$, where $\mathbf{x}_{i\ell}$ is the $\mathbf{x}$-value associated with $y_{i\ell}$. The resulting BLUP estimator of $\overline{Y}_i$ is given by

$$\tilde{\overline{Y}}_i^H = f_i\overline{y}_i + (1 - f_i)(\overline{\mathbf{x}}_{ir}^T\tilde{\boldsymbol{\beta}} + \tilde{v}_i) = f_i\overline{y}_i + (1 - f_i)\tilde{\overline{y}}_{ir}^H, \tag{7.1.21}$$

where $\overline{\mathbf{x}}_{ir}$ is the mean of nonsampled values $\mathbf{x}_{i\ell}, \ell = n_i + 1, \ldots, N_i$ from area i, and

$$\tilde{\overline{y}}_{ir}^H = \gamma_i[\overline{y}_{ia} + (\overline{\mathbf{x}}_{ir} - \overline{\mathbf{x}}_{ia})^T)\tilde{\boldsymbol{\beta}}] + (1 - \gamma_i)\overline{\mathbf{x}}_{ir}^T\tilde{\boldsymbol{\beta}}. \tag{7.1.22}$$

The BLUP property of $\tilde{\overline{Y}}_i^H$ is easily established by showing that Cov($\mathbf{b}^T\mathbf{y}$, $\tilde{\overline{Y}}_i^H - \overline{Y}_i) = 0$ for every zero linear function $\mathbf{b}^T\mathbf{y}$, that is, for every $\mathbf{b} \neq \mathbf{0}$ with $E(\mathbf{b}^T\mathbf{y}) = 0$ (Stukel 1991, Chapter 2). The BLUP property of $\tilde{\overline{Y}}_i^H$ also follows from the general results of Royall (1976).

Replacing σ_v^2 and σ_e^2 by estimators $\hat{\sigma}_v^2$ and $\hat{\sigma}_e^2$ in $\tilde{\overline{y}}_{ir}^H$ given in (7.1.22), we obtain an empirical version $\hat{\overline{y}}_{ir}^H$ of $\tilde{\overline{y}}_{ir}^H$, and an EBLUP estimator of $\overline{Y}_i$ is obtained by substituting $\hat{\overline{y}}_{ir}^H$ for $\tilde{\overline{y}}_{ir}^H$ in (7.1.21) :

$$\hat{\overline{Y}}_i^H = f_i\overline{y}_i + (1 - f_i)\hat{\overline{y}}_{ir}^H. \tag{7.1.23}$$

If $k_{ij} = 1$ for all (i,j), the estimator $\hat{\overline{Y}}_i^H$ may be also expressed as

$$\hat{\overline{Y}}_i^H = \overline{\mathbf{X}}_i^T \hat{\boldsymbol{\beta}} + \{f_i + (1 - f_i)\gamma_i\}(\overline{y}_i - \overline{\mathbf{x}}_i^T \hat{\boldsymbol{\beta}}), \tag{7.1.24}$$

noting that $\overline{\mathbf{x}}_{ir} = (N_i\overline{\mathbf{X}}_i - n_i\overline{\mathbf{x}}_i)/(N_i - n_i)$. It follows from (7.1.24) that the EBLUP estimator of $\overline{Y}_i$ can be computed without the knowledge of the membership of each population unit (i,j) to the sampled or nonsampled parts from ith area. It is also clear from (7.1.24) that $\hat{\overline{Y}}_i^H$ approaches $\hat{\mu}_i^H$ given in (7.1.6) when the sampling fraction, f_i, is negligible. In practice, sampling fractions f_i are often small for most of the areas.

7.2 MSE ESTIMATION

7.2.1 Unconditional MSE of EBLUP

Consider the EBLUP estimator (7.1.16) of the parameter $\mu_i = \overline{\mathbf{X}}_i^T \boldsymbol{\beta} + v_i$ under the nested error linear regression model (7.1.2). Under regularity conditions (i) and (ii) and normality of the errors v_i and e_{ij}, a second-order approximation of the MSE is obtained by applying (5.3.9). For the mentioned model and parameters, this approximation reduces to

$$\text{MSE}(\hat{\mu}_i^H) \approx g_{1i}(\sigma_v^2, \sigma_e^2) + g_{2i}(\sigma_v^2, \sigma_e^2) + g_{3i}(\sigma_v^2, \sigma_e^2), \tag{7.2.1}$$

where $g_{1i}(\sigma_v^2, \sigma_e^2)$ and $g_{2i}(\sigma_v^2, \sigma_e^2)$ are given by (7.1.13) and (7.1.14), respectively. Furthermore,

$$g_{3i}(\sigma_v^2, \sigma_e^2) = a_{i\cdot}^{-2}(\sigma_v^2 + \sigma_e^2/a_{i\cdot})^{-3} h(\sigma_v^2, \sigma_e^2) \tag{7.2.2}$$

with

$$h(\sigma_v^2, \sigma_e^2) = \sigma_e^4 \overline{V}_{vv}(\boldsymbol{\delta}) + \sigma_v^4 \overline{V}_{ee}(\boldsymbol{\delta}) - 2\sigma_e^2\sigma_v^2 \overline{V}_{ve}(\boldsymbol{\delta}), \tag{7.2.3}$$

where $\boldsymbol{\delta} = (\sigma_v^2, \sigma_e^2)^T$, $\overline{V}_{ee}(\boldsymbol{\delta})$, and $\overline{V}_{vv}(\boldsymbol{\delta})$ are, respectively, the asymptotic variances of the estimators $\hat{\sigma}_e^2$ and $\hat{\sigma}_v^2$ and $\overline{V}_{ve}(\boldsymbol{\delta})$ is the asymptotic covariance of $\hat{\sigma}_v^2$ and $\hat{\sigma}_e^2$.

If the method of fitting-of-constants is used, then $\hat{\boldsymbol{\delta}} = (\hat{\sigma}_v^2, \hat{\sigma}_e^2)^T = (\hat{\sigma}_{vm}^2, \hat{\sigma}_{em}^2)^T$ with asymptotic variances

$$\overline{V}_{vvm}(\boldsymbol{\delta}) = 2\eta_1^{-2}[\nu_1^{-1}(n - p - \nu_1)(n - p)\sigma_e^4 + \eta_2\sigma_v^4 + 2\eta_1\sigma_e^2\sigma_v^2], \tag{7.2.4}$$

$$\overline{V}_{eem}(\boldsymbol{\delta}) = 2\nu_1^{-1}\sigma_e^4, \tag{7.2.5}$$

and asymptotic covariance

$$\overline{V}_{evm}(\boldsymbol{\delta}) = \overline{V}_{vem}(\boldsymbol{\delta}) = -2\eta_1^{-1}\nu_1^{-1}(n - p - \nu_1)\sigma_e^4, \tag{7.2.6}$$

where η_1 is given by (7.1.18), ν_1 is defined below (7.1.17) and

$$\eta_2 = \sum_{i=1}^{m} a_{i\cdot}^2(1 - 2a_{i\cdot}\bar{\mathbf{x}}_{ia}^T\mathbf{A}_1^{-1}\bar{\mathbf{x}}_{ia}) + \mathrm{tr}\left[\left(\mathbf{A}_1^{-1}\sum_{i=1}^{m} a_{i\cdot}^2\bar{\mathbf{x}}_{ia}\bar{\mathbf{x}}_{ia}^T\right)^2\right] \tag{7.2.7}$$

with $\mathbf{A}_1 = \sum_{i=1}^{m}\sum_{j=1}^{n_i} a_{ij}\mathbf{x}_{ij}\mathbf{x}_{ij}^T$ (Stukel 1991, Chapter 2). If the errors e_{ij} have equal variance σ_e^2 (i.e., with $k_{ij} = 1$ and $a_{ij} = k_{ij}^{-2} = 1$), then (7.2.4)–(7.2.6) reduce to the formulae given by Prasad and Rao (1990).

If ML or REML is used to estimate σ_v^2 and σ_e^2, then $\hat{\boldsymbol{\delta}} = \hat{\boldsymbol{\delta}}_{\mathrm{ML}}$ or $\hat{\boldsymbol{\delta}} = \hat{\boldsymbol{\delta}}_{\mathrm{RE}}$ with asymptotic covariance matrix $\overline{V}(\boldsymbol{\delta}) = \overline{V}_{\mathrm{ML}}(\boldsymbol{\delta}) = \overline{V}_{\mathrm{RE}}(\boldsymbol{\delta})$ given by the inverse of the 2×2 information matrix $\mathcal{I}(\boldsymbol{\delta})$ with diagonal elements

$$\mathcal{I}_{vv}(\boldsymbol{\delta}) = \frac{1}{2}\sum_{i=1}^{m} \mathrm{tr}[(\mathbf{V}_i^{-1}\partial\mathbf{V}_i/\partial\sigma_v^2)^2],$$

$$\mathcal{I}_{ee}(\boldsymbol{\delta}) = \frac{1}{2}\sum_{i=1}^{m} \mathrm{tr}[(\mathbf{V}_i^{-1}\partial\mathbf{V}_i/\partial\sigma_e^2)^2]$$

and off-diagonal elements

$$\mathcal{I}_{ve}(\boldsymbol{\delta}) = \mathcal{I}_{ev}(\boldsymbol{\delta}) = \frac{1}{2}\sum_{i=1}^{m} \mathrm{tr}[\mathbf{V}_i^{-1}(\partial\mathbf{V}_i/\partial\sigma_v^2)\mathbf{V}_i^{-1}(\partial\mathbf{V}_i/\partial\sigma_e^2)],$$

where

$$\partial\mathbf{V}_i/\partial\sigma_v^2 = \mathbf{1}_{n_i}\mathbf{1}_{n_i}^T; \quad \partial\mathbf{V}_i/\partial\sigma_e^2 = \sigma_e^{-2}\mathbf{R}_i.$$

Using the formula (7.1.4) for $\mathbf{V}_i^{-1}$ we get, after simplification,

$$\mathcal{I}_{vv}(\boldsymbol{\delta}) = \frac{1}{2}\sum_{i=1}^{m} a_{i\cdot}^2\alpha_i^{-2}, \tag{7.2.8}$$

$$\mathcal{I}_{ee}(\boldsymbol{\delta}) = \frac{1}{2}\sum_{i=1}^{m}[(n_i - 1)\sigma_e^{-4} + \alpha_i^{-2}] \tag{7.2.9}$$

and

$$\mathcal{I}_{ve}(\boldsymbol{\delta}) = \frac{1}{2}\sum_{i=1}^{m} a_{i\cdot}\alpha_i^{-2}, \tag{7.2.10}$$

where

$$\alpha_i = \sigma_e^2 + a_{i\cdot}\sigma_v^2. \tag{7.2.11}$$

In the special case of equal error variances (i.e., with $k_{ij} = 1$), (7.2.8)–(7.2.10) reduce to the formulae given by Datta and Lahiri (2000).

7.2.2 Unconditional MSE Estimators

The expression for the true MSE given in (7.2.1) can be estimated by (5.3.11) when variance components are estimated either by REML or by the fitting-of-constants method. Under regularity conditions (i) and (ii) and normality of the errors v_i and e_{ij} in the nested error linear regression model (7.1.2), the MSE estimator (5.3.11) is second-order unbiased and it reduces to

$$\text{mse}(\hat{\mu}_i^H) = g_{1i}(\hat{\sigma}_v^2, \hat{\sigma}_e^2) + g_{2i}(\hat{\sigma}_v^2, \hat{\sigma}_e^2) + 2g_{3i}(\hat{\sigma}_v^2, \hat{\sigma}_e^2), \tag{7.2.12}$$

where $\hat{\boldsymbol{\delta}} = (\hat{\sigma}_v^2, \hat{\sigma}_e^2)^T$ is chosen as $\hat{\boldsymbol{\delta}}_{\text{RE}} = (\hat{\sigma}_{v\text{RE}}^2, \hat{\sigma}_{e\text{RE}}^2)^T$ or $\hat{\boldsymbol{\delta}}_m = (\hat{\sigma}_{vm}^2, \hat{\sigma}_{em}^2)^T$. The corresponding area-specific versions, $\text{mse}_1(\hat{\mu}_i^H)$ and $\text{mse}_2(\hat{\mu}_i^H)$, are obtained from (5.3.15) and (5.3.16), by evaluating $g_{3i}^*(\boldsymbol{\delta}, \mathbf{y}_i)$ given by (5.3.14). After some simplification, we get

$$g_{3i}^*(\boldsymbol{\delta}, \mathbf{y}_i) = a_{i\cdot}^{-2}(\sigma_v^2 + \sigma_e^2/a_{i\cdot})^{-4} h(\sigma_v^2, \sigma_e^2)(\bar{y}_{ia} - \bar{\mathbf{x}}_{ia}^T \tilde{\boldsymbol{\beta}})^2. \tag{7.2.13}$$

Hence, the two area-specific versions are given by

$$\text{mse}_1(\hat{\mu}_i^H) = g_{1i}(\hat{\sigma}_v^2, \hat{\sigma}_e^2) + g_{2i}(\hat{\sigma}_v^2, \hat{\sigma}_e^2) + 2g_{3i}^*(\hat{\boldsymbol{\delta}}, \mathbf{y}_i) \tag{7.2.14}$$

and

$$\text{mse}_2(\hat{\mu}_i^H) = g_{1i}(\hat{\sigma}_v^2, \hat{\sigma}_e^2) + g_{2i}(\hat{\sigma}_v^2, \hat{\sigma}_e^2) + g_{3i}(\hat{\sigma}_v^2, \hat{\sigma}_e^2) + g_{3i}^*(\hat{\boldsymbol{\delta}}, \mathbf{y}_i). \tag{7.2.15}$$

Battese et al. (1988) proposed an alternative second-order unbiased estimator of $\text{MSE}(\hat{\mu}_i^H)$, based on a method of moment estimators of σ_v^2 and σ_e^2. This estimator uses the idea behind Beale's nearly unbiased ratio estimator of a total (see Cochran 1977, p. 176).

For the ML estimator $\hat{\boldsymbol{\delta}}_{\text{ML}}$, the MSE estimator (5.3.12) is applicable and in this case it reduces to

$$\text{mse}(\hat{\mu}_i^H) = g_{1i}(\hat{\boldsymbol{\delta}}) - \mathbf{b}^T(\hat{\boldsymbol{\delta}}; \boldsymbol{\delta})\nabla g_{1i}(\hat{\boldsymbol{\delta}}) + g_{2i}(\hat{\boldsymbol{\delta}}) + 2g_{3i}(\hat{\boldsymbol{\delta}}), \tag{7.2.16}$$

where $\hat{\boldsymbol{\delta}} = \hat{\boldsymbol{\delta}}_{\text{ML}}$ and

$$\nabla g_{1i}(\hat{\boldsymbol{\delta}}) = \alpha_i^{-2}(\sigma_e^4, \sigma_v^4 a_{i\cdot})^T.$$

The bias term $\mathbf{b}(\hat{\boldsymbol{\delta}}; \boldsymbol{\delta})$ for the ML estimator $\hat{\boldsymbol{\delta}} = \hat{\boldsymbol{\delta}}_{\text{ML}}$ is obtained from (5.3.13) as

$$\mathbf{b}(\hat{\boldsymbol{\delta}}_{\text{ML}}; \boldsymbol{\delta}) = \frac{1}{2}\mathcal{I}^{-1}(\boldsymbol{\delta})\left\{\text{tr}\left[\left(\sum_{i=1}^m \mathbf{X}_i^T \mathbf{V}_i^{-1}\mathbf{X}_i\right)^{-1} \sum_{i=1}^m \mathbf{X}_i^T(\partial \mathbf{V}_i^{-1}/\partial \sigma_v^2)\mathbf{X}_i\right],\right.$$

$$\left.\text{tr}\left[\left(\sum_{i=1}^m \mathbf{X}_i^T \mathbf{V}_i^{-1}\mathbf{X}_i\right)^{-1} \sum_{i=1}^m \mathbf{X}_i^T(\partial \mathbf{V}_i^{-1}/\partial \sigma_e^2)\mathbf{X}_i\right]\right\}^T, \tag{7.2.17}$$

where $\mathcal{I}(\boldsymbol{\delta})$ is the 2×2 information matrix with elements given by (7.2.8)–(7.2.10), and $\mathbf{X}_i^T\mathbf{V}_i^{-1}\mathbf{X}_i$ is spelled out in (7.1.9). Furthermore, it follows from (7.1.9) that

$$\mathbf{X}_i^T(\partial\mathbf{V}_i^{-1}/\partial\sigma_v^2)\mathbf{X}_i = -\alpha_i^{-2}a_{i\cdot}^2\overline{\mathbf{x}}_{ia}\overline{\mathbf{x}}_{ia}^T \tag{7.2.18}$$

and

$$\mathbf{X}_i^T(\partial\mathbf{V}_i^{-1}/\partial\sigma_e^2)\mathbf{X}_i = -\sigma_e^{-4}\sum_{j=1}^{n_i} a_{ij}\mathbf{x}_{ij}\mathbf{x}_{ij}^T + \alpha_i^{-2}\sigma_e^{-4}\sigma_v^2(\sigma_e^2+\alpha_i)a_{i\cdot}^2\overline{\mathbf{x}}_{ia}\overline{\mathbf{x}}_{ia}^T. \tag{7.2.19}$$

7.2.3 *MSE Estimation: Nonnegligible Sampling Fractions

We now turn to MSE estimation of the EBLUP $\hat{\overline{Y}}_i^H$ in the case of nonnegligible sampling fraction f_i. Let $\mu_{ir} = \overline{\mathbf{X}}_{ir}^T\boldsymbol{\beta} + v_i$ be the true mean and $\overline{e}_{ir}$ the mean of the errors $e_{i\ell}$, for nonsampled units $\ell = n_i+1,\ldots,N_i$ in area i. Noting that

$$\hat{\overline{Y}}_i^H - \overline{Y}_i = (1-f_i)[(\hat{\overline{y}}_{ir}^H - \mu_{ir}) - \overline{e}_{ir}],$$

the MSE of $\hat{\overline{Y}}_i^H$ is given by

$$\text{MSE}(\hat{\overline{Y}}_i^H) = E(\hat{\overline{Y}}_i^H - \overline{Y}_i)^2 = (1-f_i)^2E(\hat{\overline{y}}_{ir}^H - \mu_{ir})^2 + (N_i^{-2}\mathbf{k}_{ir}^T\mathbf{k}_{ir})\sigma_e^2, \tag{7.2.20}$$

where $\mathbf{k}_{ir}$ is the known vector of $k_{i\ell}$-values for nonsampled units $\ell = n_i+1,\ldots,N_i$.

Now noting that $\hat{\overline{y}}_{ir}^H$ is the EBLUP estimator of μ_{ir}, we can use (7.2.1) to get a second-order approximation to $E(\hat{\overline{y}}_{ir}^H - \mu_{ir})^2$ as

$$E(\hat{\overline{y}}_{ir}^H - \mu_{ir})^2 \approx g_{1i}(\sigma_v^2,\sigma_e^2) + \tilde{g}_{2i}(\sigma_v^2,\sigma_e^2) + g_{3i}(\sigma_v^2,\sigma_e^2), \tag{7.2.21}$$

where $\tilde{g}_{2i}(\sigma_v^2,\sigma_e^2)$ is obtained from $g_{2i}(\sigma_v^2,\sigma_e^2)$ by changing $\overline{\mathbf{X}}_i$ to $\overline{\mathbf{x}}_{ir}$. Substituting (7.2.21) in (7.2.20) gives a second-order approximation to $\text{MSE}(\hat{\overline{Y}}_i^H)$.

Similarly, an estimator of $\text{MSE}(\hat{\overline{Y}}_i^H)$, unbiased up to $o(m^{-1})$ terms, is given by

$$\text{mse}(\hat{\overline{Y}}_i^H) = (1-f_i)^2\text{mse}(\hat{\overline{y}}_{ir}^H) + (N_i^{-2}\mathbf{k}_{ir}^T\mathbf{k}_{ir})\hat{\sigma}_e^2, \tag{7.2.22}$$

where

$$\text{mse}(\hat{\overline{y}}_{ir}^H) = g_{1i}(\hat{\sigma}_v^2,\hat{\sigma}_e^2) + \tilde{g}_{2i}(\hat{\sigma}_v^2,\hat{\sigma}_e^2) + 2g_{3i}(\hat{\sigma}_v^2,\hat{\sigma}_e^2) \tag{7.2.23}$$

and $(\hat{\sigma}_v^2,\hat{\sigma}_e^2)$ are the REML or the fitting-of-constants estimators of (σ_v^2,σ_e^2). Area-specific versions can be obtained similarly, using (7.2.14) and (7.2.15). From the last term in (7.2.22), it follows that the MSE estimator of $\hat{\overline{Y}}_i^H$ requires the knowledge of the $k_{i\ell}$-values for the nonsampled units $\ell = n_i+1,\ldots,N_i$ from area i, unless $k_{i\ell} = 1, \ell = n_i+1,\ldots,N_i$, in which case the last term reduces to $N_i^{-2}(N_i-n_i)\hat{\sigma}_e^2$.

Small area EBLUP estimators $\hat{\bar{Y}}_i^H$ are often aggregated to obtain an estimator for a larger area. To obtain the MSE of the larger area estimator, we also need the mean cross-product error (MCPE) of the estimators $\hat{\bar{Y}}_i^H$ for two different areas $i \neq t$ belonging to the larger area. It is easy to see that the MCPE correct up to $o(m^{-1})$ terms is given by

$$\text{MCPE}(\hat{\bar{Y}}_i^H, \hat{\bar{Y}}_t^H) \approx (\bar{\mathbf{X}}_{ir} - \gamma_i \bar{\mathbf{x}}_{ia})^T \left(\sum_{i=1}^{m} A_i \right)^{-1} (\bar{\mathbf{X}}_{ir} - \gamma_i \bar{\mathbf{x}}_{ia}) =: g_{2it}(\sigma_v^2, \sigma_e^2). \quad (7.2.24)$$

It follows from (7.2.24) that an estimator of MCPE is given by

$$\text{mcpe}(\hat{\bar{Y}}_i^H, \hat{\bar{Y}}_t^H) = g_{2it}(\hat{\sigma}_v^2, \hat{\sigma}_e^2), \quad (7.2.25)$$

which is unbiased up to $o(m^{-1})$ terms (Militino, Ugarte, and Goicoa 2007).

7.2.4 *Bootstrap MSE Estimation

We first study parametric bootstrap estimation of MSE($\hat{\mu}_i^H$) and MSE($\hat{\bar{Y}}_i^H$), assuming that $v_i \stackrel{\text{iid}}{\sim} N(0, \sigma_v^2)$, independent of $e_{ij} \stackrel{\text{iid}}{\sim} N(0, k_{ij}^2 \sigma_e^2), j = 1, \ldots, n_i, i = 1, \ldots, m$. The normality assumption will then be relaxed, following Hall and Maiti (2006a).

The parametric bootstrap procedure generates bootstrap sample data $\{(y_{ij}^*, \mathbf{x}_{ij}); j = 1, \ldots, n_i,\ i = 1, \ldots, m\}$ as $y_{ij}^* = \mathbf{x}_{ij}^T \hat{\boldsymbol{\beta}} + v_i^* + e_{ij}^*$, where v_i^* is generated from $N(0, \hat{\sigma}_v^2)$ and e_{ij}^* from $N(0, k_{ij}^2 \hat{\sigma}_e^2), j = 1, \ldots, n_i$, for specified estimators $\hat{\boldsymbol{\beta}}, \hat{\sigma}_v^2$, and $\hat{\sigma}_e^2$, assuming that $\hat{\sigma}_v^2 > 0$. Let $\mu_i^* = \bar{\mathbf{X}}_i^T \hat{\boldsymbol{\beta}} + v_i^*$ be the bootstrap version of the target parameter $\mu_i = \bar{\mathbf{X}}_i^T \boldsymbol{\beta} + v_i$. Then using the bootstrap data, bootstrap version $\hat{\mu}_{i*}^H$ of the EBLUP estimator $\hat{\mu}_i^H$ is obtained as $\hat{\mu}_{i*}^H = \bar{\mathbf{X}}_i^T \hat{\boldsymbol{\beta}}^* + \hat{v}_i^*$, where $\hat{\boldsymbol{\beta}}^*$ and $\hat{v}_i^*$ are calculated the same way as $\hat{\boldsymbol{\beta}}$ and $\hat{v}_i$, but using the above bootstrap sample data. The theoretical bootstrap estimator of MSE($\hat{\mu}_i^H$) is given by $\text{mse}_B(\hat{\mu}_i^H) = E_*(\hat{\mu}_{i*}^H - \mu_i^*)^2$. This bootstrap estimator is approximated by Monte Carlo simulation, in which the above steps are repeated a large number, B, of times. In this way, we obtain B replicates $\mu_i^*(1), \ldots, \mu_i^*(B)$ of the bootstrap true value μ_i^*, where $\mu_{i*}(b) = \bar{\mathbf{X}}_i^T \hat{\boldsymbol{\beta}} + v_i^*(b)$, together with B replicates $\hat{\mu}_{i*}^H(1), \ldots, \hat{\mu}_{i*}^H(B)$ of the EBLUP estimate $\hat{\mu}_{i*}^H$, where $\hat{\mu}_{i*}^H(b) = \bar{\mathbf{X}}_i^T \hat{\boldsymbol{\beta}}^*(b) + \hat{v}_i^*(b)$, and $v_i^*(b), \hat{\boldsymbol{\beta}}^*(b)$, and $\hat{v}_i^*(b)$ are the bth replicates of $v_i^*, \hat{\boldsymbol{\beta}}^*$, and $\hat{v}_i^*$, respectively. A first-order unbiased bootstrap MSE estimator of $\hat{\mu}_i^H$ is then given by

$$\text{mse}_{B1}(\hat{\mu}_i^H) = B^{-1} \sum_{b=1}^{B} [\hat{\mu}_{i*}^H(b) - \mu_{i*}(b)]^2. \quad (7.2.26)$$

Applying the method of imitation to $\text{mse}_B(\hat{\mu}_i^H) = E_*(\hat{\mu}_{i*}^H - \mu_{i*})^2$, it follows that

$$\text{mse}_B(\hat{\mu}_i^H) \approx g_{1i}(\hat{\sigma}_v^2) + g_{2i}(\hat{\sigma}_v^2) + g_{3i}(\hat{\sigma}_v^2). \quad (7.2.27)$$

Now noting that $\text{mse}_{B1}(\hat{\mu}_i^H)$ is a Monte Carlo approximation to $\text{mse}_B(\hat{\mu}_i^H)$, for large B and comparing (7.2.27) to the second-order unbiased MSE estimator (7.2.12), valid for REML and the method of fitting-of-constants, it follows that $\text{mse}_{B1}(\hat{\mu}_i^H)$ is not second-order unbiased. A double-bootstrap method can be used to rectify this problem (Hall and Maiti 2006a). Alternatively, hybrid methods, similar to those in Section 6.2.4, depending on $g_{1i}(\hat{\sigma}_v^2)$ and $g_{2i}(\hat{\sigma}_v^2)$, may be used to obtain second-order unbiased bootstrap estimators of $\text{MSE}(\hat{\mu}_i^H)$.

For the case of nonnegligible sampling fraction f_i, González-Manteiga et al. (2008a) proposed a first-order unbiased bootstrap MSE estimator and also studied hybrid bootstrap MSE estimators that are second-order unbiased. Let $\overline{Y}_i^* = \overline{\mathbf{X}}_i^T\hat{\boldsymbol{\beta}} + v_i^* + \overline{E}_i^*$ be the bootstrap population mean, where $\overline{E}_i^*$ is the mean of N_i bootstrap values e_{ij}^* generated from $N(0, k_{ij}^2\hat{\sigma}_e^2)$ or equivalently $\overline{E}_i^*$ is generated from $N[0, (N_i^{-2}\sum_{j=1}^{N_i} k_{ij}^2)\hat{\sigma}_e^2]$. Let $\hat{\overline{Y}}_{i*}^H$ be the bootstrap version of the EBLUP estimator $\hat{\overline{Y}}_i^H$ obtained from the bootstrap sample data. The above process is repeated a large number, B, of times to obtain B replicates $\overline{Y}_i^*(1), \ldots, \overline{Y}_i^*(B)$ of the true bootstrap mean $\overline{Y}_i^*$ together with B replicates $\hat{\overline{Y}}_{i*}^H(1), \ldots, \hat{\overline{Y}}_{i*}^H(B)$ of the EBLUP estimator $\hat{\overline{Y}}_{i*}^H$. Then, a first-order unbiased bootstrap MSE estimator of $\hat{\overline{Y}}_i^H$ is given by

$$\text{mse}_{B1}(\hat{\overline{Y}}_i^H) = B^{-1}\sum_{b=1}^{B}[\hat{\overline{Y}}_{i*}^H(b) - \overline{Y}_{i*}(b)]^2. \tag{7.2.28}$$

If $\hat{\sigma}_v^2 = 0$, then we proceed as in Section 6.2.4, that is, we generate bootstrap values y_{ij}^* from $N[\mathbf{x}_{ij}^T\tilde{\boldsymbol{\beta}}(0), \hat{\sigma}_e^2]$ and then use the bootstrap data $\{(y_{ij}^*, \mathbf{x}_{ij}; j = 1, \ldots, n_i,\ i = 1, \ldots, m\}$ to calculate the regression-synthetic estimator $\hat{\mu}_{i*}^{\text{RS}} = \overline{\mathbf{X}}_i^T\tilde{\boldsymbol{\beta}}^*(0)$, where $\tilde{\boldsymbol{\beta}}^*(0)$ is obtained the same as $\tilde{\boldsymbol{\beta}}(0)$ but using the bootstrap sample data. Repeating the process B times, a first-order bootstrap MSE estimator is given by

$$\text{mse}_{B1}(\hat{\mu}_i^H) = B^{-1}\sum_{b=1}^{B}[\hat{\mu}_{i*}^{\text{RS}}(b) - \mu_{i*}]^2, \tag{7.2.29}$$

where $\mu_i^* = \overline{\mathbf{X}}_i^T\tilde{\boldsymbol{\beta}}(0)$. A similar method can be used to handle the case of nonnegligible sampling fraction f_i.

We now turn to the case of nonparametric bootstrap MSE estimation by relaxing the normality assumption on v_i and e_{ij} and noting that the EBLUP $\hat{\mu}_i^H$ does not require normality if a moment method is used to estimate σ_v^2 and σ_e^2. Hall and Maiti (2006a) used the fitting-of-constants method to estimate σ_v^2 and σ_e^2 and noted that the MSE of $\hat{\mu}_i^H$ depends only on σ_v^2, σ_e^2 and the fourth moments $\delta_v = E_m(v_i^4)$ and $\delta_e = E_m(\tilde{e}_{ij}^4)$ if normality assumption is relaxed, where $\tilde{e}_{ij} = k_{ij}^{-1}e_{ij} \overset{\text{iid}}{\sim} (0, \sigma_e^2)$. They also noted that it is analytically difficult to obtain second-order unbiased MSE estimators through the linearization method in this case.

Hall and Maiti (2006a) used a matched-moment bootstrap method to get around the difficulty with the linearization method. First, moment estimators $(\hat{\sigma}_v^2, \hat{\sigma}_e^2)$ of (σ_v^2, σ_e^2) and $(\hat{\delta}_v, \hat{\delta}_e)$ of (δ_v, δ_e) are obtained. Then, bootstrap values v_i^* and $\tilde{e}_{ij}^*$ are generated from distributions $D(\hat{\sigma}_v^2, \hat{\delta}_v)$ and $D(\hat{\sigma}_e^2, \hat{\delta}_e)$, respectively, where $D(\xi_2, \xi_4)$ denotes the distribution of a random variable Z for which $E(Z) = 0, E(Z^2) = \xi_2$, and $E(Z^4) = \xi_4$ for specified $\xi_2, \xi_4 > 0$ with $\xi_2^2 \leq \xi_4$. The proposed method implicitly assumes that $\hat{\sigma}_v^2 > 0$. A simple way of generating values from $D(\xi_2, \xi_4)$ is to let $Z = \xi_2^{1/2}\tilde{Z}$, where $\tilde{Z}$ has the three-point distribution

$$P(\tilde{Z} = 0) = 1 - p, \quad P(\tilde{Z} = \mp p^{-1/2}) = p/2 \tag{7.2.30}$$

with $p = \xi_2^2/\xi_4$.

Bootstrap data are obtained by letting $y_{ij}^* = \mathbf{x}_{ij}^T\hat{\boldsymbol{\beta}} + v_i^* + k_{ij}\tilde{e}_{ij}^*$ and then we proceed as before to get the bootstrap versions $\hat{\mu}_{i*}^H$ and μ_i^* and the bootstrap MSE estimator, $\text{mse}_{B1}(\hat{\mu}_i^H)$, given by (7.2.26). A bias-corrected MSE estimator of $\hat{\mu}_i^H$ may be obtained by using a double-bootstrap method. This is accomplished by drawing v_i^{**} and $\tilde{e}_{ij}^{**}$ from $D(\hat{\sigma}_v^{2*}, \hat{\delta}_v^*)$ and $D(\hat{\sigma}_e^{2*}, \hat{\delta}_e^*)$, respectively, and then letting $y_{ij}^{**} = \mathbf{x}_{ij}^T\hat{\boldsymbol{\beta}}^* + v_i^{**} + k_{ij}\tilde{e}_{ij}^{**}$, where $(\hat{\boldsymbol{\beta}}^*, \hat{\sigma}_v^{2*}, \hat{\delta}_v^*, \hat{\sigma}_e^{2*}, \hat{\delta}_e^*)$ are the parameter estimates obtained from the first-phase bootstrap data $\{(y_{ij}^*, \mathbf{x}_{ij}); j = 1, \ldots, n_i, \; i = 1, \ldots, m\}$. The second-phase bootstrap sample data $\{(y_{ij}^{**}, \mathbf{x}_{ij}); j = 1, \ldots, n_i, \; i = 1, \ldots, m\}$ are used to calculate the bootstrap versions $\hat{\mu}_{i**}^H$ and $\mu_{i**} = \overline{\mathbf{X}}_i^T\hat{\boldsymbol{\beta}}^* + v_i^{**}$. In practice, we select C bootstrap replicates from each first-phase bootstrap replicate b and calculate

$$\text{mse}_{BC1}(\hat{\mu}_i^H) = B^{-1}C^{-1}\sum_{b=1}^{B}\sum_{c=1}^{C}[\hat{\mu}_{i**}^H(bc) - \mu_{i**}(bc)]^2. \tag{7.2.31}$$

A bias-corrected MSE estimator is then given by

$$\text{mse}_{BC2}(\hat{\mu}_i^H) = 2\,\text{mse}_{B1}(\hat{\mu}_i^H) - \text{mse}_{BC1}(\hat{\mu}_i^H), \tag{7.2.32}$$

which is second-order unbiased.

The bias-corrected MSE estimator (7.2.32) may take negative values and Hall and Maiti (2006a) proposed modifications ensuring positive MSE estimators that are second-order unbiased. Another possible modification proposed by Hall and Maiti (2006b) and Pfeffermann (2013) is given by

$$\text{mse}_{BC2}(\text{mod}) = \begin{cases} \text{mse}_{BC2} & \text{if } \text{mse}_{B1} \geq \text{mse}_{BC1}; \\ \text{mse}_{B1}\exp\left[\dfrac{\text{mse}_{B1} - \text{mse}_{BC1}}{\text{mse}_{BC2}}\right] & \text{if } \text{mse}_{B1} < \text{mse}_{BC1}. \end{cases} \tag{7.2.33}$$

Implementation of double-bootstrap MSE estimation involves several practical issues: (i) Sensitivity of MSE estimator to the choice of B and C. (ii) For some of the first-phase bootstrap replicates, b, the estimate of σ_v^2, denoted $\hat{\sigma}_v^{2*}(b)$, may be zero even when $\hat{\sigma}_v^2 > 0$. How to generate second-phase bootstrap samples in those cases? Earlier, we suggested a modification for single-phase bootstrap when $\hat{\sigma}_v^2 = 0$. It may be possible to use a similar modification if $\hat{\sigma}_v^{2*}(b) = 0$.

Regarding the choice of B and C, for a given number of bootstrap replicates $R = BC + B = B(C+1)$, Erciulescu and Fuller (2014) demonstrated that the choice $C = 1$ and $B = R/2$ leads to smaller bootstrap error than other choices of B and C. This result implies that one should select a single second-phase bootstrap sample for each first-phase bootstrap sample.

7.3 *APPLICATIONS

We now provide some details of the application in Example 4.3.1, Chapter 4, and describe another application on estimation of area under olive trees.

Example 7.3.1. County Crop Areas. Battese, Harter, and Fuller (1988) applied the nested error linear regression model with equal error variances (i.e., $k_{ij} = 1$) to estimate area under corn and area under soybeans for each of $m = 12$ counties in north-central Iowa, using farm interview data in conjunction with LANDSAT satellite data. Each county was divided into area segments, and the areas under corn and soybeans were ascertained for a sample of segments by interviewing farm operators. The number of sample segments, n_i, in a county ranged from 1 to 5 ($\sum_{i=1}^{m} n_i = n = 37$), while the total number of segments in a county ranged from 394 to 687. Because of negligible sample fractions n_i/N_i, the small area means $\overline{Y}_i$ are taken as $\mu_i = \overline{\mathbf{X}}_i^T \boldsymbol{\beta} + v_i$.

Unit level auxiliary data $\{x_{1ij}, x_{2ij}\}$ in the form of the number of pixels classified as corn, x_{1ij}, and the number of pixels classified as soybeans, x_{2ij}, were also obtained for all the area segments, including the sample segments, in each county, using the LANDSAT satellite readings. In this application, y_{ij} denotes the number of hectares of corn (soybeans) in the jth sample area segment from the ith county. Table 1 in Battese, Harter, and Fuller (1988) reports the values of $n_i, N_i, \{y_{ij}, x_{1ij}, x_{2ij}\}$ as well as the area means $\overline{X}_{1i}$ and $\overline{X}_{2i}$. The sample data for the second sample segment in Hardin county was deleted from the estimation of means μ_i because the corn area for that segment looked erroneous in a preliminary analysis. It is interesting to note that model diagnostics, based on the local influence approach (see Section 5.4.2), identified this sample data point as possibly erroneous (Hartless, Booth, and Littell 2000).

Battese et al. (1988) used model (7.1.2) with $\mathbf{x}_{ij} = (1, x_{1ij}, x_{2ij})^T$ and normally distributed errors v_i and e_{ij} with common variances σ_v^2 and σ_e^2. Estimates of σ_v^2 and σ_e^2 were obtained as $\hat{\sigma}_e^2 = \hat{\sigma}_{em}^2 = 150$, $\hat{\sigma}_v^2 = 140$ for corn and $\hat{\sigma}_e^2 = \hat{\sigma}_{em}^2 = 195$, $\hat{\sigma}_v^2 = 272$ for soybeans. The estimate $\hat{\sigma}_v^2$ is slightly different from $\hat{\sigma}_{vm}^2$, the estimate based on the fitting-of-constants method. The regression coefficients associated with x_{1ij} and x_{2ij} were both significant for the corn model, but only the coefficient associated with x_{2ij} was significant for the soybeans model. The between-county variance, σ_v^2, was significant at the 10% level for corn and at the 1% level for soybeans.

Battese et al. (1988) reported some methods for validating the assumed model with $\mathbf{x}_{ij} = (1, x_{1ij}, x_{2ij})^T$. First, they introduced quadratic terms x_{1ij}^2 and x_{2ij}^2 into the model and tested the null hypothesis that the regression coefficients of the quadratic terms are zero. The null hypothesis was not rejected at the 5% level. Second, to test the normality of the error terms v_i and e_{ij}, the transformed residuals $(y_{ij} - \hat{\tau}_i\bar{y}_i) - (\mathbf{x}_{ij} - \hat{\tau}_i\bar{\mathbf{x}}_i)^T\hat{\boldsymbol{\beta}}$ were computed, where $\hat{\tau}_i = 1 - (1 - \hat{\gamma}_i)^{1/2}$. Under the null hypothesis, the transformed residuals are approximately iid $N(0, \sigma_e^2)$. Hence, standard methods for checking normality can be applied to the transformed residuals. In particular, the well-known Shapiro–Wilk W statistic applied to the transformed residuals gave values of 0.985 and 0.957 for corn and soybeans, respectively, yielding p-values (probabilities of getting values less than those observed under normality) of 0.921 and 0.299, respectively (Shapiro and Wilk 1965). It may be noted that small values of W correspond to departures from normality. Although $n = 37$ is not large, the p-values suggest no evidence against the hypothesis of normality. Normal quantile–quantile (q–q) plots of transformed residuals also indicated no evidence against normality for both corn and soybeans. Jiang, Lahiri, and Wu (2001) proposed an alternative test of normality (or specified distributions) of the error terms v_i and e_{ij}, based on a chi-squared statistic with estimated cell frequencies. To check for the normality of the small area effects v_i and to detect outlier v_i's, a normal probability plot of EBLUP estimates $\hat{v}_i$, divided by their standardized errors, may be examined (see Section 5.4.2).

Table 7.1 reports the EBLUP estimates, $\hat{\mu}_i^H$, of small area means, μ_i, for corn and soybeans using $\hat{\sigma}_{em}^2$ and $\hat{\sigma}_v^2$. Estimated standard errors of the EBLUP estimates and the survey regression estimates, $\hat{\mu}_i^{\text{SR}} = \bar{y}_i + (\overline{\mathbf{X}}_i - \bar{\mathbf{x}}_i)^T\hat{\boldsymbol{\beta}}$, denoted by $s(\hat{\mu}_i^H)$ and $s(\hat{\mu}_i^{\text{SR}})$, are also given. It is clear from Table 7.1 that the ratio of the estimated standard error of the EBLUP estimate to that of the survey regression estimate decreases from 0.97 to 0.77 as the number of sample area segments, n_i, decreases from 5 to 1. The reduction in the estimated standard error is considerable when $n_i \leq 3$.

The EBLUP estimates, $\hat{\mu}_i^H$, were adjusted (calibrated) to agree with the survey regression estimate for the entire area covering the 12 counties. The latter estimate is given by $\hat{\mu}^{\text{SR}} = \sum_{\ell=1}^{12} W_\ell \hat{\mu}_\ell^{\text{SR}}$, where W_ℓ is the proportion of population area segments that belong to ℓth area, and the overall population mean $\mu = \sum_{\ell=1}^{12} W_\ell \mu_\ell$. The estimator $\hat{\mu}^{\text{SR}}$ is approximately design-unbiased for μ under simple random sampling within areas and its design standard error is relatively small. The adjusted EBLUP estimates are given by

$$\hat{\mu}_i^H(a) = \hat{\mu}_i^H + \hat{p}_i^{(1)}(\hat{\mu}^{\text{SR}} - \hat{\mu}^H), \tag{7.3.1}$$

TABLE 7.1 EBLUP Estimates of County Means and Estimated Standard Errors of EBLUP and Survey Regression Estimates

		Corn			Soybeans		
County	n_i	$\hat{\mu}_i^H$	$s(\hat{\mu}_i^H)$	$s(\hat{\mu}_i^{\text{SR}})$	$\hat{\mu}_i^H$	$s(\hat{\mu}_i^H)$	$s(\hat{\mu}_i^{\text{SR}})$
Cerro Gordo	1	122.2	9.6	13.7	77.8	12.0	15.6
Hamilton	1	126.3	9.5	12.9	94.8	11.8	14.8
Worth	1	106.2	9.3	12.4	86.9	11.5	14.2
Humboldt	2	108.0	8.1	9.7	79.7	9.7	11.1
Franklin	3	145.0	6.5	7.1	65.2	7.6	8.1
Pocahontas	3	112.6	6.6	7.2	113.8	7.7	8.2
Winnebago	3	112.4	6.6	7.2	98.5	7.7	8.3
Wright	3	122.1	6.7	7.3	112.8	7.8	8.4
Webster	4	115.8	5.8	6.1	109.6	6.7	7.0
Hancock	5	124.3	5.3	5.7	101.0	6.2	6.5
Kossuth	5	106.3	5.2	5.5	119.9	6.1	6.3
Hardin	5	143.6	5.7	6.1	74.9	6.6	6.9

Source: Adapted from Table 2 in Battese et al. (1988).

where $\hat{\mu}^H = \sum_{\ell=1}^{12} W_\ell \hat{\mu}_\ell^H$ and

$$\hat{p}_i^{(1)} = \left[\sum_{\ell=1}^{12} W_\ell^2 \text{mse}(\hat{\mu}_\ell^H)\right]^{-1} W_i \ \text{mse}(\hat{\mu}_i^H).$$

It follows from (7.3.1) that $\sum_{\ell=1}^{12} W_i \hat{\mu}_i^H(a) = \hat{\mu}^{\text{SR}}$. A drawback of $\hat{\mu}_i^H(a)$ is that it depends on the MSE estimates $\text{mse}(\hat{\mu}_\ell^H)$. Pfeffermann and Barnard (1991) proposed an "optimal" adjustment of the form (7.3.1) with $\hat{p}_i^{(1)}$ changed to

$$\hat{p}_i^{(2)} = [\text{mse}(\hat{\mu}^H)]^{-1} \text{mcpe}(\hat{\mu}_i^H, \hat{\mu}^H), \tag{7.3.2}$$

where $\text{mcpe}(\hat{\mu}_i^H, \hat{\mu}^H)$ is an estimate of the MCPE of $\hat{\mu}_i^H$ and $\hat{\mu}^H$, $\text{MCPE}(\hat{\mu}_i^H, \hat{\mu}^H) = E(\hat{\mu}_i^H - \mu_i)(\hat{\mu}^H - \mu)$. The term $\hat{p}_i^{(2)}$ involves estimates of $\text{MCPE}(\hat{\mu}_i^H, \hat{\mu}_\ell^H), \ell \neq i$. Following Isaki, Tsay, and Fuller (2000) and Wang (2000), a simple adjustment compared to (7.3.1) and (7.3.2) is obtained by changing $\hat{p}_i^{(1)}$ in (7.3.1) to

$$\hat{p}_i^{(3)} = \left[\sum_{\ell=1}^{12} W_\ell^2 \text{mse}(\hat{\mu}_\ell^{\text{SR}})\right]^{-1} W_i \ \text{mse}(\hat{\mu}_i^{\text{SR}}), \tag{7.3.3}$$

An alternative is to use the simple ratio adjustment, which is given by (7.3.1) with $\hat{p}_i^{(1)}$ changed to

$$\hat{p}_i^{(4)} = \hat{\mu}_i^H / \hat{\mu}^H. \tag{7.3.4}$$

Mantel, Singh, and Bureau (1993) conducted an empirical study on the performance of the adjustment estimators $\hat{p}_i^{(1)}, \hat{p}_i^{(2)}$, and $\hat{p}_i^{(4)}$, using a synthetic population based on Statistics Canada's Survey of Employment, Payroll and Hours. Their results indicated that the simple ratio adjustment, $\hat{p}_i^{(4)}$, often performs better than the more complex adjustments, possibly due to instability of mse and mcpe-terms involved in $\hat{p}_i^{(1)}$ and $\hat{p}_i^{(2)}$.

Example 7.3.2. Simulation Study. Rao and Choudhry (1995) studied the relative performance of some direct and indirect estimators, using real and synthetic populations. For the real population, a sample of 1,678 unincorporated tax filers (units) from the province of Nova Scotia, Canada, divided into 18 census divisions, was treated as the overall population. In each census division, units were further classified into four mutually exclusive industry groups. The objective was to estimate the total wages and salaries (Y_i) for each nonempty census division by industry group i (small areas of interest). Here, we focus on the industry group "construction" with 496 units and average census division size equal to 27.5. Gross business income, available for all the units, was used as an auxiliary variable (x). The overall correlation coefficient between y and x for the construction group was 0.64.

To make comparisons between estimators under customary repeated sampling, $R = 500$ samples, each of size $n = 149$, from the overall population of $N = 1,678$ units were selected by SRS. From each simulated sample, the following estimators were calculated: (i) Post-stratified estimator: $\text{PST} = N_i\bar{y}_i$ if $n_i \geq 1$ and $\text{PST} = 0$ if $n_i = 0$, where N_i and n_i are the population and sample sizes in the ith area (n_i is a random variable). (ii) Ratio-synthetic estimator: $\text{SYN} = (\bar{y}/\bar{x})X_i$, where $\bar{y}$ and $\bar{x}$ are the overall sample means in the industry group and X_i is the x-total for the ith area. (iii) Sample size-dependent estimator: $\text{SSD} = \psi_i(D)(\text{PST}) + [1 - \psi_i(D)](\text{SYN})$ with $\psi_i(D) = 1$ if $n_i/n \geq N_i/N$ and $\text{SSD} = (n_i/n)(N_i/N)^{-1}$ otherwise. (iv) EBLUP estimator, $\hat{Y}_i^H = N_i\hat{\bar{Y}}_i^H$, where $\hat{\bar{Y}}_i^H$ is given by (7.1.21) and based on the nested error regression model (7.1.2) with $\mathbf{x}_{ij}^T\boldsymbol{\beta} = \beta x_{ij}$ and $k_{ij} = x_{ij}^{1/2}$, and using the fitting-of-constants estimators $\hat{\sigma}_{vm}^2$ and $\hat{\sigma}_{em}^2$. To examine the aptness of this model, the model was fitted to the 496 population pairs (y_{ij}, x_{ij}) from the construction group and the standardized EBLUP residuals $(y_{ij} - \hat{\beta}x_{ij} - \hat{v}_i)/(\hat{\sigma}_e x_{ij}^{1/2})$ were examined. A plot of these residuals against the x_{ij}'s indicated a reasonable but not good fit in the sense that the plot revealed an upward shift with several values larger than 1.0 but none below -1.0. Several variations of the model, including a model with an intercept term, did not lead to better fits.

For each estimator, average absolute relative bias ($\overline{\text{ARB}}$), average relative efficiency ($\overline{\text{EFF}}$), and average absolute relative error ($\overline{\text{ARE}}$) were calculated as follows:

$$\overline{\text{ARB}} = \frac{1}{m}\sum_{i=1}^{m}\left|\frac{1}{500}\sum_{r=1}^{500}(\text{est}_r/Y_i - 1)\right|,$$

$$\overline{\text{EFF}} = [\overline{\text{MSE}}(\text{PST})/\overline{\text{MSE}}(\text{est})]^{1/2},$$

$$\overline{\text{ARE}} = \frac{1}{m}\sum_{i=1}^{m}\frac{1}{500}\sum_{r=1}^{500}|\text{est}_r/Y_i - 1|,$$

where the average is taken over $m = 18$ census divisions in the industry group. Here est_r denotes the value of the estimator, est, for the rth simulated sample $(r = 1, 2, \ldots, 500)$, Y_i is the true area total, and

$$\overline{\text{MSE}}(\text{est}) = \frac{1}{m}\sum_{i=1}^{m}\frac{1}{500}\sum_{r=1}^{500}(\text{est}_r - Y_i)^2;$$

$\overline{\text{MSE}}$(PST) is obtained by changing est_r to PST_r, the value of the post-stratified estimator for the rth simulated sample. Note that $\overline{\text{ARB}}$ measures the bias of an estimator, whereas both $\overline{\text{EFF}}$ and $\overline{\text{ARE}}$ measure the accuracy of an estimator.

Table 7.2 reports the percentage values of $\overline{\text{ARB}}$, $\overline{\text{EFF}}$, and $\overline{\text{ARE}}$ for the construction group. It is clear from Table 7.2 that both SYN and EBLUP perform significantly better than PST and SSD in terms of $\overline{\text{EFF}}$ and $\overline{\text{ARE}}$, leading to larger $\overline{\text{EFF}}$ values and smaller $\overline{\text{ARE}}$ values. For example, $\overline{\text{EFF}}$ for the EBLUP estimator is 261.1% compared to 137.6% for SSD. In terms of $\overline{\text{ARB}}$, SYN has the largest value (15.7%) as expected, followed by the EBLUP estimator with $\overline{\text{ARB}}$=11.3%; PST and SSD have smaller $\overline{\text{ARB}}$: 5.4% and 2.9%, respectively. Overall, EBLUP is somewhat better than SYN: $\overline{\text{EFF}}$ value of 261.1% versus 232.8% and $\overline{\text{ARE}}$ value of 13.5% versus 16.5%. It is gratifying that the EBLUP estimator under the assumed model performed well, despite the problems found in the residual analysis.

The estimators were also compared under a synthetic population generated from the assumed model with the real population x-values. The parameter values $(\beta, \sigma_v^2, \sigma_e^2)$ used for generating the synthetic population were the estimates obtained by fitting the model to the real population pairs $(y_{ij}, x_{ij}) : \beta = 0.21, \sigma_v^2 = 1.58$, and $\sigma_e^2 = 1.34$.

TABLE 7.2 Unconditional Comparisons of Estimators: Real and Synthetic Population

	Estimator			
Quality Measure	PST	SYN	SSD	EBLUP
		Real Population		
$\overline{\text{ARB}}$%	5.4	15.7	2.9	11.3
$\overline{\text{EFF}}$%	100.0	232.8	137.6	261.1
$\overline{\text{ARE}}$%	32.2	16.5	24.0	13.5
		Synthetic Population		
$\overline{\text{ARB}}$%	5.6	12.5	2.4	8.4
$\overline{\text{EFF}}$%	100.0	313.3	135.8	319.1
$\overline{\text{ARE}}$%	35.0	13.2	25.9	11.8

Source: Adapted from Tables 27.1 and 27.3 in Rao and Choudhry (1995).

TABLE 7.3 Effect of Between-Area Homogeneity on the Performance of SSD and EBLUP

	Between-Area Homogeneity: θ					
Estimator	0.1	0.5	1.0	2.0	5.0	10.0
			$\overline{\text{EFF}}$%			
SSD	136.0	136.0	135.8	135.6	134.7	133.1
EBLUP	324.3	324.6	319.1	305.0	270.8	239.9
			$\overline{\text{ARE}}$%			
SSD	25.6	25.7	25.9	26.3	27.2	28.2
EBLUP	11.5	11.6	11.8	12.5	14.5	16.7

Source: Adapted from Tables 27.4 and 27.5 in Rao and Choudhry (1995).

A plot of the standardized EBLUP residuals, obtained by fitting the model to the synthetic population, showed an excellent fit as expected. Table 7.2 also reports the percentage values of $\overline{\text{ARB}}$, $\overline{\text{EFF}}$, and $\overline{\text{ARE}}$ for the synthetic population. Comparing these values to the corresponding values for the real population, it is clear that $\overline{\text{EFF}}$ increases for EBLUP and SYN, while it remains essentially unchanged for SSD. Similarly, $\overline{\text{ARE}}$ decreases for EBLUP and SYN, while it remains essentially unchanged for SSD. The value of $\overline{\text{ARB}}$ also decreases for EBLUP and SYN: 11.3% versus 8.4% for EBLUP and 15.7% versus 12.5% for SYN.

Conditional comparisons of the estimators were also made by conditioning on the realized sample sizes in the small areas. This is a more realistic approach because the domain sample sizes, n_i, are random with known distribution. To make conditional comparisons under repeated sampling, a simple random sample of size $n = 149$ was first selected to determine the sample sizes, n_i, in the small areas. Regarding the n_i's as fixed, 500 stratified random samples were then selected, treating the small areas as strata. The conditional values of $\overline{\text{ARB}}$, $\overline{\text{EFF}}$, and $\overline{\text{ARE}}$ were computed from the simulated stratified samples. The conditional performances were similar to the unconditional performances. Results were different, however, when two separate values for each quality measure were computed: one by averaging over areas with $n_i < 6$ only, and another averaging over areas with $n_i \geq 6$. In particular, $\overline{\text{EFF}}(\overline{\text{ARE}})$ for EBLUP is much larger (smaller) than the value for SSD when $n_i < 6$.

As noted in Chapter 3, Section 3.3.2, the SSD estimator does not take advantage of the between-area homogeneity, unlike the EBLUP estimator. To demonstrate this point, a series of synthetic populations was generated, using the previous parameter values, $\beta = 0.21, \sigma_v^2 = 1.58$ and $\sigma_e^2 = 1.34$, and the model $y_{ij} = \beta x_{ij} + v_i \theta^{1/2} + e_{ij} x_{ij}^{1/2}$, by varying θ from 0.1 to 10.0 ($\theta = 1$ corresponds to the previous synthetic population). Note that for a given ratio σ_v^2/σ_e^2, the between-area homogeneity increases as θ decreases. Table 7.3 reports the unconditional values of $\overline{\text{EFF}}$ and $\overline{\text{ARE}}$ for the estimators SSD and EBLUP, as θ varies from 0.1 to 10.0. It is clear from Table 7.3 that $\overline{\text{EFF}}$ and $\overline{\text{ARE}}$ for SSD remain essentially unchanged as θ increases from 0.1 to 10.0. On the other hand, $\overline{\text{EFF}}$ for EBLUP is largest when

$\theta = 0.1$ and decreases as θ increases by 10.0. Similarly, $\overline{\text{ARE}}$ for EBLUP is smallest when $\theta = 0.1$ and increases as θ increases by 10.0.

Example 7.3.3. Area Occupied by Olive Trees. Militino et al. (2006) applied the basic unit level model (7.1.1) to develop EBLUP estimates of area occupied by olive trees in $m = 8$ nonirrigated areas located in a central region of Navarra, Spain. The region was divided into segments of 4 ha each. The total number of segments in the areas, N_i, varied from 32 to 731. Simple random samples of segments were selected from each area with sample sizes n_i varying from 1 to 12. In the majority of cases, the study domain plots are smaller than the 4 ha. segments and of different size. Surface areas, s_{ij}, of segments j in each area i where olive trees are likely to be found were ascertained from cropland maps. We denote by $S_i = \sum_{j=1}^{N_i} s_{ij}$ the total surface that is likely to contain olive trees in area i, which is known for all areas. Inside each sample segment j in area i, area occupied by olive trees was observed and denoted by y_{sij}. Furthermore, satellite imagery was used to ascertain the classified olive trees, x_{ij}, in each population segment.

Because of the unequal surface areas, s_{ij}, the areas covered by olive trees y_{sij} and x_{sij} are transformed in terms of portions of the surfaces s_{ij} that are covered by olive trees. The transformed data are given by

$$y_{ij} = \frac{S_i}{N_i}\frac{y_{sij}}{s_{ij}}, \quad x_{ij} = \frac{S_i}{N_i}\frac{x_{sij}}{s_{ij}}, \quad j = 1, \ldots, n_i, \quad i = 1, \ldots, 8.$$

Note that $y_{ij} = y_{sij}$ and $x_{ij} = x_{sij}$ when $s_{ij} = s_i$ for all j. For each area i, the parameter of interest is the mean of the portions of the segment's surface covered by olive trees multiplied by the total surface S_i, which is equal to the total of the transformed variables,

$$Y_i = \sum_{j=1}^{N_i} y_{ij} = \frac{S_i}{N_i}\sum_{j=1}^{N_i}\frac{y_{sij}}{s_{ij}}, \quad i = 1, \ldots, m.$$

Model (7.1.1) with $k_{ij}^2 = n_i$ was chosen for the transformed data after comparing it to alternative unit level models, using a bootstrap test of the null hypothesis H_0 : $\sigma_v^2 = 0$ and the conditional Akaike Information Criterion (cAIC) for mixed models (Section 5.4.1).

Model (7.1.1) with $\mathbf{x}_{ij} = (1, x_{ij})^T$ was validated by checking the normality assumption using both the transformed residuals and the EBLUP residuals in the Shapiro–Wilk statistic. Box plots of EBLUP residuals for the selected model did not show any specific pattern, suggesting the adequacy of the model. Note that the unequal error variances, $n_i\sigma_e^2$, must be considered when standardizing residuals. Thus, in this case transformed residuals and standardized EBLUP residuals introduced in Example 7.3.1 are given, respectively, by

$$\hat{u}_{ij} = n_i^{-1/2}[(y_{ij} - \hat{\tau}\bar{y}_i) - (\mathbf{x}_{ij} - \hat{\tau}\bar{\mathbf{x}}_i)^T\hat{\boldsymbol{\beta}}]$$

and

$$\hat{e}_{ij} = n_i^{-1/2}\hat{\sigma}_e^{-1}(y_{ij} - \mathbf{x}_{ij}^T\hat{\boldsymbol{\beta}} - \hat{v}_i),$$

where $\hat{\tau} = 1 - (1 - \hat{\gamma})^{1/2}$ with $\hat{\gamma} = \hat{\sigma}_v^2/(\hat{\sigma}_v^2 + \hat{\sigma}_e^2)$. Method of fitting-of-constants and REML (Section 7.1.2) gave similar estimates of $\sigma_v^2, \sigma_e^2, \boldsymbol{\beta}$, and $\hat{v}_i$.

EBLUP estimates of area totals $\hat{Y}_i^H = N_i(\overline{\mathbf{X}}_i^T\hat{\boldsymbol{\beta}} + \hat{v}_i), i = 1, \ldots, 8$, were calculated as in Section 7.1. Second-order unbiased MSE estimates were calculated using (7.2.12). The coefficient of variation (CV) of the EBLUP varied from 0.12 to 0.16 for six of the areas, while for areas S39 and S34 it was as large as 0.37 and 0.39, respectively.

7.4 *OUTLIER ROBUST EBLUP ESTIMATION

7.4.1 Estimation of Area Means

The EBLUP estimator $\hat{\mu}_i^H$ of μ_i, given by (7.1.16), and $\hat{\overline{Y}}_i^H$ of $\overline{Y}_i$, given by (7.1.24), are also empirical best or Bayes (EB) estimators under normality of the random effects v_i and the errors e_{ij} (see Section 9.3.1). Although the EB estimators are "optimal" under normality, they are sensitive to outliers in the responses, y_{ij}, and can lead to considerable inflation of the MSE. We now turn to robust methods that downweight any influential observation in the data. Outliers frequently occur in business survey data with responses exhibiting few unusually large values.

We consider the general linear mixed model with block-diagonal covariance structure, given by (5.3.1); the basic nested error model (7.1.1) is a special case. Assuming normality and maximizing the joint density of $\mathbf{y} = (\mathbf{y}_1^T, \ldots, \mathbf{y}_m^T)^T$ and $\mathbf{v} = (\mathbf{v}_1^T, \ldots, \mathbf{v}_m^T)^T$ with respect to $\boldsymbol{\beta}$ and $\mathbf{v}$ lead to "mixed model" equations given by

$$\sum_{i=1}^{m} \mathbf{X}_i^T\mathbf{R}_i^{-1}(\mathbf{y}_i - \mathbf{X}_i\boldsymbol{\beta} - \mathbf{Z}_i\mathbf{v}_i) = \mathbf{0}, \tag{7.4.1}$$

$$\mathbf{Z}_i^T\mathbf{R}_i^{-1}(\mathbf{y}_i - \mathbf{X}_i\boldsymbol{\beta} - \mathbf{Z}_i\mathbf{v}_i) - \mathbf{G}_i^{-1}\mathbf{v}_i = \mathbf{0}, \quad i = 1, \ldots, m. \tag{7.4.2}$$

The solution to (7.4.1) and (7.4.2) is identical to the BLUE $\tilde{\boldsymbol{\beta}}(\boldsymbol{\delta})$ and the BLUPs $\tilde{\mathbf{v}}_i(\boldsymbol{\delta}), i = 1, \ldots, m$, given by (5.3.4) and (5.3.3), respectively, but the representations (7.4.1) and (7.4.2) are useful in developing robust estimators. Note that for the nested error model with $k_{ij} = 1$, we have $\mathbf{R}_i = \sigma_e^2\mathbf{I}_{n_i}, \mathbf{Z}_i = \mathbf{1}_{n_i}, \mathbf{G}_i = \sigma_v^2, \mathbf{v}_i = v_i$, and $\boldsymbol{\delta} = (\sigma_v^2, \sigma_e^2)^T$, where $\mathbf{I}_{n_i}$ is the $n_i \times n_i$ identity matrix and $\mathbf{1}_{n_i}$ is an $n_i \times 1$ vector of ones.

We first study symmetric outliers in the distribution of v_i or e_{ij} or both, for the nested error model (7.1.1) with $k_{ij} = 1$. For example, v_i may be generated from a t-distribution with a small degrees of freedom (say 3) or from a mixture distribution, say $0.9N(0, \sigma_v^2) + 0.1N(0, \sigma_{v1}^2)$, which means that a large proportion $p = 0.9$ of the v_i's are generated from the true distribution $N(0, \sigma_v^2)$ and the remaining small proportion

$1-p=0.1$ from the contaminated distribution $N(0,\sigma_{v1}^2)$, with σ_{v1}^2 much larger than σ_v^2. In the case of symmetric outliers, we have zero means for v_i and e_{ij}.

We first obtain robust estimators, $\hat{\boldsymbol{\beta}}_R$ and $\hat{\boldsymbol{\delta}}_R$ of $\boldsymbol{\beta}$ and $\boldsymbol{\delta}$, by solving a robust version of the maximum-likelihood equations (under normality) of $\boldsymbol{\beta}$ and $\boldsymbol{\delta}$. This is done by applying Huber's (1972) ψ-function $\psi_b(u)=u\min(1,b/|u|)$, where $b>0$ is a tuning constant, to the standardized residuals. After that, the mixed model equation (7.4.2) for $\mathbf{v}_i$ is also robustified, using again Huber's ψ-function. The robustified equation for $\mathbf{v}_i$ is given by

$$\begin{aligned}&\mathbf{Z}_i^T\mathbf{R}_i^{-1/2}\boldsymbol{\Psi}_b[\mathbf{R}_i^{-1/2}(\mathbf{y}_i-\mathbf{X}_i\boldsymbol{\beta}-\mathbf{Z}_i\mathbf{v}_i)]\\&\quad-\mathbf{G}_i^{-1/2}\boldsymbol{\Psi}_b(\mathbf{G}_i^{-1/2}\mathbf{v}_i)=0,\quad i=1,\ldots,m,\end{aligned}\tag{7.4.3}$$

where $\boldsymbol{\Psi}_b(\mathbf{u}_i)=(\psi_b(u_{i1}),\ldots,\psi_b(u_{in_i}))^T$ and the tuning constant b is commonly taken as $b=1.345$ in the robustness literature. The choice $b=\infty$ leads to (7.4.2). Substituting the resulting robust estimates $\hat{\boldsymbol{\beta}}_R$ and $\hat{\boldsymbol{\delta}}_R$ for $\boldsymbol{\beta}$ and $\boldsymbol{\delta}$ in (7.4.3), we obtain the robust estimators $\hat{v}_{iR}$ of v_i (Sinha and Rao 2009).

Robust estimators $\hat{\boldsymbol{\beta}}_R$ and $\hat{\boldsymbol{\delta}}_R$ are obtained by solving the robustified ML equations for $\boldsymbol{\beta}$ and $\boldsymbol{\delta}$, given by

$$\sum_{i=1}^m \mathbf{X}_i^T\mathbf{V}_i^{-1}\mathbf{U}_i^{1/2}\boldsymbol{\Psi}_b(\mathbf{r}_i)=0\tag{7.4.4}$$

and

$$\begin{aligned}&\sum_{i=1}^m\left\{\boldsymbol{\Psi}_b^T(\mathbf{r}_i)\mathbf{U}_i^{1/2}\mathbf{V}_i^{-1}\frac{\partial\mathbf{V}_i}{\partial\delta_\ell}\mathbf{V}_i^{-1}\mathbf{U}_i^{1/2}\boldsymbol{\Psi}_b(\mathbf{r}_i)\right.\\&\quad\left.-\mathrm{tr}\left(\mathbf{K}_i\mathbf{V}_i^{-1}\frac{\partial\mathbf{V}_i}{\partial\delta_\ell}\right)\right\}=0,\quad \ell=1,\ldots,q,\end{aligned}\tag{7.4.5}$$

where $\mathbf{r}_i=\mathbf{U}_i^{-1/2}(\mathbf{y}_i-\mathbf{X}_i\boldsymbol{\beta})=(r_{i1},\ldots,r_{in_i})^T$, in which $\mathbf{U}_i$ is a diagonal matrix with diagonal elements equal to the diagonal elements of $\mathbf{V}_i$. Furthermore, $\mathbf{K}_i=c\mathbf{I}_{n_i}$ with $c=E[\psi_b^2(u)]$, where $u\sim N(0,1)$.

Equations (7.4.4) and (7.4.5) reduce to ML equations for the choice $b=\infty$, in which case $\mathbf{U}_i^{1/2}\boldsymbol{\Psi}_b(\mathbf{r}_i)=\mathbf{y}_i-\mathbf{X}_i\boldsymbol{\beta}$ and $\mathbf{K}_i=\mathbf{I}_{n_i}$. Note that in equations (7.4.4) and (7.4.5), the ψ-function is applied to the component-wise standardized residuals $r_{ij}=(y_{ij}-\mathbf{x}_{ij}^T\boldsymbol{\beta})/(\sigma_v^2+\sigma_e^2)^{1/2}, j\in s_i$, unlike the robust ML equations proposed by Huggins (1993) and Richardson and Welsh (1995), in which the ψ-function is applied to the standardized vectors $\mathbf{V}_i^{-1/2}(\mathbf{y}_i-\mathbf{X}_i\boldsymbol{\beta})$.

Sinha and Rao (2009) proposed the Newton–Raphson (NR) method for solving (7.4.4) and (7.4.5), respectively, but the NR algorithm is subject to stability and convergence problems unless the starting values of $\boldsymbol{\beta}$ and $\boldsymbol{\delta}$ are "close" to the true values. Furthermore, it depends on the derivatives $\psi_b'(r_{ij})=\partial\psi_b(r_{ij})/\partial r_{ij}$, which takes the value 1 if $|r_{ij}|\le b$ and 0 otherwise. The zero derivatives can also cause convergence problems.

Following Anderson (1973) for the solution of ML equations, we apply a fixed-point algorithm to solve the robust ML equations (7.4.5) (given $\boldsymbol{\beta}$), for the special case of the nested error model (7.1.1) with $k_{ij} = 1$. This method avoids the calculation of derivatives $\psi_b'(r_{ij})$. Schoch (2012) used iteratively re-weighted least squares (IRWLS) to solve (7.4.4) for given $\boldsymbol{\delta}$. IRWLS is widely used in the robust estimation literature and it is more stable than the NR method.

We first spell out the fixed-point algorithm for $\boldsymbol{\delta} = (\sigma_v^2, \sigma_e^2)^T$, given $\boldsymbol{\beta}$. Noting that $\mathbf{V}_i = \sigma_v^2 \mathbf{1}_{n_i}\mathbf{1}_{n_i}^T + \sigma_e^2 \mathbf{I}_{n_i}$ and writing $\text{tr}(\mathbf{K}_i\mathbf{V}_i^{-1}\partial\mathbf{V}_i/\partial\delta_\ell)$ as $\text{tr}[\mathbf{K}_i\mathbf{V}_i^{-1}(\partial\mathbf{V}_i/\partial\delta_\ell)\mathbf{V}_i^{-1}\mathbf{V}_i]$ with $\partial\mathbf{V}_i/\partial\sigma_v^2 = \mathbf{1}_{n_i}\mathbf{1}_{n_i}^T$ and $\partial\mathbf{V}_i/\partial\sigma_e^2 = \mathbf{I}_{n_i}$, we can express (7.4.5) as a system of fixed-point equations $\mathbf{A}(\boldsymbol{\delta})\boldsymbol{\delta} = \mathbf{a}(\boldsymbol{\delta})$. Here, $\mathbf{a}(\boldsymbol{\delta}) = (a_1(\boldsymbol{\delta}), a_2(\boldsymbol{\delta}))^T$ with

$$a_1(\boldsymbol{\delta}) = \sum_{i=1}^m \boldsymbol{\Psi}^T(\mathbf{r}_i)\mathbf{U}_i^{1/2}\mathbf{V}_i^{-1}\mathbf{U}_i^{1/2}\boldsymbol{\Psi}(\mathbf{r}_i),$$

$$a_2(\boldsymbol{\delta}) = \sum_{i=1}^m \boldsymbol{\Psi}^T(\mathbf{r}_i)\mathbf{U}_i^{1/2}\mathbf{V}_i^{-1}\mathbf{1}_{n_i}\mathbf{1}_{n_i}^T\mathbf{U}_i^{1/2}\boldsymbol{\Psi}(\mathbf{r}_i)$$

and $\mathbf{A}(\boldsymbol{\delta}) = (a_{k,\ell}(\boldsymbol{\delta}))$ is a 2×2 matrix with elements

$$a_{11}(\boldsymbol{\delta}) = \sum_{i=1}^m \text{tr}(\mathbf{K}_i\mathbf{V}_i^{-1}\mathbf{I}_{n_i}\mathbf{V}_i^{-1}\mathbf{I}_{n_i}), \quad a_{12}(\boldsymbol{\delta}) = \sum_{i=1}^m \text{tr}(\mathbf{K}_i\mathbf{V}_i^{-1}\mathbf{I}_{n_i}\mathbf{V}_i^{-1}\mathbf{1}_{n_i}\mathbf{1}_{n_i}^T),$$

$$a_{21}(\boldsymbol{\delta}) = \sum_{i=1}^m \text{tr}(\mathbf{V}_i^{-1}\mathbf{1}_{n_i}\mathbf{1}_{n_i}^T\mathbf{V}_i^{-1}\mathbf{I}_{n_i}), \quad a_{22}(\boldsymbol{\delta}) = \sum_{i=1}^m \text{tr}(\mathbf{V}_i^{-1}\mathbf{1}_{n_i}\mathbf{1}_{n_i}^T\mathbf{V}_i^{-1}\mathbf{1}_{n_i}\mathbf{1}_{n_i}^T)$$

(Chatrchi 2012). The fixed-point iterations are then given by

$$\boldsymbol{\delta}^{(t+1)} = \mathbf{A}^{-1}(\boldsymbol{\delta}^{(t)})\mathbf{a}(\boldsymbol{\delta}^{(t)}), \quad t = 0, 1, \ldots \tag{7.4.6}$$

As starting values $\boldsymbol{\delta}^{(0)} = (\sigma_{v0}^2, \sigma_{e0}^2)^T$, we can take the fitting-of-constants estimators of σ_v^2 and σ_e^2 (Section 7.1.2).

We next describe the IRWLS algorithm for the solution of equation (7.4.4) for $\boldsymbol{\beta}$. By writing $\psi_b(r_{ij}) = r_{ij}[\psi_b(r_{ij})/r_{ij}]$ in (7.4.4), letting $\mathbf{W}_i$ be the diagonal matrix with diagonal elements $[\psi_b(r_{ij})/r_{ij}]^{1/2}$ and denoting $\tilde{\mathbf{X}}_i = \mathbf{W}_i\mathbf{V}_i^{-1}\mathbf{U}_i^{1/2}\mathbf{X}_i$ and $\tilde{\mathbf{y}}_i^{(t)} = \mathbf{W}_i\mathbf{V}_i^{-1}\mathbf{U}_i^{1/2}\mathbf{y}_i$, we can solve (7.4.4) iteratively as

$$\boldsymbol{\beta}_R^{(t+1)} = \left[\sum_{i=1}^m (\tilde{\mathbf{X}}_i^{(t)})^T\tilde{\mathbf{X}}_i^{(t)}\right]^{-1} \sum_{i=1}^m (\tilde{\mathbf{X}}_i^{(t)})^T\tilde{\mathbf{y}}_i^{(t)}, \tag{7.4.7}$$

for given $\boldsymbol{\delta}$. Now, solving (7.4.6) and (7.4.7) jointly, we obtain the robust estimators $\hat{\boldsymbol{\delta}}_R$ and $\hat{\boldsymbol{\beta}}_R$. Starting with $\boldsymbol{\delta}_R^{(0)}$ and $\boldsymbol{\beta}_R^{(0)}$, we obtain $\boldsymbol{\beta}_R^{(1)}$ from (7.4.7) and then substituting $\boldsymbol{\beta}_R^{(1)}$ and $\boldsymbol{\delta}_R^{(0)}$ in the right-hand side of (7.4.6), we obtain $\boldsymbol{\delta}^{(1)}$. Continuing the iterations until convergence, we obtain $\hat{\boldsymbol{\beta}}_R$ and $\hat{\boldsymbol{\delta}}_R = (\hat{\sigma}_{vR}^2, \hat{\sigma}_{eR}^2)^T$.

Based on $\hat{\boldsymbol{\beta}}_R$ and $\hat{\boldsymbol{\delta}}_R$, Sinha and Rao (2009) solved (7.4.3) iteratively by the NR method to get robust EBLUP (REBLUP) estimators $\hat{v}_{iR}$ of v_i for the nested error model (7.1.1), using the EBLUP estimates $\hat{v}_i$ as starting values. NR iterations are quite stable in getting $\hat{v}_{iR}$, unlike in the case of robust estimation of $\boldsymbol{\delta}$. Alternatively, we can directly robustify the BLUP of $\mathbf{v}_i$, given by (5.3.3), to get a different REBLUP. For the nested error model with $k_{ij}=1$, this REBLUP is given by

$$\hat{\mathbf{v}}_{iR} = \mathbf{G}_i\mathbf{Z}_i^T\mathbf{V}_i^{-1}\mathbf{U}_i^{1/2}\boldsymbol{\Psi}_b[\mathbf{U}_i^{-1/2}(\mathbf{y}_i - \mathbf{X}_i\hat{\boldsymbol{\beta}}_R)], \tag{7.4.8}$$

where $\hat{\sigma}_{vR}^2$ and $\hat{\sigma}_{eR}^2$ are substituted for σ_v^2 and σ_e^2 in $\mathbf{G}_i=\sigma_v^2$ and $\mathbf{V}_i^{-1}=\sigma_e^{-2}[\mathbf{I}_{n_i} - \gamma_i\mathbf{1}_{n_i}\mathbf{1}_{n_i}^T]$ and $\mathbf{U}_i$ is the diagonal matrix with diagonal elements all equal to $\sigma_v^2+\sigma_e^2$. A drawback of (7.4.8) is that it applies the Huber ψ-function to the combined errors $\mathbf{v}_i+\mathbf{e}_i$ unlike (7.4.3), which applies the Huber ψ-function to $\mathbf{e}_i=\mathbf{y}_i-\mathbf{X}_i\boldsymbol{\beta}-\mathbf{Z}_i\mathbf{v}_i$ and $\mathbf{v}_i$ separately.

Now using the robust estimators $\hat{\boldsymbol{\beta}}_R$ and $\hat{\mathbf{v}}_{iR}$, a REBLUP estimator of $\mu_i=\overline{\mathbf{X}}_i^T\boldsymbol{\beta}+v_i$ is given by $\hat{\mu}_i^{\mathrm{SR}}=\overline{\mathbf{X}}_i^T\hat{\boldsymbol{\beta}}_R+\hat{v}_{iR}$. Similarly, a REBLUP estimator of the area mean $\overline{Y}_i$ is obtained as

$$\hat{\overline{Y}}_i^{\mathrm{SR}} = N_i^{-1}\left(\sum_{j\in s_i} y_{ij} + \sum_{j\in r_i}\hat{y}_{ijR}\right), \tag{7.4.9}$$

where $\hat{y}_{ijR}=\mathbf{x}_{ij}^T\hat{\boldsymbol{\beta}}_R+\hat{v}_{iR}$ and s_i and r_i denote the sets of sampled and nonsampled units in ith area.

We now turn to the case of possibly nonzero means in v_i or e_{ij} or both. In this case, $\hat{\overline{Y}}_i^{\mathrm{SR}}$ may lead to a significant bias, which in turn can affect the MSE. A bias-corrected robust estimator is obtained by treating the model with errors symmetrically distributed around zero as a working model and then adding an estimator of the mean of prediction errors $\hat{e}_{ijR}=y_{ij}-\hat{y}_{ijR}$, $j\in r_i$, to the "prediction" estimator $\hat{\overline{Y}}_i^{\mathrm{SR}}$. This leads to

$$\hat{\overline{Y}}_i^{\mathrm{SR-BC}} = \hat{\overline{Y}}_i^{\mathrm{SR}} + (1-n_iN_i^{-1})n_i^{-1}\sum_{j\in s_i}\phi_i\psi_c(\hat{e}_{ijR}/\phi_i), \tag{7.4.10}$$

where ϕ_i is a robust estimator of the scale of the area i errors $\hat{e}_{ijR}, j\in s_i$, such as the median absolute deviation, and $\psi_c(\cdot)$ is the Huber ψ-function with tuning constant $c>b$ (Chambers et al. 2014). The bias-correction in (7.4.10), however, is based only on the units $j\in s_i$ and hence it can be considerably variable if n_i is very small. Note that $\hat{\overline{Y}}_i^{\mathrm{SR-BC}}$ reduces to $\hat{\overline{Y}}_i^{\mathrm{SR}}$ if $c=0$. On the other hand, if $c=\infty$ and $n_i/N_i\approx 0$, then $\hat{\overline{Y}}_i^{\mathrm{SR}}\approx\hat{\mu}_i^{\mathrm{SR}}$ and $\hat{\overline{Y}}_i^{\mathrm{SR-BC}}\approx(\overline{\mathbf{X}}_i^T\hat{\boldsymbol{\beta}}_R+\hat{v}_{iR})+(\overline{y}_i-\overline{\mathbf{X}}_i^T\hat{\boldsymbol{\beta}}_R-\hat{v}_{iR})=\overline{y}_i+(\overline{\mathbf{X}}_i-\overline{\mathbf{x}}_i)^T\hat{\boldsymbol{\beta}}_R$, which may be regarded as a robust "survey regression" estimator. The latter is essentially a direct estimator.

We now turn to fully bias-corrected robust estimators that make use of the residuals $\hat{e}_{\ell jR}=y_{\ell jR}-\hat{y}_{\ell jR}$ from other areas $\ell\neq i$ in addition to the residuals $\hat{e}_{ijR}$ from area i. Jiongo, Haziza, and Duchesne (2013) proposed two different approaches for this purpose: approach 1 follows along the lines of Chambers (1986) for constructing robust direct estimators for large areas, while approach 2 uses conditional bias to measure the influence of units in the population (Beaumont, Haziza, and Ruiz-Gazen 2013).

We first express the EBLUP estimator $\hat{\bar{Y}}_i^H$ as a weighted sum of all the sample observations $\{y_{\ell j}, \ell = 1, \ldots, m; j = 1, \ldots, n_\ell\}$:

$$\hat{\bar{Y}}_i^H = N_i^{-1} \sum_{\ell=1}^{m} \sum_{j \in s_\ell} w_{i\ell j}(\hat{\boldsymbol{\delta}}) y_{\ell j}, \tag{7.4.11}$$

where the weights $w_{i\ell j}(\hat{\boldsymbol{\delta}})$ depend on $\mathbf{X}_\ell, \mathbf{V}_\ell, \mathbf{x}_{\ell j}(j \in r_i; \ell = 1, \ldots, m)$ and σ_v^2, see Jiongo et al. (2013) for explicit formulae for $w_{iij}(\hat{\boldsymbol{\delta}})$ and $w_{i\ell j}(\hat{\boldsymbol{\delta}}), \ell \neq i$. It is interesting to note that $\hat{\bar{Y}}_i^H$ satisfies calibration to known totals $\mathbf{X}_{i+} = \sum_{j=1}^{N_i} \mathbf{x}_{ij}$ in the sense

$$\sum_{\ell=1}^{m} \sum_{j \in s_\ell} w_{i\ell j}(\hat{\boldsymbol{\delta}}) \mathbf{x}_{\ell j} = \mathbf{X}_{i+}. \tag{7.4.12}$$

Using the representation (7.4.11) and following Chambers (1986), Jiongo et al. (2013) show that $\hat{\bar{Y}}_i^H$ can be expressed in terms of $\hat{\bar{Y}}_i^{\text{SR}}$ as

$$\hat{\bar{Y}}_i^H = \hat{\bar{Y}}_i^{\text{SR}} + N_i^{-1} \sum_{j \in s_i} [w_{iij}(\hat{\boldsymbol{\delta}}) - 1] \hat{e}_{ijR} + N_i^{-1} \sum_{\ell \neq i}^{m} \sum_{j \in s_\ell} w_{i\ell j}(\hat{\boldsymbol{\delta}}) \hat{e}_{\ell j}^R$$
$$+ N_i^{-1} \sum_{\ell=1}^{m} W_{i\ell}(\hat{\boldsymbol{\delta}}) \hat{v}_{\ell R}, \tag{7.4.13}$$

where

$$W_{i\ell}(\hat{\boldsymbol{\delta}}) = \begin{cases} \sum_{j \in s_i} w_{iij}(\hat{\boldsymbol{\delta}}) - N_i & \text{if } \ell = i; \\ \sum_{j \in s_\ell} w_{i\ell j}(\hat{\boldsymbol{\delta}}) & \text{if } \ell \neq i. \end{cases} \tag{7.4.14}$$

A fully bias-corrected REBLUP is now obtained by applying Huber ψ-functions to the weighted terms in each of the three sums on the right-hand side of (7.4.13), with possibly different tuning constants c_1 and c_2. This leads to

$$\hat{\bar{Y}}_i^{\text{JHD1}} = \hat{\bar{Y}}_i^{\text{SR}} + N_i^{-1} \sum_{j \in s_i} \psi_{c_1}\{[w_{iij}(\hat{\boldsymbol{\delta}}) - 1] \hat{e}_{ijR}\}$$
$$+ N_i^{-1} \sum_{\ell \neq i}^{m} \sum_{j \in s_\ell} \psi_{c_1}\{w_{i\ell j}(\hat{\boldsymbol{\delta}}) \hat{e}_{\ell j}^R\} + N_i^{-1} \sum_{\ell=1}^{m} \psi_{c_2}[W_{i\ell}(\hat{\boldsymbol{\delta}}) \hat{v}_{\ell R}]. \tag{7.4.15}$$

Note that $\hat{\overline{Y}}_i^{\text{JHD1}}$ tends to $\hat{\overline{Y}}_i^{\text{SR}}$ as $c_1 \to 0$ and $c_2 \to 0$ and to $\hat{\overline{Y}}_i^{H}$ as $c_1 \to \infty$ and $c_2 \to \infty$. For the choice of c_1 and c_2, Jiongo et al. (JHD) recommend $c_1 = \alpha\, \hat{\sigma}_{eR}\, \text{med}(w_{i\ell j}(\hat{\delta}))$ and $c_2 = \alpha\, \hat{\sigma}_{vR}\, \text{med}(W_{i\ell}(\hat{\delta}))$ for some constant α; in particular, a larger α (say $\alpha = 9$) seems to control the biases, and the corresponding MSEs are small.

Under the second approach, the fully bias-corrected estimator $\hat{\overline{Y}}_i^{\text{JHD1}}$ is closely related to (7.4.15), except that $\psi_{c_2}[W_{i\ell}(\hat{\delta})\hat{v}_{\ell R}]$ is changed to $W_{i\ell}(\hat{\delta})\hat{v}_{\ell R}$, that is, the Huber ψ-function is not applied to the last term of (7.4.13).

7.4.2 MSE Estimation

Given that the underlying distributions of v_i and e_{ij} are unknown, a nonparametric bootstrap method, based on resampling from the estimated random effects $\hat{v}_i$ and the residuals $\hat{e}_{ij}$, might look plausible for the estimation of MSE of the robust estimators of $\overline{Y}_i$. However, as noted by Salibian-Barrera and Van Aelst (2008), the proportion of outliers in the bootstrap data $\{y_{ij}^*\}$ may be much higher than in the original data $\{y_{ij}\}$ and, this difficulty may lead to poor performance of the nonparametric bootstrap MSE estimators in the presence of outliers.

Sinha and Rao (2009) proposed instead to use a parametric bootstrap method based on the robust estimates $\hat{\boldsymbol{\beta}}_R$ and $\hat{\boldsymbol{\delta}}_R$ assuming normality. The motivation behind this method is that our focus is on deviations from the working assumption of normality of the v_i and the e_{ij}, and that it is natural to use robust estimates $\hat{\boldsymbol{\beta}}_R$ and $\hat{\boldsymbol{\delta}}_R$ for drawing bootstrap samples since $\text{MSE}(\hat{\overline{Y}}_i^{\text{SR}})$ is not sensitive to outliers. The parametric bootstrap method is described as follows: (i) For given robust estimates $\hat{\boldsymbol{\beta}}_R$ and $\hat{\boldsymbol{\delta}}_R = (\hat{\sigma}_{vR}^2, \hat{\sigma}_{eR}^2)^T$, generate v_i^* and e_{ij}^* from $N(0, \hat{\sigma}_{vR}^2)$ and $N(0, \hat{\sigma}_{eR}^2)$ independently to create a bootstrap sample $\{y_{ij}^*; j = 1, \ldots, n_i,\ i = 1, \ldots, m\}$, where $y_{ij}^* = \mathbf{x}_{ij}^T\hat{\boldsymbol{\beta}}_R + v_i^* + e_{ij}^*$. Compute the corresponding bootstrap area mean $\overline{Y}_{i*} = N_i^{-1}\sum_{j\in s_i} y_{ij}^* + N_i^{-1}\sum_{j\in r_i}\mathbf{x}_{ij}^T\hat{\boldsymbol{\beta}}_R + (1 - n_iN_i^{-1})(v_i^* + \overline{e}_{ir}^*)$, where $\overline{e}_{ir}^*$ is generated from $N(0, (N_i - n_i)^{-1}\hat{\sigma}_{eR}^2)$. This is equivalent to generating $y_{ij}^* = \mathbf{x}_{ij}^T\hat{\boldsymbol{\beta}}_R + v_i^* + e_{ij}^*$, for $j = 1, \ldots, N_i$ and then taking their mean $\overline{Y}_{i*} = N_i^{-1}\sum_{j=1}^{N_i} y_{ij}^*$. Compute the REBLUP $\hat{\overline{Y}}_{i*}^{\text{SR}}$ from the generated bootstrap sample. (ii) Repeat step (i) B times. Let $\hat{\overline{Y}}_{i*}^{\text{SR}}(b)$ be the robust EBLUP estimate obtained in bootstrap replicate b and $\overline{Y}_{i*}(b)$ the corresponding bootstrap mean, $b = 1, \ldots, B$. Bootstrap MSE estimator of $\hat{\overline{Y}}_i^{\text{SR}}$ is then given by

$$\text{mse}_B(\hat{\overline{Y}}_i^{\text{SR}}) = \frac{1}{B}\sum_{b=1}^{B}\left[\hat{\overline{Y}}_{i*}^{\text{SR}}(b) - \overline{Y}_{i*}(b)\right]^2. \tag{7.4.16}$$

Bootstrap MSE estimator of $\hat{\overline{Y}}_i^{\text{SR-BC}}$ is obtained similarly.

Jiongo et al. (2013) proposed an alternative parametric bootstrap MSE estimator based on generating v_i^* and e_{ij}^* from $N(0, \hat{\sigma}_v^2)$ and $N(0, \hat{\sigma}_e^2)$, where $\hat{\sigma}_v^2$ and $\hat{\sigma}_e^2$ are the nonrobust ML or REML estimators.

Chambers et al. (2014) studied conditional MSE of $\hat{\bar{Y}}_i^{\text{SR}}$ and its bias-corrected version $\hat{\bar{Y}}_i^{\text{SR-BC}}$, by conditioning on the realized values of the area effects $\{v_i; i = 1, \ldots, m\}$. Two different estimators of the conditional MSE were proposed. The first one is based on a "pseudo-linear" representation for $\hat{\bar{Y}}_i^{\text{SR}}$, and the second estimator is based on a linearization of the robustified equations (7.4.3) for v_i and (7.4.4) for $\boldsymbol{\beta}$. Similar conditional MSE estimators were calculated also for the EBLUP estimator $\hat{\bar{Y}}_i^H$. The first MSE estimator, although computationally simpler, is significantly less stable, in terms of CV, than the second MSE estimator.

Simulation results in Chambers et al. (2014) showed that, in the case of no outliers, customary second-order unbiased Prasad–Rao (PR) MSE estimator (7.2.12) of the EBLUP estimator is much more stable, in terms of relative root mean squared error (RRMSE) or CV, than the proposed conditional MSE estimators. This is true especially when the area sample sizes, n_i, are small. Under the same no outlier scenario, the bootstrap MSE estimator (7.4.16) of the REBLUP estimator $\hat{\bar{Y}}_i^{\text{SR}}$ was significantly more stable than the conditional MSE estimators. Bootstrap MSE estimator for the bias-corrected estimator $\hat{\bar{Y}}_i^{\text{SR-BC}}$ performed even better, with median RRMSE about one-third of the values for the conditional MSE estimators.

7.4.3 Simulation Results

Sinha and Rao (2009) conducted a simulation study on the performance of REBLUP estimators $\hat{\mu}_i^{\text{SR}}$ and $\hat{\bar{Y}}_i^{\text{SR}}$ relative to the EBLUP estimators $\hat{\mu}_i^H$ and $\hat{\bar{Y}}_i^H$, respectively, under two different simulation models to generate symmetric outliers in the area effects v_i only, or the errors e_{ij}, or in both. The first simulation model used mixture normal distributions of the form $(1 - p_1)N(0, \sigma_v^2) + p_1 N(0, \sigma_{v1}^2)$ for v_i and $(1 - p_2)N(0, \sigma_e^2) + p_2 N(0, \sigma_{e1}^2)$ for e_{ij} with $p_1 = p_2 = 0.1, \sigma_v^2 = \sigma_e^2 = 1$, and $\sigma_{v1}^2 = \sigma_{e1}^2 = 25$. This means that a large proportion $1 - p_1 = 0.9$ of the v_i's are generated from the underlying "true" distribution $N(0, 1)$ and the remaining small proportion $p_1 = 0.10$ are generated from the contaminated distribution $N(0, 25)$, and similarly for the e_{ij}'s. The second simulation model used t-distribution with $k = 3$ degrees of freedom for both v_i and e_{ij}. The case of no outliers in v_i or e_{ij} is also studied. They considered $m = 30$ small areas with sample sizes $n_i = 4$, $i = 1, \ldots, m$.

Results of simulations may be summarized as follows: (i) In the case of no outliers, REBLUP estimators are similar to the corresponding EBLUP estimators in terms of MSE, indicating very small loss in efficiency. (ii) In the case of outliers in the errors e_{ij}, REBLUP estimators are much more efficient than the corresponding EBLUP estimators. (iii) On the other hand, in the case of outliers only in the area effects v_i, REBLUP and EBLUP gave similar results, indicating robustness of EBLUP in this case. (iv) The bootstrap MSE estimator performs well in tracking the MSE of the REBLUP.

Simulation results in Jiongo et al. (2013) for the case of nonsymmetric outliers indicated that the fully bias-corrected robust estimators $\hat{\bar{Y}}_i^{\text{JHD1}}$ and $\hat{\bar{Y}}_i^{\text{JHD2}}$ perform well in terms of MSE unlike $\hat{Y}_i^{SR}$ and $\hat{Y}_i^{SR-BC}$; in the latter case, MSE is inflated due to significant bias.

7.5 *M-QUANTILE REGRESSION

As an alternative to modeling the between-area variation through additive area random effects in the nested error model (7.1.1) and with the purpose of obtaining robust estimators, Chambers and Tzavidis (2006) proposed M-quantile regression models. In these models, the between-area variation is incorporated through area-specific quantile-like coefficients.

For a real-valued random variable y with probability density function $f(y)$, Breckling and Chambers (1988) defined the M-quantile of order $q \in (0, 1)$ as the solution $Q_\psi(q)$ of the following equation in Q:

$$\int_{-\infty}^{\infty} \psi_q(y - Q) f(y) dy = 0, \tag{7.5.1}$$

where $\psi_q(u) = 2\psi(u)[qI(u > 0) + (1 - q)I(u \leq 0)]$ and $\psi(u)$ is a monotone nondecreasing function with $\psi(-\infty) < \psi(0) = 0 < \psi(\infty)$. For $\psi(u) = \text{sign}(u)$, $Q_\psi(q)$ reduces to the ordinary quantile of order q, while the choice $\psi(u) = u$ with $q = 1/2$ yields the expected value.

The above definition of M-quantile of order q readily extends to the conditional density $f(y|\mathbf{x})$ by replacing $f(y)$ in (7.5.1) by $f(y|\mathbf{x})$. The resulting solution is denoted as $Q_\psi(\mathbf{x}; q)$. A linear M-quantile regression model for a fixed q and specified ψ is given by $Q_\psi(\mathbf{x}; q) = \mathbf{x}^T \boldsymbol{\beta}_\psi(q)$. The standard linear regression model $E(y|\mathbf{x}) = \mathbf{x}^T \boldsymbol{\beta}$ is a special case of this model by letting $q = 1/2$ and $\psi(u) = u$. Similarly, the ordinary quantile regression (Koenker and Bassett 1978) is obtained as a special case by setting $\psi(u) = \text{sign}(u)$. In the small area estimation context, a linear M-quantile regression model is assumed for every q in $(0, 1)$, in contrast to the basic unit level model, which assumes a nested error regression model only on the conditional mean $E(y|\mathbf{x})$. Moreover, $\boldsymbol{\beta}_\psi(q)$ is assumed to be a continuous function of q for a given ψ.

In M-quantile regression, based on a sample $\{(y_\ell, \mathbf{x}_\ell); \ell = 1, \ldots, n\}$, the estimator of the M-quantile regression coefficient $\boldsymbol{\beta}_\psi(q)$ is obtained by solving the following sample estimating equation for $\boldsymbol{\beta}_\psi(q)$,

$$\sum_{\ell=1}^{n} \psi_q \left(\frac{y_\ell - \mathbf{x}_\ell^T \boldsymbol{\beta}_\psi(q)}{\tilde{s}} \right) \mathbf{x}_\ell = 0, \tag{7.5.2}$$

where $\tilde{s}$ is a suitable robust scale estimator such as the mean absolute deviation $\tilde{s} = \text{med}_{\ell=1,\ldots,n} |y_\ell - \mathbf{x}_\ell^T \boldsymbol{\beta}_\psi(q)| / 0.6745$. Using $\psi(u) = \text{sign}(u)$, we obtain the estimating equation for ordinary quantile regression, while for $q = 1/2$ and $\psi(u) = u$ the solution of (7.5.2) is the least squares estimator of $\boldsymbol{\beta}$ in the usual regression model $E(y_\ell|\mathbf{x}_\ell) = \mathbf{x}_\ell^T \boldsymbol{\beta}, \ell = 1, \ldots, n$.

Equation (7.5.2), for given ψ and q, can be solved by IRWLS. Defining the weight function $W_q(u) = \psi_q(u)/u$ for $u \neq 0$ and $W_q(0) = \psi_q'(0)$, (7.5.2) can be expressed as

$$\sum_{\ell=1}^{n} W_q\left(\frac{y_\ell - \mathbf{x}_\ell^T \boldsymbol{\beta}_\psi(q)}{\tilde{s}}\right)\mathbf{x}_\ell(y_\ell - \mathbf{x}_\ell^T \boldsymbol{\beta}_\psi(q)) = 0. \tag{7.5.3}$$

Solving for the $\boldsymbol{\beta}_\psi(q)$ within the second parenthesis, we obtain the IRWLS updating equation, given by

$$\hat{\boldsymbol{\beta}}_\psi^{(k+1)}(q) = \left[\sum_{j=1}^{n} W_q\left(\frac{y_\ell - \mathbf{x}_\ell^T \hat{\boldsymbol{\beta}}_\psi^{(k)}(q)}{\tilde{s}^{(k)}}\right)\mathbf{x}_\ell \mathbf{x}_\ell^T\right]^{-1}$$
$$\times \sum_{\ell=1}^{n} W_q\left(\frac{y_\ell - \mathbf{x}_\ell^T \hat{\boldsymbol{\beta}}_\psi^{(k)}(q)}{\tilde{s}^{(k)}}\right)\mathbf{x}_\ell y_\ell. \tag{7.5.4}$$

The resulting solution is denoted by $\hat{\boldsymbol{\beta}}_\psi(q)$ and the associated conditional M-quantile by $\hat{Q}_\psi(\mathbf{x}; q) = \mathbf{x}^T \hat{\boldsymbol{\beta}}_\psi(q)$. The main reason for considering M-quantiles instead of the usual quantiles is that a continuous monotone ψ function ensures that the IRWLS algorithm (7.5.4) converges to a unique solution (Kokic et al. 1997). Moreover, selecting a bounded ψ ensures robustness in the sense of bounded influence function. An example of a bounded and continuous ψ-function is Huber's Proposal 2, given by $\psi(u) = uI(|u| \le c) + c \text{ sign}(u)I(|u| > c)$.

Chambers and Tzavidis (2006) proposed small area estimators based on a specific M-quantile regression model that accommodates area effects in a different way. They defined the M-quantile coefficient q_ℓ associated with $(y_\ell, \mathbf{x}_\ell)$ for the population unit ℓ as the value q_ℓ such that $Q_\psi(\mathbf{x}_\ell; q_\ell) = \mathbf{x}_\ell^T \boldsymbol{\beta}_\psi(q_\ell) = y_\ell$. Furthermore, they argued that if there is a hierarchical structure in the distribution of the population values y_ℓ given $\mathbf{x}_\ell$ with between-area and within-area variability, then units in the same area should have similar M-quantile coefficients. Using this argument, they assumed the population model

$$Q_\psi(\mathbf{x}_\ell; \theta_i) = \mathbf{x}_\ell^T \boldsymbol{\beta}_\psi(\theta_i), \quad \ell \in U_i, \quad i = 1, \ldots, m, \tag{7.5.5}$$

where U_i is the set of N_i population units from ith area and $\theta_i = N_i^{-1}\sum_{\ell \in U_i} q_\ell$ is the mean of the M-quantile coefficients q_ℓ for the units ℓ in area i.

In practice, θ_i for each area i needs to be estimated from the sample. This is done by solving $\hat{Q}_\psi(\mathbf{x}_\ell; q) = y_\ell$ for q to get $\hat{q}_\ell$ for each $\ell = 1, \ldots, n$, and then taking $\hat{\theta}_i := n_i^{-1}\sum_{\ell \in s_i} \hat{q}_\ell$ as the estimator of θ_i. The values $\hat{q}_\ell$ are determined by calculating $\hat{Q}_\psi(\mathbf{x}_\ell; q), \ell = 1, \ldots, n$, for a fine grid of q-values in the (0, 1) interval, and selecting the value $\hat{q}_\ell$ such that $\hat{Q}_\psi(\mathbf{x}_\ell; \hat{q}_\ell) = y_\ell$ by linear interpolation. For each $\hat{\theta}_i$, the corresponding $\hat{\boldsymbol{\beta}}_\psi(\hat{\theta}_i)$ is obtained by IRWLS applied to the sample data. The M-quantile predictor of a nonsample unit $y_\ell, \ell \in U_i - s_i = r_i$, is taken as $\hat{y}_\ell^{\text{MQ}} = \mathbf{x}_\ell^T \hat{\boldsymbol{\beta}}_\psi(\hat{\theta}_i)$ using the assumed population model (7.5.5). Finally, the resulting predictor of ith area mean $\overline{Y}_i = N_i^{-1}\sum_{\ell\in U_i} y_\ell$ is given by

$$\hat{\overline{Y}}_i^{\text{MQ}} = N_i^{-1}\left(\sum_{\ell\in s_i} y_\ell + \sum_{\ell\in r_i} \hat{y}_\ell^{\text{MQ}}\right). \tag{7.5.6}$$

Estimator (7.5.6) does not minimize any sound alternative criterion to the mean squared error (MSE), unlike the EBLUP under the assumed unit level model, which minimizes the MSE among the linear and unbiased estimators. Furthermore, it appears that uniform consistency of the estimators $\hat{\boldsymbol{\beta}}_\psi(q)$ is needed in order to justify the method theoretically, where uniform consistency means that $\sup_{0<q<1} |\hat{\boldsymbol{\beta}}_\psi(q) - \boldsymbol{\beta}_\psi(q)| \rightarrow_p 0$. Also, note that $\hat{\boldsymbol{\beta}}_\psi(\hat{\theta}_i)$ can be significantly different from $\hat{\boldsymbol{\beta}}_\psi(\theta_i)$ if n_i is small because $\hat{\theta}_i$ may not be a precise estimator of θ_i.

Note that under this approach, estimating the m area means $\overline{Y}_1, \ldots, \overline{Y}_m$ requires the estimation of m separate M-quantile regression models, one for each $\hat{\theta}_i$. A problem of fitting models for several M-quantile coefficients using the same sample data $(y_1, \mathbf{x}_1), \ldots, (y_n, \mathbf{x}_n)$, is that the regression planes for $\theta_k < \theta_i$ might cross each other. Adjustments of the fitted regression planes to force monotonicity have been studied, see Chambers (2005), Koenker (1984), and He (1997).

For the estimation of $\text{MSE}(\hat{\overline{Y}}_i^{\text{MQ}})$, Chambers and Tzavidis (2006) expressed $\hat{\overline{Y}}_i^{\text{MQ}}$ in (7.5.6) as

$$\hat{\overline{Y}}_i^{\text{MQ}} = N_i^{-1}\left(\sum_{\ell\in s_i} y_\ell + \sum_{\ell\in s} u_{i\ell} y_\ell\right), \tag{7.5.7}$$

where

$$u_{i\ell} = \left(\sum_{\ell\in r_i} \mathbf{x}_\ell\right)^T \left[\sum_{\ell=1}^{n} W_{\hat{\theta}_i}\left(\frac{y_\ell - \mathbf{x}_\ell^T \hat{\boldsymbol{\beta}}_\psi(\hat{\theta}_i)}{\tilde{s}}\right)\mathbf{x}_\ell \mathbf{x}_\ell^T\right]^{-1}$$
$$\times \sum_{\ell=1}^{n} W_{\hat{\theta}_i}\left(\frac{y_\ell - \mathbf{x}_\ell^T \hat{\boldsymbol{\beta}}_\psi(\hat{\theta}_i)}{\tilde{s}}\right)\mathbf{x}_\ell .$$

Subtracting the true mean $\overline{Y}_i = N_i^{-1}\left(\sum_{\ell\in s_i} y_\ell + \sum_{\ell\in r_i} y_\ell\right)$ from (7.5.7), we obtain

$$\hat{\overline{Y}}_i^{\text{MQ}} - \overline{Y}_i = \left(\sum_{\ell\in s} u_{i\ell} y_\ell - \sum_{\ell\in r_i} y_\ell\right).$$

Assuming that all population values $y_\ell, j = 1, \ldots, N$, are uncorrelated, ignoring the randomness of $\hat{\theta}_i$, and regarding $u_{i\ell}$ as nonstochastic for all ℓ, Chambers and Tzavidis (2006) obtained

$$V_i := V(\hat{\overline{Y}}_i^{\text{MQ}} - \overline{Y}_i) = N_i^{-2}\left[\sum_{\ell\in s} u_{i\ell}^2 V(y_\ell) + \sum_{\ell\in r_i} V(y_\ell)\right].$$

Then, V_i is estimated using the residuals from an M-quantile fit with $q = 0.5$. The proposed estimator of V_i is then given by

$$\hat{V}_i = N_i^{-2}\left[\sum_{\ell\in s} u_{i\ell}^2\left(y_\ell - \mathbf{x}_\ell^T\hat{\boldsymbol{\beta}}_\psi(0.5)\right)^2 + \frac{N_i - n_i}{n_i - 1}\sum_{\ell\in s_i}(y_\ell - \mathbf{x}_\ell^T\hat{\boldsymbol{\beta}}_\psi(0.5))^2\right]. \quad (7.5.8)$$

The estimator (7.5.8) uses residuals $y_\ell - \mathbf{x}_\ell^T\hat{\boldsymbol{\beta}}_\psi(0.5)$ that are not specific to the area from which the unit $\ell \in s$ is drawn. Alternatively, one could use a conditional approach based on the residuals $y_\ell - \mathbf{x}_\ell^T\hat{\boldsymbol{\beta}}_\psi(\hat{\theta}_i)$ if unit ℓ is drawn from area i.

Concerning the bias of $\hat{\overline{Y}}_i^{\text{MQ}}$, defining $w_{i\ell} = I(\ell \in s_i) + u_{i\ell}$, where $I(\ell \in s_i) = 1$ if $\ell \in s_i$ and $I(\ell \in s_i) = 0$ otherwise, we have $\hat{\overline{Y}}_i^{\text{MQ}} = N_i^{-1}\sum_{\ell\in s} w_{i\ell} y_\ell$ and consequently $\hat{\overline{Y}}_i^{\text{MQ}} - \overline{Y}_i = N_i^{-1}(\sum_{\ell\in s} w_{i\ell} y_\ell - \sum_{\ell\in U_i} y_\ell)$. Assuming that $E(y_\ell|\mathbf{x}_\ell) = \mathbf{x}_\ell^T\boldsymbol{\beta}_i, \ell \in U_i, i = 1, \ldots, m$, and regarding again $w_{i\ell}$ as nonstochastic, the bias of $\hat{\overline{Y}}_i^{\text{MQ}}$ with respect to the assumed model is given by

$$B_i := E(\hat{\overline{Y}}_i^{\text{MQ}} - \overline{Y}_i) = N_i^{-1}\left(\sum_{k=1}^{m}\sum_{\ell\in s_k} w_{i\ell}\mathbf{x}_\ell^T\boldsymbol{\beta}_k - \sum_{\ell\in U_i}\mathbf{x}_\ell^T\boldsymbol{\beta}_i\right).$$

Then, B_i is estimated as

$$\hat{B}_i = N_i^{-1}\left[\sum_{k=1}^{m}\sum_{\ell\in s_k} w_{i\ell}\mathbf{x}_\ell^T\hat{\boldsymbol{\beta}}_\psi(\hat{\theta}_k) - \sum_{\ell\in U_i}\mathbf{x}_\ell^T\hat{\boldsymbol{\beta}}_\psi(\hat{\theta}_i)\right].$$

Finally, an estimator of $M_i := \text{MSE}(\hat{\overline{Y}}_i^{\text{MQ}})$ is taken as

$$\hat{M}_i = \hat{V}_i + \hat{B}_i^2. \quad (7.5.9)$$

A limitation of $\hat{V}_i$ or $\hat{M}_i$ is that the variability due to estimating $\theta_i = N_i^{-1}\sum_{\ell\in U_i} q_\ell$ by $\hat{\theta}_i = n_i^{-1}\sum_{\ell\in s_i}\hat{q}_\ell$ is not accounted for (Fabrizi, Salvati, and Pratesi 2012). Moreover, $\hat{V}_i$ can be highly unstable if n_i is very small because the last term of (7.5.8) depends only on the residuals for area i.

A bias-adjusted M-quantile estimator of $\overline{Y}_i$ is given by

$$\hat{\overline{Y}}_i^{\text{MQA}} = N_i^{-1}\left[\sum_{\ell\in s_i} y_\ell + \sum_{\ell\in r_i}\hat{y}_\ell^{\text{MQ}} + \frac{N_i - n_i}{n_i}\sum_{\ell\in s_i}(y_\ell - \hat{y}_\ell^{\text{MQ}})\right],$$

see Tzavidis, Marchetti, and Chambers (2010). A drawback of $\hat{\overline{Y}}_i^{\text{MQA}}$ is that the bias adjustment is based on the mean of the residuals $y_\ell - \hat{y}_\ell^{\text{MQ}}$ from area i, which could be unstable if n_i is very small. The estimator $\hat{\overline{Y}}_i^{\text{MQA}}$ may be motivated by expressing $\overline{Y}_i$ as

$$\overline{Y}_i = N_i^{-1}\left(\sum_{\ell\in s_i} y_\ell + \sum_{\ell\in r_i} y_\ell\right) = N_i^{-1}\left[\sum_{\ell\in s_i} y_\ell + \sum_{\ell\in r_i} \hat{y}_\ell^{\text{MQ}} + \sum_{\ell\in r_i}(y_\ell - \hat{y}_\ell^{\text{MQ}})\right]$$

and then estimating the last term using the area-specific estimator $(N_i - n_i)n_i^{-1}\sum_{\ell\in s_i}(y_\ell - \hat{y}_\ell^{\text{MQ}})$. Tzavidis et al. (2010) proposed nonparametric bootstrap estimators of MSE$(\hat{\overline{Y}}_i^{\text{MQ}})$ and MSE$(\hat{\overline{Y}}_i^{\text{MQA}})$.

Turning to nonsampled areas, it is not clear how one should proceed to construct M-quantile estimators of the means $\overline{Y}_i, i = m+1, \ldots, M$, where M is the number of areas in the population and only the first m areas are sampled. One possible solution is to consider $\theta_i = 0.5$ and use the synthetic estimator

$$\hat{\overline{Y}}_{iS}^{\text{MQ}} = N_i^{-1} \sum_{\ell\in U_i} \mathbf{x}_\ell^T \hat{\boldsymbol{\beta}}_\psi(0.5), \tag{7.5.10}$$

(see Pfeffermann 2013).

Fabrizi et al. (2012) studied modified M-quantile estimators of small area means that ensure benchmarking to reliable direct estimator for a large area covering the specified small areas.

Example 7.5.1. Simulation Study. Chambers, Chandra, and Tzavidis (2011) conducted a simulation study on the relative performance of the EBLUP, M-quantile (MQ), and a model-based direct estimator (MBDE) of area mean $\overline{Y}_i$ under the basic unit level model (4.3.1) with $k_{ij} = 1$. The MBDE estimator of the area mean $\overline{Y}_i$ is obtained by, first, expressing the EBLUP of the population total Y under the same unit level model as $\hat{Y}^H = \sum_{k\in s} w_k(\hat{\delta})y_k$, and then using the weights of sample units j from area i to define the area-specific direct estimator $\hat{\overline{Y}}_i^{\text{MBD}} = \sum_{j\in s_i} w_j(\hat{\delta})y_j / \sum_{j\in s_i} w_j(\hat{\delta})$ (see Chandra and Chambers 2009). They considered two different scenarios: (i) $\sigma_v^2 = 10, \sigma_e^2 = 94.09$; (ii) $\sigma_v^2 = 43.02, \sigma_e^2 = 94.09$. The number of areas is $m = 50$ and the area sample sizes are taken equal for all areas, $n_i = 5$ or $n_i = 15$. The simulation results based on median relative root MSE (RRMSE) show that the EBLUP leads to large reduction in median RRMSE relative to MQ and MBDE when the basic unit level model holds and $n_i = 5$. In particular, for $n_i = 5$, the median RRMSE of EBLUP under scenario 1 is equal to 0.53 compared to 0.81 for MQ and 1.13 for MBDE. Under scenario 2, the respective values are 0.69 for EBLUP, 0.81 for MQ, and 1.13 for MBDE. However, as n_i increases to $n_i = 15$, the performance for MBDE and MQ relative to EBLUP improves; for example, under scenario 1, median RRMSE of EBLUP is 0.35 compared to 0.41 for MQ and 0.55 for MBDE. Note that the MBDE estimator is mainly based on the area-specific data and hence it is less efficient than EBLUP, which borrows strength across areas.

7.6 *PRACTICAL ISSUES

7.6.1 Unknown Heteroscedastic Error Variances

The basic unit level model (7.1.1) assumes that the unit errors e_{ij} have means 0 and variances $\sigma_{eij}^2 = k_{ij}^2\sigma_e^2$ with known k_{ij}; in particular, the case with $k_{ij} = 1$ is often studied. In this section, we relax the assumption on σ_{eij}^2 in different ways by letting $\sigma_{eij}^2 = \sigma_{ei}^2$, where $\sigma_{ei}^2, i = 1, \ldots, m$ are unknown.

Kleffe and Rao (1992) studied the mean model, $y_{ij} = \mu + v_i + e_{ij}$, assuming that $v_i \overset{\text{iid}}{\sim} (0, \sigma_v^2), e_{ij}|\sigma_{ei}^2 \overset{\text{ind}}{\sim} (0, \sigma_{ei}^2)$, and $\sigma_{ei}^2 \overset{\text{iid}}{\sim} (\sigma_e^2, \delta_e)$. Under this random error variance model, the EBLUP estimator of the ith area mean, $\mu_i = \mu + v_i$, for the balanced case $n_i = \overline{n}, i = 1, \ldots, m$, is identical to the EBLUP obtained under the customary assumption $\sigma_{ei}^2 = \sigma_e^2$ (i.e., $\delta_e = 0$). However, the MSE of the EBLUP under the random error variances model is different from the MSE under the equal variance model. Kleffe and Rao (1992) derived second-order approximations to the MSE and its unbiased estimator up to $o(m^{-1})$ terms, for large m, assuming normality of v_i and $e_{ij}|\sigma_{ei}^2$ but no distributional assumption on σ_{ei}^2 (see Rao 2003a, Section 8.7). Arora and Lahiri (1997) studied the regression case with $\mathbf{x}_{ij} = \mathbf{x}_i$ and known parameters $\boldsymbol{\beta}$ and σ_v^2. Under normality of v_i and $e_{ij}|\sigma_{ei}^2$ and a specified distribution on σ_{ei}^2, they showed that the Bayes (best) estimator of μ_i is more efficient than the BLUP estimator of Kleffe and Rao (1992). Note that the BLUP estimator makes no distributional assumptions unlike the Bayes estimator.

Jiang and Nguyen (2012) proposed a different approach to accommodate unequal error variances. They considered the heteroscedastic unit level model given by

$$y_{ij} = \mathbf{x}_{ij}^T\boldsymbol{\beta} + u_i + t_{ij}, \tag{7.6.1}$$

where $u_i = \sigma_i v_i$ with $v_i \overset{\text{iid}}{\sim} N(0, \sigma_v^2)$ and $t_{ij} = \sigma_i e_{ij}$ with $e_{ij} \overset{\text{iid}}{\sim} N(0, \sigma_e^2)$, and $\sigma_i^2, i = 1, \ldots, m$ are unknown. The basic unit level model (7.1.1) with $k_{ij} = 1$ is a special case of model (7.6.1) with $\sigma_i = 1$. The heteroscedastic model (7.6.1) ensures that the ratio $\tau_i = V_m(u_i)/V_m(t_{ij})$ is constant for all i, since $\tau_i = \sigma_v^2/\sigma_e^2 =: \tau$. The new model can also be motivated by noting that the standardized observations $z_{ij} = (y_{ij} - \mathbf{x}_{ij}^T\boldsymbol{\beta})/\sigma_i$ have mean zero and constant variance if $E_m(y_{ij}) = \mathbf{x}_{ij}^T\boldsymbol{\beta}$ and $E_m(y_{ij} - \mathbf{x}_{ij}^T\boldsymbol{\beta})^2 \propto \sigma_i^2$, and then letting z_{ij} satisfy the basic unit level model with zero mean $z_{ij} = v_i + e_{ij}$, where $v_i \overset{\text{iid}}{\sim} N(0, \sigma_v^2)$ and $e_{ij} \overset{\text{iid}}{\sim} N(0, \sigma_e^2)$.

The BLUP estimator of the area mean $\mu_i = \overline{\mathbf{X}}_i^T\boldsymbol{\beta} + u_i$ under the heteroscedastic model (7.6.1) is identical to the BLUP estimator under the basic unit level model (7.1.1) with $k_{ij} = 1$ when the model parameters are known, which is given by

$$\tilde{\mu}_i^{\text{JN}} = \gamma_i[\overline{y}_i + (\overline{\mathbf{X}}_i - \overline{\mathbf{x}}_i)^T\boldsymbol{\beta}] + (1 - \gamma_i)\overline{\mathbf{X}}_i^T\boldsymbol{\beta}, \tag{7.6.2}$$

where $\gamma_i = \sigma_v^2/(\sigma_v^2 + \sigma_e^2/n_i) = n_i\tau/(n_i\tau + 1)$. Note that the estimator $\tilde{\mu}_i^{\text{JN}}$ in (7.6.2) does not depend on the unknown error variances σ_i^2; the only unknown parameters are $\boldsymbol{\beta}$ and τ. In fact, the profile likelihood of model (7.6.1) obtained by replacing σ_i^2 by its ML estimator in the likelihood depends only on σ_v^2 and σ_e^2 through the ratio τ. Thus, σ_v^2 and σ_e^2 are not identifiable in model (7.6.1). However, this is not a problem since τ is identifiable and only τ needs to be estimated in (7.6.2).

Jiang and Nguyen (JN) showed that if the heteroscedastic model (7.6.1) is the true model, the ML estimators of $\boldsymbol{\beta}$ and τ, obtained by assuming the basic unit level model (7.1.1) with $k_{ij} = 1$, are inconsistent. As a result, the EBLUP obtained by plugging these ML estimates in (7.6.2) is not valid. Therefore, JN derived the ML estimators, $\hat{\boldsymbol{\beta}}_{\text{JN}}$ and $\hat{\tau}_{\text{JN}}$, of $\boldsymbol{\beta}$ and τ under the heteroscedastic model (7.6.2), and proved consistency of these estimators as $m \to \infty$, assuming that $n_i > 2$ for all i. We denote as $\hat{\mu}_i^{\text{JN}}$ the resulting EBLUP estimator of μ_i, obtained by replacing $\boldsymbol{\beta}$ by $\hat{\boldsymbol{\beta}}_{\text{JN}}$ and τ by $\hat{\tau}_{\text{JN}}$ in (7.6.2).

A difficulty with the estimator $\hat{\mu}_i^{\text{JN}}$ is that its MSE depends on σ_i^2 in addition to $\boldsymbol{\beta}$ and τ. Furthermore, it is not possible to consistently estimate $\sigma_i^2, i = 1, \ldots, m$, if they are completely unknown because n_i is assumed to be small. To get around this difficulty, JN considered the following additional assumptions:

(A1) $\sigma_1^2, \ldots, \sigma_m^2$ are random variables with $E_m(\sigma_i^2) = \phi_t, i \in A_t, t = 1, \ldots, q$ for a known partition of the set of sample areas, $A = \{1, \ldots, m\}$, into subsets $A_1, \ldots, A_q$, where $1 \le q < m$ with q small.

(A2) Conditional on $\sigma_1^2, \ldots, \sigma_m^2$, the observations obey model (7.6.1).

(A3) The area vectors $\mathbf{y}_i = (y_{i1}, \ldots, y_{in_i})^T, i = 1, \ldots, m$ are marginally independent.

Using the resampling method of Jiang, Lahiri, and Wan (2002), JN obtained a jackknife estimator of MSE($\hat{\mu}_i^{\text{JN}}$) that is second-order unbiased under the assumptions (A1)–(A3); see Section 9.2.2 for a description of the jackknife method. In a simulation study, the jackknife MSE estimator performed well in terms of relative bias (RB) for $m = 50$, but for $m = 20$ the RB ranged from 10% to 37%, which means that it can be fairly large for smaller m.

7.6.2 Pseudo-EBLUP Estimation

The EBLUP estimator $\hat{\mu}_i^H$ under the unit level model (7.1.1) with $k_{ij} = 1$ does not make use of the design weights, w_{ij}, attached to the sampled elements $(i,j), j = 1, \ldots, n_i, i = 1, \ldots, m$. As a result, it is not design-consistent unless the sampling design is self-weighting within areas, that is, $w_{ij} = w_i$ for all j. On the other hand, the EBLUP estimator under the area level model is design-consistent.

As noted by Kott (1990) and Prasad and Rao (1999), it is appealing to survey practitioners to use design-consistent, model-based estimators because such estimators provide protection against model failures as the small area sample size, n_i, increases.

Note that n_i could be moderately large for some of the areas under consideration, in which case design-consistency becomes relevant.

In this section, pseudo-EBLUP estimators that depend on the design weights and satisfy the design-consistency property are developed for the special case of equal error variances, that is, $k_{ij} = 1$ for all (i,j). These estimators satisfy the benchmarking property without any adjustment in the sense that they agree with a direct survey regression estimator when aggregated over the areas i (You and Rao 2002a).

First, consider the survey-weighted area level model obtained taking a weighted average with weights $\tilde{w}_{ij} = w_{ij}/\sum_{k=1}^{n_i} w_{ij}$ in the unit level model:

$$\bar{y}_{iw} = \sum_{j=1}^{n_i} \tilde{w}_{ij} y_{ij} = \sum_{j=1}^{n_i} \tilde{w}_{ij}(\mathbf{x}_{ij}^T \boldsymbol{\beta} + v_i + e_{ij})$$
$$= \bar{\mathbf{x}}_{iw}^T \boldsymbol{\beta} + v_i + \bar{e}_{iw}. \tag{7.6.3}$$

where $\bar{\mathbf{x}}_{iw} = \sum_{j=1}^{n_i} \tilde{w}_{ij}\mathbf{x}_{ij}$ and $\bar{e}_{iw} = \sum_{j=1}^{n_i} \tilde{w}_{ij} e_{ij}$. If the model (7.1.1) holds for the sample data, then $E(\bar{e}_{iw}) = 0$ and $V(\bar{e}_{iw}) = \sigma_e^2 \delta_{iw}$, where $\delta_{iw} = \sum_{j=1}^{n_i} \tilde{w}_{ij}^2$. The BLUP estimator of $\mu_i = \overline{\mathbf{X}}_i^T \boldsymbol{\beta} + v_i$ obtained under the aggregated model (7.6.3) is given by

$$\tilde{\mu}_{iw}^H = \overline{\mathbf{X}}_i^T \boldsymbol{\beta} + \gamma_{iw}(\bar{y}_{iw} - \bar{\mathbf{x}}_{iw}^T \boldsymbol{\beta}), \tag{7.6.4}$$

where $\gamma_{iw} = \sigma_v^2/(\sigma_v^2 + \sigma_e^2 \delta_{iw})$. This estimator depends on $\boldsymbol{\beta}, \sigma_e^2$ and σ_v^2. We estimate the variance components, σ_e^2 and σ_v^2, from the unit level model using the fitting-of-constants or the REML method (Section 7.1.2).

To estimate the regression parameter $\boldsymbol{\beta}$, we first obtain the BLUP estimator of v_i under the aggregated model (7.6.3), given $(\boldsymbol{\beta}, \sigma_e^2, \sigma_v^2)$, as

$$\tilde{v}_{iw}(\boldsymbol{\beta}, \sigma_e^2, \sigma_v^2) = \gamma_{iw}(\bar{y}_{iw} - \bar{\mathbf{x}}_{iw}^T \boldsymbol{\beta}). \tag{7.6.5}$$

We then solve the following design-weighted estimating equations for $\boldsymbol{\beta}$:

$$\sum_{i=1}^{m} \sum_{j=1}^{n_i} w_{ij}\mathbf{x}_{ij}[y_{ij} - \mathbf{x}_{ij}^T \boldsymbol{\beta} - \tilde{v}_{iw}(\boldsymbol{\beta}, \sigma_e^2, \sigma_v^2)] = \mathbf{0}. \tag{7.6.6}$$

The solution of (7.6.6) is given by

$$\tilde{\boldsymbol{\beta}}_w(\sigma_v^2, \sigma_e^2) = \left[\sum_{i=1}^{m} \sum_{j=1}^{n_i} w_{ij}\mathbf{x}_{ij}(\mathbf{x}_{ij} - \gamma_{iw}\bar{\mathbf{x}}_{iw})^T \right]^{-1}$$
$$\times \left[\sum_{i=1}^{m} \sum_{j=1}^{n_i} w_{ij}(\mathbf{x}_{ij} - \gamma_{iw}\bar{\mathbf{x}}_{iw}) y_{ij} \right]. \tag{7.6.7}$$

Given σ_v^2 and σ_e^2, the estimator $\tilde{\boldsymbol{\beta}}_w(\sigma_v^2, \sigma_e^2)$ is model-unbiased for $\boldsymbol{\beta}$. Now replacing σ_v^2 and σ_e^2 by the estimators $\hat{\sigma}_v^2$ and $\hat{\sigma}_e^2$, we obtain a design-weighted estimator of $\boldsymbol{\beta}$ given by $\hat{\boldsymbol{\beta}}_w = \tilde{\boldsymbol{\beta}}_w(\hat{\sigma}_v^2, \hat{\sigma}_e^2)$.

A pseudo-EBLUP estimator of μ_i is obtained from (7.6.4) by replacing $(\boldsymbol{\beta}, \sigma_e^2, \sigma_v^2)$ by $(\hat{\boldsymbol{\beta}}_w, \hat{\sigma}_e^2, \hat{\sigma}_v^2)$, that is,

$$\hat{\mu}_{iw}^H = \overline{\mathbf{X}}_i^T \hat{\boldsymbol{\beta}}_w + \hat{\gamma}_{iw}(\overline{y}_{iw} - \overline{\mathbf{x}}_{iw}^T \hat{\boldsymbol{\beta}}_w), \tag{7.6.8}$$

where $\hat{\gamma}_{iw} = \hat{\sigma}_v^2/(\hat{\sigma}_v^2 + \delta_{iw}\hat{\sigma}_e^2)$. The estimator $\hat{\mu}_{iw}^H$ automatically satisfies the benchmarking property when aggregated over i, assuming that the weights are calibrated to agree with the known sizes N_i, that is, $w_{i\cdot} = N_i$, where $w_{i\cdot} = \sum_{j=1}^{n_i} w_{ij}$ and that the unit level model includes the intercept term, that is, $x_{ij1} = 1$. Specifically, under those assumptions, $\sum_{i=1}^m N_i \hat{\mu}_{iw}^H$ equals the direct survey regression estimator $\hat{Y}_w + (\mathbf{X} - \hat{\mathbf{X}}_w)^T \hat{\boldsymbol{\beta}}_w$, where $\hat{Y}_w = \sum_{i=1}^m w_{i\cdot}\overline{y}_{iw} = \sum_{i=1}^m \sum_{j=1}^{n_i} w_{ij} y_{ij}$ and $\hat{\mathbf{X}}_w = \sum_{i=1}^m w_{i\cdot}\overline{\mathbf{x}}_{iw} = \sum_{i=1}^m \sum_{j=1}^{n_i} w_{ij}\mathbf{x}_{ij}$ are the direct estimators of the overall totals Y and $\mathbf{X}$, respectively. To prove this property, consider the first equation of (7.6.6) corresponding to the intercept term $x_{ij1} = 1$ and substitute $\hat{\boldsymbol{\beta}}_w$ and $\hat{v}_{iw} = \tilde{v}_{iw}(\hat{\boldsymbol{\beta}}_w, \hat{\sigma}_v^2, \hat{\sigma}_e^2)$ for $\boldsymbol{\beta}$ and $\tilde{v}_{iw}(\boldsymbol{\beta}, \sigma_v^2, \sigma_e^2)$ to get $\sum_{i=1}^m \sum_{j=1}^{n_i} w_{ij}(y_{ij} - \mathbf{x}_{ij}^T \hat{\boldsymbol{\beta}}_w - \hat{v}_{iw}) = 0$ or, equivalently,

$$\sum_{i=1}^m N_i \hat{v}_{iw} = \hat{Y}_w - \hat{\mathbf{X}}_w^T \hat{\boldsymbol{\beta}}_w. \tag{7.6.9}$$

Also, noting that $\hat{\mu}_{iw}^H = \overline{\mathbf{X}}_i^T \hat{\boldsymbol{\beta}}_w + \hat{v}_{iw}$ and using (7.6.9), we obtain

$$\sum_{i=1}^m N_i \hat{\mu}_{iw}^H = \mathbf{X}^T \hat{\boldsymbol{\beta}}_w + \sum_{i=1}^m N_i \hat{v}_{iw} = \hat{Y}_w + (\mathbf{X} - \hat{\mathbf{X}}_w)^T \hat{\boldsymbol{\beta}}_w.$$

Thus, the pseudo-EBLUP estimator, $\hat{\mu}_{iw}^H$, satisfies the benchmarking property without any adjustment, unlike the EBLUP estimator $\hat{\mu}_i^H$.

Under normality of the errors v_i and e_{ij}, You and Rao (2002a) derived an estimator of $\text{MSE}(\hat{\mu}_{iw}^H) = E(\hat{\mu}_{iw}^H - \mu_i)^2$ given by

$$\text{mse}(\hat{\mu}_{iw}^H) = g_{1iw}(\hat{\sigma}_v^2, \hat{\sigma}_e^2) + g_{2iw}(\hat{\sigma}_v^2, \hat{\sigma}_e^2) + 2g_{3iw}(\hat{\sigma}_v^2, \hat{\sigma}_e^2), \tag{7.6.10}$$

where

$$g_{1iw}(\sigma_v^2, \sigma_e^2) = \gamma_{iw}\delta_{iw}\sigma_e^2, \tag{7.6.11}$$

$$g_{2iw}(\sigma_v^2, \sigma_e^2) = (\overline{\mathbf{X}}_i - \gamma_{iw}\overline{\mathbf{x}}_{iw})^T \boldsymbol{\Phi}_w(\sigma_v^2, \sigma_e^2)(\overline{\mathbf{X}}_i - \gamma_{iw}\overline{\mathbf{x}}_{iw}), \tag{7.6.12}$$

and

$$g_{3iw}(\sigma_v^2, \sigma_e^2) = \gamma_{iw}(1 - \gamma_{iw})^2 \sigma_v^{-2}\sigma_e^{-2} h(\sigma_v^2, \sigma_e^2), \tag{7.6.13}$$

where $h(\sigma_v^2, \sigma_e^2)$ is given by (7.2.3). Furthermore, $\Phi_w(\sigma_v^2, \sigma_e^2)$ is the variance–covariance matrix of $\tilde{\boldsymbol{\beta}}_w(\sigma_v^2, \sigma_e^2)$, given by

$$\Phi_w(\sigma_v^2, \sigma_e^2) = \left(\sum_{i=1}^{m}\sum_{j=1}^{n_i}\mathbf{x}_{ij}\mathbf{z}_{ij}^T\right)^{-1}\left(\sum_{i=1}^{m}\sum_{j=1}^{n_i}\mathbf{z}_{ij}\mathbf{z}_{ij}^T\right)\left[\left(\sum_{i=1}^{m}\sum_{j=1}^{n_i}\mathbf{x}_{ij}\mathbf{z}_{ij}^T\right)^{-1}\right]^T\sigma_e^2$$
$$+\left(\sum_{i=1}^{m}\sum_{j=1}^{n_i}\mathbf{x}_{ij}\mathbf{z}_{ij}^T\right)^{-1}\left[\sum_{i=1}^{m}\left(\sum_{j=1}^{n_i}\mathbf{z}_{ij}\right)\left(\sum_{j=1}^{n_i}\mathbf{z}_{ij}\right)^T\right]$$
$$\times\left[\left(\sum_{i=1}^{m}\sum_{j=1}^{n_i}\mathbf{x}_{ij}\mathbf{z}_{ij}^T\right)^{-1}\right]^T\sigma_v^2, \tag{7.6.14}$$

where $\mathbf{z}_{ij} = w_{ij}(\mathbf{x}_{ij} - \gamma_{iw}\bar{\mathbf{x}}_{iw})$. The MSE estimator (7.6.10) is applicable for the fitting-of-constants or REML estimators of σ_v^2 and σ_e^2. However, the cross-product term $E[(\tilde{\mu}_{iw}^H - \mu_i)(\hat{\mu}_{iw}^H - \tilde{\mu}_{iw}^H)]$ may not be zero in the survey-weighted case, and the MSE estimator (7.6.10) ignores this term. As a result, the bias of (7.6.10) contains a term of order $O(m^{-1})$ and hence it is not second-order unbiased.

Torabi and Rao (2010) obtained an approximation to the cross-product term and then used that approximation to obtain a second-order unbiased estimator of $\text{MSE}(\hat{\mu}_{iw}^H)$. We refer the reader to Torabi and Rao (2010) for further details.

Although the pseudo-EBLUP estimator is derived under the assumption of noninformative sampling, empirical results seem to indicate that it is robust to informative sampling in the sense that it is not much affected in terms of bias, unlike the EBLUP estimator. Estimation under informative sampling is studied in Section 7.6.3.

Example 7.6.1. County Corn Crop Areas. You and Rao (2002a) applied the pseudo-EBLUP method to the county crop areas data from Battese, Harter, and Fuller (1988) (see Example 7.3.1). SRS within areas is assumed so that $w_{ij} = w_i = N_i/n_i$ for all sample elements j in area i. They obtained EBLUP and pseudo-EBLUP estimators under the model

$$y_{ij} = \beta_0 + \beta_1 x_{ij1} + \beta_2 x_{ij2} + v_i + e_{ij}, \quad j = 1, \ldots, n_i, \quad i = 1, \ldots, m.$$

In this case, the EBLUP estimator is also design-consistent, and the pseudo-EBLUP estimator has the same form as the EBLUP estimator except for the resulting estimator of $\boldsymbol{\beta} = (\beta_0, \beta_1, \beta_2)^T$.

Table 7.4 reports EBLUP and pseudo-EBLUP estimates of mean hectares of corn for the 12 counties, together with their associated standard errors (square root of estimated MSEs). The fitting-of-constants method is used to estimate σ_v^2 and σ_e^2. EBLUP estimates and the associated standard errors differ slightly from those reported in Table 7.1 because of differences in the method of estimating σ_v^2 and σ_e^2, as noted in Example 7.3.1.

TABLE 7.4 EBLUP and Pseudo-EBLUP Estimates and Associated Standard Errors (s.e.): County Corn Crop Areas

		EBLUP		Pseudo-EBLUP	
County	n_i	Estimate	s.e.	Estimate	s.e.
		Corn			
Cerro Gordo	1	122.2	9.7	120.5	9.9
Hamilton	1	126.2	9.6	125.2	9.7
Worth	1	106.8	9.5	106.4	9.6
Humboldt	2	108.5	8.1	107.4	8.3
Franklin	3	144.2	6.6	143.7	6.6
Pocahontas	3	112.1	6.6	111.5	6.6
Winnebago	3	112.8	6.6	112.1	6.6
Wright	3	122.0	6.7	121.3	6.8
Webster	4	115.3	5.8	115.0	5.8
Hancock	5	124.4	5.4	124.5	5.4
Kossuth	5	106.9	5.3	106.6	5.3
Hardin	5	143.0	5.6	143.5	5.8

Source: Adapted from Table 1 in You and Rao (2002a).

It is clear from Table 7.4 that the pseudo-EBLUP method compares favorably to the "optimal" EBLUP method, leading to a slight loss in efficiency. The loss in efficiency is due to the weighted method of estimating $\boldsymbol{\beta}$, but the standard errors of $\hat{\boldsymbol{\beta}}_w$ compare favorably to those of $\hat{\boldsymbol{\beta}}$. For example, the standard errors of $\hat{\beta}_1$ and $\hat{\beta}_2$ are, respectively, 0.050 and 0.056, compared to 0.054 and 0.062 in the case of $\hat{\beta}_{1w}$ and $\hat{\beta}_{2w}$.

The pseudo-EBLUP estimates satisfy the benchmarking property, in the sense that $\sum_{i=1}^{m} N_i \hat{\mu}_{iw}^H$ equals the survey regression estimate of the overall total, $\hat{Y}_w + (\mathbf{X} - \hat{\mathbf{X}}_w)^T \hat{\boldsymbol{\beta}}_w$, which turned out to be equal to 815,025.2 ha.

Example 7.6.2. Simulation Results. Torabi and Rao (2010) conducted a simulation study on the relative performance of the Torabi–Rao (TR) MSE estimator of $\hat{\mu}_i^H$ and the You–Rao estimator (7.6.10) in terms of relative bias (RB). They also obtained single- and double-bootstrap parametric MSE estimators. They considered two different scenarios for the sampling weights w_{ij}. In scenario 1, the within-area variability of the weights is smaller than in the case of scenario 2. Their results indicate that under scenario 1 the YR MSE estimator performs similarly as the second-order unbiased TR estimator in terms of RB. On the other hand, under scenario 2, the YR estimator could lead to considerable overestimation of MSE, whereas the TR estimator performs well in terms of RB (<5%). Furthermore, the single-bootstrap MSE estimator, which is only first-order unbiased, leads to significant underestimation under both scenarios, whereas the double-bootstrap MSE estimator performs well, although leading to consistent but moderate underestimation (<10%) across areas.

7.6.3 Informative Sampling

In Section 4.3, we studied the effect of informative sampling within areas under the basic unit level model, assuming that all the areas are sampled ($m = M$). In particular, we showed that the sample values $\{(y_{ij}, \mathbf{x}_{ij}); j \in s_i,\ i = 1, \ldots, m\}$ do not obey the assumed population model (4.3.3) if the sample selection probabilities are related to the outcome values. In this section, we consider two alternative approaches to estimation of small area means under informative sampling. The first approach, due to Pfeffermann and Sverchkov (2007), obtains "optimal" estimators of area means under a particular model identified for the sample data, assuming a model for the sampling weights w_{ij}. These estimators are obtained by correcting the EBLUP estimators for the sample selection bias. The second approach, due to Verret, Rao, and Hidiroglou (2014), accounts for the sample selection bias by using a suitable function of unit selection probability as an additional auxiliary variable in the sample model. Standard EBLUP estimators and the associated MSE estimators are obtained under the augmented model without requiring new theory.

Pfeffermann–Sverchkov (PS) Approach For simplicity, we assume that all the areas are sampled ($m = M$) but the PS approach extends to the case of sampled and nonsampled areas ($m < M$). The sample model is assumed to be a nested error model

$$y_{ij} = \mathbf{x}_{ij}^T \boldsymbol{\alpha} + u_i + h_{ij}, \quad j \in s_i, \quad i = 1, \ldots, m, \tag{7.6.15}$$

where $u_i \overset{\text{iid}}{\sim} N(0, \sigma_u^2)$ and $h_{ij} | j \in s_i \overset{\text{iid}}{\sim} N(0, \sigma_h^2)$. Let $\pi_{j|i}$ be the inclusion probability for the unit j in area i. The sampling weights $w_{ij} = \pi_{j|i}^{-1}$ for $j \in s_i$ are assumed to be random with conditional expectation

$$\begin{aligned} E_{s_i}(w_{ij} | \mathbf{x}_{ij}, y_{ij}, u_i) &= E_{s_i}(w_{ij} | \mathbf{x}_{ij}, y_{ij}) \\ &= k_i \exp(\mathbf{x}_{ij}^T \mathbf{a} + b\, y_{ij}). \end{aligned} \tag{7.6.16}$$

Here, E_{s_i} denotes expectation with respect to the distribution of w_{ij} for a unit j in the sample from area i ($j \in s_i$), that is, let $f_P(w_{ij} | \mathbf{x}_{ij}, y_{ij})$ denote the probability density function (pdf) of w_{ij} for population unit j within area i. Then, the sample pdf of w_{ij} is defined as $f_{s_i}(w_{ij} | \mathbf{x}_{ij}, y_{ij}) := f_P(w_{ij} | \mathbf{x}_{ij}, y_{ij}, a_{ij} = 1)$, where a_{ij} is the sample indicator variable with $a_{ij} = 1$ iff $j \in s_i$. Section 7.8.1 shows that for a sampling design with fixed within-area sample size n_i, k_i is given by

$$k_i = N_i n_i^{-1} \left[N_i^{-1} \sum_{j=1}^{N_i} \exp(-\mathbf{x}_{ij}^T \mathbf{a} - b\, y_{ij}) \right]. \tag{7.6.17}$$

For large N_i, the mean in the square brackets in (7.6.17) may be approximated by its population expectation, say a constant c_i, so that $k_i \approx c_i\, N_i n_i^{-1}$. Model (7.6.15) may be identified from the sample data $D_s = \{(y_{ij}, \mathbf{x}_{ij}); j \in s_i,\ i = 1, \ldots, m\}$. Similarly, model

(7.6.16) for the weights can be identified from the sample data $\{(w_{ij}, y_{ij}, \mathbf{x}_{ij}); j \in s_i,\ i = 1, \ldots, m\}$.

No specific form for the population model is assumed other than saying that the random area effects u_i are generated independently from a pdf $f_P(u_i)$ and that the values $y_{ij}, j = 1, \ldots, N_i$, are generated independently from the pdf $f_P(y_{ij}|\mathbf{x}_{ij}, u_i)$, conditional on $\mathbf{x}_{ij}$ and u_i. The pdf of the sampled y_{ij} given $\mathbf{x}_{ij}$ and u_i ($j \in s_i$) and the pdf of the nonsampled $y_{i\ell}$ given $\mathbf{x}_{i\ell}$ and u_i ($j \in r_i$) are, respectively, defined as

$$f_{s_i}(y_{ij}|\mathbf{x}_{ij}, u_i) := f_P(y_{ij}|\mathbf{x}_{ij}, u_i, a_{ij} = 1) \tag{7.6.18}$$

and

$$f_{r_i}(y_{i\ell}|\mathbf{x}_{i\ell}, u_i) := f_P(y_{i\ell}|\mathbf{x}_{i\ell}, u_i, a_{ij} = 0). \tag{7.6.19}$$

It follows from (7.6.18) and (7.6.19) that

$$f_{r_i}(y_{i\ell}|\mathbf{x}_{i\ell}, u_i) = \frac{E_{s_i}[(w_{i\ell} - 1)|y_{i\ell}, \mathbf{x}_{i\ell}, u_i] f_{s_i}(y_{i\ell}|\mathbf{x}_{i\ell}, u_i)}{E_{s_i}[(w_{i\ell} - 1)|\mathbf{x}_{i\ell}, u_i]}. \tag{7.6.20}$$

Proof of (7.6.20) is given in Section 7.8.2. Now using the sample model (7.6.15) for $y_{i\ell}$ and the sample model (7.6.16) for $w_{i\ell}$ in (7.6.20), we get after simplification

$$E_{r_i}(y_{i\ell}|\mathbf{x}_{i\ell}, u_i) = u_{i\ell} + \frac{\lambda_{i\ell}}{\lambda_{i\ell} - 1} b\, \sigma_h^2, \tag{7.6.21}$$

where $u_{i\ell} = \mathbf{x}_{i\ell}^T \boldsymbol{\alpha} + u_i$, $\lambda_{i\ell} = k_i \exp(b^2\sigma_h^2/2 + \mathbf{x}_{i\ell}^T \mathbf{a} + b u_{i\ell}) = E_{s_i}(w_{i\ell}|\mathbf{x}_{i\ell}, u_i)$.

The best prediction estimator of the area mean $\overline{Y}_i$ is then given by

$$\tilde{\overline{Y}}_i^{\text{PS}} = E_P(\overline{Y}_i|D_s) = N_i^{-1} \left\{ \sum_{j \in s_i} y_{ij} + \sum_{\ell \in r_i} E\left[E_{r_i}(y_{i\ell}|\mathbf{x}_{i\ell}, u_i)|D_s\right] \right\}. \tag{7.6.22}$$

Now, noting that $u_i|D_s \sim N(\tilde{u}_i, n_i^{-1}\sigma_h^2 \tilde{\gamma}_i)$ under the sample model (7.6.15), where $\tilde{u}_i = \tilde{u}_i(\boldsymbol{\alpha}, \sigma_u^2, \sigma_h^2) = \tilde{\gamma}_i(\overline{y}_i - \overline{\mathbf{X}}_i^T \boldsymbol{\alpha})$ with $\overline{y}_i$ and $\overline{\mathbf{x}}_i$ denoting the area sample means and $\tilde{\gamma}_i = \sigma_u^2/(\sigma_u^2 + n_i^{-1}\sigma_h^2)$, and using (7.6.21) we get

$$E[E_{r_i}(y_{i\ell}|\mathbf{x}_{i\ell}, u_i)|D_s] = \mathbf{x}_{i\ell}^T \boldsymbol{\alpha} + \tilde{u}_i + b\, \sigma_h^2 E[(1 - \lambda_{i\ell}^{-1})^{-1}|D_s]. \tag{7.6.23}$$

No closed-form expression for the expectation in the last term of (7.6.23) is available. PS approximate this expectation by 1 in the case of a small sampling fraction $f_i = n_i/N_i$, noting that $\lambda_{i\ell}$ is typically much larger than 1 in this case. Therefore, (7.6.23) is approximated by

$$E[E_{r_i}(y_{i\ell}|\mathbf{x}_{i\ell}, u_i)|D_s] = \mathbf{x}_{i\ell}^T \boldsymbol{\alpha} + \tilde{u}_i + b\, \sigma_h^2. \tag{7.6.24}$$

Now substituting (7.6.24) in (7.6.22), $\tilde{\overline{Y}}_i^{\text{PS}}$ reduces to

$$\tilde{\overline{Y}}_i^{\text{PS}} = (1-f_i)(\overline{\mathbf{X}}_i^T\boldsymbol{\alpha} + \tilde{u}_i) + f_i[\overline{y}_i + (\overline{\mathbf{X}}_i - \overline{\mathbf{x}}_i)^T\boldsymbol{\alpha}] + (1-f_i)b\,\sigma_h^2. \tag{7.6.25}$$

The estimator (7.6.25) depends on the model parameters $\boldsymbol{\alpha}, \sigma_u^2, \sigma_h^2$, and b. Estimators $\hat{\boldsymbol{\alpha}}, \hat{\sigma}_u^2$, and $\hat{\sigma}_h^2$ are obtained by fitting the sample model (7.6.15) for y_{ij}. The estimator $\hat{b}$ of b is obtained by fitting the sample model for w_{ij}. Substituting these estimators in (7.6.25), we get the empirical PS estimator $\hat{\overline{Y}}_i^{\text{PS}}$. But note that estimators $(\hat{\boldsymbol{\alpha}}, \hat{\sigma}_u^2, \hat{\sigma}_h^2)^T$ and $\hat{u}_i = \tilde{u}_i(\hat{\boldsymbol{\alpha}}, \hat{\sigma}_u^2, \hat{\sigma}_h^2)$ obtained under the assumed unit level sample model (7.6.15) are identical to estimators $(\hat{\boldsymbol{\beta}}, \hat{\sigma}_v^2, \hat{\sigma}_e^2)^T$ and $\hat{v}_i = \tilde{v}_i(\hat{\boldsymbol{\beta}}, \hat{\sigma}_v^2, \hat{\sigma}_e^2)$ obtained by assuming that the population nested error model (4.3.1) with $k_{ij} = 1$ for all (i,j) holds also for the sample. Therefore, we can express $\tilde{\overline{Y}}_i^{\text{PS}}$ as

$$\tilde{\overline{Y}}_i^{\text{PS}} = \hat{\overline{Y}}_i^{H} + (1-f_i)\hat{b}\,\hat{\sigma}_e^2. \tag{7.6.26}$$

The last term in (7.6.26) corrects for the sample selection bias. Note that (7.6.26) does not depend on k_i and a in the assumed sample model for the weights w_{ij}. PS proposed a bootstrap method to estimate $\text{MSE}(\hat{\overline{Y}}_i^{\text{PS}})$.

PS also proposed a test of informativeness of the sample selection within the areas. They suggested a two-step estimation procedure. If the test indicates that sampling is not informative, then use the standard EBLUP $\hat{\overline{Y}}_i^{H}$. Otherwise, use the PS estimator $\tilde{\overline{Y}}_i^{\text{PS}}$. A simple test is obtained by modeling w_{ij} as $w_{ij} = \gamma_{0i} + \boldsymbol{\gamma}_{1i}^T\mathbf{x}_{ij} + \gamma_{2i}y_{ij} + \epsilon_{ij}$ with $\epsilon_{ij} \overset{\text{iid}}{\sim} N(0, \sigma_\epsilon^2)$ and testing $H_0 : \gamma_{2i} = 0, \quad i = 1, \ldots, m$. Using ordinary least squares (OLS), the test statistics are given by $F_i = \hat{\gamma}_{2i}^2/s^2(\hat{\gamma}_{2i}), i = 1, \ldots, m$, where $\hat{\gamma}_{2i}$ and $s(\hat{\gamma}_{2i})$ are the OLS estimator of γ_{2i} and its standard error. Under $H_0, F_i \sim F(1, n_i - p - 2)$ provided $n_i > p + 2$, where p is the dimension of $\mathbf{x}_{ij}$. Assuming the same sampling design within all the areas, a combined test is based on $F_{\max} = \max\{F_i; i = 1, \ldots, m\}$.

PS also derived the optimal predictors for nonsampled areas when $m < M$. Furthermore, the PS method extends to general models fitted to the sample data, including generalized linear mixed model.

Augmented Sample Model Approach We focus on sampling methods with inclusion probabilities $\pi_{j|i}$ proportional to size measures b_{ij} attached to units, that is, $\pi_{j|i} = n_i p_{j|i}$, where $p_{j|i} = b_{ij}/b_{i+}$ are the unit selection probabilities, with $b_{i+} = \sum_{j=1}^{N_i} b_{ij}$. A sampling design with $\pi_{j|i} = n_i p_{j|i}$ is ensured using the Rao–Sampford method (Rao 1965, Sampford 1967). Assuming that all the areas are sampled ($m = M$), we consider the following augmented sample model:

$$y_{ij} = \mathbf{x}_{ij}^T\boldsymbol{\beta}_0 + \delta_0 g(p_{j|i}) + v_{0i} + e_{0ij}, \quad j = 1, \ldots, n_i, \quad i = 1, \ldots, m, \tag{7.6.27}$$

where $g_{ij} = g(p_{j|i})$ is a suitably defined function of $p_{j|i}$, $v_{0i} \overset{\text{iid}}{\sim} N(0, \sigma^2_{0v})$ and independent of $e_{0ij} \overset{\text{iid}}{\sim} N(0, \sigma^2_{0e})$. The sample model (7.6.18) can be identified after fitting model (7.6.27) to sample data $\{(y_{ij}, \mathbf{x}_{ij}, p_{j|i}); j \in s_i,\ i = 1, \ldots, m\}$ for different choices of the function $g(\cdot)$ and checking the adequacy of the fitted models. For example, residuals r_{ij} from the fitted model without the augmenting variable g_{ij} may be plotted against g_{ij} to select the suitable $g(\cdot)$. The identified model will also hold for the population (Skinner 1994, Rao 2003a, Section 5.3), that is, $f_{s_i}(y_{ij}|\mathbf{x}_{ij}, p_{j|i}, u_i) = f_P(y_{ij}|\mathbf{x}_{ij}, p_{j|i}, u_i)$. Possible choices of $g(p_{j|i})$ are $p_{j|i}$, $\log\ (p_{j|i})$, w_{ij}, and $n_i w_{ij} = p^{-1}_{j|i}$.

Using the augmented model, we obtain the EBLUP estimator of the area mean $\overline{Y}_i$, or its approximation $\mu_{0i} = \overline{\mathbf{X}}_i^T \beta_0 + \overline{G}_i \delta_0 + v_{0i}$ when the area sampling fraction f_i is small, where $\overline{G}_i = N_i^{-1} \sum_{j=1}^{N_i} g_{ij}$ is the mean of the population values g_{ij} from area i. The EBLUP of $\overline{Y}_i$ or μ_{0i} requires the knowledge of $\overline{G}_i$. However, the choice $g_{ij} = p_{j|i}$ gives $\overline{G}_i = 1$ and the choice $g_{ij} = n_i w_{j|i}$ gives $\overline{G}_i = n_i \overline{W}_i$, where $\overline{W}_i$ is the population mean of the weights $w_{j|i}$ in area i, which is often known in practice.

The EBLUP estimator of μ_{0i} under the augmented model (7.6.27) is then given by

$$\hat{\mu}^H_{i(a)} = \hat{\gamma}_{0i}\overline{y}_i + (\overline{\mathbf{X}}_i - \hat{\gamma}_{0i}\overline{\mathbf{x}}_i)^T \hat{\beta}_0 + (\overline{G}_i - \hat{\gamma}_{0i}\overline{g}_i)\hat{\delta}_0, \tag{7.6.28}$$

where $\hat{\gamma}_{0i} = \hat{\sigma}^2_{0v}/(\hat{\sigma}^2_{0v} + \hat{\sigma}^2_{0e}/n_i)$, $(\hat{\beta}^T_0, \hat{\gamma}_0)$ is the WLS estimator of (β^T_0, γ_0), and $\overline{g}_i = n_i^{-1} \sum_{j \in s_i} g_{ij}$. Similarly, the EBLUP estimator of $\overline{Y}_i$ is given by

$$\hat{\overline{Y}}^H_{i(a)} = (1 - f_i)\hat{\mu}^H_{i(a)} + f_i[\overline{y}_i + (\overline{\mathbf{X}}_i - \overline{\mathbf{x}}_i)^T \hat{\beta}_0 + (\overline{G}_i - \overline{g}_i)\hat{\delta}_0]. \tag{7.6.29}$$

Second-order unbiased MSE estimators for $\hat{\mu}^H_{i(a)}$ and $\hat{\overline{Y}}^H_{i(a)}$ are obtained by suitably modifying the existing MSE estimators presented in Section 7.2.2. Note that we only need to apply existing formulae to the augmented sample model (7.6.27) to get the EBLUP estimators and their associated MSE estimators. No new software is needed and no model for the weights w_{ij} is assumed, unlike in the PS approach.

Simulation Results Verret, Rao, and Hidiroglou (2014) used a design-model (pm) approach to study the relative performances of $\hat{\overline{Y}}^{\text{PS}}_i$, $\hat{\overline{Y}}^H_{i(a)}$, and the pseudo-EBLUP estimator given by (7.6.8) but denoted here $\hat{\mu}^{\text{YR}}_i$, and obtained from the model without the augmenting variable g_{ij}. The process of generating N population units according to a specified model and then selecting a sample according to a specified sampling design was repeated $R = 1{,}000$ times. For each simulation run, the population values y_{ij} for $m = 99$ areas and $N_i = 100$ units within each area i were generated from the unit level model $y_{ij} = \beta_0 + \beta_1 x_{ij} + v_i + e_{ij}$, where $\beta_0 = 1, \beta_1 = 1, v_i \overset{\text{iid}}{\sim} N(0, \sigma^2_v = 0.5)$ and independent of $e_{ij} \overset{\text{iid}}{\sim} N(0, \sigma^2_e = 2)$. The population x_{ij}-values were generated from a gamma distribution with mean 10 and variance 50 and then held fixed over the simulation runs. Unequal sample sizes within areas were chosen by letting $n_i = 5$

TABLE 7.5 Average Absolute Bias ($\overline{\text{AB}}$), Average Root Mean Squared Error ($\overline{\text{RMSE}}$) of Estimators, and Percent Average Absolute Relative Bias ($\overline{\text{ARB}}$) of MSE Estimators

Measure	EBLUP					Pseudo-EBLUP					
	no	g_{ij}				no	g_{ij}				PS
	g_{ij}	$p_{j\|i}$	$n_i w_{ij}$	w_{ij}	$\log p_{j\|i}$	g_{ij}	$p_{j\|i}$	$n_i w_{ij}$	w_{ij}	$\log p_{j\|i}$	
$\overline{\text{AB}}$	0.456	0.042	0.004	0.131	0.003	0.044	0.007	0.004	0.044	0.003	0.033
$\overline{\text{RMSE}}$	0.617	0.151	0.147	0.242	0.101	0.442	0.157	0.156	0.207	0.106	0.416
$\%\overline{\text{ARB}}$	53.1	3.7	6.7	62.6	6.9	3.8	4.1	5.2	39.6	6.7	—

Source: Adapted from Table 1 in Verret, Rao, and Hidiroglou (2014).

for the first 33 areas, $n_i = 7$ for the next 33 areas, and $n_i = 9$ for the remaining 33 areas.

Size measures b_{ij} used in the simulation are given by

$$b_{ij} = \exp\left[\frac{1}{3}\left\{-\frac{y_{ij} - \beta_0 - \beta_1 x_{ij}}{\sigma_e} + \frac{\delta_{ij}}{5}\right\}\right], \qquad (7.6.30)$$

where $\delta_{ij} \overset{\text{iid}}{\sim} N(0, 1)$. Size measures (7.6.30), reflecting informative sampling, are equivalent to those used by PS in their simulation study, and satisfy the sample model (7.6.16) for the weights $w_{ij} = \pi_{j|i}^{-1}$. Samples of specified within-area sizes n_i were selected in each simulation run using the Rao–Sampford method on the size measures b_{ij}, ensuring $\pi_{j|i} = n_i b_{ij}/b_{i+} = n_i p_{j|i}$.

Table 7.5 reports the average absolute bias ($\overline{\text{AB}}$) and average root MSE ($\overline{\text{RMSE}}$) of each estimator of the area means, and the average absolute relative bias ($\overline{\text{ARB}}$) of the associated MSE estimator. For MSE estimation, $\mu_{0i} = \overline{\mathbf{X}}_i^T \boldsymbol{\beta}_0 + \overline{G}_i \delta_0 + v_{0i}$ is taken as the true area mean for the augmented model and $\mu_i = \overline{\mathbf{X}}_i^T \boldsymbol{\beta} + v_i$ for the model without g_{ij} (see Verret et al. 2014 for details). Table 7.5 reports the values of $\overline{\text{AB}}, \overline{\text{RMSE}}$, and $\overline{\text{ARB}}$ for four different choices of g_{ij}, namely $p_{j|i}, w_{ij}, n_i w_{ij}$, and $\log\,(p_{j|i})$. Pseudo-EBLUP under the augmented model, denoted $\hat{\mu}_{i(a)}^{\text{YR}}$, is also included in Table 7.5. Bootstrap MSE estimator for $\hat{\overline{Y}}_i^{\text{PS}}$ was not included in the study.

Results from Table 7.5 may be summarized as follows: (i) $\overline{\text{AB}}$ of the EBLUP estimator $\hat{\overline{Y}}_i^H$ under the model without the augmenting variable g_{ij} is large ($\overline{\text{AB}} = 0.456$) relative to the EBLUP $\hat{\overline{Y}}_{i(a)}^H$ under the augmented model for all the four choices of g_{ij}. Also, the choice $g_{ij} = w_{ij}$ leads to larger $\overline{\text{AB}}$ relative to the other three choices of g_{ij}. Pseudo-EBLUP, $\hat{\mu}_i^{\text{YR}}$, under the model without g_{ij} performs well even though it is derived by assuming noninformative sampling. Augmented model pseudo-EBLUP $\hat{\mu}_{i(a)}^{\text{YR}}$ leads to significant reduction in $\overline{\text{AB}}$ as in the case of augmented model EBLUP. The PS estimator performs well in terms of $\overline{\text{AB}}$ relative to the augmented model estimators with $\overline{\text{AB}} = 0.033$. (ii) $\overline{\text{RMSE}}$ of the EBLUP $\hat{\overline{Y}}_i^H$ is large due to large $\overline{\text{AB}}$, followed by $\hat{\mu}_i^{\text{YR}}$ and $\hat{\overline{Y}}_i^{\text{PS}}$: 0.617 compared with 0.442 and 0.416. On the other hand, augmented model EBLUPs perform significantly better than $\hat{\mu}_i^{\text{YR}}$ and $\hat{\overline{Y}}_i^{\text{PS}}$ in terms of $\overline{\text{RMSE}}$; for example, $\overline{\text{RMSE}} = 0.151$ for the choice $g_{ij} = p_{j|i}$. Among the four choices of g_{ij}, the choice $g_{ij} = w_{ij}$ gives the largest $\overline{\text{RMSE}}$ (= 0.242), perhaps due to w_{ij} depending on n_i unlike the other three choices. $\overline{\text{RMSE}}$ values of $\hat{\mu}_i^H$ and $\hat{\mu}_{i(a)}^H$ (not reported here) are practically the same as the corresponding values for $\hat{\overline{Y}}_i^H$ and $\hat{\overline{Y}}_{i(a)}^H$. (iii) MSE estimator of the EBLUP estimator $\hat{\mu}_i^H$, mse($\hat{\mu}_i^H$), exhibits large $\overline{\text{ARB}}$ (= 53.1%) compared to 3.8% for mse($\hat{\mu}_i^{\text{YR}}$). The choice $g_{ij} = w_{ij}$ leads to large $\overline{\text{ARB}}$ for mse($\hat{\mu}_{i(a)}^H$) and mse($\hat{\mu}_{i(a)}^{\text{YR}}$) compared to the other three choices not depending on n_i: 62.6% and 39.6% compared to less than 7%.

Empirical results reported in Table 7.5 suggest that the augmented model approach performs well in terms of the three criteria, $\overline{\text{AB}}$ and $\overline{\text{RMSE}}$ of estimator and $\overline{\text{ARB}}$ of MSE estimator, except for the choice $g_{ij} = w_{ij}$. The PS approach also performs well with respect to $\overline{\text{AB}}$ but leads to somewhat larger $\overline{\text{RMSE}}$ in this simulation study. The customary pseudo-EBLUP, $\hat{\mu}_i^{\text{YR}}$, also performs well compared to the PS estimator $\hat{\overline{Y}}_i^{\text{PS}}$ in terms of $\overline{\text{AB}}$ and $\overline{\text{RMSE}}$. Stefan (2005) also noted the good performance of $\hat{\mu}_i^{\text{YR}}$ under informative sampling.

7.6.4 Measurement Error in Area-Level Covariate

Estimation of Area Means We consider the case of a nested error model with a single area level covariate, x_i, subject to measurement error. The population model in this case is given by

$$y_{ij} = \beta_0 + \beta_1 x_i + v_i + e_{ij}, \quad j = 1, \ldots, N_i, \quad i = 1, \ldots, m, \tag{7.6.31}$$

where area effects $v_i \overset{\text{iid}}{\sim} N(0, \sigma_v^2)$ are independent of errors $e_{ij} \overset{\text{iid}}{\sim} N(0, \sigma_e^2)$. We assume that a random sample of size n_i is selected from area i and the associated y-values are denoted as $y_{ij}, j = 1, \ldots, n_i, i = 1, \ldots, m$. Then, the BLUP of area mean $\overline{Y}_i$, for known x_i and model parameters $\boldsymbol{\beta} = (\beta_0, \beta_1)^T$ and $\boldsymbol{\delta} = (\sigma_v^2, \sigma_e^2)^T$, is given by

$$\tilde{\overline{Y}}_i^B = f_i \overline{y}_i + (1 - f_i)[\beta_0 + \beta_1 x_i + \gamma_i(\overline{y}_i - \beta_0 - \beta_1 x_i)], \tag{7.6.32}$$

where $f_i = n_i/N_i, \overline{y}_i$ is the ith area sample mean and $\gamma_i = \sigma_v^2(\sigma_v^2 + \sigma_e^2/n_i)^{-1}$. The estimator (7.6.32) is also the best prediction estimator under normality of v_i and e_{ij}.

We first estimate the unknown x_i, for given model parameters, assuming that independent measurements $\{x_{i\ell}, \ell = 1, \ldots, t_i\}$ on x_i are also available. For example, x_i might refer to the true soil nitrogen level in area i, and it is measured by taking t_i soil samples independent of crop yields $y_{i1}, \ldots, y_{in_i}$. In particular, we assume that

$$x_{i\ell} = x_i + b_{i\ell}, \quad \ell = 1, \ldots, t_i, \quad i = 1, \ldots, m, \tag{7.6.33}$$

where $b_{i\ell} \overset{\text{iid}}{\sim} (0, \sigma_b^2)$ are independent of v_i and e_{ij}. The model (7.6.31) together with (7.6.33) is called a functional measurement error model with fixed x_i (Fuller 1987). If x_i is assumed to be random, then we have a structural measurement error model. Fuller and Harter (1987), Ghosh, Sinha, and Kim (2006), and Torabi, Datta, and Rao (2009) studied EB estimation of $\overline{Y}_i$ in the structural case, assuming normality. In this section, we focus on the functional case.

To estimate x_i, we note that both $\overline{y}_i$ and $\overline{x}_i = t_i^{-1} \sum_{\ell=1}^{m} x_{i\ell}$ provide information on x_i. Hence, we minimize

$$Q(x_i) = \frac{(\overline{y}_i - \beta_0 - \beta_1 x_i)^2}{\sigma_v^2 + \sigma_e^2/n_i} + \frac{(\overline{x}_i - x_i)^2}{\sigma_b^2/t_i}$$

with respect to x_i, noting that $V(\overline{y}_i) = \sigma_v^2 + \sigma_e^2/n_i$ and $V(\overline{x}_i) = \sigma_b^2/t_i$ (Arima, Datta, and Liseo 2015). The resulting "optimal" estimator of x_i is given by

$$\tilde{x}_i = \overline{x}_i + \frac{\beta_1 \sigma_b^2 t_i^{-1} (\overline{y}_i - \beta_0 - \beta_1 \overline{x}_i)}{\sigma_v^2 + \sigma_e^2 n_i^{-1} + \beta_1^2 \sigma_b^2 t_i^{-1}}. \quad (7.6.34)$$

Now substituting $\tilde{x}_i$ for x_i in the BLUP given in (7.6.32), we obtain the pseudo-best (PB) estimator

$$\tilde{\overline{Y}}_i^{\text{PB}} = \overline{y}_i - (1 - f_i)(1 - \alpha_i)(\overline{y}_i - \beta_0 - \beta_1 \overline{x}_i), \quad (7.6.35)$$

where

$$\alpha_i = (\sigma_v^2 + \beta_1^2 \sigma_b^2 t_i^{-1})/(\sigma_v^2 + \sigma_e^2 n_i^{-1} + \beta_1^2 \sigma_b^2 t_i^{-1}). \quad (7.6.36)$$

For the special case $n_i = t_i$, $\tilde{\overline{Y}}_i^{\text{PB}}$ reduces to the estimator proposed by Datta, Rao, and Torabi (2010). Earlier Ghosh and Sinha (2007) obtained another PB estimator by estimating x_i by $\overline{x}_i$. This estimator is less efficient than (7.6.35) because it uses the nonoptimal estimator $\overline{x}_i$.

In practice, we need to estimate the model parameters in (7.6.35) to get an empirical PB (EPB) estimator of $\overline{Y}_i$. We adopt the method of moments for this purpose. Using the sample y-data and x-data, σ_e^2 and σ_b^2 are estimated, respectively, by

$$\hat{\sigma}_e^2 = \left(\sum_{i=1}^{m} n_i - m \right)^{-1} \sum_{i=1}^{m} \sum_{j=1}^{n_i} (y_{ij} - \overline{y}_i)^2 \quad (7.6.37)$$

and

$$\hat{\sigma}_b^2 = \left(\sum_{i=1}^{m} t_i - m \right)^{-1} \sum_{i=1}^{m} \sum_{\ell=1}^{t_i} (x_{i\ell} - \overline{x}_i)^2. \quad (7.6.38)$$

Ghosh and Sinha (2007) gave a simple moment method for estimating $\boldsymbol{\beta} = (\beta_0, \beta_1)^T$ and σ_v^2 for the special case of $n_i = t_i$. It is not clear how to extend their method to the case of general t_i, not necessarily equal to n_i. Arima et al. (2015) used an alternative method by first reducing the model from sufficiency principle and then appealing to the method of Ybarra and Lohr (2008), mentioned in Section 6.4.4. The reduced model is given by $\overline{y}_i = \beta_0 + \beta_1 x_i + v_i + \overline{e}_i$ and $\overline{x}_i = x_i + \overline{b}_i$, with $V(\overline{e}_i) = \sigma_e^2/n_i$ and $V(\overline{b}_i) = \sigma_b^2/t_i$. In the notation of Section 6.4.4, we have $\hat{\theta}_i = \overline{y}_i, \psi_i = \hat{\sigma}_e^2/n_i, \hat{\mathbf{z}}_i = (1, \overline{x}_i)^T$ and

$$\mathbf{C}_i = \begin{pmatrix} 0 & 0 \\ 0 & \hat{\sigma}_b^2/t_i \end{pmatrix}.$$

Substituting these values in (6.4.27) and (6.4.26) leads to the desired estimators of $\boldsymbol{\beta}$ and σ_v^2, denoted here $\hat{\boldsymbol{\beta}}$ and $\hat{\sigma}_v^2$. The EPB estimator, $\hat{\overline{Y}}_i^{PB}$, is now obtained from (7.6.35) by replacing $\boldsymbol{\beta} = (\beta_0, \beta_1)^T, \sigma_v^2, \sigma_e^2$, and σ_b^2 by $\hat{\boldsymbol{\beta}} = (\hat{\beta}_0, \hat{\beta}_1)^T, \hat{\sigma}_v^2, \hat{\sigma}_e^2$, and $\hat{\sigma}_b^2$, respectively, where $\hat{\sigma}_e^2$ and $\hat{\sigma}_b^2$ are given by (7.6.37) and (7.6.38).

MSE Estimation Datta et al. (2010) used the jackknife method of Section 9.2.2 to estimate MSE($\hat{\overline{Y}}_i^{PB}$), for the special case of $t_i = n_i$. However, the cross-product term $M_{12i} = 2E(\tilde{\overline{Y}}_i^{PB} - \overline{Y}_i)(\hat{\overline{Y}}_i^{PB} - \tilde{\overline{Y}}_i^{PB})$ in the decomposition of MSE($\hat{\overline{Y}}_i^{PB}$) needs to be ignored in order to apply the jackknife method. The jackknife method can also be applied to the case of general t_i not necessarily equal to n_i.

Simulation results reported in Datta et al. (2010) indicate that the EPB of Ghosh and Sinha (2007), based on $\overline{x}_i$, is significantly less efficient than the EPB based on the optimal estimator $\tilde{x}_i$, as expected. Furthermore, the estimator $\hat{\overline{Y}}_{i,na}^{PB}$, obtained by ignoring the measurement errors, exhibits similar efficiency as the EPB estimator $\hat{\overline{Y}}_i^{PB}$. Decomposition of simulated MSE of the naive estimator $\hat{\overline{Y}}_{i,na}^{PB}$ indicated that the associated cross-product term tends to be negative and its absolute value is not small relative to the sum of the other two terms in the decomposition, unlike in the case of $\hat{\overline{Y}}_i^{PB}$. As a result, the jackknife method can lead to serious overestimation of MSE($\hat{\overline{Y}}_{i,na}^{PB}$) because it ignores the cross-product term.

7.6.5 Model Misspecification

In Section 6.4.7, Chapter 6, we have studied misspecification of the linking model in the basic area level model. We showed that the method of Observed Best Prediction (OBP) can be robust to misspecification of the assumed linking model, unlike the customary EBLUP. Jiang, Nguyen, and Rao (2014) applied OBP to the nested error model and studied its performance under misspecification of either the mean function or the variance of the unit error e_{ij}, or both, assuming SRS within areas. Moreover, the focus is on the performance of OBP relative to the EBLUP for the special case of equal error variances σ_e^2 (with $k_{ij} = 1$ for all (i,j)) and all areas assumed to be sampled $(m = M)$.

The BLUP of the area mean $\overline{Y}_i = N_i^{-1} \sum_{j=1}^{N_i} y_{ij}$ when the parameters $\boldsymbol{\beta}$ and $\boldsymbol{\delta} = (\sigma_v^2, \sigma_e^2)^T$ are known may be expressed as

$$\tilde{\overline{Y}}_i^B = \overline{\mathbf{X}}_i^T \boldsymbol{\beta} + a_i(\boldsymbol{\delta})(\overline{y}_i - \overline{\mathbf{x}}_i^T \boldsymbol{\beta}), \tag{7.6.39}$$

where $\overline{y}_i$ and $\overline{\mathbf{x}}_i$ are the area sample means and

$$a_i(\boldsymbol{\delta}) = (1 - f_i)\gamma_i + f_i, \tag{7.6.40}$$

where $\gamma_i = \sigma_v^2/(\sigma_v^2 + n_i^{-1}\sigma_e^2)$ and $f_i = n_i/N_i$. Now noting that the design expectation of $\tilde{\overline{Y}}_i^B$ under SRS is given by

$$E_p(\tilde{\overline{Y}}_i^B) = (1 - f_i)(1 - \gamma_i)\overline{\mathbf{X}}_i^T \boldsymbol{\beta} + a_i(\boldsymbol{\delta})\overline{Y}_i, \tag{7.6.41}$$

the design MSE of $\tilde{\overline{Y}}_i^B$ is given by

$$\begin{aligned}
\text{MSE}_p(\tilde{\overline{Y}}_i^B) &= E_p(\tilde{\overline{Y}}_i^B - \overline{Y}_i)^2 \\
&= E_p\{(\tilde{\overline{Y}}_i^B)^2 - 2(1 - f_i)(1 - \gamma_i)\overline{y}_i(\overline{\mathbf{X}}_i^T \boldsymbol{\beta}) + [1 - 2a_i(\boldsymbol{\delta})]\hat{\overline{Y}}_i^{2\text{D}}\} \\
&= E_p\{Q_i(\boldsymbol{\beta}, \boldsymbol{\delta})\},
\end{aligned} \tag{7.6.42}$$

where $\hat{\overline{Y}}_i^{2\text{D}}$ is a design-unbiased estimator of $\overline{Y}_i^2$. By expressing $\overline{Y}_i^2$ in the form $\overline{Y}_i^2 = N_i^{-2}(\sum_{j=1}^{N_i} y_{ij}^2 + \sum_{j=1}^{N_i} \sum_{\ell \neq j} y_{ij} y_{i\ell})$ and assuming $n_i > 1$ for all $i = 1, \ldots, m$, an unbiased estimator of $\overline{Y}_i^2$ under SRS is given by

$$\hat{\overline{Y}}_i^{2\text{D}} = N_i^{-2} \left(\sum_{j=1}^{n_i} \pi_{j|i}^{-1} y_{ij}^2 + \sum_{j=1}^{n_i} \sum_{\substack{\ell=1 \\ \ell \neq j}}^{n_i} \pi_{j\ell|i}^{-1} y_{ij} y_{i\ell} \right), \tag{7.6.43}$$

where $\pi_{j|i} = n_i/N_i$ is the first-order inclusion probability and $\pi_{j\ell|i} = n_i(n_i - 1)/[N_i(N_i - 1)]$ is the second-order inclusion probability. Expression (7.6.43) clearly shows that $\hat{\overline{Y}}_i^{2\text{D}}$ is nonnegative if all y_{ij}'s are nonnegative. After simplification, (7.6.43) reduces to

$$\hat{\overline{Y}}_i^{2\text{D}} = \frac{1}{n_i} \sum_{j=1}^{n_i} y_{ij}^2 - \frac{N_i - 1}{N_i(n_i - 1)} \sum_{j=1}^{n_i} (y_{ij} - \overline{y}_i)^2. \tag{7.6.44}$$

The total design MSE of $\tilde{\overline{\mathbf{Y}}}^B = (\tilde{\overline{Y}}_1^B, \ldots, \tilde{\overline{Y}}_m^B)^T$ is given by

$$\begin{aligned}
\text{MSE}(\tilde{\overline{\mathbf{Y}}}^B) &= \sum_{i=1}^{m} E_p(\tilde{\overline{Y}}_i^B - \overline{Y}_i)^2 & (7.6.45) \\
&= E_p[Q(\boldsymbol{\beta}, \boldsymbol{\delta})], & (7.6.46)
\end{aligned}$$

where $Q(\boldsymbol{\beta}, \boldsymbol{\delta}) = \sum_{i=1}^{m} Q_i(\boldsymbol{\beta}, \boldsymbol{\delta})$. The Best Predictive Estimators (BPEs), $\hat{\boldsymbol{\beta}}_{\text{BPE}}$ and $\hat{\boldsymbol{\delta}}_{\text{BPE}}$, are then obtained by dropping E_p in (7.6.46) and then minimizing $Q(\boldsymbol{\beta}, \boldsymbol{\delta})$ with respect to $\boldsymbol{\beta}$ and $\boldsymbol{\delta}$. The OBP estimator of $\overline{Y}_i$, denoted $\hat{\overline{Y}}_i^{\text{OBP}}$, is obtained from (7.6.39) by substituting $\hat{\boldsymbol{\beta}}_{\text{BPE}}$ for $\boldsymbol{\beta}$ and $\hat{\boldsymbol{\delta}}_{\text{BPE}}$ for $\boldsymbol{\delta}$.

As in the case of the area level model, estimation of the area-specific $\text{MSE}_p(\hat{\overline{Y}}_i^{\text{OBP}})$ encounters difficulties because the assumed linking model is not used. Jiang et al. (2014) proposed a simple nonparametric bootstrap MSE estimator. We simply draw a simple random sample $\{(y_{ij}^*, \mathbf{x}_{ij}^*); j = 1, \ldots, n_i\}$ with replacement from the sample data $\{(y_{ij}, \mathbf{x}_{ij}); j = 1, \ldots, n_i\}$, from each area i independently, and then use the bootstrap data to calculate the OBP estimator $\hat{\overline{Y}}_{i*}^{\text{OBP}}, i = 1, \ldots, m$. Repeat the bootstrapping a large number, B, of times to obtain B replicates $\hat{\overline{Y}}_{i*}^{\text{OBP}}(1), \ldots, \hat{\overline{Y}}_{i*}^{\text{OBP}}(B)$ of $\hat{\overline{Y}}_{i*}^{\text{OBP}}$. Bootstrap estimator of $\text{MSE}_p(\hat{\overline{Y}}_i^{\text{OBP}})$ is then taken as

$$\text{mse}_B(\hat{\overline{Y}}_i^{\text{OBP}}) = B^{-1} \sum_{b=1}^{B} [\hat{\overline{Y}}_{i*}^{\text{OBP}}(b) - \overline{y}_i]^2, \tag{7.6.47}$$

where $\overline{y}_i = n_i^{-1} \sum_{j=1}^{n_i} y_{ij}$ is the sample mean for area i. The simple MSE estimator (7.6.47) is appealing, and simulation results on the relative bias (RB) indicated good performance in terms of RB. However, (7.6.47) may be somewhat unstable because it is estimating the (conditional) design MSE of $\hat{\overline{Y}}_i^{\text{OBP}}$, similar to the estimation of design MSE studied in Section 6.2.7 for the area level model.

In a simulation study, Jiang et al. (2014) studied the case where the assumed unit level model is misspecified in terms of both the mean function and the error variance and generated population data $\{(y_{ij}, \mathbf{x}_{ij}); j = 1, \ldots, N_i,\ i = 1, \ldots, m\}$ from the true model. By drawing repeated simple random samples of size $n_i = 5$ from each area i, they calculated the total empirical design MSE, of both the EBLUP under the assumed model and the OBP estimators. Their results showed that the design MSE of the EBLUP can be significantly larger than that of the OBP. For example, with $m = 400$, the total design MSE of OBP is 0.39 compared to 4.045 for the EBLUP, leading to a 930% increase in the total design MSE, when the assumed model has a zero intercept term while the intercept term is equal to 1.0 under the true model that generated the finite population. Surprisingly, OBP also performed better than EBLUP in terms of area-specific design MSE when m is large (≥ 100).

In practice, the assumed model may not be grossly misspecified as in the above case. Under such scenarios, OBP is only slightly better than the EBLUP. If the assumed model is indeed correct, then the percent increase in total design MSE of OBP over that of EBLUP is very small, especially for large m, suggesting that the OBP is fairly robust in terms of total design MSE.

7.6.6 Semi-parametric Nested Error Model: EBLUP

In the basic unit level model (7.1.1), we assumed a parametric mean function $m(\mathbf{x}, \boldsymbol{\beta}) = \mathbf{x}^T \boldsymbol{\beta}$. In this section we relax this assumption, using an unspecified mean function. For simplicity, we focus on the special case of a continuous scalar covariate, x. The unspecified mean function $m_0(x)$ is assumed to be approximated sufficiently well by a suitable spline basis. We focus on a truncated polynomial spline basis:

$$m_0(x) \approx m(x, \boldsymbol{\alpha}, \mathbf{u}) = \mathbf{x}^T\boldsymbol{\alpha} + \mathbf{w}^T\mathbf{u}. \tag{7.6.48}$$

Here, $\mathbf{x} = (1, x, \ldots, x^h)^T$ and $\mathbf{w} = ((x - q_1)_+^h, \ldots, (x - q_K)_+^h)^T$ form the truncated polynomial basis, where $(x - q)_+ = \max(0, x - q)$, h (≥ 1) is the degree of the spline, and $q_1 < \cdots < q_K$ is a set of fixed knots (Ruppert, Wand, and Carroll 2003). Finally, $\boldsymbol{\alpha} = (\alpha_0, \ldots, \alpha_h)^T$ and $\mathbf{u} = (u_1, \ldots, u_K)^T$ are, respectively, the coefficient vectors corresponding to the parametric and spline portions of the model. A linear spline is obtained by letting $h = 1$ in (7.6.48). Alternative bases, such as B-spline basis, are also used to approximate $m_0(x)$ (see Ugarte et al. 2009).

Regarding the choice of number of knots, K, and the location of the knots, $q_1, \ldots, q_K$, Ruppert, Wand, and Carroll (2003, Section 5.5) recommend the following: (i) One needs "enough" knots to ensure sufficient flexibility to fit the data, but after that additional knots do not change the fit of the model. (ii) Place the knots at the quantiles of the unique sample x-values, which give equal or nearly equal number of x-values between the knots.

In the special case of a regression model $y_{ij} = m(x_{ij}) + \epsilon_{ij}$ without random area effects, a penalized least squares estimator of $\boldsymbol{\gamma} = (\boldsymbol{\alpha}^T, \mathbf{u}^T)^T$ is obtained by minimizing $\sum_{i=1}^m \sum_{j=1}^{n_i} [y_{ij} - m(x_{ij}, \boldsymbol{\alpha}, \mathbf{u})]^2 + \lambda^2\boldsymbol{\gamma}^T\boldsymbol{\gamma}$ with respect to $\boldsymbol{\alpha}$ and $\mathbf{u}$, where λ is a smoothing parameter. The resulting ridge-regression type estimator of $\boldsymbol{\gamma}$ depends on λ. Several approaches to determining λ have been proposed, but a popular approach uses a mixed model formulation of (7.6.48) by treating $\mathbf{u}$ as a vector of random effects, with mean $\mathbf{0}$ and covariance matrix $\sigma_u^2\mathbf{I}_K$. Ruppert et al. (2003) showed that the resulting BLUE of $\boldsymbol{\beta}$ and BLUP of $\mathbf{u}$ are equivalent to the penalized least squares estimator with $\lambda^2 = \sigma_\epsilon^2/\sigma_u^2$, where σ_ϵ^2 is the variance of ϵ_{ij}. The parameter λ^2 may be estimated using ML or REML method.

Opsomer et al. (2008) extended the P-spline mixed model approach to the nested error model. The resulting model for a scalar x may be written as

$$y_{ij} = \mathbf{x}_{ij}^T\boldsymbol{\alpha} + \mathbf{w}_{ij}^T\mathbf{u} + v_i + e_{ij}, \quad j = 1, \ldots, n_i, \quad i = 1, \ldots, m, \tag{7.6.49}$$

where $\mathbf{x}_{ij} = (1, x_{ij}, \ldots, x_{ij}^h)^T$ and $\mathbf{w}_{ij} = ((x_{ij} - q_1)_+^h, \ldots, (x_{ij} - q_K)_+^h)^T =: (w_{ij1}, \ldots, w_{ijK})^T$ with specified $h \geq 1$. Additional covariates that are linearly related to y_{ij}, including categorical variables, may be included in $\mathbf{x}_{ij}$. Similarly, additional nonparametric terms in the mean function of the form $m_0(x_1) + g_0(x_2)$ can be handled by using the spline approximations to both $m_0(x_1)$ and $g_0(x_2)$ (see Ruppert et al. 2003).

The model (7.6.49) may be expressed in matrix notation as

$$\mathbf{y}_i = \mathbf{X}_i\boldsymbol{\alpha} + \mathbf{W}_i\mathbf{u} + \mathbf{Z}_i v_i + \mathbf{e}_i, \quad i = 1, \ldots, m, \tag{7.6.50}$$

where $\mathbf{y}_i = (y_{i1}, \ldots, y_{in_i})^T, \mathbf{X}_i = (\mathbf{x}_{i1}, \ldots, \mathbf{x}_{in_i})^T, \mathbf{W}_i = (\mathbf{w}_{i1}, \ldots, \mathbf{w}_{in_i})^T, \mathbf{Z}_i = \mathbf{1}_{n_i}$, and $\mathbf{u}, v_i$, and $\mathbf{e}_i = (e_{i1}, \ldots, e_{in_i})^T$ are mutually independent. Note that unlike the basic unit level model (7.1.1), model (7.6.50) does not have a block-diagonal covariance structure for $\mathbf{y}^T = (\mathbf{y}_1^T, \ldots, \mathbf{y}_n^T)$ because of the common random spline effect $\mathbf{u}$ across the areas i. Therefore, we express the m separate models in (7.6.50) as a single mixed model:

$$\mathbf{y} = \mathbf{X}\boldsymbol{\alpha} + \mathbf{W}\mathbf{u} + \mathbf{Z}\mathbf{v} + \mathbf{e}, \tag{7.6.51}$$

where $\mathbf{y} = \text{col}_{1\le i\le m}(\mathbf{y}_i), \mathbf{X} = \text{col}_{1\le i\le m}(\mathbf{X}_i), \mathbf{W} = \text{col}_{1\le i\le m}(\mathbf{W}_i), \mathbf{Z} = \text{diag}_{1\le i\le m}(\mathbf{Z}_i)$, $\mathbf{u} \sim (\mathbf{0}, \sigma_u^2\mathbf{I}_K), \mathbf{v} = (v_1, \ldots, v_m)^T \sim (\mathbf{0}, \sigma_v^2\mathbf{I}_m)$, and $\mathbf{e} = \text{col}_{1\le i\le m}(\mathbf{e}_i) \sim (\mathbf{0}, \sigma_e^2\mathbf{I}_n)$, for $n = \sum_{i=1}^m n_i$.

***Estimation of* $\boldsymbol{\alpha}, \mathbf{u}$, *and* $\mathbf{v}$** The BLUE of $\boldsymbol{\alpha}$ and the BLUP estimators of $\mathbf{u}$ and $\mathbf{v}$ for fixed $\boldsymbol{\delta} = (\sigma_u^2, \sigma_v^2, \sigma_e^2)^T$ are obtained by solving the "mixed model" equations

$$\begin{pmatrix} \frac{\mathbf{X}^T\mathbf{X}}{\sigma_e^2} & \frac{\mathbf{X}^T\mathbf{W}}{\sigma_e^2} & \frac{\mathbf{X}^T\mathbf{Z}}{\sigma_e^2} \\ \frac{\mathbf{W}^T\mathbf{X}}{\sigma_e^2} & \frac{\mathbf{I}_K}{\sigma_u^2} + \frac{\mathbf{W}^T\mathbf{W}}{\sigma_e^2} & \frac{\mathbf{W}^T\mathbf{Z}}{\sigma_e^2} \\ \frac{\mathbf{Z}^T\mathbf{X}}{\sigma_e^2} & \frac{\mathbf{Z}^T\mathbf{W}}{\sigma_e^2} & \frac{\mathbf{I}_m}{\sigma_v^2} + \frac{\mathbf{Z}^T\mathbf{Z}}{\sigma_e^2} \end{pmatrix} \begin{pmatrix} \boldsymbol{\alpha} \\ \mathbf{u} \\ \mathbf{v} \end{pmatrix} = \begin{pmatrix} \frac{\mathbf{X}^T\mathbf{y}}{\sigma_e^2} \\ \frac{\mathbf{W}^T\mathbf{y}}{\sigma_e^2} \\ \frac{\mathbf{Z}^T\mathbf{y}}{\sigma_e^2} \end{pmatrix}. \tag{7.6.52}$$

Denoting by $\mathbf{T} = (\mathbf{T}_{ij})_{1\le i,j\le 3}$ the inverse of the partitioned matrix on the left-hand side of (7.6.52), the solution to (7.6.52) may be written as

$$\begin{pmatrix} \tilde{\boldsymbol{\alpha}} \\ \tilde{\mathbf{u}} \\ \tilde{\mathbf{v}} \end{pmatrix} = \begin{pmatrix} \mathbf{T}_{11} & \mathbf{T}_{12} & \mathbf{T}_{13} \\ \mathbf{T}_{21} & \mathbf{T}_{22} & \mathbf{T}_{23} \\ \mathbf{T}_{31} & \mathbf{T}_{32} & \mathbf{T}_{33} \end{pmatrix} \begin{pmatrix} \frac{\mathbf{X}^T\mathbf{y}}{\sigma_e^2} \\ \frac{\mathbf{W}^T\mathbf{y}}{\sigma_e^2} \\ \frac{\mathbf{Z}^T\mathbf{y}}{\sigma_e^2} \end{pmatrix} \equiv \begin{pmatrix} \mathbf{Q}_1^T\mathbf{y} \\ \mathbf{Q}_2^T\mathbf{y} \\ \mathbf{Q}_3^T\mathbf{y} \end{pmatrix}. \tag{7.6.53}$$

The solution (7.6.53) is equivalent to equations (4)–(6) of Opsomer et al. (2008). However, (7.6.53) requires the inversion of a fixed matrix of size $(p + K + m) \times (p + K + m)$, unlike (4)–(6) of Opsomer et al. (2008) that require the inversion of the $n \times n$ covariance matrix, $\mathbf{V}$, of $\mathbf{y}$, where $n = \sum_{i=1}^m n_i$ is the total sample size.

It follows from (7.6.53) that $\tilde{\boldsymbol{\alpha}} = (\tilde{\alpha}_0, \tilde{\alpha}_1, \ldots, \tilde{\alpha}_h)^T, \tilde{\mathbf{u}} = (\tilde{u}_1, \ldots, \tilde{u}_K)^T$, and $\tilde{\mathbf{v}} = (\tilde{v}_1, \ldots, \tilde{v}_m)^T$ are unbiased in the sense that $E_m(\tilde{\boldsymbol{\alpha}}) = \boldsymbol{\alpha}$, $E_m(\tilde{\mathbf{u}} - \mathbf{u}) = \mathbf{0}$, and $E_m(\tilde{\mathbf{v}} - \mathbf{v}) = \mathbf{0}$. We now use (7.6.53) to develop method of moments estimators of σ_u^2, σ_v^2, and σ_e^2 (Rao, Sinha, and Dumitrescu 2014). It can be shown that

$$E\left(\sum_{k=1}^K \tilde{u}_k^2\right) = \text{tr}[E(\tilde{\mathbf{u}}\tilde{\mathbf{u}}^T)] = \sigma_u^2(K - t_2), \tag{7.6.54}$$

where $t_2 = \text{tr}(\mathbf{T}_{22})/\sigma_u^2$ and

$$E\left(\sum_{i=1}^{m} \tilde{v}_i^2\right) = \text{tr}[E(\tilde{\mathbf{v}}\tilde{\mathbf{v}}^T)] = \sigma_v^2(m - t_3), \tag{7.6.55}$$

where $t_3 = \text{tr}(\mathbf{T}_{33})/\sigma_v^2$. Furthermore, letting $\tilde{\mathbf{e}} = \mathbf{y} - \mathbf{X}\tilde{\boldsymbol{\alpha}} - \mathbf{W}\tilde{\mathbf{u}} - \mathbf{Z}\tilde{\mathbf{v}}$ with elements $\tilde{e}_{ij}$, we have

$$E\left(\sum_{i=1}^{m}\sum_{j=1}^{n_i} \tilde{e}_{ij}^2\right) - \text{tr}[E(\tilde{\mathbf{e}}\tilde{\mathbf{e}}^T)] = \sigma_e^2(n - t_4), \tag{7.6.56}$$

where $t_4 = \text{tr}(\mathbf{MTM}^T)/\sigma_e^2$ and $\mathbf{M} = (\mathbf{X}, \mathbf{W}, \mathbf{Z})$. It now follows from equations (7.6.54)–(7.6.56) that σ_u^2, σ_v^2, and σ_e^2 may be estimated iteratively from the following fixed-point equations

$$\sigma_u^2 = \sum_{k=1}^{K} \frac{\tilde{u}_k^2}{K - t_2}, \quad \sigma_v^2 = \sum_{i=1}^{m} \frac{\tilde{v}_i^2}{m - t_3}, \quad \sigma_e^2 = \sum_{i=1}^{m}\sum_{j=1}^{n_i} \frac{\tilde{e}_{ij}^2}{n - t_4}, \tag{7.6.57}$$

where t_4 can be expressed as $t_4 = p + (K - t_2) + (m - t_3)$. Equations (7.6.57) are solved iteratively as follows: (i) Compute the three summation terms in (7.6.57) using $\tilde{\mathbf{u}}, \tilde{\mathbf{v}}$, and $\tilde{\mathbf{e}}$ obtained from (7.6.53) with a starting value $\boldsymbol{\delta}^{(0)} = (\sigma_{u0}^2, \sigma_{v0}^2, \sigma_{e0}^2)^T$ of $\boldsymbol{\delta} = (\sigma_u^2, \sigma_v^2, \sigma_e^2)^T$ to get $\boldsymbol{\delta}^{(1)}$. (ii) Using $\boldsymbol{\delta}^{(1)}$, compute the next updated estimate $\boldsymbol{\delta}^{(2)}$ from (7.6.57) and so on until convergence, leading to the estimate $\hat{\boldsymbol{\delta}} = (\hat{\sigma}_u^2, \hat{\sigma}_v^2, \hat{\sigma}_e^2)^T$. Substituting $\hat{\boldsymbol{\delta}}$ for $\boldsymbol{\delta}$ into (7.6.53), we obtain the empirical BLUE (EBLUE) of $\boldsymbol{\alpha}$ and the empirical BLUP (EBLUP) estimators of $\mathbf{u}$ and $\mathbf{v}$, denoted by $\hat{\boldsymbol{\alpha}}, \hat{\mathbf{u}}$, and $\hat{\mathbf{v}}$, respectively.

Note that normality of $\mathbf{u}, \mathbf{v}$, and $\mathbf{e}$ is not assumed in deriving $\hat{\boldsymbol{\delta}}$, but under normality REML equations for σ_u^2, σ_v^2, and σ_e^2, given by Fellner (1986), have the same form as (7.6.57). General asymptotic theory for linear mixed models (Jiang 1996) shows that the REML estimator $\hat{\boldsymbol{\delta}}$ is consistent for $\boldsymbol{\delta}$ under the spline mixed model, provided the number of knots increases with the sample size n.

EBLUP of Area Mean $\overline{Y}_i$ Under the P-spline nested error model (7.6.50), the EBLUP of the area mean $\overline{Y}_i$ is given by

$$\hat{\overline{Y}}_i^H = N_i^{-1}\left(\sum_{j\in s_i} y_{ij} + \sum_{j\in r_i} \hat{y}_{ij}\right), \tag{7.6.58}$$

where s_i and $r_i = U_i - s_i$ denote, respectively, the set of sample units and the set of nonsample units from area i, and $\hat{y}_{ij} = \mathbf{x}_{ij}^T\hat{\boldsymbol{\alpha}} + \mathbf{w}_{ij}^T\hat{\mathbf{u}} + \hat{v}_i$ is the predictor of y_{ij} for $j \in r_i$. If the sampling fraction n_i/N_i is negligible, then $\hat{\overline{Y}}_i^H$ may be approximated by $\hat{\mu}_i^H = \overline{\mathbf{X}}_i^T\hat{\boldsymbol{\alpha}} + \overline{\mathbf{W}}_i^T\hat{\mathbf{u}} + \hat{v}_i = N_i^{-1}\sum_{j=1}^{N_i}\hat{y}_{ij}$, where the observed y_{ij} for $j \in s_i$ is replaced by its predictor $\hat{y}_{ij}$.

We now turn to the case of nonsampled areas $\ell = m+1,\dots,M$. Assuming that the P-spline mixed model holds also for them, we can use a synthetic predictor $\hat{y}_{\ell jS} = \mathbf{x}_{\ell j}^T\hat{\boldsymbol{\alpha}} + \mathbf{w}_{\ell j}^T\hat{\mathbf{u}}$, for each unit $y_{\ell j}, j = 1,\dots,N_\ell$. The resulting synthetic predictor of nonsampled area mean $\overline{Y}_\ell$ is given by

$$\hat{\overline{Y}}_\ell = N_\ell^{-1}\sum_{j=1}^{N_\ell}\hat{y}_{\ell jS} = \overline{\mathbf{X}}_\ell^T\hat{\boldsymbol{\alpha}} + \overline{\mathbf{W}}_\ell^T\hat{\mathbf{u}}. \tag{7.6.59}$$

MSE Estimation The mixed model formulation of penalized splines is a "convenient fiction to estimate smoothing parameters," and randomness of the spline coefficient vector $\mathbf{u}$ is only a device used to model curvature (Ruppert, Wang, and Carroll 2003, p. 138). Therefore, inferences should be conditional on $\mathbf{u}$. Rao et al. (2014) proposed a conditional parametric bootstrap method of estimating MSE, assuming normality.

The conditional bootstrap method generates v_i^* and e_{ij}^* from $N(0,\hat{\sigma}_v^2)$ and $N(0,\hat{\sigma}_e^2)$, respectively, and calculates the bootstrap population values $y_{ij}^* = \mathbf{x}_{ij}^T\hat{\boldsymbol{\alpha}} + \mathbf{w}_{ij}^T\hat{\mathbf{u}} + v_i^* + e_{ij}^*, j = 1,\dots,N_i, i = 1,\dots,m,$ for which the bootstrap population mean is $\overline{Y}_i^* = N_i^{-1}\sum_{j\in U_i} y_{ij}^*$. Then, using the corresponding sample data $\{(y_{ij}^*, \mathbf{x}_{ij}, \mathbf{w}_{ij}); j\in s_i,\ i = 1,\dots,m\}$, bootstrap estimates $\hat{\boldsymbol{\alpha}}^*, \hat{\mathbf{u}}^*$, and $\hat{v}_i^*$ and the corresponding predicted values $\hat{y}_{ij}^* = \mathbf{x}_{ij}^T\hat{\boldsymbol{\alpha}}^* + \mathbf{w}_{ij}^T\hat{\mathbf{u}}^* + \hat{v}_i^*$ for $j\in r_i$ are obtained. The resulting bootstrap EBLUP estimator of $\overline{Y}_i$ is $\hat{\overline{Y}}_i^{H*} = N_i^{-1}(\sum_{j\in s_i} y_{ij}^* + \sum_{j\in r_i}\hat{y}_{ij}^*)$.

Performing the above bootstrap operations a large number, B, of times, a conditional bootstrap estimator of $\text{MSE}(\hat{\overline{Y}}_i^H)$ is given by

$$\text{mse}_B(\hat{\overline{Y}}_i^H) = B^{-1}\sum_{b=1}^{B}[\hat{\overline{Y}}_i^{H*}(b) - \overline{Y}_i^*(b)]^2, \tag{7.6.60}$$

where $\hat{\overline{Y}}_i^{H*}(b)$ and $\overline{Y}_i^*(b)$ are the values of $\hat{\overline{Y}}_i^H$ and $\overline{Y}_i^*$ for the b-th bootstrap replicate, following the bootstrap MSE estimation method for finite populations (González-Manteiga et al. 2008a).

In the unconditional parametric bootstrap method, bootstrap response values are generated as $y_{ij}^* = \mathbf{x}_{ij}^T\hat{\boldsymbol{\beta}} + \mathbf{w}_{ij}^T\mathbf{u}^* + v_i^* + e_{ij}^*$, where $\mathbf{u}^*$ is generated from $N(0,\hat{\sigma}_u^2 I_K)$. Opsomer et al. (2008) proposed an unconditional nonparametric bootstrap method. This method can be changed to the conditional case by generating only v_i^* and e_{ij}^* in the method of Opsomer et al. (2008).

Opsomer et al. (2008) developed likelihood ratio tests, based on their bootstrap method, to test for the presence of small area effects ($H_0 : \sigma_v^2 = 0$) and the presence of spline effect ($H_0 : \sigma_u^2 = 0$).

7.6.7 Semi-parametric Nested Error Model: REBLUP

Estimation of Area Means We now study robust estimation of area means for the P-spline nested error model (7.6.49) in the presence of outliers either in the area effects v_i, or in the errors e_{ij}, or in both. The approach used in Section 7.4 for the nested error model runs into difficulties in the context of the model (7.6.49). In particular, it involves the Huber ψ-functions $\boldsymbol{\Psi}_b(\mathbf{r}) = \boldsymbol{\Psi}_b[\mathbf{U}^{-1/2}(\mathbf{y} - \mathbf{X}\boldsymbol{\alpha})]$, where $\mathbf{U}$ is a diagonal matrix, whose diagonal elements are those of the covariance matrix $\mathbf{V}$. For the model (7.6.49), majority of the elements of $\mathbf{r}$ will be large in absolute value because the spline term $\mathbf{Wu}$ is not included in $\mathbf{r}$. This in turn makes the corresponding elements of the derivative of $\boldsymbol{\Psi}_b(\mathbf{r})$ equal to zero, leading to difficulties in implementing the robust ML method.

Rao, Sinha, and Dumitrescu (2014) used an alternative approach that avoids the use of $\boldsymbol{\Psi}_b(\mathbf{r})$. This approach uses an iterative algorithm to get the REBLUP estimators of $\mathbf{u}$ and $\mathbf{v}$ and robust estimators of $\boldsymbol{\alpha}, \sigma_u^2, \sigma_v^2$, and σ_e^2 simultaneously, following Fellner (1986). First, robust mixed model equations are obtained from (7.6.52) by changing $\mathbf{y}$ to

$$\tilde{\mathbf{y}} = \mathbf{X}\boldsymbol{\alpha} + \mathbf{Wu} + \mathbf{Zv} + \sigma_e \boldsymbol{\Psi}_b(\sigma_e^{-1}\mathbf{e}) \tag{7.6.61}$$

on the right-hand side and then adding the term

$$\tilde{\mathbf{0}}_v = \mathbf{v} - \sigma_v \boldsymbol{\Psi}_b(\sigma_v^{-1}\mathbf{v}) \tag{7.6.62}$$

to $\mathbf{Z}^T\tilde{\mathbf{y}}/\sigma_e^2$. For given $\boldsymbol{\delta} = (\sigma_u^2, \sigma_v^2, \sigma_e^2)^T$, denote the resulting solution of these equations by $(\tilde{\boldsymbol{\alpha}}_F^T, \tilde{\mathbf{u}}_F^T, \tilde{\mathbf{v}}_F^T)^T$. Second, the fixed-point equations for σ_v^2 and σ_e^2 in (7.6.57) are robustified as

$$\sigma_v^2 = \sigma_v^2 \sum_{i=1}^{m} \frac{\psi_b^2(\sigma_v^{-1}\tilde{v}_{iF})}{(m - t_3)h}, \quad \sigma_e^2 = \sigma_e^2 \sum_{i=1}^{m}\sum_{j=1}^{n_i} \frac{\psi_b^2(\sigma_e^{-1}\tilde{e}_{ijF})}{(n - t_4)h}, \tag{7.6.63}$$

where $h = E[\psi_b^2(Z)]$ with $Z \sim N(0, 1)$, $\tilde{v}_{iF}$ is the ith element of $\tilde{\mathbf{v}}_F$, and $\tilde{e}_{ijF} = y_{ij} - \mathbf{x}_{ij}^T\tilde{\boldsymbol{\alpha}}_F - \mathbf{w}_{ij}^T\tilde{\mathbf{u}}_F - \tilde{v}_{iF}$. The fixed-point equation for σ_u^2 in (7.6.57) is not robustified because $\mathbf{u}$ is not affected by outliers:

$$\sigma_u^2 = \sum_{k=1}^{K} \frac{\tilde{u}_{kF}^2}{K - t_2}, \tag{7.6.64}$$

where $\tilde{u}_{kF}$ is the kth element of $\tilde{\mathbf{u}}_F$. The iterative algorithm of Fellner yielding REBLUP estimates of $\mathbf{u}$ and $\mathbf{v}$ and robust estimators of $\boldsymbol{\alpha}, \sigma_u^2$, and σ_e^2 simultaneously is now described in the following steps:

(1) Given a starting value $\boldsymbol{\delta}^{(0)}$ of $\boldsymbol{\delta}$, solve the mixed model equations (7.6.53) to obtain $\boldsymbol{\alpha}^{(0)}, \mathbf{u}^{(0)} = (u_1^{(0)}, \ldots, u_K^{(0)})^T$, and $\mathbf{v}^{(0)} = (v_1^{(0)}, \ldots, v_m^{(0)})^T$. Compute $\mathbf{e}^{(0)} = \mathbf{y} - \mathbf{X}\boldsymbol{\alpha}^{(0)} - \mathbf{W}\mathbf{u}^{(0)} - \mathbf{Z}\mathbf{v}^{(0)}$, with elements $e_{ij}^{(0)}$.

(2) Substitute the values $\boldsymbol{\delta}^{(0)}, u_k^{(0)}, v_i^{(0)}$, and $e_{ij}^{(0)}$ from (1) in place of $\boldsymbol{\delta}, \tilde{u}_{kF}, \tilde{v}_{iF}$, and $\tilde{e}_{ijF}$ on the right-hand sides of (7.6.63) and (7.6.64) to get an updated estimate $\boldsymbol{\delta}^{(1)}$ of $\boldsymbol{\delta}$.

(3) Compute the pseudo-values $\tilde{\mathbf{y}}^{(1)}$ and $\tilde{\mathbf{0}}_v^{(1)}$ from (7.6.61) and (7.6.62) using $\boldsymbol{\alpha}^{(0)}, \mathbf{u}^{(0)}, \mathbf{v}^{(0)}$, and $\mathbf{e}^{(0)}$ for $\boldsymbol{\alpha}, \mathbf{u}, \mathbf{v}$, and $\mathbf{e}$, respectively, and $\boldsymbol{\delta}^{(1)}$ for $\boldsymbol{\delta}$.

(4) Solve the robust mixed model equations (7.6.53) using $\boldsymbol{\delta}^{(1)}$ and the corresponding pseudo-values $\tilde{\mathbf{y}}^{(1)}$ and $\tilde{\mathbf{0}}_v^{(1)}$ obtained in (3). This in turn leads to new values $\boldsymbol{\alpha}^{(1)}, \mathbf{u}^{(1)}, \mathbf{v}^{(1)}$, and $\mathbf{e}^{(1)} = \mathbf{y} - \mathbf{X}\boldsymbol{\alpha}^{(1)} - \mathbf{W}\mathbf{u}^{(1)} - \mathbf{Z}\mathbf{v}^{(1)}$.

(5) Repeat steps (2)–(4) using the new values of $\boldsymbol{\alpha}, \boldsymbol{\delta}, \mathbf{u}$, and $\mathbf{v}$ until convergence, leading to $\hat{\boldsymbol{\alpha}}_F, \hat{\boldsymbol{\delta}}_F, \hat{\mathbf{u}}_F$, and $\hat{\mathbf{v}}_F$.

Let $\hat{y}_{ijF} = \mathbf{x}_{ij}^T\hat{\boldsymbol{\alpha}}_F + \mathbf{w}_{ij}^T\hat{\mathbf{u}}_F + \hat{v}_{iF}$ be the predicted value of y_{ij}, for $j \in r_i$. The spline-REBLUP estimator of $\overline{Y}_i$ is then given by

$$\hat{\overline{Y}}_i^{\text{RSD}} = N_i^{-1}\left(\sum_{j\in s_i} y_{ij} + \sum_{j\in r_i} \hat{y}_{ijF}\right). \tag{7.6.65}$$

MSE Estimation The MSE of the spline-REBLUP estimator $\hat{\overline{Y}}_i^{\text{RSD}}$ can be estimated following the Sinha–Rao parametric bootstrap method introduced in Section 7.4 (b) under model (7.6.49) conditional on $\mathbf{u}$. For this, we generate $\mathbf{v}^*$ and $\mathbf{e}^*$ from $N(\mathbf{0}, \hat{\sigma}_{vF}^2\mathbf{I}_m)$ and $N(\mathbf{0}, \hat{\sigma}_{eF}^2\mathbf{I}_n)$, respectively, and obtain bootstrap y-values, $y_{ij}^* = \mathbf{x}_{ij}^T\hat{\boldsymbol{\alpha}}_F + \mathbf{w}_{ij}^T\hat{\mathbf{u}}_F + v_i^* + e_{ij}^*, j = 1, \ldots, N_i, i = 1, \ldots, m$. Let $\overline{Y}_i^* = N_i^{-1}\sum_{j\in U_i} y_{ij}^*$ be the bootstrap population mean for area i. Using the corresponding bootstrap sample data $\{(y_{ij}^*, \mathbf{x}_{ij}); j \in s_i,\ i = 1, \ldots, m\}$, calculate bootstrap BLUE $\hat{\boldsymbol{\alpha}}^*$ and EBLUP estimates $\hat{\mathbf{u}}^*$ and $\hat{\mathbf{v}}^*$, and obtain predicted values $\hat{y}_{ij}^* = \mathbf{x}_{ij}^T\hat{\boldsymbol{\alpha}}^* + \mathbf{w}_{ij}^T\hat{\mathbf{u}}^* + \hat{v}_i^*$ for $j \in r_i$. The resulting bootstrap estimate of $\overline{Y}_i$ is given by $\hat{\overline{Y}}_i^{H*} = N_i^{-1}(\sum_{j\in s_i} y_{ij}^* + \sum_{j\in r_i} \hat{y}_{ij}^*)$. We used the EBLUP estimators rather than the REBLUP estimators because the bootstrap sample is free of outliers. However, the use of REBLUP also gave similar results. Bootstrap computations are greatly reduced by using the first option.

The above bootstrap operations are performed a large number of times, B, leading to B replicates of true values and corresponding EBLUP estimators, $\overline{Y}_i^*(b)$ and $\hat{\overline{Y}}_i^{H*}(b), b = 1, \ldots, B$. A conditional bootstrap estimator of $\text{MSE}(\hat{\overline{Y}}_i^{\text{RSD}})$ is then given by

$$\text{mse}_B(\hat{\overline{Y}}_i^{\text{RSD}}) = B^{-1}\sum_{b=1}^{B}[\hat{\overline{Y}}_i^{*H}(b) - \overline{Y}_i^*(b)]^2. \tag{7.6.66}$$

Simulation Results Rao et al. (2014) reported results of a limited simulation study on the performance of the spline-REBLUP estimator $\hat{\overline{Y}}_i^{\text{RSD}}$. The true model for generating samples for the simulation study is given by

$$y_{ij} = m_0(x_{ij}) + v_i + e_{ij}, \quad j = 1, \ldots, 40; \quad i = 1, \ldots, 40$$

and three different choices of $m_0(x)$: $1 + x$ (linear), $1 + x + x^2$ (quadratic), and $1 - x + 0.5\exp(x)$ (exponential 1) used by Breidt, Claeskens, and Opsomer (2005). Furthermore, the contaminated normal models introduced in Section 7.4 (c) are used to generate the random effects v_i and the unit errors e_{ij}. Samples are drawn with equal area sizes $n_i = 4$. A linear P-spline ($h = 1$) with $K = 20$ knots is used to approximate $m_0(x)$.

Results of the simulation study may be summarized as follows: (i) Spline-REBLUP estimator is much more efficient than the corresponding spline-EBLUP in the case of contamination in $\mathbf{e}$ only, or both $\mathbf{v}$ and $\mathbf{e}$, but only slightly more efficient in the case of contamination in $\mathbf{v}$ only. (ii) Sinha–Rao REBLUP estimator (corresponding to $K = 0$) leads to significant increase in MSE over the spline-REBLUP with $K = 20$ when the true model $m_0(x)$ is not linear. On the other hand, the increase in MSE of spline-REBLUP over the REBLUP is minimal when the true model is linear. (iii) The bootstrap MSE estimator (7.6.66) tracks well the MSE of $\hat{\overline{Y}}_i^{\text{RSD}}$. (iv) In the case of no outliers, results for the spline-EBLUP of Opsomer et al. (2008) relative to the EBLUP are similar to (i)–(iii). Also, the loss in efficiency of spline-REBLUP over the spline-EBLUP in this case is minimal, suggesting robustness of spline-REBLUP.

7.7 *SOFTWARE

The EBLUP estimates of area means $\overline{Y}_i$ given in (7.1.21), based on the basic unit level model (7.1.1) with $k_{ij} = 1$, can be computed using the function `eblupBHF` of the R package `sae`. The first-order unbiased parametric bootstrap MSE estimators (7.2.28) can be obtained using function `pbmseBHF`. Either ML or REML (default) fitting method can be specified in the argument `method` of these two functions. The calls to these functions are as follows:

```
eblupBHF(formula, dom, selectdom, meanxpop, popnsize,
    method = "REML", data)
pbmseBHF(formula, dom, selectdom, meanxpop, popnsize, B = 200,
    method = "REML", data)
```

In the argument `formula` of the two functions, we must specify the regression equation for the fixed (nonrandom) part of the model as a usual R regression formula. The left-hand side of the regression formula is the vector of individual values of the target variable for the sample, and the right-hand side contains the auxiliary variables for the sample separated by + and with an intercept included by default. In the argument `dom`, we must include a vector or factor with the domain codes for the sample elements. The population sizes of the domains, N_i, must be specified in the argument `popnsize`, in the form of a data frame including the domain codes in the first column. Similarly, the population means of the auxiliary variables must be included in the argument `meanxpop` as a data frame with the domain codes in the first column. Through the argument `selectdom`, we can select a subset of the areas for estimation. By default it considers the set of unique area codes in `dom`. For MSE estimation using the parametric bootstrap procedure, we must also choose the number of bootstrap replicates `B`.

The function `eblupBHF` returns a list of two objects: the vector `eblup` containing the EBLUP estimates for the domains, and the list `fit` containing the results of the fitting method. The function `pbmseBHF` returns a list with two objects: the first one is called `est` and contains the EBLUP estimates in `eblup` and a summary of the fitting procedure in `fit`. The second one is `mse` and contains a data frame with the estimated MSEs for each selected area. The list `fit` contains a summary of the fitting procedure (`summary`), which is the same summary as that of a linear mixed model fitting using function `lme` from R package `nlme`, the vector of estimated fixed effects (`fixed`), the vector of estimated random effects (`random`), the estimated random effects variance σ_v^2 (`refvar`), the loglikelihood (`loglike`), and the vector of raw residuals (`residuals`). Example 7.7.1 illustrates the use of the function `pbmseBHF`.

Example 7.7.1. County Crop Areas, with R. Consider the application of Battese et al. (1988) described in Example 7.3.1. We are going to illustrate how to obtain EBLUP estimates of county means of corn crop hectares using the basic unit level model with the total number of pixels in each county that are classified as corn and soybeans from the LANDSAT images, $\{x_{1ij}, x_{2ij}\}$, as auxiliary variables. In this application, the counties act as domains and the sample segments as units.

The data from Battese et al. (1988) are included in two predefined R data sets called `cornsoybean` and `cornsoybeanmeans`. The data set `cornsoybean` contains county codes (`County`), reported hectares of corn from the survey in each sample segment within each county (`CornHec`), reported hectares of soy beans from the survey in each sample segment within county (`SoyBeansHec`), number of pixels classified as corn from satellite data (`CornPix`), and number of pixels classified as soy beans from satellite data (`SoyBeansPix`).

The second data set, `cornsoybeanmeans`, contains the number of sample segments in the county or sample sizes (`SampSegments`), the number of population segments in the county or population sizes (`PopnSegments`), the county means of the number of corn pixels per segment (`MeanCornPixPerSeg`), and the number of soybeans pixels per segment (`MeanSoyBeansPixPerSeg`).

Here, the response variable is `CornHec` and the auxiliary variables are `CornPix` and `SoyBeansPix`. First, we load the two data sets:

```
R> data("cornsoybean")
R> data("cornsoybeanmeans")
```

Now we create a data frame with the true county means of the auxiliary variables called `Xmean`. We create another data frame with the county population sizes, called `Popn`. In these two data frames, the first column must contain the county codes. This column helps to identify the counties in the case that they were arranged differently in `Xmean` and `Popn`.

```
R> Xmean <- data.frame(cornsoybeanmeans[, c("CountyIndex",
+             "MeanCornPixPerSeg","MeanSoyBeansPixPerSeg")])
R> Popn <- data.frame(cornsoybeanmeans[, c("CountyIndex",
+            "PopnSegments")])
```

Now we call the function that returns EBLUP estimates of the means of corn crop area and also parametric bootstrap MSE estimates with `B=200` bootstrap replicates for all the counties, without removing any data point. We consider the default fitting method, REML.

```
R> set.seed(123)
R> BHF <- pbmseBHF(CornHec ~ CornPix + SoyBeansPix, dom = County,
+ meanxpop = Xmean, popnsize = Popn, B = 200, data = cornsoybean)

Bootstrap procedure with B = 200 iterations starts.
b = 1
...
b = 200
```

We construct a data frame with sample sizes, EBLUP estimates, and square root MSE estimates for each county, called `rescorn`, and print the obtained results. Note that the output of function `pbmseBHF` has been put in the object called `BHF`. This output includes the data frame `mse` with several columns, one of which is also called `mse` and contains the estimated MSEs of the EBLUP estimators. Similarly, the data frame `eblup` contains one column with the EBLUP estimates that is also called `eblup`.

```
R> sqrtmse.BHF <- sqrt(BHF$mse$mse)
R> rescorn <- data.frame(CountyIndex = BHF$est$eblup[, "domain"],
+  CountyName = cornsoybeanmeans[, "CountyName"],
+  BHF$est$eblup[, c("sampsize", "eblup")],sqrtmse=sqrtmse.BHF)
R> print(rescorn, row.names = FALSE)

CountyIndex CountyName sampsize    eblup  sqrtmse
          1 CerroGordo        1 122.5825 9.219717
          2   Hamilton        1 123.5274 8.640519
          3      Worth        1 113.0343 8.653569
          4   Humboldt        2 114.9901 8.020325
          5   Franklin        3 137.2659 8.319501
          6 Pocahontas        3 108.9807 7.131786
          7  Winnebago        3 116.4839 6.852539
          8     Wright        3 122.7711 7.029139
          9    Webster        4 111.5648 6.787479
         10    Hancock        5 124.1565 6.245703
         11    Kossuth        5 112.4626 6.157172
         12     Hardin        6 131.2515 5.955844
```

In Battese et al. (1988), observation num. 33, the second measurement in the county Hardin, was considered as an outlier and removed from the data. Here, this observation has not been removed and for that reason results are not the same. However, we can plot leverage measures s_{ii} versus scaled EBLUP residuals $\tilde{e}_i^2/\tilde{\mathbf{e}}^T\mathbf{e}$ as suggested by Zewotir and Galpin (2007) (see Section 5.4.2). For this, first we extract residuals and fitted variances from the results of the model fit.

```
R> resid<-BHF$est$fit$residuals
R> sigmav2est<-BHF$est$fit$refvar
R> sigmae2est<-BHF$est$fit$errorvar
```

We then obtain the design matrices of the fixed effects and the random effects. We also calculate the sample size, the number of covariates, and the number of areas.

```
R> X<-model.matrix(CornHec~CornPix+SoyBeansPix,data=cornsoybean)
R> n<-dim(X)[1]
R> p<-dim(X)[2]
R> Z<-model.matrix(CornHec~as.factor(County)-1,data=cornsoybean)
R> m<-dim(Z)[2]
```

Next, we compute the matrix **S**, whose diagonal elements s_{ii} measure the leverage effect of the observations.

```
R> In<-diag(rep(1,n))
R> V<-sigmav2est*Z\%*\%t(Z)+sigmae2est*In
R> Vinv<-solve(V)
R> P<-Vinv-Vinv\%*\%X\%*\%solve(t(X)\%*\%Vinv\%*\%X)\%*\%t(X)\%*\%Vinv
R> S<-sigmae2est*P
```

We calculate the scaled residuals that will be plotted.

```
R> resid2s<-resid^2/sum(resid^2)
```

We can now plot leverage measures against scaled residuals. We include the cutoff point for leverages proposed by Zewotir and Galpin (2007) and given by $(1-2p/n)\{1-[\hat{\sigma}_e^2/\hat{\sigma}_v^2+n/(2m)]^{-1}\}$. For scaled residuals, we draw the cutoff at mean+2 SD. We identify the observations that are out of the cutoffs.

```
R> plot(resid2s,diag(S),main="",xlab="ei^2/e^Te", ylab="sii")
R> cutoffs<-(1-2*p/n)*(1-1/(sigmae2est/sigmav2est+n/(2*m)))
R> cutoffresid2s<-mean(resid2s)+2*sd(resid2s)
R> abline(h=cutoffs)
R> abline(v=cutoffresid2s)
R> identify(resid2s,diag(S))
```

The resulting plot is shown in Figure 7.1. Observation num. 33 turns out to be a clear outlier, so we remove it from the data set. We obtain again EBLUP estimates and parametric bootstrap MSE estimates.

```
R> corn2<-cornsoybean[-33,]
R> BHF2<-pbmseBHF(corn2$CornHec~corn2$CornPix+corn2$SoyBeansPix,
+  dom=corn2$County,meanxpop=Xmean, popnsize=Popn,B=200)
```

Finally, we print the new results with observation num. 33 removed.

```
R> sqrtmse.BHF <- sqrt(BHF2$mse$mse)
R> rescorn2 <- data.frame(CountyIndex = BHF2$est$eblup[, "domain"],
+  CountyName = cornsoybeanmeans[, "CountyName"],
+  BHF2$est$eblup[, c("sampsize", "eblup")], sqrtmse=sqrtmse.BHF)
R> print(rescorn2, row.names = FALSE)
```

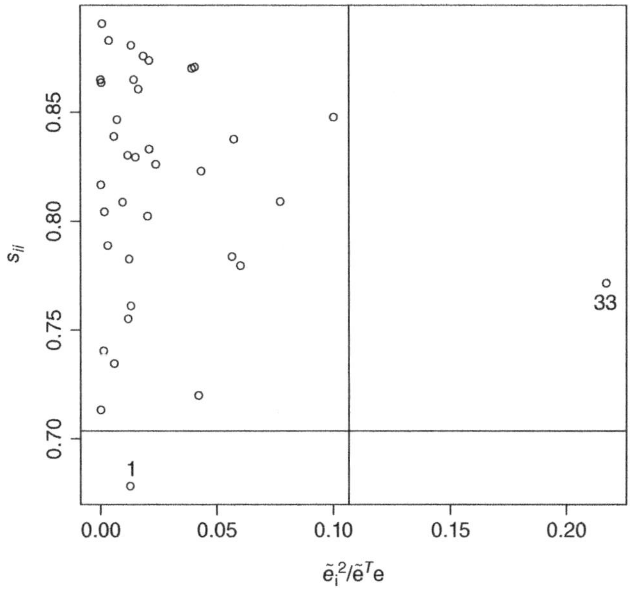

Figure 7.1 Leverage measures s_{ii} versus scaled squared residuals.

```
CountyIndex CountyName sampsize    eblup  sqrtmse
          1 CerroGordo        1 122.1954 9.684096
          2   Hamilton        1 126.2280 9.742037
          3      Worth        1 106.6638 8.854989
          4   Humboldt        2 108.4222 8.094216
          5   Franklin        3 144.3071 6.200801
          6 Pocahontas        3 112.1586 6.610806
          7  Winnebago        3 112.7801 6.380548
          8     Wright        3 122.0020 6.236414
          9    Webster        4 115.3438 5.371345
         10    Hancock        5 124.4144 5.320323
         11    Kossuth        5 106.8883 5.799831
         12     Hardin        5 143.0312 5.351844
```

Results now look pretty similar to those obtained by Battese et al. (1988).

7.8 *PROOFS

7.8.1 Derivation of (7.6.17)

We have

$$\begin{aligned} f_{s_i}(w_{ij}|\mathbf{x}_{ij}, y_{ij}) &:= f_P(w_{ij}|\mathbf{x}_{ij}, y_{ij}, a_{ij} = 1) \\ &= \frac{\Pr(a_{ij} = 1|w_{ij}, \mathbf{x}_{ij}, y_{ij}) f_P(w_{ij}|\mathbf{x}_{ij}, y_{ij})}{\Pr(a_{ij} = 1|\mathbf{x}_{ij}, y_{ij})}. \end{aligned} \tag{7.8.1}$$

Now,

$$\Pr(a_{ij}=1|w_{ij},\mathbf{x}_{ij},y_{ij}) = E_P[\Pr(a_{ij}=1|w_{ij},\mathbf{x}_{ij},y_{ij},\pi_{j|i})]$$
$$= E_P(\pi_{j|i}|w_{ij},\mathbf{x}_{ij},y_{ij}) = \pi_{j|i} = w_{ij}^{-1} \tag{7.8.2}$$

and similarly

$$\Pr(a_{ij}=1|\mathbf{x}_{ij},y_{ij}) = E_P(\pi_{j|i}|\mathbf{x}_{ij},y_{ij}). \tag{7.8.3}$$

It follows from (7.8.1), (7.8.2), and (7.8.3) that

$$f_{s_i}(w_{ij}|\mathbf{x}_{ij},y_{ij}) = w_{ij}^{-1}\frac{f_P(w_{ij}|\mathbf{x}_{ij},y_{ij})}{E_P(\pi_{j|i}|\mathbf{x}_{ij},y_{ij})},$$

which implies

$$E_{s_i}(w_{ij}|\mathbf{x}_{ij},y_{ij}) = [E_P(\pi_{j|i}|\mathbf{x}_{ij},y_{ij}]^{-1}. \tag{7.8.4}$$

Therefore, using (7.8.4) and (7.6.16), we obtain

$$E_P(\pi_{j|i}|\mathbf{x}_{ij},y_{ij}) = k_i^{-1}[\exp(\mathbf{x}_{ij}^T\mathbf{a} + b\ y_{ij})]^{-1}. \tag{7.8.5}$$

Adding (7.8.5) for $j=1,\ldots,N_i$, noting that $\sum_{j=1}^{N_i}\pi_{j|i}=n_i$ for a sampling design with fixed within-area sample size n_i and solving for k_i, it follows that k_i is given by (7.6.17).

7.8.2 Proof of (7.6.20)

Using (7.6.19), we have

$$f_{r_i}(y_{i\ell}|\mathbf{x}_{i\ell},u_i) = \Pr(a_{i\ell}=0|y_{i\ell},\mathbf{x}_{i\ell},u_i)\frac{f_P(y_{i\ell}|\mathbf{x}_{i\ell},u_i)}{\Pr(a_{i\ell}=0|\mathbf{x}_{i\ell},u_i)}. \tag{7.8.6}$$

Furthermore,

$$\Pr(a_{i\ell}=0|y_{i\ell},\mathbf{x}_{i\ell},u_i) = 1-\Pr(a_{i\ell}=1|y_{i\ell},\mathbf{x}_{i\ell},u_i)$$
$$= 1-E_P(\pi_{\ell|i}|y_{i\ell},\mathbf{x}_{i\ell},u_i)$$

and similarly $\Pr(a_{i\ell}=0|\mathbf{x}_{i\ell},u_i) = 1-E_P(\pi_{\ell|i}|\mathbf{x}_{i\ell},u_i)$. Hence, (7.8.6) may be expressed as

$$f_{r_i}(y_{i\ell}|\mathbf{x}_{i\ell},u_i) = \frac{1-E_P(\pi_{\ell|i}|y_{i\ell},\mathbf{x}_{i\ell},u_i)}{1-E_P(\pi_{\ell|i}|\mathbf{x}_{i\ell},u_i)}f_P(y_{i\ell}|\mathbf{x}_{i\ell},u_i). \tag{7.8.7}$$

Also, from (7.6.18), we have

$$f_P(y_{i\ell}|\mathbf{x}_{i\ell}, u_i) = E_P(\pi_{\ell|i}|\mathbf{x}_{i\ell}, u_i)\frac{f_{s_i}(y_{i\ell}|\mathbf{x}_{i\ell}, u_i)}{E_P(\pi_{\ell|i}|y_{i\ell}, \mathbf{x}_{i\ell}, u_i)}, \quad (7.8.8)$$

and similar to (7.8.4),

$$E_{s_i}(w_{i\ell}|\mathbf{x}_{i\ell}, u_i) = [E_P(\pi_{\ell|i}|\mathbf{x}_{i\ell}, u_i]^{-1}. \quad (7.8.9)$$

Now replacing (7.8.8) in (7.8.7), and using (7.8.4) and (7.8.9), we get (7.6.20).

8

EBLUP: EXTENSIONS

In Chapters 6 and 7, we presented a detailed account of the empirical best linear prediction (EBLUP) theory for small area estimation under the basic area and unit level models. In this chapter, we provide a brief account of the EBLUP method for various extensions of these models. Multivariate area level models are studied in Section 8.1. Area level models with correlated sampling errors are investigated in Section 8.2. Time series and cross-sectional models are studied in Section 8.3. Models that include spatial correlations among the random small area effects are considered in Section 8.4. Section 8.5 deals with twofold subarea models, designed for estimating at both area and subarea levels when covariates are available at the subarea level. A multivariate extension of the basic unit level model is studied in Section 8.6. Twofold nested error regression models, appropriate for two-stage sampling within areas, are studied in Section 8.7. Two-level models that effectively integrate the use of unit level and area level covariates are used in Section 8.8 for EBLUP estimation of small area means.

8.1 *MULTIVARIATE FAY–HERRIOT MODEL

The multivariate Fay–Herriot (FH) model was introduced in (4.4.3). This model can also be expressed as a special case of the general model (5.3.1) with block-diagonal covariance structure, by setting

$$\mathbf{y}_i = \hat{\boldsymbol{\theta}}_i, \quad \mathbf{X}_i = \mathbf{Z}_i, \quad \mathbf{Z}_i = \mathbf{I}_r, \quad \mathbf{v}_i = \mathbf{v}_i, \quad \mathbf{e}_i = \mathbf{e}_i, \quad \boldsymbol{\beta} = (\boldsymbol{\beta}_1, \ldots, \boldsymbol{\beta}_p)^T.$$

Small Area Estimation, Second Edition. J. N. K. Rao and Isabel Molina.

Here, $\hat{\boldsymbol{\theta}}_i = (\hat{\theta}_{i1}, \ldots, \hat{\theta}_{ir})^T$ is the vector of direct estimators for the r target parameters $\boldsymbol{\theta}_i = (\theta_{i1}, \ldots, \theta_{ir})^T$ from the ith area, and $\mathbf{e}_i = (e_{i1}, \ldots, e_{ir})^T$ is the $r \times 1$ vector of sampling errors for that area. Furthermore, in this case, the covariance matrices of $\mathbf{v}_i$ and $\mathbf{e}_i$ are, respectively, given by

$$\mathbf{G}_i = \boldsymbol{\Sigma}_v, \quad \mathbf{R}_i = \boldsymbol{\Psi}_i.$$

Consequently, the covariance matrix of $\hat{\boldsymbol{\theta}}_i$ is given by

$$\mathbf{V}_i = \boldsymbol{\Psi}_i + \boldsymbol{\Sigma}_v.$$

Also, the vector of target parameters is in this case $\boldsymbol{\mu}_i = \boldsymbol{\theta}_i = \mathbf{L}_i\boldsymbol{\beta} + \mathbf{M}_i\mathbf{v}_i$ with $\mathbf{L}_i = \mathbf{Z}_i$ and $\mathbf{M}_i = \mathbf{I}_r$.

Making the above substitutions in the general formula (5.3.19), we get the best linear unbiased predictor (BLUP) estimator of $\boldsymbol{\theta}_i$, given by

$$\tilde{\boldsymbol{\theta}}_i^H = \boldsymbol{\Sigma}_v(\boldsymbol{\Psi}_i + \boldsymbol{\Sigma}_v)^{-1}\hat{\boldsymbol{\theta}}_i + \boldsymbol{\Psi}_i(\boldsymbol{\Psi}_i + \boldsymbol{\Sigma}_v)^{-1}\mathbf{Z}_i\tilde{\boldsymbol{\beta}}, \tag{8.1.1}$$

which is a natural extension of the univariate BLUP estimator (6.1.3), where $\tilde{\boldsymbol{\beta}}$ is the extended weighted least squares estimator

$$\tilde{\boldsymbol{\beta}} = \left[\sum_{i=1}^m \mathbf{Z}_i^T(\boldsymbol{\Psi}_i + \boldsymbol{\Sigma}_v)^{-1}\mathbf{Z}_i\right]^{-1} \left[\sum_{i=1}^m \mathbf{Z}_i^T(\boldsymbol{\Psi}_i + \boldsymbol{\Sigma}_v)^{-1}\hat{\boldsymbol{\theta}}_i\right]. \tag{8.1.2}$$

The mean squared error (MSE) of $\tilde{\boldsymbol{\theta}}_i^H$, which equals the covariance matrix of $\tilde{\boldsymbol{\theta}}_i^H - \boldsymbol{\theta}_i$, follows from the general formula (5.3.20) by noting that $\boldsymbol{\Sigma}_v\mathbf{V}_i^{-1}\boldsymbol{\Psi}_i = (\boldsymbol{\Psi}_i^{-1} + \boldsymbol{\Sigma}_v^{-1})^{-1}$, and is given by

$$\begin{aligned}\text{MSE}(\tilde{\boldsymbol{\theta}}_i^H) &= E(\tilde{\boldsymbol{\theta}}_i^H - \boldsymbol{\theta}_i)(\tilde{\boldsymbol{\theta}}_i^H - \boldsymbol{\theta}_i)^T \\ &= (\boldsymbol{\Psi}_i^{-1} + \boldsymbol{\Sigma}_v^{-1})^{-1} + (\boldsymbol{\Psi}_i^{-1} + \boldsymbol{\Sigma}_v^{-1})^{-1}\boldsymbol{\Sigma}_v^{-1}\mathbf{Z}_i \\ &\quad \times \left[\sum_{i=1}^m \mathbf{Z}_i^T(\boldsymbol{\Psi}_i + \boldsymbol{\Sigma}_v)^{-1}\mathbf{Z}_i\right]^{-1}\mathbf{Z}_i^T\boldsymbol{\Sigma}_v^{-1}(\boldsymbol{\Psi}_i^{-1} + \boldsymbol{\Sigma}_v^{-1})^{-1}.\end{aligned} \tag{8.1.3}$$

Formula (8.1.3) reduces to (6.1.7) in the univariate case, $r = 1$. The first term on the right-hand side of (8.1.3) is $[O(1)]_{r\times r}$, whereas the second term, due to estimating $\boldsymbol{\beta}$, is $[O(m^{-1})]_{r\times r}$ for large m. The diagonal elements of $\text{MSE}(\tilde{\boldsymbol{\theta}}_i^H)$ represent the MSEs of the components $\tilde{\theta}_{ij}^H (j = 1, \ldots, r)$ of $\tilde{\boldsymbol{\theta}}_i^H$. By taking advantage of the correlations between the components $\hat{\theta}_{ij}$ of the direct estimator $\hat{\boldsymbol{\theta}}_i$, the multivariate model leads to more efficient estimators compared to those based on separate univariate models for each $j = 1, \ldots, r$.

The BLUP estimator depends on the covariance matrix of $\mathbf{v}_i, \boldsymbol{\Sigma}_v$, which is typically unknown. Substituting an estimator $\hat{\boldsymbol{\Sigma}}_v$ for $\boldsymbol{\Sigma}_v$ in (8.1.1), we obtain the EBLUP estimator $\hat{\boldsymbol{\theta}}_i^H$ of $\boldsymbol{\theta}_i$. For example, one could use the restricted maximum-likelihood (REML) method to estimate $\boldsymbol{\Sigma}_v$.

González-Manteiga et al. (2008b) studied a multivariate FH model with a common area effect for the r small area parameters, that is, with $\mathbf{v}_i = \mathbf{1}_r v_i$. Using the method of moments and the fitting-of-constants method to estimate the unknown model parameters, they derived a second-order approximation to the MSE matrix of the EBLUP estimator, $\text{MSE}(\hat{\boldsymbol{\theta}}_i^H)$. Using this approximation, a second-order unbiased estimator of $\text{MSE}(\hat{\boldsymbol{\theta}}_i^H)$ is derived. A parametric bootstrap MSE estimator is also studied. Simulation results with $r = 2$ small area parameters, under a model with correlation between sampling errors $\text{Corr}(e_{i1}, e_{i2}) = \rho_e$ varying from -0.5 to 0.5, showed that the EBLUP of the parameter θ_{ij} whose direct estimator $\hat{\theta}_{ij}$ has larger sampling variance achieves larger reduction in MSE, and the reduction is as large as 60% when $\rho_e = -0.5$ and decreases to about 30% for $\rho_e = 0.5$.

Example 8.1.1. Median Income. Datta, Fay, and Ghosh (1991) applied the multivariate FH model with $r = 2$ to estimate the current median income, θ_{i1}, of four-person families in each of the American states, $i = 1, \ldots, 51$. In this application, the second small area parameter was taken as $\theta_{i2} = 3/4$(median income of five-person families in state i)+$1/4$(median income of three-person families in state i); see Example 4.4.3, Chapter 4, for details on the auxiliary variables $\mathbf{z}_{ij}$ used in this study. Direct survey estimates, $\hat{\boldsymbol{\theta}}_i$, and associated covariance matrices, $\boldsymbol{\Psi}_i$, were obtained from the 1979 Current Population Survey data.

Datta et al. (1991) made an external evaluation of the EBLUP estimates, $\hat{\theta}_{i1}^H$, and the direct estimates, $\hat{\theta}_{i1}$, by comparing them to "true" values, θ_{i1}, available from the 1980 census data. The comparison was based on the absolute relative error (ARE) averaged over the states, defined as $\overline{\text{ARE}} = \frac{1}{51}\sum_{i=1}^{51} |\text{est}_i - \theta_{i1}|/\theta_{i1}$. In terms of $\overline{\text{ARE}}$, the EBLUP estimates, $\hat{\theta}_{i1}^H$, outperformed the direct estimates, $\hat{\theta}_{i1}$, obtaining an $\overline{\text{ARE}}$ of 2% compared to $\overline{\text{ARE}}$ of 5% for the direct estimates.

8.2 CORRELATED SAMPLING ERRORS

The FH model with correlated sampling errors is given by (4.4.5). It is a special case of the general linear mixed model (5.2.1) with unrestricted (but known) error covariance matrix $\mathbf{R} = \boldsymbol{\Psi}$, and with $\mathbf{Z} = \mathbf{I}, \mathbf{G} = \sigma_v^2 \mathbf{I}_m, \mathbf{X} = \mathbf{Z}$, and $\mathbf{y} = \hat{\boldsymbol{\theta}} = (\hat{\theta}_1, \ldots, \hat{\theta}_m)^T$. Replacing these matrices in the general BLUP estimator (5.2.9) together with $\mathbf{L} = \mathbf{Z}$ and $\mathbf{M} = \mathbf{I}$, we get the BLUP estimator of $\boldsymbol{\theta} = (\theta_1, \ldots, \theta_m)^T$, given by

$$\tilde{\boldsymbol{\theta}}^H = \mathbf{Z}\tilde{\boldsymbol{\beta}} + \sigma_v^2 \mathbf{V}^{-1}(\hat{\boldsymbol{\theta}} - \mathbf{Z}\tilde{\boldsymbol{\beta}}), \tag{8.2.1}$$

where

$$\tilde{\boldsymbol{\beta}} = (\mathbf{Z}^T \mathbf{V}^{-1} \mathbf{Z})^{-1} (\mathbf{Z}^T \mathbf{V}^{-1} \hat{\boldsymbol{\theta}})$$

and

$$\mathbf{V} = \boldsymbol{\Psi} + \sigma_v^2\mathbf{I}.$$

The EBLUP estimator $\hat{\boldsymbol{\theta}}^H$ is obtained by assuming known $\boldsymbol{\Psi} = (\psi_{i\ell})$ and substituting an estimator of $\hat{\sigma}_v^2$ for σ_v^2 in (8.2.1) .

Datta et al. (1992) obtained an approximation to the covariance matrix of $\hat{\boldsymbol{\theta}}^H - \boldsymbol{\theta}$, under normality of the errors $\mathbf{v}$ and $\mathbf{e}$, as

$$\text{MSE}(\hat{\boldsymbol{\theta}}^H) \approx \mathbf{G}_1(\sigma_v^2) + \mathbf{G}_2(\sigma_v^2) + \mathbf{G}_3(\sigma_v^2), \tag{8.2.2}$$

where

$$\mathbf{G}_1(\sigma_v^2) = \boldsymbol{\Psi} - \boldsymbol{\Psi}\mathbf{V}^{-1}\boldsymbol{\Psi},$$
$$\mathbf{G}_2(\sigma_v^2) = \boldsymbol{\Psi}\mathbf{V}^{-1}\mathbf{Z}(\mathbf{Z}^T\mathbf{V}^{-1}\mathbf{Z})^{-1}\mathbf{Z}^T\mathbf{V}^{-1}\boldsymbol{\Psi}$$

and

$$\mathbf{G}_3(\sigma_v^2) = \boldsymbol{\Psi}\mathbf{K}^3\boldsymbol{\Psi}\overline{V}(\hat{\sigma}_v^2),$$

where

$$\mathbf{K} = \mathbf{V}^{-1} - \mathbf{V}^{-1}\mathbf{Z}(\mathbf{Z}^T\mathbf{V}^{-1}\mathbf{Z})^{-1}\mathbf{Z}^T\mathbf{V}^{-1}$$

and $\overline{V}(\hat{\sigma}_v^2)$ is the asymptotic variance of the estimator $\hat{\sigma}_v^2$ of σ_v^2. If $\hat{\sigma}_v^2$ is chosen as the REML estimator, then

$$\overline{V}(\hat{\sigma}_v^2) = 2/\text{tr}(\mathbf{V}^{-2}).$$

Based on the approximation (8.2.2), an estimator of $\text{MSE}(\hat{\boldsymbol{\theta}}^H)$ is obtained along the lines of (5.3.11) as

$$\text{mse}(\hat{\boldsymbol{\theta}}^H) = \mathbf{G}_1(\hat{\sigma}_v^2) + \mathbf{G}_2(\hat{\sigma}_v^2) + 2\mathbf{G}_3(\hat{\sigma}_v^2). \tag{8.2.3}$$

In practice, ψ_{ii} is taken as a smoothed estimator of the sampling variance of the direct estimator $\hat{\theta}_i$, but the question of how to obtain smoothed estimators $\psi_{i\ell}$ of the sampling covariances between $\hat{\theta}_i$ and $\hat{\theta}_\ell, i \neq \ell$, remains. Isaki, Tsay, and Fuller (2000) addressed this issue by proposing the following sampling error covariance matrix based on a convex combination of a diagonal matrix $\hat{\boldsymbol{\Psi}}_d = \text{diag}(\psi_{ii}, i = 1, \ldots, m)$ with the smoothed variances and the original design-based estimator of the full covariance matrix $\hat{\boldsymbol{\Psi}} = (\hat{\psi}_{i\ell})$:

$$\hat{\boldsymbol{\Psi}}_\phi = \phi\hat{\boldsymbol{\Psi}}_d + (1 - \phi)\hat{\boldsymbol{\Psi}}, \tag{8.2.4}$$

where $\phi \in (0, 1)$ is a prespecified constant. We denote the EBLUP estimator of $\boldsymbol{\theta}$ with $\boldsymbol{\Psi}$ replaced by $\hat{\boldsymbol{\Psi}}_\phi$ as $\hat{\boldsymbol{\theta}}^H(\phi)$. Isaki et al. (2000) adjusted $\hat{\boldsymbol{\theta}}^H(\phi)$ to agree with a direct estimator for the entire area covering the small areas. This adjustment is similar to (7.3.1), but more complex because of correlated sampling errors. For the special case of $\phi = 1$, it uses multipliers similar to $\hat{p}_i^{(3)}$ given by (7.3.3). Isaki et al.

(2000) also obtained an estimator of the covariance matrix of $\hat{\boldsymbol{\theta}}_a^H(\boldsymbol{\phi}) - \boldsymbol{\theta}$, where $\hat{\boldsymbol{\theta}}_a^H(\boldsymbol{\phi})$ denotes the adjusted EBLUP estimator of $\boldsymbol{\theta}$. This covariance estimator is based on additional assumptions: (i) correlation between $\hat{\sigma}_v^2$ and $\hat{\boldsymbol{\Psi}}$ is negligible; (ii) $\hat{\boldsymbol{\Psi}}$ is distributed as a "Wishart" matrix with fairly large degrees of freedom. Note that the Wishart distribution is the multivariate version of the chi-squared distribution.

Example 8.2.1. U.S. Census Undercoverage. The adjustment of undercounts in the 1980 census attracted a lot of attention and controversy. In 1980, the U.S. Census Bureau expended vast sums of money and intellectual resources on improving coverage in the 1980 census enumeration. After the enumeration, however, several large states and cities claimed that the census had undercounted their populations. In fact, New York State filed a lawsuit against the Census Bureau in 1980, demanding that the Bureau adjusts its count for New York State. E.P. Ericksen, J.B. Kadane, and some other prominent statisticians appeared as the plaintiff's expert witnesses in New York State's lawsuit. They proposed EBLUP state estimates of census undercounts, $\theta_i = (T_i - C_i)/T_i$, based on a state level model, using dual-system estimates $\hat{\theta}_i$ of θ_i obtained from the 1980 post-enumeration survey (PES) (Ericksen and Kadane 1985). Here, T_i and C_i denote the true count and the census count for the ith state, respectively. Freedman and Navidi (1986) criticized the Ericksen–Kadane proposal for not validating their model fully and for not making their assumptions explicit. They also raised other technical issues, including the effect of large biases and large sampling errors in the PES estimates, $\hat{\theta}_i$. Cressie (1992) modeled the adjustment factors $\theta_i = T_i/C_i$ and proposed improved state level models.

In 1990, dual-system PES estimates of adjustment factors $\theta_i = T_i/C_i$ were produced for 1392 subdivisions (poststrata) of the total population. The PES sample contained approximately 377,000 persons in roughly 5,200 sample blocks. The poststrata were defined on the basis of geographical divisions of the country, tenure (owners or renters of homes), size of place, race, sex, and age. A poststratum level model was developed, using the dual-system estimates, $\hat{\theta}_i$, and poststratum level covariates $\mathbf{z}_i$. The PES variance estimates, $\hat{\psi}_{ii}$, were smoothed through a variance function model to obtain ψ_{ii}, while the original estimated correlations were used with the smoothed standard errors to obtain partially smoothed estimates of the sampling covariances, $\psi_{i\ell}, i \neq \ell$. Treating the resulting covariance matrix as the true $\boldsymbol{\Psi}$, EBLUP estimates, $\hat{\boldsymbol{\theta}}^H$, were obtained, using (8.2.1) with σ_v^2 changed to $\hat{\sigma}_v^2$. The EBLUP estimates were then ratio-adjusted to agree with regional total population estimates derived from the direct estimates of regional adjustment factors. Finally, the ratio-adjusted EBLUP adjustment factors were applied to block level census counts and to produce synthetic estimates of true block counts. Note that synthetic estimates are used only at the block level, in contrast to state level modeling, which entails synthetic estimation at all levels of geography lower than the state. Isaki, Huang, and Tsay (1991) provided a detailed account of EBLUP estimation of the poststratum adjustment factors $\theta_i = T_i/C_i$.

Datta et al. (1992) conducted an evaluation of the above methodology, prior to the 1990 census, using the 1988 Census Dress Rehearsal Data from test sites in Missouri. A post-enumeration survey was also conducted as part of the 1988 study to produce direct estimates of the adjustment factors. Datta et al. (1992) used a best subset selection procedure, based on the well-known Mallow's C_p criterion, to select z-variables from a set of 22 possible explanatory variables. Both $\mathbf{y}$ and $\mathbf{Z}$ were first transformed to ensure approximately iid errors, and the best subset selection procedure was then applied to the transformed data $\{\hat{\mathbf{V}}^{-1/2}\mathbf{y}, \hat{\mathbf{V}}^{-1/2}\mathbf{Z}\}$, where $\hat{\mathbf{V}}$ is an estimate of the covariance matrix of the combined error vector, $\mathbf{v}+\mathbf{e}$. Datta et al. (1992) computed the EBLUP estimates and associated standard errors, using the estimator (8.2.1) and the MSE estimator (8.2.3) . Their study revealed that the EBLUP estimates of adjustment factors outperformed the direct estimates for every poststratum.

The Census Bureau recommended the use of 1990 census-adjusted counts based on the above EBLUP methodology. However, Fay (1992) identified some serious deficiencies in the recommended EBLUP estimates. In particular, he noted that some of the regression coefficients in the model were very sensitive to the choice of estimated covariance matrix used to construct $\hat{\boldsymbol{\beta}}$. He attributed this difficulty to an unstable estimate of $\boldsymbol{\Psi}$ caused by direct estimates that were zero for many of the sample blocks. Subsequently, the Secretary of Commerce decided against using the proposed adjusted counts.

Isaki, Tsay, and Fuller (2000) investigated alternative EBLUP estimates from the 1990 census and PES data, using $\hat{\boldsymbol{\Psi}}_\phi$ given by (8.2.4) . The proposed method is less influenced by the variability of the estimated covariance matrix compared to the previous method. It was applied to a new set of 357 poststrata composed of 51 poststratum groups, each of which was subdivided into 7 age-sex categories (see Example 4.4.2, Chapter 4). Adjusted EBLUP estimates $\hat{\theta}_{ia}^H(\phi)$ of $\theta_i = T_i/C_i$ and associated standard errors, as described before, were computed for each of $m = 336$ poststrata obtained by eliminating Asian and American Indian data. The poststratum level model contained 21 explanatory variables including the intercept. The ratio of the standard error of $\hat{\theta}_{ia}^H(\phi)$ to $\hat{\theta}_{ia}^H(0.6)$, averaged over $i = 1, \ldots, 336$, was computed for selected values of ϕ including $\phi = 0$ and 1. Those average ratios revealed that $\phi = 0.5$ or 0.6 is a good choice. The standard errors of adjusted EBLUP estimates $\hat{\theta}_{ia}^H(0.6)$ of census adjustment factors $\theta_i = T_i/C_i$, were much smaller than those of the corresponding direct estimates, $\hat{\theta}_i$, with average estimated MSE efficiency of the adjusted EBLUP estimates (with $\phi = 0.6$) about 400% relative to the direct estimates $\hat{\theta}_i$. Moreover, the set of adjusted EBLUP estimates contained fewer extreme estimates than the set of direct estimates.

8.3 TIME SERIES AND CROSS-SECTIONAL MODELS

8.3.1 *Rao–Yu Model

The Rao–Yu model given by (4.4.6) and (4.4.7) with AR(1) or random walk model on the errors u_{it} provides an extension of the basic area level model (4.2.5) to handle time series and cross-sectional data. Let $\hat{\boldsymbol{\theta}}_i = (\hat{\theta}_{i1}, \ldots, \hat{\theta}_{iT})^T$ be the vector of available direct estimators in area i for the T time points, $i = 1, \ldots, m$. We can write the model in the general form (5.3.1) with block diagonal covariance structure, by considering

$$\mathbf{y}_i = \hat{\boldsymbol{\theta}}_i, \quad \mathbf{X}_i = (\mathbf{z}_{i1}, \ldots, \mathbf{z}_{iT})^T, \quad \mathbf{Z}_i = (\mathbf{1}_T, \mathbf{I}_T),$$
$$\mathbf{v}_i^T = (v_i, \mathbf{u}_i^T), \quad \mathbf{e}_i = (\mathbf{e}_{i1}, \ldots, \mathbf{e}_{iT})^T$$

and $\boldsymbol{\beta} = (\beta_1, \ldots, \beta_p)^T$, where $\mathbf{1}_T$ is the $T \times 1$ vector of ones, and $\mathbf{I}_T$ is the identity matrix of order T. Furthermore,

$$\mathbf{G}_i = \begin{bmatrix} \sigma_v^2 & \mathbf{0}^T \\ \mathbf{0} & \sigma^2 \boldsymbol{\Lambda}_T \end{bmatrix}, \quad \mathbf{R}_i = \boldsymbol{\Psi}_i,$$

where $\boldsymbol{\Lambda}_i = \boldsymbol{\Lambda}_T = (\lambda_{t,s})$ is the $T \times T$ covariance matrix of $\mathbf{u}_i = (u_{i1}, \ldots, u_{iT})^T$ with (t, s)th element given by $\lambda_{t,s} = \rho^{|t-s|}/(1 - \rho^2)$ for the AR(1) model $u_{it} = \rho u_{i,t-1} + \epsilon_{it}$, $|\rho| \ll 1$. For the random walk model $u_{it} = u_{i,t-1} + \epsilon_{it}$, the (t, s)th element of the covariance matrix $\boldsymbol{\Lambda}_T$ is given by $\lambda_{t,s} = \min(t, s)$.

We may write $\mathbf{V}_i$ as

$$\mathbf{V}_i = \boldsymbol{\Psi}_i + \sigma^2 \boldsymbol{\Lambda}_T + \sigma_v^2 \mathbf{J}_T \tag{8.3.1}$$

with $\mathbf{J}_T = \mathbf{1}_T \mathbf{1}_T^T$ denoting a $T \times T$ matrix of ones. Furthermore, the small area parameters θ_{iT} for the current time point T may be expressed as $\theta_{iT} = \mu_i = \mathbf{l}_i^T \boldsymbol{\beta} + \mathbf{m}_i^T \mathbf{v}_i$ with $\mathbf{l}_i = \mathbf{z}_{iT}$ and $\mathbf{m}_i = (1, 0, \ldots, 0, 1)^T$.

The random walk model is applicable for a finite T by specifying an initial value, u_{i0}, of the u_{it}-series, say $u_{i0} = 0$. In the latter case, Diallo (2014) noted that it is not necessary to force $\rho = 1$ as in the random walk model or confine to $|\rho| < 1$ as in the AR(1) model of Rao and Yu (RY). He postulated a model on $\{u_{it}\}$ with arbitrary ρ, given by

$$u_{it} = \rho u_{i,t-1} + \epsilon_{it}, \quad t \geq 1, \quad u_{i0} = 0, \tag{8.3.2}$$

where $\epsilon_{it} \overset{\text{iid}}{\sim} N(0, \sigma^2)$. Model (8.3.2) is applicable for small or moderate T and it is adequate for the purpose of predicting the current θ_{iT}. Under model (8.3.2), the covariance matrix $\mathbf{V}_i$ is obtained from (8.3.1) by noting that

$$\lambda_{t,t} = V(u_{it}) = \sigma^2 \sum_{\ell=1}^{t} \rho^{2(\ell-1)}, \tag{8.3.3}$$

$$\lambda_{t,s} = \text{Cov}(u_{it}, u_{is}) = \sigma^2 \rho^{|t-s|} \sum_{\ell=1}^{\min(t,s)} \rho^{2(\ell-1)}. \tag{8.3.4}$$

Making the above substitution in the general BLUP formula (5.3.2), we get the BLUP estimator of θ_{iT} as

$$\tilde{\theta}_{iT}^H = \mathbf{z}_{iT}^T \tilde{\boldsymbol{\beta}} + (\sigma_v^2 \mathbf{1}_T + \sigma^2 \boldsymbol{\lambda}_T)^T \mathbf{V}_i^{-1} (\hat{\boldsymbol{\theta}}_i - \mathbf{X}_i \tilde{\boldsymbol{\beta}}), \tag{8.3.5}$$

where λ_T^T is the Tth row of $\mathbf{\Lambda}_T$ and

$$\tilde{\beta} = \left(\sum_{i=1}^{m} \mathbf{X}_i^T \mathbf{V}_i^{-1} \mathbf{X}_i\right)^{-1} \left(\sum_{i=1}^{m} \mathbf{X}_i^T \mathbf{V}_i^{-1} \hat{\theta}_i\right).$$

The BLUP estimator (8.3.5) may also be expressed as a weighted combination of the direct estimator of $\hat{\theta}_{iT}$, the regression-synthetic estimator $\mathbf{z}_{iT}^T \tilde{\beta}$ and the residuals $\hat{\theta}_{it} - \mathbf{z}_{it}^T \tilde{\beta}, t = 1, \ldots, T-1$, as follows:

$$\tilde{\theta}_{iT}^H = w_{iT}^* \hat{\theta}_{iT} + (1 - w_{iT}^*) \mathbf{z}_{iT}^T \tilde{\beta} + \sum_{t=1}^{T-1} w_{it}^* (\hat{\theta}_{it} - \mathbf{z}_{it}^T \tilde{\beta}), \qquad (8.3.6)$$

where

$$(w_{i1}^*, \ldots, w_{iT}^*) = (\sigma_v^2 \mathbf{1}_T + \sigma^2 \lambda_T)^T \mathbf{V}_i^{-1}, \quad i = 1, \ldots, m.$$

The BLUP estimator (8.3.5) of θ_{iT} depends on unknown model parameters σ_v^2, σ^2, and ρ in the case of the Rao–Yu model and the Diallo model, and only on σ_v^2 and σ^2 in the case of the random walk model. Datta et al. (2002) obtained REML estimators of σ_v^2 and σ^2 under the random walk model, whereas Diallo (2014) studied maximum likelihood (ML) estimation of σ_v^2, σ^2, and ρ under the general model (8.3.2) for u_{it}. Estimation of ρ under the Rao–Yu model encounters difficulties when the true ρ is close to 1, because of the term $1 - \rho^2$ appearing in the denominator of $\mathbf{\Lambda}_T$.

The EBLUP estimator $\hat{\theta}_{iT}^H$ is obtained from (8.3.6) by substituting suitable estimators for the model parameters. Diallo (2014) derived a second-order approximation to $\text{MSE}(\hat{\theta}_{iT}^H)$ and a second-order unbiased estimator of $\text{MSE}(\hat{\theta}_{iT}^H)$ for large m and fixed T, under ML estimation of model parameters.

Example 8.3.1. Median Income. Datta, Lahiri, and Maiti (2002) applied the Rao–Yu model (4.4.6) and (4.4.7), Chapter 4, with random walk errors u_{it}, to estimate median income of four-person families for the 50 American states and the District of Columbia. They used the Current Population Survey (CPS) data for 9 years (1981–1989) to produce EBLUP estimates $\hat{\theta}_{iT}^H$ of the median income, θ_{iT}, for 1989 ($i = 1, \ldots, 51$). The estimates for 1989 enabled them to conduct an external evaluation, by comparing the estimates to 1990 census estimates for 1989. The sampling error model is given by $\hat{\theta}_{it} = \theta_{it} + e_{it}$ with sampling covariance matrix $\mathbf{\Psi}_i$ for each area i estimated from the CPS data, using the jackknife method and some smoothing techniques. The linking model is given by $\theta_{it} = \beta_0 + \beta_1 z_{it1} + v_i + u_{it}$, where z_{it1} is the "adjusted" census median income for year t in area i (see Example 4.4.3, Chapter 4). The EBLUP estimates were calculated using both REML and ML estimates of model parameters σ^2 and σ_v^2.

Average absolute relative error ($\overline{\text{ARE}}$) values for the CPS direct estimates and the EBLUP estimates were calculated by treating the census values as the true values θ_{iT}. Note that $\overline{\text{ARE}} = \frac{1}{51} \sum_{i=1}^{51} |\text{est}_{iT} - \theta_{iT}|/\theta_{iT}$. They obtained the following values:

TABLE 8.1 Distribution of Coefficient of Variation (%)

	Coefficient of Variation		
Estimate	2–4%	4–6%	≥ 6%
CPS	6	7	38
EBLUP (REML)	49	2	0
EBLUP (ML)	49	2	0

Source: Adapted from Table 1 in Datta, Lahiri, and Maiti (2002).

$\overline{\text{ARE}}$(CPS) = 7.3%, $\overline{\text{ARE}}$ (EBLUP) = 2.9% using REML, and $\overline{\text{ARE}}$(EBLUP) = 2.8% using ML. Thus, the EBLUP estimates outperformed the CPS estimates in terms of $\overline{\text{ARE}}$.

Datta, Lahiri, and Maiti (2002) also calculated the coefficient of variation (CV) of the EBLUP estimates, using the Tth diagonal element of $\boldsymbol{\Psi}_i$ for the CPS direct estimate $\hat{\theta}_{iT}$, and MSE estimates of the EBLUP based on REML and ML estimates of model parameters. The distribution of CVs is given in Table 8.1.

It is clear from Table 8.1 that EBLUP (ML) and EBLUP (REML) estimates outperformed the CPS direct estimates by bringing the CV down to the range 2–4% for 49 areas, while the CV of CPS estimates is at least 6% for 38 areas.

Example 8.3.2. Poverty Proportions in Spain. Esteban et al. (2012) applied the Rao–Yu model given by (4.4.6) and (4.4.7) with AR(1) errors u_{it} to estimate poverty rates for the Spanish provinces by gender, using data from the EU-SILC from years 2004–2006 ($T = 3$). They used the generalized variance function (GVF) approach to smooth the sampling variances and treated the smoothed estimates as known ψ_{it}. They compared the efficiencies of the EBLUP estimator under the Rao–Yu model relative to the direct estimator and to the EBLUP estimator under the model assuming $\rho = 0$. This comparison is based on the second-order unbiased MSE estimators for the two EBLUP estimators under their respective models. The REML estimator of ρ turned out to be 0.68. Authors conclude that the EBLUP under the Rao–Yu model performed the best among the three estimators for this data set. This conclusion is supported by simulation results for $\rho = 0.75$.

8.3.2 State-Space Models

The area level model given by (4.4.13)–(4.4.15), Chapter 4, allows the model coefficients to vary both cross-sectionally (across areas) and over time. This model is a special case of the general state-space model, which may be expressed in the form

$$\mathbf{y}_t = \mathbf{Z}_t\boldsymbol{\alpha}_t + \boldsymbol{\varepsilon}_t; \quad E(\boldsymbol{\varepsilon}_t) = \mathbf{0}, \quad E(\boldsymbol{\varepsilon}_t\boldsymbol{\varepsilon}_t^T) = \boldsymbol{\Sigma}_t \tag{8.3.7}$$

$$\boldsymbol{\alpha}_t = \mathbf{H}\boldsymbol{\alpha}_{t-1} + \mathbf{A}\boldsymbol{\eta}_t; \quad E(\boldsymbol{\eta}_t) = \mathbf{0}, \quad E(\boldsymbol{\eta}_t\boldsymbol{\eta}_t^T) = \boldsymbol{\Gamma} \tag{8.3.8}$$

where $\boldsymbol{\varepsilon}_t$ and $\boldsymbol{\eta}_t$ are uncorrelated contemporaneously and over time. This model is a special case of the general linear mixed model but the state-space form permits updating of the estimates over time using the Kalman filter equations (8.3.10) and (8.3.11) given below. It also allows smoothing past estimates as new data becomes available, using an appropriate smoothing algorithm. The vector $\boldsymbol{\alpha}_t$ is known as the *state vector*, equation for $\boldsymbol{\alpha}_t$ in (8.3.8) as the *transition equation* and equation (8.3.7) for $\mathbf{y}_t$ as the *measurement equation*.

Let $\tilde{\boldsymbol{\alpha}}_{t-1}$ be the BLUP estimator of $\boldsymbol{\alpha}_{t-1}$ based on all data observed up to time $t-1$, so that $\tilde{\boldsymbol{\alpha}}_{t|t-1} = \mathbf{H}\tilde{\boldsymbol{\alpha}}_{t-1}$ is the BLUP of $\boldsymbol{\alpha}_t$ at time $t-1$. Furthermore, let $\mathbf{P}_{t-1} = E(\tilde{\boldsymbol{\alpha}}_{t-1} - \boldsymbol{\alpha}_{t-1})(\tilde{\boldsymbol{\alpha}}_{t-1} - \boldsymbol{\alpha}_{t-1})^T$ be the covariance matrix of the prediction errors at time $t-1$. Then, the covariance matrix of the prediction errors $\tilde{\boldsymbol{\alpha}}_{t|t-1} - \boldsymbol{\alpha}_t$ is given by

$$\mathbf{P}_{t|t-1} = \mathbf{H}\mathbf{P}_{t-1}\mathbf{H}^T + \mathbf{A}\boldsymbol{\Gamma}\mathbf{A}^T.$$

This result readily follows from (8.3.8). At time t, the predictor of $\boldsymbol{\alpha}_t$ and its covariance matrix are updated using the new data $(\mathbf{y}_t, \mathbf{Z}_t)$. We have

$$\mathbf{y}_t - \mathbf{Z}_t\tilde{\boldsymbol{\alpha}}_{t|t-1} = \mathbf{Z}_t(\boldsymbol{\alpha}_t - \tilde{\boldsymbol{\alpha}}_{t|t-1}) + \boldsymbol{\varepsilon}_t, \tag{8.3.9}$$

which has the linear mixed model form (5.2.1) with $\mathbf{y} = \mathbf{y}_t - \mathbf{Z}_t\tilde{\boldsymbol{\alpha}}_{t|t-1}, \mathbf{Z} = \mathbf{Z}_t, \mathbf{v} = \boldsymbol{\alpha}_t - \tilde{\boldsymbol{\alpha}}_{t|t-1}, \mathbf{G} = \mathbf{P}_{t|t-1}, \mathbf{X}\boldsymbol{\beta}$ absent, and $\mathbf{V} = \mathbf{F}_t$, where

$$\mathbf{F}_t = \mathbf{Z}_t\mathbf{P}_{t|t-1}\mathbf{Z}_t^T + \boldsymbol{\Sigma}_t.$$

Therefore, the BLUP estimator $\tilde{\mathbf{v}} = \mathbf{G}\mathbf{Z}^T\mathbf{V}^{-1}\mathbf{y}$ reduces in this case to

$$\tilde{\boldsymbol{\alpha}}_t = \tilde{\boldsymbol{\alpha}}_{t|t-1} + \mathbf{P}_{t|t-1}\mathbf{Z}_t^T\mathbf{F}_t^{-1}(\mathbf{y}_t - \mathbf{Z}_t\tilde{\boldsymbol{\alpha}}_{t|t-1}). \tag{8.3.10}$$

Furthermore, it follows from (5.2.12) that the covariance matrix of the prediction errors $\tilde{\mathbf{v}} - \mathbf{v}$ is $\mathbf{G} - \mathbf{G}\mathbf{Z}^T\mathbf{V}^{-1}\mathbf{Z}\mathbf{G}$, which in the case of the prediction errors $\tilde{\boldsymbol{\alpha}}_t - \boldsymbol{\alpha}_t$ reduces to

$$\mathbf{P}_t = \mathbf{P}_{t|t-1} - \mathbf{P}_{t|t-1}\mathbf{Z}_t^T\mathbf{F}_t^{-1}\mathbf{Z}_t\mathbf{P}_{t|t-1}. \tag{8.3.11}$$

The general state-space model, (8.3.7) and (8.3.8), covers both area level and unit level models.

To illustrate the Kalman filter, consider the area level model

$$\hat{\theta}_{it} = \mathbf{z}_{it}^T\boldsymbol{\beta}_{it} + e_{it} \tag{8.3.12}$$

with the coefficients β_{itj} obeying the random walk model:

$$\beta_{itj} = \beta_{i,t-1,j} + v_{itj}, \quad j = 1, \ldots, p. \tag{8.3.13}$$

The sampling errors, e_{it}, for each area i, are assumed to be serially uncorrelated with mean 0 and variance ψ_{it}. Furthermore, the model errors, v_{itj}, for each area i, are also uncorrelated over time, with mean zero and covariances $E_m(v_{itj}v_{it\ell}) = \sigma_{vj\ell}$, and for areas $i \neq h, E_m(v_{itj}v_{htj}) = \rho_j\sigma_{vjj}$ and $E_m(v_{itj}v_{ht\ell}) = 0, j \neq \ell$. In this case,

$\mathbf{y}_t = \hat{\boldsymbol{\theta}}_t = (\hat{\theta}_{1t}, \ldots, \hat{\theta}_{mt})^T, \mathbf{z}_{it}^T \boldsymbol{\alpha}_{it} = \mathbf{z}_{it}^T \boldsymbol{\beta}_{it} = \theta_{it}, \mathbf{H} = \mathbf{I}, \mathbf{A} = \mathbf{I}$. The BLUP estimator of the realized vector of parameters at time t, $\boldsymbol{\theta}_t = (\theta_{1t}, \ldots, \theta_{mt})^T = \mathbf{Z}_t \boldsymbol{\alpha}_t$, is $\tilde{\boldsymbol{\theta}}_t^H = (\tilde{\theta}_{1t}^H, \ldots, \tilde{\theta}_{mt}^H)^T = \mathbf{Z}_t \tilde{\boldsymbol{\alpha}}_t$, where $\mathbf{Z}_t$ is block diagonal with the blocks defined by the rows $\mathbf{z}_{it}^T, \boldsymbol{\alpha}_t = (\boldsymbol{\beta}_{1t}^T, \ldots, \boldsymbol{\beta}_{mt}^T)^T$ and $\tilde{\boldsymbol{\alpha}}_t$ is obtained from the Kalman filter equation (8.3.10) . After simplification, it can be shown that the BLUP estimator $\tilde{\theta}_{it}^H$ is a weighted combination of the direct estimator $\hat{\theta}_{it}$, the regression-synthetic estimator $\mathbf{z}_{it}^T \tilde{\boldsymbol{\beta}}_{it|t-1}$, where $\tilde{\boldsymbol{\beta}}_{it|t-1}$ is the BLUP estimator of $\boldsymbol{\beta}_{it} = \boldsymbol{\alpha}_{it}$ at time $t-1$, and a third component based on the "adjustment factors" $\hat{\theta}_{ht} - \mathbf{z}_{ht}^T \tilde{\boldsymbol{\beta}}_{ht|t-1}$ for areas $h \neq i$ (Pfeffermann and Burck 1990). This third component vanishes when $\rho_j = 0$ for all $j = 1, \ldots, p$, that is, when the spatial correlations, ρ_j, are absent. In this case, the BLUP estimator $\tilde{\theta}_{it}^H$ has the familiar FH form (6.1.3) with the weight attached to the direct estimator decreasing as the sampling variance ψ_{it} increases.

To implement the recursive calculations (8.3.10) and (8.3.11), we need to specify a mean $\tilde{\boldsymbol{\alpha}}_0$ and a covariance matrix $\mathbf{P}_0$ of the initial state vector $\boldsymbol{\alpha}_0$. When the transition equation is nonstationary, as in the case of the random walk model (8.3.13), the values $\tilde{\boldsymbol{\alpha}}_0 = \mathbf{0}$ and $\mathbf{P}_0 = \kappa \mathbf{I}$ could be used, where κ is a large positive constant. This choice corresponds to a diffuse or noninformative prior distribution. For the stationary case, $\tilde{\boldsymbol{\alpha}}_0$ and $\mathbf{P}_0$ are taken as the unconditional mean and covariance matrix of $\boldsymbol{\alpha}_t$. We refer the reader to Harvey (1990, Section 3.3.4) for a careful discussion on the initial conditions.

The BLUP estimator $\tilde{\boldsymbol{\alpha}}_t$ involves unknown parameters $\boldsymbol{\delta}$ that specify the covariance matrices $\boldsymbol{\Sigma}_t$ and $\boldsymbol{\Gamma}$. The vector $\boldsymbol{\delta}$ may also contain unknown elements of the transition matrix. Assuming normality of the errors $\boldsymbol{\epsilon}_t$ and $\boldsymbol{\eta}_t$, ML estimators of these parameters can be obtained by expressing the loglikelihood in the *prediction error decomposition* form:

$$\log\ L = \text{const.} - \frac{1}{2} \sum_{t=1}^{T} \log\ |\mathbf{F}_t| - \frac{1}{2} \sum_{t=1}^{T} (\mathbf{y}_t - \tilde{\mathbf{y}}_{t|t-1})^T \mathbf{F}_t^{-1} (\mathbf{y}_t - \tilde{\mathbf{y}}_{t|t-1}), \quad (8.3.14)$$

where $\tilde{\mathbf{y}}_{t|t-1} = \mathbf{Z}_t \tilde{\boldsymbol{\alpha}}_{t|t-1}$ is the BLUP of $\mathbf{y}_t$ at time $t-1$ and $\mathbf{y}_t - \tilde{\mathbf{y}}_{t|t-1}$ is the vector of prediction errors. The representation (8.3.14) follows by writing L as

$$L = \prod_{t=1}^{T} p(\mathbf{y}_t | \mathbf{Y}_{t-1}),$$

where $p(\mathbf{y}_t | \mathbf{Y}_{t-1})$ is the conditional distribution of $\mathbf{y}_t$ given $\mathbf{Y}_{t-1} = \{\mathbf{y}_{t-1}, \ldots, \mathbf{y}_1\}$, and noting that $p(\mathbf{y}_t | \mathbf{Y}_{t-1})$ is normal with mean $\tilde{\mathbf{y}}_{t|t-1}$ and covariance matrix $\mathbf{F}_t$. The ML estimator of $\boldsymbol{\delta}$ and the parameters of the initial specification, if any, can be obtained by using the Fisher-scoring algorithm with a variable step length (Pfeffermann and Burck 1990). The updating equations of this iterative algorithm are

$$\boldsymbol{\delta}^{(a+1)} = \boldsymbol{\delta}^{(a)} + r_i [\mathcal{I}(\boldsymbol{\delta}^{(a)})]^{-1} \mathbf{s}(\boldsymbol{\delta}^{(a)}).$$

Here, $\boldsymbol{\delta}^{(a)}$ is the value of the estimator at the ath iteration, $\mathbf{s}(\boldsymbol{\delta})$ is the score vector consisting of the elements $\partial \log L/\partial \delta_i$, $\mathcal{I}(\boldsymbol{\delta})$ is the information matrix and r_i is a variable step length introduced to ensure that the likelihood is nondecreasing, that is, $L(\boldsymbol{\delta}^{(a+1)}) \geq L(\boldsymbol{\delta}^{(a)})$ at each iteration a. The method of moments can also be used to estimate the model parameters $\boldsymbol{\delta}$ without the normality assumption. Singh, Mantel, and Thomas (1994) presented such estimators for special cases, using a moment method analogous to the Fay and Herriot (1979) method for cross-sectional data.

The EBLUP estimator $\hat{\boldsymbol{\alpha}}_t$ is obtained by substituting an estimator $\hat{\boldsymbol{\delta}}$ for $\boldsymbol{\delta}$ in $\tilde{\boldsymbol{\alpha}}_t$. The resulting EBLUP estimator of $\boldsymbol{\theta}_t = \mathbf{Z}_t\boldsymbol{\alpha}_t$ is given by

$$\hat{\boldsymbol{\theta}}_t^H = \mathbf{Z}_t\hat{\boldsymbol{\alpha}}_t. \tag{8.3.15}$$

A naive covariance matrix of the prediction errors $\hat{\boldsymbol{\alpha}}_t - \boldsymbol{\alpha}_t$ is obtained by substituting $\hat{\boldsymbol{\delta}}$ for $\boldsymbol{\delta}$ in the formula (8.3.11) for $\mathbf{P}_t$, ignoring the variability associated with $\hat{\boldsymbol{\delta}}$. The resulting estimated covariance matrix of $\hat{\boldsymbol{\theta}}_t^H - \boldsymbol{\theta}_t$ is

$$\text{mse}_N(\hat{\boldsymbol{\theta}}_t^H) = \mathbf{Z}_t\hat{\mathbf{P}}_t\mathbf{Z}_t^T, \tag{8.3.16}$$

where $\hat{\mathbf{P}}_t$ is the estimator of $\mathbf{P}_t$. We refer the reader to Harvey (1990, Chapter 3) for a detailed account of the state-space model and the Kalman filter.

Methods of estimating the MSE of the EBLUP estimator $\hat{\theta}_{it}^H$ that account for the variability associated with $\hat{\boldsymbol{\delta}}$ have also been studied. Under normality of the errors,

$$\text{MSE}(\hat{\theta}_{it}^H) = \text{MSE}(\tilde{\theta}_{it}^H) + E(\hat{\theta}_{it}^H - \tilde{\theta}_{it}^H)^2, \tag{8.3.17}$$

where

$$\text{MSE}(\tilde{\theta}_{it}^H) = \mathbf{z}_{it}^T\mathbf{P}_t\mathbf{z}_{it}. \tag{8.3.18}$$

A naive estimator of $\text{MSE}(\tilde{\theta}_{it}^H)$ is given by

$$\text{mse}_N(\tilde{\theta}_{it}^H) = \mathbf{z}_{it}^T\mathbf{P}_t(\hat{\boldsymbol{\delta}})\mathbf{z}_{it}. \tag{8.3.19}$$

To approximate the last term in (8.3.17), $g_{3ij}(\boldsymbol{\delta}) = E(\hat{\theta}_{it}^H - \tilde{\theta}_{it}^H)^2$, Ansley and Kohn (1986) expanded $\hat{\theta}_{it}^H = \tilde{\theta}_{it}^H(\hat{\boldsymbol{\delta}})$, where $\hat{\boldsymbol{\delta}}$ is the ML estimator of $\boldsymbol{\delta}$, around $\boldsymbol{\delta}$ as follows:

$$g_{3ij}(\boldsymbol{\delta}) \approx [\partial\tilde{\theta}_{it}(\boldsymbol{\delta})/\partial\boldsymbol{\delta}]^T\mathcal{I}^{-1}(\boldsymbol{\delta})\partial\tilde{\theta}_{it}(\boldsymbol{\delta})/\partial\boldsymbol{\delta}, \tag{8.3.20}$$

where $\mathcal{I}(\hat{\boldsymbol{\delta}})$ is information matrix. Ansley and Kohn used $\text{mse}_N(\tilde{\theta}_{it}^H) + g_{3ij}(\hat{\boldsymbol{\delta}})$ as an estimator of $\text{MSE}(\hat{\theta}_{it}^H)$, but the bias of $\text{mse}_N(\tilde{\theta}_{it}^H)$ is of the same order as $g_{3ij}(\boldsymbol{\delta})$. As a result, the MSE estimator of Ansley and Kohn is not second-order unbiased.

Pfeffermann and Tiller (2005) proposed an estimator of MSE based on the parametric bootstrap. Assuming normality of the errors, their method consists of the following steps:

(1) Generate a large number, B, of bootstrap series $\{\hat{\theta}_{it}^{(b)}\}, b = 1, \ldots, B$ from the model fitted to the original series, with model parameters $\boldsymbol{\delta} = \hat{\boldsymbol{\delta}}$.

(2) Re-estimate $\boldsymbol{\delta}$ for each of the generated series using the same method as used for estimating $\boldsymbol{\delta}$ from the original series $\{\hat{\theta}_{it}\}$. Denote the bootstrap estimators as $\hat{\boldsymbol{\delta}}^{(b)}, b = 1, \ldots, B$.

(3) Estimate $g_{3ij}(\boldsymbol{\delta}) = E(\hat{\theta}_{it}^H - \tilde{\theta}_{it}^H)^2$ as

$$g_{3ij}^B = \frac{1}{B}\sum_{b=1}^{B}[\tilde{\theta}_{it}^H(\hat{\boldsymbol{\delta}}^{(b)}) - \tilde{\theta}_{it}^H(\hat{\boldsymbol{\delta}})]^2, \tag{8.3.21}$$

noting that $\tilde{\theta}_{it}^H = \tilde{\theta}_{it}^H(\boldsymbol{\delta})$ and $\hat{\theta}_{it}^H = \tilde{\theta}_{it}^H(\hat{\boldsymbol{\delta}})$.

(4) Correct $\text{mse}_N(\tilde{\theta}_{it}^H)$ for its bias as follows:

$$\text{mse}_B(\tilde{\theta}_{it}^H) = 2\ \text{mse}_N(\tilde{\theta}_{it}^H) - \frac{1}{B}\sum_{b=1}^{B}[\mathbf{z}_{it}^T\mathbf{P}_t(\hat{\boldsymbol{\delta}}^{(b)})\mathbf{z}_{it}]. \tag{8.3.22}$$

(5) The bootstrap estimator of $\text{MSE}(\hat{\theta}_{it}^H)$ is given by

$$\text{mse}_B(\hat{\theta}_{it}^H) = \text{mse}_B(\tilde{\theta}_{it}^H) + g_{3ij}^B.$$

Pfeffermann and Tiller (2005) established the unbiasedness of the bootstrap MSE estimator to second-order terms, under specified regularity conditions. The bootstrap MSE estimator is similar to the jackknife MSE estimator studied in Section 9.2.2.

The U.S. Bureau of Labor Statistics (BLS) uses area level state-space models for the prediction of all the major employment and unemployment estimates in the 50 states and the District of Columbia. The CPS direct estimates over time are used to produce the model-based estimates, which are "time indirect" in the terminology of Section 1.1 (Tiller 1982). The models used in this application account for seasonal effects, trend, a covariate, and sampling error correlations over time. Pfeffermann, Feder, and Signorelli (1998) applied a similar state-space model to labor force data from Australia.

In the context of BLS small area estimates, produced by fitting a separate state-space model within each area, Pfeffermann and Tiller (2006) developed benchmarked small area estimates that agree with a reliable CPS direct estimate when aggregated over a group of areas.

Example 8.3.3. Simulation Study. Singh, Mantel, and Thomas (1994) conducted a simulation study to compare the EBLUP estimator at current time T based on the state-space model, the EBLUP estimator based on FH model (FH), a synthetic (SYN) estimator, and a sample size dependent (SSD) estimator based only on the cross-sectional data for time T. For this, they generated a pseudo-population from Statistics Canada's biannual farm survey data from Quebec over six time points

(June 1988, January 1989, …, January 1991). The farm survey used dual frames, namely a list frame and an area frame, but the pseudo-population was generated from the list frame survey data only. The farm units in the list frame were replicated proportional to their sampling weights to form a pseudo-population of $N = 10,362$ farm units. The parameter of interest, θ_{iT}, is the total number of cattle and calves for each crop district $i = 1, \ldots, 12$ (small areas) in Quebec for the current period T (January 1991).

The pseudo-population was stratified into four take-some and one take-all strata using 1986 Agricultural Census count data on cattle and calves as the stratification variable. Independent stratified random samples were generated for each occasion from the pseudo-population, and the estimates SYN, SSD, FH, and EBLUP based on the state-space model, were computed from each simulated sample r.

Let θ_i^c and $\theta_\cdot^c = \sum_{i=1}^{12} \theta_i^c$ be the 1986 census counts of cattle and calves in crop district i and in the population, and let $\hat{\theta}_{\cdot T}$ be the direct estimate of $\theta_{\cdot T} = \sum_{i=1}^{12} \theta_{iT}$ obtained from the farm survey. The ratio-synthetic estimate $z_{iT} = (\hat{\theta}_{\cdot T}/\theta_\cdot^c)\theta_i^c$ of the parameter θ_{iT} for area i at time T is used to derive the SYN estimate as follows:

$$\text{SYN}_{iT} = \hat{\beta}_1 + \hat{\beta}_2 z_{iT},$$

where $\hat{\beta}_1$ and $\hat{\beta}_2$ are the least squares estimates of β_1 and β_2 obtained fitting the model $\hat{\theta}_{iT} = \beta_1 + \beta_2 z_{iT} + e_{iT}$, where $\hat{\theta}_{iT}$ is the direct estimate of θ_{iT} and $e_{iT} \overset{\text{iid}}{\sim} (0, \psi_T), i = 1, \ldots, m$. SSD is a weighted combination of $\hat{\theta}_{iT}$ and SYN, that is, SSD $= \phi_{iT}(\text{S1})\hat{\theta}_{iT} + [1 - \phi_{iT}(\text{S1})]\text{SYN}_{iT}$, where $\phi_{iT}(\text{S1})$ is given by (3.3.8) with $\delta = 1.0$. FH is given by (6.1.12), Chapter 6, where the model variance, σ_{vT}^2, is estimated by the FH method of moments, see (6.1.14), Chapter 6. Finally, EBLUP estimator is obtained under the state-space model $\hat{\theta}_{it} = \beta_{t1} + \beta_{t2} z_{it} + a_{it}$, with $\beta_{tj} = \beta_{t-1,j} + v_{tj}, j = 1, 2$ and $a_{it} = a_{i,t-1} + \epsilon_{it}$. Here, the errors v_{tj} and a_{it} are assumed to be uncorrelated with mean 0 and variances σ_{vj}^2 and σ_a^2, respectively. The parameters σ_{vj}^2 and σ_a^2 were estimated by a method similar to the FH moment method for cross-sectional data.

Table 8.2 presents the simulated values of absolute relative bias and relative root MSE (or CV) of the estimates averaged over $m = 12$ small areas within specified size groups, denoted by $\overline{\text{ARB}}$ and $\overline{\text{RRMSE}}$. It is clear from Table 8.2 that EBLUP estimator based on the state-space model leads to significant reduction in $\overline{\text{RRMSE}}$ relative to FH which, in turn, performs better than SSD. In terms of $\overline{\text{ARB}}$, again EBLUP estimator based on the state-space model is better than SYN and FH. As expected, SSD has the smallest $\overline{\text{ARB}}$.

8.4 *SPATIAL MODELS

In real applications, direct estimators from neighboring areas may be correlated. When the area level auxiliary variables in the FH model do not explain the spatial correlation sufficiently well, including in the model spatially correlated area effects v_i might increase the efficiency of the final small area estimators. Spatial FH models were introduced in Section 4.4.4. In the spatial FH model, the covariance

TABLE 8.2 Average Absolute Relative Bias ($\overline{\text{ARB}}$) and Average Relative Root MSE ($\overline{\text{RRMSE}}$) of SYN, SSD, FH, and EBLUP (State-Space)

	Estimator			
Size Group	SYN	SSD	FH	EBLUP
		$\overline{\text{ARB}}$ %		
Small	8.7	2.9	5.2	3.9
Medium	14.1	1.7	9.5	8.0
Large	6.3	0.6	4.9	4.2
		$\overline{\text{RRMSE}}$ %		
Small	15.4	19.4	16.1	12.0
Medium	15.9	18.5	15.1	12.7
Large	9.8	14.9	10.0	8.4

Source: Adapted from Table 4 in Singh, Mantel, and Thomas (1994).

matrix $\mathbf{G} = \sigma_v^2 \mathbf{I}_m$ of the FH model is changed to $\mathbf{G} = \mathbf{\Gamma}(\boldsymbol{\delta})$, where $\mathbf{\Gamma}(\boldsymbol{\delta})$ may take different forms depending on the particular spatial model assumed for $\mathbf{v}$. Examples are $\mathbf{\Gamma}(\boldsymbol{\delta}) = \sigma_v^2(\mathbf{I} - \rho\mathbf{Q})^{-1}\mathbf{B}$ with $\boldsymbol{\delta} = (\rho, \sigma_v^2)^T$ for the CAR model (4.4.16), $\mathbf{G} = \mathbf{G}(\boldsymbol{\delta}) = \sigma_u^2[(\mathbf{I} - \phi\mathbf{W})(\mathbf{I} - \phi\mathbf{W})^T]^{-1}$, with $\boldsymbol{\delta} = (\phi, \sigma_u^2)^T$ for the simultaneously autoregressive process (SAR) model (4.4.18), or alternatively $\mathbf{\Gamma}(\boldsymbol{\delta}) = \sigma_v^2(\delta_1\mathbf{I} + \delta_2\mathbf{D})$ or $\mathbf{\Gamma}(\boldsymbol{\delta}) = \sigma_v^2[\delta_1\mathbf{I} + \delta_2\mathbf{D}(\delta_3)]$, as used in the geostatistics literature, for $\mathbf{D}$ and $\mathbf{D}(\delta_3)$ defined in Section 4.4.4.

Spatial models described above are special cases of the general linear mixed model (5.2.1). Therefore, the BLUP estimator of $\theta_i = \mathbf{z}_i^T\boldsymbol{\beta} + b_i v_i$ can be obtained from the general formula (5.2.4), for a specified vector of parameters $\boldsymbol{\delta}$. In practice, $\boldsymbol{\delta}$ is unknown and we need to replace it by an estimator $\hat{\boldsymbol{\delta}}$ to obtain the EBLUP estimator of θ_i. ML or REML estimators of $\boldsymbol{\delta}$ may be obtained from the general results in Section 5.2.4. Cressie and Chan (1989) discussed ML estimation of $\boldsymbol{\delta}$ for spatial models.

Again, since spatial models are special cases of (5.2.1), the MSE of the BLUP is given by (5.2.11). For the EBLUP estimator, the decomposition (5.2.29), given by Kackar and Harville (1984), still holds provided $\hat{\boldsymbol{\delta}}$ is odd and translation invariant, which is true for ML and REML estimators. However, a rigorous second-order approximation for the term $E[t(\hat{\boldsymbol{\delta}}) - t(\boldsymbol{\delta})]^2$ cannot be obtained from the general framework of Das, Jiang, and Rao (2004) because of the presence of correlation among observations from different areas. Petrucci and Salvati (2006) and Singh, Shukla, and Kundu (2005) directly applied the general linearization approach (5.2.30)–(5.2.32) to the spatial model with area effects following the SAR model (4.4.18), and gave a heuristic MSE approximation given by (5.2.34) with $\boldsymbol{\delta} = (\phi, \sigma_u^2)^T$. Petrucci and Salvati (2006, 2008) used also the MSE estimators (5.2.39) or (5.2.40), depending on whether $\boldsymbol{\delta} = (\phi, \sigma_u^2)^T$ is estimated by REML or ML, respectively. Singh et al. (2005) obtained different MSE estimators obtained by subtracting an additional term from (5.2.39) or (5.2.40). Proofs of results in Singh et al. (2005) assume that the number of small areas, m, is finite.

Singh et al. (2005) used the state-space model approach to develop EBLUP estimators for spatio-temporal models. They also studied the relative performance of spatial and spatio-temporal models on monthly data on per capita consumer expenditure from India. Marhuenda, Molina, and Morales (2013) studied EBLUP estimation under spatiotemporal models obtained by assuming that the area effects in the Rao–Yu model (4.4.6) and (4.4.7) follow a SAR model. Their simulation showed that, in the model with temporal correlation, including additionally spatial correlation in the area effects leads to negligible gains in efficiency. This empirical result suggests that including spatial correlations in a temporal model may not be necessary in practice.

Parametric and nonparametric bootstrap procedures for estimation of the MSE of the EBLUP under the spatial model with SAR random effects were proposed by Molina, Salvati, and Pratesi (2008). Both naive and bias-corrected bootstrap MSE estimates are provided. Marhuenda et al. (2013) extended the parametric bootstrap method to the spatiotemporal case based on the Rao–Yu model. The order of bias of the bootstrap MSE estimators under spatial models remains to be studied.

Schmid and Münnich (2014) extended the robust EBLUP (REBLUP) theory of Sinha and Rao (2009) to spatial linear mixed models. A simulation study was also conducted to study the relative performance of the spatial REBLUP (SREBLUP) and the REBLUP under both spatial and nonspatial settings.

Chandra, Salvati, and Chambers (2007) studied the unit level model (4.3.1) with spatially correlated random effects following a SAR model, similar to the spatial FH model. They developed EBLUP estimators and associated MSE estimators using the Taylor linearization approach. They conducted simulations based on data from the ISTAT farm structure survey in Tuscany and the U.S. Environmental Monitoring and Assessment Program survey.

Example 8.4.1. U.S. Census Undercount. As noted in Example 4.4.6, Chapter 4, Cressie (1989) modeled the census adjustment factor $\theta_i = T_i/C_i$ for state i, using the basic area level model, $\hat{\theta}_i = \mathbf{z}_i^T\boldsymbol{\beta} + b_i v_i + e_i$, with $b_i = 1/\sqrt{C_i}(i = 1, \ldots, 51)$. Cressie (1991) extended this model by allowing spatial correlations among the state effects v_i. The covariance matrix of the vector with elements $b_i v_i$ is given by $\boldsymbol{\Gamma}(\boldsymbol{\delta}) = \sigma_v^2(\mathbf{I} - \rho\mathbf{Q})^{-1}\mathbf{B}$, where $\mathbf{B} = \text{diag}(b_1^2, \ldots, b_m^2)$ and the elements of $\mathbf{Q}$ are given by $q_{ii} = 0, q_{i\ell} = \sqrt{C_\ell/C_i}$ if the "distance" between the ith and the ℓth states, $d_{i\ell}$, is less than or equal to 700 miles, and $q_{i\ell} = 0$ otherwise ($i \neq \ell$); see Example 4.4.6. Dual-system PES estimates, $\hat{\theta}_i$, were used as direct estimates in the spatial FH model.

A subset of eight explanatory variables was selected following the variable selection method of Ericksen, Kadane, and Tukey (1989), but using weighted least squares (WLS) with census counts, C_i, as weights. This procedure essentially ignores sampling errors, e_i, and spatial dependence, and fits the model $\hat{\theta}_i = \mathbf{z}_i^T\boldsymbol{\beta} + v_i/\sqrt{C_i}$ by WLS to select the z-variables. The variable selection method is as follows: (i) Consider all subsets of two, three, and four variables. (ii) Consider all regression equations in which each WLS regression coefficient is at least twice its standard error. (iii) Select the regression equation that minimizes the residual mean sum of squares. This method yielded the following variables: z_2= minority percentage, z_3= education (percentage of persons aged 25 and older who have not graduated from high school), and the constant term $z_1 = 1$.

Using $\mathbf{z} = (z_1, z_2, z_3)^T$ and assuming normality, ML estimates of $\boldsymbol{\beta}, \sigma_v^2$, and ρ and the resulting EBLUP estimates were calculated. Denote these estimates with no restrictions on $\boldsymbol{\beta}$ and ρ as $\hat{\theta}_i^H(2,2)$. To study the effect of spatial correlation, the following models with the variable z_3 either deliberately omitted or not omitted, and spatial correlation coefficient with either set equal to zero or not, were also considered: model (1,1) with $\beta_3 = 0$ and $\rho = 0$; model (1,2) with $\beta_3 = 0$; model (2,1) with $\rho = 0$. Estimate of σ_v^2 for each of the four models was used to summarize the model fit, similarly as in Example 6.1.1. The following estimates were obtained for each respective model: $\hat{\sigma}_v^2(1,1) = 109, \hat{\sigma}_v^2(1,2) = 47.32, \hat{\sigma}_v^2(2,1) = 18.39$, and $\hat{\sigma}_v^2(2,2) = 0$. It is interesting to note that the spatial model (1,2) with omitted variable has a smaller $\hat{\sigma}_v^2$ than the nonspatial model (2,1) with the variable included. This result suggests that spatial models can compensate for omitted variables.

The difference between the "raw" adjustment factor $\hat{\theta}_i$ and no adjustment was decomposed into differences that correspond to the effects of smoothing, spatial modeling, omitting the variable z_3, and adjusting:

$$\hat{\theta}_i - 1 = [\hat{\theta}_i - \hat{\theta}_i^H(1,1)] + [\hat{\theta}_i^H(1,1) - \hat{\theta}_i^H(1,2)] \\ +[\hat{\theta}_i^H(1,2) - \hat{\theta}_i^H(2,2)] + [\hat{\theta}_i^H(2,2) - 1].$$

The components of $\hat{\theta}_i - 1$ were calculated for each state i and summarized as weighted sum of squares (WSS) with weights C_i; for example, $\sum_{i=1}^{51} C_i[\hat{\theta}_i - \hat{\theta}_i^H(1,1)]^2$ for the first component $\hat{\theta}_i - \hat{\theta}_i^H(1,1)$. The WSS-values obtained for these four components were 27,838, 2618, 4248, and 28,602, respectively. Large WSS-values for the first and the fourth components suggest that the smoothing of raw adjustment factors and the adjustment of census counts are both important. On the other hand, small WSS-values for the second and third components indicate that the omission of an explanatory variable or the spatial correlation has relatively smaller effect on the smoothed adjustment factors.

8.5 *TWO-FOLD SUBAREA LEVEL MODELS

The twofold subarea model given by (4.4.20) provides an extension of the basic area level model to handle estimation of the means in areas and subareas simultaneously. This model is a special case of the general linear mixed model with block diagonal covariance structure (5.3.1), where

$$\mathbf{y}_i = (\hat{\theta}_{i1}, \ldots, \hat{\theta}_{i,n_i})^T, \quad \mathbf{e}_i = (e_{i1}, \ldots, e_{i,n_i})^T,$$

$\mathbf{X}_i$ is the $n_i \times p$ matrix with rows $\mathbf{z}_{ij}^T, \mathbf{Z}_i = (\mathbf{1}_{n_i} | \mathbf{I}_{n_i})$, and $\mathbf{v}_i = (v_i, \mathbf{u}_i^T)^T$ with $\mathbf{u}_i = (u_{i1}, \ldots, u_{i,n_i})^T$.

Making the above substitutions, noting that

$$\mathbf{V}_i = \sigma_v^2 \mathbf{1}_{n_i} \mathbf{1}_{n_i}^T + \text{diag}(\sigma_u^2 + \psi_{i1}, \ldots, \sigma_u^2 + \psi_{i,n_i}),$$

and appealing to the general formula (5.3.3) for $\mathbf{v}_i$, we obtain the BLUP estimators of v_i and u_{ij}, given by

$$\tilde{v}_i(\boldsymbol{\delta}) = \gamma_i \left[\hat{\theta}_{i\gamma} - \bar{\mathbf{z}}_{i\gamma}^T \tilde{\boldsymbol{\beta}}(\boldsymbol{\delta}) \right], \tag{8.5.1}$$

$$\tilde{u}_{ij}(\boldsymbol{\delta}) = \gamma_{ij} \left[\hat{\theta}_{ij} - \mathbf{z}_{ij}^T \tilde{\boldsymbol{\beta}}(\boldsymbol{\delta}) \right] - \gamma_i \gamma_{ij} \left[\hat{\theta}_{i\gamma} - \bar{\mathbf{z}}_{i\gamma}^T \tilde{\boldsymbol{\beta}}(\boldsymbol{\delta}) \right]. \tag{8.5.2}$$

Here, $\tilde{\boldsymbol{\beta}}(\boldsymbol{\delta})$ is the WLS estimator of $\boldsymbol{\beta}, \gamma_{ij} = \sigma_u^2/(\sigma_u^2 + \psi_{ij}), \gamma_{i\cdot} = \sum_{j=1}^{n_i} \gamma_{ij}, \gamma_i = \sigma_v^2/(\sigma_v^2 + \sigma_u^2/\gamma_{i\cdot}), \hat{\theta}_{i\gamma} = \gamma_{i\cdot}^{-1} \sum_{j=1}^{n_i} \hat{\theta}_{ij}$, and $\bar{\mathbf{z}}_{i\gamma} = \gamma_{i\cdot}^{-1} \sum_{j=1}^{n_i} \mathbf{z}_{ij}$ (see Torabi and Rao (2014) for details).

It follows that the BLUP estimator of θ_{ij} for a sampled subarea is given by

$$\tilde{\theta}_{ij}^H(\boldsymbol{\delta}) = \mathbf{z}_{ij}^T \tilde{\boldsymbol{\beta}}(\boldsymbol{\delta}) + \tilde{v}_i(\boldsymbol{\delta}) + \tilde{u}_{ij}(\boldsymbol{\delta}), \quad j = 1, \ldots, n_i, \quad i = 1, \ldots, m, \tag{8.5.3}$$

where $\tilde{v}_i(\boldsymbol{\delta})$ and $\tilde{u}_{ij}(\boldsymbol{\delta})$ are obtained from (8.5.1) and (8.5.2). For nonsampled subareas $k = n_i + 1, \ldots, N_i$, we use the quasi-BLUP estimators

$$\tilde{\theta}_{ik}^H(\boldsymbol{\delta}) = \mathbf{z}_{ik}^T \tilde{\boldsymbol{\beta}}(\boldsymbol{\delta}) + \tilde{v}_i(\boldsymbol{\delta}), \quad k = n_i + 1, \ldots, N_i, \quad i = 1, \ldots, m. \tag{8.5.4}$$

The BLUP estimator of the area mean θ_i is then given by

$$\tilde{\theta}_i^H(\boldsymbol{\delta}) = N_{i\cdot}^{-1} \left[\sum_{j=1}^{n_i} N_{ij} \tilde{\theta}_{ij}^H(\boldsymbol{\delta}) + \sum_{k=n_i+1}^{N_i} N_{ik} \tilde{\theta}_{ik}^H(\boldsymbol{\delta}) \right]. \tag{8.5.5}$$

Note that the subarea estimators (8.5.3) and (8.5.4) automatically benchmark to the corresponding area estimators $\tilde{\theta}_i^H(\boldsymbol{\delta})$ in (8.5.5) . If N_i is large, then the mean subarea effect within area i is $\overline{U}_i = N_{i\cdot}^{-1} \sum_{j=1}^{N_i} N_{ij} u_{ij} \approx 0$ and $\theta_i \approx \overline{\mathbf{Z}}_i^T \boldsymbol{\beta} + v_i$, where $\overline{\mathbf{Z}}_i = N_{i\cdot}^{-1} \sum_{j=1}^{N_i} N_{ij} \mathbf{z}_{ij}$. In this case, the BLUP estimator of θ_i may be taken as $\tilde{\theta}_i^H(\boldsymbol{\delta}) = \overline{\mathbf{Z}}_i^T \tilde{\boldsymbol{\beta}}(\boldsymbol{\delta}) + \tilde{v}_i(\boldsymbol{\delta})$.

The BLUP estimators depend on the unknown model parameters $\boldsymbol{\delta}$. Substitution of suitable estimators $\hat{\boldsymbol{\delta}}$ in the BLUP estimators leads to the EBLUP estimators $\hat{\theta}_i^H = \tilde{\theta}_i^H(\hat{\boldsymbol{\delta}})$ for areas, $\hat{\theta}_{ij}^H = \tilde{\theta}_{ij}^H(\hat{\boldsymbol{\delta}})$ for sampled subareas $j = 1, \ldots, n_i$ and $\hat{\theta}_{ik}^H = \tilde{\theta}_{ik}^H(\hat{\boldsymbol{\delta}})$ for nonsampled subareas $k = n_i + 1, \ldots, N_i$. Torabi and Rao (2014) extended the iterative moment method of Fay and Herriot (1979), described in Section 6.1.2, to estimate $\boldsymbol{\delta}$ in the twofold subarea model (4.4.20). They also derived a second-order approximation to the MSEs of the EBLUP estimators and, using this approximation, second-order unbiased estimators of the MSEs are obtained.

Torabi and Rao (2014) conducted a simulation study on the performance of the subarea EBLUP estimators. Their simulation results suggest that, for sampled subareas, the EBLUP estimators $\hat{\theta}_{ij}^H$ lead to modest efficiency gains compared to the corresponding EBLUP estimators under the subarea level FH model $\hat{\theta}_{ij} = \mathbf{z}_{ij}^T\boldsymbol{\beta} + u_{ij} + e_{ij}$. On the other hand, for nonsampled subareas, the EBLUP estimators $\hat{\theta}_{ik}^H$ are significantly more efficient than the corresponding regression-synthetic estimators under the FH model.

8.6 *MULTIVARIATE NESTED ERROR REGRESSION MODEL

The sample part of the multivariate nested error regression model (4.5.1) can be expressed as a special case of the general linear mixed model (5.3.1) with block-diagonal covariance structure, taking

$$\mathbf{y}_i^T = (\mathbf{y}_{i1}^T, \ldots, \mathbf{y}_{in_i}^T), \quad \mathbf{e}_i^T = (\mathbf{e}_{i1}^T, \ldots, \mathbf{e}_{in_i}^T),$$
$$\mathbf{v}_i = \mathbf{v}_i, \quad \boldsymbol{\beta} = \text{vec}(\mathbf{B}),$$
$$\mathbf{X}_i^T = [(\mathbf{I}_r \otimes \mathbf{x}_{i1}^T)^T, \ldots, (\mathbf{I}_r \otimes \mathbf{x}_{in_i}^T)^T],$$
$$\mathbf{Z}_i = \mathbf{1}_{n_i} \otimes \mathbf{I}_r, \quad i = 1, \ldots, m.$$

Here, the operator $\otimes$ denotes the direct product, $\mathbf{1}_{n_i}$ is the $n_i \times 1$ vector of ones, vec($\mathbf{B}$) is the $pr \times 1$ vector obtained by listing the columns of the $r \times p$ matrix $\mathbf{B}$ one beneath the other beginning with the first column, and n_i is the number of sample multivariate observations $\mathbf{y}_{ij}$. Furthermore, in this model we have

$$\mathbf{G}_i = \boldsymbol{\Sigma}_v, \quad \mathbf{R}_i = \mathbf{I}_{n_i} \otimes \boldsymbol{\Sigma}_e$$

so that

$$\mathbf{V}_i = (\mathbf{J}_{n_i} \otimes \boldsymbol{\Sigma}_v) + (\mathbf{I}_{n_i} \otimes \boldsymbol{\Sigma}_e),$$

where $\mathbf{J}_{n_i} = \mathbf{1}_{n_i}\mathbf{1}_{n_i}^T$ and $\mathbf{I}_{n_i}$ is the identity matrix of order n_i. The parameters of interest are the small area mean vectors $\overline{\mathbf{Y}}_i \approx \mathbf{B}\overline{\mathbf{X}}_i + \mathbf{v}_i = (\mathbf{I}_r \otimes \overline{\mathbf{X}}_i^T)\boldsymbol{\beta} + \mathbf{v}_i =: \boldsymbol{\mu}_i$ if the population size N_i is large. Therefore, the target parameter vector has the form $\boldsymbol{\mu}_i = \mathbf{L}_i\boldsymbol{\beta} + \mathbf{M}_i\mathbf{v}_i$ with $\mathbf{L}_i = \mathbf{I}_r \otimes \overline{\mathbf{X}}_i^T$ and $\mathbf{M}_i = \mathbf{I}_r$.

Making the above substitutions in the general formula (5.3.19), we get the BLUP estimator $\tilde{\boldsymbol{\mu}}_i^H$ of $\boldsymbol{\mu}_i = \mathbf{B}^T\overline{\mathbf{X}}_i + \mathbf{v}_i$. The EBLUP estimator $\hat{\boldsymbol{\mu}}_i^H$ is obtained by substituting suitable estimators $\hat{\boldsymbol{\Sigma}}_v$ and $\hat{\boldsymbol{\Sigma}}_e$ for $\boldsymbol{\Sigma}_v$ and $\boldsymbol{\Sigma}_e$ in (5.3.19).

A method of moments was used by Fuller and Harter (1987) to estimate $\boldsymbol{\Sigma}_v$ and $\boldsymbol{\Sigma}_e$. Alternatively, ML or REML methods can be used, using the general results of Section (5.2.4). Second-order unbiased MSE estimators may also be obtained from the general results for block-diagonal covariance structures given in Section 5.3.2 (see Datta, Day, and Basawa 1999).

Lohr and Prasad (2003) considered sampling on two occasions with partial sample overlap and developed small area estimators by combining information from the two samples. They used a multivariate nested error model to derive EBLUP estimators of the area means on the second occasion, and associated second-order unbiased MSE estimators. Significant gains in efficiency, relative to EBLUP estimators based only on the second occasion sample, may be achieved if the measurements on the matched portion of the sample are highly correlated. The method of Lohr and Prasad (2003) may also be used to combine information from separate surveys that overlap in some of the areas.

Baíllo and Molina (2009) studied a particular case of the multivariate nested error regression model with $\mathbf{v}_i = \mathbf{1}_r v_i$, similar to the special case of the multivariate FH model considered by González-Manteiga et al. (2008b). Baíllo and Molina (2009) derived a second-order unbiased MSE estimator of the EBLUP estimator of the area mean vector $\boldsymbol{\mu}_i = \overline{\mathbf{X}}_i^T \boldsymbol{\beta} + \mathbf{1}_r v_i$. A simulation study based on the 2006 Spanish Labor Force Survey (SLFS) data was conducted. They considered a model with $r = 2$ response variables explained by the same set of auxiliary variables. Simulation results showed that the efficiency gains relative to the univariate modeling are considerable for the response variable that is less explained by the auxiliary variables in the univariate modeling. Thus, in this simulation study, bivariate modeling is borrowing strength from the first response variable to predict the area means of the second response variable.

Example 8.6.1. Simulation Study. Datta, Day, and Basawa (1999) used simulation to compare the efficiencies of EBLUP estimators under bivariate ($r = 2$) and univariate nested error regression models. They estimated the model parameters using the LANDSAT data of Battese, Harter, and Fuller (1988), and treated $\hat{\mathbf{B}}$ and the diagonal elements $\hat{\sigma}_{vii}$ of $\hat{\boldsymbol{\Sigma}}_v$ as the true values, $\mathbf{B}$ and σ_{vii}. The vectors $\mathbf{y}_{ij}$ and $\mathbf{x}_{ij} = (1, x_{ij1}, x_{ij2})^T$ are as described in Example 7.3.1. Using selected values of $r_1 = \sigma_{v11}/\sigma_{e11}$ and $r_2 = \sigma_{v22}/\sigma_{e22}$, and correlation coefficients, ρ_v and ρ_e, associated with $\boldsymbol{\Sigma}_v$ and $\boldsymbol{\Sigma}_e$, respectively, they generated several pairs $(\boldsymbol{\Sigma}_v, \boldsymbol{\Sigma}_e)$. From each parameter setting, 100 data sets were generated under the bivariate nested error regression model, and MSE estimates of EBLUP estimates (univariate and bivariate) for the $m = 12$ counties (small areas) were computed from each data set. Based on the average values of MSE estimates over the 100 data sets for each parameter setting, they computed the estimated reduction in MSE of the bivariate method relative to the univariate method for each of the two response variables (corn and soybeans). Their results suggest that little improvement over the univariate method is achieved if ρ_v and ρ_e have the same sign, while ρ_v and ρ_e with opposite signs and large absolute values lead to large improvements in efficiency. It may be noted that the reduction in MSE in this simulation does not depend on the choice of $\mathbf{B}$ and $(\sigma_{v11}, \sigma_{v22})$.

8.7 TWO-FOLD NESTED ERROR REGRESSION MODEL

In Section 4.5.2, we described a twofold nested error regression model appropriate for two-stage sampling within areas, that is, a sample, s_i, of m_i primary units (clusters) is selected from the ith area and, if the jth cluster is sampled, then a subsample, s_{ij}, of n_{ij} elements is selected from it and the associated values $(y_{ij\ell}, \mathbf{x}_{ij\ell}), \ell = 1, \ldots, n_{ij}$ are observed. We assume that the sample obeys the twofold model given by (4.5.2), that is,

$$y_{ij\ell} = \mathbf{x}_{ij\ell}^T \boldsymbol{\beta} + v_i + u_{ij} + e_{ij\ell}, \quad \ell = 1, \ldots, n_{ij}, \quad j = 1, \ldots, m_i, \quad i = 1, \ldots, m, \tag{8.7.1}$$

where $v_i \overset{\text{iid}}{\sim} (0, \sigma_v^2)$, $u_{ij} \overset{\text{iid}}{\sim} (0, \sigma_u^2)$, and $e_{ij\ell} = k_{ij\ell}\tilde{e}_{ij\ell}$ with $\tilde{e}_{ij\ell} \overset{\text{iid}}{\sim} (0, \sigma_e^2)$, and known constants $k_{ij\ell}$. Furthermore, $\{v_i\}, \{u_{ij}\}$ and $\{e_{ij\ell}\}$ are mutually independent.

The twofold model (8.7.1) is a special case of the general linear mixed model (5.3.1) with block-diagonal covariance structure. Letting $\text{col}_{1\le i\le t}(\mathbf{a}_i) = (\mathbf{a}_1^T, \ldots, \mathbf{a}_t^T)^T$ and $\text{diag}_{1\le i\le t}(\mathbf{A}_i) = \text{blockdiag}(\mathbf{A}_1, \ldots, \mathbf{A}_t)$, we have

$$\begin{aligned}
&\mathbf{y}_i = \text{col}_{1\le j\le m_i}(\mathbf{y}_{ij}), \quad \mathbf{y}_{ij} = \text{col}_{1\le \ell\le n_{ij}}(y_{ij\ell}),\\
&\mathbf{X}_i = \text{col}_{1\le j\le m_i}(\text{col}_{1\le \ell\le n_{ij}}(\mathbf{x}_{ij\ell}^T)), \quad \mathbf{Z}_i = (\mathbf{z}_i|\mathbf{Z}_{2i}),\\
&\mathbf{v}_i = \begin{pmatrix} v_i \\ \mathbf{v}_{2i} \end{pmatrix}, \quad \boldsymbol{\beta} = (\beta_1, \ldots, \beta_p)^T,
\end{aligned}$$

and $\mathbf{e}_i$ is obtained from $\mathbf{y}_i$ by changing $y_{ij\ell}$ to $e_{ij\ell}$, where

$$\begin{aligned}
&\mathbf{z}_i = \text{col}_{1\le j\le m_i}(\text{col}_{1\le k\le n_{ij}}(1)) = \text{col}_{1\le j\le m_i}(\mathbf{z}_{ij}),\\
&\mathbf{Z}_{2i} = \text{diag}_{1\le j\le m_i}(\mathbf{z}_{ij}), \quad \mathbf{v}_{2i} = \text{col}_{1\le j\le m_i}(u_{ij}).
\end{aligned}$$

Furthermore,

$$\mathbf{G}_i = \begin{bmatrix} \sigma_v^2 & \mathbf{0}^T \\ \mathbf{0} & \sigma_u^2 \mathbf{I}_{m_i} \end{bmatrix}, \quad \mathbf{R}_i = \text{diag}_{1\le j\le m_i}(\mathbf{R}_{ij})$$

and $\mathbf{R}_{ij} = \text{diag}_{1\le \ell\le n_{ij}}(\sigma_e^2 k_{ij\ell}^2)$, where $\mathbf{I}_b$ is the identity matrix of order b and $\mathbf{0}$ is a vector of zeros.

Let us define $a_{ij\ell} = k_{ij\ell}^{-2}, a_{ij\cdot} = \sum_{\ell=1}^{n_{ij}} a_{ij\ell}, \bar{y}_{ija} = \sum_{\ell=1}^{n_{ij}} a_{ij\ell} y_{ij\ell}/a_{ij\cdot}, \bar{\mathbf{x}}_{ija} = \sum_{\ell=1}^{n_{ij}} a_{ij\ell}\mathbf{x}_{ij\ell}/a_{ij\cdot}$,

$$\gamma_{ij} = \frac{\sigma_u^2}{\sigma_u^2 + \sigma_e^2/a_{ij\cdot}}, \quad \gamma_i = \frac{\sigma_v^2}{\sigma_v^2 + \sigma_u^2/(\sum_{j=1}^{m_i} \gamma_{ij})}, \tag{8.7.2}$$

$\bar{y}_{i\gamma} = (\sum_{j=1}^{m_i} \gamma_{ij})^{-1} \sum_{j=1}^{m_i} \gamma_{ij}\bar{y}_{ija}$ and $\bar{\mathbf{x}}_{i\gamma} = (\sum_{j=1}^{m_i} \gamma_{ij})^{-1} \sum_{j=1}^{m_i} \gamma_{ij}\bar{\mathbf{x}}_{ija}$. Using an explicit form for $\mathbf{V}_i^{-1}$, the general formula (5.3.4) for $\tilde{\boldsymbol{\beta}}$ reduces to

$$\tilde{\boldsymbol{\beta}} = \mathbf{A}^{-1}\mathbf{b}, \tag{8.7.3}$$

where

$$\sigma_e^2\mathbf{A} = \sum_{i=1}^{m}\sum_{j=1}^{m_i}\sum_{\ell=1}^{n_{ij}} a_{ij\ell}\mathbf{x}_{ij\ell}\mathbf{x}_{ij\ell}^T - \sum_{i=1}^{m}\sum_{j=1}^{m_i} a_{ij\cdot}\gamma_{ij}\overline{\mathbf{x}}_{ija}\overline{\mathbf{x}}_{ija}^T$$
$$- (\sigma_e^2/\sigma_u^2)\sum_{i=1}^{m}\gamma_i\left(\sum_{j=1}^{m_i}\gamma_{ij}\right)\overline{\mathbf{x}}_{i\gamma}\overline{\mathbf{x}}_{i\gamma}^T$$

and

$$\sigma_e^2\mathbf{b} = \sum_{i=1}^{m}\sum_{j=1}^{m_i}\sum_{\ell=1}^{n_{ij}} a_{ij\ell}\mathbf{x}_{ij\ell}y_{ij\ell} - \sum_{i=1}^{m}\sum_{j=1}^{m_i} a_{ij\cdot}\gamma_{ij}\overline{\mathbf{x}}_{ija}\overline{y}_{ija}$$
$$- (\sigma_e^2/\sigma_u^2)\sum_{i=1}^{m}\gamma_i\left(\sum_{j=1}^{m_i}\gamma_{ij}\right)\overline{\mathbf{x}}_{i\gamma}\overline{y}_{i\gamma},$$

see Stukel and Rao (1999). Furthermore, the general formula (5.3.3) for the BLUP estimator of $\mathbf{v}_i$ reduces to $\tilde{\mathbf{v}}_i = (\tilde{v}_i, \tilde{\mathbf{v}}_{2i}^T)^T$, where

$$\tilde{v}_i = \gamma_i(\overline{y}_{i\gamma} - \overline{\mathbf{x}}_{i\gamma}^T\tilde{\boldsymbol{\beta}}) \tag{8.7.4}$$

and the jth element of $\tilde{\mathbf{v}}_{2i}$ is

$$\tilde{u}_{ij} = \gamma_{ij}(\overline{y}_{ija} - \overline{\mathbf{x}}_{ija}^T\tilde{\boldsymbol{\beta}}) - \gamma_i\gamma_{ij}(\overline{y}_{i\gamma} - \overline{\mathbf{x}}_{i\gamma}^T\tilde{\boldsymbol{\beta}}). \tag{8.7.5}$$

Note that $\tilde{\boldsymbol{\beta}}$ and $\tilde{v}_i$ involve the inversion of a $p \times p$ matrix only, where p is the dimension of $\mathbf{x}_{ij\ell}$.

The BLUP estimator of the small area mean $\overline{Y}_i$ is then given by

$$\tilde{\overline{Y}}_i^H = \frac{1}{N_i}\left[\sum_{j\in s_i}\sum_{k\in s_{ij}} y_{ijk} + \sum_{j\in s_i}\left(\sum_{l\in r_{ij}}\tilde{y}_{ij\ell}\right) + \sum_{j\in r_i}\left(\sum_{l=1}^{N_{ij}}\tilde{y}_{ij\ell}\right)\right], \tag{8.7.6}$$

where r_{ij} is the set of nonsampled units in the sampled cluster (i,j), r_i is the set of nonsampled clusters from area i, N_{ij} is the population size of cluster j from area i $(j = 1, \ldots, M_i)$, and $N_i = \sum_{j=1}^{M_i} N_{ij}$ is the population size of area i. Furthermore, for a unit ℓ in a sampled cluster, the predicted value is

$$\tilde{y}_{ij\ell} = \mathbf{x}_{ij\ell}^T\tilde{\boldsymbol{\beta}} + \tilde{v}_i + \tilde{u}_{ij}, \quad j \in s_i \quad l \in r_{ij}, \tag{8.7.7}$$

whereas for a unit ℓ in nonsampled cluster, we have

$$\tilde{y}_{ij\ell} = \mathbf{x}_{ij\ell}^T\tilde{\boldsymbol{\beta}} + \tilde{v}_i, \quad j \in r_i, \quad \ell = 1, \ldots, N_{ij}. \tag{8.7.8}$$

The BLUP property of $\tilde{\overline{Y}}_i^H$ is established by showing that $\text{Cov}(\mathbf{a}^T\mathbf{y}, \tilde{\overline{Y}}_i^H - \overline{Y}_i) = 0$ for every zero function $\mathbf{a}^T\mathbf{y} = \sum_{i=1}^m \mathbf{a}_i^T\mathbf{y}_i$, that is, $E(\mathbf{a}^T\mathbf{y}) = 0$; see Stukel (1991, Chapter 3). The form (8.7.6) for $\tilde{\overline{Y}}_i^H$ shows that $\tilde{\overline{Y}}_i^H$ is readily obtained from $\tilde{\boldsymbol{\beta}}$, $\tilde{v}_i$ and $\tilde{u}_{ij}$ given by (8.7.3), (8.7.4), and (8.7.5), respectively.

The BLUP estimator $\tilde{\overline{Y}}_i^H$ depends on the variance components $\boldsymbol{\delta} = (\sigma_v^2, \sigma_u^2, \sigma_e^2)^T$. Stukel and Rao (1997) extended the simple transformation method of Section 7.1.2 to obtain a moment estimator, $\hat{\boldsymbol{\delta}}_m$, of $\boldsymbol{\delta}$. Fuller and Battese (1973) proposed a similar extension for the special case of $k_{ij\ell} = 1$ for all (i, j, l). The transformation method is identical to the fitting-of-constants method, but it is computationally less cumbersome and more stable than the latter method, as noted in Section 7.1.2. One could also use the REML estimator $\hat{\boldsymbol{\delta}}_{\text{RE}}$. Replacing $\boldsymbol{\delta}$ in (8.7.6) by an estimator $\hat{\boldsymbol{\delta}}$, we get the EBLUP estimator $\hat{\overline{Y}}_i^H$.

If the population size of area i, N_i, is large, then $\overline{Y}_i$ may be approximated as $\overline{Y}_i \approx \mu_i = \overline{\mathbf{X}}_i^T\boldsymbol{\beta} + v_i$, as noted in Section 4.5.2, where $\overline{\mathbf{X}}_i$ is the vector of known population x-means. In this case, the BLUP estimator of μ_i is given by

$$\tilde{\mu}_i^H = \gamma_i[\overline{y}_{i\gamma} + (\overline{\mathbf{X}}_i - \overline{\mathbf{x}}_{i\gamma})^T\tilde{\boldsymbol{\beta}}] + (1-\gamma_i)\overline{\mathbf{X}}_i^T\tilde{\boldsymbol{\beta}}. \tag{8.7.9}$$

The form (8.7.9) shows that the BLUP estimator of μ_i is a weighted average of the survey regression estimator $\overline{y}_{i\gamma} + (\overline{\mathbf{X}}_i - \overline{\mathbf{x}}_{i\gamma})^T\tilde{\boldsymbol{\beta}}$ and the regression-synthetic estimator $\overline{\mathbf{X}}_i^T\tilde{\boldsymbol{\beta}}$. Note that $\tilde{\mu}_i^H$ requires only the population means $\overline{\mathbf{X}}_i$ and not the individual x-values. The MSE formula (5.3.5) reduces to $\text{MSE}(\tilde{\mu}_i^H) = g_{1i}(\boldsymbol{\delta}) + g_{2i}(\boldsymbol{\delta})$ with

$$g_{1i}(\boldsymbol{\delta}) = \sigma_v^2(1-\gamma_i) \tag{8.7.10}$$

and

$$g_{2i}(\boldsymbol{\delta}) = (\overline{\mathbf{X}}_i - \gamma_i\overline{\mathbf{x}}_{i\gamma})^T\mathbf{A}^{-1}(\overline{\mathbf{X}}_i - \gamma_i\overline{\mathbf{x}}_{i\gamma}). \tag{8.7.11}$$

The first term, $g_{1i}(\boldsymbol{\delta})$, is of order $O(1)$ while the second term, $g_{2i}(\boldsymbol{\delta})$, due to estimating $\boldsymbol{\beta}$, is of order $O(m^{-1})$ for large m.

The EBLUP estimator $\hat{\mu}_i^H$ is obtained by replacing $\boldsymbol{\delta}$ by $\hat{\boldsymbol{\delta}}$ in (8.7.9) . Applying the general formula (5.3.11), under normality of the errors v_i, u_{ij}, and $e_{ij\ell}$, a second-order unbiased MSE estimator is given by

$$\text{mse}(\hat{\mu}_i^H) = g_{1i}(\hat{\boldsymbol{\delta}}) + g_{2i}(\hat{\boldsymbol{\delta}}) + 2g_{3i}(\hat{\boldsymbol{\delta}}), \tag{8.7.12}$$

where

$$g_{3i}(\boldsymbol{\delta}) = \sigma_u^{-2}\sigma_v^4(1-\gamma_i)^2\left\{(1-\gamma_i)\sum_{j=1}^{m_i}\gamma_{ij}\right.$$

$$\times \overline{V}\left[(\sigma_v^{-2}\hat{\sigma}_v^2 - \sigma_e^{-2}\hat{\sigma}_e^2)\left(\sum_{j=1}^{m_i}\gamma_{ij}\right)^{-1}\sum_{j=1}^{m_i}\gamma_{ij}^2(\sigma_u^{-2}\hat{\sigma}_u^2 - \sigma_e^{-2}\hat{\sigma}_e^2)\right]$$

$$+\left(\sum_{j=1}^{m_i}\gamma_{ij}\right)^{-1}\left[\left(\sum_{j=1}^{m_i}\gamma_{ij}^3\right)\sum_{j=1}^{m_i}\gamma_{ij}-\left(\sum_{j=1}^{m_i}\gamma_{ij}^2\right)^2\right]$$

$$\times\ \overline{V}(\sigma_u^{-2}\hat{\sigma}_u^2-\sigma_e^{-2}\hat{\sigma}_e^2)\Bigg\} \tag{8.7.13}$$

and $\overline{V}$ denotes the asymptotic variance (Stukel and Rao 1999). Formula (8.7.12) is valid for $\hat{\boldsymbol{\delta}}_{\text{RE}}$ and $\hat{\boldsymbol{\delta}}_m$. The asymptotic covariance matrix of $\hat{\boldsymbol{\delta}}_m$ is given in Stukel and Rao (1999), while the asymptotic covariance matrix of $\hat{\boldsymbol{\delta}}_{\text{RE}}$ may be obtained from the general results in Section (5.2.4). Area-specific versions of (8.7.12) may be obtained from (5.3.14)–(5.3.16).

Example 8.7.1. Simulation Study. Stukel and Rao (1999) conducted a simulation study to analyze the performance of $\text{mse}(\hat{\mu}_i^H)$ in terms of relative bias (RB). To this end, they considered a twofold nested error model without auxiliary variables, given by $y_{ij\ell}=\beta+v_i+u_{ij}+e_{ij\ell}$ with $m=30$ areas, $m_i=\overline{m}=2$ clusters sampled within each area, $n_{ij}=\overline{n}=2$ units drawn from cluster $(i,j), k_{ij\ell}=1$ for all i, j, and ℓ, and normally distributed errors. In this balanced case, fitting-of-constants estimators, $\hat{\boldsymbol{\delta}}_m=(\hat{\sigma}_v^2,\hat{\sigma}_u^2,\hat{\sigma}_e^2)^T$, reduce to the simple analysis of variance (ANOVA) estimators $\hat{\sigma}_v^2=\max(0,\tilde{\sigma}_v^2), \hat{\sigma}_u^2=\max(0,\tilde{\sigma}_u^2)$, and $\hat{\sigma}_e^2=\tilde{\sigma}_e^2=[m\overline{m}(\overline{n}-1)]^{-1}\sum_{i=1}^m\sum_{j=1}^{m_i}\sum_{\ell=1}^{n_{ij}}(y_{ij\ell}-\overline{y}_{ij\cdot})^2$, where $\tilde{\sigma}_u^2=[m(\overline{m}-1)]^{-1}\sum_{i=1}^m\sum_{j=1}^{m_i}(\overline{y}_{ij\cdot}-\overline{y}_{i\cdot\cdot})^2-\hat{\sigma}_e^2, \tilde{\sigma}_v^2=(m-1)^{-1}\sum_{i=1}^m(\overline{y}_{i\cdot\cdot}-\overline{y}_{\cdot\cdot\cdot})^2-\tilde{\sigma}_u^2/\overline{m}-\tilde{\sigma}_e^2/(\overline{m}\overline{n}), \overline{y}_{ij\cdot}=\sum_{\ell=1}^{n_{ij}}y_{ij\ell}/\overline{n}$, and $\overline{y}_{i\cdot\cdot}=\sum_{j=1}^{m_i}\overline{y}_{ij\cdot}/\overline{m}$. Note that asymptotically, $\hat{\boldsymbol{\delta}}=(\hat{\sigma}_v^2,\hat{\sigma}_u^2,\hat{\sigma}_e^2)^T$ has the same covariance matrix as $\tilde{\boldsymbol{\delta}}=(\tilde{\sigma}_v^2,\tilde{\sigma}_u^2,\tilde{\sigma}_e^2)^T$; the latter may be obtained from Searle, Casella, and McCulloch (1992, p. 148).

Simulated values of RB were computed using independent data sets, each of size 2000, generated from the model with $\beta=0, \sigma_e^2=300$ and selected values, 0.1, 0.2, 0.5, 1.0, 2.0 of the ratios σ_u^2/σ_e^2 and σ_v^2/σ_e^2, leading to 25 combinations $(\sigma_v^2/\sigma_e^2,\sigma_u^2/\sigma_e^2)$. The RB values were then averaged over the areas i for each combination $(\sigma_v^2/\sigma_e^2,\sigma_u^2/\sigma_e^2)$; note that the theoretical RB of $\text{mse}(\hat{\mu}_i^H)$ does not depend on i under the balanced twofold model. The average RB values suggest the following properties of $\text{mse}(\hat{\mu}_i^H)$: (i) RB depends on both σ_v^2/σ_e^2 and σ_u^2/σ_e^2. (ii) RB is small ($<10\%$) in a parameter region of interest where $\sigma_v^2/\sigma_e^2\geq\sigma_u^2/\sigma_e^2$ and $\sigma_v^2/\sigma_e^2\geq 0.5$. However, RB is quite large (i.e., significant overestimation) when σ_v^2/σ_e^2 is small (≤ 0.2), especially when σ_v^2/σ_e^2 is significantly lower than σ_u^2/σ_e^2. On the other hand, the naive MSE estimator that ignores the variability in $\hat{\boldsymbol{\delta}}$ almost always leads to significant underestimation, except for very few cases of the parameter region under consideration, where the underestimation is mild. Note that the REML estimator $\hat{\boldsymbol{\delta}}_{\text{RE}}$ is not identical to the ANOVA estimator $\hat{\boldsymbol{\delta}}$ in the balanced twofold model because REML uses a different form of truncation to handle negative values of $\tilde{\sigma}_v^2$ and $\tilde{\sigma}_u^2$ (see Searle et al. 1992, p. 149).

8.8 *TWO-LEVEL MODEL

The two-level model introduced in Section 4.5.3 effectively integrates unit and area level covariates into a single model (4.5.8). We assume that the sample obeys the same two-level model, that is,

$$\mathbf{y}_i = \tilde{\mathbf{X}}_i \tilde{\mathbf{Z}}_i \boldsymbol{\alpha} + \tilde{\mathbf{X}}_i \mathbf{v}_i + \mathbf{e}_i; \quad i = 1, \ldots, m, \tag{8.8.1}$$

where $(\mathbf{y}_i, \tilde{\mathbf{X}}_i, \mathbf{e}_i)$ correspond to the sample part of the population model (4.5.8); we use $\tilde{\mathbf{X}}_i$ here to denote the sample part of $\mathbf{X}_i^P$ to avoid confusion with matrix $\mathbf{X}_i$ for the general linear mixed model (see below). The model (8.8.1) is a special case of the general linear mixed model (5.3.1) with block-diagonal covariance structure, setting

$$\begin{aligned} y_i &= \mathbf{y}_i = \text{col}_{1\le j\le n_i}(y_{ij}), \quad \mathbf{X}_i = \tilde{\mathbf{X}}_i \tilde{\mathbf{Z}}_i, \quad \mathbf{Z}_i = \tilde{\mathbf{X}}_i, \\ \boldsymbol{\beta} &= \boldsymbol{\alpha}, \quad \mathbf{v}_i = \mathbf{v}_i, \quad \mathbf{e}_i = \mathbf{e}_i, \\ \mathbf{G}_i &= \boldsymbol{\Sigma}_v, \quad \mathbf{R}_i = \sigma_e^2 \text{diag}_{1\le j\le n_i}(k_{ij}^2), \end{aligned}$$

where n_i is the sample size in the ith area ($i = 1, \ldots, m$) and the k_{ij}'s are known constants.

If N_i is large, we can express the mean $\overline{Y}_i$ as $\overline{\mathbf{X}}_i^T \mathbf{Z}_i \boldsymbol{\alpha} + \overline{\mathbf{X}}_i^T \mathbf{v}_i$, where $\overline{\mathbf{X}}_i$ is the vector of known population x-means. It now follows that the BLUP estimator, $\tilde{\mu}_i^H$, and its MSE are readily obtained from (5.3.2) and (5.3.5) using $\mathbf{l}_i^T = \overline{\mathbf{X}}_i^T \mathbf{Z}_i$ and $\mathbf{m}_i^T = \overline{\mathbf{X}}_i^T$.

The BLUP estimator, $\tilde{\mu}_i^H$, depends on a vector of unknown parameters $\boldsymbol{\delta}$, composed by the $p(p+1)/2$ distinct elements of the $p \times p$ matrix $\boldsymbol{\Sigma}_v$ and the error variance σ_e^2. Moura and Holt (1999) used restricted iterative generalized least squares (RIGLS) to estimate $\boldsymbol{\delta}$. Goldstein (1989) proposed the same method for multilevel models, which is equivalent to REML under normality of the errors. The EBLUP estimator, $\hat{\mu}_i^H$, is obtained from the BLUP estimator by replacing $\boldsymbol{\delta}$ by the RIGLS estimator $\hat{\boldsymbol{\delta}}$. A second-order unbiased estimator of MSE($\hat{\mu}_i^H$) is obtained from the general formula (5.3.11).

When the sampling fraction, $f_i = n_i/N_i$, is not small, the EBLUP estimator of $\overline{Y}_i$ and its MSE estimator are obtained along the lines of Section 7.1.3. We refer the readers to Moura and Holt (1999) for details.

Example 8.8.1. Simulation Study. Moura and Holt (1999) investigated the properties of alternative estimators, using data from 38,740 households in $m = 140$ enumeration districts (small areas) in one county in Brazil. The data consisted of values for each household (unit) of the response variable y = income of the household head, the covariates x_1 = number of rooms in the house (1–11+), and x_2 = educational attainment of the household head (ordinal scale of 0–5) centered around the means $(\overline{x}_{1i}, \overline{x}_{2i})$, as well as values of an area level covariate, z = number of cars per household, for each area. A two-level model of the form

$$y_{ij} = \beta_{0i} + \beta_{1i}x_{1ij} + \beta_{2i}x_{2ij} + e_{ij}, \tag{8.8.2}$$

$$\beta_{0i} = \alpha_{00} + v_{0i}, \beta_{1i} = \alpha_{10} + \alpha_{11}z_i + v_{1i}, \beta_{2i} = \alpha_{20} + \alpha_{21}z_i + v_{2i}, \tag{8.8.3}$$

with diagonal covariance matrix $\mathbf{\Sigma}_v$, was fitted to the data to estimate the elements of $\boldsymbol{\delta}$, namely, $\sigma_{v0}^2, \sigma_{v1}^2, \sigma_{v2}^2$, and σ_e^2 and the regression parameters $\alpha_{00}, \alpha_{10}, \alpha_{11}, \alpha_{20}$, and α_{21}. Note that z_i is not used in modeling the random intercept term, β_{0i}.

To simulate samples from the two-level model, a 10% random sample of (x_1, x_2)-values was selected from each area and treated as fixed with average area sample size $\overline{n} = 28$. Treating the parameter estimates as true values, a realization, $\boldsymbol{\beta}_i^{(1)}$, of $\boldsymbol{\beta}_i = (\beta_{0i}, \beta_{1i}, \beta_{2i})^T$ was first generated from (8.8.3), assuming normality of the errors $\mathbf{v}_i = (v_{0i}, v_{1i}, v_{2i})^T$. This, in turn, leads to a realization of μ_i, namely, $\mu_i^{(1)} = \beta_{0i}^{(1)} + \beta_{1i}^{(1)}\overline{X}_{1i} + \beta_{2i}^{(1)}\overline{X}_{2i}$, where $\overline{X}_{1i}$ and $\overline{X}_{2i}$ are the known population x-means. At the second stage, y-values were generated as $y_{ij}^{(1)} = \beta_{0i}^{(1)} + \beta_{1i}^{(1)}x_{1ij} + \beta_{2i}^{(1)}x_{2ij} + e_{ij}^{(1)}$, where $e_{ij}^{(1)} \overset{\text{iid}}{\sim} N(0, \sigma_e^2)$, yielding a simulated sample $\{y_{ij}^{(1)}, x_{1ij}, x_{2ij}\}$. This two-stage process was repeated $R = 5,000$ times, and from each simulated sample $\{y_{ij}^{(r)}, x_{1ij}, x_{2ij}\}$, the EBLUP estimate $\hat{\mu}_i^{H(r)}$ under the two-level model and the corresponding EBLUP estimate, $\hat{\mu}_{i*}^{H(r)}$, under the model without the covariate z_i, were computed. Simulated values of MSE and ARE of $\hat{\mu}_i^H$ and $\hat{\mu}_{i*}^H$ were then computed as

$$\text{MSE}_i = \frac{1}{R}\sum_{r=1}^{R}(\text{est}_i^{(r)} - \mu_i^{(r)})^2, \quad \text{ARE}_i = \frac{1}{R}\sum_{r=1}^{R}|\text{est}_i^{(r)} - \mu_i^{(r)}|/\mu_i^{(r)},$$

where $\text{est}_i^{(r)}$ denotes the value of an estimator of $\mu_i^{(r)}$ for the rth simulated sample. Denote the average values of MSE and ARE over i by $\overline{\text{MSE}}$ and $\overline{\text{ARE}}$. Efficiency of $\hat{\mu}_i^H$ relative to $\hat{\mu}_{i*}^H$ was measured by $R_1 = \overline{\text{MSE}}(\hat{\mu}_{i*}^H)/\overline{\text{MSE}}(\hat{\mu}_i^H)$ and $R_2 = \overline{\text{ARE}}(\hat{\mu}_{i*}^H)/\overline{\text{ARE}}(\hat{\mu}_i^H)$. The resulting values were $R_1 = 110.3\%$ and $R_2 = 125.4\%$. These values suggest that the introduction of the area level covariate, z_i, leads to moderate improvement in efficiency. It may be noted that area level covariates, z_i, related to the random parameters, $\boldsymbol{\beta}_i$, are more difficult to find in practice than unit level covariates, $\mathbf{x}_{ij}$, related to y_{ij}.

Lehtonen and Veijanen (1999) proposed a new model-assisted generalized regression (GREG) estimator of μ_i (or $\overline{Y}_i$), which under simple random sampling within areas is given as

$$\hat{\mu}_i^{\text{LV}} = \overline{y}_i + (\overline{\mathbf{X}}_i - \overline{\mathbf{x}}_i)^T(\tilde{\mathbf{Z}}_i\hat{\boldsymbol{\beta}} + \hat{\mathbf{v}}_i), \tag{8.8.4}$$

where $\overline{y}_i$ and $\overline{\mathbf{x}}_i$ are sample means, and $\overline{\mathbf{X}}_i$ is the population mean for area i. The estimator (8.8.4) is obtained by expressing $\overline{Y}_i$ as $N_i^{-1}[\sum_{j=1}^{N_i}\hat{y}_{ij} + \sum_{j=1}^{N_i}(y_{ij} - \hat{y}_{ij})]$ and then estimating it by

$$\hat{\mu}_i^{\mathrm{LV}} = \frac{1}{N_i}\left[\sum_{j=1}^{N_i} \hat{y}_{ij} + \sum_{j\in s_i} \frac{N_i}{n_i}(y_{ij} - \hat{y}_{ij})\right], \qquad (8.8.5)$$

where $\hat{y}_{ij} = \mathbf{x}_{ij}^T(\tilde{\mathbf{Z}}_i\hat{\boldsymbol{\beta}} + \hat{\mathbf{v}}_i)$ is the EBLUP predictor of y_{ij} under the assumed two-level model, and the last term in (8.8.5) is a design-unbiased estimator of the total prediction error for area i. The estimator $\hat{\mu}_i^{\mathrm{LV}}$ readily extends to general sampling designs within areas by replacing in (8.8.5) N_i/n_i by the design weight w_{ij} for the specified design.

Torabi and Rao (2008) used the above simulation setup to compare the relative performance of the EBLUP estimator based on the two-level model and the new GREG estimator under both model-based and design-based setups. For the latter, they simulated a synthetic finite population from the Brazilian sample data. Simulation results indicated that the new GREG performs slightly better than the customary GREG estimator in terms of average MSE and average ARE. On the other hand, the EBLUP estimator is substantially more efficient than the new GREG and the customary GREG due to borrowing strength across areas.

8.9 *MODELS FOR MULTINOMIAL COUNTS

Now consider the case of estimating the proportions $p_{ij1}, \ldots, p_{ijK}$ of individuals in the K categories of a qualitative variable, in a certain population group j within area i ($j = 1, \ldots, n_i, i = 1, \ldots, m$), where $\sum_{k=1}^K p_{ijk} = 1$. Let $(y_{ij1}, \ldots, y_{ijK})$ be the vector of observed counts of individuals in the K categories in group j within area i. We assume that $(y_{ij1}, \ldots, y_{ijK})^T \sim \text{multinomial}(m_{ij}; p_{ij1}, \ldots, p_{ijK})$, where $\sum_{k=1}^K y_{ijk} = m_{ij}$. The probabilities for all the areas are linked through the logistic mixed model

$$\log\,(p_{ijk}/p_{ijK}) = \mathbf{x}_{ij}^T\boldsymbol{\beta}_k + v_{ik}, \quad k = 1, \ldots, K-1, \quad j = 1, \ldots, n_i, \quad i = 1, \ldots, m, \qquad (8.9.1)$$

where $(v_{i1}, \ldots, v_{i,K-1})^T \overset{\text{iid}}{\sim} N_{K-1}(\mathbf{0}, \boldsymbol{\Sigma}_v)$ are vectors of random effects for area i. In the special case of $m_{ij} = 1$, we have $y_{ijk} \in \{0, 1\}, k = 1, \ldots, K$, such that $\sum_{k=1}^K y_{ijk} = 1$. Furthermore, letting $K = 2$, we obtain the logistic mixed model (4.6.1).

Example 8.9.1. Labor Force Estimates. Molina, Saei, and Lombardía (2007) considered the multinomial logit model (8.9.1) with common area effects for all groups k, that is, with $v_{ik} = v_i$ for all k and $v_i \overset{\text{iid}}{\sim} N(0, \sigma_v^2)$, to estimate labor force characteristics in each of $m = 406$ unitary authority and local authority areas (UALAD) from the United Kingdom, treating the UALADs as small areas. The categories of the variable of interest are unemployed ($k = 1$), employed ($k = 2$), and not in labor force ($k = 3$), and the available data was in the form of counts $(y_{ij1}, y_{ij2}, y_{ij3})$ within gender–age group j ($j = 1, \ldots, 6$) for each UALAD i. The covariates $\mathbf{x}_{ij}$ considered in the model were the log-proportion of registered unemployed and 22 dummy indicators including gender–age categories and regions covering the small areas.

The multinomial logit mixed model (8.9.1) was fitted using a two-stage iterative algorithm based on the penalized quasi-likelihood (PQL) method (Breslow and Clayton 1993) for estimating the regression coefficients $\boldsymbol{\beta} = (\boldsymbol{\beta}_1^T, \boldsymbol{\beta}_2^T)^T$ and the random effects $\mathbf{v} = (v_1, \ldots, v_m)^T$, and approximate ML or REML for estimation of σ_v^2. This combined algorithm was introduced by Schall (1991) and used by Saei and Chambers (2003) for fitting generalized linear mixed models (GLMM) in the small area estimation context. PQL estimates jointly the regression coefficients $\boldsymbol{\beta}$ and the random effects $\mathbf{v}$, given a fixed value of σ_v^2, by maximizing the joint loglikelihood of the observations $(y_{ij1}, \ldots, y_{ijK})^T$ and $\mathbf{v}$ instead of maximizing the marginal loglikelihood of the data. In the case that $(y_{ij1}, \ldots, y_{ijK})^T$ are multivariate normal, PQL estimates of $\boldsymbol{\beta}$ and $\mathbf{v}$ coincide with the ML estimator of $\boldsymbol{\beta}$ and the BLUP of $\mathbf{v}$. Note that the joint loglikelihood has a closed form in the case of a GLMM, unlike the marginal loglikelihood. In a second stage, for $\boldsymbol{\beta}$ and $\mathbf{v}$ fixed, Schall (1991) applied a linearization of the model and a normal approximation of the marginal likelihood to obtain approximate ML or REML estimates of the variance components (in this case, σ_v^2). Iterating this two-stage procedure, we obtain estimates of all required quantities, namely, $\boldsymbol{\beta}, \mathbf{v}$, and the variance components. Let $\hat{\boldsymbol{\beta}}_1, \hat{\boldsymbol{\beta}}_2$ and $\hat{v}_i$ be the estimated regression coefficients and area effects. The probabilities are then estimated as

$$\hat{p}_{ijk} = \frac{\exp(\mathbf{x}_{ij}^T \hat{\boldsymbol{\beta}}_k + \hat{v}_i)}{1 + \sum_{k=1}^2 \exp(\mathbf{x}_{ij}^T \hat{\boldsymbol{\beta}}_k + \hat{v}_i)}, \qquad k = 1, 2,$$

and $\hat{p}_{ij3} = 1 - \sum_{k=1}^2 \hat{p}_{ijk}$.

Letting now $(y_{ij1}^r, y_{ij2}^r, y_{ij3}^r)^T$ be the vector of nonsample counts and $m_{ij}^r = \sum_{k=1}^K y_{ijk}^r$ the total count of individuals in gender–age class j within UALAD i, the area total of individuals in each category $T_{ik} = \sum_{j=1}^6 (y_{ijk} + y_{ijk}^r)$ is estimated as $\hat{T}_{ik} = \sum_{j=1}^6 (y_{ijk} + \hat{y}_{ijk}^r)$, $k = 1, 2, 3$, where $\hat{y}_{ijk}^r = m_{ij}^r \hat{p}_{ijk}$, and the unemployment rate $R_i = T_{i1}/(T_{i1} + T_{i2})$ is estimated as $\hat{R}_i = \hat{T}_{i1}/(\hat{T}_{i1} + \hat{T}_{i2})$. A parametric bootstrap method was proposed for MSE estimation.

López-Vizcaíno, Lombardía, and Morales (2013) considered the model (8.9.1) but at the area level ($n_i = 1$ for all i) and with independent category-specific random effects $v_{ik}, k = 1, \ldots, K - 1$, that is, with diagonal $\boldsymbol{\Sigma}_v$. They applied this model to estimate labor force characteristics in 51 counties from the region of Galicia by gender, using data from the Spanish Labor Force Survey (LFS) for that region. López-Vizcaíno, Lombardía, and Morales (2014) extended this model by including, independently for each category k and area i, a vector of time-correlated random effects following an AR(1) process.

8.10 *EBLUP FOR VECTORS OF AREA PROPORTIONS

Berg and Fuller (2014) considered a generalized multivariate area level model for estimating vectors of proportions $\mathbf{p}_i = (p_{1i}, \ldots, p_{Ri})^T$ associated with a multicategory response variable in each area i. This leads to a two-way table of proportions $p_{ri}, r = 1, \ldots, R, i = 1, \ldots, m$ with $\sum_{r=1}^{R} p_{ri} = 1$ for each i. For example, the Canadian LFS reports direct estimates, $\hat{\mathbf{p}}_i = (\hat{p}_{1i}, \ldots, \hat{p}_{Ri})^T$, of proportions in the two-way table defined by occupation classes (r) crossed with the Canadian provinces (i). The objective is to improve on the direct estimators by borrowing strength across provinces and occupation classes.

The proposed area level model consists of the sampling model

$$\hat{\mathbf{p}}_i = \mathbf{p}_i + \mathbf{e}_i, \quad i = 1, \ldots, m, \tag{8.10.1}$$

and the linking model

$$\mathbf{p}_i = \mathbf{p}_{Ti} + \mathbf{v}_i, \quad i = 1, \ldots, m, \tag{8.10.2}$$

where $\mathbf{p}_{Ti} = (p_{T,1i}, \ldots, p_{T,Ri})^T$ with $p_{T,ri}$ a specified function of covariates, $\mathbf{z}_{1i}, \ldots, \mathbf{z}_{Ri}$, satisfying $0 \le p_{T,ri} \le 1$ and $\sum_{r=1}^{R} p_{T,ri} = 1$. Here, the $\mathbf{z}_{ri}$'s are area level covariate vectors, the $\mathbf{e}_i$'s are sampling errors with $E(\mathbf{e}_i|\mathbf{p}_i) = \mathbf{0}$ and conditional variance $V(\mathbf{e}_i|\mathbf{p}_i)$, and the $\mathbf{v}_i$ are model errors with mean $\mathbf{0}$ and covariance matrix $\mathbf{\Sigma}_{vv,i}$. The unconditional covariance matrix of $\mathbf{e}_i$ is $V(\mathbf{e}_i) = E[V(\mathbf{e}_i|\mathbf{p}_i)] =: \mathbf{\Sigma}_{ee,i}$. One possible choice of $p_{T,ri}$ is the logistic function given by

$$p_{T,ri} = \frac{\exp(\mathbf{z}_{ri}^T\boldsymbol{\beta})}{1 + \sum_{k=2}^{R} \exp(\mathbf{z}_{ki}^T\boldsymbol{\beta})}, \quad r = 2, \ldots, R, \tag{8.10.3}$$

and $p_{T,1i} = 1 - \sum_{r=2}^{R} p_{T,ri}$. Note that (8.10.3) may be expressed equivalently as $\log\,(p_{T,ri}/p_{T,1i}) = \mathbf{z}_{ri}^T\boldsymbol{\beta}, r = 2, \ldots, R$.

Since $\sum_{r=1}^{R} p_{ri} = 1$, we drop the first category and denote the vector of remaining proportions as $\mathbf{p}_i^{(1)} = (p_{2i}, \ldots, p_{Ri})^T$. Similarly, we denote $\mathbf{e}_i^{(1)}, \hat{\mathbf{p}}_i^{(1)}$, and $\mathbf{p}_{T,i}^{(1)}$ as the respective vectors $\mathbf{e}_i, \hat{\mathbf{p}}_i$, and $\mathbf{p}_{T,i}$ without the first element. Let $\mathbf{\Sigma}_{vv,i}^{(1)} = V(\mathbf{p}_i^{(1)})$ and $\mathbf{\Sigma}_{ee,i}^{(1)}$ be the unconditional covariance matrix of $\mathbf{e}_i^{(1)}$. Then, Berg and Fuller (2014) proved that the BLUP of $\mathbf{p}_i^{(1)}$ is given by

$$\tilde{\mathbf{p}}_i^{(1)\mathrm{BF}} = \mathbf{p}_{T,i}^{(1)} + \mathbf{\Sigma}_{vv,i}^{(1)}(\mathbf{\Sigma}_{vv,i}^{(1)} + \mathbf{\Sigma}_{ee,i}^{(1)})^{-1}(\hat{\mathbf{p}}_i^{(1)} - \mathbf{p}_{T,i}^{(1)}). \tag{8.10.4}$$

The proportion for the first category is then estimated in terms of the BLUP estimators for the other proportions $\tilde{\mathbf{p}}_i^{(1)\mathrm{BF}} = (\tilde{p}_{2i}^{\mathrm{BF}}, \ldots, \tilde{p}_{Ri}^{\mathrm{BF}})^T$ as $\tilde{p}_{1i}^{\mathrm{BF}} = 1 - \sum_{r=1}^{R} \tilde{p}_{ri}^{\mathrm{BF}}$. The estimators $\tilde{p}_{ri}^{\mathrm{BF}}$ sum to one for each area i, but they may not remain between zero and one. Modified estimators p_{ri}^* that satisfy the constraints $\delta_{ri} \le p_{ri}^* \le 1 - \delta_{ri}$ and $\sum_{r=1}^{R} p_{ri}^* = 1$ for specified δ_{ri} (≥ 0) are obtained by minimizing a distance function between $\tilde{\mathbf{p}}_i^{(1)\mathrm{BF}}$ and $\mathbf{p}_i^{(1)*} = (p_{2i}^*, \ldots, p_{Ri}^*)^T$ subject to the constraints. The BLUP estimators are invariant to the choice of the dropped category in (8.10.4) .

The estimator (8.10.4) depends on $\boldsymbol{\beta}$ and the matrices $\mathbf{\Sigma}_{vv,i}^{(1)}$ and $\mathbf{\Sigma}_{ee,i}^{(1)}$, but it is greatly simplified by assuming that

$$\mathbf{\Sigma}_{vv,i}^{(1)} = \delta[\mathrm{diag}(\mathbf{p}_{T,i}^{(1)}) - \mathbf{p}_{T,i}^{(1)}(\mathbf{p}_{T,i}^{(1)})^T] \tag{8.10.5}$$

with $\delta \geq 0$. Note that (8.10.5) is a multiple of a multinomial covariance matrix. Furthermore, we assume that

$$V(\mathbf{e}_i|\mathbf{p}_i) = n_i^{-1}d_i[\text{diag}(\mathbf{p}_i) - \mathbf{p}_i\mathbf{p}_i^T], \tag{8.10.6}$$

where d_i is a measure of average design effect for area i (Rao and Scott 1981) and n_i is the area sample size. Then (8.10.5) and (8.10.6) lead to

$$\boldsymbol{\Sigma}_{ee,i} = V(\mathbf{e}_i) = E[V(\mathbf{e}_i|\mathbf{p}_i)] = n_i^{-1}c_i[\text{diag}(\mathbf{p}_{T,i}) - \mathbf{p}_{T,i}(\mathbf{p}_{T,i})^T], \tag{8.10.7}$$

where $c_i = d_i(1-\delta)$. The parameter c_i may be interpreted as an average of generalized design effects (Rao and Scott 1981). It now follows from (8.10.4), (8.10.5), and (8.10.7) that $\hat{p}_{ri}^{\text{BF}}$ reduces to

$$\tilde{p}_{ri}^{\text{BF}} = p_{T,ri} + \gamma_i(\hat{p}_{ri} - p_{T,ri}), \quad r = 2, \ldots, R, \tag{8.10.8}$$

where $\gamma_i = \delta/(\delta + c_i n_i^{-1})$. Note that in this case it holds $0 \leq \hat{p}_{ri}^{\text{BF}} \leq 1$ and $\sum_{r=1}^R \hat{p}_{ri}^{\text{BF}} = 1$ because γ_i does not depend on the category r.

Berg and Fuller (2014) used generalized least squares to obtain the estimators $\hat{\delta}$ and $\hat{\boldsymbol{\beta}}$ of the model parameters δ and $\boldsymbol{\beta}$ in (8.10.8) . Substituting $\hat{\delta}$ and $\hat{\boldsymbol{\beta}}$ for δ and $\boldsymbol{\beta}$ in (8.10.8), we obtain an EBLUP estimator $\hat{p}_{ri}^{\text{BF}}$ of the proportion p_{ri} ($r = 1, \ldots, R, i = 1, \ldots, m$). Berg and Fuller (2014) also studied MSE estimation using a bootstrap method. They applied the new method to estimate occupation proportions, p_{ri}, for each province r in Canada, using direct estimates from the LFS and auxiliary data, $\mathbf{z}_{ri}$, from the previous census (Hidiroglou and Patak 2009). They also compared their method to structure preserving estimation (SPREE) estimates (Section 3.2.6).

8.11 *SOFTWARE

EBLUP estimates of area parameters $\theta_i, i = 1, \ldots, m$, based on the spatial model with SAR area effects, can be computed using the function `eblupSFH` of the R package `sae`. The analytical MSE estimators of Singh et al. (2005) can be obtained using function `mseSFH` of the same package. Instead, parametric or nonparametric bootstrap MSE estimates proposed by Molina et al. (2008) are provided by functions `pbmseSFH` and `npbmseSFH`, respectively. Either ML or REML (default) fitting method can be specified in the argument `method` of these functions. The calls to the functions are as follows:

```
eblupSFH(formula, vardir, proxmat, method = "REML",
    MAXITER = 100, PRECISION = 0.0001, data)
mseSFH(formula, vardir, proxmat, method = "REML",
    MAXITER = 100, PRECISION = 0.0001, data)
pbmseSFH(formula, vardir, proxmat, B = 100, method = "REML",
    MAXITER = 100, PRECISION = 0.0001, data)
npbmseSFH(formula, vardir, proxmat, B = 100, method = "REML",
```

```
MAXITER = 100, PRECISION = 0.0001, data)
```

In the argument `formula`, we must specify the regression equation for the fixed (nonrandom) part of the model as a usual R regression formula. The left-hand side of the regression formula is the vector of direct estimates $\hat{\theta}_i$ of the target means θ_i, and the right-hand side contains the area level auxiliary variables included in $\mathbf{z}_i$ separated by + and with an intercept included by default. In the argument `vardir`, we must include the vector of sampling variances ψ_i of direct estimators, in `proxmat` we must give the $m \times m$ matrix of proximities between all pairs of areas $\mathbf{W}$ and, finally, in argument `B` of the two functions implementing bootstrap procedures, we must specify the desired number of bootstrap replicates.

The function `eblupSFH` returns a list of two objects: the vector `eblup`, containing the EBLUP estimates for the areas, and the list `fit`, containing the results of the fitting method. The functions `mseSFH`, `pbmseSFH`, and `npbmseSFH` also return a list with two objects: the first object, `est`, contains the EBLUP estimates in the vector `eblup`, and a summary of the fitting procedure in the list `fit`. The second object, `mse`, is the vector with the estimated MSEs of the EBLUP estimators for each area in the case of function `mseSFH`. In functions `pbmseSFH` and `npbmseSFH`, the object `mse` is a data frame with two columns containing, respectively, naive and bias-corrected bootstrap MSE estimates (see Molina et al. 2008). In all the functions, the list `fit` contains a summary of the fitting procedure, which includes the specified fitting method (`method`), the logical value `convergence=TRUE` if the Fisher-scoring algorithm has converged in the specified maximum number of iterations (`MAXITER`), the total number of iterations of Fisher-scoring algorithm until convergence (`iterations`), a data frame with the required output concerning the estimated regression coefficients (`estcoef`), the estimated random effects variance σ_u^2 (`refvar`), the estimated spatial autocorrelation parameter ϕ (`spatialcorr`), and finally a vector (`goodness`) with three goodness-of-fit measures: loglikelihood, AIC, and BIC.

Example 8.11.1. EBLUP under a Spatial FH Model, with R. Here we illustrate how to obtain EBLUP estimates of $\theta_i, i = 1, \ldots, m$, under a spatial model with random effects following a SAR process, together with MSE estimates using the functions `mseSFH` and `npmseSFH`. We consider the predefined R data set `grapesprox`, which contains synthetic data on grape production for $m = 274$ municipalities in the region of Tuscany (Italy). The data set contains the variables `grapehect`, direct estimates of the mean surface area (in hectares) used for production of grape for each municipality, `area`, agrarian surface area (in hectares) used for production, `workdays`, average number of working days in the reference year, and `var`, sampling variance of the direct estimators for each municipality. We calculate EBLUP estimates of mean surface area used for production of grapes, based on a spatial FH model with SAR area effects and `area` and `workdays` as auxiliary variables. We supplement the EBLUP estimates with the analytical MSE estimates derived by Singh et al. (2005) and the nonparametric bootstrap estimates of Molina et al. (2008).

The predefined R data set `grapesprox` contains the proximity matrix $\mathbf{W} = (w_{i\ell})$ representing the neighborhood structure of the municipalities in Tuscany. The elements of this matrix are $w_{i\ell} = \tilde{w}_{i\ell} / \sum_{k=1}^{m} \tilde{w}_{ik}$, where $\tilde{w}_{i\ell} = 1$ if $\ell \neq i$ and ℓth municipality shares a boundary with i th municipality, and $\tilde{w}_{i\ell} = 0$ otherwise.

First load the package `sae` and the two required data sets:

```
R> library(sae)
R> data("grapes")
R> data("grapesprox")
```

Next, obtain EBLUP estimators based on REML fitting method (default), accompanied by the analytical MSE estimates given by Singh et al. (2005):

```
R> SFH <- mseSFH(grapehect ~ area + workdays - 1, var, grapesprox,
+                    data = grapes)
```

The two considered auxiliary variables are significant, and the estimated model parameters σ_u^2 and ϕ, returned, respectively, writing `SFH$fit$refvar` and `SFH$fit$spatialcorr`, are $\hat{\sigma}_u^2 = 71.189$ and $\hat{\phi} = 0.583$.

Now compare the obtained analytical MSE estimates with the naive and bias-corrected nonparametric bootstrap MSE estimates by calling function `npbmseSFH`:

```
R> SFH.npb <- npbmseSFH(grapehect ~ area + workdays, var, grapesprox,
+                        B=100,data=grapes)
```

Plot analytical MSE estimates against naive and bias-corrected nonparametric bootstrap MSE estimates

```
R> xmin<-min(SFH$mse,SFH.npb$mse[,1])
R> xmax<-max(SFH$mse,SFH.npb$mse[,1])
R> plot(SFH$mse,SFH.npb$mse[,1],xlim=c(xmin,xmax),ylim=c(xmin,xmax),
+  xlab="Analytical",ylab="Naive nonparametric bootstrap")
R> abline(a=0,b=1)

R> xmin<-min(SFH$mse,SFH.npb$mse[,2])
R> xmax<-max(SFH$mse,SFH.npb$mse[,2])
R> plot(SFH$mse,SFH.npb$mse[,2],xlim=c(xmin,xmax),ylim=c(xmin,xmax),
+  xlab="Analytical",ylab="Bias-corrected nonparametric bootstrap")
abline(a=0,b=1)
```

Naive nonparametric bootstrap against analytical MSE estimates are displayed in Figure 8.1a, whereas Figure 8.1b plots bias-corrected MSE estimates against analytical estimates. Figure 8.1 shows that the naive nonparametric bootstrap MSE estimates clearly differ from the analytical estimates. On the other hand, the bias-corrected nonparametric bootstrap estimates are very similar to the analytical estimates obtained by Singh et al. (2005), except for a slight deviation for the areas with largest MSE estimates. This agreement gives some credibility to these two sets of MSE estimates.

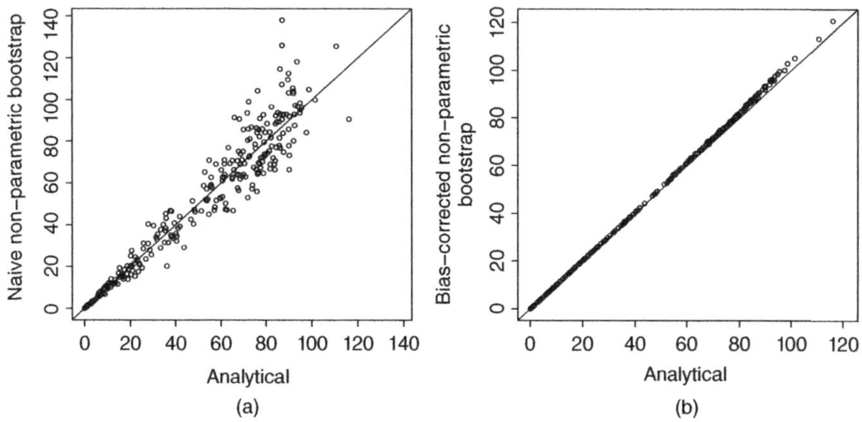

Figure 8.1 Naive Nonparametric Bootstrap MSE Estimates Against Analytical MSE Estimates (a). Bias-corrected Nonparametric Bootstrap MSE Estimates Against Analytical MSE Estimates (b).

Now considering the bias-corrected nonparametric bootstrap MSE estimates, calculate the CVs and arrange results in a data frame:

```
R> cv.SFH <- 100*sqrt(SFH.npb$mse$msebc)/SFH$est$eblup
R> grapes.est <- data.frame(dir = grapes$grapehect,
+                 cv.dir = 100 * abs(sqrt(grapes$var)/grapes$grapehect),
+                 eblup.SFH = SFH$est$eblup,cv.SFH)
```

Sort results by increasing CVs of direct estimators:

```
R> sortedgrapes<-grapes.est[order(grapes.est$cv.dir),]
```

Then, plot the sorted EBLUP estimates together with direct estimates:

```
R> plot(sortedgrapes$dir,type="n",xlab="area",ylab="Estimate")
R> points(sortedgrapes$dir,type="b",col=3,lwd=2,pch=1)
R> points(sortedgrapes$eblup.SFH,type="b",col=4,lwd=2,pch=4)
R> legend(1,350,legend=c("Direct","EBLUP"),ncol=2,col=c(3,4),
+  lwd=rep(2,2),pch=c(1,4))
```

Plot also the CVs of EBLUP and direct estimates:

```
R> plot(sortedgrapes$cv.dir,type="n",xlab="area",ylab="CV",
   ylim=c(0,400))
R> points(sortedgrapes$cv.dir,type="b",col=3,lwd=2,pch=1)
R> points(sortedgrapes$cv.SFH,type="b",col=4,lwd=2,pch=4)
R> legend(1,400,legend=c("CV(Direct)","CV(EBLUP)"),col=c(3,4),
+  lwd=rep(2,2),pch=c(1,4))
```

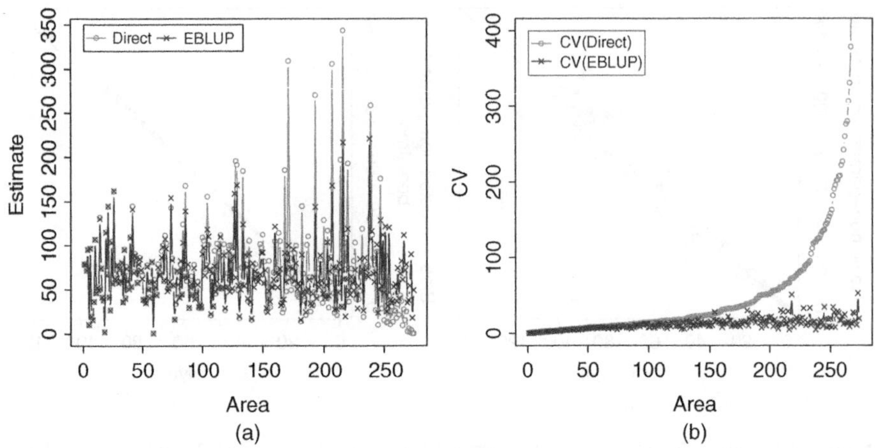

Figure 8.2 EBLUP Estimates, Based on the Spatial FH Model with SAR Random Effects, and Direct Estimates of Mean Surface Area Used for Production of Grapes for Each Municipality (a). CVs of EBLUP Estimates and of Direct Estimates for Each Municipality (b). Municipalities are Sorted by Increasing CVs of Direct Estimates.

Figure 8.2a shows EBLUP and direct estimates of mean surface area used for grape production for each municipality and Figure 8.2b shows the CVs of EBLUP and direct estimates, with municipalities sorted by increasing CVs of direct estimates. Figure 8.2a shows that EBLUP estimates are more stable than direct estimates. Moreover, the CVs of EBLUP estimates are smaller than those of direct estimates for most municipalities, and the reduction in CV is considerable in the municipalities where direct estimates are very inefficient (on the right side of the plots).

9

EMPIRICAL BAYES (EB) METHOD

9.1 INTRODUCTION

The empirical best linear unbiased prediction (EBLUP) method, studied in Chapters 5–8, is applicable to linear mixed models that cover many applications of small area estimation. Normality of random effects and errors is not needed for point estimation, but normality is used for getting accurate MSE estimators. The MSE estimator for the basic area level model remains valid under nonnormality of the random effects, v_i (see Section 6.2.1), but normality is generally needed.

Linear mixed models are designed for continuous variables, y, but they are not suitable for handling binary or count data. In Section 4.6, we proposed suitable models for binary and count data; in particular, logistic regression models with random area effects for binary data, and log-linear models with random effects for count data. Empirical Bayes (EB) and hierarchical Bayes (HB) methods are applicable more generally in the sense of handling models for binary and count data as well as normal linear mixed models. In the latter case, EB and EBLUP estimators are identical.

In this chapter, we study the EB method in the context of small area estimation. The EB approach may be summarized as follows: (i) Using the conditional density, $f(\mathbf{y}|\boldsymbol{\mu}, \boldsymbol{\lambda}_1)$, of $\mathbf{y}$ given $\boldsymbol{\mu}$ and the density $f(\boldsymbol{\mu}|\boldsymbol{\lambda}_2)$ of $\boldsymbol{\mu}$, obtain the posterior density, $f(\boldsymbol{\mu}|\mathbf{y}, \boldsymbol{\lambda})$, of the small area (random) parameters of interest, $\boldsymbol{\mu}$, given the data $\mathbf{y}$, where $\boldsymbol{\lambda} = (\boldsymbol{\lambda}_1^T, \boldsymbol{\lambda}_2^T)^T$ denotes the vector of model parameters. (ii) Estimate the model parameters, $\boldsymbol{\lambda}$, from the marginal density, $f(\mathbf{y}|\boldsymbol{\lambda})$, of $\mathbf{y}$. (iii) Use the estimated posterior density, $f(\boldsymbol{\mu}|\mathbf{y}, \hat{\boldsymbol{\lambda}})$, for making inferences about $\boldsymbol{\mu}$, where $\hat{\boldsymbol{\lambda}}$ is an estimator of $\boldsymbol{\lambda}$.

Small Area Estimation, Second Edition. J. N. K. Rao and Isabel Molina.

The density of $\boldsymbol{\mu}$ is often interpreted as prior density on $\boldsymbol{\mu}$, but it is actually a part of the postulated model on $(\mathbf{y}, \boldsymbol{\mu})$ and it can be validated from the data, unlike subjective priors on model parameters, $\boldsymbol{\lambda}$, used in the HB approach. In this sense the EB approach is essentially frequentist, and EB inferences refer to averaging over the joint distribution of $\mathbf{y}$ and $\boldsymbol{\mu}$. Sometimes, a prior density is chosen for $\boldsymbol{\lambda}$ but used only to derive estimators and associated measures of variability as well as confidence intervals (CIs) with good frequentist properties (Morris 1983b).

In the parametric empirical Bayes (PEB) approach, a parametric form, $f(\boldsymbol{\mu}|\boldsymbol{\lambda}_2)$, is assumed for the density of $\boldsymbol{\mu}$. On the other hand, nonparametric empirical Bayes (NPEB) methods do not specify the form of the (prior) distribution of $\boldsymbol{\mu}$. Nonparametric maximum likelihood is used to estimate the (prior) distribution of $\boldsymbol{\mu}$ (Laird 1978). Semi-nonparametric (SNP) representations of the density of $\boldsymbol{\mu}$ are also used in making EB inferences. For example, Zhang and Davidian (2001) studied linear mixed models with block-diagonal covariance structures by approximating the density of the random effects by a SNP representation. Their representation includes normality as a special case, and it provides flexibility in capturing nonnormality through a user-chosen tuning parameter. In this chapter, we focus on PEB approach to small area estimation. We refer the reader to Maritz and Lwin (1989) and Carlin and Louis (2008) for excellent accounts of the EB methodology.

The basic area level model (6.1.1) with normal random effects, v_i, is used in Section 9.2 to introduce the EB methodology. A jackknife method of MSE estimation is given in Section 9.2.2. This method is applicable more generally, as shown in subsequent sections. Inferences based on the estimated posterior density of the small area parameters, θ_i, do not account for the variability in the estimators of model parameters $(\boldsymbol{\beta}, \sigma_v^2)$. Methods that account for the variability are studied in Section 9.2.3. CI estimation is addressed in Section 9.2.4. Section 9.3 provides extensions to linear mixed models with a block-diagonal covariance structure. Section 9.4 describes EB estimation of general nonlinear parameters under the basic unit level model and an application to poverty mapping. The case of binary data is studied in Section 9.5. Applications to disease mapping using count data are given in Section 9.6. Section 9.7 describes design-weighted EB estimation under exponential family models. The EB (or EBLUP) estimator, $\hat{\theta}_i^{\text{EB}}$, may not perform well in estimating the histogram of the θ_i's or in ranking them; Section 9.8 proposes constrained EB and other estimators that address this problem. Finally, empirical linear Bayes (ELB) and empirical constrained linear Bayes (ECLB) methods are studied in Sections 9.9 and 9.10, respectively. These methods avoid distributional assumptions. Section 9.11 describes R software for EB estimation of general area parameters under the basic unit level model.

9.2 BASIC AREA LEVEL MODEL

Assuming normality, the basic area level model (6.1.1) may be expressed as a two-stage hierarchical model: (i) $\hat{\theta}_i|\theta_i \overset{\text{ind}}{\sim} N(\theta_i, \psi_i), \quad i = 1, \ldots, m$; (ii) $\theta_i \overset{\text{ind}}{\sim} N(\mathbf{z}_i^T\boldsymbol{\beta}, b_i^2\sigma_v^2), \quad i = 1, \ldots, m$, where $\boldsymbol{\beta}$ is the $p \times 1$ vector of regression parameters. In the Bayesian framework, the model parameters $\boldsymbol{\beta}$ and σ_v^2 are random, and the two-stage hierarchical model is called the conditionally independent hierarchical model (CIHM) because the pairs $(\hat{\theta}_i, \theta_i)$ are independent across areas i, conditionally on $\boldsymbol{\beta}$ and σ_v^2 (Kass and Steffey 1989).

9.2.1 EB Estimator

The "optimal" estimator of the realized value of θ_i is given by the conditional expectation of θ_i given $\hat{\theta}_i$, $\boldsymbol{\beta}$, and σ_v^2:

$$E(\theta_i|\hat{\theta}_i, \boldsymbol{\beta}, \sigma_v^2) = \hat{\theta}_i^B = \gamma_i \hat{\theta}_i + (1 - \gamma_i)\mathbf{z}_i^T \boldsymbol{\beta}, \tag{9.2.1}$$

where $\gamma_i = b_i^2 \sigma_v^2 / (b_i^2 \sigma_v^2 + \psi_i)$. The result (9.2.1) follows from the posterior (or conditional) distribution of θ_i given $\hat{\theta}_i$, $\boldsymbol{\beta}$, and σ_v^2, which is given by

$$\theta_i|\hat{\theta}_i, \boldsymbol{\beta}, \sigma_v^2 \overset{\text{ind}}{\sim} N(\hat{\theta}_i^B, g_{1i}(\sigma_v^2) = \gamma_i \psi_i). \tag{9.2.2}$$

The estimator $\hat{\theta}_i^B = \hat{\theta}_i^B(\boldsymbol{\beta}, \sigma_v^2)$ is the "Bayes" estimator under squared error loss and it is optimal in the sense that its MSE, $\text{MSE}(\hat{\theta}_i^B) = E(\hat{\theta}_i^B - \theta_i)^2$, is smaller than the MSE of any other estimator of θ_i, linear or nonlinear in the $\hat{\theta}_i$'s (see Section 6.2.3, Chapter 6). It may be more appropriate to name $\hat{\theta}_i^B$ as the best prediction (BP) estimator of θ_i because it is obtained from the conditional distribution (9.2.1) without assuming a prior distribution on the model parameters (Jiang, Lahiri, and Wan 2002).

The Bayes estimator $\hat{\theta}_i^B$ depends on the model parameters $\boldsymbol{\beta}$ and σ_v^2, which are estimated from the marginal distribution given by $\hat{\theta}_i \overset{\text{ind}}{\sim} N(\mathbf{z}_i^T \boldsymbol{\beta}, b_i^2 \sigma_v^2 + \psi_i)$, using ML or REML. Denoting the resulting estimators as $\hat{\boldsymbol{\beta}}$ and $\hat{\sigma}_v^2$, we obtain the EB (or the empirical best prediction (EBP)) estimator of θ_i from $\hat{\theta}_i^B$ by substituting $\hat{\boldsymbol{\beta}}$ for $\boldsymbol{\beta}$ and $\hat{\sigma}_v^2$ for σ_v^2:

$$\hat{\theta}_i^{\text{EB}} = \hat{\theta}_i^B(\hat{\boldsymbol{\beta}}, \hat{\sigma}_v^2) = \hat{\gamma}_i \hat{\theta}_i + (1 - \hat{\gamma}_i)\mathbf{z}_i^T \hat{\boldsymbol{\beta}}. \tag{9.2.3}$$

The EB estimator, $\hat{\theta}_i^{\text{EB}}$, is identical to the EBLUP estimator $\hat{\theta}_i^H$ given by (6.1.12). Note that $\hat{\theta}_i^{\text{EB}}$ is also the mean of the estimated posterior density, $f(\theta_i|\hat{\boldsymbol{\theta}}, \hat{\boldsymbol{\beta}}, \hat{\sigma}_v^2)$, of θ_i given the data $\hat{\boldsymbol{\theta}} = (\hat{\theta}_1, \ldots, \hat{\theta}_m)^T$, which is $N(\hat{\theta}_i^{\text{EB}}, \hat{\gamma}_i \psi_i)$.

For the special case of equal sampling variances $\psi_i = \psi$ and $b_i = 1$ for all i, Morris (1983b) studied the use of an unbiased estimator of $1 - \gamma_i = 1 - \gamma$ given by

$$1 - \gamma^* = \psi(m - p - 2)/S, \tag{9.2.4}$$

where $S = \sum_{i=1}^m (\hat{\theta}_i - \mathbf{z}_i^T \hat{\boldsymbol{\beta}}_{\text{LS}})^2$ and $\hat{\boldsymbol{\beta}}_{\text{LS}}$ is the least squares estimator of $\boldsymbol{\beta}$. The resulting EB estimator

$$\hat{\theta}_i^{\text{EB}} = \gamma^* \hat{\theta}_i + (1 - \gamma^*)\mathbf{z}_i^T \hat{\boldsymbol{\beta}}_{\text{LS}} \tag{9.2.5}$$

is identical to the James–Stein estimator, studied in Section 3.4.2. Note that (9.2.4) may be expressed as

$$1 - \gamma^* = [(m - p - 2)/(m - p)]\psi/(\psi + \tilde{\sigma}_v^2), \tag{9.2.6}$$

where $\tilde{\sigma}_v^2 = S/(m-p) - \psi$ is the unbiased moment estimator of σ_v^2. Morris (1983b) used the REML estimator $\hat{\sigma}_v^2 = \max(0, \tilde{\sigma}_v^2)$ in (9.2.6), instead of the unbiased estimator $\tilde{\sigma}_v^2$, to ensure that $1 - \gamma^* < 1$. The multiplying constant $(m-p-2)/(m-p)$ offsets the positive bias introduced by the substitution of a nearly unbiased estimator $\hat{\sigma}_v^2$ into $1 - \gamma = \psi/(\psi + \sigma_v^2)$; note that $1 - \gamma$ is a convex nonlinear function of σ_v^2 so that $E[\psi/(\psi + \tilde{\sigma}_v^2)] > 1 - \gamma$ by Jensen's inequality.

For the case of unequal sampling variances ψ_i with $b_i = 1$, Morris (1983b) used the multiplying constant $(m-p-2)/(m-p)$ in the EB estimator (9.2.3); that is, $1 - \hat{\gamma}_i$ is replaced by

$$1 - \gamma_i^* = [(m-p-2)/(m-p)]\psi_i/(\psi_i + \hat{\sigma}_v^2), \tag{9.2.7}$$

where $\hat{\sigma}_v^2$ is the REML estimator of σ_v^2. He also proposed an alternative estimator of σ_v^2 that is similar to the Fay–Herriot (FH) moment estimator (Section 6.1.2). This estimator is obtained by solving, iteratively for σ_v^2, the equation

$$\sigma_v^2 = \left(\sum_{i=1}^m \alpha_i\right)^{-1} \sum_{i=1}^m \alpha_i \left[\frac{m}{m-p}(\hat{\theta}_i - \mathbf{z}_i^T \tilde{\boldsymbol{\beta}})^2 - \psi_i\right], \tag{9.2.8}$$

where $\tilde{\boldsymbol{\beta}} = \tilde{\boldsymbol{\beta}}(\sigma_v^2)$ is the weighted least squares estimator of $\boldsymbol{\beta}$ and $\alpha_i = 1/(\sigma_v^2 + \psi_i)$. To avoid a negative solution, $\tilde{\sigma}_v^2$, we take $\hat{\sigma}_v^2 = \max(\tilde{\sigma}_v^2, 0)$. If α_i is replaced by α_i^2 in (9.2.8), then the resulting solution is approximately equal to the REML estimator of σ_v^2.

An advantage of EB (or EBP) is that it can be applied to find the EB estimator of any function $\phi_i = h(\theta_i)$; in particular, $\overline{Y}_i = g^{-1}(\theta_i) = h(\theta_i)$. The EB estimator is obtained from the Bayes estimator $\hat{\phi}_i^B = E(\phi_i|\hat{\theta}_i, \boldsymbol{\beta}, \sigma_v^2)$ by substituting $(\hat{\boldsymbol{\beta}}, \hat{\sigma}_v^2)$ for $(\boldsymbol{\beta}, \sigma_v^2)$. The computation of the EB estimator $\hat{\phi}_i^{\text{EB}}$ might require the use of Monte Carlo or numerical integration. For example, $\{\theta_i^{(r)}, r = 1, \ldots, R\}$ can be simulated from the estimated posterior density, namely, $N(\hat{\theta}_i^{\text{EB}}, \hat{\gamma}_i\psi_i)$, to obtain a Monte Carlo approximation:

$$\hat{\phi}_i^{\text{EB}} \approx \frac{1}{R}\sum_{r=1}^R h(\theta_i^{(r)}). \tag{9.2.9}$$

The computation of (9.2.9) can be simplified by rewriting (9.2.9) as

$$\hat{\phi}_i^{\text{EB}} \approx \frac{1}{R}\sum_{r=1}^R h(\hat{\theta}_i^{\text{EB}} + z_i^{(r)}\sqrt{\hat{\gamma}_i\psi_i}), \tag{9.2.10}$$

where $\{z_i^{(r)}, r = 1, \ldots, R\}$ are generated from $N(0, 1)$.

The approximation (9.2.10) will be accurate if the number of simulated samples, R, is large. Note that we used $h(\hat{\theta}_i^H) = h(\hat{\theta}_i^{\text{EB}})$ in Section 6.2.3 and remarked that the estimator $h(\hat{\theta}_i^H)$ does not retain the optimality of $\hat{\theta}_i^H$.

9.2.2 MSE Estimation

The results in Section 6.2.1 on the estimation of MSE of the EBLUP estimator, $\hat{\theta}_i^H$, are applicable to the EB estimator $\hat{\theta}_i^{\text{EB}}$ because $\hat{\theta}_i^H$ and $\hat{\theta}_i^{\text{EB}}$ are identical under normality. Also, the area-specific estimator (6.2.5) or (6.2.6) may be used as an estimator of the conditional MSE, $\text{MSE}_c(\hat{\theta}_i^{\text{EB}}) = E[(\hat{\theta}_i^{\text{EB}} - \theta_i)^2|\hat{\theta}_i]$, where the expectation is conditional on the observed $\hat{\theta}_i$ for the ith area (see Section 6.2.7). As noted in Section 6.2.1, the MSE estimators are second-order unbiased, that is, their bias is of lower order than m^{-1}, for large m.

Jiang, Lahiri, and Wan (2002) proposed a jackknife method of estimating the MSE of EB estimators. This method is more general than the Taylor linearization methods of Section 6.2.1 for $\hat{\theta}_i^H = \hat{\theta}_i^{\text{EB}}$, in the sense that it can also be applied to MSE estimation under models for binary and count data (see Sections 9.5 and 9.6). We illustrate its use here for estimating $\text{MSE}(\hat{\theta}_i^{\text{EB}})$.

Let us first decompose $\text{MSE}(\hat{\theta}_i^{\text{EB}})$ as follows:

$$\text{MSE}(\hat{\theta}_i^{\text{EB}}) = E(\hat{\theta}_i^{\text{EB}} - \hat{\theta}_i^B)^2 + E(\hat{\theta}_i^B - \theta_i)^2 \tag{9.2.11}$$

$$= E(\hat{\theta}_i^{\text{EB}} - \hat{\theta}_i^B)^2 + g_{1i}(\sigma_v^2)$$

$$=: M_{2i} + M_{1i}, \tag{9.2.12}$$

where the expectation is over the joint distribution of $(\hat{\theta}_i, \theta_i)$, $i = 1, \ldots, m$; see Section 9.12 for a proof of (9.2.11). Note that $\hat{\theta}_i^{\text{EB}}$ depends on all the data, $\hat{\boldsymbol{\theta}} = (\hat{\theta}_1, \ldots, \hat{\theta}_m)^T$, through the estimators $\hat{\boldsymbol{\beta}}$ and $\hat{\sigma}_v^2$.

The jackknife steps for estimating the two terms, M_{2i} and M_{1i}, in (9.2.12) are as follows. Let $\hat{\theta}_i^{\text{EB}} = k_i(\hat{\theta}_i, \hat{\boldsymbol{\beta}}, \hat{\sigma}_v^2)$ be the EB estimator of θ_i expressed as a function of the direct estimator $\hat{\theta}_i$ and the parameter estimators $\hat{\boldsymbol{\beta}}$ and $\hat{\sigma}_v^2$.

Step 1. Calculate the delete-ℓ estimators $\hat{\boldsymbol{\beta}}_{-\ell}$ and $\hat{\sigma}_{v,-\ell}^2$ by deleting the ℓth area data set $(\hat{\theta}_\ell, \mathbf{z}_\ell)$ from the full data set $\{(\hat{\theta}_i, \mathbf{z}_i); i = 1, \ldots, m\}$. This calculation is done for each ℓ to get m estimators of $\boldsymbol{\beta}$ and σ_v^2, $\{(\hat{\boldsymbol{\beta}}_{-\ell}, \hat{\sigma}_{v,-\ell}^2); \ell = 1, \ldots, m\}$ which, in turn, provide m estimators of θ_i, $\{\hat{\theta}_{i,-\ell}^{\text{EB}}; \ell = 1, \ldots, m\}$, where $\hat{\theta}_{i,-\ell}^{\text{EB}} = k_i(\hat{\theta}_i, \hat{\boldsymbol{\beta}}_{-\ell}, \hat{\sigma}_{v,-\ell}^2)$.

Step 2. Calculate the estimator of M_{2i} as

$$\hat{M}_{2i} = \frac{m-1}{m} \sum_{\ell=1}^{m} (\hat{\theta}_{i,-\ell}^{\text{EB}} - \hat{\theta}_i^{\text{EB}})^2. \tag{9.2.13}$$

Step 3. Calculate the estimator of M_{1i} as

$$\hat{M}_{1i} = g_{1i}(\hat{\sigma}_v^2) - \frac{m-1}{m} \sum_{\ell=1}^{m} [g_{1i}(\hat{\sigma}_{v,-\ell}^2) - g_{1i}(\hat{\sigma}_v^2)]. \tag{9.2.14}$$

The estimator $\hat{M}_{1i}$ corrects the bias of $g_{1i}(\hat{\sigma}_v^2)$.

Step 4. Calculate the jackknife estimator of MSE($\hat{\theta}_i^{\text{EB}}$) as

$$\text{mse}_J(\hat{\theta}_i^{\text{EB}}) = \hat{M}_{1i} + \hat{M}_{2i}. \tag{9.2.15}$$

Note that $\hat{M}_{1i}$ estimates the MSE when the model parameters are known, and $\hat{M}_{2i}$ estimates the additional variability due to estimating the model parameters. The jackknife estimator of MSE, given by (9.2.15), is also second-order unbiased. The proof of this result is highly technical, and we refer the reader to Jiang, Lahiri, and Wan (2002) for details.

The jackknife method is applicable to ML, REML, or moment estimators of the model parameters. It is computer-intensive compared to the MSE estimators studied in Section 6.2.1. In the case of ML or REML, computation of $\{(\hat{\boldsymbol{\beta}}_{-\ell}, \hat{\sigma}^2_{v,-\ell}); \ell = 1, \ldots, m\}$ may be simplified by performing only a single step of the Newton–Raphson algorithm using $(\hat{\boldsymbol{\beta}}, \hat{\sigma}_v^2)$ as the starting values of $\boldsymbol{\beta}$ and σ_v^2. However, properties of the resulting simplified jackknife MSE estimator are not known.

The jackknife method is applicable to the EB estimator of any function $\phi_i = h(\theta_i)$; in particular, $\overline{Y}_i = g^{-1}(\theta_i) = h(\theta_i)$. However, the computation of $\text{mse}_J(\hat{\phi}_i^{\text{EB}}) = \hat{M}_{1i} + \hat{M}_{2i}$ might require repeated Monte Carlo or numerical integration to obtain $\hat{M}_{2i} = (m-1)m^{-1}\sum_{\ell=1}^m (\hat{\phi}_{i,-\ell}^{\text{EB}} - \hat{\phi}_i^{\text{EB}})^2$ and the bias-corrected estimator, $\hat{M}_{1i}$, of $E(\hat{\phi}_i^B - \phi_i)^2$.

Example 9.2.1. Visits to Doctor's Office. Jiang et al. (2001) applied the jackknife method to data from the U.S. National Health Interview Survey (NHIS). The objective here is to estimate the proportion of individuals who did not visit a doctor's office during the previous 12 months for all the 50 states and the District of Columbia (regarded as small areas). Direct NHIS estimates, $\hat{P}_i$, of the proportions P_i are not reliable for the smaller states. The arcsine transformation was used to stabilize the variances of the direct estimates. We take $\hat{\theta}_i = \arcsin\sqrt{\hat{P}_i}$ and assume that $\hat{\theta}_i \overset{\text{ind}}{\sim} (\theta_i, \psi_i = \hat{D}_i/(4n_i))$, where $\hat{D}_i$ is the estimated design effect (deff) of $\hat{P}_i$ and n_i is the sample size from the ith area. Note that $V(\hat{\theta}_i) \approx D_i/(4n_i)$, where $D_i = V_i[P_i(1-P_i)/n_i]^{-1}$ is the population deff of $\hat{P}_i$ and $V_i = V(\hat{P}_i)$. The estimated deff $\hat{D}_i$ was calculated as $\hat{D}_i = \hat{V}_i[\hat{P}_i(1-\hat{P}_i)/n_i]^{-1}$, where $\hat{V}_i$ is the NHIS variance estimate of $\hat{P}_i$.

For the covariate selection, the largest 15 states with small ψ_i-values were chosen. This permitted the use of standard selection methods for linear regression models because the basic area level model for those states may be treated as $\hat{\theta}_i \approx \mathbf{z}_i^T\boldsymbol{\beta} + v_i$ with $v_i \overset{\text{iid}}{\sim} (0, \sigma_v^2)$. Based on C_p, R^2, and adjusted R^2 criteria used in SAS, the following covariates were selected: $z_1 = 1, z_2 =$1995 Bachelor's degree completion for population aged 25+, and $z_3 =$1995 health insurance coverage. The simple moment estimator $\hat{\sigma}_{vs}^2 = \max(\tilde{\sigma}_{vs}^2, 0)$ was used to estimate σ_v^2, where $\tilde{\sigma}_{vs}^2$ is given by (6.1.15) with $m = 51$ and $p = 3$. The resulting weights $\hat{\gamma}_i = \hat{\sigma}_v^2/(\hat{\sigma}_v^2 + \psi_i)$ varied from 0.09 (South Dakota) to 0.95 (California). Using the estimated weights $\hat{\gamma}_i$, the EB (EBLUP) estimates $\hat{\theta}_i^{\text{EB}}$ were computed for each of the small areas, which in turn provided the estimates of the proportions P_i as $\tilde{P}_i^{\text{EB}} = \sin^2(\hat{\theta}_i^{\text{EB}})$. Note that $\tilde{P}_i^{\text{EB}}$ is not equal to the true EB estimate $\hat{P}_i^{\text{EB}} = E(P_i|\hat{\theta}_i, \hat{\boldsymbol{\beta}}, \hat{\sigma}_v^2)$, but it simplifies the computations.

MSE estimates of the estimated proportions $\tilde{P}_i^{\text{EB}}$ were computed from the jackknife estimates $\text{mse}_J(\hat{\theta}_i^{\text{EB}})$ given by (9.2.15). Using Taylor linearization, we have $\text{mse}_J(\tilde{P}_i^{\text{EB}}) \approx 4\tilde{P}_i^{\text{EB}}(1 - \tilde{P}_i^{\text{EB}})\text{mse}_J(\hat{\theta}_i^{\text{EB}}) =: s_J^2(\tilde{P}_i^{\text{EB}})$. Performance of $\tilde{P}_i^{\text{EB}}$ relative to the direct estimate $\hat{P}_i$ was measured by the percent improvement, $\text{PI}_i = 100[s(\hat{P}_i) - s_J(\tilde{P}_i^{\text{EB}})]/s(\hat{P}_i)$, where $s^2(\hat{P}_i) = \hat{V}_i$, the NHIS variance estimate of $\hat{P}_i$. Values of PI_i indicated that the improvement is quite substantial (30% to 55%) for small states (e.g., South Dakota and Vermont). On the other hand, the improvement is small for large states (e.g., California and Texas), as expected. Note that $\hat{\theta}_i^{\text{EB}}$ gives more weight, $\hat{\gamma}_i$, to the direct estimate $\hat{\theta}_i$ for large states (e.g., $\hat{\gamma}_i = 0.95$ for California and $\hat{\gamma}_i = 0.94$ for Texas).

9.2.3 Approximation to Posterior Variance

In Section 9.2.2, we studied MSE estimation, but alternative measures of variability associated with $\hat{\theta}_i^{\text{EB}}$ have also been proposed. Those measures essentially use a HB approach and provide approximations to the posterior variance of θ_i, denoted by $V(\theta_i|\hat{\boldsymbol{\theta}})$, based on a prior distribution on the model parameters $\boldsymbol{\beta}$ and σ_v^2.

If the model parameters $\boldsymbol{\beta}$ and σ_v^2 are given, then the posterior (or conditional) distribution $f(\boldsymbol{\theta}|\hat{\boldsymbol{\theta}})$ is completely known and it provides a basis for inference on $\boldsymbol{\theta} = (\theta_1, \ldots, \theta_m)^T$. In particular, the Bayes estimator $\hat{\theta}_i^B = E(\theta_i|\hat{\theta}_i, \boldsymbol{\beta}, \sigma_v^2)$ is used to estimate the realized value of θ_i, and the posterior variance $V(\theta_i|\hat{\theta}_i, \boldsymbol{\beta}, \sigma_v^2) = g_{1i}(\sigma_v^2) = \gamma_i\psi_i$ is used to measure the variability associated with θ_i. The posterior variance is identical to $\text{MSE}(\hat{\theta}_i^B)$, and $\hat{\theta}_i^B$ is the BP estimator of θ_i. Therefore, the frequentist approach agrees with the Bayesian approach for the basic area level model when the model parameters are known. This agreement also holds for the general linear mixed model, but not necessarily for nonlinear models with random effects.

In practice, the model parameters are not known and the EB approach uses the marginal distribution of the $\hat{\theta}_i$'s, namely, $\hat{\theta}_i \overset{\text{ind}}{\sim} N(\mathbf{z}_i^T\boldsymbol{\beta}, b_i^2\sigma_v^2 + \psi_i)$, to estimate $\boldsymbol{\beta}$ and σ_v^2. A naive EB approach uses the estimated posterior density of θ_i, namely, $N(\hat{\theta}_i^{\text{EB}}, \hat{\gamma}_i\psi_i)$, to make inferences on θ_i. In particular, θ_i is estimated by $\hat{\theta}_i^{\text{EB}}$, and the estimated posterior variance $g_{1i}(\hat{\sigma}_v^2) = \hat{\gamma}_i\psi_i$ is used as a measure of variability. The use of $g_{1i}(\hat{\sigma}_v^2)$, however, leads to severe underestimation of $\text{MSE}(\hat{\theta}_i^{\text{EB}})$. Note that the naive EB approach treats $\boldsymbol{\beta}$ and σ_v^2 as fixed, unknown parameters, so no prior distributions are involved. However, if we adopt an HB approach by treating the model parameters $\boldsymbol{\beta}$ and σ_v^2 as random with a prior density $f(\boldsymbol{\beta}, \sigma_v^2)$, then the posterior mean $\hat{\theta}_i^{\text{HB}} = E(\theta_i|\hat{\boldsymbol{\theta}})$ is used as an estimator of θ_i and the posterior variance $V(\theta_i|\hat{\boldsymbol{\theta}})$ as a measure of variability associated with $\hat{\theta}_i^{\text{HB}}$. We can express $E(\theta_i|\hat{\boldsymbol{\theta}})$ as

$$E(\theta_i|\hat{\boldsymbol{\theta}}) = E_{\boldsymbol{\beta},\sigma_v^2}[E(\theta_i|\hat{\theta}_i, \boldsymbol{\beta}, \sigma_v^2)] = E_{\boldsymbol{\beta},\sigma_v^2}[\hat{\theta}_i^B(\boldsymbol{\beta}, \sigma_v^2)] \tag{9.2.16}$$

and $V(\theta_i|\hat{\boldsymbol{\theta}})$ as

$$V(\theta_i|\hat{\boldsymbol{\theta}}) = E_{\boldsymbol{\beta},\sigma_v^2}[V(\theta_i|\hat{\theta}_i,\boldsymbol{\beta},\sigma_v^2)] + V_{\boldsymbol{\beta},\sigma_v^2}[E(\theta_i|\hat{\theta}_i,\boldsymbol{\beta},\sigma_v^2)]$$
$$= E_{\boldsymbol{\beta},\sigma_v^2}[g_{1i}(\sigma_v^2)] + V_{\boldsymbol{\beta},\sigma_v^2}[\hat{\theta}_i^B(\boldsymbol{\beta},\sigma_v^2)], \quad (9.2.17)$$

where $E_{\boldsymbol{\beta},\sigma_v^2}$ and $V_{\boldsymbol{\beta},\sigma_v^2}$, respectively, denote the expectation and the variance with respect to the posterior distribution of $\boldsymbol{\beta}$ and σ_v^2, given $\hat{\boldsymbol{\theta}}$, that is, $f(\boldsymbol{\beta},\sigma_v^2|\hat{\boldsymbol{\theta}})$.

For large m, we can approximate the last term of (9.2.16) by $\hat{\theta}_i^B(\hat{\boldsymbol{\beta}},\hat{\sigma}_v^2) = \hat{\theta}_i^{\text{EB}}$, where $\hat{\boldsymbol{\beta}}$ and $\hat{\sigma}_v^2$ are ML (REML) estimators. More precisely, we have $E(\theta_i|\hat{\boldsymbol{\theta}}) = \hat{\theta}_i^B(\hat{\boldsymbol{\beta}},\hat{\sigma}_v^2)[1+O(m^{-1})]$, regardless of the prior $f(\boldsymbol{\beta},\sigma_v^2)$. Hence, the EB estimator $\hat{\theta}_i^{\text{EB}}$ tracks $E(\theta_i|\hat{\boldsymbol{\theta}})$ well. However, the naive EB measure of variability, $g_{1i}(\hat{\sigma}_v^2)$, provides only a first-order approximation to the first variance term $E_{\boldsymbol{\beta},\sigma_v^2}[g_{1i}(\sigma_v^2)]$ on the right-hand side of (9.2.17). The second variance term, $V_{\boldsymbol{\beta},\sigma_v^2}[\hat{\theta}_i^B(\boldsymbol{\beta},\sigma_v^2)]$, accounts for the uncertainty about the model parameters, and the naive EB approach ignores this uncertainty. As a result, the naive EB approach can lead to severe underestimation of the true posterior variance, $V(\theta_i|\hat{\boldsymbol{\theta}})$. Note that the naive measure of variability, $g_{1i}(\hat{\sigma}_v^2)$, also underestimates $\text{MSE}(\hat{\theta}_i^{\text{EB}})$.

The HB approach may be used to evaluate the posterior mean, $E(\theta_i|\hat{\boldsymbol{\theta}})$, and the posterior variance, $V(\theta_i|\hat{\boldsymbol{\theta}})$, exactly, for any specified prior on the model parameters $\boldsymbol{\beta}$ and σ_v^2. Moreover, it can handle complex small area models (see Chapter 10). Typically, "improper" priors that reflect lack of information on the model parameters are used in the HB calculations; for example, $f(\boldsymbol{\beta},\sigma_v^2) \propto 1$ may be used for the basic area level model. In the EB context, two methods of approximating the posterior variance, regardless of the prior, have been proposed. The first method, based on bootstrap resampling, attempts to account for the underestimation of $V(\theta_i|\hat{\boldsymbol{\theta}})$ imitating the decomposition (9.2.17) of the posterior variance $V(\theta_i|\hat{\boldsymbol{\theta}})$. In the bootstrap method, a large number, B, of independent samples $\{\hat{\theta}_i^*(b); b = 1, \ldots, B\}$ are first drawn from the estimated marginal distribution, $N(\mathbf{z}_i^T\hat{\boldsymbol{\beta}}, b_i^2\hat{\sigma}_v^2 + \psi_i), i = 1, \ldots, m$. Estimates $\{\hat{\boldsymbol{\beta}}^*(b), \hat{\sigma}_v^{*2}(b)\}$ and EB estimates $\hat{\theta}_i^{*\text{EB}}(b)$ are then computed from the bootstrap data $\{(\hat{\theta}_i^*(b), \mathbf{z}_i); i = 1, \ldots, m\}$. This leads to the following approximation to the posterior variance:

$$V_{\text{LL}}(\theta_i|\hat{\boldsymbol{\theta}}) = \frac{1}{B}\sum_{b=1}^{B} g_{1i}(\hat{\sigma}_v^{*2}(b)) + \frac{1}{B}\sum_{b=1}^{B}[\hat{\theta}_i^{*\text{EB}}(b) - \hat{\theta}_i^{*\text{EB}}(\cdot)]^2, \quad (9.2.18)$$

where $\hat{\theta}_i^{*\text{EB}}(\cdot) = B^{-1}\sum_{b=1}^{B}\hat{\theta}_i^{*\text{EB}}(b)$ (see Laird and Louis 1987). Note that (9.2.18) has the same form as the decomposition (9.2.17) of the posterior variance. The last term of (9.2.18) is designed to account for the uncertainty about the model parameters. The bootstrap method uses $\hat{\theta}_i^{\text{EB}}$ as the estimator of θ_i and $V_{\text{LL}}(\theta_i|\hat{\boldsymbol{\theta}})$ as a measure of its variability.

Butar and Lahiri (2003) studied the performance of the bootstrap measure, $V_{\text{LL}}(\theta_i|\hat{\boldsymbol{\theta}})$, as an estimator of $\text{MSE}(\hat{\theta}_i^{\text{EB}})$. They obtained an analytical approximation to (9.2.18) by letting $B \to \infty$. Under REML estimation, we have

$$V_{\rm LL}(\theta_i|\hat{\boldsymbol{\theta}}) \approx g_{1i}(\hat{\sigma}_v^2) + g_{2i}(\hat{\sigma}_v^2) + g_{3i}^*(\hat{\sigma}_v^2, \hat{\theta}_i) =: \tilde{V}_{\rm LL}(\theta_i|\hat{\boldsymbol{\theta}}), \tag{9.2.19}$$

for large m, where $g_{2i}(\sigma_v^2)$ and $g_{3i}^*(\hat{\sigma}_v^2, \hat{\theta}_i)$ are given by (6.1.9) and (6.2.4), respectively. Comparing (9.2.19) to the second-order unbiased MSE estimator (6.2.6), it follows that $V_{\rm LL}(\theta_i|\hat{\boldsymbol{\theta}})$ is not second-order unbiased for MSE($\hat{\theta}_i^{\rm EB}$). In particular,

$$E[\tilde{V}_{\rm LL}(\theta_i|\hat{\boldsymbol{\theta}})] = {\rm MSE}(\hat{\theta}_i) - g_{3i}(\sigma_v^2). \tag{9.2.20}$$

A bias-corrected MSE estimator is therefore given by

$${\rm mse}_{\rm BL}(\hat{\theta}_i^{\rm EB}) = \tilde{V}_{\rm LL}(\theta_i|\hat{\boldsymbol{\theta}}) + g_{3i}(\hat{\sigma}_v^2). \tag{9.2.21}$$

This estimator is second-order unbiased and also area-specific. It is identical to ${\rm mse}_2(\hat{\theta}_i^H)$ given by (6.2.6).

In the second method, due to Kass and Steffey (1989), $\hat{\theta}_i^{\rm EB}$ is taken as the estimator of θ_i, noting that

$$E(\theta_i|\hat{\boldsymbol{\theta}}) = \hat{\theta}_i^{\rm EB}[1 + O(m^{-1})], \tag{9.2.22}$$

but a positive correction term is added to the estimated posterior variance, $g_{1i}(\hat{\sigma}_v^2)$, to account for the underestimation of $V(\theta_i|\hat{\boldsymbol{\theta}})$. This positive term depends on the information matrix and the partial derivatives of $\hat{\theta}_i^B$ evaluated at the ML (REML) estimates $\hat{\boldsymbol{\beta}}$ and $\hat{\sigma}_v^2$. The following first-order approximation to $V(\theta_i|\hat{\boldsymbol{\theta}})$ is obtained, noting that the information matrix for $\boldsymbol{\beta}$ and σ_v^2 is block diagonal:

$$\begin{aligned} V_{\rm KS}(\theta_i|\hat{\boldsymbol{\theta}}) = {} & g_{1i}(\hat{\sigma}_v^2) + (\partial\hat{\theta}_i^B/\partial\boldsymbol{\beta})^T\overline{V}(\hat{\boldsymbol{\beta}})(\partial\hat{\theta}_i^B/\partial\boldsymbol{\beta})|_{\boldsymbol{\beta}=\hat{\boldsymbol{\beta}},\sigma_v^2=\hat{\sigma}_v^2} \\ & + (\partial\hat{\theta}_i^B/\partial\sigma_v^2)^2\overline{V}(\hat{\sigma}_v^2)|_{\boldsymbol{\beta}=\hat{\boldsymbol{\beta}},\sigma_v^2=\hat{\sigma}_v^2}, \end{aligned} \tag{9.2.23}$$

where $\overline{V}(\hat{\boldsymbol{\beta}}) = [\sum_{i=1}^m \mathbf{z}_i\mathbf{z}_i^T/(\psi_i + \sigma_v^2 b_i^2)]^{-1}$ is the asymptotic covariance matrix of $\hat{\boldsymbol{\beta}}$, $\overline{V}(\hat{\sigma}_v^2) = [\mathcal{I}(\sigma_v^2)]^{-1}$ is the asymptotic variance of $\hat{\sigma}_v^2$ with $\mathcal{I}(\sigma_v^2)$ given by (6.1.17),

$$\frac{\partial\hat{\theta}_i^B}{\partial\boldsymbol{\beta}} = (1 - \gamma_i)\mathbf{z}_i \tag{9.2.24}$$

and

$$\frac{\partial\hat{\theta}_i^B}{\partial\sigma_v^2} = \frac{\psi_i b_i^2}{(\psi_i + \sigma_v^2 b_i^2)^2}(\hat{\theta}_i - \mathbf{z}_i^T\boldsymbol{\beta}). \tag{9.2.25}$$

After simplification, (9.2.23) reduces to

$$V_{\rm KS}(\theta_i|\hat{\boldsymbol{\theta}}) = g_{1i}(\hat{\sigma}_v^2) + g_{2i}(\sigma_v^2) + g_{3i}^*(\sigma_v^2, \hat{\theta}_i), \tag{9.2.26}$$

which is identical to $\tilde{V}_{LL}(\theta_i|\hat{\theta})$. Therefore, $V_{KS}(\theta_i|\hat{\theta})$ is also not second-order unbiased for MSE($\hat{\theta}_i$). Kass and Steffey (1989) have given a more accurate approximation to $V(\theta_i|\hat{\theta})$. This approximation ensures that the neglected terms are of lower order than m^{-1}, but it depends on the prior density, $f(\boldsymbol{\beta}, \sigma_v^2)$, unlike the first-order approximation (9.2.23). If a prior needs to be specified, then it may be better to use the HB approach (see Chapter 10) because it is free of asymptotic approximations. Singh, Stukel, and Pfeffermann (1998) studied Kass and Steffey's (1989) approximations in the context of small area estimation.

Kass and Steffey's (1989) method is applicable to general functions $\phi_i = h(\theta_i)$, but the calculations might require the use of Monte Carlo or numerical integration. Similarly as in (9.2.22), we have

$$E(\phi_i|\hat{\theta}) = \hat{\phi}_i^{\text{EB}}[1 + O(m^{-1})] \tag{9.2.27}$$

and the first-order approximation to $V(\phi_i|\hat{\theta})$ is then given by

$$\begin{aligned} V_{KS}(\phi_i|\hat{\theta}) = & \, V(\phi_i|\hat{\theta}, \hat{\boldsymbol{\beta}}, \hat{\sigma}_v^2) + (\partial\hat{\phi}_i^B/\partial\boldsymbol{\beta})^T\overline{V}(\hat{\boldsymbol{\beta}})(\partial\hat{\phi}_i^B/\partial\boldsymbol{\beta})|_{\boldsymbol{\beta}=\hat{\boldsymbol{\beta}},\sigma_v^2=\hat{\sigma}_v^2} \\ & + [(\partial\hat{\phi}_i^B/\partial\sigma_v^2)^2\overline{V}(\hat{\sigma}_v^2)]|_{\boldsymbol{\beta}=\hat{\boldsymbol{\beta}},\sigma_v^2=\hat{\sigma}_v^2}, \end{aligned} \tag{9.2.28}$$

where $\hat{\phi}_i^B = h(\hat{\theta}_i^B)$ and $V(\phi_i|\hat{\theta}, \hat{\boldsymbol{\beta}}, \hat{\sigma}_v^2)$ denotes $V(\phi_i|\hat{\theta}, \boldsymbol{\beta}, \sigma_v^2)$ evaluated at $\boldsymbol{\beta} = \hat{\boldsymbol{\beta}}$ and $\sigma_v^2 = \hat{\sigma}_v^2$.

Morris (1983a) studied the basic area level model without auxiliary information ($\mathbf{z}_i = 1$ for all i) and in the case of equal sampling variances, $\psi_i = \psi$ for all i, with known ψ. This model may be expressed as $\hat{\theta}_i|\theta_i \overset{\text{ind}}{\sim} N(\theta_i, \psi)$ and $\theta_i|\mu, \sigma_v^2 \overset{\text{iid}}{\sim} N(\mu, \sigma_v^2)$, $i = 1, \ldots, m$. The Bayes estimator (9.2.1) reduces in this case to

$$\hat{\theta}_i^B = \gamma\hat{\theta}_i + (1 - \gamma)\mu, \tag{9.2.29}$$

where $\gamma = \sigma_v^2/(\sigma_v^2 + \psi)$. We first obtain the HB estimator, $\hat{\theta}_i^{\text{HB}}(\sigma_v^2)$, for a given σ_v^2, assuming that $\mu \sim$ uniform$(-\infty, \infty)$ to reflect the absence of prior information on μ; that is, $f(\mu) =$ constant. It is easy to verify that

$$\mu|\hat{\theta}, \sigma_v^2 \sim N(\hat{\theta}_\cdot, m^{-1}(\sigma_v^2 + \psi)), \tag{9.2.30}$$

where $\hat{\theta}_\cdot = \sum_{i=1}^m \hat{\theta}_i/m$ (see Section 9.12.2). It now follows from (9.2.29) that

$$\begin{aligned} \hat{\theta}_i^{\text{HB}}(\sigma_v^2) &= \gamma\hat{\theta}_i + (1 - \gamma)E(\mu|\hat{\theta}, \sigma_v^2) \\ &= \hat{\theta}_i - (1 - \gamma)(\hat{\theta}_i - \hat{\theta}_\cdot). \end{aligned} \tag{9.2.31}$$

The HB estimator $\hat{\theta}_i^{\text{HB}}(\sigma_v^2)$ is identical to the BLUP estimator $\tilde{\theta}_i^H$.

The posterior variance of θ_i given $\hat{\boldsymbol{\theta}}$ and σ_v^2 is given by

$$\begin{aligned} V(\theta_i|\hat{\boldsymbol{\theta}},\sigma_v^2) &= E_\mu[V(\theta_i|\hat{\theta}_i,\mu,\sigma_v^2)] + V_\mu[E(\theta_i|\hat{\theta}_i,\mu,\sigma_v^2)] \\ &= E_\mu[g_1(\sigma_v^2)] + V_\mu(\hat{\theta}_i^B), \end{aligned} \tag{9.2.32}$$

where E_μ and V_μ, respectively, denote the expectation and the variance with respect to the posterior distribution of μ given $\hat{\boldsymbol{\theta}}$ and σ_v^2, and

$$g_1(\sigma_v^2) = \gamma\psi = \psi - (1-\gamma)\psi. \tag{9.2.33}$$

It now follows from (9.2.29), (9.2.30), and (9.2.32) that

$$V(\theta_i|\hat{\boldsymbol{\theta}},\sigma_v^2) = g_1(\sigma_v^2) + g_2(\sigma_v^2), \tag{9.2.34}$$

where

$$g_2(\sigma_v^2) = (1-\gamma)^2(\sigma_v^2+\psi)/m = (1-\gamma)\psi/m. \tag{9.2.35}$$

The conditional posterior variance (9.2.34) is identical to the MSE of BLUP estimator.

To take account of the uncertainty associated with σ_v^2, Morris (1983a) assumed that $\sigma_v^2 \sim$ uniform$[0,\infty)$; that is, $f(\sigma_v^2)$ =constant. The resulting posterior density, $f(\sigma_v^2|\hat{\boldsymbol{\theta}})$, however, does not yield closed-form expressions for $E[(1-\gamma)|\hat{\boldsymbol{\theta}})]$ and $E[(1-\gamma)^2|\hat{\boldsymbol{\theta}}]$. The latter terms are needed in the evaluation of

$$\hat{\theta}_i^{\text{HB}} = E_{\sigma_v^2}[\hat{\theta}_i^{\text{HB}}(\sigma_v^2)] \tag{9.2.36}$$

and the posterior variance

$$V(\theta_i|\hat{\boldsymbol{\theta}}) = E_{\sigma_v^2}[V(\theta_i|\hat{\boldsymbol{\theta}},\sigma_v^2)] + V_{\sigma_v^2}[\hat{\theta}_i^{\text{HB}}(\sigma_v^2)], \tag{9.2.37}$$

where $E_{\sigma_v^2}$ and $V_{\sigma_v^2}$, respectively, denote the expectation and the variance with respect to $f(\sigma_v^2|\hat{\boldsymbol{\theta}})$. Morris (1983a) evaluated $\hat{\theta}_i^{\text{HB}}$ and $V(\theta_i|\hat{\boldsymbol{\theta}})$ numerically.

Morris (1983a) also obtained closed-form approximations to $E[(1-\gamma)|\hat{\boldsymbol{\theta}}]$ and $E[(1-\gamma)^2|\hat{\boldsymbol{\theta}}]$ by expressing them as ratios of definite integrals over the range $[0,1)$ and then replacing $\int_0^1$ by $\int_0^\infty$. This method is equivalent to assuming that $\sigma_v^2+\psi$ is uniform on $[0,\infty)$. We have

$$E_{\sigma_v^2}[(1-\gamma)|\hat{\boldsymbol{\theta}}] \approx \psi(m-3)/S = 1-\gamma^*, \tag{9.2.38}$$

where $S = \sum_{i=1}^m (\hat{\theta}_i - \hat{\theta}_\cdot)^2$. Hence, it follows from (9.2.31) and (9.2.36) that

$$\hat{\theta}_i^{\text{HB}} \approx \gamma^*\hat{\theta}_i + (1-\gamma^*)\hat{\theta}_\cdot\,, \tag{9.2.39}$$

which is identical to $\hat{\theta}_i^{\text{EB}}$ given by (9.2.5) with $\mathbf{z}_i = 1$ for all i. Turning to the approximation to the posterior variance $V(\theta_i|\hat{\boldsymbol{\theta}})$, we first note that

$$E_{\sigma_v^2}[(1-\gamma)^2|\hat{\boldsymbol{\theta}}] \approx 2\psi^2(m-3)/S^2 = 2(1-\gamma^*)^2/(m-3). \tag{9.2.40}$$

It now follows from (9.2.31), (9.2.33), (9.2.34), and (9.2.37) that $V(\theta_i|\hat{\boldsymbol{\theta}}) \approx V_M(\theta_i|\hat{\boldsymbol{\theta}})$, where

$$V_M(\theta_i|\hat{\boldsymbol{\theta}}) = \psi - \frac{m-1}{m}(1-\gamma^*)\psi + \frac{2(1-\gamma^*)^2}{m-3}(\hat{\theta}_i - \hat{\theta}_\cdot)^2 \tag{9.2.41}$$

$$= \psi\gamma^* + \frac{(1-\gamma^*)\psi}{m} + \frac{2(1-\gamma^*)^2}{m-3}(\hat{\theta}_i - \hat{\theta}_\cdot)^2. \tag{9.2.42}$$

Morris (1983a) obtained (9.2.41). The alternative form (9.2.42) shows that $V_M(\theta_i|\hat{\boldsymbol{\theta}})$ is asymptotically equivalent to $\text{mse}_2(\hat{\theta}_i^H)$ up to terms of order m^{-1}, noting that

$$\psi\gamma^* = \psi\hat{\gamma} + \frac{2}{m-1}(1-\hat{\gamma})\psi \tag{9.2.43}$$

and

$$\text{mse}_2(\hat{\theta}_i^H) = \psi\hat{\gamma} + \frac{(1-\hat{\gamma})\psi}{m} + \frac{2(1-\hat{\gamma})\psi}{m} + \frac{2(1-\hat{\gamma})^2}{m}(\hat{\theta}_i - \hat{\theta}_\cdot)^2 \tag{9.2.44}$$

$$= g_1(\hat{\sigma}_v^2) + g_2(\hat{\sigma}_v^2) + g_3(\hat{\sigma}_v^2) + g_{3i}^*(\hat{\sigma}_v^2, \hat{\theta}_i),$$

where $1-\hat{\gamma} = \psi(m-1)/S$ and $\hat{\sigma}_v^2$ is the REML estimator of σ_v^2. This result shows that the frequentist inference, based on the MSE estimator, and the HB inference, based on the approximation to posterior variance, agree up to terms of order m^{-1}. Neglected terms in this comparison are of lower order than m^{-1}.

The approximation (9.2.41) to the posterior variance extends to the regression model, $\theta_i = \mathbf{z}_i^T\boldsymbol{\beta} + v_i$, with equal sampling variances $\psi_i = \psi$ (Morris 1983b). We have

$$V_M(\theta_i|\hat{\boldsymbol{\theta}}) \approx \psi - \frac{m-p}{m}(1-\gamma^*)\psi + \frac{2(1-\gamma^*)^2}{m-p-2}(\hat{\theta}_i - \mathbf{z}_i^T\hat{\boldsymbol{\beta}}_{\text{LS}})^2, \tag{9.2.45}$$

where $1-\gamma^*$ is given by (9.2.6). Note that the approximation to $\hat{\theta}_i^{\text{HB}}$ is identical to the EB estimator given by (9.2.5). Morris (1983b) also proposed an extension to the case of unequal sampling variances ψ_i but with $b_i = 1$. This extension is obtained from (9.2.45) by changing ψ to ψ_i, $1-\gamma^*$ to $1-\gamma_i^*$ given by (9.2.7), $\hat{\boldsymbol{\beta}}_{\text{LS}}$ to the weighted least squares estimator $\hat{\boldsymbol{\beta}}$, and finally multiplying the last term of (9.2.45) by the factor $(\overline{\psi}_w + \hat{\sigma}_v^2)/(\psi_i + \hat{\sigma}_v^2)$, where $\overline{\psi}_w = \sum_{i=1}^m (\hat{\sigma}_v^2 + \psi_i)^{-1}\psi_i / \sum_{i=1}^m (\hat{\sigma}_v^2 + \psi_i)^{-1}$ and $\hat{\sigma}_v^2$ is either the REML estimator or the estimator obtained by solving the moment equation (9.2.8) iteratively. Note that the adjustment factor $(\overline{\psi}_w + \hat{\sigma}_v^2)/(\psi_i + \hat{\sigma}_v^2)$

reduces to 1 in the case of equal sampling variances, $\psi_i = \psi$. Theoretical justification of this extension remains to be studied.

Example 9.2.2. Simulation Study. Jiang, Lahiri, and Wan (2002) reported simulation results on the relative performance of estimators of MSE($\hat{\theta}_i^{\text{EB}}$) under the simple mean model $\hat{\theta}_i|\theta_i \stackrel{\text{ind}}{\sim} N(\theta_i, \psi)$ and $\theta_i \stackrel{\text{iid}}{\sim} N(\mu, \sigma_v^2), i = 1, \ldots, m$, where $\hat{\theta}_i^{\text{EB}}$ is given by (9.2.39), the approximation of Morris (1983a) to $\hat{\theta}_i^{\text{HB}}$. The estimators studied include (i) the Prasad–Rao estimator $\text{mse}_{\text{PR}}(\hat{\theta}_i^{\text{EB}}) = g_1(\hat{\sigma}_v^2) + g_2(\hat{\sigma}_v^2) + 2g_3(\hat{\sigma}_v^2)$, where $g_1(\sigma_v^2)$ and $g_2(\sigma_v^2)$ are given by (9.2.33) and (9.2.35), respectively, and $g_3(\sigma_v^2) = 2\psi(1-\gamma)/m$; (ii) the area-specific estimator $\text{mse}_2(\hat{\theta}_i^{\text{EB}})$, which is equivalent to $\text{mse}_{\text{BL}}(\hat{\theta}_i^{\text{EB}})$ given by (9.2.21); (iii) the approximation to Laird and Louis bootstrap, $\tilde{V}_{\text{LL}}(\theta_i|\hat{\boldsymbol{\theta}})$, given by (9.2.19); (iv) the jackknife estimator $\text{mse}_J(\hat{\theta}_i^{\text{EB}})$; (v) the Morris estimator, $V_M(\theta_i|\hat{\boldsymbol{\theta}})$, given by (9.2.45), and (vi) the naive estimator $\text{mse}_N(\hat{\theta}_i^{\text{EB}}) = g_1(\hat{\sigma}_v^2) + g_2(\hat{\sigma}_v^2)$, which ignores the variability associated with $\hat{\sigma}_v^2$. Average relative bias, $\overline{\text{RB}}$, was used as the criterion for comparison of the estimators, where $\overline{\text{RB}} = m^{-1}\sum_{i=1}^m \text{RB}_i, \text{RB}_i = [E(\text{mse}_i) - \text{MSE}_i]/\text{MSE}_i$, and mse_i denotes an estimator of MSE for the ith area and $\text{MSE}_i = \text{MSE}(\hat{\theta}_i^{\text{EB}})$. Note that $\text{mse}_{\text{PR}}(\hat{\theta}_i^{\text{EB}})$, $\text{mse}_{\text{BL}}(\hat{\theta}_i^{\text{EB}})$, $V_M(\theta_i|\hat{\boldsymbol{\theta}})$, and $\text{mse}_J(\hat{\theta}_i^{\text{EB}})$ are second-order unbiased unlike $\tilde{V}_{\text{LL}}(\theta_i|\hat{\boldsymbol{\theta}})$.

One million samples were simulated from the model by letting $\mu = 0$ without loss of generality, $\sigma_v^2 = \psi = 1$ and $m = 30, 60$ and 90. Values of $\overline{\text{RB}}$ calculated from the simulated samples are reported in Table 9.1. As expected, the naive estimator $\text{mse}_N(\hat{\theta}_i^{\text{EB}})$ and, to a lesser extent, the Laird and Louis estimator $\tilde{V}_{\text{LL}}(\theta_i|\hat{\boldsymbol{\theta}})$ lead to underestimation of MSE. The remaining estimators are nearly unbiased with $\overline{\text{RB}}$ less than 1%. The performance of all the estimators improves as m increases, that is, $\overline{\text{RB}}$ decreases as m increases.

9.2.4 *EB Confidence Intervals

We now turn to EB CIs on the individual small area parameters, θ_i, under the basic area level model (i) $\hat{\theta}_i|\theta_i \stackrel{\text{ind}}{\sim} N(\theta_i, \psi_i)$ and (ii) $\theta_i \stackrel{\text{ind}}{\sim} N(\mathbf{z}_i^T\boldsymbol{\beta}, \sigma_v^2), i = 1, \ldots, m$. We define

TABLE 9.1 Percent Average Relative Bias ($\overline{\text{RB}}$) of MSE Estimators

MSE Estimators	$m = 30$	$m = 60$	$m = 90$
mse_N	−8.4	−4.8	−3.2
$\tilde{V}_{\text{LL}}(\theta_i\|\hat{\boldsymbol{\theta}})$	−3.8	−2.9	−2.0
mse_J	0.6	0.2	0.1
$V_M(\theta_i\|\hat{\boldsymbol{\theta}})$	0.7	−0.1	0.0
mse_{BL}	0.1	−0.2	−0.1
mse_{PR}	0.7	−0.1	0.0

Source: Adapted from Table 1 in Jiang, Lahiri, and Wan (2002).

a $(1-\alpha)$-level CI on θ_i as $I_i(\alpha)$ such that

$$P[\theta_i \in I_i(\alpha)] = 1-\alpha \tag{9.2.46}$$

for all possible values of $\boldsymbol{\beta}$ and σ_v^2, where $P(\cdot)$ refers to the assumed model. A normal theory EB interval is given by

$$I_i^{\text{NT}}(\alpha) = [\hat{\theta}_i^{\text{EB}} - z_{\alpha/2}\{s(\hat{\theta}_i^{\text{EB}})\}, \hat{\theta}_i^{\text{EB}} + z_{\alpha/2}\{s(\hat{\theta}_i^{\text{EB}})\}], \tag{9.2.47}$$

where $z_{\alpha/2}$ is the upper $(\alpha/2)$-point of $N(0,1)$ and $s^2(\hat{\theta}_i^{\text{EB}})$ is a second-order unbiased estimator of $\text{MSE}(\hat{\theta}_i^{\text{EB}})$. The interval $I_i^{\text{NT}}(\alpha)$, however, is only first-order correct in the sense of $P[\theta_i \in I_i^{\text{NT}}(\alpha)] = 1-\alpha+O(m^{-1})$ (see Diao et al. 2014).

It is desirable to construct more accurate second-order correct intervals $I_i(\alpha)$ that satisfy $P[\theta_i \in I_i(\alpha)] = 1-\alpha+o(m^{-1})$. Section 9.2.4 of Rao (2003a) reviews second-order correct intervals for the special case of equal sampling variances $\psi_i = \psi$ for all i. Here, we provide a brief account of recent methods of constructing second-order correct intervals for the general case of unequal ψ_i.

(i) *Parametric Bootstrap Intervals*

In Section 6.2.4, bootstrap data $\{(\hat{\theta}_i^*, \mathbf{z}_i); i = 1, \ldots, m\}$ are generated by first drawing θ_i^* from $N(\mathbf{z}_i^T\hat{\boldsymbol{\beta}}, \hat{\sigma}_v^2)$ and then drawing $\hat{\theta}_i^*$ from $N(\theta_i^*, \psi_i)$. Here, we use the bootstrap data to approximate the distribution of

$$t = (\hat{\theta}_i^{\text{EB}} - \theta_i)/[g_{1i}(\hat{\sigma}_v^2)]^{1/2}, \tag{9.2.48}$$

where $g_{1i}(\sigma_v^2) = \gamma_i\psi_i$ with $\gamma_i = \sigma_v^2/(\sigma_v^2+\psi_i)$. Denote the EB estimator of θ_i and the estimator of σ_v^2 obtained from the bootstrap data by $\hat{\theta}_{i*}^{\text{EB}}$ and $\hat{\sigma}_{v*}^2$, respectively, and the corresponding value of t by

$$t^* = (\hat{\theta}_{i*}^{\text{EB}} - \theta_i^*)/[g_{1i}(\hat{\sigma}_{v*}^2)]^{1/2}. \tag{9.2.49}$$

When $\hat{\sigma}_v^2$ computed from the data or $\hat{\sigma}_{v*}^2$ in any bootstrap sample is zero, we set it to a small threshold such as 0.01. Alternatively, any of the strictly positive estimators of σ_v^2 given in Section 6.4.2 may be used.

We approximate the distribution of t by the known bootstrap distribution of t^* and calculate quantiles q_1 and q_2 $(q_1 < q_2)$ satisfying $P_*(q_1 \le t^* \le q_2) = 1-\alpha$, where P_* refers to the bootstrap distribution. A bootstrap interval on θ_i is then obtained from $q_1 \le t \le q_2$ as

$$I_i^{\text{CLL}}(\alpha) = [\hat{\theta}_i^{\text{EB}} - q_2\{g_{1i}(\hat{\sigma}_v^2)\}^{1/2}, \hat{\theta}_i^{\text{EB}} - q_1\{g_{1i}(\hat{\sigma}_v^2)\}^{1/2}]. \tag{9.2.50}$$

The choice of q_1 and q_2 may be based on equal tail probabilities of size $\alpha/2$ or shortest length interval. In practice, we generate a large number, B, of t^*-values, denoted by $t_i^*(1), \ldots, t_i^*(B)$, and determine q_1 and q_2 from the resulting empirical bootstrap distribution. The number of bootstrap samples, B, for obtaining the quantiles q_1 and q_2 should be much larger than the value of B used for bootstrap MSE estimation.

Chatterjee, Lahiri, and Li (2008) proved that, under regularity assumptions,

$$P[\theta_i \in I_i^{\text{CLL}}(\alpha)] = 1 - \alpha + O(m^{-3/2}), \tag{9.2.51}$$

which establishes second-order accuracy of $I_i^{\text{CLL}}(\alpha)$ in terms of coverage. Chatterjee, Lahiri, and Li (CLL) have, in fact, extended the result to construct bootstrap intervals under a general linear mixed model (Section 5.2). Proof of (9.2.51) is highly technical and we refer the reader to Chatterjee et al. (2008) for details.

If $\theta_i = g(\overline{Y}_i)$ is a one-to-one function of the area mean $\overline{Y}_i$, then the corresponding second-order correct bootstrap interval for the area mean $\overline{Y}_i$ is given by $[g^{-1}(c_1), g^{-1}(c_2)]$, where $c_1 = \hat{\theta}_i^{\text{EB}} - q_2\{g_{1i}(\hat{\sigma}_v^2)\}^{1/2}$ and $c_2 = \hat{\theta}_i^{\text{EB}} - q_1\{g_{1i}(\hat{\sigma}_v^2)\}^{1/2}$. For example, if $\theta_i = \log(\overline{Y}_i)$, then $g^{-1}(c_1) = \exp(c_1)$ and $g^{-1}(c_2) = \exp(c_2)$.

Hall and Maiti (2006b) proposed a different parametric bootstrap method to construct CIs on θ_i. We first note that a $(1-\alpha)$-level CI on θ_i, without using the direct estimator $\hat{\theta}_i$, is given by $(\mathbf{z}_i^T\boldsymbol{\beta} - z_{\alpha/2}\,\sigma_v, \mathbf{z}_i^T\boldsymbol{\beta} + z_{\alpha/2}\,\sigma_v)$ if $\boldsymbol{\beta}$ and σ_v are known. Now replacing $\boldsymbol{\beta}$ and σ_v by their estimators $\hat{\boldsymbol{\beta}}$ and $\hat{\sigma}_v$, we get the estimated interval $(\mathbf{z}_i^T\hat{\boldsymbol{\beta}} - z_{\alpha/2}\,\hat{\sigma}_v, \mathbf{z}_i^T\hat{\boldsymbol{\beta}} + z_{\alpha/2}\,\hat{\sigma}_v)$ with coverage probability deviating from the nominal $1-\alpha$ by terms of order m^{-1}. To achieve higher order accuracy, parametric bootstrap is used to find points $b_{\alpha/2}$ and $b_{1-\alpha/2}$ such that $P_*(\theta_i^* \le \mathbf{z}_i^T\hat{\boldsymbol{\beta}} + b_{\alpha/2}\,\hat{\sigma}_v^*) = \alpha/2$ and $P_*(\theta_i^* \le \mathbf{z}_i^T\hat{\boldsymbol{\beta}} + b_{1-\alpha/2}\,\hat{\sigma}_v^*) = 1-\alpha/2$. The resulting bootstrap interval for θ_i, given by $I_i^{\text{HM}}(\alpha) = (\mathbf{z}_i^T\hat{\boldsymbol{\beta}} - b_{\alpha/2}\,\hat{\sigma}_v, \mathbf{z}_i^T\hat{\boldsymbol{\beta}} + b_{1-\alpha/2}\,\hat{\sigma}_v)$, achieves second-order accuracy in the sense $P[\theta_i \in I_i^{\text{HM}}(\alpha)] = 1 - \alpha + o(m^{-1})$, as shown by Hall and Maiti (2006b). However, the Hall–Maiti (HM) bootstrap interval is not centered around the reported point estimator, $\hat{\theta}_i^{\text{EB}}$, unlike the CLL bootstrap interval. On the other hand, note that $I_i^{\text{HM}}(\alpha)$ is based on the regression-synthetic estimator, $\mathbf{z}_i^T\hat{\boldsymbol{\beta}}$, of θ_i, and therefore it can be used for nonsampled areas ℓ for which only $\mathbf{z}_\ell^T\hat{\boldsymbol{\beta}}$ is available for estimating $\theta_\ell, \ell = m+1, \ldots, M$.

Example 9.2.3. Simulation study. Chatterjee et al. (2008) reported the results of a limited simulation study, based on $R = 10,000$ simulation runs, on the relative performance of the normal theory (NT) interval (9.2.47) and the bootstrap interval (9.2.50) in terms of CI coverage and length. The setup used in Example 6.2.1 was employed for this purpose. It consists of the FH model with $\mathbf{z}_i^T\boldsymbol{\beta} = \mu$, and five groups of areas with three areas in each group ($m = 15$). Two different patterns for the sampling variances ψ_i with equal ψ_i's within each group were used: (a) 0.2, 0.4, 0.5, 0.6, 4.0; and (b) 0.4, 0.8, 1.0, 1.2, 8.0. The ratio $\psi_i/(\sigma_v^2 + \psi_i)$ is made equal for the two patterns by choosing $\sigma_v^2 = 1$ for pattern (a) and $\sigma_v^2 = 2$ for pattern (b).

The MSE estimator $s^2(\hat{\theta}_i^{\text{EB}})$ used in the NT interval (9.2.47) is based on the second-order correct MSE estimator (6.2.7), based on the FH moment estimator $\hat{\sigma}_{vm}^2$. The latter estimator of σ_v^2 is also used in the bootstrap interval (9.2.50) based on $B = 1,000$ bootstrap replicates. The NT interval (9.2.47) consistently led to undercoverage for pattern (b) with coverage ranging from 0.84 to 0.90 compared to nominal 0.95. Undercoverage of the NT interval is less severe for pattern (a) with smaller variability: 0.90 to 0.95. On the other hand, the performance of the bootstrap interval (9.2.50) remains stable over the two patterns (a) and (b), with coverage close to the nominal 0.95. In terms of average length, the shortest length bootstrap interval is slightly better than the bootstrap interval based on equal tail probabilities.

(ii) *Closed-Form CIs*

We now study closed-form second-order correct CI that avoids bootstrap calibration and is computationally simpler. Using an expansion of the coverage probability $P[\hat{\theta}_i^{\text{EB}} - z\, s(\hat{\theta}_i^{\text{EB}}) \le \theta_i \le \hat{\theta}_i^{\text{EB}} + z\, s(\hat{\theta}_i^{\text{EB}})]$, Diao et al. (2014) showed that the customary choice $z = z_{\alpha/2}$ leads to a first-order correct interval. However, an adjusted choice $z = z_i(\hat{\sigma}^2_{v\text{RE}})$ gives second-order accuracy in the case of REML estimator $\hat{\sigma}^2_{v\text{RE}}$, where

$$z_i(\hat{\sigma}^2_{v\text{RE}}) = z_{\alpha/2} + \frac{(z^3_{\alpha/2} + z_{\alpha/2})g_{3i}(\hat{\sigma}^2_{v\text{RE}})\psi_i^2}{8k_i^2(\hat{\sigma}^2_{v\text{RE}})(\hat{\sigma}^2_{v\text{RE}} + \psi_i)}. \tag{9.2.52}$$

Here, $k_i(\hat{\sigma}^2_{v\text{RE}}) = g_{1i}(\hat{\sigma}^2_{v\text{RE}}) + g_{2i}(\hat{\sigma}^2_{v\text{RE}})$ and g_{1i}, g_{2i}, and g_{3i} are as defined in Chapter 6. For the special case $\psi_i = \psi$, (9.2.52) reduces to

$$z_i(\hat{\sigma}^2_{v\text{RE}}) = z_{\alpha/2} + \frac{(z^3_{\alpha/2} + z_{\alpha/2})\psi^2}{4m(\hat{\sigma}^2_{v\text{RE}} + \psi/m)^2}, \tag{9.2.53}$$

noting that $k_i(\sigma_v^2) = \sigma_v^2\psi/(\sigma_v^2 + \psi) + \psi^2/[m(\sigma_v^2 + \psi)]$ and $g_{3i}(\sigma_v^2) = 2\psi^2/[m(\sigma_v^2 + \psi)]$. Proof of (9.2.52) is highly technical and we refer the reader to Diao et al. (2014) for details. Result (9.2.52) requires that σ_v^2/ψ_i values are bounded away from zero.

Simulation results under pattern (a) of Chatterjee et al. (2008) showed that the Diao et al. method, based on equal tail probabilities, is comparable to the Chatterjee et al. method in terms of coverage accuracy. Diao et al. (2014) also studied coverage accuracy for the nonnormal case by generating the sampling errors e_i and the random effects v_i from shifted chi-squared distributions with mean 0 and variance ψ_i and mean 0 and variance σ_v^2, respectively. Simulation results indicated that the interval based on (9.2.52) performs better than the normality-based bootstrap interval (9.2.50) in terms of coverage accuracy for that nonnormal case.

Yoshimori and Lahiri (2014c) introduced a new adjusted residual (restricted) maximum-likelihood method to produce second-order correct CIs for the means θ_i. This method is based on the estimator of σ_v^2 maximizing the adjusted residual likelihood $L_i(\sigma_v^2) = h_i(\sigma_v^2)L_R(\sigma_v^2)$ for a specified function $h_i(\sigma_v^2)$, where $L_R(\sigma_v^2)$ is the residual likelihood of σ_v^2 (Section 5.2.4). Note that the choice $h_i(\sigma_v^2) = \sigma_v^2$ is studied in Section 6.4.2 in the context of getting positive estimators of σ_v^2. Denote the resulting estimator of σ_v^2 by $\hat{\sigma}^2_{vh_i}$ and the NT interval based on the pivotal quantity (9.2.48) by

$$I_{ih_i}(\alpha) = [\hat{\theta}_i^{\text{EB}} - z_{\alpha/2}\{g_{1i}(\hat{\sigma}^2_{vh_i})\}^{1/2}, \hat{\theta}_i^{\text{EB}} + z_{\alpha/2}\{g_{1i}(\hat{\sigma}^2_{vh_i})\}^{1/2}]. \tag{9.2.54}$$

Yoshimori and Lahiri (YL) showed that the coverage error term associated with (9.2.54) is $O(m^{-1})$ and depends on $h_i(\sigma_v^2)$. They chose the adjustment function $h_i(\sigma_v^2) = \bar{h}_i(\sigma_v^2)$ such that the $O(m^{-1})$ term vanishes at $\bar{h}_i(\sigma_v^2)$. Then, they found the corresponding estimator of σ_v^2 that maximizes the adjusted likelihood $\bar{L}_i = \bar{h}_i(\sigma_v^2)L_R(\sigma_v^2)$. Denote this estimator by $\hat{\sigma}^2_{v\bar{h}_i}$ and the corresponding $\hat{\theta}_i^{\text{EB}}$ obtained from $\hat{\sigma}^2_{v\bar{h}_i}$ by $\hat{\theta}_i^{\text{EB}}(\hat{\sigma}^2_{v\bar{h}_i})$. The resulting second-order correct CI on θ_i is given by

$$I_i^{\text{YL}}(\alpha) = \left[\hat{\theta}_i^{\text{EB}}(\hat{\sigma}^2_{v\bar{h}_i}) - z_{\alpha/2}\{g_{1i}(\hat{\sigma}^2_{v\bar{h}_i})\}^{1/2}, \hat{\theta}_i^{\text{EB}}(\hat{\sigma}^2_{v\bar{h}_i}) + z_{\alpha/2}\{g_{1i}(\hat{\sigma}^2_{v\bar{h}_i})\}^{1/2}\right]. \quad (9.2.55)$$

The coverage error associated with $I_i^{\text{YL}}(\alpha)$ is $O(m^{-3/2})$. Proofs for establishing the second-order correct property of $I_i^{\text{YL}}(\alpha)$ are highly technical, and we refer the reader to Yoshimori and Lahiri (2014c) for details. Note that the EB estimator $\hat{\theta}_i^{\text{EB}}(\hat{\sigma}^2_{v\bar{h}_i})$ is different from $\hat{\theta}_i^{\text{EB}}(\hat{\sigma}^2_{v\text{RE}})$. Yoshimori and Lahiri (2014c) derived a second-order unbiased estimator of MSE associated with $\hat{\theta}_i^{\text{EB}}(\hat{\sigma}^2_{v\bar{h}_i})$. However, they recommended the use of $\hat{\theta}_i^{\text{EB}}$ and estimator of MSE($\hat{\theta}_i^{\text{EB}}$) based on the positive estimator of σ_v^2 proposed in Yoshimori and Lahiri (2014a), see Section 6.4.2. Simulation results reported in Yoshimori and Lahiri (2014c) indicated that the interval $I_i^{\text{YL}}(\alpha)$ is comparable to the bootstrap interval $I_i^{\text{CLL}}(\alpha)$ in terms of coverage accuracy. Concerning length, intervals $I_i^{\text{YL}}(\alpha)$ are designed to have smaller length than the intervals for the direct estimators. In simulations, the bootstrap intervals $I_i^{\text{CLL}}(\alpha)$ also showed average lengths smaller than those of direct estimators, but currently there is no formal theory supporting this fact.

Note that the closed-form intervals are at present available only for the basic area level model, unlike the parametric bootstrap method, which can be extended to general linear mixed models including the basic unit level model (Chatterjee et al. 2008).

Dass et al. (2012) studied CI estimation for the case of unknown sampling variances ψ_i in the FH model. They assumed simple random sampling within areas and considered that $\hat{\theta}_i = \bar{y}_i$ is the mean of n_i observations $y_{ij} \overset{\text{iid}}{\sim} N(\theta_i, \sigma_i^2)$, similar to Rivest and Vandal (2003), see Section 6.4.1. Further, they assume $\theta_i \sim N(\mathbf{z}_i^T\boldsymbol{\beta}, \sigma_v^2)$, $[(n_i - 1)s_i^2/\sigma_i^2]|\sigma_i^2 \sim \mathcal{X}^2_{n_i-1}$ and $\sigma_i^{-2} \overset{\text{iid}}{\sim} G(a, b), a > 0, b > 0$, independently for $i = 1, \ldots, m$. Under this setup, they obtained improved EB estimators of the area means θ_i as well as smoothed estimators of the sampling variances $\psi_i = \sigma_i^2/n_i$, assuming that the data $\{(\bar{y}_i, s_i^2, \mathbf{z}_i); i = 1, \ldots, m\}$ are available. Dass et al. (2012) also developed CIs for the means θ_i using a decision theory approach. Note that no prior distribution on the model parameters $\boldsymbol{\beta}, \sigma_v^2$, a and b is assumed, unlike in the HB approach. The proposed model, however, assumes random sampling variances σ_i^2, which in turn implies that $\psi_i = \sigma_i^2/n_i$ is also random. An HB approach by You and Chapman (2006), under the above variant of the FH model, is outlined in Section 10.3.4.

(iii) *Population-Specific Simultaneous Coverage*

In sections (i) and (ii), we studied unconditional coverage of CIs under the FH model, assuming normality of random effects v_i and sampling errors e_i. It is of practical interest to study the population-specific (or design-based) coverage of model-based CIs, by conditioning on $\boldsymbol{\theta} = (\theta_1, \ldots, \theta_m)^T$ and referring only to the sampling model $\hat{\theta}_i|\theta_i \overset{\text{ind}}{\sim} N(\theta_i, \psi_i), i = 1, \ldots, m$.

Zhang (2007) studied design-based coverages of CIs for the special case of known model parameters $\boldsymbol{\beta}$ and σ_v^2. In this case, the $(1-\alpha)$-level model-based interval on θ_i is given by

$$I_i(\alpha) = [\hat{\theta}_i^B - z_{\alpha/2}\{g_{1i}(\sigma_v^2)\}^{1/2}, \hat{\theta}_i^B + z_{\alpha/2}\{g_{1i}(\sigma_v^2)\}^{1/2}] \tag{9.2.56}$$

The interval $I_i(\alpha)$ has exact coverage of $1-\alpha$ under the assumed FH model, where $\hat{\theta}_i^B = \mathbf{z}_i^T\boldsymbol{\beta} + \gamma_i(\hat{\theta}_i - \mathbf{z}_i^T\boldsymbol{\beta})$ is the BP estimator of θ_i and $g_{1i}(\sigma_v^2) = \gamma_i\psi_i = (1-\gamma_i)\sigma_v^2$ as earlier. Letting $A_i = A_i(\alpha) = 1$ iff $\theta_i \in I_i(\alpha)$, model coverage of $I_i(\alpha)$ may be expressed as $\Delta_i(\alpha) = E(A_i) = 1-\alpha$, where the expectation E is with respect to both the sampling model and the linking model.

Let $\delta_i(\alpha) = E_p(A_i)$ denote the conditional design coverage of the interval $I_i(\alpha)$, where E_p is the expectation with respect to the sampling model conditional on θ_i (or v_i). Also, suppose that $\sigma_v^2/\psi_i \approx 0$. In this case, the interval $I_i(\alpha)$ for $\theta_i = \mathbf{z}_i^T\boldsymbol{\beta} + v_i$ reduces to the following interval for v_i:

$$I_i(\alpha) = (-z_{\alpha/2}\, \sigma_v \le v_i \le z_{\alpha/2}\, \sigma_v). \tag{9.2.57}$$

It follows from (9.2.57) that $\delta_i(\alpha) = 1$ if $|v_i| \le z_{\alpha/2}\sigma_v$ and $\delta_i(\alpha) = 0$ otherwise. Hence, the area-specific interval $I_i(\alpha)$ leads to degenerate design coverage; that is, the design coverage of $I_i(\alpha)$ is uncontrollable.

To get around the above difficulty with area-specific design coverage, Zhang (2007) proposed the use of simultaneous design coverage $\delta(\alpha) = m^{-1}\sum_{i=1}^m \delta_i(\alpha)$, which summarizes all the area-specific design coverages in a single number. The proposed coverage $\delta(\alpha)$ may be interpreted as the expected proportion of parameters θ_i covered by the set of intervals $I_i(\alpha)$ under the sampling model conditional on $\boldsymbol{\theta}$.

Zhang (2007) showed that $\delta(\alpha)$ converges in model probability to the nominal level $1-\alpha$ as $m \to \infty$. That is, the design-based simultaneous coverage of model-based intervals $I_i(\alpha)$ is close to the nominal level, given a large number of areas. This result follows by noting that $E_m[\delta_i(\alpha)] = 1-\alpha$, $V_m[\delta_i(\alpha)] = V(A_i) - E_m V_p(A_i) \le V(A_i) = \alpha(1-\alpha)$, $Cov_m[\delta_i(\alpha), \delta_t(\alpha)] = 0, i \ne t$, and then appealing to the Law of Large Numbers (LLN), where V denotes the total variance and V_p the variance with respect to the sampling model. The covariance result follows by noting that

$$\text{Cov}_m[\delta_i(\alpha), \delta_t(\alpha)] = \text{Cov}(A_i, A_t) = 0, \quad i \ne t,$$

since A_i is a function of (v_i, e_i), which is independent of A_t, a function of (v_t, e_t) under the FH model, where Cov denotes the total covariance.

The above result on simultaneous coverage is applicable to intervals in Section (ii) for the case of unknown $\boldsymbol{\beta}$ and σ_v^2, provided $E(A_i) \approx 1-\alpha$ as $m \to \infty$. We can apply the above ideas to show that the design expectation of the average of the model-based MSE estimators, $\text{mse}(\hat{\theta}_i^B) = g_{1i}(\sigma_v^2) = \gamma_i\psi_i$, converges in model probability to the average of the design MSEs, $\text{MSE}_p(\hat{\theta}_i^B) = E_p(\hat{\theta}_i - \theta_i)^2$, as $m \to \infty$. That is, the average design bias of the estimators $\text{mse}(\hat{\theta}_i^B), i = 1, \ldots, m$, converges in model probability to zero, as $m \to \infty$. This result follows by expressing the average design bias as

$$\frac{1}{m}\sum_{i=1}^{m}\{E_p[\text{mse}(\hat{\theta}_i^B)] - \text{MSE}_p(\hat{\theta}_i^B)\}$$

$$= -\frac{1}{m}\sum_{i=1}^{m}(1-\gamma_i)^2(v_i^2 - \sigma_v^2) =: -\frac{1}{m}\sum_{i=1}^{m}u_i = -\bar{u},$$

and then appealing to the LLN, noting that $E_m(u_i) = 0$ and that $V(\bar{u}) = m^{-1}\left[m^{-1}\sum_{i=1}^{m}(1-\gamma_i)^2 V(v_i^2)\right] \to 0$, as $m \to \infty$.

9.3 LINEAR MIXED MODELS

9.3.1 EB Estimation of $\mu_i = \mathbf{l}_i^T\boldsymbol{\beta} + \mathbf{m}_i^T\mathbf{v}_i$

Assuming normality, the linear mixed model (5.3.1) with block-diagonal covariance structure may be expressed as

$$\mathbf{y}_i|\mathbf{v}_i \overset{\text{ind}}{\sim} N(\mathbf{X}_i\boldsymbol{\beta} + \mathbf{Z}_i\mathbf{v}_i, \mathbf{R}_i) \tag{9.3.1}$$

with

$$\mathbf{v}_i \overset{\text{ind}}{\sim} N(\mathbf{0}, \mathbf{G}_i), \quad i = 1, \ldots, m,$$

where $\mathbf{G}_i$ and $\mathbf{R}_i$ depend on a vector of variance parameters $\boldsymbol{\delta}$. The Bayes or BP estimator of realized $\mu_i = \mathbf{l}_i^T\boldsymbol{\beta} + \mathbf{m}_i^T\mathbf{v}_i$ is given by the conditional expectation of μ_i given $\mathbf{y}_i$, $\boldsymbol{\beta}$, and $\boldsymbol{\delta}$:

$$\hat{\mu}_i^B = \hat{\mu}_i^B(\boldsymbol{\beta}, \boldsymbol{\delta}) := E(\mu_i|\mathbf{y}_i, \boldsymbol{\beta}, \boldsymbol{\delta}) = \mathbf{l}_i^T\boldsymbol{\beta} + \mathbf{m}_i^T\hat{\mathbf{v}}_i^B, \tag{9.3.2}$$

where

$$\hat{\mathbf{v}}_i^B = E(\mathbf{v}_i|\mathbf{y}_i, \boldsymbol{\beta}, \boldsymbol{\delta}) = \mathbf{G}_i\mathbf{Z}_i^T\mathbf{V}_i^{-1}(\mathbf{y}_i - \mathbf{X}_i\boldsymbol{\beta}) \tag{9.3.3}$$

and $\mathbf{V}_i = \mathbf{R}_i + \mathbf{Z}_i\mathbf{G}_i\mathbf{Z}_i^T$. The result (9.3.2) follows from the posterior (or conditional) distribution of μ_i given $\mathbf{y}_i$:

$$\mu_i|\mathbf{y}_i, \boldsymbol{\beta}, \boldsymbol{\delta} \overset{\text{ind}}{\sim} N(\hat{\mu}_i^B, g_{1i}(\boldsymbol{\delta})), \tag{9.3.4}$$

where $g_{1i}(\boldsymbol{\delta})$ is given by (5.3.6).

The estimator $\hat{\mu}_i^B$ depends on the model parameters $\boldsymbol{\beta}$ and $\boldsymbol{\delta}$, which are estimated from the marginal distribution

$$\mathbf{y}_i \overset{\text{ind}}{\sim} N(\mathbf{X}_i\boldsymbol{\beta}, \mathbf{V}_i), \quad i = 1, \ldots, m \tag{9.3.5}$$

using ML or REML. Denoting the estimators as $\hat{\beta}$ and $\hat{\delta}$, we obtain the EB or the EBP estimator of μ_i from $\hat{\mu}_i^B$ by substituting $\hat{\beta}$ for β and $\hat{\delta}$ for δ:

$$\hat{\mu}_i^{\text{EB}} = \hat{\mu}_i^B(\hat{\beta}, \hat{\delta}) = \mathbf{l}_i^T\hat{\beta} + \mathbf{m}_i^T\hat{\mathbf{v}}_i^B(\hat{\beta}, \hat{\delta}). \tag{9.3.6}$$

The EB estimator $\hat{\mu}_i^{\text{EB}}$ is identical to the EBLUP estimator $\hat{\mu}_i^H$ given in (5.3.8). Note that $\hat{\mu}_i^{\text{EB}}$ is also the mean of the estimated posterior density, $f(\mu_i|\mathbf{y}_i, \hat{\beta}, \hat{\delta})$, of μ_i, which is $N[\hat{\mu}_i^{\text{EB}}, g_{1i}(\hat{\delta})]$.

9.3.2 MSE Estimation

The results in Section 5.3.2 on the estimation of the MSE of the EBLUP estimator $\hat{\mu}_i^H$ are applicable to the EB estimator $\hat{\mu}_i^{\text{EB}}$ because $\hat{\mu}_i^H$ and $\hat{\mu}_i^{\text{EB}}$ are identical under normality. Also, the area-specific estimators (5.3.15)–(5.3.18) may be used as estimators of the conditional MSE, $\text{MSE}_c(\hat{\mu}_i^{\text{EB}}) = E[(\hat{\mu}_i^{\text{EB}} - \mu_i)^2|\mathbf{y}_i]$, where the expectation is conditional on the observed $\mathbf{y}_i$ for the ith area. As noted in Section 5.3.2, the MSE estimators are second-order unbiased.

The jackknife MSE estimator, (9.2.15), for the basic area level model, readily extends to the linear mixed model with block-diagonal covariance structure. We calculate the delete-ℓ estimators $\hat{\beta}_{-\ell}$ and $\hat{\delta}_{-\ell}$ by deleting the ℓth area data set $(\mathbf{y}_\ell, \mathbf{X}_\ell, \mathbf{Z}_\ell)$ from the full data set. This calculation is done for each ℓ to get m estimators $\{(\hat{\beta}_{-\ell}, \hat{\delta}_{-\ell}); \ell = 1, \ldots, m\}$ which, in turn, provide m estimators of μ_i, $\{\hat{\mu}_{-\ell}^{\text{EB}}; \ell = 1, \ldots, m\}$, where $\hat{\mu}_{-\ell}^{\text{EB}}$ is obtained from $\hat{\mu}_i^B = k_i(\mathbf{y}_i, \beta, \delta)$ by changing β and δ to $\hat{\beta}_{-\ell}$ and $\hat{\delta}_{-\ell}$, respectively. The jackknife estimator is given by

$$\text{mse}_J(\hat{\mu}_i^B) = \hat{M}_{1i} + \hat{M}_{2i} \tag{9.3.7}$$

with

$$\hat{M}_{1i} = g_{1i}(\hat{\delta}) - \frac{m-1}{m}\sum_{\ell=1}^{m}[g_{1i}(\hat{\delta}_{-\ell}) - g_{1i}(\hat{\delta})] \tag{9.3.8}$$

and

$$\hat{M}_{2i} = \frac{m-1}{m}\sum_{\ell=1}^{m}(\hat{\mu}_{i,-\ell}^{\text{EB}} - \hat{\mu}_i^{\text{EB}})^2, \tag{9.3.9}$$

where $g_{1i}(\delta)$ is given by (5.3.6). The jackknife MSE estimator, $\text{mse}_J(\hat{\mu}_i^B)$, is also second-order unbiased (Jiang, Lahiri, and Wan 2002).

9.3.3 Approximations to the Posterior Variance

As noted in Section 9.2.3, the naive EB approach uses the estimated posterior density, $f(\mu_i|\mathbf{y}_i, \hat{\beta}, \hat{\delta})$ to make inference on μ_i. In practice, μ_i is estimated by $\hat{\mu}_i^{\text{EB}}$ and the estimated posterior variance $V(\mu_i|\mathbf{y}_i, \hat{\beta}, \hat{\delta}) = g_{1i}(\hat{\delta})$ is used as a measure of variability.

But the naive measure, $g_{1i}(\hat{\boldsymbol{\delta}})$, can lead to severe underestimation of the true posterior variance $V(\mu_i|\mathbf{y})$ as well as of $\text{MSE}(\hat{\mu}_i^{\text{EB}})$.

The bootstrap method of Laird and Louis (1987) readily extends to the linear mixed model. Bootstrap data $\{\mathbf{y}_i^*(b), \mathbf{X}_i, \mathbf{Z}_i; i = 1, \ldots, m\}$ are first generated from the estimated marginal distribution, $N(\mathbf{X}_i\hat{\boldsymbol{\beta}}, \hat{\mathbf{V}}_i)$, for $b = 1, \ldots, B$. Estimates $\{\hat{\boldsymbol{\beta}}(b), \hat{\boldsymbol{\delta}}(b)\}$ and EB estimates $\hat{\mu}_i^{*\text{EB}}(b)$ are then computed from the bootstrap data. The bootstrap approximation to the posterior variance is given by

$$V_{\text{LL}}(\mu_i|\mathbf{y}) = \frac{1}{B}\sum_{b=1}^{B} g_{1i}(\hat{\boldsymbol{\delta}}(b)) + \frac{1}{B}\sum_{b=1}^{B} [\hat{\mu}_i^{*\text{EB}}(b) - \hat{\mu}_i^{*\text{EB}}(\cdot)]^2, \tag{9.3.10}$$

where $\hat{\mu}_i^{*\text{EB}}(\cdot) = B^{-1}\sum_{b=1}^{B} \hat{\mu}_i^{*\text{EB}}(b)$.

By deriving an approximation, $\tilde{V}_{\text{LL}}(\mu_i|\mathbf{y})$ of $V_{\text{LL}}(\mu_i|\mathbf{y})$, as $B \to \infty$, Butar and Lahiri (2003) showed that $V_{\text{LL}}(\mu_i|\mathbf{y})$ is not second-order unbiased for $\text{MSE}(\hat{\mu}_i^{\text{EB}})$. Then, they proposed a bias-corrected MSE estimator

$$\text{mse}_{\text{BL}}(\hat{\mu}_i^{\text{EB}}) = \tilde{V}_{\text{LL}}(\mu_i|\mathbf{y}) + g_{3i}(\hat{\boldsymbol{\delta}}), \tag{9.3.11}$$

where $g_{3i}(\boldsymbol{\delta})$ is given by 5.3.10 and

$$\tilde{V}_{\text{LL}}(\mu_i|\mathbf{y}) = g_{1i}(\hat{\boldsymbol{\delta}}) + g_{2i}(\hat{\boldsymbol{\delta}}) + g_{3i}^*(\hat{\boldsymbol{\delta}}, \mathbf{y}_i), \tag{9.3.12}$$

where $g_{2i}(\boldsymbol{\delta})$ and $g_{3i}^*(\boldsymbol{\delta}, \mathbf{y}_i)$ are given by (5.3.7) and (5.3.14), respectively. It now follows from (9.3.11) and (9.3.12) that $\text{mse}_{\text{BL}}(\hat{\mu}_i^{\text{EB}})$ is identical to the area-specific MSE estimator given by (5.3.16).

Kass and Steffey (1989) obtained a first-order approximation to $V(\mu_i|\mathbf{y})$, denoted here $V_{\text{KS}}(\mu_i|\mathbf{y})$. After simplification, $V_{\text{KS}}(\mu_i|\mathbf{y})$ reduces to $\tilde{V}_{\text{LL}}(\mu_i|\mathbf{y})$ given by (9.3.12). Therefore, $V_{\text{KS}}(\mu_i|\mathbf{y})$ is also not second-order unbiased for $\text{MSE}(\hat{\mu}_i^{\text{EB}})$. A second-order approximation to $V(\mu_i|\mathbf{y})$ ensures that the neglected terms are of lower order than m^{-1}, but it depends on the prior density, $f(\boldsymbol{\beta}, \boldsymbol{\delta})$, unlike the first-order approximation.

9.4 *EB ESTIMATION OF GENERAL FINITE POPULATION PARAMETERS

This section deals with EB estimation of general (possibly nonlinear) parameters of a finite population. The finite population P contains N units and a sample s of size n is drawn from P. We denote by $r = P - s$ the sample complement of size $N - n$. We denote by $\mathbf{y}^P$ the vector with the unit values of the target variable in the population, which is assumed to be random with a given joint probability distribution. We denote by $\mathbf{y}_s$ the subvector of $\mathbf{y}^P$ corresponding to the sample units, by $\mathbf{y}_r$ the subvector of nonsampled units and consider without loss of generality that the first n elements of $\mathbf{y}$ are the sample elements, that is, $\mathbf{y}^P = (\mathbf{y}_s^T, \mathbf{y}_r^T)^T$. The target parameter is a real measurable function $\tau = h(\mathbf{y}^P)$ of the population vector $\mathbf{y}^P$ that we want to estimate using the available sample data $\mathbf{y}_s$.

9.4.1 BP Estimator Under a Finite Population

Let $\hat{\tau}$ denote an estimator of τ based on $\mathbf{y}_s$. The MSE of $\hat{\tau}$ is defined as $\text{MSE}(\hat{\tau}) = E(\hat{\tau} - \tau)^2$, where E denotes expectation with respect to the joint distribution of $\mathbf{y}^P$. The BP estimator of τ is the function of $\mathbf{y}_s$ with minimum MSE. Let us define $\tau^0 = E_{\mathbf{y}_r}(\tau|\mathbf{y}_s)$, where the expectation is now taken with respect to the conditional distribution of $\mathbf{y}_r$ given $\mathbf{y}_s$. Note that τ^0 is a function of the sample data $\mathbf{y}_s$ and model parameters. We can write

$$\begin{aligned}\text{MSE}(\hat{\tau}) &= E(\hat{\tau} - \tau^0 + \tau^0 - \tau)^2 \\ &= E(\hat{\tau} - \tau^0)^2 + 2\,E[(\hat{\tau} - \tau^0)(\tau^0 - \tau)] + E(\tau^0 - \tau)^2.\end{aligned}$$

In this expression, the last term does not depend on $\hat{\tau}$. For the second term, using the Law of Iterated Expectations, we get

$$\begin{aligned}E[(\hat{\tau} - \tau^0)(\tau^0 - \tau)] &= E_{\mathbf{y}_s}\{E_{\mathbf{y}_r}[(\hat{\tau} - \tau^0)(\tau^0 - \tau)|\mathbf{y}_s]\} \\ &= E_{\mathbf{y}_s}\{(\hat{\tau} - \tau^0)[\tau^0 - E_{\mathbf{y}_r}(\tau|\mathbf{y}_s)]\} = 0.\end{aligned}$$

Thus, the BP estimator of τ is the minimizer $\hat{\tau}$ of $E[(\hat{\tau} - \tau^0)^2]$. Since this quantity is nonnegative and its minimum value of zero is obtained at $\hat{\tau} = \tau^0$, the BP estimator of τ is given by

$$\hat{\tau}^B = \tau^0 = E_{\mathbf{y}_r}(\tau|\mathbf{y}_s). \tag{9.4.1}$$

Note that the BP estimator is unbiased in the sense that $E(\hat{\tau}^B - \tau) = 0$ because

$$E_{\mathbf{y}_s}(\hat{\tau}^B) = E_{\mathbf{y}_s}[E_{\mathbf{y}_r}(\tau|\mathbf{y}_s)] = E(\tau).$$

Typically, the distribution of $\mathbf{y}^P$ depends on an unknown parameter vector $\boldsymbol{\lambda}$, which can be estimated using the sample data $\mathbf{y}_s$. Then, the EB (or EBP) estimator of τ, denoted $\hat{\tau}^{\text{EB}}$, is given by (9.4.1), with the expectation taken with respect to the distribution of $\mathbf{y}_r|\mathbf{y}_s$ with $\boldsymbol{\lambda}$ replaced by an estimator $\hat{\boldsymbol{\lambda}}$. The EB estimator is not exactly unbiased, but the bias coming from the estimation of $\boldsymbol{\lambda}$ is negligible for large samples, provided that $\hat{\boldsymbol{\lambda}}$ is consistent for $\boldsymbol{\lambda}$.

9.4.2 EB Estimation Under the Basic Unit Level Model

Suppose now that the population contains m areas (or subpopulations) $P_1, \ldots, P_m$ of sizes $N_1, \ldots, N_m$. Let s_i be a sample of size n_i drawn from P_i and $r_i = P_i - s_i$ the sample complement, $i = 1, \ldots, m$. We assume that the value y_{ij} of a target variable for jth unit in ith area follows the basic unit level model (4.3.1). In this section, the target parameter is $\tau_i = h(\mathbf{y}_i^P)$, where $\mathbf{y}_i^P = (\mathbf{y}_{is}^T, \mathbf{y}_{ir}^T)^T$ is the vector with variables y_{ij} for sample and nonsampled units from area i. Applying (9.4.1) to $\tau_i = h(\mathbf{y}_i^P)$ and noting that $\mathbf{y}_1^P, \ldots, \mathbf{y}_m^P$ are independent under the assumed model, the BP estimator of τ_i is given by

$$\hat{\tau}_i^B = E_{\mathbf{y}_{ir}}[h(\mathbf{y}_i^P)|\mathbf{y}_{is}] = \int_{\mathbb{R}^{N_i-n_i}} h(\mathbf{y}_i^P) f(\mathbf{y}_{ir}|\mathbf{y}_{is})\, d\mathbf{y}_{ir}, \tag{9.4.2}$$

where $f(\mathbf{y}_{ir}|\mathbf{y}_{is})$ is the joint density of $\mathbf{y}_{ir}$ given the observed data vector $\mathbf{y}_{is}$. For the special case of the area mean $\tau_i = \overline{Y}_i$, if N_i is large, then by (4.3.1) and the LLN, $\tau_i \approx \overline{\mathbf{X}}_i^T\boldsymbol{\beta} + v_i$, which is a special case of the mixed effect $\mu_i = \mathbf{l}_i^T\boldsymbol{\beta} + \mathbf{m}_i^T\mathbf{v}_i$, studied in Section 9.3.1.

For area parameters $\tau_i = h(\mathbf{y}_i^P)$ with nonlinear $h(\cdot)$, we might not be able to calculate analytically the expectation in (9.4.2). Molina and Rao (2010) proposed to approximate this expectation by Monte Carlo. For this, first generate A replicates $\{\mathbf{y}_{ir}^{(a)}; a = 1, \ldots, A\}$ of $\mathbf{y}_{ir}$ from the conditional distribution $f(\mathbf{y}_{ir}|\mathbf{y}_{is})$. Then, attach the vector of sample elements $\mathbf{y}_{is}$ to $\mathbf{y}_{ir}^{(a)}$ as $\mathbf{y}_i^{(a)} = ((\mathbf{y}_{ir}^{(a)})^T, \mathbf{y}_{is}^T)^T$. With the area vector $\mathbf{y}_i^{(a)}$, calculate the target quantity $\tau_i^{(a)} = h(\mathbf{y}_i^{(a)})$ for each a and then average over the A replicates, that is, take

$$\hat{\tau}_i^B \approx \frac{1}{A}\sum_{a=1}^{A} \tau_i^{(a)}. \tag{9.4.3}$$

Let $\mathbf{y}_i^P = (y_{i1}, \ldots, y_{iN_i})^T, \mathbf{X}_i^P = (\mathbf{x}_{i1}, \ldots, \mathbf{x}_{iN_i})^T, i = 1, \ldots, m, \mathbf{I}_k$ the identity matrix of order k, and $\mathbf{1}_k$ a $k \times 1$ vector of ones. We assume the unit level model (4.3.1) with normality of v_i and e_{ij}, so that

$$\mathbf{y}_i^P \overset{\text{ind}}{\sim} N(\mathbf{X}_i^P\boldsymbol{\beta}, \mathbf{V}_i^P), \quad i = 1, \ldots, m, \tag{9.4.4}$$

where $\mathbf{V}_i^P = \sigma_v^2\mathbf{1}_{N_i}\mathbf{1}_{N_i}^T + \sigma_e^2\text{diag}_{1\le j\le N_i}(k_{ij}^2), i = 1, \ldots, m$. Consider the decomposition into sample and nonsampled elements of $\mathbf{X}_i^P$ and of the elements of the covariance matrix,

$$\mathbf{X}_i^P = \begin{pmatrix}\mathbf{X}_{is}\\ \mathbf{X}_{ir}\end{pmatrix}, \quad \mathbf{V}_i^P = \begin{pmatrix}\mathbf{V}_{is} & \mathbf{V}_{isr}\\ \mathbf{V}_{irs} & \mathbf{V}_{ir}\end{pmatrix}.$$

By the normality assumption in (9.4.4), we have that $\mathbf{y}_{ir}|\mathbf{y}_{is} \sim N(\boldsymbol{\mu}_{ir|s}, \mathbf{V}_{ir|s})$, where the conditional mean vector and covariance matrix are given by

$$\boldsymbol{\mu}_{ir|s} = \mathbf{X}_{ir}\boldsymbol{\beta} + \gamma_i(\overline{y}_{ia} - \overline{\mathbf{x}}_{ia}^T\boldsymbol{\beta})\mathbf{1}_{N_i-n_i}, \tag{9.4.5}$$

$$\mathbf{V}_{ir|s} = \sigma_v^2(1-\gamma_i)\mathbf{1}_{N_i-n_i}\mathbf{1}_{N_i-n_i}^T + \sigma_e^2\text{diag}_{j\in r_i}(k_{ij}^2), \tag{9.4.6}$$

where $\gamma_i = \sigma_v^2(\sigma_v^2 + \sigma_e^2/a_{i\cdot})^{-1}$, with $a_{i\cdot} = \sum_{j\in s_i} a_{ij}$ and $a_{ij} = k_{ij}^{-2}$, and where $\overline{y}_{ia}$ and $\overline{\mathbf{x}}_{ia}$ are defined in (7.1.7). Note that the application of the Monte Carlo approximation (9.4.3) involves simulation of m multivariate normal vectors $\mathbf{y}_{ir}$ of sizes $N_i - n_i, i = 1, \ldots, m$, from the conditional distribution of $\mathbf{y}_{ir}|\mathbf{y}_{is}$. Then this process has to be repeated a large number of times A, which may be computationally unfeasible for $N_i - n_i$ large. This can be avoided by noting that the conditional covariance matrix

$\mathbf{V}_{ir|s}$, given by (9.4.6), corresponds to the covariance matrix of a random vector $\mathbf{y}_{ir}$ generated from the model

$$\mathbf{y}_{ir} = \boldsymbol{\mu}_{ir|s} + u_i \mathbf{1}_{N_i - n_i} + \boldsymbol{\epsilon}_{ir}, \tag{9.4.7}$$

with new random effects u_i and errors $\boldsymbol{\epsilon}_{ir}$ that are independent and satisfy

$$u_i \sim N(0, \sigma_v^2(1-\gamma_i)) \quad \text{and} \quad \boldsymbol{\epsilon}_{ir} \sim N(\mathbf{0}_{N_i-n_i}, \sigma_e^2 \text{diag}_{j \in r_i}(k_{ij}^2));$$

see Molina and Rao (2010). Using model (9.4.7), instead of generating a multivariate normal vector of size $N_i - n_i$, we just need to generate $1 + N_i - n_i$ independent univariate normal variables $u_i \overset{\text{ind}}{\sim} N(0, \sigma_v^2(1-\gamma_i))$ and $\epsilon_{ij} \overset{\text{iid}}{\sim} N(0, \sigma_e^2 k_{ij}^2)$, for $j \in r_i$. Then, we obtain the corresponding nonsampled elements $y_{ij}, j \in r_i$, from (9.4.7) using as means the corresponding elements of $\boldsymbol{\mu}_{ir|s}$ given by (9.4.5).

In practice, the model parameters $\boldsymbol{\lambda} = (\boldsymbol{\beta}^T, \sigma_v^2, \sigma_e^2)^T$ are replaced by consistent estimates $\hat{\boldsymbol{\lambda}} = (\hat{\boldsymbol{\beta}}^T, \hat{\sigma}_v^2, \hat{\sigma}_e^2)^T$, such as the ML or REML estimates, and then the variables y_{ij} are generated from (9.4.7) with $\boldsymbol{\lambda}$ replaced by $\hat{\boldsymbol{\lambda}}$, leading to the Monte Carlo approximation of the EB estimator $\hat{\tau}_i^{\text{EB}}$ of τ_i.

For nonsampled areas $i = m+1, \ldots, M$, we generate $y_{ij}^{(a)}$ for $j = 1, \ldots, N_i$ as $y_{ij}^{(a)} = \mathbf{x}_{ij}^T \hat{\boldsymbol{\beta}} + v_i^{(a)} + \epsilon_{ij}^{(a)}$, where $v_i^{(a)} \overset{\text{iid}}{\sim} N(0, \hat{\sigma}_v^2)$ and $\epsilon_{ij}^{(a)} \overset{\text{ind}}{\sim} N(0, \hat{\sigma}_e^2 k_{ij}^2)$ and $\epsilon_{ij}^{(a)}$ is independent of $v_i^{(a)}$. The EB estimator of τ_i for a nonsampled area i is then given by

$$\hat{\tau}_i^{\text{EB}} = \frac{1}{A} \sum_{a=1}^{A} h(\mathbf{y}_i^{P(a)}), \tag{9.4.8}$$

where $\mathbf{y}_i^{P(a)} = (y_{i1}^{(a)}, \ldots, y_{iN_i}^{(a)})^T$, for $i = m+1, \ldots, M$.

For very large populations and/or computationally complex indicators such as those that require sorting all population elements, the EB method described in Section 9.4.2 might be computationally unfeasible. For those cases, a faster version of the EB method was proposed by Ferretti and Molina (2012). This method is based on replacing in the Monte Carlo approximation (9.4.3), the true value of the parameter for ath Monte Carlo population $\tau_i^{(a)}$ by a design-based estimator $\hat{\tau}_i^{\text{DB}(a)}$ of $\tau_i^{(a)}$ based on a sample $s_i^{(a)}$ drawn from the population units P_i in area i, independently for each $a = 1, \ldots, A$. In particular, we can select simple random samples, $s_i^{(a)}$, from each area i. Then the values of the auxiliary variables corresponding to the units drawn in $s_i^{(a)}$ are taken: $\mathbf{x}_{ij}, j \in s_i^{(a)}$. Using those values, the corresponding responses $y_{ij}, j \in s_i^{(a)}$, are generated for $i = 1, \ldots, m$ from (9.4.7). Denoting the vector containing those generated sample values as $\mathbf{y}_s^{(a)}$, calculate the design-based estimator of the poverty indicator, $\hat{\tau}_i^{\text{DB}(a)}$. Finally, the fast EB estimator is given by

$$\hat{\tau}_i^{\text{FEB}} \approx \frac{1}{A} \sum_{a=1}^{A} \hat{\tau}_i^{\text{DB}(a)}. \tag{9.4.9}$$

In the fast EB estimator (9.4.9), only the response variables corresponding to the sample units $j \in s_i^{(a)}$ need to be generated from (9.4.11) for each Monte Carlo replicate a, avoiding the generation of the full population of responses. Another advantage of this variation of the EB method is that the identification of the sample units in the population register from where the auxiliary variables are obtained is not necessary; linking the units in the sample file to the population register is often not possible. Simulation results not reported here indicate that the fast EB method loses little efficiency as compared with the EB estimator described in Section 9.4.2.

Diallo and Rao (2014) extended the Molina–Rao normality-based EB results by allowing the random effects v_i and/or errors e_{ij} to follow skew normal (SN) family of distributions. This extended SN family includes the normal distribution as a special case.

9.4.3 FGT Poverty Measures

Nonlinear area parameters of special relevance are poverty indicators. Poverty can be measured in many different ways but, when measured in terms of a quantitative welfare variable such as income or expenditure, popular poverty indicators are the members of the FGT family (Foster, Greer, and Thorbecke, 1984). Let E_{ij} be the welfare measure for jth individual within ith area, and let z be a (fixed) poverty line defined for the actual population. The FGT family of poverty indicators for area i is given by $F_{\alpha i} = N_i^{-1} \sum_{j=1}^{N_i} F_{\alpha ij}$, where

$$F_{\alpha ij} = \left(\frac{z - E_{ij}}{z}\right)^{\alpha} I(E_{ij} < z), \quad \alpha \geq 0, \quad j = 1, \ldots, N_i. \tag{9.4.10}$$

For $\alpha = 0$, F_{0i} reduces to the proportion of individuals with income below the poverty line, which is called poverty incidence or at risk of poverty rate. For $\alpha = 1$, F_{1i} is the average of the relative distances of the individuals' income to the poverty line, which is known as poverty gap. Thus, the poverty incidence F_{0i} measures the frequency of poverty, whereas the poverty gap F_{1i} measures the degree or intensity of poverty.

The distribution of the welfare variables E_{ij} involved in $F_{\alpha i}$ is seldom normal due to the typical right skewness of economic variables. However, after some transformation, such as $\log(E_{ij} + c)$ for given c, the resulting distribution might be approximately normal. Molina and Rao (2010) assumed that $y_{ij} = T(E_{ij})$ follows a normal distribution, where $T(\cdot)$ is a one-to-one transformation. Then $F_{\alpha i}$, given by (9.4.10), may be expressed as a function of $\mathbf{y}_i^P = (y_{i1}, \ldots, y_{iN_i})^T$ as follows:

$$F_{\alpha i} = \frac{1}{N_i} \sum_{j=1}^{N_i} \left\{\frac{z - T^{-1}(y_{ij})}{z}\right\}^{\alpha} I\{T^{-1}(y_{ij}) < z\} =: h_\alpha(\mathbf{y}_i^P), \quad \alpha \geq 0.$$

The EB estimator of $\tau_i = F_{\alpha i} = h_\alpha(\mathbf{y}_i^P)$ is then obtained by Monte Carlo approximation as in (9.4.3).

In some cases, it may not be possible to link the sample file to the population register from where the auxiliary variables are obtained. Then we simulate the census values $y_{ij}^{(a)}$, for $j = 1, \ldots, N_i$, by changing (9.4.7) to

$$\mathbf{y}_i^P = \boldsymbol{\mu}_{i|s}^P + u_i \mathbf{1}_{N_i} + \boldsymbol{\epsilon}_i^P, \tag{9.4.11}$$

and generating from (9.4.11), where

$$\boldsymbol{\mu}_{i|s}^P = \mathbf{X}_i^P \boldsymbol{\beta} + \gamma_i(\bar{y}_{ia} - \bar{\mathbf{x}}_{ia}^T \boldsymbol{\beta}) \mathbf{1}_{N_i}$$

and $\boldsymbol{\epsilon}_i^P \overset{\text{ind}}{\sim} N(\mathbf{0}_{N_i}, \sigma_e^2 \text{diag}_{1 \le j \le N_i}(k_{ij}^2))$. The EB estimator is then calculated from (9.4.8) by using the above simulated census values $\mathbf{y}_i^{P(a)}, a = 1, \ldots, A$.

9.4.4 Parametric Bootstrap for MSE Estimation

Estimators of the MSE of EB estimators $\hat{\tau}_i^{\text{EB}}$ of τ_i can be obtained using the parametric bootstrap for finite populations introduced in Section 7.2.4. This method proceeds as follows:

(1) Fit the basic unit level model (4.3.1) by ML, REML, or a moments method to obtain model parameter estimators $\hat{\boldsymbol{\beta}}$, $\hat{\sigma}_v^2$ and $\hat{\sigma}_e^2$.

(2) Generate bootstrap domain effects as $v_i^* \overset{\text{iid}}{\sim} N(0, \hat{\sigma}_v^2), i = 1, \ldots, m$.

(3) Generate, independently of $v_1^*, \ldots, v_m^*$, unit errors as

$$e_{ij}^* \overset{\text{ind}}{\sim} N(0, \hat{\sigma}_e^2 k_{ij}^2), \quad j = 1, \ldots, N_i, \quad i = 1, \ldots, m.$$

(4) Generate a bootstrap population of response variables from the model

$$y_{ij}^* = \mathbf{x}_{ij}{}' \hat{\boldsymbol{\beta}} + v_i^* + e_{ij}^*, \quad j = 1, \ldots, N_i, \quad i = 1, \ldots, m.$$

(5) Let $\mathbf{y}_i^{P*} = (y_{i1}^*, \ldots, y_{iN_i}^*)^T$ denote the vector of generated bootstrap response variables for area i. Calculate target quantities for the bootstrap population as $\tau_i^* = h(\mathbf{y}_i^{P*}), i = 1, \ldots, m$.

(6) Let $\mathbf{y}_s^*$ be the vector whose elements are the generated y_{ij}^* with indices contained in the sample s. Fit the model to the bootstrap sample data $\{(y_{ij}^*, \mathbf{x}_{ij}); j \in s_i, i = 1, \ldots, m\}$ and obtain bootstrap model parameter estimators, denoted $\hat{\sigma}_v^{2*}, \hat{\sigma}_e^{2*}$, and $\hat{\boldsymbol{\beta}}^*$.

(7) Obtain the bootstrap EB estimator of τ_i through the Monte Carlo approximation, denoted $\hat{\tau}_i^{\text{EB}*}, i = 1, \ldots, m$.

(8) Repeat steps (2)–(7) a large number of times B. Let $\tau_i^*(b)$ be true value and $\hat{\tau}_i^{\text{EB}*}(b)$ the EB estimator obtained in bth replicate of the bootstrap procedure, $b = 1, \ldots, B$.

(9) The bootstrap MSE estimator of $\hat{\tau}_i^{\text{EB}}$ is given by

$$\text{mse}_B(\hat{\tau}_i^{\text{EB}}) = B^{-1} \sum_{b=1}^{B} [\hat{\tau}_i^{\text{EB}*}(b) - \tau_i^*(b)]^2. \tag{9.4.12}$$

A similar parametric bootstrap approach can be used to get a bootstrap MSE estimator for the fast EB estimator $\hat{\tau}_i^{\text{FEB}}$.

9.4.5 ELL Estimation

The method of Elbers, Lanjouw, and Lanjouw (2003), called ELL method, assumes a nested error model on the transformed welfare variables but using random cluster effects, where the clusters are based on the sampling design and may be different from the small areas. In fact, the small areas are not specified in advance. Then ELL estimators of area parameters, τ_i, are computed by applying a nonparametric bootstrap method similar to the parametric bootstrap procedure described in Section 9.4.4. To make their method comparable with the EB method described in Section 9.4.2, here we consider that the clusters are the same as areas and assume normality of v_i and e_{ij}. ELL estimator of τ_i under the above setup is obtained as follows:

(1) With the original sample data $\mathbf{y}_s$, fit model (4.3.1). Let $\hat{\boldsymbol{\beta}}, \hat{\sigma}_v^2$, and $\hat{\sigma}_e^2$ be the resulting estimators of $\boldsymbol{\beta}, \sigma_v^2$, and σ_e^2.

(2) Generate bootstrap area/cluster effects as $v_i^* \stackrel{\text{iid}}{\sim} N(0, \hat{\sigma}_v^2), i = 1, \ldots, m$.

(3) Independently of the cluster effects, generate bootstrap model errors as

$$e_{ij}^* \stackrel{\text{iid}}{\sim} N(0, \hat{\sigma}_e^2 k_{ij}^2), \quad j = 1, \ldots, N_i, \quad i = 1, \ldots, m.$$

(4) Construct a population vector $\mathbf{y}^{P*} = ((\mathbf{y}_1^{P*})^T, \ldots, (\mathbf{y}_m^{P*})^T)^T$ from the bootstrap model

$$y_{ij}^* = \mathbf{x}_{ij}'\hat{\boldsymbol{\beta}} + v_i^* + e_{ij}^*, \quad j = 1, \ldots, N_i, \quad i = 1, \ldots, m. \tag{9.4.13}$$

(5) Calculate the target area quantities for the generated bootstrap population, $\tau_i^* = h(\mathbf{y}_i^{P*}), i = 1, \ldots, m$.

(6) The ELL estimator of τ_i is then given by the bootstrap expectation

$$\hat{\tau}_i^{\text{ELL}} = E_*(\tau_i^*).$$

The MSE estimator of $\hat{\tau}_i^{\text{ELL}}$ obtained by the ELL method is the bootstrap variance of τ_i^*, that is,

$$\text{mse}(\hat{\tau}_i^{\text{ELL}}) = E_*[\tau_i^* - E_*(\tau_i^*)]^2,$$

where E_* denotes the expectation with respect to bootstrap model (9.4.13) given the sample data. Note that $E_*(\tau_i^*)$ is tracking $E(\tau_i)$ and $E_*[\tau_i^* - E_*(\tau_i^*)]^2$ is tracking $E[\tau_i - E(\tau_i)]^2$. In practice, ELL estimators are obtained from a Monte Carlo approximation by generating a large number, A, of population vectors $\mathbf{y}^{P*(a)} = ((\mathbf{y}_1^{P*(a)})^T, \ldots, (\mathbf{y}_m^{P*(a)})^T)^T, a = 1, \ldots, A$, from model (9.4.13), calculating the bootstrap area parameters for each population a in the form $\tau_i^{*(a)} = h(\mathbf{y}_i^{P*(a)}), i = 1, \ldots, m$, and then averaging over the A populations. This leads to

$$\hat{\tau}_i^{\text{ELL}} \approx \frac{1}{A}\sum_{a=1}^{A} \tau_i^{*(a)}, \quad \text{mse}(\hat{\tau}_i^{\text{ELL}}) \approx \frac{1}{A}\sum_{a=1}^{A} (\tau_i^{*(a)} - \hat{\tau}_i^{\text{ELL}})^2. \tag{9.4.14}$$

Note that, in contrast to the EB method described in Section 9.4.2, the population vectors $\mathbf{y}^{P*(a)}$ in ELL method are generated from the marginal distribution of the model responses instead of the conditional distribution given sample data $\mathbf{y}_s$, and they do not contain the observed sample data. Thus, if the model parameters were known, the ELL method as described here would not be using the sample data at all. In fact, it is easy to see that for $\tau_i = \overline{Y}_i$, the ELL estimator is equal to the synthetic estimator $\hat{\tau}_i^{\text{ELL}} = \overline{\mathbf{X}}_i^T \hat{\boldsymbol{\beta}}$, which is a poor estimator when area effects v_i are present (i.e., σ_v^2 is significant). However, EB and ELL estimators coincide for areas with zero sample size or when data do not show significant area effects, that is, when the available covariates explain all the between-area variability.

9.4.6 Simulation Experiments

Molina and Rao (2010) carried out a simulation study to analyze the performance of the EB method to estimate area poverty incidences and poverty gaps. Populations of size $N = 20,000$ composed of $m = 80$ areas with $N_i = 250$ elements in each area $i = 1, \ldots, m$ were generated from model (4.3.1) with $k_{ij} = 1$. The transformation $T(\cdot)$ of the welfare variables E_{ij}, defined in Section 9.4.2, is taken as $T(E_{ij}) = \log(E_{ij}) = y_{ij}$.

As auxiliary variables in the model, we considered two binary variables x_1 and x_2 apart from the intercept. These binary variables were simulated from Bernoulli distributions with probabilities $p_{1i} = 0.3 + 0.5\ i/m$ and $p_{2i} = 0.2, i = 1, \ldots, m$. Independent simple random samples s_i without replacement are drawn from each area i with area sample sizes $n_i = 50, i = 1, \ldots, m$. Variables x_1 and x_2 for the population units and sample indices were held fixed over all Monte Carlo simulations. The regression coefficients were taken as $\boldsymbol{\beta} = (3, 0.03, -0.04)^T$.

The random area effects variance was taken as $\sigma_u^2 = (0.15)^2$ and the model error variance as $\sigma_e^2 = (0.5)^2$. The poverty line was fixed at $z = 12$, which is roughly 0.6 times the median of the welfare variables E_{ij} for a population generated as mentioned earlier. Hence, the poverty incidence for the simulated populations is approximately 16%.

A total of $K = 10{,}000$ population vectors $\mathbf{y}^{P(k)} = ((\mathbf{y}_1^{P(k)})^T, \ldots, (\mathbf{y}_m^{P(k)})^T)^T$ were generated from the true model described earlier. For each population vector $\mathbf{y}^{P(k)}$, true values of poverty incidence and gap $F_{\alpha i}^{(k)}, \alpha = 0, 1$ were calculated, along with direct, EB, and ELL estimators, where the direct estimator of $F_{\alpha i}$ is the area sample mean of $F_{\alpha ij}, j \in s_i$. Then, means over Monte Carlo populations $k = 1, \ldots, K$ of true values of poverty indicators were computed as

$$E(F_{\alpha i}) = \frac{1}{K}\sum_{k=1}^{K} F_{\alpha i}^{(k)}, \quad \alpha = 0, 1, \quad i = 1, \ldots, m.$$

Biases were computed for the direct estimator $\hat{F}_{\alpha i}$, the EB estimator $\hat{F}_{\alpha i}^{\text{EB}}$, and the ELL estimator $\hat{F}_{\alpha i}^{\text{ELL}}$ as $E(\hat{F}_{\alpha i}) - E(F_{\alpha i})$, $E(\hat{F}_{\alpha i}^{\text{EB}}) - E(F_{\alpha i})$, and $E(\hat{F}_{\alpha i}^{\text{ELL}}) - E(F_{\alpha i})$. The MSEs over Monte Carlo populations of the three estimators were also computed as $E(\hat{F}_{\alpha i} - F_{\alpha i})^2$, $E(\hat{F}_{\alpha i}^{\text{EB}} - F_{\alpha i})^2$, and $E(\hat{F}_{\alpha i}^{\text{ELL}} - F_{\alpha i})^2$.

Figure 9.1 reports the model biases and MSEs of the three estimators of the poverty gap for each area. Figure 9.1a shows that the EB estimator has the smallest absolute biases followed by the ELL estimator, although in terms of bias all estimators perform reasonably well. However, in terms of MSE, Figure 9.1b shows that the EB estimator is significantly more efficient than the ELL and the direct estimators. In this simulation study, the auxiliary variables are not very informative, and, due to this, ELL estimators turn out to be even less efficient than direct estimators. Conclusions for the poverty incidence are similar but plots are not reported here.

Turning to MSE estimation, the parametric bootstrap procedure described in Section 9.4.4 was implemented with $B = 500$ replicates and the results are plotted in Figure 9.2 for the poverty gap. The number of Monte Carlo simulations was $K = 500$, and the true values of the MSEs were independently computed with $K = 50,000$ Monte Carlo replicates. Figure 9.2 shows that the bootstrap MSE

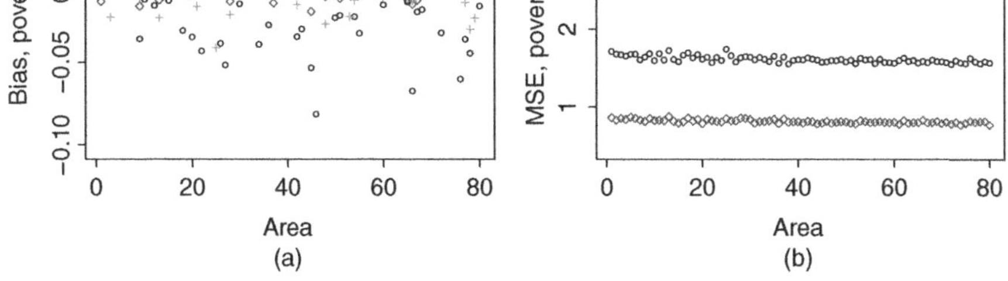

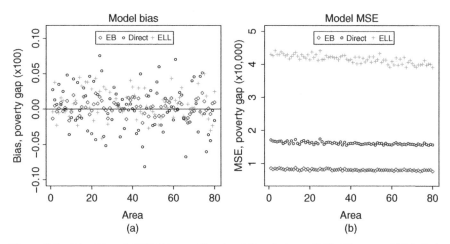

Figure 9.1 Bias (a) and MSE (b) over Simulated Populations of EB, Direct, and ELL Estimates of Percent Poverty Gap 100 F_{1i} for Each Area i. *Source*: Adapted from Molina and Rao (2010).

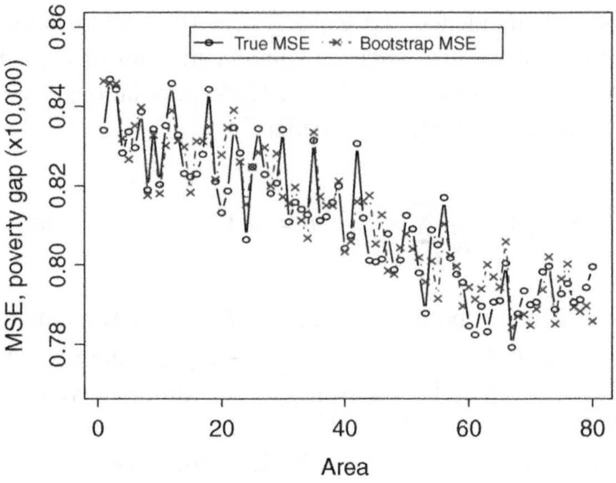

Figure 9.2 True MSEs of EB Estimators of Percent Poverty Gap and Average of Bootstrap MSE Estimators Obtained with $B = 500$ for Each Area i. *Source*: Adapted from Molina and Rao (2010).

estimator tracks the pattern of the true MSE values. Similar results were obtained for the poverty incidence.

An additional simulation experiment was carried out to study the performance of the estimators under repeated sampling from a fixed population. For this, a fixed population was generated exactly as described earlier. Keeping this population fixed, $K = 1000$ samples were drawn independently from the population using simple random sampling without replacement within each area. For each sample, the three estimators of the poverty incidences and gaps were computed, namely, direct, EB, and ELL estimators.

Figure 9.3 displays the design bias and the design MSE of the estimators of the poverty gap for each area. As expected, direct estimator shows practically zero bias, whereas the EB estimators show small biases, and the largest biases are for ELL estimators. Concerning MSE, ELL estimators show small MSEs for some of the areas but very large for other areas. In contrast, the MSEs of EB and direct estimators are small for all areas. Surprisingly, the design MSEs of the EB estimators are even smaller than those of the direct estimators for most of the areas.

In the above simulation study, $L = 50$ and $A = 50$ were used for EB and ELL methods, respectively. A limited comparison of EB estimators for $L = 50$ with the corresponding values for $L = 1000$ showed that the choice $L = 50$ gives fairly accurate results. In practice, however, when dealing with a given sample data set, it is advisable to use $L \geq 200$.

9.5 BINARY DATA

In this section, we study unit level models for binary responses, y_{ij}, that is, $y_{ij} = 1$ or 0. In this case, linear mixed models are not suitable and alternative models have

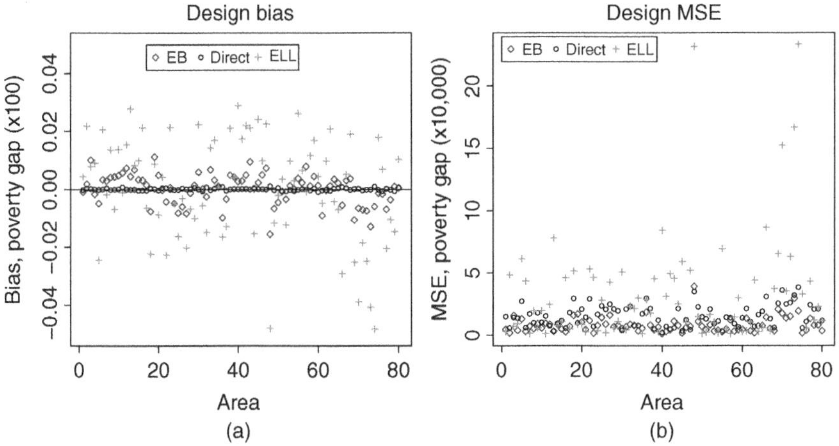

Figure 9.3 Bias (a) and MSE (b) of EB, Direct and ELL Estimators of the Percent Poverty Gap 100 F_{1i} for Each Area i under Design-Based Simulations. *Source*: Adapted from Molina and Rao (2010).

been proposed. If all the covariates $\mathbf{x}_{ij}$ associated with y_{ij} are area-specific, that is, $\mathbf{x}_{ij} = \mathbf{x}_i$, then we can transform the sample area proportions, $\hat{p}_i = \sum_{j=1}^{n_i} y_{ij}/n_i = y_i/n_i$, using an arcsine transformation, as in Example 9.2.1, and reduce the model to an area level model. However, the resulting transformed estimators, $\hat{\theta}_i$, may not satisfy the sampling model with zero mean sampling errors if n_i is small. Also, the estimator of the proportion, p_i, obtained from $\hat{\theta}_i^{\text{EB}}$ is not equal to the true EB estimator $\hat{p}_i^{\text{EB}}$ due to the transformation.

Unit level modeling avoids the above difficulties, and EB estimators of proportions may be obtained directly for the general case of unit-specific covariates. In Section 9.5.1, we study the special case of no covariates. A generalized linear mixed model is used in Section 9.5.2 to handle covariates.

9.5.1 *Case of No Covariates

We assume a two-stage model on the sample observations y_{ij}, $j = 1, \ldots, n_i, i = 1, \ldots, m$. In the first stage, we assume that $y_{ij}|p_i \overset{\text{iid}}{\sim} \text{Bernoulli}(p_i), i = 1, \ldots, m$. A model linking the p_i's is assumed in the second stage; in particular, $p_i \overset{\text{iid}}{\sim} \text{beta}(\alpha, \beta)$, with $\alpha > 0$ and $\beta > 0$, where beta(α, β) denotes the beta distribution with parameters α and β, that is,

$$f(p_i|\alpha,\beta) = \frac{\Gamma(\alpha+\beta)}{\Gamma(\alpha)\Gamma(\beta)} p_i^{\alpha-1}(1-p_i)^{\beta-1}, \quad \alpha > 0, \beta > 0, \tag{9.5.1}$$

where $\Gamma(\cdot)$ is the gamma function. We reduce $\mathbf{y}_i = (y_{i1}, \ldots, y_{in_i})^T$ to the sample total $y_i = \sum_{j=1}^{n_i} y_{ij}$, noting that y_i is a minimal sufficient statistic for the first-stage model.

We note that $y_i|p_i \overset{\text{ind}}{\sim}$ binomial (n_i, p_i), that is,

$$f(y_i|p_i) = \binom{n_i}{y_i} p_i^{y_i}(1-p_i)^{n_i-y_i}. \tag{9.5.2}$$

It follows from (9.5.1) and (9.5.2) that $p_i|y_i, \alpha, \beta \overset{\text{ind}}{\sim} \text{beta}(y_i+\alpha, n_i-y_i+\beta)$. Therefore, the Bayes estimator of p_i and the posterior variance of p_i are given by

$$\hat{p}_i^B(\alpha,\beta) = E(p_i|y_i,\alpha,\beta) = (y_i+\alpha)/(n_i+\alpha+\beta) \tag{9.5.3}$$

and

$$V(p_i|y_i,\alpha,\beta) = \frac{(y_i+\alpha)(n_i-y_i+\beta)}{(n_i+\alpha+\beta+1)(n_i+\alpha+\beta)^2}. \tag{9.5.4}$$

Note that the linking distribution, $f(p_i|\alpha,\beta)$, is a "conjugate prior" in the sense that the posterior (or conditional) distribution, $f(p_i|y_i,\alpha,\beta)$, has the same form as the prior distribution.

We obtain estimators of the model parameters from the marginal distribution, given by $y_i|\alpha,\beta \overset{\text{ind}}{\sim}$ beta-binomial, with pdf

$$f(y_i|\alpha,\beta) = \binom{n_i}{y_i} \frac{\Gamma(\alpha+y_i)\Gamma(\beta+n_i-y_i)}{\Gamma(\alpha+\beta+n_i)} \frac{\Gamma(\alpha+\beta)}{\Gamma(\alpha)\Gamma(\beta)}. \tag{9.5.5}$$

Maximum-likelihood (ML) estimators, $\hat{\alpha}_{\text{ML}}$ and $\hat{\beta}_{\text{ML}}$, may be obtained by maximizing the loglikelihood, given by

$$\begin{aligned} l(\alpha,\beta) = \text{const} + \sum_{i=1}^{m} \Bigg[& \sum_{h=0}^{y_i-1} \log(\alpha+h) + \sum_{h=0}^{n_i-y_i-1} \log(\beta+h) \\ & - \sum_{h=0}^{n_i-1} \log(\alpha+\beta+h) \Bigg], \end{aligned} \tag{9.5.6}$$

where $\sum_{h=0}^{y_i-1}\log(\alpha+h)$ is taken as zero if $y_i=0$ and $\sum_{h=0}^{n_i-y_i-1}\log(\beta+h)$ is taken as zero if $y_i=n_i$. A convenient representation is in terms of the mean $E(y_{ij})=\mu=\alpha/(\alpha+\beta)$ and $\tau=1/(\alpha+\beta)$, which is related to the intraclass correlation $\rho=\text{Corr}(y_{ij},y_{ik})=1/(\alpha+\beta+1)$ for $j\neq k$. In terms of μ and τ, the loglikelihood (9.5.6) takes the form

$$\begin{aligned} l(\mu,\tau) = \text{const} + \sum_{i=1}^{m} \Bigg[& \sum_{h=0}^{y_i-1} \log(\mu+h\tau) + \sum_{h=0}^{n_i-y_i-1} \log(1-\mu+h\tau) \\ & - \sum_{h=0}^{n_i-1} \log(1+h\tau) \Bigg]. \end{aligned} \tag{9.5.7}$$

Closed-form expressions for $\hat{\alpha}_{\text{ML}}$ and $\hat{\beta}_{\text{ML}}$ (or $\hat{\mu}_{\text{ML}}$ and $\hat{\tau}_{\text{ML}}$) do not exist, but the ML estimates may be obtained by the Newton–Raphson method or some other iterative method. McCulloch and Searle (2001, Section 2.6) have given the asymptotic covariance matrices of $(\hat{\alpha}_{\text{ML}}, \hat{\beta}_{\text{ML}})$, and $(\hat{\mu}_{\text{ML}}, \hat{\tau}_{\text{ML}})$. Substituting $\hat{\alpha}_{\text{ML}}$ and $\hat{\beta}_{\text{ML}}$ in (9.5.3) and (9.5.4), we get the EB estimator of p_i and the estimated posterior variance.

We can also use simple method of moments estimators of α and β. We equate the weighted sample mean $\hat{p} = \sum_{i=1}^{m}(n_i/n)\hat{p}_i$ and the weighted sample variance $s_p^2 = \sum_{i=1}^{m}(n_i/n)(\hat{p}_i - \hat{p})^2$ to their expected values and solve the resulting moment equations for α and β, where $n = \sum_{i=1}^{m} n_i$. This leads to moment estimators, $\hat{\alpha}$ and $\hat{\beta}$, given by

$$\hat{\alpha}/(\hat{\alpha} + \hat{\beta}) = \hat{p} \tag{9.5.8}$$

and

$$\frac{1}{\hat{\alpha} + \hat{\beta} + 1} = \frac{ns_p^2 - \hat{p}(1 - \hat{p})(m - 1)}{\hat{p}(1 - \hat{p})[n - \sum_{i=1}^{m} n_i^2/n - (m - 1)]}; \tag{9.5.9}$$

see Kleinman (1973).

We substitute the moment estimators $\hat{\alpha}$ and $\hat{\beta}$ into (9.5.3) to get an EB estimator of p_i as

$$\hat{p}_i^{\text{EB}} = \hat{p}_i^B(\hat{\alpha}, \hat{\beta}) = \hat{\gamma}_i\hat{p}_i + (1 - \hat{\gamma}_i)\hat{p} = k_i(\hat{p}_i, \hat{\alpha}, \hat{\beta}), \tag{9.5.10}$$

where $\hat{\gamma}_i = n_i/(n_i + \hat{\alpha} + \hat{\beta})$. Note that $\hat{p}_i^{\text{EB}}$ is a weighted average of the direct estimator $\hat{p}_i$ and the synthetic estimator $\hat{p}$, and more weight is given to $\hat{p}_i$ as the ith area sample size, n_i, increases. It is therefore similar to the FH estimator for the basic area level model, but the weight $\hat{\gamma}_i$ avoids the assumption of known sampling variance of $\hat{p}_i$. The estimator $\hat{p}_i^{\text{EB}}$ is nearly unbiased for p_i in the sense that its bias, $E(\hat{p}_i^{\text{EB}} - p_i)$, is of order m^{-1}, for large m.

A naive EB approach uses $\hat{p}_i^{\text{EB}}$ as the estimator of realized p_i, and its variability is measured by the estimated posterior variance $V(p_i|y_i, \alpha, \beta) =: g_{1i}(\alpha, \beta, y_i)$, evaluated at $\alpha = \hat{\alpha}$ and $\beta = \hat{\beta}$. However, $g_{1i}(\hat{\alpha}, \hat{\beta}, y_i)$ can lead to severe underestimation of $\text{MSE}(\hat{p}_i^{\text{EB}})$ because it ignores the variability associated with $\hat{\alpha}$ and $\hat{\beta}$. Note that $\text{MSE}(\hat{p}_i^{\text{EB}}) = E[g_{1i}(\alpha, \beta, y_i)] + E(\hat{p}_i^{\text{EB}} - \hat{p}_i^B)^2 = M_{1i} + M_{2i}$, say.

The jackknife method may be applied to the estimation of $\text{MSE}(\hat{p}_i^{\text{EB}})$. We have $\hat{p}_i^{\text{EB}} = k_i(\hat{p}_i, \hat{\alpha}, \hat{\beta}), \hat{p}_{i,-\ell}^{\text{EB}} = k_i(\hat{p}_i, \hat{\alpha}_{-\ell}, \hat{\beta}_{-\ell})$ and the estimator of M_{2i} is taken as

$$\hat{M}_{2i} = \frac{m-1}{m}\sum_{\ell=1}^{m}(\hat{p}_{i,-\ell}^{\text{EB}} - \hat{p}_i^{\text{EB}})^2, \tag{9.5.11}$$

where $\hat{\alpha}_{-\ell}$ and $\hat{\beta}_{-\ell}$ are the delete$-\ell$ moment estimators obtained from $\{(\hat{p}_i, n_i); i = 1, \ldots, \ell - 1, \ell + 1, \ldots, m\}$. Furthermore, we propose to estimate M_{1i} by

$$\hat{M}_{1i} = g_{1i}(\hat{\alpha}, \hat{\beta}, y_i) - \sum_{\substack{\ell=1 \\ \ell \neq i}}^{m}[g_{1i}(\hat{\alpha}_{-\ell}, \hat{\beta}_{-\ell}, y_i) - g_{1i}(\hat{\alpha}, \hat{\beta}, y_i)] \tag{9.5.12}$$

(Lohr and Rao 2009). The jackknife estimator of M_{1i} given in Rao (2003a, p. 199) is not second-order unbiased, unlike (9.5.12), as shown by Lohr and Rao (2009). The jackknife estimator of MSE($\hat{p}_i^{\text{EB}}$) is then given by

$$\text{mse}_J(\hat{p}_i^{\text{EB}}) = \hat{M}_{1i} + \hat{M}_{2i}. \qquad (9.5.13)$$

This estimator is second-order unbiased for MSE($\hat{p}_i^{\text{EB}}$).

Note that the leading term $g_{1i}(\hat{\alpha}, \hat{\beta}, y_i)$ of $\text{mse}_J(\hat{p}_i^{\text{EB}})$ is area-specific in the sense that it depends on y_i. Our method of estimating M_{1i} differs from the unconditional method used by Jiang, Lahiri, and Wan (2002). They evaluated $E[g_{1i}(\alpha, \beta, y_i)] = \tilde{g}_{1i}(\alpha, \beta)$ using the marginal distribution of y_i, and then used $\tilde{g}_{1i}(\hat{\alpha}, \hat{\beta})$ in (9.5.12) in place of $g_{1i}(\hat{\alpha}, \hat{\beta}, y_i)$ to get an estimator of M_{1i}. This estimator is computationally more complex than $\hat{M}_{1i}$ and its leading term, $\tilde{g}_{1i}(\hat{\alpha}, \hat{\beta})$, is not area-specific, unlike $g_{1i}(\hat{\alpha}, \hat{\beta}, y_i)$.

The jackknife estimator (9.5.13) can also be used to estimate the conditional MSE of $\hat{p}_i^{\text{EB}}$ given by $\text{MSE}_c(\hat{p}_i^{\text{EB}}) = E[(\hat{p}_i^{\text{EB}} - p_i)^2 | y_i]$, which depends on y_i, unlike in the case of linear mixed models. Lohr and Rao (2009) showed that $\text{mse}_J(\hat{p}_i^{\text{EB}})$ is conditionally second-order unbiased in the sense that its conditional bias is of order $o_p(m^{-1})$ in probability.

Alternative two-stage models have also been proposed. The first-stage model is not changed, but the second-stage linking model is changed to either (i) $\text{logit}(p_i) = \log[p_i/(1-p_i)] \overset{\text{iid}}{\sim} N(\mu, \sigma^2)$ or (ii) $\Phi^{-1}(p_i) \overset{\text{iid}}{\sim} N(\mu, \sigma^2)$, where $\Phi(\cdot)$ is the cumulative distribution function (CDF) of a $N(0, 1)$ variable. The models (i) and (ii) are called logit-normal and probit-normal models, respectively. Implementation of EB is more complicated for these alternative models because no closed-form expressions for the Bayes estimator and the posterior variance of p_i exist.

For the logit-normal model, the Bayes estimator of p_i can be expressed as a ratio of single-dimensional integrals. Writing p_i as $p_i = h_1(\mu + \sigma z_i)$, where $h_1(a) = e^a/(1 + e^a)$ and $z_i \sim N(0, 1)$, we obtain $\hat{p}_i^B(\mu, \sigma) = E(p_i | y_i, \mu, \sigma)$ from the conditional distribution of z_i given y_i. We have

$$\hat{p}_i^B(\mu, \sigma) = A(y_i, \mu, \sigma)/B(y_i, \mu, \sigma), \qquad (9.5.14)$$

where

$$A(y_i, \mu, \sigma) = E[h_1(\mu + \sigma z) \exp\{h_2(y_i, \mu + \sigma z)\}] \qquad (9.5.15)$$

and

$$B(y_i, \mu, \sigma) = E[\exp\{h_2(y_i, \mu + \sigma z)\}], \qquad (9.5.16)$$

where $h_2(y_i, a) = a y_i - n_i \log(1 + e^a)$ and the expectation is with respect to $z \sim N(0, 1)$; see McCulloch and Searle (2001, p. 67). We can evaluate (9.5.15) and (9.5.16) by simulating samples from $N(0, 1)$. Alternatively, numerical integration, as sketched below, can be used.

The loglikelihood, $l(\mu, \sigma)$, for the logit-normal model may be written as

$$l(\mu, \sigma) = \text{const} + \sum_{i=1}^{m} \log[B(y_i, \mu, \sigma)], \tag{9.5.17}$$

where $B(y_i, \mu, \sigma)$ is given by (9.5.16). Each of the single-dimensional integrals in (9.5.15)–(9.5.16) is of the form $E[a(z)]$ for a given function $a(z)$, which may be approximated by a finite sum of the form $\sum_{k=1}^{d} b_k a(z_k)/\sqrt{\pi}$, using Gauss–Hermite quadrature (McCulloch and Searle 2001, Section 10.3). The weights b_k and the evaluation points z_k for a specified value of d can be calculated using mathematical software. For ML estimation of μ and $\sigma, d = 20$ usually provides good approximations. Derivatives of $l(\mu, \sigma)$, needed in Newton–Raphson type methods for calculating ML estimates, can be approximated in a similar manner.

Using ML estimators $\hat{\mu}$ and $\hat{\sigma}$, we obtain an EB estimator of p_i as $\hat{p}_i^{\text{EB}} = \hat{p}_i^{\text{B}}(\hat{\mu}, \hat{\sigma})$. The posterior variance, $V(p_i|y_i, \mu, \sigma)$, may also be expressed in terms of expectation over $z \sim N(0, 1)$, noting that $V(p_i|y_i, \mu, \sigma) = E(p_i^2|y_i, \mu, \sigma) - [\hat{p}_i^B(\mu, \sigma)]^2$. Denoting $V(p_i|y_i, \mu, \sigma) = g_{1i}(\mu, \sigma, y_i)$, the jackknife method can be applied to obtain a second-order unbiased estimator of $\text{MSE}(\hat{p}_i^{\text{EB}})$. We obtain the jackknife estimator, $\text{mse}_J(\hat{p}_i^{\text{EB}})$, from (9.5.13) by substituting $\hat{p}_i^{\text{EB}} = k_i(y_i, \hat{\mu}, \hat{\sigma})$ and $\hat{p}_{i,-\ell}^{\text{EB}} = k_i(y_i, \hat{\mu}_{-\ell}, \hat{\sigma}_{-\ell})$ in (9.5.11) and $g_{1i}(\hat{\mu}, \hat{\sigma}, y_i)$ and $g_{1i}(\hat{\mu}_{-\ell}, \hat{\sigma}_{-\ell}, y_i)$ in (9.5.12), where $\hat{\mu}_{-\ell}$ and $\hat{\sigma}_{-\ell}$ are the delete$-\ell$ ML estimators obtained from $\{(y_i, n_i), i = 1, \ldots, \ell - 1, \ell + 1, \ldots, m\}$.

Computation of $\text{mse}_J(\hat{p}_i^{\text{EB}})$ using ML estimators is very cumbersome. However, computations may be simplified by using moment estimators of μ and σ obtained by equating the weighted mean $\hat{p}$ and the weighted variance s_p^2 to their expected values, as in the beta-binomial model, and solving the resulting equations for μ and σ. The expected values involve the marginal moments $E(y_{ij}) = E(p_i)$ and $E(y_{ij}y_{ik}) = E(p_i^2), j \neq k$, which can be calculated by numerical or Monte Carlo integration, using $E(p_i) = E[h_1(\mu + \sigma z)]$ and $E(p_i^2) = E[h_1^2(\mu + \sigma z)]$. Jiang (1998) equated $\sum_{i=1}^{m} y_i = n\hat{p}$ and $\sum_{i=1}^{m}(\sum_{j \neq k} y_{ij}y_{ik}) = \sum_{i=1}^{m}(y_i^2 - y_i) = \sum_{i=1}^{m} n_i^2\hat{p}_i^2 - n\hat{p}$ to their expected values to obtain different moment estimators, that is, solved equations

$$\sum_{i=1}^{m} y_i = nE[h_1(\mu + \sigma z)], \quad \sum_{i=1}^{m}(y_i^2 - y_i) = \left[\sum_{i=1}^{m} n_i(n_i - 1)\right] E[h_1^2(\mu + \sigma z)] \tag{9.5.18}$$

for μ and σ to get the moment estimators, and established asymptotic consistency as $m \to \infty$. Jiang and Zhang (2001) proposed more efficient moment estimators, using a two-step procedure. In the first step, (9.5.18) is solved for μ and σ, and then used in the second step to produce two "optimal" weighted combinations of $(\sum_{i=1}^{m} y_i, y_1^2 - y_1, \ldots, y_m^2 - y_m)$. Note that the first step uses $(1, 0, \ldots, 0)$ and $(0, 1, \ldots, 1)$ as the weights. The optimal weights involve the derivatives of $E[h_1(\mu + \sigma z)]$ and $E[h_1^2(\mu + \sigma z)]$ with respect to μ and σ as well as the covariance matrix of $(\sum_{i=1}^{m} y_i, y_1^2 - y_1, \ldots, y_m^2 - y_m)$, which depend on μ and σ. Replacing μ and σ by the first-step estimators, estimated weighted combinations are obtained.

The second-step moment estimators are then obtained by equating the estimated weighted combinations to their expectations, treating the estimated weights as fixed. Simulation studies indicated considerable overall gain in efficiency over the first-step estimators in the case of unequal n_i's. Note that the optimal weights reduce to first-step weights in the balanced case, $n_i = \overline{n}$. Jiang and Zhang (2001) established the asymptotic consistency of the second-step moment estimators.

Jiang and Lahiri (2001) used $\hat{p}_i^{\text{EB}}$ based on the first-step moment estimators and obtained a Taylor expansion estimator of $\text{MSE}(\hat{p}_i^{\text{EB}})$, similar to the second-order unbiased MSE estimators of Chapter 5 for the linear mixed model. This MSE estimator is second-order unbiased, as in the case of the jackknife MSE estimator.

Lahiri et al. (2007) proposed a bootstrap-based perturbation method to estimate $\text{MSE}(\hat{p}_i^{\text{EB}}) = M_i(\boldsymbol{\delta}) = M_{1i}(\boldsymbol{\delta}) + M_{2i}(\boldsymbol{\delta})$ under a given two-stage model, where $\boldsymbol{\delta}$ is the vector of model parameters. Their estimator of $M_i(\boldsymbol{\delta})$ does not require the explicit evaluation of $M_{1i}(\boldsymbol{\delta})$ at $\boldsymbol{\delta} = \hat{\boldsymbol{\delta}}$ using the marginal distribution of y_i, unlike the Jiang et al. (2002) estimator. Moreover, it is always positive and second-order unbiased. We refer the reader to Lahiri et al. (2007) for technical details, which are quite complex. It may be noted that the proposed method is applicable to general two-stage area level models.

9.5.2 Models with Covariates

The logit-normal model of Section 9.5.1 readily extends to the case of covariates. In the first stage, we assume that $y_{ij}|p_{ij} \overset{\text{ind}}{\sim} \text{Bernoulli}(p_{ij})$ for $j = 1, \ldots, N_i, i = 1, \ldots, m$. The probabilities p_{ij} are linked in the second stage by assuming a logistic regression model with random area effects, $\text{logit}(p_{ij}) = \mathbf{x}_{ij}^T\boldsymbol{\beta} + v_i$, where $v_i \overset{\text{iid}}{\sim} N(0, \sigma_v^2)$ and $\mathbf{x}_{ij}$ is the vector of fixed covariates. The two-stage model belongs to the class of generalized linear mixed models and is called a logistic linear mixed model. The target parameters are the small area proportions $P_i = \overline{Y}_i = \sum_j y_{ij}/N_i, i = 1, \ldots, m$. As in the case of the basic unit level model, we assume that the model holds for the sample $\{(y_{ij}, \mathbf{x}_{ij}); j \in s_i, i = 1, \ldots, m\}$, where s_i is the sample of size n_i from the ith area.

We express the area proportion P_i as $P_i = f_i\overline{y}_i + (1 - f_i)\overline{y}_{ir}$, where $f_i = n_i/N_i, \overline{y}_i$ is the sample mean (proportion) and $\overline{y}_{ir} = \sum_{\ell \in r_i} y_{i\ell}/(N_i - n_i)$ is the mean of the nonsampled units r_i in the ith area. Now noting that $E(y_{i\ell}|p_{i\ell}, \mathbf{y}_i, \boldsymbol{\beta}, \sigma_v) = p_{i\ell}$, for $\ell \in r_i$, the Bayes estimator of $\overline{y}_{ir}$ is given by $\hat{p}_{i(c)}^B = E(p_{i(c)}|\mathbf{y}_i, \boldsymbol{\beta}, \sigma_v)$, where $p_{i(c)} = \sum_{\ell \in r_i} p_{i\ell}/(N_i - n_i)$ and $\mathbf{y}_i$ is the vector of sample y-values from the ith area. Therefore, the Bayes estimator of P_i may be expressed as

$$\hat{P}_i^B = \hat{P}_i^B(\boldsymbol{\beta}, \sigma_v) = f_i\overline{y}_i + (1 - f_i)\hat{p}_{i(c)}^B. \tag{9.5.19}$$

If the sampling fraction f_i is negligible, we can express $\hat{P}_i^B$ as

$$\hat{P}_i^B \approx \frac{1}{N_i}E\left(\sum_{\ell=1}^{N_i} p_{i\ell}|\mathbf{y}_i, \boldsymbol{\beta}, \sigma_v\right). \tag{9.5.20}$$

The posterior variance of P_i reduces to

$$V(P_i|\mathbf{y}_i, \boldsymbol{\beta}, \sigma_v) = (1-f_i)^2 E(\bar{y}_{ir} - \hat{p}_{i(c)}^B)^2$$
$$= N_i^{-2}\left\{E\left[\sum_{\ell\in r_i} p_{i\ell}(1-p_{i\ell})|\mathbf{y}_i, \boldsymbol{\beta}, \sigma_v)\right] + V\left[\sum_{\ell\in r_i} p_{i\ell}|\mathbf{y}_i, \boldsymbol{\beta}, \sigma_v\right]\right\}; \tag{9.5.21}$$

see Malec et al. (1997). Note that (9.5.21) involves expectations of the form $E(\sum_{\ell\in r_i} p_{i\ell}^2|\mathbf{y}_i, \boldsymbol{\beta}, \sigma_v)$ and $E[(\sum_{\ell\in r_i} p_{i\ell})^2|\mathbf{y}_i, \boldsymbol{\beta}, \sigma_v]$, as well as the expectation $E(\sum_{\ell\in r_i} p_{i\ell}|\mathbf{y}_i, \boldsymbol{\beta}, \sigma_v) = (N_i - n_i)\hat{p}_{ic}^B$. No closed-form expressions for these expectations exist. However, we can express the expectations as ratios of single-dimensional integrals, similar to (9.5.14). For example, writing $\sum_{\ell\in r_i} p_{i\ell}^2$ as a function of $z \sim N(0, 1)$, we can write

$$E\left(\sum_{\ell\in r_i} p_{i\ell}^2|\mathbf{y}_i, \boldsymbol{\beta}, \sigma_v\right) = \frac{E\left\{\left(\sum_{\ell\in r_i} p_{i\ell}^2\right)\exp\left[h_i\left(\sum_{j\in s_i}\mathbf{x}_{ij}^T y_{ij}, y_i, \sigma z, \boldsymbol{\beta}\right)\right]\right\}}{E\left\{\exp\left[h_i\left(\sum_{j\in s_i}\mathbf{x}_{ij}^T y_{ij}, y_i, \sigma z, \boldsymbol{\beta}\right)\right]\right\}}, \tag{9.5.22}$$

where $y_i = \sum_{j\in s_i} y_{ij}$,

$$h_i\left(\sum_{j\in s_i}\mathbf{x}_{ij}^T y_{ij}, y_i, \sigma z, \boldsymbol{\beta}\right) = \left(\sum_{j\in s_i}\mathbf{x}_{ij}^T y_{ij}\right)\boldsymbol{\beta} + (\sigma z)y_i - \sum_{j\in s_i}\log[1+\exp(\mathbf{x}_{ij}^T\boldsymbol{\beta} + \sigma z)] \tag{9.5.23}$$

and the expectation is with respect to $z \sim N(0, 1)$. Note that (9.5.23) reduces to $h_2(y_i, \mu + \sigma z)$ for the logit-normal model without covariates, that is, with $\mathbf{x}_{ij}^T\boldsymbol{\beta} = \mu$ for all i and j.

ML estimation of model parameters, $\boldsymbol{\beta}$ and σ_v, for the logistic linear mixed model and other generalized linear mixed models has received considerable attention in recent years. Methods proposed include numerical quadrature, EM algorithm, Markov chain Monte Carlo (MCMC), and stochastic approximation. We refer the readers to McCulloch and Searle, (2001, Section 10.4) for details of the algorithms. Simpler methods, called penalized quasi-likelihood (PQL), based on maximizing the joint distribution of $\mathbf{y} = (\mathbf{y}_1^T, \ldots, \mathbf{y}_m^T)^T$ and $\mathbf{v} = (v_1, \ldots, v_m)^T$ with respect to $\boldsymbol{\beta}$ and $\mathbf{v}$ have also been proposed. For the special case of linear mixed models under normality, PQL leads to "mixed model" equations 5.2.8, whose solution is identical to the BLUE of $\boldsymbol{\beta}$ and the BLUP of $\mathbf{v}$. However, for the logistic linear mixed model, the PQL estimator of $\boldsymbol{\beta}$ is asymptotically biased (as $m \to \infty$) and hence inconsistent, unlike the ML estimator of $\boldsymbol{\beta}$.

Consistent moment estimators of $\boldsymbol{\beta}$ and σ_v may be obtained by equating $\sum_{i=1}^m \sum_{j\in s_i}\mathbf{x}_{ij}y_{ij}$ and $\sum_{i=1}^m(\sum_{j\neq k} y_{ij}y_{ik})$ to their expectations and then solving the equations for $\boldsymbol{\beta}$ and σ_v. The expectations depend on $E(y_{ij}) = E(p_{ij}) = E[(h_1(\mathbf{x}_{ij}^T\boldsymbol{\beta} + \sigma z)]$ and $E(y_{ij}y_{ik}) = E(p_{ij}p_{ik}) = E[h_1(\mathbf{x}_{ij}^T\boldsymbol{\beta} + \sigma z)h_1(\mathbf{x}_{ik}^T\boldsymbol{\beta} + \sigma z)]$. The two-step method of Section 9.5.1 can be extended to get more efficient moment estimators.

Sutradhar (2004) proposed a quasi-likelihood (QL) method for the estimation of $\boldsymbol{\beta}$ and σ_v^2. It requires the covariance matrix of the vectors $\mathbf{y}_i$ and $\mathbf{u}_i = (y_{i1}y_{i2}, \ldots, y_{ij}y_{ik}, \ldots, y_{i(n_i-1)}y_{n_i})^T$. This covariance matrix involves the evaluation of third- and fourth-order expectations of the form $E(p_{ij}\, p_{ik}\, p_{i\ell})$ and $E(p_{ij}\, p_{ik}\, p_{i\ell}\, p_{it})$ in addition to $E(p_{ij})$ and $E(p_{ij}^2)$. These expectations can be evaluated by simulating samples from $N(0, 1)$ or by one-dimensional numerical integration, as outlined in Section 9.5.1.

Sutradhar (2004) made a numerical comparison of the QL method and the one-step moment method in terms of asymptotic variance, assuming the model $\text{logit}(p_{ij}) = \beta x_i + v_i$, with $v_i \overset{\text{iid}}{\sim} N(0, \sigma_v^2)$ and equal sample sizes $n_i = \overline{n}$, where x_i is an area level covariate. The QL method turned out to be significantly more efficient in estimating σ_v^2 than the first-step moment method. Relative efficiency of QL with respect to the two-step method has not been studied. Note that the equivalence of the first-step and the second-step moment estimators in the case of equal n_i holds only for the special case of a mean model $\text{logit}(p_{ij}) = \mu + v_i$.

Using either ML or moment estimators $\hat{\boldsymbol{\beta}}$ and $\hat{\sigma}_v$, we obtain an EB estimator of the ith area proportion P_i as $\hat{P}_i^{\text{EB}} = \hat{P}_i^{B}(\hat{\boldsymbol{\beta}}, \hat{\sigma}_v)$. The estimator $\hat{P}_i^{\text{EB}}$ is nearly unbiased for P_i in the sense that its bias, $E(\hat{P}_i^{\text{EB}} - \hat{P}_i^{\text{B}})$, is of order m^{-1} for large m. A naive EB approach uses $\hat{P}_i^{\text{EB}}$ as the estimator of the realized proportion P_i and its variability is measured by the estimated posterior variance $V(P_i|\mathbf{y}_i, \hat{\boldsymbol{\beta}}, \hat{\sigma}_v) = g_{1i}(\hat{\boldsymbol{\beta}}, \hat{\sigma}_v, \mathbf{y}_i)$, using (9.5.21). However, the estimated posterior variance ignores the variability associated with $\hat{\boldsymbol{\beta}}$ and $\hat{\sigma}_v$.

The bootstrap method of estimating MSE, described in Section 7.2.4 for the basic unit level model, is applicable to $\text{MSE}(\hat{P}_i^{\text{EB}})$, in particular the one-step bootstrap estimator (7.2.26) and the bias-corrected estimators (7.2.32) or (7.2.33) based on double bootstrap. Pfeffermann and Correa (2012) proposed an alternative empirical bias-correction method and applied it to the estimation of $\text{MSE}(\hat{P}_i^{\text{EB}})$. This method involves (i) drawing several plausible values of the model parameter vector $\boldsymbol{\delta} = (\boldsymbol{\beta}^T, \sigma_v^2)^T$, say $\boldsymbol{\delta}_1, \ldots, \boldsymbol{\delta}_T$; (ii) generating a pseudo-original sample from the model corresponding to each $\boldsymbol{\delta}_t$; and (iii) generating a large number of bootstrap samples from each pseudo-original sample generated in step (i). The method makes use of the one-step bootstrap estimators computed from the original sample and the pseudo-original samples to construct an empirical bootstrap bias-corrected MSE estimator. We refer the reader to Pfeffermann–Correa (PC) paper for further details of the method. PC studied the performance of the proposed method in a simulation study, using the linking model $\text{logit}(p_{ij}) = \mathbf{x}_{ij}^T\boldsymbol{\beta} + v_i$ with $v_i \overset{\text{iid}}{\sim} N(0, \sigma_v^2), i = 1, \ldots, m$. In terms of average absolute relative bias ($\overline{\text{ARB}}$), the PC estimator of MSE performed similar to the double-bootstrap MSE estimator, with $\overline{\text{ARB}}$ values of 5.5% and 4.15%, respectively. On the other hand, the one-step bootstrap MSE estimator led to consistent underestimation (as expected). The PC estimator performed significantly better than the double-bootstrap estimator in terms of average coefficient of variation (CV), with CV of 26% compared to about 50% for the double bootstrap estimator.

It is clear from the foregoing account that the implementation of EB for the logistic linear mixed model is quite cumbersome computationally. Approximate EB methods that are computationally simpler have been proposed in the literature, but those methods are not asymptotically valid as $m \to \infty$. As a result, the approximate methods might perform poorly in practice, especially for small sample sizes, n_i. We give a brief account of an approximate method proposed by MacGibbon and Tomberlin (1989), based on approximating the posterior distribution $f(\boldsymbol{\beta}, \mathbf{v}|\mathbf{y}, \sigma_v^2)$ by a multivariate normal distribution, assuming a flat prior on $\boldsymbol{\beta}$, that is, $f(\boldsymbol{\beta}) =$ constant. We have

$$f(\boldsymbol{\beta}, \mathbf{v}|\mathbf{y}, \sigma_v^2) \propto \left[\prod_{i,j} p_{ij}^{y_{ij}}(1-p_{ij})^{1-y_{ij}}\right] \exp\left(-\sum_{i=1}^{m} v_i^2/2\sigma_v^2\right), \tag{9.5.24}$$

where $\text{logit}(p_{ij}) = \mathbf{x}_{ij}^T\boldsymbol{\beta} + v_i$. The right-hand side (9.5.24) is approximated by a multivariate normal having its mean at the mode of (9.5.24) and covariance matrix equal to the inverse of the information matrix evaluated at the mode. The elements of the information matrix are given by the second derivatives of $\log f(\boldsymbol{\beta}, \mathbf{v}|\mathbf{y}, \sigma_v^2)$ with respect to the elements of $\boldsymbol{\beta}$ and $\mathbf{v}$, multiplied by -1. The mode $(\boldsymbol{\beta}^*, \mathbf{v}^*)$ is obtained by maximizing $\log f(\boldsymbol{\beta}, \mathbf{v}|\mathbf{y}, \sigma_v^2)$ with respect to $\boldsymbol{\beta}$ and $\mathbf{v}$, using the Newton–Raphson algorithm, and then substituting the ML estimator of σ_v^2 obtained from the EM algorithm. An EB estimator of P_i is then taken as

$$\hat{P}_{i*}^{\text{EB}} = \frac{1}{N_i}\left(n_i\bar{y}_i + \sum_{j\in r_i} p_{ij}^*\right), \tag{9.5.25}$$

where

$$p_{ij}^* = \exp(\mathbf{x}_{ij}^T\boldsymbol{\beta}^* + v_i^*)/[1+\exp(\mathbf{x}_{ij}^T\boldsymbol{\beta}^* + v_i^*)]. \tag{9.5.26}$$

Unlike the EB estimator $\hat{P}_i^{\text{EB}} = \hat{P}_i^B(\hat{\boldsymbol{\beta}}, \hat{\sigma}_v)$, the estimator $\hat{P}_{i*}^{\text{EB}}$ is not nearly unbiased for P_i. Its bias is of order n_i^{-1}, and hence it may not perform well when n_i is small; v_i^* is a biased estimator of v_i to the same order.

Farrell, MacGibbon, and Tomberlin (1997a) used a bootstrap measure of accuracy of $\hat{P}_{i*}^{\text{EB}}$, similar to the Laird and Louis (1987) parametric bootstrap (also called type III bootstrap) to account for the sampling variability of the estimates of $\boldsymbol{\beta}$ and σ_v^2. Simulation results, based on $m = 20$ and $n_i = 50$, indicated good performance of the proposed EB estimator $\hat{P}_{i*}^{\text{EB}}$. Farrell, MacGibbon, and Tomberlin (1997b) relaxed the assumption of normality on the v_i's, using the following linking model: $\text{logit}(p_{ij}) = \mathbf{x}_{ij}^T\boldsymbol{\beta} + v_i$ and area effects v_i are iid following an unspecified distribution. They used a nonparametric ML method, proposed by Laird (1978), to obtain an EB estimator of P_i, similar in form to (9.5.25). They also used the type II bootstrap of Laird and Louis (1987) to obtain a bootstrap measure of accuracy.

The EB estimators $\hat{P}_i^{\text{EB}}$ and $\hat{P}_{i*}^{\text{EB}}$ require the knowledge of individual $\mathbf{x}$-values in the population, unlike the EB estimator for the basic unit level model, which depends only on the population mean $\overline{\mathbf{X}}_i = N_i^{-1} \sum_{j=1}^{N_i} \mathbf{x}_{ij}$. This is a practical drawback of logistic linear mixed models and other nonlinear models because microdata for all individuals in a small area may not be available. Farrell, MacGibbon, and Tomberlin (1997c) obtained an approximation to $\hat{P}_{i*}^{\text{EB}}$ depending only on the mean $\overline{\mathbf{X}}_i$ and the cross-product matrix $\sum_{j=1}^{N_i} \mathbf{x}_{ij}\mathbf{x}_{ij}^T$ with elements $\sum_{j=1}^{N_i} x_{ija}x_{ijb}, a, b = 1, \ldots, p$. It uses a second-order multivariate Taylor expansion of p_{ij}^* around $\overline{\mathbf{x}}_{ir}$, the mean vector of the nonsampled elements $\mathbf{x}_{ij}, j \in r_i$. However, the deviations $\mathbf{x}_{ij} - \overline{\mathbf{x}}_{ir}$ in the expansion are of order $O(1)$, and hence terms involving higher powers of the deviations are not necessarily negligible. Farrell et al. (1997c) investigated the accuracy of the second-order approximation for various distributions of $\mathbf{x}_{ij}^T\boldsymbol{\beta} + v_i$ and provided some guidelines.

9.6 DISEASE MAPPING

Mapping of small area mortality (or incidence) rates of diseases such as cancer is a widely used tool in public health research. Such maps permit the analysis of the geographical variation in the rates of diseases, which may be useful in formulating and assessing etiological hypotheses, resource allocation, and identifying areas of unusually high-risk warranting intervention. Examples of disease rates studied in the literature include lip cancer rates in the 56 counties (small areas) of Scotland (Clayton and Kaldor 1987), incidence of leukemia in 281 census tracts (small areas) of upstate New York (Datta, Ghosh, and Waller 2000), stomach cancer mortality rates in Missouri cities (small areas) for males aged 47–64 years (Tsutakawa, Shoop, and Marienfeld 1985), all cancer mortality rates for white males in health service areas (small areas) of the United States (Nandram, Sedransk, and Pickle 1999), prostate cancer rates in Scottish counties (Langford et al. 1999), and infant mortality rates for local health areas (small areas) in British Columbia, Canada (Dean and MacNab 2001). We refer the reader to the October 2000 issue of *Statistics in Medicine* for a review of methods used in disease mapping. In disease mapping, data on event counts and related auxiliary variables are typically obtained from administrative sources, and sampling is not used.

Suppose that the country (or the large region) used for disease mapping is divided into m nonoverlapping small areas. Let θ_i be the unknown relative risk (RR) in the ith area. A direct (or crude) estimator of θ_i is given by the standardized mortality ratio (SMR), $\hat{\theta}_i = y_i/e_i$, where y_i and e_i denote, respectively, the observed and expected number of deaths (cases) over a given period in area i, for $i = 1, \ldots, m$. The expected counts e_i are calculated as

$$e_i = n_i \left(\sum_{i=1}^{m} y_i / \sum_{i=1}^{m} n_i \right), \tag{9.6.1}$$

where n_i is the number of person-years at risk in the ith area, and then treated as fixed. Some authors use mortality (event) rates, τ_i, as parameters, instead of RRs, and a crude estimator of τ_i is then given by $\hat{\tau}_i = y_i/n_i$. The two approaches, however, are equivalent because the factor $\sum_{i=1}^m y_i / \sum_{i=1}^m n_i$ is treated as a constant.

A common assumption in disease mapping is that $y_i|\theta_i \overset{\text{ind}}{\sim} \text{Poisson}(e_i\theta_i)$. Under this assumption, the ML estimator of θ_i is the SMR, $\hat{\theta}_i = y_i/e_i$. However, a map of crude rates $\{\hat{\theta}_i\}$ can badly distort the geographical distribution of disease incidence or mortality because it tends to be dominated by areas of low population, e_i, exhibiting extreme SMR's that are least reliable. Note that $V(\hat{\theta}_i) = \theta_i/e_i$ is large if e_i is small.

EB or HB methods provide reliable estimators of RR by borrowing strength across areas. As a result, maps based on EB or HB estimates are more reliable than crude maps. In this section, we give a brief account of EB methods based on simple linking models. Various extensions have been proposed in the literature, including bivariate models and models exhibiting spatial correlation.

9.6.1 Poisson–Gamma Model

We first study a two-stage model for count data $\{y_i\}$ similar to the beta-binomial model for binary data. In the first stage, we assume that $y_i \overset{\text{ind}}{\sim} \text{Poisson}(e_i\theta_i), i = 1, \ldots, m$. A "conjugate" model linking the RRs θ_i is assumed in the second stage: $\theta_i \overset{\text{iid}}{\sim} \text{gamma}(\nu, \alpha)$, where gamma$(\nu, \alpha)$ denotes the gamma distribution with shape parameter $\nu(> 0)$ and scale parameter $\alpha(> 0)$. Then,

$$f(\theta_i|\alpha, \nu) = \frac{\alpha^\nu}{\Gamma(\nu)} e^{-\alpha\theta_i} \theta_i^{\nu-1} \tag{9.6.2}$$

and

$$E(\theta_i) = \nu/\alpha = \mu, \quad V(\theta_i) = \nu/\alpha^2. \tag{9.6.3}$$

Noting that $\theta_i|y_i, \alpha, \nu \overset{\text{ind}}{\sim} \text{gamma}(y_i + \nu, e_i + \alpha)$, the Bayes estimator of θ_i and the posterior variance of θ_i are obtained from (9.6.3) by changing α to $e_i + \alpha$ and ν to $y_i + \nu$, that is,

$$\hat{\theta}_i^B(\alpha, \nu) = E(\theta_i|y_i, \alpha, \nu) = (y_i + \nu)/(e_i + \alpha) \tag{9.6.4}$$

and

$$V(\theta_i|y_i, \alpha, \nu) = g_{1i}(\alpha, \nu, y_i) = (y_i + \nu)/(e_i + \alpha)^2. \tag{9.6.5}$$

We can obtain ML estimators of α and ν from the marginal distribution, $y_i|\alpha, \nu \overset{\text{iid}}{\sim}$ negative binomial, using the loglikelihood

$$l(\alpha, \nu) = \sum_{i=1}^m \left[\sum_{h=0}^{y_i-1} \log(\nu + h) + \nu \log(\alpha) - (y_i + \nu) \log(e_i + \alpha) \right]. \tag{9.6.6}$$

Closed-form expressions for $\hat{\alpha}_{\text{ML}}$ and $\hat{\nu}_{\text{ML}}$ do not exist. Marshall (1991) obtained simple moment estimators by equating the weighted sample mean $\hat{\theta}_{e\cdot} = m^{-1} \sum_{\ell=1}^m (e_\ell/e_\cdot)\hat{\theta}_\ell$ and the weighted sample variance $s_e^2 = m^{-1} \sum_{i=1}^m (e_i/e_\cdot)(\hat{\theta}_i - \hat{\theta}_{e\cdot})^2$

to their expected values and then solving the resulting moment equations for α and ν, where $e_{\cdot} = \sum_{i=1}^{m} e_i/m$. This leads to moment estimators, $\hat{\alpha}$ and $\hat{\nu}$, given by

$$\hat{\nu}/\hat{\alpha} = \hat{\theta}_{e\cdot} \tag{9.6.7}$$

and

$$\hat{\nu}/\hat{\alpha}^2 = s_e^2 - \hat{\theta}_{e\cdot}/e_{\cdot}. \tag{9.6.8}$$

Lahiri and Maiti (2002) provided more efficient moment estimators. The moment estimators may also be used as starting values for ML iterations.

We substitute the moment estimators $\hat{\alpha}$ and $\hat{\nu}$ into (9.6.4) to get an EB estimator of θ_i as

$$\hat{\theta}_i^{\text{EB}} = \hat{\theta}_i^B(\hat{\alpha}, \hat{\nu}) = \hat{\gamma}_i \hat{\theta}_i + (1 - \hat{\gamma}_i)\hat{\theta}_{e\cdot}, \tag{9.6.9}$$

where $\hat{\gamma}_i = e_i/(e_i + \hat{\alpha})$. Note that $\hat{\theta}_i^{\text{EB}}$ is a weighted average of the direct estimator (SMR) $\hat{\theta}_i$ and the synthetic estimator $\hat{\theta}_{e\cdot}$, and more weight is given to $\hat{\theta}_i$ as the ith area expected deaths, e_i, increases. If $s_e^2 < \hat{\theta}_{e\cdot}/e_{\cdot}$, then $\hat{\theta}_i^{\text{EB}}$ is taken as the synthetic estimator $\hat{\theta}_{e\cdot}$. The EB estimator is nearly unbiased for θ_i in the sense that its bias is of order m^{-1}, for large m.

As in the binary case, the jackknife method may be used to obtain a second-order unbiased estimator of $\text{MSE}(\hat{\theta}_i^{\text{EB}})$. We obtain the jackknife estimator, $\text{mse}_J(\hat{\theta}_i^{\text{EB}})$ from (9.5.13) by substituting $\hat{\theta}_i^{\text{EB}} = k_i(y_i, \hat{\alpha}, \hat{\nu})$ for $\hat{p}_i^{\text{EB}}$ and $\hat{\theta}_{i,-\ell}^{\text{EB}} = k_i(y_i, \hat{\alpha}_{-\ell}, \hat{\nu}_{-\ell})$ for $\hat{p}_{i,-\ell}^{\text{EB}}$ in (9.5.11) and using $g_{1i}(\hat{\alpha}, \hat{\nu}, y_i)$ and $g_{1i}(\hat{\alpha}_{-\ell}, \hat{\nu}_{-\ell}, y_i)$ in (9.5.12), where $\hat{\alpha}_{-\ell}$ and $\hat{\nu}_{-\ell}$ are the delete$-\ell$ moment estimators obtained from $\{(y_i, e_i); i = 1, \ldots, \ell - 1, \ell + 1, \ldots, m\}$. Note that $\text{mse}_J(\hat{\theta}_i^{\text{EB}})$ is area-specific in the sense that it depends on y_i. Lahiri and Maiti (2002) obtained a Taylor expansion estimator of MSE, using a parametric bootstrap to estimate the covariance matrix of $(\hat{\alpha}, \hat{\nu})$.

EB estimators of RRs θ_i can be improved by extending the linking gamma model on the θ_i's to allow for area level covariates, $\mathbf{z}_i$, such as degree of urbanization of areas. Clayton and Kaldor (1987) allowed varying scale parameters, α_i, and assumed a log-linear model on $E(\theta_i) = \nu/\alpha_i$ given by $\log[E(\theta_i)] = \mathbf{z}_i^T \boldsymbol{\beta}$. EB estimation for this extension is implemented by changing α to α_i in (9.6.4) and (9.6.5) and using ML or moment estimators of ν and $\boldsymbol{\beta}$. Christiansen and Morris (1997) studied the Poisson-gamma regression model in detail, and proposed accurate approximations to the posterior mean and the posterior variance of θ_i. The posterior mean approximation is used as the EB estimator and the posterior variance approximation as a measure of its variability.

9.6.2 Log-normal Models

Log-normal two-stage models have also been proposed. The first-stage model is not changed, but the second-stage linking model is changed to $\xi_i = \log(\theta_i) \overset{\text{iid}}{\sim} N(\mu, \sigma^2)$, $i = 1, \ldots, m$ in the case of no covariates. As in the case of logit-normal models, implementation of EB is more complicated for the log-normal model because no closed-form expression for the Bayes estimator, $\hat{\theta}_i^B(\mu, \sigma^2)$, and the posterior variance, $V(\theta_i|y_i, \mu, \sigma^2)$, exist. Clayton and Kaldor (1987) approximated the joint posterior density of $\boldsymbol{\xi} = (\xi_1, \ldots, \xi_m)^T, f(\boldsymbol{\xi}|\mathbf{y}, \mu, \sigma^2)$, for $\mathbf{y} = (y_1, \ldots, y_m)^T$, by a multivariate normal distribution, which gives an explicit approximation to the BP estimator $\hat{\xi}_i^B$ of ξ_i. ML estimators of model parameters μ and σ^2 were obtained using the EM algorithm, and then used in the approximate formula for $\hat{\xi}_i^B$ to get EB estimators $\hat{\xi}_i^{\text{EB}}$ of ξ_i and $\hat{\theta}_i^{\text{EB}} = \exp(\hat{\xi}_i^{\text{EB}})$ of θ_i. The EB estimator $\hat{\theta}_i^{\text{EB}}$, however, is not nearly unbiased for θ_i. We can employ numerical integration, as done in Section 9.5.1, to get nearly unbiased EB estimators, but we omit details here. Moment estimators of μ and σ (Jiang and Zhang 2001) may be used to simplify the calculation of a jackknife estimator of $\text{MSE}(\hat{\theta}_i^{\text{EB}})$.

The above basic log-normal model readily extends to the case of covariates, $\mathbf{z}_i$ by considering the linking model $\xi_i = \log(\theta_i) \overset{\text{ind}}{\sim} N(\mathbf{z}_i^T\boldsymbol{\beta}, \sigma^2)$. Also, the basic model can be extended to allow spatial correlations; mortality data sets often exhibit significant spatial relationships between the log RRs, $\xi_i = \log(\theta_i)$. A simple conditional autoregression (CAR) normal model on $\boldsymbol{\xi}$ assumes that $\boldsymbol{\xi}$ is multivariate normal, with

$$E(\xi_i|\xi_\ell, \ell \neq i) = \mu + \rho \sum_{\substack{\ell=1 \\ \ell \neq i}}^{m} q_{i\ell}(\xi_\ell - \mu), \tag{9.6.10}$$

$$V(\xi_i|\xi_\ell, \ell \neq i) = \sigma^2, \tag{9.6.11}$$

where ρ is the correlation parameter and $\mathbf{Q} = (q_{i\ell})$ is the "adjacency" matrix of the map, with elements $q_{i\ell} = 1$ if i and ℓ are adjacent areas and $q_{i\ell} = 0$ otherwise. It follows from Besag (1974) that $\boldsymbol{\xi}$ is multivariate normal with mean $\boldsymbol{\mu} = \mu\mathbf{1}$ and covariance matrix $\boldsymbol{\Sigma} = \sigma^2(\mathbf{I} - \rho\mathbf{Q}^{-1})$, where ρ is bounded above by the inverse of the largest eigenvalue of $\mathbf{Q}$. Clayton and Kaldor (1987) approximated the posterior density, $f(\boldsymbol{\xi}|\mathbf{y}, \mu, \sigma^2, \rho)$, similar to the log-normal case.

The assumption (9.6.11) of a constant conditional variance for the ξ_i's results in the conditional mean (9.6.10) proportional to the sum, rather than the mean, of the neighboring ξ_i's. Clayton and Bernardinelli (1992) proposed an alternative joint density of the ξ_i's given by

$$f(\boldsymbol{\xi}) \propto (\sigma^2)^{-m/2} \exp\left[-\frac{1}{2\sigma^2}\sum_{i=1}^{m}\sum_{\substack{\ell=1 \\ \ell \neq i}}^{m}(\xi_i - \xi_\ell)^2 q_{i\ell}\right]. \tag{9.6.12}$$

This specification leads to

$$E(\xi_i|\xi_\ell, \ell \neq i) = \left(\sum_{\ell=1}^{m} q_{i\ell}\right)^{-1} \sum_{\ell=1}^{m} q_{i\ell}\xi_\ell \tag{9.6.13}$$

and

$$V(\xi_i|\xi_\ell, \ell \neq i) = \sigma^2 \left(\sum_{\ell=1}^{m} q_{i\ell} \right)^{-1}. \tag{9.6.14}$$

Note that the conditional variance is now inversely proportional to $\sum_{\ell=1}^{m} q_{i\ell}$, the number of neighbors of area i, and the conditional mean is equal to the mean of the neighboring values ξ_ℓ. In the context of disease mapping, the alternative specification may be more appropriate.

Example 9.6.1. Lip Cancer. Clayton and Kaldor (1987) applied EB estimation to data on observed cases, y_i, and expected cases, e_i, of lip cancer registered during the period 1975–1980 in each of 56 counties (small areas) of Scotland. They reported the SMR, the EB estimate of θ_i based on the Poisson-gamma model, denoted $\hat{\theta}_i^{\text{EB}}(1)$, and the approximate EB estimates of θ_i based on the log-normal model and the CAR-normal model, denoted $\hat{\theta}_i^{\text{EB}}(2)$ and $\hat{\theta}_i^{\text{EB}}(3)$, for each of the 56 counties (all values multiplied by 100). The SMR values varied between 0 and 652, while the EB estimates showed considerably less variability across counties, as expected: $\hat{\theta}_i^{\text{EB}}(1)$ varied between 31 and 422 (with CV=0.78) and $\hat{\theta}_i^{\text{EB}}(2)$ varied between 34 and 495 (with CV=0.85), suggesting little difference between the two sets of EB estimates. Ranks of EB estimates differed little from the corresponding ranks of the SMRs for most counties, despite less variability exhibited by the EB estimates.

Turning to the CAR-normal model, the adjacency matrix, $\mathbf{Q}$, was specified by listing adjacent counties for each county i. The ML estimate of ρ was 0.174 compared to the upper bound of 0.175, suggesting a high degree of spatial relationship in the data set. Most of the CAR estimates, $\hat{\theta}_i^{\text{EB}}(3)$, differed little from the corresponding estimates $\hat{\theta}_i^{\text{EB}}(1)$ and $\hat{\theta}_i^{\text{EB}}(2)$ based on the independence assumption. Counties with few cases, y_i, and SMRs differing appreciably from adjacent counties are the only counties affected substantially by spatial correlation. For example, county number 24 with $y_{24} = 7$ is adjacent to several low-risk counties, and the CAR estimate $\hat{\theta}_{24}^{\text{EB}}(3) = 83.5$ is substantially smaller than $\hat{\theta}_{24}^{\text{EB}}(1) = 127.7$ and $\hat{\theta}_{24}^{\text{EB}}(2) = 123.6$, based on the independence assumption.

9.6.3 Extensions

Various extensions of the disease mapping models studied in Sections 9.6.1 and 9.6.2 have been proposed in the recent literature. DeSouza (1992) proposed a two-stage, bivariate logit-normal model to study joint RRs (or mortality rates), θ_{1i} and θ_{2i}, of two cancer sites (e.g., lung and large bowel cancers), or two groups (e.g., lung cancer in males and females) over several geographical areas. Denote the observed and expected number of deaths at the two sites as (y_{1i}, y_{2i}) and (e_{1i}, e_{2i}), respectively, for the ith area ($i = 1, \ldots, m$). The first stage assumes that $(y_{1i}, y_{2i})|(\theta_{1i}, \theta_{2i}) \overset{\text{ind}}{\sim} \text{Poisson}(e_{1i}\theta_{1i}) * \text{Poisson}(e_{2i}\theta_{2i}), i = 1, \ldots, m$, where $*$ denotes that $f(y_{1i}, y_{2i}|\theta_{1i}, \theta_{2i}) = f(y_{1i}|\theta_{1i})f(y_{2i}|\theta_{2i})$. The joint risks $(\theta_{1i}, \theta_{2i})$ are linked in the second stage by assuming that the vectors $(\text{logit}(\theta_{1i}), \text{logit}(\theta_{2i}))$ are independent, bivariate normal with means μ_1, μ_2; standard deviations σ_1 and σ_2; and correlation ρ, denoted $N(\mu_1, \mu_2, \sigma_1, \sigma_2, \rho)$. Bayes estimators of θ_{1i} and θ_{2i} involve double integrals, which may be calculated numerically using Gauss–Hermite quadrature. EB estimators are obtained by substituting ML estimators of model parameters in the Bayes estimators. DeSouza (1992) applied the bivariate EB method to two different data sets consisting of cancer mortality rates in 115 counties of the state of Missouri during 1972–1981: (i) lung and large bowel cancers; (ii) lung cancer in males and females. The EB estimates based on the bivariate model lead to improved efficiency for each site (group) compared to the EB estimates based on the univariate logit-normal model, because of significant correlation; $\hat{\rho} = 0.54$ for data set (i) and $\hat{\rho} = 0.76$ for data set (ii). Kass and Steffey's (1989) first-order approximation to the posterior variance was used as a measure of variability of the EB estimates.

Kim, Sun, and Tsutakawa (2001) extended the bivariate model by introducing spatial correlations (via CAR) and covariates into the model. They used a HB approach instead of the EB approach. They applied the bivariate spatial model to male and female lung cancer mortality in the State of Missouri, and constructed disease maps of male and female lung cancer mortality rates by age group and time period.

Dean and MacNab (2001) extended the Poisson-gamma model to handle nested data structures, such as a hierarchical health administrative structure consisting of health districts, i, in the first level and local health areas, j, within districts in the second level ($j = 1, \ldots, n_i, i = 1, \ldots, m$). The data consist of incidence or mortality counts, y_{ij}, and the corresponding population at risk counts, n_{ij}. Dean and MacNab (2001) derived EB estimators of the local health area rates, θ_{ij}, using a nested error Poisson-gamma model. The Bayes estimator of θ_{ij} is a weighted combination of the crude local area rate, y_{ij}/n_{ij}, the correspond crude district rate $y_{i\cdot}/n_{i\cdot}$, and the overall rate $y_{\cdot\cdot}/n_{\cdot\cdot}$, where $y_{i\cdot} = \sum_{j=1}^{n_i} y_{ij}$ and $y_{\cdot\cdot} = \sum_{i=1}^{m} y_{i\cdot}$, and $(n_{i\cdot}, n_{\cdot\cdot})$ similarly defined. Dean and MacNab (2001) used the Kass and Steffey (1989) first-order approximation to posterior variance as a measure of variability. They applied the nested error model to infant mortality data from the province of British Columbia, Canada.

9.7 *DESIGN-WEIGHTED EB ESTIMATION: EXPONENTIAL FAMILY MODELS

In Section 7.6.2, we studied pseudo-EBLUP estimation of area means under the basic unit level model (7.1.1) with $k_{ij} = 1$. The pseudo-EBLUP estimator $\hat{\mu}_{iw}^{H}$, given by (7.6.8), takes account of the design weights, w_{ij}, and in turn leads to a design-consistent estimator of the area mean μ_i. Ghosh and Maiti (2004) extended the pseudo-EBLUP approach to exponential family models. Their results are applicable to the case of area level covariates $\mathbf{x}_i$, and the basic data available to the user consist of the weighted area means $\bar{y}_{iw} = \sum_{j=1}^{n_i} \tilde{w}_{ij} y_{ij}, i = 1, \ldots, m$ but not the unit level values, y_{ij}.

For a given θ_i, we assume that the $y_{ij}, j = 1, \ldots, n_i$, are independently and identically distributed with probability density function $f(y_{ij}|\theta_i)$ belonging to the natural exponential family with canonical parameter θ_i and quadratic variance function. Specifically,

$$f(y_{ij}|\theta_i) = \exp\{\theta_i y_{ij} - a(\theta_i) + b(y_{ij})\} \tag{9.7.1}$$

with

$$E(y_{ij}|\theta_i) = a'(\theta_i) = \nu_i, \quad V(y_{ij}|\theta_i) = a''(\theta_i) = c(\nu_i), \tag{9.7.2}$$

where $c(\nu_i) = d_0 + d_1\nu_i + d_2\nu_i^2$. This family covers the Bernouilli distribution $(d_0 = 0, d_1 = 1, d_2 = -1)$ and the Poisson distribution $(d_0 = d_2 = 0, d_1 = 1)$. In the special case of $y_{ij} \sim \text{Bernouilli}(p_i)$, we have $\nu_i = p_i, \theta_i = \log[p_i(1-p_i)]$ and $a(\theta_i) = -\log(1-p_i)$, where $p_i = e^{\theta_i}/(1+e^{\theta_i})$. The canonical parameters θ_i are assumed to obey the conjugate family with density function

$$f(\theta_i) = C(\lambda, h_i)\exp\{\lambda[h_i\theta_i - a(\theta_i)]\}, \tag{9.7.3}$$

where $h_i = a'(\mathbf{z}_i^T\boldsymbol{\beta})$ is the canonical link function and $\lambda > \max(0, d_2)$. The mean and variance of ν_i are given by

$$E(\nu_i) = h_i, \quad V(\nu_i) = c(h_i)(\lambda - d_2)^{-1} \tag{9.7.4}$$

(Morris 1983c). It now follows from (9.7.2) and (9.7.4) that

$$E(\bar{y}_{iw}) = h_i, \quad V(\bar{y}_{iw}) = c(h_i)(\lambda - d_2)^{-1}(1 + \lambda\delta_{2i}) =: \phi_i c(h_i), \tag{9.7.5}$$

and

$$\text{Cov}(\bar{y}_{iw}, \nu_i) = c(h_i)(\lambda - d_2)^{-1}, \tag{9.7.6}$$

where $\delta_{2i} = \sum_{j=1}^{n_i} \tilde{w}_{ij}^2$. The BLUP estimator of ν_i based on $\bar{y}_{iw}$ is given by

$$\tilde{\nu}_{iw}^{\text{GM}} = h_i + \frac{\text{Cov}(\bar{y}_{iw}, \nu_i)}{V(\bar{y}_{iw})}(\bar{y}_{iw} - h_i) \tag{9.7.7}$$

$$= r_{iw}\bar{y}_{iw} + (1 - r_{iw})h_i, \tag{9.7.8}$$

where $r_{iw} = (1 + \lambda\delta_{2i})^{-1}$. Expression (9.7.8) follows from (9.7.5) and (9.7.6). Note that (9.7.8) requires the knowledge of δ_{2i}.

The BLUP estimator (9.7.8) depends on the unknown parameters $\boldsymbol{\beta}$ and λ. The marginal distribution of $\bar{y}_{iw}$ is not tractable, and hence it cannot be used to estimate $\boldsymbol{\beta}$ and λ. Therefore, Ghosh and Maiti (2004) proposed an estimating function approach by combining the elementary unbiased estimating functions $\mathbf{u}_i = (u_{i1}, u_{i2})^T, i = 1, \ldots, m$, where $u_{1i} = \bar{y}_{iw} - h_i$ and $u_{2i} = (\bar{y}_{iw} - h_i)^2 - \phi_i c(h_i)$ with $E(u_{1i}) = 0$ and $E(u_{2i}) = 0$. The resulting optimal estimating equations (Godambe and Thompson 1989) require the evaluation of third and fourth moments of $\bar{y}_{iw}$, which depend on $\delta_{3i} = \sum_{j=1}^{n_i} \tilde{w}_{ij}^3$ and $\delta_{4i} = \sum_{j=1}^{n_i} \tilde{w}_{ij}^4$. Estimators $\hat{\boldsymbol{\beta}}$ and $\hat{\lambda}$ are obtained by solving the optimal estimating equations iteratively. We refer the reader to Ghosh and Maiti (2004) for details.

Substituting $\hat{\boldsymbol{\beta}}$ and $\hat{\lambda}$ for $\boldsymbol{\beta}$ and λ in (9.7.8), we obtain an EBLUP estimator of v_i as

$$\hat{v}_{iw}^{\text{GM}} = \hat{r}_{iw}\bar{y}_{iw} + (1 - \hat{r}_{iw})\hat{h}_i, \quad (9.7.9)$$

where $\hat{r}_{iw} = (1 + \hat{\lambda}\delta_{2i})^{-1}$ and $\hat{h}_i = a'(\mathbf{z}_i^T\hat{\boldsymbol{\beta}}), i = 1, \ldots, m$.

Ghosh and Maiti (2004) also obtained a second-order approximation to MSE($\hat{v}_{iw}^{\text{GM}}$), using the linearization method. Using this approximation, a second-order unbiased MSE estimator is also obtained.

Example 9.7.1. Poverty Proportions. Ghosh and Maiti (2004) applied the EBLUP estimator (9.7.9) to $m = 39$ county poverty proportions $\hat{p}_{iw} = \bar{y}_{iw}$ of poor school-age children of a certain U.S. state for the year 1989, where y_{ij} is 1 or 0 according as the jth sample child in county i is poor or not poor. The estimates $\bar{y}_{iw}$ and $\sum_{j=1}^{n_i} \tilde{w}_{ij}^t, t = 2, 3, 4$ from the March supplement of the Current Population Survey were provided by the Census Bureau. The responses y_{ij} follow Bernouilli($p_i = v_i$), and the link function is taken as $\log[h_i/(1 - h_i)] = \beta_0 + \beta_1 z_{1i} + \beta_2 z_{2i} + \beta_3 z_{3i} + \beta_4 z_{4i} = \mathbf{z}_i^T\boldsymbol{\beta}$, where the covariates $z_{1i}, \ldots z_{4i}$ for county i are taken as z_{1i} = log(proportion of child exemptions reported by families in poverty on tax returns), z_{2i} = log(proportion of people receiving food stamps), z_{3i} = log(proportion of child exemptions on tax returns), and z_{4i} = log(proportion of poor school-age children estimated from the previous census.

The estimates $\hat{p}_{iw} = \bar{y}_{iw}$ and $\hat{p}_{iw}^{\text{GM}} = \hat{v}_{iw}^{\text{GM}}$ were compared to the estimate $\hat{p}_{iw}^{\text{FH}}$ obtained under the FH model $\hat{\theta}_i = \log(\bar{y}_{iw}) = \mathbf{z}_i^T\boldsymbol{\beta} + v_i + e_i$ with $v_i \overset{\text{iid}}{\sim} N(0, \sigma_v^2)$ and $e_i \overset{\text{ind}}{\sim} N(0, \psi_i = 9.97/n_i)$; the known sampling variance is based on the value 9.97 supplied by the Census Bureau. Note that the above model is different from the poverty counts model of the Census Bureau (e.g., 6.1.2). Note also that the counties with zero poor school-age children cause difficulty with the FH model because $\bar{y}_{iw} = 0$ for those counties. On the other hand, this is not a problem with the models (9.7.1) and (9.7.3) used to derive $\hat{p}_{iw}^{\text{GM}}$.

An external evaluation was conducted by comparing the three estimates for 1989 to the corresponding 1990 long-form census estimate for 1989, in terms of absolute relative error (ARE) averaged over the counties. Average ARE values reported are 0.54 for the direct estimates $\bar{y}_{iw}$, 0.66 for the FH estimates $\hat{p}_{iw}^{\text{FH}}$, and 0.47 for the Ghosh–Maiti estimates $\hat{p}_{iw}^{\text{GM}}$. These values suggest that the Ghosh–Maiti estimator performs somewhat better than the other two estimators.

9.8 TRIPLE-GOAL ESTIMATION

We have focused so far on the estimation of area-specific parameters (means, RRs, etc.), but in some applications the main objective is to produce an ensemble of parameter estimates whose distribution is in some sense close enough to the distribution of area-specific parameters, θ_i. For example, Spjøtvoll and Thomsen (1987) were interested in finding how 100 municipalities in Norway were distributed according to proportions of persons in the labor force. By comparing with the actual distribution in their example, they showed that the EB estimates, $\hat{\theta}_i^{\text{EB}}$, under a simple area level model distort the distribution by overshrinking toward the synthetic component $\hat{\theta}_{\cdot} = \sum_{i=1}^{m} \hat{\theta}_i/m$. In particular, the variability of the EB estimates was smaller than the variability of the θ_i's. On the other hand, the set of direct estimates $\{\hat{\theta}_i\}$ were overdispersed in the sense of variability larger than the variability of the θ_i's.

We are also often interested in the ranks of the θ_i's (e.g., ranks of schools, hospitals or geographical areas) or in identifying domains (areas) with extreme θ_i's. Ideally, it is desirable to construct a set of "triple-goal" estimates that can produce good ranks, a good histogram, and good area-specific estimates. However, simultaneous optimization is not feasible, and it is necessary to seek a compromise set that can strike an effective balance between the three goals (Shen and Louis 1998).

9.8.1 Constrained EB

Consider a two-stage model of the form $\hat{\theta}_i|\theta_i \overset{\text{ind}}{\sim} f(\hat{\theta}_i|\theta_i, \lambda_1)$ and $\theta_i \overset{\text{iid}}{\sim} f(\theta_i|\lambda_2)$, $i = 1, \ldots, m$, where $\lambda = (\lambda_1^T, \lambda_2^T)$ is the vector of model parameters. The set of direct estimators $\{\hat{\theta}_i\}$ are generally overdispersed under this model. For example, consider the simple model $\hat{\theta}_i = \theta_i + e_i$ with $e_i \overset{\text{iid}}{\sim} (0, \psi)$ independent of $\theta_i \overset{\text{iid}}{\sim} (\mu, \sigma_v^2), i = 1, \ldots, m$. Noting that $\hat{\theta}_i \overset{\text{iid}}{\sim} (\mu, \psi + \sigma_v^2)$, it immediately follows that

$$E\left[\frac{1}{m-1}\sum_{i=1}^{m}(\hat{\theta}_i - \hat{\theta}_{\cdot})^2\right] = \psi + \sigma_v^2 > \sigma_v^2 = E\left[\frac{1}{m-1}\sum_{i=1}^{m}(\theta_i - \theta_{\cdot})^2\right], \quad (9.8.1)$$

where $\hat{\theta}_{\cdot} = \sum_{i=1}^{m} \hat{\theta}_i/m$ and $\theta_{\cdot} = \sum_{i=1}^{m} \theta_i/m$. On the other hand, if $\hat{\theta}_i^{\text{B}} = E(\theta_i|\hat{\theta}_i, \lambda)$ denotes the Bayes estimator of θ_i under squared error, the set of Bayes estimators $\{\hat{\theta}_i^B\}$ exhibit underdispersion under the two-stage model $\hat{\theta}_i|\theta_i \overset{\text{ind}}{\sim} f(\hat{\theta}_i|\theta_i, \lambda_1)$ and $\theta_i \overset{\text{iid}}{\sim} f(\theta_i|\lambda_2)$. Specifically, we have

$$E\left[\frac{1}{m-1}\sum_{i=1}^{m}(\theta_i - \theta_{\cdot})^2|\hat{\theta}\right] = \frac{1}{m-1}\sum_{i=1}^{m} V(\theta_i - \theta_{\cdot}|\hat{\theta}) + \frac{1}{m-1}\sum_{i=1}^{m}(\hat{\theta}_i^B - \hat{\theta}_{\cdot}^B)^2 \quad (9.8.2)$$

$$> \frac{1}{m-1}\sum_{i=1}^{m}(\hat{\theta}_i^B - \hat{\theta}_{\cdot}^B)^2, \quad (9.8.3)$$

where $\hat{\theta} = (\hat{\theta}_1, \ldots, \hat{\theta}_m)^T, \hat{\theta}_{\cdot}^{\text{B}} = \sum_{i=1}^{m} \hat{\theta}_i^{\text{B}}/m$, and the dependence on λ is suppressed for simplicity. It follows from (9.8.3) that $E[\sum_{i=1}^{m}(\theta_i - \theta_{\cdot})^2] > E[\sum_{i=1}^{m}(\hat{\theta}_i^B - \hat{\theta}_{\cdot}^B)^2]$. However, note that $\{\hat{\theta}_i^B\}$ match the ensemble mean because $\hat{\theta}_{\cdot}^B = E(\theta_{\cdot}|\hat{\theta})$ which, in turn, implies $E(\hat{\theta}_{\cdot}^B) = E(\theta_{\cdot})$.

We can match the ensemble variance by finding the estimators $t_1, \ldots, t_m$ that minimize the posterior expected squared error loss $E[\sum_{i=1}^m (\theta_i - t_i)^2 | \hat{\theta}]$ subject to the constraints

$$t_{.} = \hat{\theta}^B_{.} \tag{9.8.4}$$

$$\frac{1}{m-1}\sum_{i=1}^m (t_i - t_{.})^2 = E\left[\frac{1}{m-1}\sum_{i=1}^m (\theta_i - \theta_{.})^2 | \hat{\boldsymbol{\theta}}\right], \tag{9.8.5}$$

where $t_{.} = \sum_{i=1}^m t_i/m$. Using Lagrange multipliers, we obtain the constrained Bayes (CB) estimators $\{\hat{\theta}_i^{\text{CB}}\}$ as the solution to the minimization problem, where

$$t_{i,\,\text{opt}} = \hat{\theta}_i^{\text{CB}} = \hat{\theta}^B_{.} + a(\hat{\boldsymbol{\theta}}, \boldsymbol{\lambda})(\hat{\theta}_i^B - \hat{\theta}^B_{.}) \tag{9.8.6}$$

with

$$a(\hat{\boldsymbol{\theta}}, \boldsymbol{\lambda}) = \left\{1 + \frac{(1/m)\sum_{i=1}^m V(\theta_i|\hat{\theta}_i, \boldsymbol{\lambda})}{[1/(m-1)]\sum_{i=1}^m (\hat{\theta}_i^B - \hat{\theta}^B_{.})^2}\right\}^{1/2}. \tag{9.8.7}$$

Louis (1984) derived the CB estimator under normality. Ghosh (1992b) obtained (9.8.6) for arbitrary distributions. A proof of (9.8.6) is given in Section 9.12.3. It follows from (9.8.6) that $\sum_{i=1}^m (\hat{\theta}_i^{\text{CB}} - \hat{\theta}^{\text{CB}}_{.})^2 > \sum_{i=1}^m (\hat{\theta}_i^B - \hat{\theta}^B_{.})^2$ because the term $a(\hat{\boldsymbol{\theta}}, \boldsymbol{\lambda})$ in (9.8.6) is greater than 1 and $\hat{\theta}^{\text{CB}}_{.} = \hat{\theta}^B_{.}$, that is, the variability of $\{\hat{\theta}_i^{\text{CB}}\}$ is larger than that of $\{\hat{\theta}_i^B\}$. Note that the constraint (9.8.5) implies that $E[\sum_{i=1}^m (\hat{\theta}_i^{\text{CB}} - \hat{\theta}^{\text{CB}}_{.})^2] = E[\sum_{i=1}^m (\theta_i - \theta_{.})^2]$, so that $\{\hat{\theta}_i^{\text{CB}}\}$ matches the variability of $\{\theta_i\}$.

The CB estimator $\hat{\theta}_i^{\text{CB}}$ is a function of the set of posterior variances $\{V(\theta_i|\hat{\boldsymbol{\theta}}, \boldsymbol{\lambda})\}$ and the set of Bayes estimators $\{\hat{\theta}_i^B = E(\theta_i|\hat{\boldsymbol{\theta}}, \boldsymbol{\lambda})\}$. Replacing the model parameters $\boldsymbol{\lambda}$ by suitable estimators $\hat{\boldsymbol{\lambda}}$, we obtain an empirical CB (ECB) estimator $\hat{\theta}_i^{\text{ECB}} = \hat{\theta}_i^{\text{CB}}(\hat{\boldsymbol{\lambda}})$.

Example 9.8.1. Simple Model. We illustrate the calculation of $\hat{\theta}_i^{\text{CB}}$ for the simple model $\hat{\theta}_i = \theta_i + e_i$, with $e_i \overset{\text{iid}}{\sim} N(0, \psi)$ and independent of $\theta_i \overset{\text{iid}}{\sim} N(\mu, \sigma_v^2)$. The Bayes estimator is given by $\hat{\theta}_i^B = \gamma\hat{\theta}_i + (1-\gamma)\mu$, where $\gamma = \sigma_v^2/(\sigma_v^2 + \psi)$. Furthermore, $V(\theta_i|\hat{\theta}_i, \boldsymbol{\lambda}) = \gamma\psi, \hat{\theta}^B_{.} = \gamma\hat{\theta}_{.} + (1-\gamma)\mu$ and $\sum_{i=1}^m (\hat{\theta}_i^B - \hat{\theta}^B_{.})^2 = \gamma^2 \sum_{i=1}^m (\hat{\theta}_i - \hat{\theta}_{.})^2$. Hence,

$$\hat{\theta}_i^{\text{CB}} = [\gamma\hat{\theta}_{.} + (1-\gamma)\mu] + \left\{1 + \frac{\psi/\gamma}{[1/(m-1)]\sum_{i=1}^m (\hat{\theta}_i - \hat{\theta}_{.})^2}\right\}^{1/2} \gamma(\hat{\theta}_i - \hat{\theta}_{.}). \tag{9.8.8}$$

Noting that $\hat{\theta}_i \overset{\text{iid}}{\sim} N(\mu, \psi + \sigma_v^2)$, it follows that $\hat{\theta}_{.}$ and $(m-1)^{-1}\sum_{i=1}^m (\hat{\theta}_i - \hat{\theta}_{.})^2$ converge in probability to μ and $\psi + \sigma_v^2 = \psi/(1-\gamma)$, respectively, as $m \to \infty$. Hence,

$$\hat{\theta}_i^{\text{CB}} \approx \gamma^{1/2}\hat{\theta}_i + (1-\gamma^{1/2})\mu. \tag{9.8.9}$$

It follows from (9.8.9) that the weight attached to the direct estimator is larger than the weight used by the Bayes estimator, and the shrinkage toward the synthetic component μ is reduced. Assuming normality, Ghosh (1992b) proved that the total MSE, $\sum_{i=1}^{m} \text{MSE}(\hat{\theta}_i^{\text{CB}}) = \sum_{i=1}^{m} E(\hat{\theta}_i^{\text{CB}} - \theta_i)^2$, of the CB estimators is smaller than the total MSE of the direct estimators if $m \geq 4$. Hence, the CB estimators perform better than the direct estimators, but are less efficient than the Bayes estimators.

Shen and Louis (1998) studied the performance of CB estimators for exponential families $f(\hat{\theta}_i|\theta_i, \lambda_1)$ with conjugate $f(\theta_i|\lambda_2)$. They showed that, for large m, CB estimators are always more efficient than the direct estimators in terms of total MSE. Further, the maximum loss in efficiency relative to the Bayes estimators is 24%. Note that the exponential families cover many commonly used distributions, including the binomial, Poisson, normal, and gamma distributions.

9.8.2 Histogram

The empirical distribution function based on the CB estimators is given by $F_m^{\text{CB}}(t) = m^{-1} \sum_{i=1}^{m} I(\hat{\theta}_i^{\text{CB}} \leq t), -\infty < t < \infty$, where $I(\theta_i \leq t) = 1$ if $\theta_i \leq t$ and $I(\theta_i \leq t) = 0$ otherwise. The estimator $F_m^{\text{CB}}(t)$ is generally not consistent for the true distribution of the θ_i's as $m \to \infty$, though the CB approach matches the first and second moments. As a result, $F_m^{\text{CB}}(t)$ can perform poorly as an estimator of $F_m(t) = m^{-1} \sum_{i=1}^{m} I(\theta_i \leq t)$.

An "optimal" estimator of $F_m(t)$ is obtained by minimizing the posterior expected integrated squared error loss $E[\int \{A(t) - F_m(t)\}^2 dt|\hat{\theta}]$. The optimal $A(\cdot)$ is given by

$$A_{\text{opt}}(t) = \overline{F}_m(t) = \frac{1}{m} \sum_{i=1}^{m} P(\theta_i \leq t|\hat{\theta}_i). \tag{9.8.10}$$

Adding the constraint that $A(\cdot)$ is a discrete distribution with at most m mass points, the optimal estimator $\hat{F}_m(\cdot)$ is discrete with mass $1/m$ at $\hat{U}_\ell = \overline{F}_m^{-1}(\frac{2\ell-1}{2m}), \ell = 1, \ldots, m$; see Shen and Louis (1998) for a proof. Note that $\hat{U}_\ell$ depends on model parameters, λ, and an EB version of $\hat{U}_\ell$ is obtained by substituting a suitable estimator $\hat{\lambda}$ for λ.

9.8.3 Ranks

How good are the ranks based on the BP estimators $\hat{\theta}_i^B$ compared to those based on the true (realized but unobservable) values θ_i? In the context of BLUP estimation under the linear mixed model, Portnoy (1982) showed that ranking based on the BLUP estimators is "optimal" in the sense of maximizing the probability of correctly ranking with respect to the true values θ_i. Also, the ranks based on the $\hat{\theta}_i^B$'s often agree with the ranks based on the direct estimators $\hat{\theta}_i$, and it follows from (9.8.6) that the ranks based on the CB estimators $\hat{\theta}_i^{\text{CB}}$ are always identical to the ranks based on the $\hat{\theta}_i^B$'s.

Let $R(i)$ be the rank of the true θ_i, that is, $R(i) = \sum_{\ell=1}^{m} I(\theta_i \geq \theta_\ell)$. Then the "optimal" estimator of $R(i)$ that minimizes the expected posterior squared error loss $E[\sum_{i=1}^{m} (Q(i) - R(i))^2 | \hat{\theta}]$ is given by the Bayes estimator

$$Q_{\text{opt}}(i) = \tilde{R}^B(i) = E[R(i)|\hat{\theta}] = \sum_{\ell=1}^{m} P(\theta_i \geq \theta_\ell | \hat{\theta}). \qquad (9.8.11)$$

Generally, the estimators $\tilde{R}^B(i)$ are not integers, so we rank the $\tilde{R}^B(i)$ to produce integer ranks $\hat{R}^B(i)$ = rank of $\tilde{R}^B(i)$ in the set $\{\tilde{R}^B(1), \ldots, \tilde{R}^B(m)\}$.

Shen and Louis (1998) proposed $\hat{\theta}_i^{\text{TG}} = \hat{U}_{\hat{R}^B(i)}$ as a compromise triple-goal estimator of the realized θ_i. The set $\{\hat{\theta}_i^{\text{TG}}\}$ is optimal for estimating the distribution $F_m(t)$ as well as the ranks $\{R(i)\}$. Simulation results indicate that the proposed method performs better than the CB method and achieves the three inferential goals.

9.9 EMPIRICAL LINEAR BAYES

EB methods studied so far are based on distributional assumptions on $\hat{\theta}_i|\theta_i$ and θ_i. Empirical linear Bayes (ELB) methods avoid distributional assumptions by specifying only the first and second moments, but confining to the linear class of estimators, as in the case of EBLUP for the linear mixed models. Maritz and Lwin (1989) provide an excellent account of linear Bayes (LB) methods.

9.9.1 LB Estimation

We assume a two-stage model of the form $\hat{\theta}_i|\theta_i \overset{\text{ind}}{\sim} (\theta_i, \psi_i(\theta_i))$ and $\theta_i \overset{\text{ind}}{\sim} (\mu_i, \sigma_i^2)$, $i = 1, \ldots, m$. Then, we have $\hat{\theta}_i \overset{\text{ind}}{\sim} (\mu_i, \psi_i + \sigma_i^2)$ unconditionally, where $\psi_i = E[\psi_i(\theta_i)]$. This result follows by noting that $E(\hat{\theta}_i) = E[E(\hat{\theta}_i|\theta_i)] = E(\theta_i) = \mu_i$ and $V(\hat{\theta}_i) = E[V(\hat{\theta}_i|\theta_i)] + V[E(\hat{\theta}_i|\theta_i)] = E[\psi_i(\theta_i)] + V(\theta_i) = \psi_i + \sigma_i^2$. We consider a linear class of estimators of the realized θ_i of the form $a_i\hat{\theta}_i + b_i$ and then minimize the unconditional MSE, $E(a_i\hat{\theta}_i + b_i - \theta_i)^2$ with respect to the constants a_i and b_i. The optimal estimator, called the LB estimator, is given by

$$\hat{\theta}_i^{\text{LB}} = \mu_i + \gamma_i(\hat{\theta}_i - \mu_i) = \gamma_i\hat{\theta}_i + (1 - \gamma_i)\mu_i, \qquad (9.9.1)$$

where $\gamma_i = \sigma_i^2/(\psi_i + \sigma_i^2)$ (see Griffin and Krutchkoff 1971 and Hartigan 1969). A proof of (9.9.1) is given in Section 9.12.4. The LB estimator (9.9.1) involves $2m$ parameters $(\mu_i, \sigma_i^2), i = 1, \ldots, m$. In practice, we need to assume that μ_i and σ_i^2 depend on a fixed set of parameters λ in order to "borrow strength". The MSE of $\hat{\theta}_i^{\text{LB}}$ is given by

$$\text{MSE}(\hat{\theta}_i^{\text{LB}}) = E(\hat{\theta}_i^{\text{LB}} - \theta_i)^2 = \gamma_i\psi_i. \qquad (9.9.2)$$

We estimate λ by the method of moments and use the estimator $\hat{\lambda}$ in (9.9.1) to obtain the ELB estimator, given by

$$\hat{\theta}_i^{\text{ELB}} = \hat{\gamma}_i\hat{\theta}_i + (1 - \hat{\gamma}_i)\hat{\mu}_i, \tag{9.9.3}$$

where $\hat{\mu}_i = \mu_i(\hat{\lambda})$ and $\hat{\gamma}_i = \gamma_i(\hat{\lambda})$. A naive estimator of $\text{MSE}(\hat{\theta}_i^{\text{ELB}})$ is obtained as

$$\text{mse}_N(\hat{\theta}_i^{\text{ELB}}) = \hat{\gamma}_i\hat{\psi}_i, \tag{9.9.4}$$

where $\hat{\psi}_i = \psi_i(\hat{\lambda})$. But the naive estimator underestimates the MSE because it ignores the variability associated with $\hat{\lambda}$. It is difficult to find approximately unbiased MSE estimators without further assumptions. In general, $\text{MSE}(\hat{\theta}_i^{\text{ELB}}) \neq \text{MSE}(\hat{\theta}_i^{\text{LB}}) + E(\hat{\theta}_i^{\text{ELB}} - \hat{\theta}_i^{\text{LB}})^2$ because of the nonzero covariance term $E(\hat{\theta}_i^{\text{LB}} - \theta_i)(\hat{\theta}_i^{\text{ELB}} - \hat{\theta}_i^{\text{LB}})$. As a result, the jackknife method is not applicable here without further assumptions.

To illustrate ELB estimation, suppose that θ_i is the RR and $\hat{\theta}_i = y_i/e_i$ is the SMR for the ith area. We assume a two-stage model of the form $E(\hat{\theta}_i|\theta_i) = \theta_i, \psi_i(\theta_i) = V(\hat{\theta}_i|\theta_i) = \theta_i/e_i$ and $E(\theta_i) = \mu_i = \mu$, $V(\theta_i) = \sigma_i^2 = \sigma^2$. The LB estimator is given by (9.9.1) with $\mu_i = \mu$ and $\gamma_i = \sigma^2/(\sigma^2 + \mu/e_i)$. Note that the conditional first and second moments of y_i are identical to the Poisson moments. We obtain moment estimators of μ and σ^2 by equating the weighted sample mean $\hat{\theta}_{e.} = m^{-1}\sum_{i=1}^m (e_i/e.)\hat{\theta}_i$ and the weighted variance, $s_e^2 = m^{-1}\sum_{i=1}^m (e_i/e.)(\hat{\theta}_i - \hat{\theta}_{e.})^2$ to their expected values, as in the case of the Poisson-gamma model (Section 9.6.1). This leads to the moment estimators

$$\hat{\mu} = \hat{\theta}_{e.}, \quad \hat{\sigma}^2 = s_e^2 - \hat{\theta}_{e.}/e.. \tag{9.9.5}$$

Example 9.9.1. Infant Mortality Rates. Marshall (1991) obtained ELB estimates of infant mortality rates in $m = 167$ census area units (CAUs) of Auckland, New Zealand, for the period 1977–1985. For this application, we change $(\theta_i, \hat{\theta}_i, e_i)$ to $(\tau_i, \hat{\tau}_i, n_i)$ and let $E(\tau_i) = \mu$, $V(\tau_i) = \sigma^2$, where τ_i is the mortality rate, $\hat{\tau}_i = y_i/n_i$ is the crude rate, n_i is the number of person-years at risk, and y_i is the number of deaths in the ith area. The n_i for a CAU was taken as nine times its recorded population in the 1981 census. This n_i-value should be a good approximation to the true n_i, because 1981 is the midpoint of the study period 1977–1985. "Global" ELB estimates of the τ_i's were obtained from (9.9.3), using the moment estimator based on (9.9.5). These estimators shrink the crude rates $\hat{\tau}_i$ toward the overall mean $\hat{\mu} = 2.63$ deaths per thousand. Marshall (1991) also obtained "local" estimates by defining the neighbors of each CAU to be those sharing a common boundary; the smallest neighborhood contained 3 CAUs and the largest 13 CAUs. A local estimate of τ_i was obtained by using local estimates $\hat{\mu}_{(i)}$ and $\hat{\sigma}_{(i)}^2$ of μ and σ^2 for each CAU i. The estimates $\hat{\mu}_{(i)}$ and $\hat{\sigma}_{(i)}^2$ were obtained from (9.9.5) using only the neighborhood areas of i to calculate the weighted sample mean and variance. This heuristic method of local smoothing is similar to smoothing based on spatial modeling of the areas. Marshall (1991) did not report standard errors of the estimates. For a global estimate, the naive standard error based on (9.9.4) should be adequate because $m = 167$ is large, but it may be inadequate as a measure of variability for local estimates based on 3–13 CAUs.

Karunamuni and Zhang (2003) studied LB and ELB estimation of finite population small area means, $\overline{Y}_i = \sum_{j=1}^{N_i} y_{ij}/N_i$, for unit level models, without covariates. Suppose that the population model is a two-stage model of the form $y_{ij}|\theta_i \overset{\text{ind}}{\sim} (\theta_i, \mu_2(\theta_i))$ for each i $(j = 1, \ldots, N_i)$ and $\theta_i \overset{\text{iid}}{\sim} (\mu, \sigma_v^2), \sigma_e^2 = E[\mu_2(\theta_i)]$. We assume that the model holds for the sample $\{y_{ij}; j = 1, \ldots, n_i, i = 1, \ldots, m\}$. We consider a linear class of estimators of the realized $\overline{Y}_i$ of the form $a_i\overline{y}_i + b_i$, where $\overline{y}_i$ is the ith area sample mean. Minimizing the unconditional MSE, $E(a_i\overline{y}_i + b_i - \overline{Y}_i)^2$, with respect to the constants a_i and b_i, the "optimal" LB estimator of $\overline{Y}_i$ is obtained as

$$\hat{\overline{Y}}_i^{\text{LB}} = f_i\overline{y}_i + (1 - f_i)[\gamma_i\overline{y}_i + (1 - \gamma_i)\mu], \tag{9.9.6}$$

where $\gamma_i = \sigma_v^2/(\sigma_v^2 + \sigma_e^2/n_i)$ and $f_i = n_i/N_i$. If we replace μ by the BLUE estimator $\tilde{\mu} = \sum_{i=1}^m \gamma_i\overline{y}_i / \sum_{i=1}^m \gamma_i$, we get a first-step ELB estimator similar to the BLUP estimator for the basic unit level model (7.1.1) without covariates and $k_{ij} = 1$. Ghosh and Lahiri (1987) used ANOVA-type estimators of σ_e^2 and σ_v^2 given by

$$\hat{\sigma}_e^2 = \sum_{i=1}^{m}\sum_{j=1}^{n_i} (y_{ij} - \overline{y}_i)^2/(n - m) =: s_w^2 \tag{9.9.7}$$

where $n = \sum_{i=1}^m n_i$ and

$$\hat{\sigma}_v^2 = \max\{0, (s_b^2 - s_w^2)(m - 1)g^{-1}\}, \tag{9.9.8}$$

where $s_b^2 = \sum_{i=1}^m n_i(\overline{y}_i - \overline{y}_.)^2/(m - 1)$, with $\overline{y}_. = \sum_{i=1}^m n_i\overline{y}_i/n$ and $g = n - \sum_{i=1}^m n_i^2/n$. Replacing σ_e^2 and σ_v^2 by $\hat{\sigma}_e^2$ and $\hat{\sigma}_v^2$ in the first-step estimator, we get an ELB estimator of $\overline{Y}_i$. Note that the first-step estimator depends only on the ratio $\tau = \sigma_v^2/\sigma_e^2$. Then, instead of $\hat{\tau} = \hat{\sigma}_v^2/\hat{\sigma}^2$, we can use an approximately unbiased estimator of τ along the lines of Morris (1983b):

$$\tau^* = \max\left[0, \left\{\frac{(m - 1)s_b^2}{(m - 3)s_w^2} - 1\right\}(m - 1)g^{-1}\right]. \tag{9.9.9}$$

The multiplier $(m - 1)/(m - 3)$ corrects the bias of $\hat{\tau}$. The modified ELB estimator is obtained from the first-step estimator by substituting τ^* for τ, and using $\hat{\mu} = \sum_{i=1}^m \gamma_i^*\overline{y}_i / \sum_{i=1}^m \gamma_i^*$ if $\tau^* \neq 0$ and $\hat{\mu} = \sum_{i=1}^m n_i\overline{y}_i/n$ if $\tau^* = 0$.

The MSE of $\hat{\overline{Y}}_i^{\text{LB}}$ is given by

$$\begin{aligned}\text{MSE}(\hat{\overline{Y}}_i^{\text{LB}}) &= E(\hat{\overline{Y}}_i^{\text{LB}} - \overline{Y}_i)^2 \\ &= (1 - f_i)^2 g_{1i}(\sigma_v^2, \sigma_e^2) + \frac{1}{N_i}(1 - f_i)\sigma_e^2 \qquad (9.9.10) \\ &\approx (1 - f_i)^2 g_{1i}(\sigma_v^2, \sigma_e^2), \qquad (9.9.11)\end{aligned}$$

where

$$g_{1i}(\sigma_v^2, \sigma_e^2) = \gamma_i \sigma_e^2 / n_i \tag{9.9.12}$$

and $\gamma_i = \sigma_v^2/(\sigma_v^2 + \sigma_e^2/n_i)$. The last term in (9.9.10) is negligible if N_i is large, leading to the approximation (9.9.11). Note that (9.9.10) has the same form as the MSE of the Bayes estimator of $\overline{Y}_i$ under the basic unit level model with normal effects v_i and normal errors e_{ij} with equal variance σ_e^2. A naive estimator of MSE($\hat{\overline{Y}}_i^{\text{ELB}}$) is obtained by substituting $(\hat{\sigma}_v^2, \hat{\sigma}_e^2)$ for (σ_v^2, σ_e^2) in (9.9.10), but it underestimates MSE($\hat{\overline{Y}}_i^{\text{ELB}}$) because it ignores the variability associated with $(\hat{\sigma}_v^2, \hat{\sigma}_e^2)$. Again, it is difficult to find approximately unbiased estimators of MSE without further assumptions.

9.9.2 Posterior Linearity

A different approach, called linear EB (LEB), has also been used to estimate the small area means $\overline{Y}_i$ under the two-stage unit level model studied in Section 9.9.1 (Ghosh and Lahiri 1987). The basic assumption underlying this approach is the posterior linearity condition (Goldstein 1975):

$$E(\theta_i | \mathbf{y}_i, \lambda) = a_i \overline{y}_i + b_i, \tag{9.9.13}$$

where $\mathbf{y}_i$ is the $n_i \times 1$ vector of sample observations from the ith area and $\lambda = (\mu, \sigma_v^2, \sigma_e^2)^T$. This condition holds for a variety of distributions on the θ_i's. However, the distribution of the θ_i's becomes a conjugate family if the conditional distribution of y_{ij} given θ_i belongs to the exponential family. For example, the model $y_{ij}|\theta_i \overset{\text{iid}}{\sim} \text{Bernoulli}(\theta_i)$ with posterior linearity (9.9.13) implies that $\theta_i \sim \text{beta}(\alpha, \beta)$, for $\alpha > 0, \beta > 0$. The LEB approach assumes posterior linearity and uses the two-stage model on the y_{ij}'s without distributional assumptions. It leads to an estimator of $\overline{Y}_i$ identical to the estimator under the LB approach, namely, $\hat{\overline{Y}}_i^{\text{LB}}$ given by (9.9.6). The two approaches are therefore similar, but the ELB approach is more appealing as it avoids further assumptions on the two-stage model by confining to the linear class of estimators.

We briefly outline the LEB approach. Using the two-stage model and the posterior linearity condition (9.9.13), it follows from Section 9.9.1 that the optimal a_i and b_i that minimize $E(\theta_i - a_i\overline{y}_i - b_i)^2$ are given by $a_i^* = \gamma_i$ and $b_i^* = (1-\gamma_i)\mu$, where $\gamma_i = \sigma_v^2/(\sigma_v^2 + \sigma_e^2/n_i)$ as in Section 9.9.1. Therefore, the Bayes estimator under posterior linearity is given by

$$E(\theta_i | \mathbf{y}_i, \lambda) = \gamma_i \overline{y}_i + (1-\gamma_i)\mu, \tag{9.9.14}$$

which is identical to the LB estimator. Using (9.9.14), we get

$$E(\overline{Y}_i | \mathbf{y}_i, \lambda) = f_i \overline{y}_i + (1-f_i)[\gamma_i \overline{y}_i + (1-\gamma_i)\mu], \tag{9.9.15}$$

noting that for any unit j in the nonsampled set r_i, it holds that $E(y_{ij}|\mathbf{y}_i, \lambda) = E[E(y_{ij}|\theta_i, y_i, \lambda)|\mathbf{y}_i, \lambda] = E(\theta_i|\mathbf{y}_i, \lambda)$; see Ghosh and Lahiri (1987). It follows from (9.9.15) that $E(\overline{Y}_i|\mathbf{y}_i, \lambda)$ is identical to the LB estimator $\hat{\overline{Y}}_i^{LB}$ given by (9.9.6). Substituting moment estimators $\hat{\lambda}$ in (9.9.15), we get the LEB estimator, $\hat{\overline{Y}}_i^{LEB} = E(\overline{Y}_i|\mathbf{y}_i, \hat{\lambda})$, which is identical to the ELB estimator $\hat{\overline{Y}}_i^{ELB}$.

It appears that nothing is gained over the ELB approach by making the posterior linearity assumption and then using the LEB approach. However, it permits the use of the jackknife to obtain an approximately unbiased MSE estimator, unlike the ELB approach, by making use of the orthogonal decomposition

$$\begin{aligned}\text{MSE}(\hat{\overline{Y}}_i^{LEB}) &= E(\hat{\overline{Y}}_i^{LB} - \overline{Y}_i)^2 + E(\hat{\overline{Y}}_i^{LEB} - \hat{\overline{Y}}_i^{LB})^2\\ &= \tilde{g}_{1i}(\sigma_v^2, \sigma_e^2) + E(\hat{\overline{Y}}_i^{LEB} - \hat{\overline{Y}}_i^{LB})^2 = M_{1i} + M_{2i},\end{aligned} \tag{9.9.16}$$

where $\tilde{g}_{1i}(\sigma_v^2, \sigma_e^2)$ is given by (9.9.10). We obtain a jackknife estimator of the last term in (9.9.16) as

$$\hat{M}_{2i} = \frac{m-1}{m}\sum_{\ell=1}^{m}(\hat{\overline{Y}}_{i,-\ell}^{LEB} - \hat{\overline{Y}}_i^{LEB})^2, \tag{9.9.17}$$

where $\hat{\overline{Y}}_{i,-\ell}^{LEB} = E(\overline{Y}_i|\mathbf{y}_i, \hat{\lambda}_{-\ell})$ is obtained from $\hat{\overline{Y}}_i^{LEB}$ by substituting the delete$-\ell$ estimators $\hat{\lambda}_{-\ell}$ for $\hat{\lambda}$. The leading term, M_{1i}, is estimated as

$$\hat{M}_{1i} = \tilde{g}_{1i}(\hat{\sigma}_v^2, \hat{\sigma}_e^2) - \frac{m-1}{m}\sum_{\ell=1}^{m}[\tilde{g}_{1i}(\hat{\sigma}_{v,-\ell}^2, \hat{\sigma}_{e,-\ell}^2) - \tilde{g}_{1i}(\hat{\sigma}_v^2, \hat{\sigma}_e^2)]. \tag{9.9.18}$$

The jackknife MSE estimator is then given by

$$\text{mse}_J(\hat{\overline{Y}}_i^{LEB}) = \hat{M}_{1i} + \hat{M}_{2i}. \tag{9.9.19}$$

Results for the infinite population case are obtained by letting $f_i = 0$.

The above results are applicable to two-stage models without covariates. Raghunathan (1993) proposed two-stage models with area level covariates, by specifying only the first and second moments (see Section 4.6.5). He obtained a quasi-EB estimator of the small area parameter, θ_i, using the following steps: (i) Based on the mean and variance specifications, obtain the conditional quasi-posterior density of θ_i given the data and model parameters λ. (ii) Evaluate the quasi-Bayes estimator and the conditional quasi-posterior variance by numerical integration, using the density from step (i). (iii) Use a generalized EM (GEM) algorithm to get quasi-ML estimator $\hat{\lambda}$ of the model parameters. (iv) Replace λ by $\hat{\lambda}$ in the quasi-Bayes estimator to obtain the quasi-EB estimator of θ_i. Raghunathan (1993) also proposed a jackknife method of estimating the MSE of the quasi-EB estimator, taking account of the variability of $\hat{\lambda}$. This method is different from the jackknife method of Jiang, Lahiri, and Wan (2002), and its asymptotic properties have not been studied.

9.10 CONSTRAINED LB

Consider the two-stage model (i) $\hat{\theta}_i|\theta_i \overset{\text{ind}}{\sim} (\theta_i, \psi_i(\theta_i))$ and (ii) $\theta_i \overset{\text{iid}}{\sim} (\mu, \sigma^2)$ and $\psi_i = E(\psi_i(\theta_i))$. We consider a linear class of estimators of the realized θ_i of the form $a_i\hat{\theta}_i + b_i$ and determine the constants a_i and b_i to match the mean μ and the variance σ^2 of θ_i:

$$E(a_i\hat{\theta}_i + b_i) = \mu, \tag{9.10.1}$$

$$E(a_i\hat{\theta}_i + b_i - \mu)^2 = \sigma^2. \tag{9.10.2}$$

Noting that $E(\hat{\theta}_i) = \mu$, we get $b_i = \mu - (1 - a_i)$ from (9.10.1), and (9.10.2) reduces to

$$a_i^2 E(\hat{\theta}_i - \mu)^2 = a_i^2(\sigma^2 + \psi_i) = \sigma^2 \tag{9.10.3}$$

or $a_i = \gamma_i^{1/2}$, where $\gamma_i = \sigma^2/(\sigma^2 + \psi_i)$. The resulting estimator is a constrained LB (CLB) estimator:

$$\hat{\theta}_i^{\text{CLB}} = \gamma_i^{1/2}\hat{\theta}_i + (1 - \gamma_i^{1/2})\mu \tag{9.10.4}$$

(Spjøtvoll and Thomsen 1987). Method of moments may be used to estimate the model parameters. The resulting estimator is an empirical CLB estimator. Note that $\hat{\theta}_i^{\text{LB}}$ attaches a smaller weight, γ_i, to the direct estimator $\hat{\theta}_i$ which leads to over-shrinking toward μ compared to $\hat{\theta}_i^{\text{CLB}}$.

As an example, consider the model for binary data: $\hat{\theta}_i|\theta_i \overset{\text{ind}}{\sim} (\theta_i, \theta_i(1-\theta_i)/n_i)$ and $\theta_i \overset{\text{iid}}{\sim} (\mu, \sigma^2)$, where $\hat{\theta}_i$ is the sample proportion and n_i is the sample size in the ith area. Noting that $E[\theta_i(1 - \theta_i)] = \mu(1 - \mu) - \sigma^2$, we get

$$\gamma_i = \sigma^2\left[\left(1 - \frac{1}{n_i}\right)\sigma^2 + \frac{\mu(1-\mu)}{n_i}\right]^{-1}. \tag{9.10.5}$$

Spjøtvoll and Thomsen (1987) used empirical CLB to study the distribution of $m = 100$ municipalities in Norway with respect to the proportion of persons not in the labor force. A comparison with the actual distribution in their example showed that the CLB method tracked the actual distribution much better than the LB method.

The above CLB method ignores the simultaneous aspect of the problem, by considering each area separately. However, the CLB estimator (9.10.4) is similar to the CB estimator, $\hat{\theta}_i^{\text{CB}}$. In fact, as shown in (9.8.9), $\hat{\theta}_i^{\text{CB}} \approx \hat{\theta}_i^{\text{CLB}}$ under the simple model $\hat{\theta}_i = \theta_i + e_i$ with $e_i \overset{\text{iid}}{\sim} N(0, \psi)$ independent of $\theta_i \overset{\text{iid}}{\sim} N(\mu, \sigma_v^2)$.

To take account of the simultaneous aspect of the problem, we can use the method of Section 9.8.1 assuming posterior linearity $E(\theta_i|\hat{\theta}_i) = a_i\hat{\theta}_i + b_i$. We minimize the posterior expected squared error loss under the two-stage model subject to the constraints (9.8.4) and (9.8.5) on the ensemble mean and variance, respectively. The resulting CLB estimator is equal to (9.8.6) with $\hat{\theta}_i^B$ and $V(\theta_i|\hat{\theta}_i, \lambda)$ changed to $\hat{\theta}_i^{\text{LB}} =$

$\gamma_i\hat{\theta}_i + (1-\gamma_i)\mu$ and $E[(\theta_i - \hat{\theta}_i^{\text{LB}})^2|\hat{\theta}_i, \lambda]$, respectively. This estimator avoids distributional assumptions. Lahiri (1990) called this method "adjusted" Bayes estimation.

The posterior variance term $E[(\theta_i - \hat{\theta}_i^{\text{LB}})^2|\hat{\theta}_i, \lambda]$ cannot be calculated without additional assumptions; Lahiri (1990) and Ghosh (1992b) incorrectly stated that it is equal to $\gamma_i\psi_i$. However, for large m, we can approximate $m^{-1}\sum_{i=1}^m E[(\theta_i - \hat{\theta}_i^{\text{LB}})^2|\hat{\theta}_i, \lambda]$ by its expectation $m^{-1}\sum_{i=1}^m E[(\theta_i - \hat{\theta}_i^{\text{LB}})^2|\lambda] = m^{-1}\sum_{i=1}^m \gamma_i\psi_i$. Using this approximation, we get the following CLB estimator:

$$\hat{\theta}_i^{\text{CLB}}(1) = \hat{\theta}_\cdot^{\text{LB}} + a^*(\hat{\boldsymbol{\theta}}, \lambda)(\hat{\theta}_i^{\text{LB}} - \hat{\theta}_\cdot^{\text{LB}}), \tag{9.10.6}$$

where

$$a^*(\hat{\boldsymbol{\theta}}, \lambda) = \left\{1 + \frac{(1/m)\sum_{i=1}^m \gamma_i\psi_i}{[1/(m-1)]\sum_{i=1}^m(\hat{\theta}_i^{\text{LB}} - \hat{\theta}_\cdot^{\text{LB}})^2}\right\}^{1/2}. \tag{9.10.7}$$

In general, $\hat{\theta}_i^{\text{CLB}}(1)$ differs from $\hat{\theta}_i^{\text{CLB}}$ given by (9.10.4), but for the special case of equal ψ_i, we get $\hat{\theta}_i^{\text{CLB}}(1) \approx \hat{\theta}_i^{\text{CLB}}$ for large m, similar to the result (9.8.9).

The CLB estimator (9.10.6) depends on the unknown model parameters $\lambda = (\mu, \sigma^2)^T$. Replacing λ by moment estimators $\hat{\lambda}$, we obtain an empirical CLB estimator.

9.11 *SOFTWARE

As already mentioned, under the basic area level model (6.1.1) with normality, the EB estimator of θ_i given in (9.2.3) equals the EBLUP of θ_i. Section 6.5 describes the functions of the R package `sae` that calculate EBLUP estimates and analytical MSE estimates based on the basic area level model. Similarly, equivalence of EB and EBLUP occurs when estimating a small area mean $\overline{Y}_i$ under the basic unit level model (7.1.1) with $k_{ij} = 1$ and with normality. The corresponding functions of the R package `sae` are described in Section 7.7.

For estimation of general nonlinear parameters $\tau_i = h(\mathbf{y}_i^P)$ under the basic unit level model (7.1.1) with $k_{ij} = 1$, EB estimates together with parametric bootstrap MSE estimates can be obtained using functions `ebBHF()` and `pbmseebBHF()`. The calls to these functions are as follows:

```
ebBHF(formula, dom, selectdom, Xnonsample, MC = 100,
    data, transform = "BoxCox", lambda = 0, constant = 0, indicator)
pbmseebBHF(formula, dom, selectdom, Xnonsample, B = 100, MC = 100,
    data, transform = "BoxCox", lambda = 0, constant = 0, indicator)
```

These functions assume that the target variable for jth individual in ith area is E_{ij}, but the response (or dependent) variable in the basic unit level model is $y_{ij} = T(E_{ij})$, where $T(\cdot)$ is a one-to-one transformation. A typical example of this situation is when

E_{ij} is income or expenditure, and we consider the model for $y_{ij} = \log(E_{ij} + c)$. The target area parameter is $\tau_i = h(\{y_{ij}; j = 1, \ldots, N_i\}) = h(\{T(E_{ij}); j = 1, \ldots, N_i\})$.

Thus, on the left-hand side of `formula`, we must write the name of the vector containing the sample data on the original target variables E_{ij}. On the right-hand side of `formula`, we must write the auxiliary variables considered in the basic unit level model separated by "+". Note that an intercept is assumed by default as in any usual R formula. The target variables E_{ij}, placed on the left-hand side of `formula`, can be transformed by first adding a constant to it through the argument `constant`, and then applying a transformation chosen from either the Box–Cox or power families through the argument `transform`. The value of the parameter for the chosen family of transformations can be specified in the argument `lambda`. This parameter is set by default to 0, which corresponds to log transformation for both families. Setting `lambda=1` means no transformation. Note that these functions fit the basic unit level model for the transformed variables $y_{ij} = T(E_{ij})$. The target parameter τ_i, expressed as an R function of the original target variables E_{ij}, that is, as $h(\{T(E_{ij}); j = 1, \ldots, N_i\})$, must be specified in argument `indicator`. For example, if we want to estimate the median of $\{E_{ij}; j = 1, \ldots, N_i\}$, we just need to specify `indicator=median`.

The vector with the area codes for sample observations must be specified in `dom`. The functions allow to select a subset of the areas for estimation, just specifying the vector of selected (unique) area codes in `selectdom`. Moreover, to generate the nonsample vectors and apply the Monte Carlo approximation to the EB estimator of τ_i as given in (9.4.3), we need the values of the auxiliary variables for all nonsampled units. A matrix or data frame with those values must be specified in `Xnonsample`. Additionally, the desired number of Monte Carlo samples can be specified in `MC`.

The function `ebBHF()` returns a list of two objects: the first one is called `eb` and contains the EB estimates for each selected area, and the second one is called `fit` and contains a summary of the model-fitting results. The function `pbmseebBHF()` obtains the parametric bootstrap MSE estimates together with EB estimates. It delivers a list with two objects. The first one, `est`, is another list containing itself two objects: the results of the point estimation process (`eb`) and a summary of the fitting results (`fit`). The second object, called `mse`, contains a data frame with the estimated MSEs for each selected area.

Example 9.11.1. Poverty Mapping, with R. In this example, we estimate poverty incidences in Spanish provinces (areas) using the predefined data set `incomedata`, which contains synthetic unit level data on income E_{ij} and other sociological variables for a sample of individuals, together with province identifiers. The following variables from the data set `incomedata` will be used: province name (`provlab`), province code (`prov`), income (`income`), sampling weight (`weight`), education level (`educ`), labor status (`labor`), and finally the indicators of each of the categories of `educ` and `labor`.

We calculate EB estimates of province poverty incidences based on the basic unit level model (with $k_{ij} = 1$), for a transformation of the variable `income` and the categories of education level and of labor status as auxiliary variables. The EB method described in Section 9.4 requires (at least approximately) normality of the

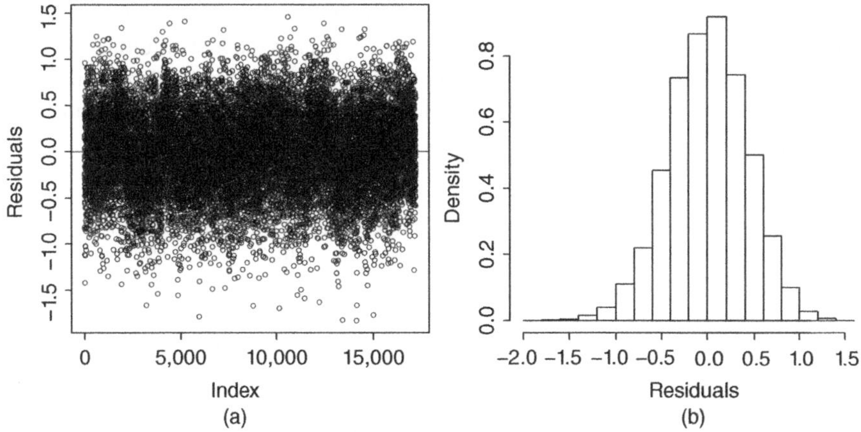

Figure 9.4 Index Plot of Residuals (a) and Histogram of Residuals (b) from the Fitting of the Basic Unit Level Model with Response Variable log(income+constant).

response variable in the model. The histogram of `income` is severely right-skewed, but log(`income` + 3, 500) leads to model residuals with an approximately symmetric distribution (see Figure 9.4).

The poverty incidence for province i is obtained by taking $\alpha = 0$ in the FGT poverty indicator $F_{\alpha i}$ given in (9.4.10), that is, F_{0i}. In R, it can be calculated as the province mean of the indicator variable taking value 1 when the person's income is below the poverty line z and 0 otherwise. The poverty line is defined as 0.6 the median income, which turns out to be `z = 6,557.143`. We first create the function `povertyincidence()` defining the target parameter $\tau_i = F_{0i}$ as function of `y=income`:

```
R > z <- 6,557.143
R > povertyincidence <- function(y) {
+   result <- mean(y < z)
+   return (result)
+ }
```

Although sample data from all the provinces will be used to fit the basic unit level model, to save computation time in this example, we compute EB estimates and corresponding MSE estimates only for the five provinces with the smallest sample sizes. For these selected provinces, the data set `Xoutsamp` contains the values of several auxiliary variables (age groups, nationality, education levels, labor status) for each nonsampled individual. We read the two required data sets `incomedata` and `Xoutsamp`:

```
R> data("incomedata")
R> data("Xoutsamp")
```

Now we create the vector with the codes for the five selected provinces. These codes are taken as the unique province codes that appear in the first column of `Xoutsamp` data set. Then we construct the data frame `Xoutsamp_AuxVar`, containing the values of the selected auxiliary variables (education levels and labor status) for nonsampled individuals from these five selected provinces:

```
R> provincecodes <- unique(Xoutsamp$domain)
R> provincelabels <- unique(incomedata$provlab)[provincecodes]
R> Xoutsamp_AuxVar <- Xoutsamp[,c("domain","educ1","educ3","labor1",
+   "labor2")]
```

Next, we use the function `ebBHF` to calculate EB estimates of the poverty incidences under the basic unit level model for log(`income+constant`) for the five selected provinces specified in the argument `selectdom`. The function `povertyincidence()` defining the target parameter is specified in `indicator`:

```
R> set.seed(123)
R> EB <- ebBHF(income~ educ1 + educ3 + labor1 + labor2, dom = prov,
+ selectdom = provincecodes, Xnonsample = Xoutsamp_AuxVar, MC = 50,
+ constant = 3,500, indicator = povertyincidence, data = incomedata)
```

The list `fit` of the output gives a summary of the fitting results. For example, we can see whether auxiliary variables are significant as follows:

```
R> EB$fit$summary

Linear mixed-effects model fit by REML
 Data: NULL
       AIC       BIC    logLik
  18,980.72 19,034.99 -9,483.361

Random effects:
 Formula: ~1 | as.factor(dom)
        (Intercept)  Residual
StdDev:  0.09436138 0.4179426

Fixed effects: ys ~ -1 + Xs
                   Value   Std.Error     DF  t-value p-value
Xs(Intercept)   9.505176 0.014384770 17,143 660.7805       0
Xseduc1        -0.124043 0.007281270 17,143 -17.0359       0
Xseduc3         0.291927 0.010366323 17,143  28.1611       0
Xslabor1        0.145985 0.006915979 17,143  21.1084       0
Xslabor2       -0.081624 0.017082634 17,143  -4.7782       0
 Correlation:
        Xs(In) Xsedc1 Xsedc3 Xslbr1
Xseduc1  -0.212
Xseduc3  -0.070  0.206
Xslabor1 -0.199  0.128 -0.228
Xslabor2 -0.079  0.039 -0.039  0.168
```

```
Standardized Within-Group Residuals:
       Min          Q1         Med          Q3         Max
-4.2201202 -0.6617181  0.0203607  0.6881828  3.5797393

Number of Observations: 17,199
Number of Groups: 52
```

Checking model assumptions is crucial in the EB method since the optimality properties of the EB estimates depend on the extent to which those model assumptions are true. To detect departures from normality of the transformed income, we can display the usual residual plots. The following R commands draw an index plot and a histogram of residuals:

```
R> plot(EB$fit$residuals, xlab = "Index", ylab = "Residuals",
+   cex.axis = 1.5, cex.lab = 1.5)
R> hist(EB$fit$residuals, prob = TRUE, xlab = "Residuals",
+   main = "", cex.axis = 1.5, cex.lab = 1.5)
```

Figure 9.4 displays the two mentioned plots, which show no evidence of serious model departure.

Finally, we compute parametric bootstrap MSE estimates and calculate CVs of EB estimates. This process might be slow for large number of bootstrap or Monte Carlo iterations `B` and `MC`, respectively, large sample size or large number of auxiliary variables. Note that function `pbmseebBHF()` gives again the EB estimates:

```
R> set.seed(123)
R> pbmse.EB <- pbmseebBHF(income ~ educ1 + educ3 + labor1 + labor2,
+ dom = prov, selectdom = provincecodes, Xnonsample = Xoutsamp_AuxVar,
+ B = 200, MC = 50, constant = 3,500, indicator = povertyincidence,
+ data = incomedata)
R> pbcv.EB <- 100*sqrt(pbmse.EB$mse$mse)/abs(pbmse.EB$est$eb$eb)
```

In the data frame `results.EB`, we collect the sample sizes, EB estimates, and CVs of the poverty incidence for the five selected provinces:

```
R> results.EB <-data.frame(ProvinceIndex=pbmse.EB$est$eb$domain,
+ ProvinceName = provincelabels,
+ SampleSize = pbmse.EB$est$eb$sampsize,
+ EB = pbmse.EB$est$eb$eb, cv.EB = pbcv.EB)
R> results.EB
  ProvinceIndex ProvinceName SampleSize        EB    cv.EB
1            42        Soria         20 0.2104329 21.06776
2             5        Avila         58 0.1749877 19.49466
3            34     Palencia         72 0.2329916 11.57829
4            44       Teruel         72 0.2786618 11.89621
5            40      Segovia         58 0.2627178 13.21378
```

Finally, we obtain direct estimates for comparison:

```
R> poor <- as.integer(incomedata$income < z)
R> DIR <- direct(y = poor, dom = incomedata$provlab,
+ sweight = incomedata$weight, domsize = Popn)
R> results.DIR<-DIR[DIR$Domain %in% c("Soria","Avila","Palencia",
+ "Teruel","Segovia"),c("Domain","Direct","CV")]
     Domain     Direct       CV
5     Avila 0.05512200 46.35946
34 Palencia 0.30166074 23.80085
40  Segovia 0.22262002 25.33449
42    Soria 0.02541207 99.97815
44   Teruel 0.27364239 24.57017
```

We see that CVs of the EB estimates are much smaller than those of the direct estimates for the same five provinces.

9.12 PROOFS

9.12.1 Proof of (9.2.11)

Let us decompose $(\hat{\theta}_i^{\text{EB}} - \theta_i)^2$ as

$$(\hat{\theta}_i^{\text{EB}} - \hat{\theta}_i^B + \hat{\theta}_i^B - \theta_i)^2 = (\hat{\theta}_i^{\text{EB}} - \hat{\theta}_i^B)^2 + (\hat{\theta}_i^B - \theta_i)^2 + 2(\hat{\theta}_i^{\text{EB}} - \hat{\theta}_i^B)(\hat{\theta}_i^B - \theta_i), \tag{9.12.1}$$

Now taking expectation in (9.12.1), using the Law of Iterated Expectations $E = E_{\hat{\theta}} E_{\theta|\hat{\theta}}$, where $E_{\theta|\hat{\theta}}$ is the expectation over the conditional distribution of θ given $\hat{\theta}$ and $E_{\hat{\theta}}$ is the expectation over the marginal distribution of $\hat{\theta}$, and noting that

$$E_{\theta|\hat{\theta}}[(\hat{\theta}_i^{\text{EB}} - \hat{\theta}_i^B)(\hat{\theta}_i^B - \theta_i)] = (\hat{\theta}_i^{\text{EB}} - \hat{\theta}_i^B)[E_{\theta_i|\hat{\theta}_i}(\hat{\theta}_i^B - \theta_i)] = 0, \tag{9.12.2}$$

we get the decomposition (9.2.11).

9.12.2 Proof of (9.2.30)

We note that $\hat{\theta}_i|\mu, \sigma_v^2 \overset{\text{iid}}{\sim} N(\mu, \sigma_v^2 + \psi), i = 1, \ldots, m$ and $f(\mu) = \text{constant}$. Therefore, by Bayes theorem,

$$\begin{aligned} f(\mu|\hat{\theta}, \sigma_v^2) &\propto f(\hat{\theta}|\mu, \sigma_v^2) f(\mu) \\ &\propto \exp\left\{-\frac{1}{2}\sum_{i=1}^m (\hat{\theta}_i - \mu)^2/(\sigma_v^2 + \psi)\right\} \\ &\propto \exp\left\{-\frac{m}{2}\sum_{i=1}^m (\mu - \hat{\theta}_\cdot)^2/(\sigma_v^2 + \psi)\right\}, \end{aligned}$$

which is the kernel of a $N[\hat{\theta}_\cdot, m^{-1}(\sigma_v^2 + \psi)]$ distribution. Hence, we get (9.2.30).

9.12.3 Proof of (9.8.6)

The posterior expected squared error loss $E[\sum_{i=1}^{m}(\theta_i - t_i)^2|\hat{\theta}]$ may be written as

$$E\left[\sum_{i=1}^{m}(\theta_i - t_i)^2|\hat{\theta},\lambda\right] = E\left[\sum_{i=1}^{m}(\theta_i - \hat{\theta}_i^B)^2|\hat{\theta},\lambda\right] + \sum_{i=1}^{m}(\hat{\theta}_i^B - t_i)^2. \tag{9.12.3}$$

Therefore, it is sufficient to minimize the last term of (9.12.3). Using Lagrange multipliers a_1 and a_2, we minimize $\sum_{i=1}^{m}(\hat{\theta}_i^B - t_i)^2$ subject to constraints $\sum_{i=1}^{m} t_i = c_1$ and $\sum_{i=1}^{m}(t_i - t_.)^2 = c_2$ or $\sum_{i=1}^{m} t_i^2 = c_2 + c_1^2/m$, for given constants c_1 and c_2; that is, we minimize the objective function $\phi = \sum_{i=1}^{m}(\hat{\theta}_i^B - t_i)^2 - a_1(\sum_{i=1}^{m} t_i - c_1) - a_2(\sum_{i=1}^{m} t_i^2 - c_2 - c_1^2/m)$ with respect to $t_1, \ldots, t_m$. The solution is given by

$$t_{i,\,\text{opt}} = \frac{1}{1-a_2}\left(\hat{\theta}_i^B + \frac{a_1}{2}\right). \tag{9.12.4}$$

Imposing the constraints on (9.12.4), we get $a_1/2 = (1-a_2)c_1/m - \hat{\theta}_.^B$ and $(1-a_2)^2 = \sum_{i=1}^{m}(\hat{\theta}_i^B - \hat{\theta}_.^B)^2/c_2$. Now taking the positive square root of $(1-a_2)^2$, we get

$$t_{i,\,\text{opt}} = \frac{c_1}{m} + \frac{1}{1-a_2}(\hat{\theta}_i^B - \hat{\theta}_.^B) \tag{9.12.5}$$

from (9.12.4), where $1 - a_2 = [\sum_{i=1}^{m}(\hat{\theta}_i^B - \hat{\theta}_.^B)^2/c_2]^{1/2}$. The estimator (9.12.5) is valid for any specified c_1 and c_2. In particular, the constraints in (9.8.4) and (9.8.5) specify $c_1 = m\,\hat{\theta}_.^B$ and $c_2 = E[\sum_{i=1}^{m}(\theta_i - \theta_.)^2|\hat{\theta}]$. Now using (9.8.2) and these values of c_1 and c_2 in (9.12.5), we get $t_{i,\,\text{opt}} = \hat{\theta}_i^{\text{CB}}$ given by (9.8.6). This proof is due to Singh and Folsom (2001) and Shen and Louis (1998).

9.12.4 Proof of (9.9.1)

We minimize $\phi_i = E(a_i\hat{\theta}_i + b_i - \theta_i)^2$ with respect to a_i and b_i to get the optimal a_i^* and b_i^* from

$$a_i^*\mu_i + b_i^* = \mu_i, \tag{9.12.6}$$

$$E(\theta_i\hat{\theta}_i) = a_i^*E(\hat{\theta}_i^2) + b_i^*\mu_i. \tag{9.12.7}$$

Substituting for b_i^* from (9.12.6) into (9.12.7), we get $\text{Cov}(\hat{\theta}_i, \theta_i)/V(\hat{\theta}_i) = \sigma_i^2/(\psi_i + \sigma_i^2) = \gamma_i$, noting that $E(\theta_i\hat{\theta}_i) = E[\theta_i E(\hat{\theta}_i|\theta_i)] = E(\theta_i^2) = \sigma_i^2 + \mu_i^2$ and $V(\theta_i) = \psi_i + \sigma_i^2$. Hence,

$$a_i^*\hat{\theta}_i + b_i^* = \mu_i + \gamma_i(\hat{\theta}_i - \mu_i),$$

which is identical to the LB estimator $\hat{\theta}_i^{\text{LB}}$ given by (9.9.1).

10

HIERARCHICAL BAYES (HB) METHOD

10.1 INTRODUCTION

In the hierarchical Bayes (HB) approach, a subjective prior distribution $f(\boldsymbol{\lambda})$ on the model parameters $\boldsymbol{\lambda}$ is specified and the posterior distribution $f(\boldsymbol{\mu}|\mathbf{y})$ of the small area (random) parameters of interest $\boldsymbol{\mu}$, given the data $\mathbf{y}$, is obtained. The two-stage model, $f(\mathbf{y}|\boldsymbol{\mu}, \boldsymbol{\lambda}_1)$ and $f(\boldsymbol{\mu}|\boldsymbol{\lambda}_2)$, is combined with the subjective prior on $\boldsymbol{\lambda} = (\boldsymbol{\lambda}_1^T, \boldsymbol{\lambda}_2^T)^T$, using Bayes theorem, to arrive at the posterior $f(\boldsymbol{\mu}|\mathbf{y})$. Inferences are based on $f(\boldsymbol{\mu}|\mathbf{y})$; in particular, a parameter of interest, say $\phi = h(\boldsymbol{\mu})$, is estimated by its posterior mean $\hat{\phi}^{\text{HB}} = E[h(\boldsymbol{\mu})|\mathbf{y}]$, and the posterior variance $V[h(\boldsymbol{\mu})|\mathbf{y}]$ is used as a measure of precision of the estimator, provided they are finite.

The HB approach is straightforward, and HB inferences are clear-cut and "exact," but require the specification of a subjective prior $f(\boldsymbol{\lambda})$ on the model parameters $\boldsymbol{\lambda}$. Priors on $\boldsymbol{\lambda}$ may be informative or "diffuse." Informative priors are based on substantial prior information, such as previous studies judged relevant to the current data set $\mathbf{y}$. However, informative priors are seldom available in real HB applications, particularly those related to public policy. Diffuse (or noninformative or default) priors are designed to reflect lack of information about $\boldsymbol{\lambda}$. The choice of a diffuse prior is not unique, and some diffuse improper priors could lead to improper posteriors. It is therefore essential to make sure that the chosen diffuse prior, $f(\boldsymbol{\lambda})$, leads to a proper posterior $f(\boldsymbol{\mu}|\mathbf{y})$. Also, it is desirable to select a diffuse prior that leads to well-calibrated inferences in the sense of validity under the frequentist framework. In practice, the frequentist bias, $E(\hat{\phi}^{\text{HB}} - \phi)$, of the HB estimator $\hat{\phi}^{\text{HB}}$ and the relative

Small Area Estimation, Second Edition. J. N. K. Rao and Isabel Molina.

frequentist bias of the posterior variance as an estimator of mean squared error (MSE) ($\hat{\phi}^{\text{HB}}$) should be small. In addition, the frequentist coverage of an HB interval on ϕ should be close to the nominal level (Dawid 1985, Browne and Draper 2006).

Applying Bayes theorem, we have

$$f(\boldsymbol{\mu}, \boldsymbol{\lambda}|\mathbf{y}) = \frac{f(\mathbf{y}, \boldsymbol{\mu}|\boldsymbol{\lambda})f(\boldsymbol{\lambda})}{f_1(\mathbf{y})}, \tag{10.1.1}$$

where $f_1(\mathbf{y})$ is the marginal density of $\mathbf{y}$, given by

$$f_1(\mathbf{y}) = \int f(\mathbf{y}, \boldsymbol{\mu}|\boldsymbol{\lambda})f(\boldsymbol{\lambda})d\boldsymbol{\mu}d\boldsymbol{\lambda}. \tag{10.1.2}$$

The desired posterior density $f(\boldsymbol{\mu}|\mathbf{y})$ is obtained from (10.1.1) as

$$f(\boldsymbol{\mu}|\mathbf{y}) = \int f(\boldsymbol{\mu}, \boldsymbol{\lambda}|\mathbf{y})d\boldsymbol{\lambda} \tag{10.1.3}$$

$$= \int f(\boldsymbol{\mu}|\mathbf{y}, \boldsymbol{\lambda})f(\boldsymbol{\lambda}|\mathbf{y})d\boldsymbol{\lambda}. \tag{10.1.4}$$

The form (10.1.4) shows that $f(\boldsymbol{\mu}|\mathbf{y})$ is a mixture of conditional densities $f(\boldsymbol{\mu}|\mathbf{y}, \boldsymbol{\lambda})$ and $f(\boldsymbol{\lambda}|\mathbf{y})$; note that $f(\boldsymbol{\mu}|\mathbf{y}, \boldsymbol{\lambda})$ is used for empirical Bayes (EB) inferences. Because of the mixture form (10.1.4), HB is also called Bayes EB or fully Bayes.

It is clear from (10.17.1) and (10.1.3) that the evaluation of $f(\boldsymbol{\mu}|\mathbf{y})$ and associated posterior quantities such as $E[h(\boldsymbol{\mu})|\mathbf{y}]$ involves multi-dimensional integrations. However, it is often possible to perform integration analytically with respect to some of the components of $\boldsymbol{\mu}$ and $\boldsymbol{\lambda}$. If the reduced problem involves one- or two-dimensional integration, then direct numerical integration can be used to calculate the desired posterior quantities. For complex problems, however, it becomes necessary to evaluate high-dimensional integrals. Recently developed Markov chain Monte Carlo (MCMC) methods seem to overcome the computational difficulties to a large extent. These methods generate samples from the posterior distribution, and then use the simulated samples to approximate the desired posterior quantities. Software packages Bayesian inference Using Gibbs Sampling BUGS and Convergence Diagnosis and Output Analysis (CODA) implement MCMC and convergence diagnostics (see Section 10.2.4). We refer the reader to the review paper by Brooks (1998) and the book edited by Gilks, Richardson, and Spiegelhalter (1996) for details on MCMC methods used in Bayesian computations. Section 10.2 gives a brief account of MCMC methods; in particular, the Gibbs sampling algorithm and its extension, the Metropolis–Hastings (M–H) algorithm. Chib and Greenberg (1995) have provided a detailed, introductory exposition of the M–H algorithm.

10.2 MCMC METHODS

10.2.1 Markov Chain

Let $\boldsymbol{\eta} = (\boldsymbol{\mu}^T, \boldsymbol{\lambda}^T)^T$ be the vector of small area parameters $\boldsymbol{\mu}$ and model parameters $\boldsymbol{\lambda}$. In general, it is not feasible to draw independent samples from the joint posterior $f(\boldsymbol{\eta}|\mathbf{y})$ because of the intractable denominator $f_1(\mathbf{y})$. MCMC avoids this difficulty by constructing a Markov chain $\{\boldsymbol{\eta}^{(k)}, k = 0, 1, 2, \ldots\}$ such that the distribution of $\boldsymbol{\eta}^{(k)}$ converges to a unique stationary (or invariant) distribution equal to $f(\boldsymbol{\eta}|\mathbf{y})$, denoted $\pi(\boldsymbol{\eta})$. Thus, after a sufficiently large "burn-in", say, d, we can regard $\boldsymbol{\eta}^{(d+1)}, \ldots, \boldsymbol{\eta}^{(d+D)}$ as D dependent samples from the target distribution $f(\boldsymbol{\eta}|\mathbf{y})$, regardless of the starting point $\boldsymbol{\eta}^{(0)}$.

To construct a Markov chain, we need to specify a one-step transition probability (or kernel) $P(\boldsymbol{\eta}^{(k+1)}|\boldsymbol{\eta}^{(k)})$, which depends only on the current "state" $\boldsymbol{\eta}^{(k)}$ of the chain. That is, the conditional distribution of $\boldsymbol{\eta}^{(k+1)}$ given $\boldsymbol{\eta}^{(0)}, \ldots, \boldsymbol{\eta}^{(k)}$ does not depend on the "history" of the chain $\{\boldsymbol{\eta}^{(0)}, \ldots, \boldsymbol{\eta}^{(k-1)}\}$. The transition kernel must satisfy the stationarity condition:

$$\int \pi(\boldsymbol{\eta}^{(k)}) P(\boldsymbol{\eta}^{(k+1)}|\boldsymbol{\eta}^{(k)}) d\boldsymbol{\eta}^{(k)} = \pi(\boldsymbol{\eta}^{(k+1)}). \tag{10.2.1}$$

Equation (10.2.1) says that if $\boldsymbol{\eta}^{(k)}$ is from $\pi(\cdot)$, then $\boldsymbol{\eta}^{(k+1)}$ will also be from $\pi(\cdot)$. Stationarity is satisfied if the chain is "reversible":

$$\pi(\boldsymbol{\eta}^{(k)}) P(\boldsymbol{\eta}^{(k+1)}|\boldsymbol{\eta}^{(k)}) = \pi(\boldsymbol{\eta}^{(k+1)}) P(\boldsymbol{\eta}^{(k)}|\boldsymbol{\eta}^{(k+1)}). \tag{10.2.2}$$

Verification of (10.2.2) ensures that the stationary distribution of the chain generated by $P(\cdot|\cdot)$ is $\pi(\cdot)$.

It is also necessary to ensure that the distribution of $\boldsymbol{\eta}^{(k)}$ given $\boldsymbol{\eta}^{(0)}$, denoted $P^{(k)}(\boldsymbol{\eta}^{(k)}|\boldsymbol{\eta}^{(0)})$, converges to $\pi(\boldsymbol{\eta}^{(k)})$ regardless of $\boldsymbol{\eta}^{(0)}$. For this, the chain needs to be "irreducible" and "aperiodic." Irreducibility means that from all starting points $\boldsymbol{\eta}^{(0)}$, the chain will eventually reach any nonempty set in the state space with positive probability. Aperiodicity means that the chain is not permitted to oscillate between different sets in a periodic manner. For an irreducible and aperiodic chain, the ergodic theorem also holds

$$\overline{h}_D = \frac{1}{D} \sum_{k=d+1}^{d+D} h(\boldsymbol{\eta}^{(k)}) \rightarrow_p E[h(\boldsymbol{\eta})|\mathbf{y}] \tag{10.2.3}$$

as $D \rightarrow \infty$, where $\rightarrow_p$ denotes convergence in probability. Thus, for sufficiently large D, we may be able to obtain an estimator, $\overline{h}_D$, of $E[h(\boldsymbol{\eta})|\mathbf{y}]$ with adequate precision.

However, it is not easy to find a Monte Carlo standard error of $\overline{h}_D$ because of dependence in the simulated samples $\boldsymbol{\eta}^{(d+1)}, \ldots, \boldsymbol{\eta}^{(d+D)}$.

10.2.2 Gibbs Sampler

To generate the samples $\boldsymbol{\eta}^{(k)}$, we partition $\boldsymbol{\eta}$ into suitable blocks $\boldsymbol{\eta}_1, \ldots, \boldsymbol{\eta}_r$. Some of the blocks might contain only single elements, whereas others contain more than one element. For example, consider the basic area level model with $\boldsymbol{\mu} = (\theta_1, \ldots, \theta_m)^T = \boldsymbol{\theta}$ and $\boldsymbol{\lambda} = (\boldsymbol{\beta}^T, \sigma_v^2)^T$. In this case, $\boldsymbol{\eta}$ may be partitioned as $\eta_1 = \boldsymbol{\beta}, \eta_2 = \theta_1, \ldots, \eta_{m+1} = \theta_m, \eta_{m+2} = \sigma_v^2$ $(r = m + 2)$. We need the following set of Gibbs conditional distributions: $f(\boldsymbol{\eta}_1|\boldsymbol{\eta}_2, \ldots, \boldsymbol{\eta}_r, \mathbf{y})$, $f(\boldsymbol{\eta}_2|\boldsymbol{\eta}_1, \boldsymbol{\eta}_3, \ldots, \boldsymbol{\eta}_r, \mathbf{y})$, $\ldots$, $f(\boldsymbol{\eta}_r|\boldsymbol{\eta}_1, \ldots, \boldsymbol{\eta}_{r-1}, \mathbf{y})$. The Gibbs sampler uses these conditional distributions to construct a transition kernel, $P(\cdot|\cdot)$, such that the stationary distribution of the resulting Markov chain is $\pi(\boldsymbol{\eta}) = f(\boldsymbol{\eta}|\mathbf{y})$. This result follows from the fact that $f(\boldsymbol{\eta}|\mathbf{y})$ is uniquely determined by the set of Gibbs conditionals.

If a conditional distribution has a standard, closed form, such as normal or inverted gamma, then samples can be generated directly from the conditional distribution. Otherwise, alternative algorithms, such as M–H rejection sampling, may be used to generate samples from the conditional distribution. Some authors suggest using M–H also for the closed-form cases. If M–H is used only for the cases without a closed form, then the algorithm is called M–H within Gibbs.

The Gibbs sampling algorithm involves the following steps.

Step 0. Choose a starting point $\boldsymbol{\eta}^{(0)}$ with components $\boldsymbol{\eta}_1^{(0)}, \ldots, \boldsymbol{\eta}_r^{(0)}$; set $k = 0$. For example, one could use restricted maximum-likelihood (REML) or moment estimates of model parameters $\boldsymbol{\lambda}$ and EB estimates of $\boldsymbol{\mu}$ as starting values. However, the starting points can be arbitrary.

Step 1. Generate $\boldsymbol{\eta}^{(k+1)} = (\boldsymbol{\eta}_1^{(k+1)}, \ldots, \boldsymbol{\eta}_r^{(k+1)})$ as follows: Draw $\boldsymbol{\eta}_1^{(k+1)}$ from $f(\boldsymbol{\eta}_1|\boldsymbol{\eta}_2^{(k)}, \ldots, \boldsymbol{\eta}_r^{(k)}, \mathbf{y})$; $\boldsymbol{\eta}_2^{(k+1)}$ from $f(\boldsymbol{\eta}_2|\boldsymbol{\eta}_1^{(k+1)}, \boldsymbol{\eta}_3^{(k)}, \ldots, \boldsymbol{\eta}_r^{(k)}, \mathbf{y})$; $\ldots$; $\boldsymbol{\eta}_r^{(k+1)}$ from $f(\boldsymbol{\eta}_r|\boldsymbol{\eta}_1^{(k+1)}, \ldots, \boldsymbol{\eta}_{r-1}^{(k+1)}, \mathbf{y})$.

Step 2. Set $k = k + 1$ and go to Step 1.

Steps 1 and 2 constitute one cycle for each k. The sequence $\{\boldsymbol{\eta}^{(k)}\}$ generated by the Gibbs sampler is a Markov chain with stationary distribution $\pi(\boldsymbol{\eta}) = f(\boldsymbol{\eta}|\mathbf{y})$; see Gelfand and Smith (1990). Note that the one-step transition kernel is the product of the r Gibbs conditional distributions.

10.2.3 M–H Within Gibbs

If all the Gibbs conditionals have closed forms belonging to standard families, then it is straightforward to generate samples from the conditionals. If a conditional does not admit a closed form, then several methods are available for generating samples from the conditional. In particular, we have adaptive rejection sampling for univariate log-concave conditionals (Gilks and Wild 1992) and more generally M–H (Metropolis et al. 1953; Hastings 1970) within Gibbs.

Let $f(\boldsymbol{\eta}_i|\boldsymbol{\eta}_{-i}^{(k)}, \mathbf{y})$ denote the Gibbs conditional after completing the first $i-1$ drawings of the $(k+1)$th cycle of Gibbs sampling, where

$$\boldsymbol{\eta}_{-i}^{(k)} = \{\boldsymbol{\eta}_1^{(k+1)}, \ldots, \boldsymbol{\eta}_{i-1}^{(k+1)}, \boldsymbol{\eta}_{i+1}^{(k)}, \ldots, \boldsymbol{\eta}_r^{(k)}\}.$$

The M–H algorithm for generating a sample, $\boldsymbol{\eta}_i^{(k+1)}$, from $f(\boldsymbol{\eta}_i|\boldsymbol{\eta}_{-i}^{(k)}, \mathbf{y})$ involves the following steps.

Step 1. Approximate $f(\boldsymbol{\eta}_i|\boldsymbol{\eta}_{-i}^{(k)}, \mathbf{y})$ by a candidate (or proposal) density $q_i(\boldsymbol{\eta}_i|\boldsymbol{\eta}_i^{(k)}, \boldsymbol{\eta}_{-i}^{(k)})$ that is easy to sample from, such as a normal or a t distribution. The proposal density may depend on the current values $\{\boldsymbol{\eta}_i^{(k)}, \boldsymbol{\eta}_{-i}^{(k)}\}$.

Step 2. Generate a "candidate" for $\boldsymbol{\eta}_i$, say $\boldsymbol{\eta}_i^*$, from the candidate density, and u from Uniform(0,1).

Step 3. Set $\boldsymbol{\eta}_i^{(k+1)} = \boldsymbol{\eta}_i^*$ if $u \le a(\boldsymbol{\eta}_{-i}^{(k)}, \boldsymbol{\eta}_i^{(k)}, \boldsymbol{\eta}_i^*)$ and $\boldsymbol{\eta}_i^{(k+1)} = \boldsymbol{\eta}_i^{(k)}$ otherwise, where the acceptance probability $a(\boldsymbol{\eta}_{-i}^{(k)}, \boldsymbol{\eta}_i^{(k)}, \boldsymbol{\eta}_i^*)$ is given by

$$a(\boldsymbol{\eta}_{-i}^{(k)}, \boldsymbol{\eta}_i^{(k)}, \boldsymbol{\eta}_i^*) = \min\left\{\frac{f(\boldsymbol{\eta}_i^*|\boldsymbol{\eta}_{-i}^{(k)}, \mathbf{y})q_i(\boldsymbol{\eta}_i^{(k)}|\boldsymbol{\eta}_i^*, \boldsymbol{\eta}_{-i}^{(k)})}{f(\boldsymbol{\eta}_i^{(k)}|\boldsymbol{\eta}_{-i}^{(k)}, \mathbf{y})q_i(\boldsymbol{\eta}_i^*|\boldsymbol{\eta}_i^{(k)}, \boldsymbol{\eta}_{-i}^{(k)})}, 1\right\}. \tag{10.2.4}$$

Note that the acceptance probability (10.2.4) depends on the ratio $f(\boldsymbol{\eta}_i^*|\cdot)/f(\boldsymbol{\eta}_i^{(k)}|\cdot)$, so we need to know $f(\boldsymbol{\eta}_i|\cdot)$ only up to a constant of proportionality, that is, the normalizing constant cancels out in (10.2.4). Also, if the candidate density is symmetric, that is, $q_i(\boldsymbol{\eta}_i^{(k)}|\boldsymbol{\eta}_i^*, \cdot) = q_i(\boldsymbol{\eta}_i^*|\boldsymbol{\eta}_i^{(k)}, \cdot)$, then the acceptance probability reduces to

$$a(\boldsymbol{\eta}_{-i}^{(k)}, \boldsymbol{\eta}_i^{(k)}, \boldsymbol{\eta}_i^*) = \min\left\{\frac{f(\boldsymbol{\eta}_i^*|\boldsymbol{\eta}_{-i}^{(k)}, \mathbf{y})}{f(\boldsymbol{\eta}_i^{(k)}|\boldsymbol{\eta}_{-i}^{(k)}, \mathbf{y})}, 1\right\} \tag{10.2.5}$$

(Metropolis et al. 1953). The Gibbs sampler is a special case of M–H with $q_i(\boldsymbol{\eta}_i|\boldsymbol{\eta}_i^{(k)}, \boldsymbol{\eta}_{-i}^{(k)}) = f(\boldsymbol{\eta}_i|\boldsymbol{\eta}_{-i}^{(k)}, \mathbf{y})$, and the corresponding acceptance probability equals 1 so that the Gibbs candidate $\boldsymbol{\eta}_i^*$ is automatically accepted.

To improve convergence of the MCMC sampling, several useful variations of M–H have been proposed. We refer the reader to Chen, Shao, and Ibrahim (2000, Chapter 2) for details of these algorithms.

10.2.4 Posterior Quantities

The MCMC output $\{\boldsymbol{\eta}^{(k)}, k = d+1, \ldots, d+D\}$ from a single long run may be employed to compute posterior quantities of interest, noting that $\{\boldsymbol{\mu}^{(k)}\}$ is a sample from the marginal posterior $f(\boldsymbol{\mu}|\mathbf{y})$. In particular, the posterior mean (or the HB estimator) of $\phi = h(\boldsymbol{\mu})$ is estimated by the "ergodic average"

$$\hat{\phi}^{\text{HB}} = \frac{1}{D} \sum_{k=d+1}^{d+D} \phi^{(k)} = \phi^{(\cdot)}, \tag{10.2.6}$$

where $\phi^{(k)} = h(\boldsymbol{\mu}^{(k)})$. Similarly, the posterior variance of ϕ is estimated by

$$\hat{V}(\phi|\mathbf{y}) = \frac{1}{D-1} \sum_{k=d+1}^{d+D} (\phi^{(k)} - \phi^{(\cdot)})^2. \tag{10.2.7}$$

By the ergodic theorem for Markov chains, $\hat{\phi}^{\text{HB}}$ converges to $E(\phi|\mathbf{y})$ and $\hat{V}(\phi|\mathbf{y})$ to $V(\phi|\mathbf{y})$ as $D \to \infty$. However, Monte Carlo standard errors of $\hat{\phi}^{\text{HB}}$ and $\hat{V}(\phi|\mathbf{y})$ are not easy to find if the simulated values $\phi^{(d+1)}, \ldots, \phi^{(d+D)}$ are dependent.

If the $\boldsymbol{\eta}^{(k)}$ are iid vectors, as in the multiple runs method of Gelfand and Smith (1990) described in Section 10.2.5, then estimates of ϕ with reduced standard errors may be obtained if the mathematical form for the conditional expectation of ϕ given the data $\mathbf{y}$ and the model parameters $\boldsymbol{\lambda}$ is known. Suppose that the form of $E(\phi|\mathbf{y}, \boldsymbol{\lambda})$ is known. Then an improved estimate of $E(\phi|\mathbf{y})$ is given by

$$\tilde{\phi}^{\text{HB}} = \frac{1}{D} \sum_{k=d+1}^{d+D} E(\phi|\mathbf{y}, \boldsymbol{\lambda}^{(k)}), \tag{10.2.8}$$

where $\{\boldsymbol{\lambda}^{(k)}\}$ is the iid sample from the marginal posterior $f(\boldsymbol{\lambda}|\mathbf{y})$. This result readily follows by noting that

$$\begin{aligned} V(\phi|\mathbf{y}) &= E[V(\phi|\mathbf{y}, \boldsymbol{\lambda})|\mathbf{y}] + V[E(\phi|\mathbf{y}, \boldsymbol{\lambda})|\mathbf{y}] \\ &\geq V[E(\phi|\mathbf{y}, \boldsymbol{\lambda})|\mathbf{y}], \end{aligned}$$

and that the Monte Carlo variances of $\hat{\phi}^{\text{HB}}$ and $\tilde{\phi}^{\text{HB}}$ are given by $D^{-1}V(\phi|\mathbf{y})$ and $D^{-1}V[E(\phi|\mathbf{y}, \boldsymbol{\lambda})|\mathbf{y}]$, respectively. The improved estimator $\tilde{\phi}^{\text{HB}}$ is called a Rao–Blackwell estimator (Gelfand and Smith 1991) because of the similarity of the result to the well-known Rao–Blackwell theorem (see, e.g., Casella and Berger 1990, p. 316). An estimate of posterior variance associated with $\tilde{\phi}^{\text{HB}}$ is given by

$$\tilde{V}(\phi|\mathbf{y}) = \frac{1}{D} \sum_{k=d+1}^{d+D} V(\phi|\mathbf{y}, \boldsymbol{\lambda}^{(k)}) + \frac{1}{D-1} \sum_{k=d+1}^{d+D} \left[E(\phi|\mathbf{y}, \boldsymbol{\lambda}^{(k)}) - \tilde{\phi}^{\text{HB}}\right]^2. \tag{10.2.9}$$

For the case of a single run with dependent samples, (10.2.9) may not hold. However, if the D samples are "thinned" by selecting say every 5th or 10th element $\boldsymbol{\eta}^{(k)}$, then the reduced MCMC output may be approximately iid and the result (10.2.9) may hold for the thinned sample. Thinning of the chain is sometimes used to save storage space and computational time, especially when the consecutive samples $\boldsymbol{\eta}^{(k)}$ are highly correlated, necessitating a very long run, that is, very large D.

We can also estimate $(1-2\alpha)$-level posterior (or credible) intervals for ϕ from the MCMC output $\{\boldsymbol{\eta}^{(k)}\}$. For example, by setting ϕ_α equal to the αth quantile of $\{\phi^{(k)}\}$ and $\phi_{1-\alpha}$ to the $(1-\alpha)$th quantile, we obtain a "central" or "equal-tailed" posterior interval $[\phi_\alpha, \phi_{1-\alpha}]$ for ϕ such that $P(\phi_\alpha \leq \phi \leq \phi_{1-\alpha}|\mathbf{y}) = 1-2\alpha$. An alternative "interval," called the highest posterior density (HPD) "interval," is given by $A = \{\phi : f(\phi|\mathbf{y}) \geq c\}$, where c is chosen such that $P(\phi \in A|\mathbf{y}) = 1-2\alpha$; the set A may not be an interval. The HPD interval is the shortest length interval for a given credible probability, $1-2\alpha$, provided $f(\phi|\mathbf{y})$ is unimodal (see Casella and Berger 1990, p. 430). Generally, the central posterior interval is preferred to the HPD interval because it is invariant to one-to-one transformations of ϕ and is easier to compute. Also, the end points ϕ_α and $\phi_{1-\alpha}$ can be interpreted as the posterior α and $1-\alpha$ quantiles.

A computer program, called BUGS, is widely used to implement MCMC and to calculate posterior quantities from the MCMC output (Spiegelhalter et al. 1997). BUGS can handle many of the small area models studied in this chapter. It uses inverse gamma priors on variance parameters and normal priors on regression parameters. BUGS runs are monitored using a menu-driven set of S-Plus and R functions, called CODA. Convergence diagnostics and statistical and graphical output analyses of the simulated values from BUGS may be performed using CODA. The BUGS software package is freely available and can be downloaded from http://www.mrc-bsu.cam.ac.uk/software/bugs/the-bugs-project-winbugs/. Currently, there are WinBUGS version 1.4.3. for Windows and OpenBUGS version 3.2.3. for Windows and Linux. These versions include M–H within Gibbs steps to handle non-log-concave conditionals not admitting closed forms; the older version 0.5 handles only log-concave conditionals using adaptive rejection sampling (Gilks and Wild 1992). CODA is supplied with BUGS. We refer the reader to the BUGS manual for examples of typical applications. Spiegelhalter et al. (1996) presented a case study of MCMC methods using BUGS.

10.2.5 Practical Issues

The HB approach based on MCMC has several limitations, so it is important to exercise caution when implementing MCMC. We now give a brief account of some practical issues associated with MCMC.

Choice of Prior Diffuse priors $f(\boldsymbol{\lambda})$, reflecting lack of information about the model parameters $\boldsymbol{\lambda}$, are commonly used in the HB approach to small area estimation. However, if the diffuse prior is improper, that is, $\int f(\boldsymbol{\lambda})d\boldsymbol{\lambda} = \infty$, then the Gibbs sampler could lead to seemingly reasonable inferences about a nonexistent posterior $f(\boldsymbol{\mu}, \boldsymbol{\lambda}|\mathbf{y})$. This happens when the posterior is improper and yet all the Gibbs conditionals are proper (Natarajan and McCulloch 1995; Hobert and Casella 1996). To demonstrate this, consider the simple nested error model without covariates: $y_{ij} = \mu + v_i + e_{ij}$ with

$v_i \overset{\text{iid}}{\sim} N(0, \sigma_v^2)$ and independent of $e_{ij} \overset{\text{iid}}{\sim} N(0, \sigma_e^2)$. If we choose an improper prior of the form $f(\mu, \sigma_v^2, \sigma_e^2) = f(\mu)f(\sigma_v^2)f(\sigma_e^2)$ with $f(\mu) \propto 1$, $f(\sigma_v^2) \propto 1/\sigma_v^2$ and $f(\sigma_e^2) \propto 1/\sigma_e^2$, then the joint posterior of $\mu, \mathbf{v} = (v_1, \ldots, v_m)^T$, σ_v^2, and σ_e^2 is improper (Hill 1965). On the other hand, all the Gibbs conditionals are proper for this choice of prior. In particular, σ_v^{-2}|all others ~ gamma (G), or equivalently σ_v^2|all others ~ inverted gamma (IG), σ_e^{-2}|all others ~ G, μ|all others ~ Normal (N), and v_i|all others ~ N. To get around this difficulty, the BUGS software (Spiegelhalter et al. 1996) uses diffuse (or vague) proper priors of the form $\mu \sim N(0, \sigma_0^2)$, $\sigma_v^{-2} \sim G(a_0, a_0)$, and $\sigma_e^{-2} \sim G(a_0, a_0)$ as default priors, where σ_0^2 is chosen very large (say 10,000) and $a_0 (> 0)$ very small (say 0.001) to reflect lack of prior information on μ, σ_v^2, and σ_e^2. Here, $G(a, b)$ denotes a gamma distribution with shape parameter a and scale parameter b and that the variance of $G(a_0, a_0)$ is $1/a_0$, which becomes very large as $a_0 \to 0$. The posterior resulting from the above priors remains proper as $\sigma_0^2 \to \infty$, which is equivalent to choosing $f(\mu) \propto 1$, but it becomes improper as $a_0 \to 0$. Therefore, the posterior is nearly improper (or barely proper) for very small a_0, and this feature can affect the convergence of the Gibbs sampler. Alternative choices are $\sigma_v^2 \sim \text{uniform}(0, 1/a_0)$ and $\sigma_e^2 \sim \text{uniform}(0, 1/a_0)$, which avoid this difficulty in the sense that the posterior remains proper as $a_0 \to 0$.

Gelman (2006) studied the choice of prior on σ_v^2 for the simple nested error model without covariates. He noted that the posterior inferences based on the prior $\sigma_v^{-2} \sim G(a_0, a_0)$ with very small $a_0 > 0$ becomes very sensitive to the choice of a_0 for data sets in which small values of σ_v^2 are possible. He recommended the flat prior on σ_v, $f(\sigma_v) \propto 1$ if $m \geq 5$. He also suggested the use of a proper uniform$(0, A)$ prior on σ_v with sufficiently large A, say $A = 100$, if $m \geq 5$. Resulting posterior inferences are not sensitive to the choice of A and they can be implemented in WinBUGS, which requires a proper prior.

We have noted in Section 10.1 that it is desirable to choose diffuse priors that lead to well-calibrated inferences. Browne and Draper (2006) compared frequentist performances of posterior quantities under the simple nested error model and $G(a_0, a_0)$ or uniform$(0, 1/a_0)$ priors on the variance parameters. In particular, for σ_v^2 they examined the bias of the posterior mean, median, and mode and the Bayesian interval coverage in repeated sampling. All the Gibbs conditionals have closed forms here, so Gibbs sampling was used to generate samples from the posterior. They found that the posterior quantities are generally insensitive to the specific choice of a_0 around 0.001 (default setting used in BUGS with gamma prior). In terms of bias, the posterior median performed better than the posterior mean (HB estimator) for the gamma prior, while the posterior mode performed much better than the posterior mean for the uniform prior. Bayesian intervals for uniform or gamma priors did not attain nominal coverage when the number of small areas, m, and/or the variance ratio $\tau = \sigma_v^2/\sigma_e^2$ are small. Browne and Draper (2006) did not study the frequentist behavior of the posterior quantities associated with the small area means $\mu_i = \mu + v_i$. Datta, Ghosh, and Kim (2001) developed noninformative priors for the simple nested error model, called probability matching priors, and focused on the variance ratio τ. These priors ensure that the frequentist coverage of Bayesian intervals for τ approaches the nominal level asymptotically, as $m \to \infty$. In Section 10.3, we consider priors for the basic area level model that make the posterior variance of a small area mean, μ_i, approximately unbiased for the MSE of EB/HB estimators, as $m \to \infty$.

Single Run versus Multiple Runs Sections 10.2.2 and 10.2.3 discussed the generation of one single long run $\{\boldsymbol{\eta}^{(d+1)}, \ldots, \boldsymbol{\eta}^{(d+D)}\}$. One single long run may provide reliable Monte Carlo estimates of posterior quantities by choosing D sufficiently large, but it may leave a significant portion of the space generated by the posterior, $f(\boldsymbol{\eta}|\mathbf{y})$, unexplored. To avoid this problem, Gelman and Rubin (1992) and others recommended the use of multiple parallel runs with different starting points, leading to parallel samples. Good starting values are required to generate multiple runs, and it may not be easy to find such values. On the other hand, for generating one single run, REML estimates of model parameters, $\boldsymbol{\lambda}$, and EB estimates of small area parameters, $\boldsymbol{\mu}$, may work quite well as starting points. For multiple runs, Gelman and Rubin (1992) recommended generating starting points from an "overdispersed" distribution compared to the target distribution $\pi(\boldsymbol{\eta}|\mathbf{y})$. They proposed some methods for generating overdispersed starting values, but general methods that lead to good multiple starting values are not available. Note that the normalizing factor of $\pi(\boldsymbol{\eta}|\mathbf{y})$ is not known in a closed form.

Multiple runs can be wasteful because initial burn-in periods are discarded from each run, although this may not be a serious limitation if parallel processors are used to generate the parallel samples. Gelfand and Smith (1990) used many short runs, each consisting of $(d + 1)$ $\boldsymbol{\eta}$-values, and kept only the last observation from each run. This method provides independent samples $\boldsymbol{\eta}^{(d+1)}(1), \ldots, \boldsymbol{\eta}^{(d+1)}(L)$, where $\boldsymbol{\eta}^{(d+1)}(\ell)$ is the last $\boldsymbol{\eta}$-value of the ℓth run ($\ell = 1, \ldots, L$), but it discards most of the generated values. Gelman and Rubin (1992) used small L (say 10) and generated $2d$ values for each run and kept the last d values from each run, leading to L independent sets of $\boldsymbol{\eta}$-values. This approach facilitates convergence diagnostics (see later). Note that it is not necessary to generate independent samples for getting Monte Carlo estimates of posterior quantities because of the ergodic theorem for Markov chains. However, independent samples facilitate the calculation of Monte Carlo standard errors.

Proponents of one single long run argue that comparing chains can never "prove" convergence. On the other hand, multiple runs' proponents assert that comparing seemingly converged runs might disclose real differences if the runs have not yet attained stationarity.

Burn-in Length The length of burn-in, d, depends on the starting point $\boldsymbol{\eta}^{(0)}$ and the convergence rate of $P^{(k)}(\cdot)$ to the stationary distribution $\pi(\cdot)$. Convergence rates have been studied in the literature (Roberts and Rosenthal 1998), but it is not easy to use these rates to determine d. Note that the convergence rates of different MCMC algorithms may vary significantly and also depend on the target distribution $\pi(\cdot)$.

Convergence diagnostics, based on the MCMC output $\{\boldsymbol{\eta}^{(k)}\}$, are often used in practice to determine the burn-in length d. At least 13 convergence diagnostics have been proposed in the literature (see Cowles and Carlin 1996). A diagnostic measure based on multiple runs and classical analysis of variance (ANOVA) is currently popular (Gelman and Rubin 1992). Suppose $h(\boldsymbol{\eta})$ denotes a scalar summary of $\boldsymbol{\eta}$; for example, $h(\boldsymbol{\eta}) = \mu_i$, the mean of the ith small area. Denote the values of $h(\boldsymbol{\eta})$ for the ℓth run as $h_{\ell,d+1}, \ldots, h_{\ell,2d}$, $\ell = 1, \ldots, L$. Calculate the between-run variance $B = d\sum_{\ell=1}^{L}(\overline{h}_{\ell\cdot} - \overline{h}_{\cdot\cdot})^2/(L-1)$ and the within-run variance $W = \sum_{\ell=1}^{L}\sum_{k=d+1}^{2d}(h_{\ell,k} - \overline{h}_{\ell\cdot})^2/[L(d-1)]$, where $\overline{h}_{\ell\cdot} = \sum_{k=d+1}^{2d} h_{\ell k}/d$ and $\overline{h}_{\cdot\cdot} = \sum_{\ell=1}^{L}\overline{h}_{\ell\cdot}/L$. Using the two variance components B and W, we calculate an estimate of the variance of $h(\boldsymbol{\eta})$ in the target distribution as

$$\hat{V} = \frac{d-1}{d}W + \frac{1}{d}B. \tag{10.2.10}$$

Under stationarity, $\hat{V}$ is unbiased but it is an overestimate if the L points $h_{1,d+1}, \ldots, h_{L,d+1}$ are overdispersed. Using the latter property, we calculate the estimated "potential scale reduction" factor

$$\hat{R} = \hat{V}/W. \tag{10.2.11}$$

If stationarity is attained, $\hat{R}$ will be close to 1. Otherwise, $\hat{R}$ will be significantly larger than 1, suggesting a larger burn-in length d. It is necessary to calculate $\hat{R}$ for all scalar summaries of interest, for example, all small area means μ_i.

Unfortunately, most of the proposed diagnostics have shortcomings. Cowles and Carlin (1996) compared the performance of 13 convergence diagnostics, including $\hat{R}$, in two simple models and concluded that "all of the methods can fail to detect the sorts of convergence failure that they were designed to identify".

Methods of generating independent samples from the exact stationary distribution of a Markov chain, that is, $\pi(\boldsymbol{\eta}) = f(\boldsymbol{\eta}|\mathbf{y})$, have been proposed recently. These methods eliminate the difficulties noted above, but currently the algorithms are not easy to implement. We refer the reader to Casella, Lavine, and Robert (2001) for a brief introduction to the proposed methods, called perfect (or exact) samplings.

10.2.6 Model Determination

MCMC methods are also used for model determination. In particular, methods based on Bayes factors (BFs), posterior predictive densities and cross-validation predictive densities, are employed for model determination. These approaches require the specification of priors on the model parameters associated with the competing models. Typically, noninformative priors are employed with the methods based on posterior predictive and cross-validation predictive densities. Because of the dependence on priors, some authors have suggested a hybrid strategy, in which prior-free, frequentist methods are used in the model exploration phase, such as those given in Chapters 5–9 and Bayesian diffuse-prior MCMC methods for inference from the selected model (see, e.g., Browne and Draper 2006). We refer the reader to Gelfand (1996) for an excellent account of model determination procedures using MCMC methods.

Bayes Factors Suppose that M_1 and M_2 denote two competing models with associated parameters (random effects and model parameters) $\boldsymbol{\eta}_1$ and $\boldsymbol{\eta}_2$, respectively. We denote the marginal densities of the "observables" $\mathbf{y}$ under M_1 and M_2 by $f(\mathbf{y}|M_1)$ and $f(\mathbf{y}|M_2)$, respectively. Note that

$$f(\mathbf{y}|M_i) = \int f(\mathbf{y}|\boldsymbol{\eta}_i, M_i) f(\boldsymbol{\eta}_i|M_i) d\boldsymbol{\eta}_i, \tag{10.2.12}$$

where $f(\boldsymbol{\eta}_i|M_i)$ is the density of $\boldsymbol{\eta}_i$ under model $M_i (i = 1, 2)$. We denote the actual observations as $\mathbf{y}_{\text{obs}}$. The BF is defined as

$$\text{BF}_{12} = \frac{f(\mathbf{y}_{\text{obs}}|M_1)}{f(\mathbf{y}_{\text{obs}}|M_2)}, \tag{10.2.13}$$

where $\mathbf{y}_{\text{obs}}$ denotes the actual observations. It provides relative weight of evidence for M_1 compared to M_2. BF_{12} may be interpreted as the ratio of the posterior odds for M_1 versus M_2 to prior odds for M_1 versus M_2, noting that the posterior odds is $f(M_1|\mathbf{y})/f(M_2|\mathbf{y})$ and the prior odds is $f(M_1)/f(M_2)$, where

$$f(M_i|\mathbf{y}) = f(\mathbf{y}|M_i) f(M_i)/f(\mathbf{y}), \tag{10.2.14}$$

$f(M_i)$ is the prior on M_i and

$$f(\mathbf{y}) = f(\mathbf{y}|M_1) f(M_1) + f(\mathbf{y}|M_2) f(M_2). \tag{10.2.15}$$

Kass and Raftery (1995) classified the evidence against M_2 as follows: (i) BF_{12} between 1 and 3: not worth more than a bare mention; (ii) BF_{12} between 3 and 20: positive; (iii) BF_{12} between 20 and 50: strong; (iv) $\text{BF}_{12} > 50$: very strong. BF is appealing for model selection, but $f(\mathbf{y}|M_i)$ is necessarily improper if the prior on the model parameters λ_i is improper, even if the posterior $f(\boldsymbol{\eta}_i|\mathbf{y}, M_i)$ is proper. A proper prior on λ_i $(i = 1, 2)$ is needed to implement BF. However, the use of BF with diffuse proper priors usually gives bad answers (Berger and Pericchi 2001). Gelfand (1996) noted several limitations of BF, including the above-mentioned impropriety of $f(\mathbf{y}|M_i)$, and concluded that "use of the Bayes factor often seems inappropriate in real applications." Sinharay and Stern (2002) showed that the BF can be very sensitive to the choice of prior distributions for the parameters of M_1 and M_2. We refer the reader to Kass and Raftery (1995) for methods of calculating BFs.

Posterior Predictive Densities The posterior predictive density, $f(\mathbf{y}|\mathbf{y}_{\text{obs}})$, is defined as the predictive density of new independent observations, $\mathbf{y}$, under the model, given $\mathbf{y}_{\text{obs}}$. We have

$$f(\mathbf{y}|\mathbf{y}_{\text{obs}}) = \int f(\mathbf{y}|\boldsymbol{\eta}) f(\boldsymbol{\eta}|\mathbf{y}_{\text{obs}}) d\boldsymbol{\eta}, \tag{10.2.16}$$

and the marginal posterior predictive density of an element y_r of $\mathbf{y}$ is given by

$$f(y_r|\mathbf{y}_{\text{obs}}) = \int f(y_r|\boldsymbol{\eta})f(\boldsymbol{\eta}|\mathbf{y}_{\text{obs}})d\boldsymbol{\eta}. \tag{10.2.17}$$

Using the MCMC output $\{\boldsymbol{\eta}^{(k)}; k = d+1, \ldots, d+D\}$, we can draw a sample $\{\mathbf{y}^{(k)}\}$ from $f(\mathbf{y}|\mathbf{y}_{\text{obs}})$ as follows: For each $\boldsymbol{\eta}^{(k)}$, draw $\mathbf{y}^{(k)}$ from $f(\mathbf{y}|\boldsymbol{\eta}^{(k)})$. Furthermore, $\{y_r^{(k)}\}$ constitutes a sample from the marginal density $f(y_r|\mathbf{y}_{\text{obs}})$, where $y_r^{(k)}$ is the rth element of $\mathbf{y}^{(k)}$.

To check the overall fit of a proposed model to data $\mathbf{y}_{\text{obs}}$, Gelman and Meng (1996) proposed the criterion of posterior predictive p-value, noting that the hypothetical replications $\{\mathbf{y}^{(k)}\}$ should look similar to $\mathbf{y}_{\text{obs}}$ if the assumed model is reasonably accurate. Let $T(\mathbf{y}, \boldsymbol{\eta})$ denote a measure of discrepancy between $\mathbf{y}$ and $\boldsymbol{\eta}$. The posterior predictive p-value is then defined as

$$pp = P\{T(\mathbf{y}, \boldsymbol{\eta}) \geq T(\mathbf{y}_{\text{obs}}, \boldsymbol{\eta})|\mathbf{y}_{\text{obs}}\}. \tag{10.2.18}$$

The MCMC output $\{\boldsymbol{\eta}^{(k)}\}$ may be used to approximate pp by

$$\widehat{pp} = \frac{1}{D}\sum_{k=d+1}^{d+D} I\{T(\mathbf{y}^{(k)}, \boldsymbol{\eta}^{(k)}) \geq T(\mathbf{y}_{\text{obs}}, \boldsymbol{\eta}^{(k)})\}, \tag{10.2.19}$$

where $I(\cdot)$ is the indicator function taking the value 1 when its argument is true and 0 otherwise.

A limitation of the posterior predictive p-value is that it makes "double use" of the data $\mathbf{y}_{\text{obs}}$, first to generate $\{\mathbf{y}^{(k)}\}$ from $f(\mathbf{y}|\mathbf{y}_{\text{obs}})$ and then to compute $\widehat{pp}$ given by (10.2.19). This double use of the data can induce unnatural behavior, as demonstrated by Bayarri and Berger (2000). They proposed two alternative p-measures, named the partial posterior predictive p-value and the conditional predictive p-value, that attempt to avoid double use of the data. These measures, however, are more difficult to implement than the posterior predictive p-value, especially for small area models.

If the model fits $\mathbf{y}_{\text{obs}}$, then the two values $T(\mathbf{y}^{(k)}, \boldsymbol{\eta}^{(k)})$ and $T(\mathbf{y}_{\text{obs}}, \boldsymbol{\eta}^{(k)})$ should tend to be similar for most k, and $\widehat{pp}$ should be close to 0.5. Extreme value of $\widehat{pp}$ (closer to 0 or 1) suggest poor fit. It is informative to plot the realized values $T(\mathbf{y}_{\text{obs}}, \boldsymbol{\eta}^{(k)})$ versus the predictive values $T(\mathbf{y}^{(k)}, \boldsymbol{\eta}^{(k)})$. If the model is a good fit, then about half the points in the scatter plot would fall above the 45° line and the remaining half below the line. Brooks, Catchpole, and Morgan (2000) studied competing animal survival models, and used the above scatter plots for four competing models, called discrepancy plots, to select a model.

Another measure of fit is given by the posterior expected predictive deviance (EPD), $E\{\Delta(\mathbf{y}, \mathbf{y}_{\text{obs}})|\mathbf{y}_{\text{obs}}\}$, where $\Delta(\mathbf{y}, \mathbf{y}_{\text{obs}})$ is a measure of discrepancy between $\mathbf{y}_{\text{obs}}$ and $\mathbf{y}$. For count data $\{y_r; r = 1, \ldots, n\}$, we can use a chi-squared measure given by

$$\Delta(\mathbf{y}, \mathbf{y}_{\text{obs}}) = \sum_{r=1}^{n} (y_{r,\text{obs}} - y_r)^2/(y_r + 0.5). \tag{10.2.20}$$

It is a general measure of agreement. Nandram, Sedransk, and Pickle (1999) studied (10.2.20) and some other measures in the context of disease mapping (see Section 10.11). Using the predictions $\mathbf{y}^{(k)}$, the posterior EPD may be estimated as

$$\hat{E}\{\Delta(\mathbf{y},\mathbf{y}_{\text{obs}})|\mathbf{y}_{\text{obs}}\} = \frac{1}{D}\sum_{k=d+1}^{d+D}\Delta(\mathbf{y}^{(k)},\mathbf{y}_{\text{obs}}). \tag{10.2.21}$$

Note that the deviance measure (10.2.21) also makes double use of the data.

The deviance measure is useful for comparing the relative fits of candidate models. The model with the smallest deviance value may then be used to check its overall fit to the data $\mathbf{y}_{\text{obs}}$, using the posterior predictive p-value.

Cross-validation Predictive Densities Let $\mathbf{y}_{(r)}$ be the vector obtained by removing the rth element y_r from $\mathbf{y}$. The cross-validation predictive density of y_r is given by

$$f(y_r|\mathbf{y}_{(r)}) = \int f(y_r|\boldsymbol{\eta},\mathbf{y}_{(r)})f(\boldsymbol{\eta}|\mathbf{y}_{(r)})d\boldsymbol{\eta}. \tag{10.2.22}$$

This density suggests what values of y_r are likely when the model is fitted to $\mathbf{y}_{(r)}$. The actual observation $y_{r,\text{obs}}$ may be compared to the hypothetical values y_r in a variety of ways to see whether $y_{r,\text{obs}}$, for each r, supports the model. In most applications, $f(y_r|\boldsymbol{\eta},\mathbf{y}_{(r)}) = f(y_r|\boldsymbol{\eta})$, that is, y_r and $\mathbf{y}_{(r)}$ are conditionally independent given $\boldsymbol{\eta}$. Also, $f(\boldsymbol{\eta}|\mathbf{y}_{(r)})$ is usually proper if $f(\boldsymbol{\eta}|\mathbf{y})$ is proper. As a result, $f(y_r|\mathbf{y}_{(r)})$ will remain proper, unlike $f(\mathbf{y})$ used in the computation of the BF. If $f(\mathbf{y})$ is proper, then the set $\{f(y_r|\mathbf{y}_{(r)})\}$ is equivalent to $f(\mathbf{y})$ in the sense that $f(\mathbf{y})$ is uniquely determined from $\{f(y_r|\mathbf{y}_{(r)})\}$ and vice versa.

The product of the densities $f(y_r|\mathbf{y}_{(r)})$ is often used as a substitute for $f(\mathbf{y})$ to avoid an improper marginal $f(\mathbf{y})$. This leads to a pseudo-Bayes factor (PBF)

$$\text{PBF}_{12} = \prod_r \frac{f(y_{r,\text{obs}}|\mathbf{y}_{(r),\text{obs}},M_1)}{f(y_{r,\text{obs}}|\mathbf{y}_{(r),\text{obs}},M_2)}, \tag{10.2.23}$$

which is often used as a substitute for BF. However, PBF cannot be interpreted as the ratio of posterior odds to prior odds, unlike BF.

Global summary measures, such as the posterior predictive p-value (10.2.18) and the posterior expected predictive deviance (10.2.21), are useful for checking the overall fit of a model, but not helpful for discovering the reasons for poor global performance. The univariate density $f(y_r|\mathbf{y}_{(r),\text{obs}})$ can be used through a "checking" function $c(y_r, y_{r,\text{obs}})$ to see whether $y_{r,\text{obs}}$, for each r, supports the model. Gelfand (1996) proposed a variety of checking functions and calculated their expectations over $f(y_r|\mathbf{y}_{(r),\text{obs}})$. In particular, choosing

$$c_1(y_r, y_{r,\text{obs}}) = \frac{1}{2\epsilon}I[y_{r,\text{obs}} - \epsilon \le y_r \le y_{r,\text{obs}} + \epsilon], \tag{10.2.24}$$

then taking its expectation, and letting $\epsilon \to 0$, we obtain the conditional predictive ordinate (CPO):

$$\text{CPO}_r = f(y_{r,\text{obs}} | \mathbf{y}_{(r),\text{obs}}). \tag{10.2.25}$$

Models with larger CPO values provide better fit to the observed data. A plot of CPO_r versus r for each model is useful for comparing the models visually. We can easily see from a CPO plot which models are better than the others, which models are indistinguishable, which points, $y_{r,\text{obs}}$, are poorly fit under all the competing models, and so on. If two or more models are indistinguishable and provide good fits, then we should choose the simplest among the models.

We approximate CPO_r from the MCMC output $\{\boldsymbol{\eta}^{(k)}\}$ as follows:

$$\widehat{\text{CPO}}_r = \hat{f}(y_{r,\text{obs}} | \mathbf{y}_{(r),\text{obs}}) = \left[\frac{1}{D} \sum_{k=d+1}^{d+D} \frac{1}{f(y_{r,\text{obs}} | \mathbf{y}_{(r),\text{obs}}, \boldsymbol{\eta}^{(k)})} \right]^{-1}. \tag{10.2.26}$$

That is, $\widehat{\text{CPO}}_r$ is the harmonic mean of the D ordinates $f(y_{r,\text{obs}} | \mathbf{y}_{(r),\text{obs}}, \boldsymbol{\eta}^{(k)})$, $k = d + 1, \ldots, d + D$. For several small area models, y_r and $\mathbf{y}_{(r)}$ are conditionally independent given $\boldsymbol{\eta}$, that is, $f(y_{r,\text{obs}} | \mathbf{y}_{(r),\text{obs}}, \boldsymbol{\eta}^{(k)}) = f(y_{r,\text{obs}} | \boldsymbol{\eta}^{(k)})$ in (10.2.26). A proof of (10.2.26) is given in Section 10.18.

A checking function of a "residual" form is given by

$$c_2(y_r, y_{r,\text{obs}}) = y_{r,\text{obs}} - y_r \tag{10.2.27}$$

with expectation $d_{2r} = y_{r,\text{obs}} - E(y_r | \mathbf{y}_{(r),\text{obs}})$. We can use a standardized form

$$d_{2r}^* = \frac{d_{2r}}{\sqrt{V(y_r | \mathbf{y}_{(r),\text{obs}})}} \tag{10.2.28}$$

and plot d_{2r}^* versus r, similar to standard residual analysis, where $V(y_r | \mathbf{y}_{(r),\text{obs}})$ is the variance of y_r with respect to $f(y_r | \mathbf{y}_{(r),\text{obs}})$. Another choice of checking function is $c_{3r} = I(-\infty < y_r \le y_{r,\text{obs}})$ with expectation

$$d_{3r} = P(y_r \le y_{r,\text{obs}} | \mathbf{y}_{(r),\text{obs}}). \tag{10.2.29}$$

We can use d_{3r} to measure how unlikely y_r is under $f(y_r | \mathbf{y}_{(r),\text{obs}})$. If y_r is discrete, then we can use $P(y_r = y_{r,\text{obs}} | \mathbf{y}_{(r),\text{obs}})$. Yet another choice is given by $c_4(y_r, \mathbf{y}_{(r),\text{obs}}) = I(y_r \in B_r)$, where

$$B_r = \{y_r : f(y_r | \mathbf{y}_{(r),\text{obs}}) \le f(y_{r,\text{obs}} | \mathbf{y}_{(r),\text{obs}})\}. \tag{10.2.30}$$

Its expectation is equal to

$$d_{4r} = P(B_r | \mathbf{y}_{(r),\text{obs}}). \tag{10.2.31}$$

We can use d_{4r} to see how likely $f(y_r | \mathbf{y}_{(r),\text{obs}})$ will be smaller than the corresponding conditional predictive ordinate, CPO_r.

To estimate the measures d_{2r}, d_{3r}, and d_{4r}, we need to evaluate expectations of the form $E[a(y_r)|\mathbf{y}_{(r),\text{obs}}]$ for specified functions $a(y_r)$. Suppose that $f(y_r|\mathbf{y}_{(r),\text{obs}}, \boldsymbol{\eta}) = f(y_r|\boldsymbol{\eta})$. Then we can estimate the expectations as

$$\hat{E}[a(y_r)|\mathbf{y}_{(r),\text{obs}}] = \hat{f}(y_r|\mathbf{y}_{(r),\text{obs}})\frac{1}{D}\sum_{k=d+1}^{d+D}\frac{b_r(\boldsymbol{\eta}^{(k)})}{f(y_{r,\text{obs}}|\boldsymbol{\eta}^{(k)})}, \tag{10.2.32}$$

where $b_r(\boldsymbol{\eta})$ is the conditional expectation of $a(y_r)$ over y_r given $\boldsymbol{\eta}$. A proof of (10.2.32) is given by Section 10.18. If a closed-form expression for $b_r(\boldsymbol{\eta})$ is not available, then we have to draw a sample from $f(y_r|\mathbf{y}_{(r)})$ and estimate $E[a(y_r)|\mathbf{y}_{(r),\text{obs}}]$ directly. Gelfand (1996) has given a method of drawing such a sample without rerunning the MCMC sampler, using only $\mathbf{y}_{(r),\text{obs}}$.

10.3 BASIC AREA LEVEL MODEL

In this section, we apply the HB approach to the basic area level model (6.1.1), assuming a prior distribution on the model parameters $(\boldsymbol{\beta}, \sigma_v^2)$. We first consider the case of known σ_v^2 and assume a "flat" prior on $\boldsymbol{\beta}$ given by $f(\boldsymbol{\beta}) \propto 1$, and rewrite (6.1.1) as an HB model:

$$\begin{aligned}
&\text{(i)}\quad \hat{\theta}_i|\theta_i, \boldsymbol{\beta}, \sigma_v^2 \overset{\text{ind}}{\sim} N(\theta_i, \psi_i), \quad i = 1, \ldots, m,\\
&\text{(ii)}\quad \theta_i|\boldsymbol{\beta}, \sigma_v^2 \overset{\text{ind}}{\sim} N(\mathbf{z}_i^T\boldsymbol{\beta}, b_i^2\sigma_v^2), \quad i = 1, \ldots, m,\\
&\text{(iii)}\quad f(\boldsymbol{\beta}) \propto 1.
\end{aligned} \tag{10.3.1}$$

We then extend the results to the case of unknown σ_v^2 by replacing (iii) in (10.3.1) by

$$\text{(iii)}'\quad f(\boldsymbol{\beta}, \sigma_v^2) = f(\boldsymbol{\beta})f(\sigma_v^2) \propto f(\sigma_v^2), \tag{10.3.2}$$

where $f(\sigma_v^2)$ is a prior on σ_v^2.

10.3.1 Known σ_v^2

Straightforward calculations show that the posterior distribution of θ_i given the data $\hat{\boldsymbol{\theta}} = (\hat{\theta}_1, \ldots, \hat{\theta}_m)^T$ and σ_v^2, under the HB model (10.3.1), is normal with mean equal to the best linear unbiased predictor (BLUP) estimator $\tilde{\theta}_i^H$ and variance equal to $M_{1i}(\sigma_v^2)$ given by (6.1.7). Thus, the HB estimator of θ_i is

$$\tilde{\theta}_i^{\text{HB}}(\sigma_v^2) = E(\theta_i|\hat{\boldsymbol{\theta}}, \sigma_v^2) = \tilde{\theta}_i^H, \tag{10.3.3}$$

and the posterior variance of θ_i is

$$V(\theta_i|\hat{\boldsymbol{\theta}}, \sigma_v^2) = M_{1i}(\sigma_v^2) = \text{MSE}(\tilde{\theta}_i^H). \tag{10.3.4}$$

Hence, when σ_v^2 is assumed to be known and $f(\boldsymbol{\beta}) \propto 1$, the HB and BLUP approaches under normality lead to identical point estimates and measures of variability.

10.3.2 *Unknown σ_v^2: Numerical Integration

In practice, σ_v^2 is unknown and it is necessary to take account of the uncertainty about σ_v^2 by assuming a prior, $f(\sigma_v^2)$, on σ_v^2. The HB model is given by (i) and (ii) of (10.3.1) and (iii)′ given by (10.3.2). We obtain the HB estimator of θ_i as

$$\hat{\theta}_i^{\text{HB}} = E(\theta_i|\hat{\boldsymbol{\theta}}) = E_{\sigma_v^2}[\tilde{\theta}_i^{\text{HB}}(\sigma_v^2)], \tag{10.3.5}$$

where $E_{\sigma_v^2}$ denotes the expectation with respect to the posterior distribution of σ_v^2, $f(\sigma_v^2|\hat{\boldsymbol{\theta}})$. The posterior variance of θ_i is given by

$$V(\theta_i|\hat{\boldsymbol{\theta}}) = E_{\sigma_v^2}[M_{1i}(\sigma_v^2)] + V_{\sigma_v^2}[\tilde{\theta}_i^{\text{HB}}(\sigma_v^2)], \tag{10.3.6}$$

where $V_{\sigma_v^2}$ denotes the variance with respect to $f(\sigma_v^2|\hat{\boldsymbol{\theta}})$. It follows from (10.3.5) and (10.3.6) that the evaluation of $\hat{\theta}_i^{\text{HB}}$ and $V(\theta_i|\hat{\boldsymbol{\theta}})$ involves only single dimensional integrations.

The posterior $f(\sigma_v^2|\hat{\boldsymbol{\theta}})$ may be obtained from the restricted likelihood function $L_R(\sigma_v^2)$ as

$$f(\sigma_v^2|\hat{\boldsymbol{\theta}}) \propto L_R(\sigma_v^2)f(\sigma_v^2), \tag{10.3.7}$$

where

$$\log[L_R(\sigma_v^2)] = \text{const} - \frac{1}{2}\sum_{i=1}^m \log(\sigma_v^2 b_i^2 + \psi_i) - \frac{1}{2}\log\left|\sum_{i=1}^m (\sigma_v^2 b_i^2 + \psi_i)^{-1}\mathbf{z}_i\mathbf{z}_i^T\right|$$
$$- \frac{1}{2}\sum_{i=1}^m [\hat{\theta}_i - \mathbf{z}_i^T\tilde{\boldsymbol{\beta}}(\sigma_v^2)]^2/(\sigma_v^2 b_i^2 + \psi_i). \tag{10.3.8}$$

Here, $\mathbf{z}_i$ is a $p \times 1$ vector of covariates and $\tilde{\boldsymbol{\beta}}(\sigma_v^2) = \tilde{\boldsymbol{\beta}}$ is the weighted least squares estimator of $\boldsymbol{\beta}$ given in equation (6.1.5); see Harville (1977). Under a "flat" prior $f(\sigma_v^2) \propto 1$, the posterior $f(\sigma_v^2|\hat{\boldsymbol{\theta}})$ is proper provided $m > p + 2$, and proportional to $L_R(\sigma_v^2)$; note that $f(\sigma_v^2)$ is improper. More generally, for any improper prior $f(\sigma_v^2) \propto h(\sigma_v^2)$, the posterior $f(\sigma_v^2|\hat{\boldsymbol{\theta}})$ will be proper if $h(\sigma_v^2)$ is a bounded function of σ_v^2 and $m > p + 2$. Note that $h(\sigma_v^2) = 1$ for the flat prior. Morris (1983a) and Ghosh (1992a) used the flat prior $f(\sigma_v^2) \propto 1$.

The posterior mean of σ_v^2 under the flat prior $f(\sigma_v^2) \propto 1$ may be expressed as

$$\hat{\sigma}_{v\text{HB}}^2 := E(\sigma_v^2|\hat{\boldsymbol{\theta}}) = \int_0^\infty \sigma_v^2 L_R(\sigma_v^2)d\sigma_v^2 \Big/ \int_0^\infty L_R(\sigma_v^2)d\sigma_v^2. \tag{10.3.9}$$

This estimator is always unique and positive, unlike the posterior mode or the REML estimator of σ_v^2. Also, the HB estimator $\hat{\theta}_i^{\text{HB}}$ attaches a positive weight $\hat{\gamma}_i^{\text{HB}}$ to the direct estimator $\hat{\theta}_i$, where $\hat{\gamma}_i^{\text{HB}} = E[\gamma_i(\sigma_v^2)|\hat{\boldsymbol{\theta}}]$ is obtained from (10.3.9) by changing σ_v^2 to $\gamma_i(\sigma_v^2) = \gamma_i$. On the other hand, if the REML estimator $\hat{\sigma}_{v\text{RE}}^2 = 0$, the empirical best linear unbiased prediction (EBLUP) (or EB) will give a zero weight to $\hat{\theta}_i$ for all the small areas, regardless of the area sample sizes. Because of the latter difficulty, the HB estimator $\hat{\theta}_i^{\text{HB}}$ may be more appealing (Bell 1999).

The frequentist bias of $\hat{\sigma}^2_{v\text{HB}}$ can be substantial when σ_v^2 is small (Browne and Draper 2006). However, $\hat{\theta}_i^{\text{HB}}$ may not be affected by the bias of $\hat{\sigma}^2_{v\text{HB}}$ because the weighted average $a_i\hat{\theta}_i + (1-a_i)\mathbf{z}_i^T\hat{\boldsymbol{\beta}}$ is approximately unbiased for θ_i for any choice of the weight $a_i (0 \le a_i \le 1)$, provided the linking model (ii) of the HB model (10.3.1) is correctly specified.

The posterior variance (10.3.6), computed similar to (10.3.9), is used as a measure of uncertainty associated with $\hat{\theta}_i^{\text{HB}}$. As noted in Section 10.1, it is desirable to select a "matching" improper prior that leads to well-calibrated inferences. In particular, the posterior variance should be second-order unbiased for the MSE, that is, $E[V(\theta_i|\hat{\boldsymbol{\theta}})] - \text{MSE}(\hat{\theta}_i^{\text{HB}}) = o(m^{-1})$. Such a dual justification is very appealing to survey practitioners. Datta, Rao, and Smith (2005) showed that the matching prior for the special case of $b_i = 1$ is given by

$$f_i(\sigma_v^2) \propto (\sigma_v^2 + \psi_i)^2 \sum_{\ell=1}^{m} (\sigma_v^2 + \psi_\ell)^{-2}. \tag{10.3.10}$$

The proof of (10.3.10) is quite technical and we refer the reader to Datta et al. (2005) for details. The matching prior (10.3.10) is a bounded function of σ_v^2 so that the posterior is proper provided $m > p + 2$. Also, the prior (10.3.10) depends collectively on all the sampling variances for all the areas as well as on the individual sampling variance, ψ_i, of the ith area. Strictly speaking, a prior on the common parameter σ_v^2 should not vary with area i. Hence, we should use the area-specific matching prior (10.3.10) when area i is of primary interest. The same prior is also employed for the remaining areas, although one could argue in favor of using different priors for different areas.

For the balanced case, $\psi_i = \psi$ for all i, the matching prior (10.3.10) reduces to the flat prior $f(\sigma_v^2) \propto 1$ and the resulting HB inferences have dual justification. However, the use of a flat prior when the sampling variances vary significantly across areas could lead to posterior variances not tracking the MSE.

The matching prior (10.3.10) may be generalized to the case of matching a weighted average of expected posterior variances to the corresponding weighted average of MSEs in the following sense:

$$E\left[\sum_{\ell=1}^{m} a_\ell V(\theta_\ell|\hat{\boldsymbol{\theta}})\right] - \sum_{\ell=1}^{m} a_\ell \text{MSE}(\hat{\theta}_\ell^H) = o(m^{-1}), \tag{10.3.11}$$

where a_ℓ are specified weights satisfying $a_\ell \ge 0$, $\ell = 1, \ldots, m$, and $\sum_{\ell=1}^m a_\ell = 1$. For this case, Ganesh and Lahiri (2008) obtained a matching prior given by

$$f_{\text{GL}}(\sigma_v^2) \propto \frac{\sum_{\ell=1}^m (\sigma_v^2 + \psi_i)^{-2}}{\sum_{\ell=1}^m a_\ell \psi_\ell^2 (\sigma_v^2 + \psi_\ell)^{-2}}. \tag{10.3.12}$$

By letting $a_\ell = \psi_\ell^{-1} / \sum_{t=1}^m \psi_t^{-2}$, the prior (10.3.12) reduces to the flat prior $f(\sigma_v^2) \propto 1$. Average moment-matching prior is obtained by letting $a_\ell = m^{-1}$ for $\ell = 1, \ldots, m$.

The area-specific matching prior (10.3.10) is also a particular case of (10.3.12) by letting $a_i = 1$ and $a_\ell = 0$, $\ell \neq i$, $i = 1, \ldots, m$. For the special case of $\psi_\ell = \psi$ for all ℓ, the average moment-matching prior reduces to the flat prior $f(\sigma_v^2) \propto 1$.

Example 10.3.1. U.S. Poverty Counts. In Example 6.1.2, we considered EBLUP estimation of school-age children in poverty in the United States at the county and state levels. Bell (1999) used the state model to calculate maximum-likelihood (ML), REML, and HB estimates of σ_v^2 for 5 years, 1989–1993. Both ML and REML estimates are zero for the first 4 years while $\hat{\sigma}_{v\text{HB}}^2$, based on the flat prior $f(\sigma_v^2) \propto 1$, varied from 1.6 to 3.4. The resulting EBLUP estimates of poverty rates attach zero weight to the direct estimates regardless of the Current Population Survey (CPS) state sample sizes n_i (number of households). Also, the leading term, $g_{1i}(\hat{\sigma}_v^2) = \hat{\gamma}_i \psi_i$, of the MSE estimator (6.2.7) becomes zero when the estimate of σ_v^2 is zero. As a result, the contribution to (6.2.7) comes entirely from the terms $g_{2i}(\hat{\sigma}_v^2)$ and $g_{3i}(\hat{\sigma}_v^2)$, which account for the estimation of $\boldsymbol{\beta}$ and σ_v^2, respectively.

Table 10.1 reports results for four states in increasing order of the sampling variance ψ_i: California (CA), North Carolina (NC), Indiana (IN), and Mississippi (MS). This table shows the sample sizes n_i, poverty rates $\hat{\theta}_i$ as a percentage, sampling variances ψ_i, MSE estimates mse_{ML} and mse_{RE} based on ML and REML estimation of σ_v^2 using (6.2.7) and (6.2.3), and the posterior variances $V(\theta_i|\hat{\boldsymbol{\theta}})$. Naive MSE estimates, based on the formula $g_{1i}(\hat{\sigma}_v^2) + g_{2i}(\hat{\sigma}_v^2)$, are also included for ML and REML estimation ($\text{mse}_{N,\text{ML}}$ and $\text{mse}_{N,\text{RE}}$). Results for 1992 (with $\hat{\sigma}_{v\text{ML}}^2 = \hat{\sigma}_{v\text{RE}}^2 = 0$, $\hat{\sigma}_{v\text{HB}}^2 = 1.6$) are compared in Table 10.1 to those for 1993 (with $\hat{\sigma}_{v\text{ML}}^2 = 0.4$, $\hat{\sigma}_{v\text{RE}}^2 = 1.7$, $\hat{\sigma}_{v\text{HB}}^2 = 3.4$).

Comparing $\text{mse}_{N,\text{ML}}$ to mse_{ML} and $\text{mse}_{N,\text{RE}}$ to mse_{RE} given in Table 10.1, we note that the naive MSE estimates lead to significant underestimation. Here, σ_v^2 is not estimated precisely, and this fact is reflected in the contribution from $2g_{3i}(\hat{\sigma}_v^2)$. For 1992, the leading term $g_{1i}(\hat{\sigma}_v^2)$ is zero and the contribution to MSE estimates

TABLE 10.1 MSE Estimates and Posterior Variance for Four States

State	n_i	$\hat{\theta}_i$	ψ_i	$\text{mse}_{N,\text{ML}}$	mse_{ML}	$\text{mse}_{N,\text{RE}}$	mse_{RE}	$V(\theta_i\|\hat{\boldsymbol{\theta}})$
				1992				
CA	4,927	20.9	1.9	1.3	3.6	1.3	2.8	1.4
NC	2,400	23.0	5.5	0.6	2.0	0.6	1.2	2.0
IN	670	11.8	9.3	0.3	1.4	0.3	0.6	1.7
MS	796	29.6	12.4	2.8	3.8	2.8	3.0	4.0
				1993				
CA	4,639	23.8	2.3	1.5	3.2	1.6	2.2	1.7
NC	2,278	17.0	4.5	1.0	2.4	1.7	2.2	2.0
IN	650	10.3	8.5	0.8	1.9	1.8	2.2	3.0
MS	747	30.5	13.6	3.2	4.3	4.2	4.5	5.1

Source: Adapted from Tables 2 and 3 in Bell (1999).

comes entirely from $g_{2i}(\hat{\sigma}_v^2)$ and $2g_{3i}(\hat{\sigma}_v^2)$. The MSE estimates for NC and IN turned out to be smaller than the corresponding MSE estimates for CA, despite the smaller sample sizes and larger sampling variances compared to CA. The reason for this occurrence becomes evident by examining $g_{2i}(\hat{\sigma}_v^2)$ and $2g_{3i}(\hat{\sigma}_v^2)$, which reduce to $g_{2i}(\hat{\sigma}_v^2) = \mathbf{z}_i^T(\sum_{i=1}^m \mathbf{z}_i\mathbf{z}_i^T/\psi_i)^{-1}\mathbf{z}_i$ and $2g_{3i}(\hat{\sigma}_v^2) = 4/(\psi_i \sum_{i=1}^m \psi_i^{-2})$ for ML and REML when $\hat{\sigma}_v^2 = 0$. The term $2g_{3i}(\hat{\sigma}_v^2)$ for CA is larger because the sampling variance $\psi_{\text{CA}} = 1.9$, which appears in the denominator, is much smaller than $\psi_{\text{NC}} = 5.5$ and $\psi_{\text{IN}} = 9.3$. The naive MSE estimator is also smaller for NC and IN compared to CA, and in this case only the g_{2i}-term contributes. It appears from the form of $g_{2i}(\hat{\sigma}_v^2)$ that the covariates $\mathbf{z}_i$ for CA contribute to this increase.

Turning to the HB values for 1992, we see that the leading term of the posterior variance $V(\theta_i|\hat{\boldsymbol{\theta}})$ is $g_{1i}(\hat{\sigma}_{v\text{HB}}^2) = \hat{\gamma}_i^{\text{HB}}\psi_i$. Here, $\hat{\sigma}_{v\text{HB}}^2 = 1.6$ and the leading term dominates the posterior variance. As a result, the posterior variance is the smallest for CA, although the value for NC is slightly larger compared to IN, despite the larger sample size and smaller sampling variance (5.5 vs 9.3). For 1993 with nonzero ML and REML estimates of σ_v^2, mse_{RE} exhibits a trend similar to $V(\theta_i|\hat{\boldsymbol{\theta}})$, but mse_{ML} values are similar to 1992 values due to a small $\hat{\sigma}_{v\text{ML}}^2 (= 0.4)$ compared to $\hat{\sigma}_{v\text{RE}}^2 (= 1.7)$ and $\hat{\sigma}_{v\text{HB}}^2 (= 3.4)$.

The occurrences of zero estimates of σ_v^2 in the frequentist approach is problematic (Hulting and Harville 1991), but it is not clear if the HB approach based on a flat prior on σ_v^2 leads to well-calibrated inferences. We have already noted that the matching prior is different from the flat prior if the ψ_i-values vary significantly, as in the case of state CPS variance estimates with $\max(\psi_i)/\min(\psi_i)$ as large as 20.

10.3.3 Unknown σ_v^2: Gibbs Sampling

In this section, we apply Gibbs sampling to the basic area level model, given by (i) and (ii) of (10.3.1), assuming the prior (10.3.2) on $\boldsymbol{\beta}$ and σ_v^2 with $\sigma_v^{-2} \sim G(a,b)$, $a > 0, b > 0$, that is, a gamma distribution with shape parameter a and scale parameter b. Note that σ_v^2 is distributed as inverted gamma $IG(a,b)$, with $f(\sigma_v^2) \propto \exp(-b/\sigma_v^2)(1/\sigma_v^2)^{a+1}$. The positive constants a and b are set to very small values (BUGS uses $a = b = 0.001$ as the default setting). It is easy to verify that the Gibbs conditionals are then given by

$$\text{(i)}\quad [\boldsymbol{\beta}|\boldsymbol{\theta},\sigma_v^2,\hat{\boldsymbol{\theta}}] \sim N_p\left[\boldsymbol{\beta}^*, \sigma_v^2\left(\sum_{i=1}^m \tilde{\mathbf{z}}_i\tilde{\mathbf{z}}_i^T\right)^{-1}\right], \tag{10.3.13}$$

$$\text{(ii)}\quad [\theta_i|\boldsymbol{\beta},\sigma_v^2,\hat{\boldsymbol{\theta}}] \sim N[\hat{\theta}_i^B(\boldsymbol{\beta},\sigma_v^2), \gamma_i\psi_i],\quad i = 1, \ldots, m, \tag{10.3.14}$$

$$\text{(iii)}\quad [\sigma_v^{-2}|\boldsymbol{\beta},\boldsymbol{\theta},\hat{\boldsymbol{\theta}}] \sim G\left[\frac{m}{2} + a, \frac{1}{2}\sum_{i=1}^m (\tilde{\theta}_i - \tilde{\mathbf{z}}_i^T\boldsymbol{\beta})^2 + b\right], \tag{10.3.15}$$

where $\tilde{\theta}_i = \theta_i/b_i$, $\tilde{\mathbf{z}}_i = \mathbf{z}_i/b_i$, $\boldsymbol{\beta}^* = (\sum_{i=1}^m \tilde{\mathbf{z}}_i\tilde{\mathbf{z}}_i^T)^{-1}(\sum_{i=1}^m \tilde{\mathbf{z}}_i\tilde{\theta}_i)$, $N_p(\cdot)$ denotes a p-variate normal, and $\hat{\theta}_i^B(\boldsymbol{\beta},\sigma_v^2) = \gamma_i\hat{\theta}_i + (1-\gamma_i)\mathbf{z}_i^T\boldsymbol{\beta}$ is the Bayes estimator of θ_i. A proof of

(10.3.13)–(10.3.15) is sketched in Section 10.18.3. Note that all the Gibbs conditionals (i)–(iii) have closed forms, and hence the MCMC samples can be generated directly from them.

Denote the MCMC samples as $\{(\boldsymbol{\beta}^{(k)}, \boldsymbol{\theta}^{(k)}, \sigma_v^{2(k)}), k = d+1, \ldots, d+D\}$. Using (10.3.14), we can obtain Rao–Blackwell estimators of the posterior mean and the posterior variance of θ_i as

$$\hat{\theta}_i^{\text{HB}} = \frac{1}{D}\sum_{k=d+1}^{d+D} \hat{\theta}_i^B(\boldsymbol{\beta}^{(k)}, \sigma_v^{2(k)}) =: \hat{\theta}_i^B(\cdot,\cdot) \tag{10.3.16}$$

and

$$\hat{V}(\theta_i|\hat{\boldsymbol{\theta}}) = \frac{1}{D}\sum_{k=d+1}^{d+D} g_{1i}(\sigma_v^{2(k)}) + \frac{1}{D-1}\sum_{k=d+1}^{d+D}\left[\hat{\theta}_i^B(\boldsymbol{\beta}^{(k)}, \sigma_v^{2(k)}) - \hat{\theta}_i^B(\cdot,\cdot)\right]^2. \tag{10.3.17}$$

More efficient estimators may be obtained by exploiting the closed-form results of Section 10.3.1 for known σ_v^2. We have

$$\hat{\theta}_i^{\text{HB}} = \frac{1}{D}\sum_{k=d+1}^{d+D} \tilde{\theta}_i^H(\sigma_v^{2(k)}) =: \tilde{\theta}_i^H(\cdot) \tag{10.3.18}$$

and

$$\hat{V}(\theta_i|\hat{\boldsymbol{\theta}}) = \frac{1}{D}\sum_{k=d+1}^{d+D}\left[g_{1i}(\sigma_v^{2(k)}) + g_{2i}(\sigma_v^{2(k)})\right] + \frac{1}{D-1}\sum_{k=d+1}^{d+D}\left[\tilde{\theta}_i^H(\sigma_v^{2(k)}) - \tilde{\theta}_i^H(\cdot)\right]^2. \tag{10.3.19}$$

Based on the Rao–Blackwell estimator $\hat{\theta}_i^{\text{HB}}$, an estimator of the total $Y_i = g^{-1}(\theta_i)$ is given by $g^{-1}(\hat{\theta}_i^{\text{HB}})$, but it is not equal to the desired posterior mean $E(Y_i|\hat{\boldsymbol{\theta}})$. However, the marginal MCMC samples $\{Y_i^{(k)} = g^{-1}(\theta_i^{(k)}); k = d+1, \ldots, d+D\}$ can be used directly to estimate the posterior mean of Y_i as

$$\hat{Y}_i^{\text{HB}} = \frac{1}{D}\sum_{k=d+1}^{d+D} Y_i^{(k)} =: Y_i^{(\cdot)}. \tag{10.3.20}$$

Similarly, the posterior variance of Y_i is estimated as

$$\hat{V}(Y_i|\hat{\boldsymbol{\theta}}) = \frac{1}{D-1}\sum_{k=d+1}^{d+D}(Y_i^{(k)} - Y_i^{(\cdot)})^2. \tag{10.3.21}$$

If L independent runs are generated, instead of a single long run, then the posterior mean is estimated as

$$\hat{Y}_i^{\text{HB}} = \frac{1}{Ld} \sum_{\ell=1}^{L} \sum_{k=d+1}^{2d} Y_i^{(\ell k)} = \frac{1}{L} \sum_{\ell=1}^{L} Y_i^{(\ell \cdot)} =: Y_i^{(\cdot\cdot)}, \qquad (10.3.22)$$

where $Y_i^{(\ell k)}$ is the kth retained value in the ℓth run of length $2d$ with the first d burn-in iterations deleted. The posterior variance is estimated from (10.2.10):

$$\hat{V}(Y_i|\hat{\boldsymbol{\theta}}) = \frac{d-1}{d} W_i + \frac{1}{d} B_i, \qquad (10.3.23)$$

where $B_i = d \sum_{\ell=1}^{L} (Y_i^{(\ell \cdot)} - Y_i^{(\cdot\cdot)})^2/(L-1)$ is the between-run variance and $W_i = \sum_{\ell=1}^{L} \sum_{k=d+1}^{2d} (Y_i^{(\ell k)} - Y_i^{(\ell \cdot)})^2/[L(d-1)]$ is the within-run variance.

Example 10.3.2. Canadian Unemployment Rates. Example 4.4.4 mentioned the use of time series and cross-sectional models to estimate monthly unemployment rates for $m = 62$ Census Metropolitan Areas (CMAs) and Census Agglomerations (CAs) (small areas) in Canada, using time series and cross-sectional data. We study this application in Example 10.9.1. Here we consider only the cross-sectional data for the "current" month June 1999 to illustrate the Gibbs sampling HB method for the basic area level model.

Let $\hat{\theta}_i$ be the Labor Force Survey (LFS) unemployment rate for the ith area. To obtain smoothed estimates ψ_i of the sampling variances of $\hat{\theta}_i$, You, Rao, and Gambino (2003) first computed the average CV (over time), $\overline{\text{CV}}_i$, and then took $\psi_i = (\hat{\theta}_i \, \overline{\text{CV}}_i)^2$. Employment Insurance (EI) beneficiary rates were used as auxiliary variables, z_i, in the linking model $\theta_i = \beta_0 + \beta_1 z_i + v_i$, $i = 1, \ldots, m$, where $m = 62$.

Gibbs sampling was implemented using $L = 10$ parallel runs, each of length $2d = 2000$. The first $d = 1000$ "burn-in" iterations of each run were deleted. The convergence of the Gibbs sampler was monitored using the method of Gelman and Rubin (1992); see Section 10.2.5. The Gibbs sampler converged very well in terms of the estimated potential scale reduction factor $\hat{R}$ given by (10.2.11).

To check the overall fit of the model, the posterior predictive p-value, pp, was approximated from each run of the MCMC output $\{\boldsymbol{\eta}^{(\ell k)}\}, \ell = 1, \ldots, 10$, using the formula (10.2.19) with the measure of discrepancy given by $T(\hat{\boldsymbol{\theta}}, \boldsymbol{\eta}) = \sum_{i=1}^{62} (\hat{\theta}_i - \theta_i)^2/\psi_i$. The average of the $L = 10$ p-values, $\widehat{pp} = 0.59$, indicates a good fit of the model to the current cross-sectional data. Note that the hypothetical replication $\hat{\theta}_i^{(\ell k)}$, used in (10.2.19), is generated from $N(\theta_i^{(\ell k)}, \psi_i)$ for each $\theta_i^{(\ell k)}$.

Rao–Blackwell estimators (10.3.16) and (10.3.17) were used to calculate the estimated posterior mean, $\hat{\theta}_i^{\text{HB}}$, and the estimated posterior variance, $\hat{V}(\theta_i|\hat{\boldsymbol{\theta}})$, for each area i. Figure 10.1 displays the coefficient of variations (CVs) of the direct estimates $\hat{\theta}_i$ and the HB estimates $\hat{\theta}_i^{\text{HB}}$; the CV of $\hat{\theta}_i^{\text{HB}}$ is taken as $[\hat{V}(\theta_i|\hat{\boldsymbol{\theta}})]^{1/2}/\hat{\theta}_i^{\text{HB}}$. It is clear from Figure 10.1 that the HB estimates lead to significant reduction in CV, especially for the areas with smaller population sizes.

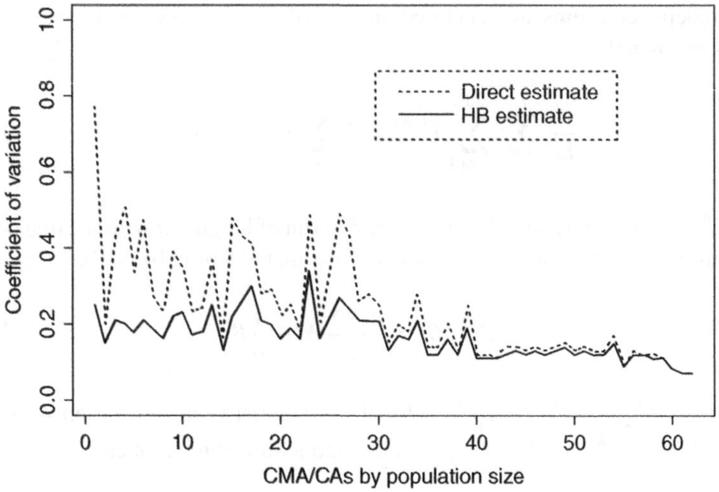

Figure 10.1 Coefficient of Variation (CV) of Direct and HB Estimates. *Source*: Adapted from Figure 3 in You, Rao, and Gambino (2003).

10.3.4 *Unknown Sampling Variances ψ_i

Suppose that simple random sampling within areas is assumed and that the available data are $D = \{(\hat{\theta}_i = \bar{y}_i, s_i^2); i = 1, \ldots, m\}$, where $s_i^2 = (n_i - 1)^{-1} \sum_{j=1}^{n_i} (y_{ij} - \bar{y}_i)^2$. The HB model is specified as (i) $\bar{y}_i|\theta_i, \sigma_i^2 \overset{\text{iid}}{\sim} N(\theta_i, \psi_i = \sigma_i^2/n_i)$; (ii) $[(n_i - 1)s_i^2/\sigma_i^2]|\sigma_i^2 \sim \chi^2_{n_i-1}$; and (iii) $\theta_i|\boldsymbol{\beta}, \sigma_v^2 \overset{\text{ind}}{\sim} N(\mathbf{z}_i^T\boldsymbol{\beta}, \sigma_v^2)$, where σ_i^2 is an unknown parameter. Priors for the model parameters are specified as $f(\boldsymbol{\beta}) \propto 1, \sigma_i^{-2} \overset{\text{iid}}{\sim} G(a_1, b_1)$, and $\sigma_v^{-2} \sim G(a_0, b_0)$ with $a_t > 0$ and $b_t > 0, t = 0, 1$.

Let $\boldsymbol{\theta} = (\theta_1, \ldots, \theta_m)^T$, $\hat{\boldsymbol{\theta}} = (\hat{\theta}_1, \ldots, \hat{\theta}_m)^T$, and $\boldsymbol{\sigma}^2 = (\sigma_1^2, \ldots, \sigma_m^2)^T$. Under the specified HB model (denoted Model 1), the Gibbs conditionals are given by

$$[\boldsymbol{\beta}|\boldsymbol{\theta}, \sigma_v^2, \boldsymbol{\sigma}^2, D] \sim N_p\left[\left(\sum_{i=1}^m \mathbf{z}_i\mathbf{z}_i^T\right)^{-1} \sum_{i=1}^m \mathbf{z}_i\hat{\theta}_i, \sigma_v^2\left(\sum_{i=1}^m \mathbf{z}_i\mathbf{z}_i^T\right)^{-1}\right], \tag{10.3.24}$$

$$[\theta_i|\boldsymbol{\beta}, \sigma_v^2, \boldsymbol{\sigma}^2, D] \sim N[\tilde{\theta}_i^B(\boldsymbol{\beta}, \sigma_v^2), \gamma_i\psi_i], \quad i = 1, \ldots, m, \tag{10.3.25}$$

$$[\sigma_i^{-2}|\boldsymbol{\beta}, \sigma_v^2, D, \hat{\boldsymbol{\theta}}] \sim G\left\{a_1 + \frac{n_i}{2}, b_1 + \frac{1}{2}\left[n_i(\bar{y}_i - \theta_i)^2 + (n_i - 1)s_i^2\right]\right\}, \quad i = 1, \ldots, m, \tag{10.3.26}$$

$$[\sigma_v^{-2}|\boldsymbol{\beta}, \boldsymbol{\sigma}^2, \boldsymbol{\theta}, D] \sim G\left[a_0 + \frac{m}{2}, b_0 + \frac{1}{2}\sum_{i=1}^m (\theta_i - \mathbf{z}_i^T\boldsymbol{\beta})^2\right], \tag{10.3.27}$$

Proof of (10.3.24)–(10.3.27) follows along the lines of Section 10.18.3.

It follows from (10.3.24)–(10.3.27) that all the Gibbs conditionals have closed forms, and hence the MCMC samples can be generated directly from the conditionals. The posterior means and variances of θ_i are obtained from the MCMC samples, as in Section 10.3.3. You and Chapman (2006) obtained the Gibbs conditionals assuming that $\psi_i^{-1} \sim G(a_i, b_i), i = 1, \ldots, m$. They also derived the Gibbs conditionals under an alternative HB model (denoted Model 2) that assumes $\psi_i = n_i^{-1}s_i^2$. They compared the posterior variances of the θ_i's under the two different HB models, using a data set with small area sample sizes n_i (ranging from 3 to 5) and another data set with larger n_i (≥ 95). In the case of small n_i, the posterior variance of θ_i under Model 1 is significantly larger than the corresponding value under Model 2 because Model 1 accounts for the uncertainty associated with $\boldsymbol{\sigma}^2$ unlike Model 2. On the other hand, for the second data set, the two posterior variances are practically the same because of much larger area sample sizes n_i. Model 2 should not be used in practice if n_i is small.

You and Chapman (2006) also studied the sensitivity of posterior inferences to the choice of a_t and b_t in the priors for σ_v^2 and σ_i^2, by varying their values from 0.0001 to 0.1. Their results indicate that the HB estimates of the area means are not sensitive to the choice of proper priors specified by small values of a_t and b_t.

Arora and Lahiri (1997) proposed an HB method of incorporating the design weights, w_{ij}, into the model. Assuming the unit level model $y_{ij}|\theta_i, \sigma_i^2 \overset{\text{iid}}{\sim} N(\theta_i, \sigma_i^2)$ and no sample selection bias, it holds that $\bar{y}_{iw}|\theta_i, \sigma_i^2 \overset{\text{ind}}{\sim} N(\theta_i, \psi_i = \delta_{iw}\sigma_i^2)$, where $\delta_{iw} = \sum_{j=1}^{n_i} \tilde{w}_{ij}$ with $\tilde{w}_{ij} = w_{ij}/\sum_{j=1}^{n_i} w_{ij}$ and $\bar{y}_{iw} = \sum_{j=1}^{n_i} \tilde{w}_{ij}y_{ij}$, as in Section 7.6.2. The HB model is then specified as (i) $\bar{y}_{iw}|\theta_i, \sigma_i^2 \overset{\text{ind}}{\sim} N(\theta_i, \delta_{iw}\sigma_i^2)$, (ii) $[(n_i - 1)s_i^2/\sigma_i^2]|\sigma_i^2 \sim \chi^2_{n_i-1}$, (iii) $\theta_i|\boldsymbol{\beta}, \sigma_v^2 \overset{\text{ind}}{\sim} N(\mathbf{z}_i^T\boldsymbol{\beta}, \sigma_v^2)$, and (iv) $\sigma_i^{-2} \overset{\text{iid}}{\sim} G(a, b)$, $a > 0$, $b > 0$. Priors for the model parameters are specified as $\boldsymbol{\beta} \sim U_p$, $\sigma_v^{-2} \sim U^+$, $a \sim U^+$, and $b \sim U^+$, where U^+ denotes a uniform distribution over a subset of $\mathbb{R}^+$ with large but finite length and U_p denotes a uniform distribution over a p-dimensional rectangle whose sides are of large but finite length. M–H within Gibbs algorithm is used to generate MCMC samples from the joint posterior distribution. We refer the reader to Arora and Lahiri (1997) for technical details. Note that the data required to implement the proposed method are $\{(\bar{y}_{iw}, s_i^2, \sum_{j=1}^{n_i} \tilde{w}_{ij}^2); i = 1, \ldots, m\}$.

Note that You and Chapman (2006) treat σ_i^2 as a model parameter, considering the prior $\sigma_i^{-2} \sim G(a_i, b_i)$ with fixed parameters a_i and b_i, unlike Arora and Lahiri's assumption of random σ_i^2 satisfying (iv) above, with random parameters a and b. Assumption of random σ_i^2 following (iv) above allows borrowing strength across both means θ_i and sampling variances σ_i^2.

10.3.5 *Spatial Model

We now study an extension of the model introduced in Section 10.3.4 by allowing spatial correlations among the area effects $\mathbf{v} = (v_1, \ldots, v_m)^T$. In particular, the spatial model is specified as (i) $\bar{y}_i|\theta_i, \sigma_i^2 \overset{\text{ind}}{\sim} N(\theta_i, \psi_i = \sigma_i^2/n_i)$; (ii) $[(n_i - 1)s_i^2/\sigma_i^2]|\sigma_i^2 \overset{\text{ind}}{\sim} \chi^2_{n_i-1}$; and (iii) $\boldsymbol{\theta} = (\theta_1, \ldots, \theta_m)^T|\boldsymbol{\beta}, \sigma_v^2 \sim N_m(\mathbf{Z}\boldsymbol{\beta}, \sigma_v^2\mathbf{D}^{-1})$, where $\mathbf{Z} = (\mathbf{z}_1, \ldots, \mathbf{z}_m)^T$ and $\mathbf{D} = \lambda\mathbf{R} + (1 - \lambda)\mathbf{I}$. Here, $\mathbf{R}$ is a neighborhood matrix with elements r_{ii} = number of neighbors of area i and $r_{i\ell} = -1$ if area ℓ is a neighbor of area i and 0 otherwise, $i \neq \ell$, and λ is a spatial autocorrelation parameter, $0 \leq \lambda \leq 1$. Priors for the model parameters $(\boldsymbol{\beta}, \lambda, \sigma_v^2, \sigma_1^2, \dots, \sigma_m^2)$ are specified as (iv) $f(\boldsymbol{\beta}) \propto 1$, $\sigma_i^{-2} \overset{\text{iid}}{\sim} G(a, b)$, $i = 1, \dots, m$, $\sigma_v^{-2} \sim G(a_0, b_0)$ and $\lambda \sim U(0, 1)$. Model component (iii) corresponds to a conditional autoregressive (CAR) model on $\mathbf{v} = (v_1, \dots, v_m)^T$.

Under the specified HB model (i)–(iv), all Gibbs conditionals excepting that of $[\lambda|\boldsymbol{\theta}, \boldsymbol{\beta}, \sigma_v^2, \boldsymbol{\sigma}^2]$ have closed-form expressions. Therefore, we can employ M–H within Gibbs algorithm to generate MCMC from the joint posterior distribution of $(\boldsymbol{\theta}, \boldsymbol{\beta}, \lambda, \sigma_v^2, \sigma_1^2, \dots, \sigma_m^2)$. Appendix A.4 of You and Zhou (2011) gives the full Gibbs conditionals assuming that $\psi_i^{-1} \sim G(a_i, b_i)$.

You and Zhou (2011) applied their HB spatial model to estimate asthma rates, θ_i, for the $m = 20$ health regions in British Columbia, Canada. For this purpose, they used data from the Canadian Community Health Survey (CCHS) and six area level auxiliary variables. Since the CCHS used a complex design, assumption (ii) of the spatial model is replaced by $(d_i\hat{\psi}_i/\psi_i)|\psi_i \overset{\text{ind}}{\sim} \chi^2_{d_i}$ and the prior on ψ_i is taken as $\psi_i^{-1} \sim G(a, b)$, where $\hat{\psi}_i$ is a direct estimator of the sampling variance of $\hat{\theta}_i$ and d_i is the degrees of freedom associated with $\hat{\psi}_i$. You and Zhou (2011) took $d_i = n_i - 1$, which ignores the design effect associated with $\hat{\psi}_i$. Results based on the CCHS data indicated that the spatial model leads to significant reduction in CV relative to the model without spatial correlation when the number of neighbors to an area increases; for example, a 21% reduction in CV is achieved when the number of neighboring areas is seven.

10.4 *UNMATCHED SAMPLING AND LINKING AREA LEVEL MODELS

As noted in Section 4.2, the assumption of zero design-bias, that is, $E_p(e_i|\theta_i) = 0$, in the sampling model $\hat{\theta}_i = \theta_i + e_i$ may not be valid if the sample size n_i in the ith area is small and $\theta_i = g(Y_i)$ is a nonlinear function of the total Y_i. A more realistic sampling model is given by

$$\hat{Y}_i = Y_i + e_i^*, \quad i = 1, \dots, m, \tag{10.4.1}$$

with $E_p(e_i^*|Y_i) = 0$, where $\hat{Y}_i$ is a design-unbiased (p-unbiased) estimator of Y_i. A generalized regression (GREG) estimator, $\hat{Y}_{i\text{GR}}$, of the form (2.4.7) may also be used as $\hat{Y}_i$, since it is approximately p-unbiased if the overall sample size is large. However, the sampling model (10.4.1) is not matched with the linking model $\theta_i = \mathbf{z}_i^T\boldsymbol{\beta} + b_i v_i$, where $v_i \overset{\text{iid}}{\sim} N(0, \sigma_v^2)$. As a result, the sampling and linking models cannot be combined to produce a basic area level model.

The HB approach readily extends to unmatched sampling and linking models (You and Rao 2002b). We write the sampling model (10.4.1) under normality as

$$\hat{Y}_i \overset{\text{ind}}{\sim} N(Y_i, \phi_i), \tag{10.4.2}$$

where the sampling variance ϕ_i is known or a known function of Y_i. We first consider the case of known ϕ_i. We combine (10.4.2) with the linking model (ii) of (10.3.1) and the prior (10.3.2) with $\sigma_v^{-2} \sim G(a,b)$. The resulting Gibbs conditionals $[\boldsymbol{\beta}|\mathbf{Y},\sigma_v^2,\hat{\mathbf{Y}}]$ and $[\sigma_v^{-2}|\boldsymbol{\beta},\mathbf{Y},\hat{\mathbf{Y}}]$ are identical to (10.3.13) and (10.3.15), respectively, where $\mathbf{Y} = (Y_1, \ldots, Y_m)^T$ and $\hat{\mathbf{Y}} = (\hat{Y}_1, \ldots, \hat{Y}_m)^T$. However, the Gibbs conditional $[Y_i|\boldsymbol{\beta},\sigma_v^2,\hat{\mathbf{Y}}]$ does not admit a closed form, unlike (10.3.14). It is easy to verify that

$$f(Y_i|\boldsymbol{\beta},\sigma_v^2,\hat{\mathbf{Y}}) \propto h(Y_i|\boldsymbol{\beta},\sigma_v^2)d(Y_i), \tag{10.4.3}$$

where $d(Y_i)$ is the normal density $N(\hat{Y}_i,\phi_i)$, and

$$h(Y_i|\boldsymbol{\beta},\sigma_v^2) \propto g'(Y_i)\exp\left\{-\frac{(\tilde{\theta}_i - \tilde{\mathbf{z}}_i^T\boldsymbol{\beta})^2}{2\sigma_v^2}\right\} \tag{10.4.4}$$

is the density of Y_i given $\boldsymbol{\beta}$ and σ_v^2 and $g'(Y_i) = \partial g(Y_i)/\partial Y_i$, where $\tilde{\theta}_i = \hat{\theta}_i/b_i$ and $\tilde{\mathbf{z}}_i = \mathbf{z}_i/b_i$. Using $d(Y_i)$ as the "candidate" density to implement M–H within Gibbs, the acceptance probability (10.2.5) reduces to

$$a(\boldsymbol{\beta}^{(k)},\sigma_v^{2(k)},Y_i^{(k)},Y_i^*) = \min\left\{\frac{h(Y_i^*|\boldsymbol{\beta}^{(k)},\sigma_v^{2(k)})}{h(Y_i^{(k)}|\boldsymbol{\beta}^{(k)},\sigma_v^{2(k)})},1\right\}, \tag{10.4.5}$$

where the candidate Y_i^* is drawn from $N(\hat{Y}_i,\phi_i)$, and $\{Y_i^{(k)},\boldsymbol{\beta}^{(k)},\sigma_v^{2(k)}\}$ are the values of Y_i, $\boldsymbol{\beta}$, and σ_v^2 after the kth cycle. The candidate Y_i^* is accepted with probability $a(\boldsymbol{\beta}^{(k)},\sigma_v^{2(k)},Y_i^{(k)},Y_i^*)$; that is, $Y_i^{(k+1)} = Y_i^*$ if Y_i^* is accepted, and $Y_i^{(k+1)} = Y_i^{(k)}$ otherwise. Repeating this updating procedure for $i = 1, \ldots, m$, we get $\mathbf{Y}^{(k+1)} = (Y_1^{(k+1)}, \ldots, Y_m^{(k+1)})^T$, noting that the Gibbs conditional of Y_i does not depend on Y_ℓ, $\ell \neq i$. Given $\mathbf{Y}^{(k+1)}$, we then generate $\boldsymbol{\beta}^{(k+1)}$ from $[\boldsymbol{\beta}|\mathbf{Y}^{(k+1)},\sigma_v^{2(k)},\hat{\mathbf{Y}}]$ and $\sigma_v^{2(k+1)}$ from $[\sigma_v^2|\mathbf{Y}^{(k+1)},\boldsymbol{\beta}^{(k+1)},\hat{\mathbf{Y}}]$ to complete the $(k+1)$th cycle $\{\mathbf{Y}^{(k+1)},\boldsymbol{\beta}^{(k+1)},\sigma_v^{2(k+1)}\}$.

If ϕ_i is a known function of Y_i, we use $h(Y_i|\boldsymbol{\beta}^{(k)},\sigma_v^{2(k)})$ to draw the candidate Y_i^*, noting that $Y_i = g^{-1}(\theta_i)$ and $\theta_i|\boldsymbol{\beta},\sigma_v^2 \sim N(\mathbf{z}_i^T\boldsymbol{\beta},b_i^2\sigma_v^2)$. In this case, the acceptance probability is given by

$$a(Y_i^{(k)},Y_i^*) = \min\left\{\frac{d(Y_i^*)}{d(Y_i^{(k)})},1\right\}. \tag{10.4.6}$$

Example 10.4.1. Canadian Census Undercoverage. In Example 6.1.3, we provided a brief account of an application of the basic area level model to estimate domain undercount in the 1991 Canadian census, using the EBLUP method. You and Rao (2002b) applied the unmatched sampling and linking models to 1991 Canadian census data, using the HB method, and estimated the number of missing persons $M_i(= Y_i)$ and the undercoverage rate $U_i = M_i/(M_i + C_i)$ for each province i $(i = 1, \ldots, 10)$, where C_i is the census count.

The sampling variances ϕ_i were estimated through a generalized variance function model of the form $V(\hat{Y}_i) \propto C_i^{\gamma}$ and then treated as known in the sampling model (10.4.2). The linking model is given by (ii) of (10.3.1) with $\theta_i = \log\ \{M_i/(M_i + C_i)\}$, $\mathbf{z}_i^T\boldsymbol{\beta} = \beta_0 + \beta_1 \log(C_i)$, and $b_i = 1$. The prior on the model parameters $(\beta_0, \beta_1, \sigma_v^2)$ is given by (10.3.2) with $\sigma_v^{-2} \sim G(0.001, 0.001)$ to reflect lack of prior information. You and Rao (2002b) simulated $L = 8$ parallel M–H within Gibbs runs independently, each of length $2d = 10,000$. The first $d = 5,000$ "burn-in" iterations of each run were deleted. Furthermore, to reduce the autocorrelation within runs, they selected every 10th iteration of the remaining 5,000 iterations of each run, leading to 500 iterations for each run. Thus, the total number of MCMC runs retained is $Ld = 4,000$. The convergence of the M–H within Gibbs sampler was monitored using the method of Gelman and Rubin (1992) described in Section 10.2.4. The MCMC sampler converged very well in terms of the potential scale reduction factor $\hat{R}$ given by (10.2.11); values of $\hat{R}$ for the 10 provinces were close to 1. We denote the marginal MCMC sample as $\{M_i^{(\ell k)}; k = d + 1, \ldots, 2d, \ell = 1, \ldots, L\}$.

To check the overall fit of the model, the posterior predictive p-value was estimated from each run of the MCMC output $\{\boldsymbol{\eta}^{(\ell k)}\}, \ell = 1, \ldots, 8$ using the formula (10.2.19) with $T(\hat{\mathbf{Y}}, \boldsymbol{\eta}) = \sum_{i=1}^m (\hat{Y}_i - Y_i)^2/\phi_i$. The average of the $L = 8$ p-values, $\widehat{pp} = 0.38$, indicates a good fit of the model. Note that the hypothetical replications $\hat{Y}_i^{(\ell k)}$, used in (10.2.18), are generated from $N(Y_i^{(\ell k)}, \phi_i)$, for each $Y_i^{(\ell k)}$.

To assess model fit at the individual province level, You and Rao (2002b) computed two statistics proposed by Daniels and Gatsonis (1999). The first statistic, given by

$$p_i = P(\hat{Y}_i < \hat{Y}_{i,\text{obs}} | \hat{\mathbf{Y}}_{\text{obs}}), \tag{10.4.7}$$

provides information on the degree of consistent overestimation or underestimation of $\hat{Y}_{i,\text{obs}}$. This statistic is similar to the cross-validation statistic (10.2.29) but uses the full predictive density $f(\hat{Y}_i|\hat{\mathbf{Y}}_{\text{obs}})$. As a result, it is computationally simpler than (10.2.29). The p_i-values, computed from the hypothetical replications $\hat{Y}_i^{(\ell k)}$, ranged from 0.28 to 0.87 with a mean of 0.51 and a median of 0.49, indicating no consistent overestimation or underestimation. The second statistic is similar to the cross-validation standardized residual (10.2.28) but uses the full predictive density and it is given by

$$d_i = [E(\hat{Y}_i|\hat{\mathbf{Y}}_{\text{obs}}) - \hat{Y}_{i,\text{obs}}]/\sqrt{V(\hat{Y}_i|\hat{\mathbf{Y}}_{\text{obs}})}, \tag{10.4.8}$$

where the expectation and variance are with respect to the full predictive density. The estimated standardized residuals, d_i, ranged from -1.13 (New Brunswick) to 0.57 (Prince Edward Island), indicating adequate model fit for individual provinces.

TABLE 10.2 1991 Canadian Census Undercount Estimates and Associated CVs

Prov	$\hat{M}_i$	CV($\hat{M}_i$)	$\hat{M}_i^{\text{HB}}$	CV($\hat{M}_i^{\text{HB}}$)	$\hat{U}_i$(%)	CV($\hat{U}_i$)	$\hat{U}_i^{\text{HB}}$(%)	CV($\hat{U}_i^{\text{HB}}$)
Nfld	11,566	0.16	10,782	0.14	1.99	0.16	1.86	0.13
PEI	1,220	0.30	1,486	0.19	0.93	0.30	1.13	0.19
NS	17,329	0.20	17,412	0.14	1.89	0.20	1.90	0.14
NB	24,280	0.14	18,948	0.17	3.25	0.13	2.55	0.17
Que	184,473	0.08	189,599	0.08	2.58	0.08	2.65	0.08
Ont	381,104	0.08	368,424	0.08	3.64	0.08	3.52	0.08
Man	20,691	0.21	21,504	0.14	1.86	0.20	1.93	0.14
Sask	18,106	0.19	18,822	0.14	1.80	0.18	1.87	0.13
Alta	51,825	0.15	55,392	0.12	2.01	0.14	2.13	0.12
BC	92,236	0.10	89,929	0.09	2.73	0.10	2.67	0.09

Source: Adapted from Table 2 in You and Rao (2002b).

The posterior mean and the posterior variance of $M_i(=Y_i)$ were estimated using (10.3.20) and (10.3.21) with $L = 8$ and $d = 500$. Table 10.2 reports the direct estimates, $\hat{M}_i$, the HB estimates, $\hat{M}_i^{\text{HB}}$, and the associated CVs. For the direct estimate, the CV is based on the design-based variance estimate, while the CV for $\hat{M}_i^{\text{HB}}$ is based on the estimated posterior variance. Table 10.2 also reports the estimates of undercoverage rates U_i, denoted $\hat{U}_i$ and $\hat{U}_i^{\text{HB}}$, and the associated CVs, where $\hat{U}_i$ is the direct estimate of U_i. The HB estimate $\hat{U}_i^{\text{HB}}$ and the associated posterior variance are obtained from (10.3.22) and (10.3.23) by changing $Y_i^{(\ell k)}$ to $U_i^{(\ell k)} = M_i^{(\ell k)}/(M_i^{(\ell k)} + C_i)$.

It is clear from Table 10.2 that the HB estimates ($\hat{M}_i^{\text{HB}}, \hat{U}_i^{\text{HB}}$) perform better than the direct estimates ($\hat{M}_i$, $\hat{U}_i$) in terms of CV, except the estimate for New Brunswick. For Ontario, Quebec, and British Columbia, the CVs of HB and direct estimates are nearly equal due to larger sample sizes in these provinces.

Note that the sampling model (10.4.1) assumes that the direct estimate $\hat{Y}_i$ is design-unbiased. In the context of census undercoverage estimation, this assumption may be restrictive because the estimate may be subject to nonsampling bias (Zaslavsky 1993). This potential bias was ignored due to lack of a reliable estimate of bias.

Example 10.4.2. Adult Literacy Proportions in the United States. In Section 8.5 we studied a twofold subarea level linking model matching to the sampling model for the subareas, and developed EBLUP estimators of subarea and area means simultaneously. We now provide an application of the HB method to estimate the state and county proportions of adults in the United States at the lowest level of literacy, using unmatched sampling model and twofold subarea level linking model for the county proportions. Mohadjer et al. (2012) used data from the National Assessment of Adult Literacy (NAAL) to produce state and county HB estimates and associated credible intervals. In this application, subarea refers to county and area to state.

The sampling model is given by $\hat{\theta}_{ij} = \theta_{ij} + e_{ij}$, where θ_{ij} is the population proportion of adults at the lowest level of literacy in county j belonging to state i, and

$\hat{\theta}_{ij}$ is the direct estimator of θ_{ij}. The sampling errors e_{ij} are assumed to be $N(0, \psi_{ij})$ with known coefficient of variation $\phi_{ij} = \psi_{ij}^{1/2}/\theta_{ij}$. In the case of NAAL, "smoothed" estimates $\tilde{\phi}_{ij}$ of ϕ_{ij} were obtained from the direct estimates $\hat{\phi}_{ij}$ and then used as proxies for the true ϕ_{ij}. Smoothed estimates $\tilde{\phi}_{ij}$ are obtained in two steps. In step 1, predicted values $\tilde{\theta}_{ij}$ of θ_{ij} were obtained by fitting the model $\text{logit}(\hat{\theta}_{ij}) = \gamma_0 + \gamma_1 z_{ij1} + \cdots + \gamma_4 z_{ij4} + \varepsilon_{ij}$, where $\varepsilon_{ij} \sim N(0, \sigma_\varepsilon^2)$ and $z_{ij1}, \ldots, z_{ij4}$ are county-level covariates (see Mohadjer et al. 2012 for details). In the second step, predicted (smoothed) values $\tilde{\phi}_{ij}$ of ϕ_{ij} are obtained by fitting the model $\log(\hat{\phi}_{ij}^2) = \eta_0 + \eta_1 \log(\tilde{\theta}_{ij}) + \eta_2 \log(1 - \tilde{\theta}_{ij}) + \eta_3 \log(n_{ij}) + \tau_{ij}$, where $\tau_{ij} \sim N(0, \sigma_\tau^2)$ and n_{ij} is the sample sizes in jth county in state i. The smoothed values $\tilde{\phi}_{ij}$ are then treated as the true values ϕ_{ij}, leading to $\psi_{ij} = \tilde{\phi}_{ij}^2 \theta_{ij}^2$.

The unmatched linking model is given by $\text{logit}(\theta_{ij}) = \mathbf{z}_{ij}^T \boldsymbol{\beta} + v_i + u_{ij}$, where $v_i \overset{\text{iid}}{\sim} N(0, \sigma_v^2)$ is the ith state random effect, $u_{ij} \overset{\text{iid}}{\sim} N(0, \sigma_u^2)$ is the random effect associated with the jth county in the ith state, and $\mathbf{z}_{ij}$ is a set of predictor variables that includes $z_{ij1}, \ldots, z_{ij4}$ used in the model for smoothing the direct estimates $\hat{\theta}_{ij}$. County-level data from the 2000 Census of Population was the primary source for selecting the predictor variables. The census data contains variables related to adult literacy skills, such as country of birth, education, age, and disabilities. Importance of selecting a proper linking model is magnified for NAAL because the NAAL sample contains only about 10% of the U.S. counties.

HB estimation of the county proportions, θ_{ij}, was implemented by assuming a flat prior for $\boldsymbol{\beta}$ and gamma priors $G(a, b)$ for σ_v^{-2} and σ_u^{-2} with small values $a = 0.001$ and $b = 0.001$. Alternative noninformative priors were also examined and the resulting HB estimators were highly correlated with those obtained from the above choice, suggesting that the final estimates are not sensitive to the choice of priors. Let $\mathbf{v}$ and $\mathbf{u}$ denote the vectors of county and state effects v_i and u_{ij}, respectively. Using $L = 3$ independent runs, $B = 27{,}000$ MCMC samples $(\boldsymbol{\beta}^{(b)}, \mathbf{v}^{(b)}, \mathbf{u}^{(b)}, \sigma_v^{2(b)}, \sigma_e^{2(b)})$, $b = 1, \ldots, B$, were generated from the joint posterior distribution of $\boldsymbol{\beta}$, $\mathbf{v}$ and $\mathbf{u}$, σ_v^2, and σ_e^2. WinBUGS software (Lunn et al. 2000), based on M–H within Gibbs, was employed to generate the MCMC samples. Model-fitting methods, outlined in Section 10.2.6, were also implemented to assess the goodness of fit of the final model.

The HB estimates (posterior means) for sampled counties are calculated as

$$\hat{\theta}_{ij}^{\text{HB}} = \frac{1}{B} \sum_{b=1}^{B} \theta_{ij}^{(b)}$$

where $\theta_{ij}^{(b)}$ is obtained from $\log(\theta_{ij}^{(b)}) = \mathbf{z}_{ij}^T \boldsymbol{\beta}^{(b)} + v_i^{(b)} + u_{ij}^{(b)}$. However, for nonsampled counties within a sampled state, the $u_{ij}^{(b)}$-values are not available. For a nonsample state, neither $v_i^{(b)}$ nor $u_{ij}^{(b)}$ are available. In the former case, $\theta_{ij}^{(b)}$ is calculated from $\log(\theta_{ij}^{(b)}) = \mathbf{z}_{ij}^T \boldsymbol{\beta}^{(b)} + v_i^{(b)} + u_{ij(RD)}^{(b)}$, where $u_{ij(RD)}^{(b)}$ is a random draw from $N(0, \sigma_u^{2(b)})$. Similarly, in the case of nonsampled state, $\theta_{ij}^{(b)}$ is calculated from $\log(\theta_{ij}^{(b)}) = \mathbf{z}_{ij}^T \boldsymbol{\beta}^{(b)}$

$+ v_{i(RD)}^{(b)} + u_{ij(RD)}^{(b)}$, where $v_{i(RD)}^{(b)}$ is a random draw from $N(0, \sigma_v^{2(b)})$. A 95% posterior (credible) interval on θ_{ij} is calculated by determining the lower 2.5% and the upper 2.5% quantiles of the B values $\theta_{ij}^{(b)}$.

State proportions are given by $\theta_{i\cdot} = \sum_{j=1}^{N_i} W_{ij}\theta_{ij}$, where N_i is the number of counties in state i and W_{ij} is the proportion of state i adult population in county j. The corresponding HB estimate is given by $\hat{\theta}_{i\cdot}^{\text{HB}} = \sum_{j=1}^{N_i} W_{ij}\hat{\theta}_{ij}^{\text{HB}}$. Credible intervals for θ_i are obtained from the B posterior values $\theta_{i\cdot}^{(b)} = \sum_{j=1}^{N_i} W_{ij}\theta_{ij}^{(b)}$, $b = 1, \ldots, B$.

Example 10.4.3. Beta Sampling Model. Liu, Lahiri, and Kalton (2014) proposed a beta sampling model to estimate small area proportions, P_i, using design-weighted estimators, p_{iw}, based on stratified simple random sampling (SRS) within each area i. In particular, $p_{iw} = \sum_{h=1}^{H_i} W_{ih}p_{ih}$, where $p_{ih} = n_{ih}^{-1}\sum_{k=1}^{n_{ih}} y_{ihk}$ is the estimator of the proportion P_{ih} associated with a binary variable y_{ihk}, $W_{ih} = N_{ih}/N_i$ is the stratum weight, $\sum_{h=1}^{H_i} N_{ih} = N_i$, and n_{ih} is the sample size in stratum $h = 1, \ldots, H_i$ within area i. The sampling model is given by

$$p_{iw}|P_i \overset{\text{ind}}{\sim} \text{Beta}(a_i, b_i), \quad i = 1, \ldots, m, \tag{10.4.9}$$

where

$$a_i = P_i\left(\frac{n_i}{\text{deff}_{iw}} - 1\right), \quad b_i = (1 - P_i)\left(\frac{n_i}{\text{deff}_{iw}} - 1\right) \tag{10.4.10}$$

and deff_{iw} is an approximation to the design effect associated with p_{iw}. In particular, assuming negligible sampling fractions $f_{ih} = n_{ih}/N_{ih}$ and $P_{ih}(1 - P_{ih}) \approx P_i(1 - P_i)$, we have

$$\text{deff}_{iw} = n_i \sum_{h=1}^{H_i} W_{ih}^2/n_{ih}, \tag{10.4.11}$$

where $n_i = \sum_{h=1}^{H_i} n_{ih}$. The resulting smoothed sampling variance ψ_i is given by

$$\psi_i = [P_i(1 - P_i)/n_i]\text{deff}_{iw}, \tag{10.4.12}$$

which is a function of the unknown P_i only because deff_{iw} is known.

The linking model is taken as

$$\text{logit}(P_i)|\boldsymbol{\beta}, \sigma_v^2 \overset{\text{ind}}{\sim} N(\mathbf{z}_i^T\boldsymbol{\beta}, \sigma_v^2). \tag{10.4.13}$$

HB inference on the proportions P_i is implemented by assuming a flat prior on $\boldsymbol{\beta}$ and gamma prior on σ_v^{-2}. MCMC samples from the joint posterior of $(P_1, \ldots, P_m, \boldsymbol{\beta}, \sigma_v^2)$ are generated using the M–H within Gibbs algorithm; Gibbs conditional of P_i does not have a closed-form expression. Liu et al. (2014) provide a WinBUGS code to implement HB inference.

Liu et al. (2014) also report design-based simulation results on the performance of the beta-logistic (BL) model relative to a normal-logistic (NL) model that assumes $p_{iw}|P_i \overset{\text{ind}}{\sim} N(P_i, \psi_i)$, both using the sampling variances ψ_i given by (10.4.12). The finite population studied consisted of records of live births weights (y) in 2002 in the 50 states of the United States and the District of Columbia (DC). The parameters of interest, P_i, are the state-level low birth weight rates, and the values of P_i ranged from 5% to 11% across the states. Simulation results indicated that the equal tail-area credible interval associated with the NL model leads to undercoverage, especially for small n_i (< 30). The BL-based interval performed better in terms of coverage. In terms of bias of the HB estimator, $\hat{P}_i^{\text{HB}}$, the NL model leads to larger bias but smaller MSE.

10.5 BASIC UNIT LEVEL MODEL

In this section, we apply the HB approach to the basic unit level model 7.1.1 with equal error variances (that is, $k_{ij} = 1$ for all i and j), assuming a prior distribution on the model parameters $(\boldsymbol{\beta}, \sigma_v^2, \sigma_e^2)$. We first consider the case of known σ_v^2 and σ_e^2, and assume a "flat" prior on $\boldsymbol{\beta}$: $f(\boldsymbol{\beta}) \propto 1$. We rewrite 7.1.1 as an HB model:

$$\text{(i)} \quad y_{ij}|\boldsymbol{\beta}, v_i, \sigma_e^2 \overset{\text{ind}}{\sim} N(\mathbf{x}_{ij}^T\boldsymbol{\beta} + v_i, \sigma_e^2), \quad j = 1, \ldots, n_i, \quad i = 1, \ldots, m,$$

$$\text{(ii)} \quad v_i|\sigma_v^2 \overset{\text{iid}}{\sim} N(0, \sigma_v^2), \quad i = 1, \ldots, m,$$

$$\text{(iii)} \quad f(\boldsymbol{\beta}) \propto 1. \tag{10.5.1}$$

We then extend the results to the case of unknown σ_v^2 and σ_e^2 by replacing (iii) in (10.5.1) by

$$\text{(iii)}' \quad f(\boldsymbol{\beta}, \sigma_v^2, \sigma_e^2) = f(\boldsymbol{\beta}) f(\sigma_v^2) f(\sigma_e^2) \propto f(\sigma_v^2) f(\sigma_e^2), \tag{10.5.2}$$

where $f(\sigma_v^2)$ and $f(\sigma_e^2)$ are the priors on σ_v^2 and σ_e^2. For simplicity, we take $\mu_i = \overline{\mathbf{X}}_i^T\boldsymbol{\beta} + v_i$ as the ith small area mean, assuming that the area population size, N_i, is large.

10.5.1 Known σ_v^2 and σ_e^2

When σ_v^2 and σ_e^2 are assumed to be known, the HB and BLUP approaches under normality lead to identical point estimates and measures of variability, assuming a flat prior on $\boldsymbol{\beta}$. This result, in fact, is valid for general linear mixed models with known variance parameters. The HB estimator of μ_i is given by

$$\tilde{\mu}_i^{\text{HB}}(\sigma_v^2, \sigma_e^2) = E(\mu_i|\mathbf{y}, \sigma_v^2, \sigma_e^2) = \tilde{\mu}_i^H, \tag{10.5.3}$$

where $\mathbf{y}$ is the vector of sample observations and $\tilde{\mu}_i^H$ is the BLUP estimator given by (7.1.6). Similarly, the posterior variance of μ_i is

$$V(\mu_i|\sigma_v^2, \sigma_e^2, \mathbf{y}) = M_{1i}(\sigma_v^2, \sigma_e^2) = \text{MSE}(\tilde{\mu}_i^H), \tag{10.5.4}$$

where $M_{1i}(\sigma_v^2, \sigma_e^2)$ is given by (7.1.12).

10.5.2 Unknown σ_v^2 and σ_e^2: Numerical Integration

In practice, σ_v^2 and σ_e^2 are unknown and it is necessary to take account of the uncertainty about σ_v^2 and σ_e^2 by assuming a prior on σ_v^2 and σ_e^2. The HB model is given by (i) and (ii) of (10.5.1) and (iii)$'$ given by (10.5.2). We obtain the HB estimator of μ_i and the posterior variance of μ_i as

$$\hat{\mu}_i^{\text{HB}} = E(\mu_i|\mathbf{y}) = E_{\sigma_v^2,\sigma_e^2}[\tilde{\mu}_i^{\text{HB}}(\sigma_v^2, \sigma_e^2)] \tag{10.5.5}$$

and

$$V(\mu_i|\mathbf{y}) = E_{\sigma_v^2,\sigma_e^2}[M_{1i}(\sigma_v^2, \sigma_e^2)] + V_{\sigma_v^2,\sigma_e^2}[\tilde{\mu}_i^{\text{HB}}(\sigma_v^2, \sigma_e^2)], \tag{10.5.6}$$

where $E_{\sigma_v^2,\sigma_e^2}$ and $V_{\sigma_v^2,\sigma_e^2}$, respectively, denote the expectation and variance with respect to the posterior distribution $f(\sigma_v^2, \sigma_e^2|\mathbf{y})$.

As in Section 10.3, the posterior $f(\sigma_v^2, \sigma_e^2|\mathbf{y})$ may be obtained from the restricted likelihood function $L_R(\sigma_v^2, \sigma_e^2)$ as

$$f(\sigma_v^2, \sigma_e^2|\mathbf{y}) \propto L_R(\sigma_v^2, \sigma_e^2)f(\sigma_v^2)f(\sigma_e^2). \tag{10.5.7}$$

Under flat priors $f(\sigma_v^2) \propto 1$ and $f(\sigma_e^2) \propto 1$, the posterior $f(\sigma_v^2, \sigma_e^2|\mathbf{y})$ is proper (subject to a mild sample size restriction) and proportional to $L_R(\sigma_v^2, \sigma_e^2)$. Evaluation of the posterior mean (10.5.5) and the posterior variance (10.5.6), using $f(\sigma_v^2, \sigma_e^2|\mathbf{y}) \propto L_R(\sigma_v^2, \sigma_e^2)$, involves two-dimensional integrations.

If we assume a diffuse gamma prior on σ_e^{-2}, that is, $\sigma_e^{-2} \sim G(a_e, b_e)$ with $a_e \geq 0$ and $b_e > 0$, then it is possible to integrate out σ_e^2 with respect to $f(\sigma_e^2|\tau_v, \mathbf{y})$, where $\tau_v = \sigma_v^2/\sigma_e^2$. The evaluation of (10.5.5) and (10.5.6) is now reduced to single-dimensional integration with respect to the posterior of τ_v, that is, $f(\tau_v|\mathbf{y})$. Datta and Ghosh (1991) expressed $f(\tau_v|\mathbf{y})$ as $f(\tau_v|\mathbf{y}) \propto h(\tau_v)$ and obtained an explicit expression for $h(\tau_v)$, assuming a gamma prior on τ_v^{-1}, that is, $\tau_v^{-1} \sim G(a_v, b_v)$ with $a_v \geq 0$ and $b_v > 0$; note that a_v is the shape parameter and b_v is the scale parameter. Datta and Ghosh (1991) applied the numerical integration method to the data on county crop areas (Example 7.3.1) studied by Battese, Harter, and Fuller (1988). They calculated the HB estimate of mean hectares of soybeans and associated standard error (square root of posterior variance) for each of the $m = 12$ counties in north-central Iowa, using flat priors on $\boldsymbol{\beta}$ and gamma priors on σ_e^{-2} and τ_v^{-1} with $a_e = a_v = 0$ and $b_e = b_v = 0.005$ to reflect lack of prior information. Datta and Ghosh (1991) actually studied the more complex case of finite population means $\overline{Y}_i$ instead of the means μ_i, but the sampling fractions n_i/N_i are negligible in Example 7.3.1, so that $\overline{Y}_i \approx \mu_i$.

10.5.3 Unknown σ_v^2 and σ_e^2: Gibbs Sampling

In this section, we apply Gibbs sampling to the basic unit level model given by (i) and (ii) of (10.5.1), assuming the prior (10.5.2) on $\boldsymbol{\beta}, \sigma_v^2$, and σ_e^2 with $\sigma_v^{-2} \sim G(a_v, b_v)$, $a_v \geq 0, b_v > 0$ and $\sigma_e^{-2} \sim G(a_e, b_e)$, $a_e \geq 0, b_e > 0$. It is easy to verify that the Gibbs conditionals are given by

$$\text{(i)}\ [\beta|\mathbf{v}, \sigma_v^2, \sigma_e^2, \mathbf{y}] \sim N_p\left[\left(\sum_{i=1}^{m}\sum_{j=1}^{n_i}\mathbf{x}_{ij}\mathbf{x}_{ij}^T\right)^{-1}\sum_{i=1}^{m}\sum_{j=1}^{n_i}\mathbf{x}_{ij}(y_{ij}-v_i),\ \sigma_e^2\left(\sum_{i=1}^{m}\sum_{j=1}^{n_i}\mathbf{x}_{ij}\mathbf{x}_{ij}^T\right)^{-1}\right], \tag{10.5.8}$$

$$\text{(ii)}\ [v_i|\beta, \sigma_v^2, \sigma_e^2, \mathbf{y}] \sim N[\gamma_i(\bar{y}_i - \bar{\mathbf{x}}_i^T\beta), g_{1i}(\sigma_v^2, \sigma_e^2) = \gamma_i\sigma_e^2/n_i],\ \ i = 1, \ldots, m, \tag{10.5.9}$$

$$\text{(iii)}\ [\sigma_e^{-2}|\beta, \mathbf{v}, \sigma_v^2, \mathbf{y}] \sim G\left[\frac{n}{2} + a_e, \frac{1}{2}\sum_{i=1}^{m}\sum_{j=1}^{n_i}(y_{ij} - \mathbf{x}_{ij}^T\beta - v_i)^2 + b_e\right], \tag{10.5.10}$$

$$\text{(iv)}\ [\sigma_v^{-2}|\beta, \mathbf{v}, \sigma_e^2, \mathbf{y}] \sim G\left(\frac{m}{2} + a_v, \frac{1}{2}\sum_{i=1}^{m}v_i^2 + b_v\right), \tag{10.5.11}$$

where $n = \sum_{i=1}^{m} n_i$, $\mathbf{v} = (v_1, \ldots, v_m)^T$, $\bar{y}_i = \sum_{j=1}^{n_i} y_{ij}/n_i$, $\bar{\mathbf{x}}_i = \sum_{j=1}^{n_i}\mathbf{x}_{ij}/n_i$, and $\gamma_i = \sigma_v^2/(\sigma_v^2 + \sigma_e^2/n_i)$. The proof of (10.5.8)–(10.5.11) follows along the lines of the proof of (10.3.13)–(10.3.15) given in Section 10.18.3. Note that all the Gibbs conditionals have closed forms and hence the MCMC samples can be generated directly from (i)–(iv). WinBUGS can be used to generate samples from the above conditionals using either default inverted gamma priors on σ_v^2 and σ_e^2 (with $a_e = b_e = 0.001$ and $a_v = b_v = 0.001$) or with specified values of a_e, b_e, a_v, and b_v. CODA can be used to perform convergence diagnostics.

Denote the MCMC samples from a single large run by $\{\boldsymbol{\beta}^{(k)}, \mathbf{v}^{(k)}, \sigma_v^{2(k)}, \sigma_e^{2(k)}, k = d+1, \ldots, d+D\}$. The marginal MCMC samples $\{\boldsymbol{\beta}^{(k)}, \mathbf{v}^{(k)}\}$ can be used directly to estimate the posterior mean of μ_i as

$$\hat{\mu}_i^{\text{HB}} = \frac{1}{D}\sum_{k=d+1}^{d+D}\mu_i^{(k)} =: \mu_i^{(\cdot)}, \tag{10.5.12}$$

where $\mu_i^{(k)} = \overline{\mathbf{X}}_i^T\beta^{(k)} + v_i^{(k)}$. Similarly, the posterior variance of μ_i is estimated as

$$V(\mu_i|\mathbf{y}) = \frac{1}{D-1}\sum_{k=d+1}^{d+D}(\mu_i^{(k)} - \mu_i^{(\cdot)})^2. \tag{10.5.13}$$

Alternatively, Rao–Blackwell estimators of the posterior mean and the posterior variance of μ_i may be used along the lines of (10.3.16) and (10.3.17):

$$\hat{\mu}_i^{\text{HB}} = \frac{1}{D} \sum_{k=d+1}^{d+D} \tilde{\mu}_i^{\text{HB}} \left(\sigma_v^{2(k)}, \sigma_e^{2(k)}\right) =: \tilde{\mu}_i^{\text{HB}}(\cdot, \cdot) \tag{10.5.14}$$

and

$$V(\mu_i|\mathbf{y}) = \frac{1}{D} \sum_{k=d+1}^{d+D} \left[g_{1i} \left(\sigma_v^{2(k)}, \sigma_e^{2(k)}\right) + g_{2i} \left(\sigma_v^{2(k)}, \sigma_e^{2(k)}\right) \right] + \frac{1}{D-1} \sum_{k=d+1}^{d+D} \left[\tilde{\mu}_i^{\text{HB}} \left(\sigma_v^{2(k)}, \sigma_e^{2(k)}\right) - \tilde{\mu}_i^{\text{HB}}(\cdot, \cdot) \right]^2. \tag{10.5.15}$$

Example 10.5.1. County Crop Areas. The Gibbs sampling method was applied to the data on area under corn for each of $m = 12$ counties in north-central Iowa, excluding the second sample segment in Hardin county (see Example 7.3.1). We fitted the nested error linear regression model, $y_{ij} = \beta_0 + \beta_1 x_{1ij} + \beta_2 x_{2ij} + v_i + e_{ij}$, for corn ($y_{ij}$) using the LANDSAT satellite readings of corn (x_{1ij}) and soybeans (x_{2ij}); see Example 7.3.1 for details. We used the BUGS program with default flat prior on $\boldsymbol{\beta}$ and gamma priors on σ_v^{-2} and σ_e^{-2} to generate samples from the joint posterior distribution of $(\boldsymbol{\beta}, \mathbf{v}, \sigma_v^2, \sigma_e^2)$. We generated a single long run of length $D = 5,000$ after discarding the first $d = 5,000$ "burn-in" iterations. We used CODA to implement convergence diagnostics and statistical and output analyses of the simulated samples.

To check the overall fit of the model, the posterior predictive value, pp, was estimated, using the formula (10.2.19) with measure of discrepancy given by $T(\mathbf{y}, \boldsymbol{\eta}) = \sum_{i=1}^{m} \sum_{j=1}^{n_i} (y_{ij} - \beta_0 - \beta_1 x_{1ij} - \beta_2 x_{2ij})^2 / (\sigma_v^2 + \sigma_e^2)$. The estimated p-value, $p = 0.51$, indicates a good fit of the model. We also fitted the simpler model $y_{ij} = \beta_0 + \beta_1 x_{1ij} + v_i + e_{ij}$, using only the corn ($x_{1ij}$) satellite reading as an auxiliary variable. The resulting posterior predictive p-value, $\widehat{pp} = 0.497$, indicates that the simpler model also fits the data well. We also compared the CPO plots for the two models using the formula (10.2.26). The CPO plots indicate that the full model with x_1 and x_2 is slightly better than the simpler model with x_1 only.

Table 10.3 gives the EBLUP estimates and associated standard errors (taken from Battese, Harter, and Fuller 1988), and the HB estimates and associated standard errors. It is clear from Table 10.3 that the EBLUP and HB estimates are similar. Standard errors are also similar except for Kossuth county (with $n_i = 5$) where the HB standard error (square root of posterior variance) is about 20% larger than the corresponding EBLUP standard error; in fact, the HB standard error is slightly larger than the corresponding sample regression (SR) standard error reported in Table 10.3. Reasons for this exception are not clear.

10.5.4 Pseudo-HB Estimation

In Section 7.6.2, we obtained a pseudo-EBLUP estimator, $\hat{\mu}_{iw}^H$, of the small area mean μ_i that makes use of survey weights w_{ij}, unlike the EBLUP estimator $\hat{\mu}_i^H$.

TABLE 10.3 EBLUP and HB Estimates and Associated Standard Errors: County Corn Areas

		Estimate		Standard Error	
County	n_i	EBLUP	HB	EBLUP	HB
Cerro Gordo	1	122.2	122.2	9.6	8.9
Hamilton	1	126.3	126.1	9.5	8.7
Worth	1	106.2	108.1	9.3	9.8
Humboldt	2	108.0	109.5	8.1	8.1
Franklin	3	145.0	142.8	6.5	7.3
Pocahontas	3	112.6	111.2	6.6	6.5
Winnebago	3	112.4	113.8	6.6	6.5
Wright	3	122.1	122.0	6.7	6.2
Webster	4	115.8	114.5	5.8	6.1
Hancock	5	124.3	124.8	5.3	5.2
Kossuth	5	106.3	108.4	5.2	6.3
Hardin	5	143.6	142.2	5.7	6.0

Source: Adapted from You and Rao (2002a) and You and Rao (2003).

Also, $\hat{\mu}_{iw}^H$ satisfies the benchmarking property without any adjustment, unlike $\hat{\mu}_i^H$. In this section, we obtain a pseudo-HB estimator, $\hat{\mu}_{iw}^{\text{HB}}$, that is analogous to the pseudo-EBLUP estimator $\hat{\mu}_{iw}^H$. The estimator $\hat{\mu}_{iw}^{\text{HB}}$ also makes use of the survey weights and satisfies the benchmarking property without any adjustment.

We make use of the unit level MCMC samples $\{\sigma_v^{2(k)}, \sigma_e^{2(k)}\}$ generated from (i) and (ii) of (10.5.1) and the prior (10.5.2) with $\sigma_v^{-2} \sim G(a_v, b_v)$ and $\sigma_e^{-2} \sim G(a_e, b_e)$, but replace $\boldsymbol{\beta}^{(k)}$ by $\boldsymbol{\beta}_w^{(k)}$ that makes use of the design-weighted estimator $\tilde{\boldsymbol{\beta}}_w(\sigma_v^2, \sigma_e^2) = \tilde{\boldsymbol{\beta}}_w$ given by (7.6.7). Now, assuming a flat prior $f(\boldsymbol{\beta}) \propto 1$ and noting that $\tilde{\boldsymbol{\beta}}_w | \boldsymbol{\beta}, \sigma_v^2, \sigma_e^2 \sim N_p(\boldsymbol{\beta}, \boldsymbol{\Phi}_w)$, we get the posterior distribution $[\boldsymbol{\beta} | \tilde{\boldsymbol{\beta}}_w, \sigma_v^2, \sigma_e^2] \sim N(\tilde{\boldsymbol{\beta}}_w, \boldsymbol{\Phi}_w)$, where $\boldsymbol{\Phi}_w = \boldsymbol{\Phi}_w(\sigma_v^2, \sigma_e^2)$ is given by (7.6.14). We calculate $\tilde{\boldsymbol{\beta}}_w^{(k)} = \tilde{\boldsymbol{\beta}}_w(\sigma_v^{2(k)}, \sigma_e^{2(k)})$ and $\boldsymbol{\Phi}_w(\sigma_v^{2(k)}, \sigma_e^{2(k)})$ using $\sigma_v^{2(k)}$ and $\sigma_e^{2(k)}$ and then generate $\boldsymbol{\beta}_w^{(k)}$ from $N_p(\tilde{\boldsymbol{\beta}}_w^{(k)}, \boldsymbol{\Phi}_w^{(k)})$.

Under the survey-weighted model 7.6.3, the conditional posterior mean $E(\mu_i | \overline{y}_{iw}, \boldsymbol{\beta}, \sigma_v^2, \sigma_e^2)$ is identical to the BLUP estimator $\tilde{\mu}_{iw}^H = \tilde{\mu}_{iw}^H(\boldsymbol{\beta}, \sigma_v^2, \sigma_e^2)$ given by (7.6.2), where $\overline{y}_{iw} = \sum_{j=1}^{n_i} w_{ij} y_{ij} / \sum_{j=1}^{n_i} w_{ij} = \sum_{j=1}^{n_i} \tilde{w}_{ij} y_{ij}$. Similarly, the conditional posterior variance $V(\mu_i | \overline{y}_{iw}, \boldsymbol{\beta}, \sigma_v^2, \sigma_e^2)$ is equal to $g_{1iw}(\sigma_v^2, \sigma_e^2)$ given by (7.6.11). Now using the generated MCMC samples $\{\boldsymbol{\beta}^{(k)}, \sigma_v^{2(k)}, \sigma_e^{2(k)}; k = d+1, \ldots, d+D\}$, we get a pseudo-HB estimator of μ_i as

$$\hat{\mu}_{iw}^{\text{HB}} = \frac{1}{D} \sum_{k=d+1}^{d+D} \tilde{\mu}_{iw} \left(\boldsymbol{\beta}^{(k)}, \sigma_v^{2(k)}, \sigma_e^{2(k)} \right)$$

$$= \overline{\mathbf{X}}_i^T \boldsymbol{\beta}_w^{(\cdot)} + \tilde{v}_{iw}^{(\cdot)}, \tag{10.5.16}$$

where $\boldsymbol{\beta}_w^{(\cdot)} = \sum_{k=d+1}^{d+D} \boldsymbol{\beta}_w^{(k)}/D$, $\tilde{v}_{iw}^{(\cdot)}$ is the average of $\tilde{v}_{iw}(\boldsymbol{\beta}^{(k)}, \sigma_v^{2(k)}, \sigma_e^{2(k)})$ over k, and $\tilde{v}_{iw}(\boldsymbol{\beta}, \sigma_v^2, \sigma_e^2)$ is given by (7.6.5). The pseudo-HB estimator (10.5.16) is design-consistent as n_i increases.

It is easy to verify that $\sum_{i=1}^m N_i \hat{\mu}_{iw}^{\text{HB}}$ benchmarks to the direct survey regression estimator $\hat{Y}_w + (\mathbf{X} - \hat{\mathbf{X}}_w)^T \boldsymbol{\beta}_w^{(\cdot)}$, where $\hat{Y}_w = \sum_{i=1}^m \sum_{j=1}^{n_i} \tilde{w}_{ij} y_{ij}$ and $\hat{\mathbf{X}}_w = \sum_{i=1}^m \sum_{j=1}^{n_i} \tilde{w}_{ij} x_{ij}$ are the direct estimators of the overall totals $Y = \sum_{i=1}^m \sum_{j=1}^{N_i} y_{ij}$ and $\mathbf{X}_+ = \sum_{i=1}^m \sum_{j=1}^{N_i} \mathbf{x}_{ij}$, respectively. Note that the direct survey regression estimator here differs from the estimator in Section 7.6.2. The latter uses $\hat{\boldsymbol{\beta}}_w = \tilde{\boldsymbol{\beta}}_w(\hat{\sigma}_v^2, \hat{\sigma}_e^2)$, but the difference between $\hat{\boldsymbol{\beta}}_w$ and $\boldsymbol{\beta}_w^{(\cdot)}$ should be very small.

A pseudo-posterior variance of μ_i is obtained as

$$\hat{V}^{\text{PHB}}(\mu_i) = \frac{1}{D} \sum_{k=d+1}^{d+D} g_{1iw}\left(\sigma_v^{2(k)}, \sigma_e^{2(k)}\right) + \frac{1}{D-1} \sum_{k=d+1}^{d+D} \left[\tilde{\mu}_{iw}^H\left(\boldsymbol{\beta}^{(k)}, \sigma_v^{2(k)}, \sigma_e^{2(k)}\right) - \hat{\mu}_{iw}^{\text{HB}}\right]^2. \quad (10.5.17)$$

The last term in (10.5.17) accounts for the uncertainty associated with $\boldsymbol{\beta}$, σ_v^2, and σ_e^2. We refer the reader to You and Rao (2003) for details of the pseudo-HB method.

Example 10.5.2. County Corn Areas. In Example 7.3.1, we applied the pseudo-EBLUP method to county corn area data from Battese et al. (1988), assuming simple random sampling within areas, that is, $w_{ij} = w_i = N_i/n_i$. We generated MCMC samples $\{\boldsymbol{\beta}_w^{(k)}, \sigma_v^{2(k)}, \sigma_e^{2(k)}\}$ from this data set, using diffuse priors on $\boldsymbol{\beta}$, σ_v^2, and σ_e^2. Table 10.4 compares the results from the pseudo-HB method to those

TABLE 10.4 Pseudo-HB and Pseudo-EBLUP Estimates and Associated Standard Errors: County Corn Areas

		Estimate		Standard Error	
County	n_i	Pseudo-HB	Pseudo-EBLUP	Pseudo-HB	Pseudo-EBLUP
Cerro Gordo	1	120.6	120.5	9.3	9.9
Hamilton	1	125.2	125.2	9.4	9.7
Worth	1	107.5	106.4	10.2	9.6
Humboldt	2	108.4	107.4	8.2	8.3
Franklin	3	142.5	143.7	7.4	6.6
Pocahontas	3	110.6	111.5	7.0	6.6
Winnebago	3	113.2	112.1	7.0	6.6
Wright	3	121.1	121.3	6.4	6.8
Webster	4	114.2	115.0	6.4	5.8
Hancock	5	124.8	124.5	5.4	5.4
Kossuth	5	108.0	106.6	6.6	5.3
Hardin	5	142.3	143.5	6.1	5.8

Source: Adapted from Table 1 in You and Rao (2003).

from the pseudo-EBLUP method (Table 10.3). It is clear from Table 10.4 that the pseudo-HB and the pseudo-EBLUP estimates are very similar. Standard errors are also similar except for Kossuth county (with $n_i = 5$), where the pseudo-HB standard error is significantly larger than the corresponding pseudo-EBLUP standard error, similar to the EBLUP and HB standard errors reported in Table 10.3.

We assumed that the basic unit level model also holds for the sample $\{(y_{ij}, \mathbf{x}_{ij}); j \in s_i, i = 1, \ldots, m\}$ in developing the pseudo-HB estimator (10.5.16) and the pseudo-posterior variance (10.5.17). If the model holds for the sample, then the HB estimator (without weights) is optimal in the sense of providing the smallest posterior variance, but it is not design-consistent. As noted in Section 7.6.2, survey practitioners prefer design-consistent estimators as a form of insurance, and the sample size, n_i, could be moderately large for some of the areas under consideration, in which case design-consistency becomes relevant.

10.6 GENERAL ANOVA MODEL

In this section, we apply the HB approach to the general ANOVA model 5.2.15, assuming a prior distribution on the model parameters $(\boldsymbol{\beta}, \boldsymbol{\delta})$, where $\boldsymbol{\delta} = (\sigma_0^2, \ldots, \sigma_r^2)^T$ is the vector of variance components. In particular, we assume that $f(\boldsymbol{\beta}, \boldsymbol{\delta}) = f(\boldsymbol{\beta}) \prod_{i=0}^r f(\sigma_i^2)$ with $f(\boldsymbol{\beta}) \propto 1$, $f(\sigma_i^2) \propto (\sigma_i^2)^{-(a_i+1)}, i = 1, \ldots, r$ and $f(\sigma_0^2) \propto (\sigma_0^2)^{-(b+1)}$ for specified values a_i and b. Letting $\sigma_0^2 = \sigma_e^2$, the HB model may be written as

$$\text{(i)}\quad \mathbf{y}|\mathbf{v}, \sigma_e^2, \boldsymbol{\beta}, \sim N\left(\mathbf{X}\boldsymbol{\beta} + \sum_{i=1}^r \mathbf{Z}_i \mathbf{v}_i, \sigma_e^2 \mathbf{I}\right),$$

$$\text{(ii)}\quad \mathbf{v}_i|\sigma_1^2, \ldots, \sigma_r^2 \overset{\text{ind}}{\sim} N_{h_i}(\mathbf{0}, \sigma_i^2 \mathbf{I}_{h_i}), \quad i = 1, \ldots, r,$$

$$\text{(iii)}\quad f(\boldsymbol{\beta}) \propto 1, \quad f(\sigma_i^2) \propto (\sigma_i^2)^{-(a_i+1)}, \quad f(\sigma_e^2) \propto (\sigma_e^2)^{-(b+1)}. \tag{10.6.1}$$

Under model (10.6.1), Hobert and Casella (1996) derived Gibbs conditionals and showed that the Gibbs conditionals are all proper if $2a_i > -h_i$ for all i and $2b > -n$. These conditions are satisfied for diffuse priors with small values of a_i and b. However, propriety of the conditionals does not imply propriety of the joint posterior $f(\boldsymbol{\beta}, \boldsymbol{\delta}, \mathbf{v}|\mathbf{y})$. In fact, many values of the vector $(a_1, \ldots, a_r, b)$ simultaneously yield proper conditionals and an improper joint posterior. It is therefore important to verify that the chosen improper joint prior yields a proper joint posterior before proceeding with Gibbs sampling. Hobert and Casella (1996) derived conditions on the constants $(a_1, \ldots, a_r, b)$ that ensure propriety of the joint posterior. Theorem 10.6.1 gives these conditions.

Theorem 10.6.1. Let $t = \text{rank}(\mathbf{P_X Z}) = \text{rank}(\mathbf{Z}^T \mathbf{P_X Z}) \leq h$, where $h = \sum_{i=1}^m h_i$, $\mathbf{Z} = (\mathbf{Z}_1, \ldots, \mathbf{Z}_r)$ and $\mathbf{P_X} = \mathbf{I}_n - \mathbf{X}(\mathbf{X}^T\mathbf{X})^{-1}\mathbf{X}^T$. Consider the following two cases: (1) $t = h$ or $r = 1$; (2) $t < h$ and $r > 1$. For case 1, the following conditions are necessary and sufficient for the propriety of the joint posterior: (a) $a_i < 0$, (b) $h_i > h - t - 2a_i$, (c) $n + 2\sum_{i=1}^{m} a_i + 2b - p > 0$. For case 2, conditions (a)–(c) are sufficient for the property of the joint posterior while necessary conditions result when (b) is replaced by (b′) $h_i > -2a_i$.

The proof of Theorem 10.6.1 is very technical and we refer the reader to Hobert and Casella (1996) for details of the proof. If we choose $a_i = -1$ for all i and $b = -1$, we get uniform (flat) priors on the variance components, that is, $f(\sigma_i^2) \propto 1$ and $f(\sigma_e^2) \propto 1$. This choice typically yields a proper joint posterior in practical applications, but not always. For example, for the balanced onefold random effects model, $y_{ij} = \mu + v_i + e_{ij}, j = 1, \ldots, \overline{n}, i = 1, \ldots, m$, a flat prior on $\sigma_v^2 (= \sigma_1^2)$ violates condition (b) if $m = 3$, noting that $r = 1$, $h_i = h = 3$, $t = \text{rank}(\mathbf{Z}^T\mathbf{P}_\mathbf{X}\mathbf{Z}) = \text{rank}(\mathbf{I}_3 - \mathbf{1}_3\mathbf{1}_3^T/3) = 2$, and $a_i = -1$. As a result, the joint posterior is improper.

PROC MIXED in SAS, Version 8.0, implements the HB method for the general ANOVA model. It uses flat priors on the variance components $\sigma_e^2, \sigma_1^2, \ldots, \sigma_r^2$, and the regression parameters $\boldsymbol{\beta}$. PRIOR option in SAS generates MCMC samples from the joint posterior of variance components and regression parameters, while RANDOM option generates samples from the marginal posterior of random effects $\mathbf{v}_1, \ldots, \mathbf{v}_r$.

The ANOVA model (10.6.1) covers the onefold and twofold nested error regression models as special cases, but not the two-level model (8.8.1), which is a special case of the general linear mixed model with a block-diagonal covariance structure (see 5.3.1). In Section 10.8, we apply the HB method to the two-level model (8.8.1).

10.7 *HB ESTIMATION OF GENERAL FINITE POPULATION PARAMETERS

The HB approach is a good alternative to the EB method described in Section 9.4 for the estimation of general finite population quantities because it avoids the use of the bootstrap for MSE estimation. Note that the EB estimator of Section 9.4 is approximated empirically by Monte Carlo simulation. This is done by generating many nonsample vectors, attaching to each of them the sample data to get the full census vectors, and then calculating the simulated census parameter in each Monte Carlo replicate. Many censuses have to be generated in that way to get an accurate Monte Carlo approximation. Moreover, in the parametric bootstrap described in Section 9.4.4, each bootstrap replicate consists of generating a census from the fitted model. Then, for each bootstrap replicate, the sample part of the census is extracted and, from that sample, the Monte Carlo approximation of the EB estimator is obtained, which implies generating again many Monte Carlo censuses from each bootstrap sample as described earlier. This approach might be computationally unfeasible for very large populations or very complex parameters such as some poverty indicators that require sorting all census values. On the other hand, HB approaches provide approximations to the whole posterior distribution of the parameter of interest, and any summary of the posterior distribution can be directly obtained. In particular, the posterior variance is used as measure of uncertainty of the HB estimator, and credible intervals can also be obtained without practically any additional effort.

10.7.1 HB Estimator under a Finite Population

Consider the general finite population parameter $\tau = h(\mathbf{y}^P)$, where $\mathbf{y}^P = (\mathbf{y}_s^T, \mathbf{y}_r^T)^T$ is the (random) vector with the values of the study variable in the sampled and non-sampled units of the population, and $h(\cdot)$ is a given measurable function. HB models typically assume a density $f(\mathbf{y}^P|\boldsymbol{\lambda})$ for the population vector $\mathbf{y}^P$ given the vector of parameters $\boldsymbol{\lambda}$, and then a prior distribution $f(\boldsymbol{\lambda})$ for $\boldsymbol{\lambda}$. The posterior distribution is then given by

$$f(\boldsymbol{\lambda}|\mathbf{y}_s) = \frac{f(\mathbf{y}_s|\boldsymbol{\lambda})f(\boldsymbol{\lambda})}{\int f(\mathbf{y}_s|\boldsymbol{\lambda})f(\boldsymbol{\lambda})d\boldsymbol{\lambda}}, \tag{10.7.1}$$

The HB estimator of $\tau = h(\mathbf{y}_s, \mathbf{y}_r)$, under squared loss, is the posterior mean given by

$$\hat{\tau}^{\text{HB}} = E(\tau|\mathbf{y}_s) = \int h(\mathbf{y}_s, \mathbf{y}_r)f(\mathbf{y}_r|\mathbf{y}_s)d\mathbf{y}_r, \tag{10.7.2}$$

where the predictive density is given by

$$f(\mathbf{y}_r|\mathbf{y}_s) = \int f(\mathbf{y}_r|\boldsymbol{\lambda})f(\boldsymbol{\lambda}|\mathbf{y}_s)d\boldsymbol{\lambda}. \tag{10.7.3}$$

Monte Carlo methods based on the generation of random values of $\boldsymbol{\lambda}$ from the posterior distribution (10.7.1), and of random values of $\mathbf{y}_r$ from the predictive distribution (10.7.3), can be used to approximate the HB estimator (10.7.2).

10.7.2 Reparameterized Basic Unit Level Model

To avoid the use of MCMC methods, Molina, Nandram, and Rao (2014) considered a reparameterization of the basic unit level model (4.3.1), by using the intraclass correlation coefficient $\rho = \sigma_v^2/(\sigma_v^2 + \sigma_e^2)$. They considered the following HB population model:

$$\begin{aligned}
&y_{ij}|v_i, \boldsymbol{\beta}, \sigma_e^2 \overset{\text{ind}}{\sim} N(\mathbf{x}_{ij}^T\boldsymbol{\beta}, \sigma_e^2 k_{ij}^2), \quad j = 1, \ldots, N_i, \quad i = 1, \ldots, M,\\
&v_i|\rho, \sigma_e^2 \overset{\text{ind}}{\sim} N\left(0, \frac{\rho}{1-\rho}\sigma_e^2\right), \quad i = 1, \ldots, M,\\
&f(\boldsymbol{\beta}, \rho, \sigma_e^2) \propto \sigma_e^{-2}, \quad \epsilon \le \rho \le 1 - \epsilon,
\end{aligned} \tag{10.7.4}$$

where $\epsilon > 0$ is a small number.

A sample s_i of size $n_i < N_i$ is drawn from each area i, for $i = 1, \ldots, m$, where the last $M - m$ areas in the population are not sampled. Let $\mathbf{y}_s = (\mathbf{y}_{1s}^T, \ldots, \mathbf{y}_{ms}^T)^T$ be the vector containing the sample data from the m sampled areas. Then, assuming that the population model (10.7.4) holds also for the sample, the posterior distribution of the vector of parameters $\boldsymbol{\lambda} = (\mathbf{v}^T, \boldsymbol{\beta}^T, \sigma_e^2, \rho)^T$, where $\mathbf{v} = (v_1, \ldots, v_m)^T$ is the vector of area effects, is given by

$$f(\boldsymbol{\lambda}|\mathbf{y}_s) = f_1\left(\mathbf{v}|\boldsymbol{\beta}, \sigma_e^2, \rho, \mathbf{y}_s\right) f_2\left(\boldsymbol{\beta}|\sigma_e^2, \rho, \mathbf{y}_s\right) f_3\left(\sigma_e^2|\rho, \mathbf{y}_s\right) f_4\left(\rho|\mathbf{y}_s\right) \quad (10.7.5)$$

The conditional densities f_1, f_2, and f_3 appearing in (10.7.5) all have simple forms, given by

$$v_i|\boldsymbol{\beta}, \sigma_e^2, \rho, \mathbf{y}_s \overset{\text{ind}}{\sim} N\left\{\lambda_i(\rho)(\bar{y}_i - \bar{\mathbf{x}}_i{}'\boldsymbol{\beta}), [1 - \lambda_i(\rho)]\frac{\rho}{1-\rho}\sigma_e^2\right\}, \quad (10.7.6)$$

$$\boldsymbol{\beta}|\sigma_e^2, \rho, \mathbf{y}_s \sim N[\hat{\boldsymbol{\beta}}(\rho), \sigma_e^2\mathbf{Q}^{-1}(\rho)], \quad (10.7.7)$$

$$\sigma_e^{-2}|\rho, \mathbf{y}_s \sim \text{Gamma}\left[\frac{n-p}{2}, \frac{\gamma(\rho)}{2}\right], \quad (10.7.8)$$

where $\lambda_i(\rho) = a_{i\cdot}[a_{i\cdot} + (1-\rho)/\rho]^{-1}$, $i = 1, \ldots, m$, and $\hat{\boldsymbol{\beta}}(\rho) = \mathbf{Q}^{-1}(\rho)\mathbf{p}(\rho)$, with

$$\mathbf{Q}(\rho) = \sum_{i=1}^{m}\sum_{j\in s_i} k_{ij}^{-1/2}(\mathbf{x}_{ij} - \bar{\mathbf{x}}_i)(\mathbf{x}_{ij} - \bar{\mathbf{x}}_i)' + \frac{1-\rho}{\rho}\sum_{i=1}^{m}\lambda_i(\rho)\,\bar{\mathbf{x}}_i\bar{\mathbf{x}}_i{}',$$

$$\mathbf{p}(\rho) = \sum_{i=1}^{m}\sum_{j\in s_i} k_{ij}^{-1/2}(\mathbf{x}_{ij} - \bar{\mathbf{x}}_i)(y_{ij} - \bar{y}_i) + \frac{1-\rho}{\rho}\sum_{i=1}^{m}\lambda_i(\rho)\bar{\mathbf{x}}_i\bar{y}_i$$

and

$$\gamma(\rho) = \sum_{i=1}^{m}\sum_{j\in s_i} k_{ij}^{-1/2}\left[y_{ij} - \bar{y}_i - (\mathbf{x}_{ij} - \bar{\mathbf{x}}_i)'\hat{\boldsymbol{\beta}}(\rho)\right]^2 + \frac{1-\rho}{\rho}\sum_{i=1}^{m}\lambda_i(\rho)\left[\bar{y}_i - \bar{\mathbf{x}}_i{}'\hat{\boldsymbol{\beta}}(\rho)\right]^2.$$

Thus, random values of v_i, $\boldsymbol{\beta}$, and σ_e^2 can be easily drawn from (10.7.6), (10.7.7), and (10.7.8). Unfortunately, f_4 does not have a simple form:

$$f_4(\rho|\mathbf{y}_s) \propto \left(\frac{1-\rho}{\rho}\right)^{m/2} |\mathbf{Q}(\rho)|^{-1/2}\gamma(\rho)^{-(n-p)/2}\prod_{i=1}^{m}\lambda_i^{1/2}(\rho), \quad \epsilon \le \rho \le 1-\epsilon. \quad (10.7.9)$$

However, since ρ is in the bounded interval $[\epsilon, 1-\epsilon]$, random values can be generated from (10.7.9) using a grid method. Thus, the considered reparameterization of the basic unit level model in terms of ρ, together with the considered priors, allows us to draw random values of $\boldsymbol{\lambda}$ from the posterior (10.7.5) avoiding the use of MCMC methods in this particular model.

10.7.3 HB Estimator of a General Area Parameter

In this section, we confine to the estimation of a general area parameter $\tau_i = h(\mathbf{y}_i^P)$, where $\mathbf{y}_i^P = (\mathbf{y}_{is}^T, \mathbf{y}_{ir}^T)^T$ is the vector with the values of the response variable for the sampled and nonsampled units from area i. By the HB model (10.7.4), given the vector of parameters $\boldsymbol{\lambda}$ that includes area effects, responses $\{y_{ij}, i \in r_i\}$ for nonsampled units are independent of sample responses $\mathbf{y}_s$, with

$$y_{ij}|\boldsymbol{\lambda} \stackrel{\text{ind}}{\sim} N(\mathbf{x}_{ij}^T\boldsymbol{\beta} + v_i, \sigma_e^2 k_{ij}^2), \quad j \in r_i, \quad i = 1, \ldots, M. \tag{10.7.10}$$

Then, the posterior predictive density of $\mathbf{y}_{ir}$ is given by

$$f(\mathbf{y}_{ir}|\mathbf{y}_s) = \int \prod_{i \in r_i} f(y_{ij}|\boldsymbol{\lambda}) f(\boldsymbol{\lambda}|\mathbf{y}_s) d\boldsymbol{\lambda}.$$

The HB estimator of the target parameter $\tau_i = h(\mathbf{y}_i^P)$ is then given by the posterior mean

$$\hat{\tau}_i^{\text{HB}} = E(\tau_i|\mathbf{y}_s) = \int h(\mathbf{y}_{is}, \mathbf{y}_{ir}) f(\mathbf{y}_{ir}|\mathbf{y}_s) d\mathbf{y}_{ir}. \tag{10.7.11}$$

A Monte Carlo approximation of (10.7.11) is obtained by first generating samples from the posterior $f(\boldsymbol{\lambda}|\mathbf{y}_s)$. For this, we first draw ρ from $f_4(\rho|\mathbf{y}_s)$, then σ_e^2 from $f_3(\sigma_e^2|\rho, \mathbf{y}_s)$, then $\boldsymbol{\beta}$ from $f_2(\boldsymbol{\beta}|\sigma_e^2, \rho, \mathbf{y}_s)$, and finally $\mathbf{v}$ from $f_1(\mathbf{v}|\boldsymbol{\beta}, \sigma_e^2, \rho, \mathbf{y}_s)$. We can repeat this procedure a large number, A, of times to get a random sample $\boldsymbol{\lambda}^{(a)}$, $a = 1, \ldots, A$, from $f(\boldsymbol{\lambda}|\mathbf{y}_s)$. Then, for each generated $\boldsymbol{\lambda}^{(a)}$, $a = 1, \ldots, A$, from $f(\boldsymbol{\lambda}|\mathbf{y}_s)$, we draw nonsample values $y_{ij}^{(a)}$, $j \in r_i$, $i = 1, \ldots, m$, from (10.7.10). Thus, for each sampled area $i = 1, \ldots, m$, we have generated a nonsample vector $\mathbf{y}_{ir}^{(a)} = \{y_{ij}^{(a)}, \quad i \in r_i\}$ and we have also the sample data $\mathbf{y}_{is}$ available. Attaching both vectors, we construct the full census vector $\mathbf{y}_i^{P(a)} = (\mathbf{y}_{is}^T, (\mathbf{y}_{ir}^{(a)})^T)^T$. For areas i with zero sample size, $i = m + 1, \ldots, M$, the whole vector $\mathbf{y}_i^{P(a)} = \mathbf{y}_{ir}^{(a)}$ is generated from (10.7.10), since in that case $r_i = P_i$. Using $\mathbf{y}_i^{P(a)}$, we compute the area parameter $\tau_i^{(a)} = h(\mathbf{y}_i^{P(a)})$, $i = 1, \ldots, M$. In this way, we obtain a random sample $\tau_i^{(a)}$, $a = 1, \ldots, A$, from the posterior density of the target parameter τ_i. Finally, the HB estimator $\hat{\tau}_i^{\text{HB}}$ and its posterior variance are approximated as

$$\hat{\tau}_i^{\text{HB}} = E(\tau_i|\mathbf{y}_s) \approx \frac{1}{A} \sum_{a=1}^{A} \tau_i^{(a)}, \quad V(\tau_i|\mathbf{y}_s) \approx \frac{1}{A} \sum_{a=1}^{A} \left(\tau_i^{(a)} - \hat{\tau}_i^{\text{HB}}\right)^2. \tag{10.7.12}$$

Other useful posterior summaries, such as credible intervals, can be computed in a straightforward manner.

Similarly, as described at the end of Section 9.4.2 for the EB method, an HB approach can be implemented analogously to the fast EB approach introduced in Ferretti and Molina (2012). For this, from each Monte Carlo population vector $\mathbf{y}_i^{P(a)}$ we draw a sample $s_i^{(a)}$ and, with this sample, we obtain a design-based estimator

$\hat{\tau}_i^{\text{DB}(a)}$ of $\tau_i^{(a)}$. Then, the fast HB estimator is given by $\hat{\tau}_i^{\text{FHB}} = H^{-1}\sum_{h=1}^{H}\hat{\tau}_i^{\text{DB}(a)}$, and its posterior variance can be approximated similarly by $H^{-1}\sum_{h=1}^{H}(\hat{\tau}_i^{\text{DB}(a)} - \hat{\tau}_i^{\text{FHB}})^2$.

In the particular case of estimating the FGT poverty measure $\tau_i = F_{\alpha i}$, from the census vector $\mathbf{y}_i^{P(a)} = (\mathbf{y}_{is}^T, (\mathbf{y}_{ir}^{(a)})^T)^T$, we calculate

$$F_{\alpha i}^{(a)} = \frac{1}{N_i}\left[\sum_{j\in s_i}\left(\frac{z - E_{ij}}{z}\right)^{\alpha} I(E_{ij} < z) + \sum_{i\in r_i}\left(\frac{z - E_{ij}^{(a)}}{z}\right)^{\alpha} I(E_{ij}^{(a)} < z)\right], \tag{10.7.13}$$

where $E_{ij} = T^{-1}(y_{ij}), j \in s_i$ and $E_{ij}^{(a)} = T^{-1}(y_{ij}^{(a)}), i \in r_i, i = 1, \ldots, M$, for the selected transformation $T(\cdot)$ of the welfare variables E_{ij}. Then we average for $a = 1, \ldots, A$ and calculate the posterior variance as in (10.7.12).

Molina, Nandram, and Rao (2014) conducted a simulation study to compare EB and HB estimates of poverty incidence and poverty gap under the frequentist setup. The setup of the simulation is exactly the same as in Molina and Rao (2010), which is also described in Section 9.4.6. Mean values of EB and HB estimates across Monte Carlo simulations turned out to be practically equal. Approximately, the same point estimates were also obtained in an application with data from the Spanish Survey on Income and Living Conditions from year 2006. Results in this application also show that posterior variances of HB estimators are of comparable magnitude to MSEs of frequentist EB estimators. Thus, HB estimates obtained from model (10.7.4) appear to have also good frequentist properties.

Example 10.7.1. Poverty Mapping in Spain. Molina and Rao (2010) and Molina, Nandram, and Rao (2014) applied EB and HB methods to unit level data from the 2006 Spanish Survey on Income and Living Conditions to estimate poverty incidences and gaps for provinces by gender. They considered the basic unit level model for the log(income+constant), where the constant was selected to achieve symmetry of the distribution of model residuals. As explanatory variables in the model, they considered the indicators of five age groups, of having Spanish nationality, of three education levels and of the labor force status (unemployed, employed, or inactive). Table 10.5 reports the CVs of EBLUP estimators based on the FH model and

TABLE 10.5 Estimated % CVs of the Direct, EB, and HB Estimators of Poverty Incidence for Selected Provinces by Gender

Province	Gender	n_i	CV(dir.)	CV(EBLUP)	CV(EB)	CV(HB)
Soria	F	17	51.9	25.4	16.6	19.8
Tarragona	M	129	24.4	20.1	14.9	12.3
Córdoba	F	230	13.0	10.5	6.2	6.9
Badajoz	M	472	8.4	7.6	3.5	4.2
Barcelona	F	1,483	9.4	7.8	6.5	4.5

Source: Adapted from Table 17.1 in Rao and Molina (2015).

of the EB and HB estimators of poverty incidence for a selected set of domains (province×gender); for the HB estimator, the CV is computed from the posterior variance. Table 10.5 shows that both EB and HB estimators have much smaller CV than the direct estimators and the EBLUP estimators based on the FH model, especially for provinces with small sample sizes; for example, Soria with a sample size of 17 females.

10.8 TWO-LEVEL MODELS

In Section 8.8, we studied EBLUP estimation for the two-level model (8.8.1). In this section, we study three different HB models, including the HB version of (8.8.1), assuming priors on the model parameters.

Model 1. The HB version of (8.8.1) with $k_{ij} = 1$ may be written as

(i) $y_{ij}|\beta_i, \sigma_e^2 \overset{\text{ind}}{\sim} N(\mathbf{x}_{ij}^T\beta_i, \sigma_e^2)$,

(ii) $\beta_i|\alpha, \Sigma_v \overset{\text{ind}}{\sim} N_p(\mathbf{Z}_i\alpha, \Sigma_v)$,

(iii) $f(\alpha, \sigma_e^2, \Sigma_v) = f(\alpha)f(\sigma_e^2)f(\Sigma_v)$, where

$$\alpha \sim N_q(\mathbf{0}, \mathbf{D}), \sigma_e^{-2} \sim G(a, b), \Sigma_v^{-1} \sim W_p(d, \mathbf{\Delta}), \tag{10.8.1}$$

and $W_p(d, \mathbf{\Delta})$ denotes a Wishart distribution with df d and scale matrix $\mathbf{\Delta}$:

$$f(\mathbf{\Sigma}_v^{-1}) \propto |\mathbf{\Sigma}_v^{-1}|^{(d-p-1)/2} \exp\left\{-\frac{1}{2}\text{tr}(\mathbf{\Delta}\mathbf{\Sigma}_v^{-1})\right\}, \qquad d \geq p. \tag{10.8.2}$$

The constants $a \geq 0, b > 0, d$, and the elements of $\mathbf{D}$ and $\mathbf{\Delta}$ are chosen to reflect lack of prior information on the model parameters. In particular, using a diagonal matrix $\mathbf{D}$ with very large diagonal elements, say 10^4, is roughly equivalent to a flat prior $f(\boldsymbol{\alpha}) \propto 1$. Similarly, $d = p, a = 0.001, b = 0.001$, and a scale matrix $\mathbf{\Delta}$ with diagonal elements equal to 1 and off diagonals equal to 0.001 may be chosen. Daniels and Kass (1999) studied alternative priors for $\mathbf{\Sigma}_v^{-1}$ and discuss the limitations of Wishart prior when the number of small areas, m, is not large.

Model 2. By relaxing the assumption of constant error variance, σ_e^2, we obtain a HB two-level model with unequal error variances:

(i) $y_{ij}|\beta_i, \sigma_{ei}^2 \overset{\text{ind}}{\sim} N(\mathbf{x}_{ij}^T\beta_i, \sigma_{ei}^2)$,

(ii) Same as in (ii) of Model 1.,

(iii) Marginal priors on $\boldsymbol{\alpha}$ and $\mathbf{\Sigma}_v^{-1}$same as in (iii) of Model 1

$$\text{and } \sigma_{ei}^{-2} \overset{\text{ind}}{\sim} G(a_i, b_i), \quad i = 1, \ldots, m. \tag{10.8.3}$$

The constants a_i and b_i are chosen as $a_i = 0.001, b_i = 0.001$ to reflect lack of prior information.

Model 3. In Section 7.6.1, we studied a simple random effects model given by $y_{ij} = \beta + v_i + e_{ij}$ with random error variance σ_{ei}^2. We also noted that the HB approach may be used to handle extensions to nested error regression models with random error variances. Here, we consider an HB version of a two-level model with random error variances:

(i) Same as in (i) of Model 2,

(ii) Same as in (ii) of Model 2,

(iii) $\sigma_{ei}^{-2} \stackrel{\text{ind}}{\sim} G(\eta, \lambda)$,

(iv) Marginal priors on $\boldsymbol{\alpha}$ and $\boldsymbol{\Sigma}_v^{-1}$same as in (iii) of Model 2,

and η and λ uniform over a large interval $(0, 10^4]$. (10.8.4)

Note that (iii) of (10.8.4) is a component of the two-level model, and priors on model parameters are introduced only in (iv) of (10.8.4). The priors on η and λ reflect vague prior knowledge on the model parameters $\eta > 0, \lambda > 0$.

You and Rao (2000) showed that all the Gibbs conditionals for Models 1–3 have closed forms. They also obtained Rao–Blackwell HB estimators of small area means $\mu_i = \overline{\mathbf{X}}_i^T \boldsymbol{\beta}_i$ and posterior variance $V(\mu_i|\mathbf{y}) = \overline{\mathbf{X}}_i^T V(\boldsymbol{\beta}_i|\mathbf{y})\overline{\mathbf{X}}_i$, where $\overline{\mathbf{X}}_i$ is the vector of population x-means for the ith area.

Example 10.8.1. Household Income. In Example 8.8.1, a two-level model with equal error variances σ_e^2 was applied to data from 38,704 households in $m = 140$ enumeration districts (small areas) in one county in Brazil. The two-level model is given by (8.8.2) and (8.8.3), where y_{ij} is the jth household's income in the ith small area, (x_{1ij}, x_{2ij}) are two unit level covariates: number of rooms in the (i,j)th household and educational attainment of head of the (i,j)th household centered around the means $(\overline{x}_{1i}, \overline{x}_{2i})$. The area level covariate z_i, the number of cars per household in the ith area, is related to the random slopes $\boldsymbol{\beta}_i$, using (8.8.3.) You and Rao (2000) used a subset of $m = 10$ small areas with $n_i = 28$ households each, obtained by simple random sampling in each area, to illustrate model selection and HB estimation.

The Gibbs sampler for the three models was implemented using the BUGS program aided by CODA for convergence diagnostics. Using priors as specified above, the Gibbs sampler for each model was first run for a "burn-in" of $d = 2,000$ iterations. Then, $D = 5,000$ more iterations were run and kept for model selection and HB estimation.

For model selection, You and Rao (2000) calculated CPO values for the three models, using (10.2.26). In particular, for Model 1,

$$\widehat{\text{CPO}}_{ij} = \left[\frac{1}{D} \sum_{k=d+1}^{d+D} \frac{1}{f(y_{ij}|\boldsymbol{\beta}_i^{(k)}, \sigma_e^{2(k)})} \right]^{-1}, \tag{10.8.5}$$

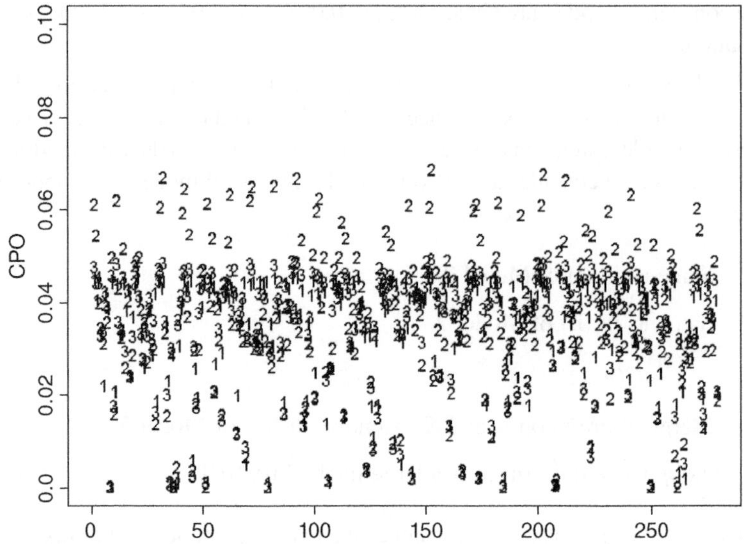

Figure 10.2 CPO Comparison Plot for Models 1–3. *Source*: Adapted from Figure 1 in You and Rao (2000).

where $\{\boldsymbol{\beta}_i^{(k)}, \sigma_e^{2(k)}, \boldsymbol{\alpha}^{(k)}, \boldsymbol{\Sigma}_v^{(k)}; k = d+1, \ldots, d+D\}$ denote the MCMC samples. For Models 2, and 3, $\sigma_e^{2(k)}$ in (10.8.5) is replaced by $\sigma_{ei}^{2(k)}$. A CPO plot for the three models is given in Figure 10.2. This plot shows that a majority of CPO-values for Model 2 were significantly larger than those for Models 1 and 3, thus indicating Model 2 as the best fitting model among the three models.

You and Rao (2000) calculated Rao–Blackwell HB estimates of small area means $\mu_i = \beta_0 + \beta_1 \overline{X}_{1i} + \beta_2 \overline{X}_{2i}$ under Model 2 and demonstrated the benefits of Rao–Blackwellization in terms of simulation standard errors.

Example 10.8.2. Korean Unemployment. Cheng, Lee, and Kim (2001) applied Models 1 and 2 to adjust direct unemployment estimates, y_{ij}, associated with the jth month survey data in the ith small area (here j refers to May, July, and December 2000). Auxiliary variables $\mathbf{x}_{ij}$ were obtained from the Economically Active Population Survey (EAPS), census, and administrative records. HB analysis for these data was conducted using WinBUGS program (Lunn et al. 2000).

A $\widehat{\text{CPO}}_{ij}$ plot for the two models showed that $\widehat{\text{CPO}}$ values for Model 2 are significantly larger in every small area than those for Model 1. Chung et al. (2003) also calculated standardized residuals, d_{ij}, similar to (10.4.8), and two other measures of fit, namely negative cross-validatory loglikelihood, $-\sum_{i=1}^{m}\sum_{j=1}^{n_i} \log\,\{\hat{f}(y_{ij,\text{obs}}|\mathbf{y}_{(ij),\text{obs}})\}$, and posterior mean deviance, $-2\sum_{i=1}^{m}\sum_{j=1}^{n_i}[D^{-1}\sum_{k=d+1}^{d+D} \log\,\{f(y_{ij,\text{obs}}|\boldsymbol{\eta}^{(k)})\}]$, where $\hat{f}(y_{ij,\text{obs}}|\mathbf{y}_{(ij),\text{obs}})$ is obtained from (10.2.26). These measures are computable directly in WinBUGS. Model 2 yielded a negative cross-validatory loglikelihood of 121.5

(compared to 188.7 for Model 1) and a posterior mean deviance of 243.0 (compared to 377.3 for Model 1), thus supporting Model 2 relative to Model 1. Using Model 2, Chung et al. (2003) calculated the posterior means and variances of the small area means $\mu_{ij} = \mathbf{x}_{ij}^T \boldsymbol{\beta}_i$.

10.9 TIME SERIES AND CROSS-SECTIONAL MODELS

In Section 8.3.1, we studied EBLUP estimation from time series and cross-sectional data using the sampling model (4.4.6) and the linking model (4.4.7): (i) $\hat{\theta}_{it} = \theta_{it} + e_{it}$; (ii) $\theta_{it} = \mathbf{z}_{it}^T \boldsymbol{\beta} + v_i + u_{it}$, where $\{u_{it}\}$ follows either an AR(1) model $u_{it} = \rho u_{i,t-1} + \epsilon_{it}$, $|\rho| < 1$, or a random walk model $u_{it} = u_{i,t-1} + \epsilon_{it}$. In this section, we give a brief account of HB estimation under this model, assuming a prior distribution on the model parameters. The HB version of the model with random walk effects u_{it} may be expressed as

$$
\begin{aligned}
&\text{(i)} \quad \hat{\boldsymbol{\theta}}_i | \boldsymbol{\theta}_i \overset{\text{ind}}{\sim} N_T(\boldsymbol{\theta}_i, \boldsymbol{\Psi}_i), \quad \text{where } \boldsymbol{\Psi}_i \text{ is the known sampling covariance} \\
&\qquad \text{matrix of } \hat{\boldsymbol{\theta}}_i = (\hat{\theta}_{i1}, \ldots, \hat{\theta}_{iT})^T, \\
&\text{(ii)} \quad \theta_{it} | \boldsymbol{\beta}, u_{it}, \sigma_v^2 \overset{\text{ind}}{\sim} N(\mathbf{z}_{it}^T \boldsymbol{\beta} + u_{it}, \sigma_v^2), \\
&\text{(iii)} \quad u_{it} | u_{i,t-1}, \sigma_\epsilon^2 \overset{\text{ind}}{\sim} N(u_{i,t-1}, \sigma_\epsilon^2), \\
&\text{(iv)} \quad f(\boldsymbol{\beta}, \sigma_v^2, \sigma^2) = f(\boldsymbol{\beta}) f(\sigma_v^2) f(\sigma^2) \quad \text{with} \\
&\qquad f(\boldsymbol{\beta}) \propto 1, \sigma_v^{-2} \sim G(a_1, b_1), \sigma^{-2} \sim G(a_2, b_2).
\end{aligned}
\tag{10.9.1}
$$

For the AR(1) case with known ρ, replace $u_{i,t-1}$ in (iii) of (10.9.1) by $\rho\, u_{i,t-1}$.

Datta et al. (1999) and You, Rao, and Gambino (2003) obtained Gibbs conditionals under the HB model (10.9.1) and showed that all the conditionals have closed forms. You et al. (2003) also obtained Rao–Blackwell estimators of the posterior mean $E(\theta_{iT}|\hat{\boldsymbol{\theta}})$ and the posterior variance $V(\theta_{iT}|\hat{\boldsymbol{\theta}})$ for the current time T. We refer the reader to the above papers for technical details.

You (2008) extended the matched random walk model (10.9.1) to unmatched models by replacing the linking model (ii) by

$$
\log(\theta_{it}) | \boldsymbol{\beta}, u_{it}, \sigma_v^2 \overset{\text{ind}}{\sim} N(\mathbf{z}_{it}^T \boldsymbol{\beta} + u_{it}, \sigma_v^2) \tag{10.9.2}
$$

and retaining (i), (iii), and (iv). In this case, the Gibbs conditional for $\boldsymbol{\theta}_i$ does not have a closed form and the M–H within Gibbs algorithm is used to generate MCMC samples from the joint posterior distribution.

Example 10.9.1. Canadian Unemployment Rates. You et al. (2003) used the HB model (10.9.1) to estimate unemployment rates, θ_{iT}, for $m = 62$ Census Agglomerations (CAs) in Canada, using Canadian LFS direct estimates $\hat{\theta}_{it}$ and auxiliary data $\mathbf{z}_{it}$.

Here, CA's are treated as small areas. They used data $\{\hat{\theta}_{it}, \mathbf{z}_{it}\}$ for $T = 6$ consecutive months, from January 1999 to June 1999, and the parameters of interest are the true unemployment rates, θ_{iT}, in June 1999 for each of the small areas. The choice $T = 6$ was motivated by the fact that the correlation between the estimates $\{\hat{\theta}_{it}\}$ and $\{\hat{\theta}_{is}\}$ $(s \neq t)$ is weak after a lag of 6 months because of the LFS sample rotation based on a 6-month cycle; each month, one-sixth of the sample is replaced.

To obtain a smoothed estimate of the sampling covariance matrix $\mathbf{\Psi}_i$ used in the model (10.9.1), You et al. (2003) first computed the average CV $(\overline{\text{CV}}_i)$ for each CA i over time and the average lag correlations, $\bar{r}_a$, over time and over all CA's. A smoothed estimate of $\mathbf{\Psi}_i$ was then obtained using those smoothed CVs and lag correlations: the tth diagonal element, ψ_{itt}, of $\mathbf{\Psi}_i$ (i.e., the sampling variance of $\hat{\theta}_{it}$) equals $[\hat{\theta}_{it}(\overline{\text{CV}}_i)]^2$ and the (t, s)th element, ψ_{its}, of $\mathbf{\Psi}_i$ (i.e., sampling covariance of $\hat{\theta}_{it}$ and $\hat{\theta}_{is}$) equals $\bar{r}_a^2(\psi_{itt}\psi_{iss})^{1/2}$ with $a = |t - s|$. The smoothed estimate of $\mathbf{\Psi}_i$ was treated as the true $\mathbf{\Psi}_i$.

You et al. (2003) used a divergence measure proposed by Laud and Ibrahim (1995) to compare the relative fits of the random walk model (10.9.1) and the corresponding AR(1) model with $\rho = 0.75$ and $\rho = 0.50$. This measure is given by $d(\hat{\boldsymbol{\theta}}_*, \hat{\boldsymbol{\theta}}_{\text{obs}}) = E[(mT)^{-1}\|\hat{\boldsymbol{\theta}}_* - \hat{\boldsymbol{\theta}}_{\text{obs}}\|^2|\hat{\boldsymbol{\theta}}_{\text{obs}}]$, where $\hat{\boldsymbol{\theta}}_{\text{obs}}$ is the (mt)-vector of direct estimates $\hat{\theta}_{it}$ and the expectation is with respect to the posterior predictive distribution $f(\hat{\boldsymbol{\theta}}_*|\hat{\boldsymbol{\theta}}_{\text{obs}})$ of a new observation $\hat{\boldsymbol{\theta}}_*$. Models yielding smaller values of this measure are preferred.

The Gibbs output with $L = 10$ parallel runs was used to generate samples $\{\boldsymbol{\theta}^{(\ell k)}; \ell = 1, \ldots, 10\}$ from the posterior distribution, $f(\boldsymbol{\theta}|\hat{\boldsymbol{\theta}}_{\text{obs}})$, where $\boldsymbol{\theta}$ is the (mT)-vector of small area parameters θ_{it}. For each $\boldsymbol{\theta}^{(\ell k)}$, a new observation $\hat{\boldsymbol{\theta}}_*^{(\ell k)}$ was then generated from $f(\hat{\boldsymbol{\theta}}|\boldsymbol{\theta}^{(\ell k)})$. The new observations $\{\hat{\boldsymbol{\theta}}_*^{(\ell k)}\}$ represent simulated samples from $f(\hat{\boldsymbol{\theta}}_*|\hat{\boldsymbol{\theta}}_{\text{obs}})$. The measure $d(\hat{\boldsymbol{\theta}}_*, \hat{\boldsymbol{\theta}}_{\text{obs}})$ was approximated by using these observations from the posterior predictive distribution. The following values of $d(\hat{\boldsymbol{\theta}}_*, \hat{\boldsymbol{\theta}}_{\text{obs}})$ were obtained: (i) 13.36 for the random walk model; (ii) 14.62 for the AR(1) model with $\rho = 0.75$; and (iii) 14.52 for the AR(1) model with $\rho = 0.5$. Based on these values, the random walk model was selected.

To check the overall fit of the random walk model, the simulated values $\{(\boldsymbol{\theta}^{(\ell k)}, \hat{\boldsymbol{\theta}}_*^{(\ell k)}); \ell = 1, \ldots, 10\}$ were employed to approximate the posterior predictive p value from each run, ℓ, using the formula (10.2.19) with measure of discrepancy given by $T(\hat{\boldsymbol{\theta}}, \boldsymbol{\eta}) = \sum_{i=1}^{62} (\hat{\boldsymbol{\theta}}_i - \boldsymbol{\theta}_i)^T \mathbf{\Psi}_i^{-1}(\hat{\boldsymbol{\theta}}_i - \boldsymbol{\theta}_i)$. The average of the $L = 10$ posterior predictive p-values, $\widehat{pp} = 0.615$, indicated a good fit of the random walk model to the time series and cross-sectional data.

Rao–Blackwell method was used to calculate the posterior mean, $\hat{\theta}_{i6}^{\text{HB}}$, and the posterior variance, $V(\theta_{i6}|\hat{\boldsymbol{\theta}})$, for each area i. Figure 10.3 displays the HB estimates under model (10.9.1), denoted HB1, the HB estimates under the Fay–Herriot model using only the current cross-sectional data, denoted HB2, and the direct LFS estimates, denoted DIRECT. It is clear from Figure 10.3 that the HB2 estimates tend to be smoother than HB1, whereas HB1 leads to moderate smoothing of the direct estimates. For the CAs with larger population sizes and therefore larger sample sizes, DIRECT and HB1 are very close to each other, whereas DIRECT differs substantially from HB1 for some smaller CAs.

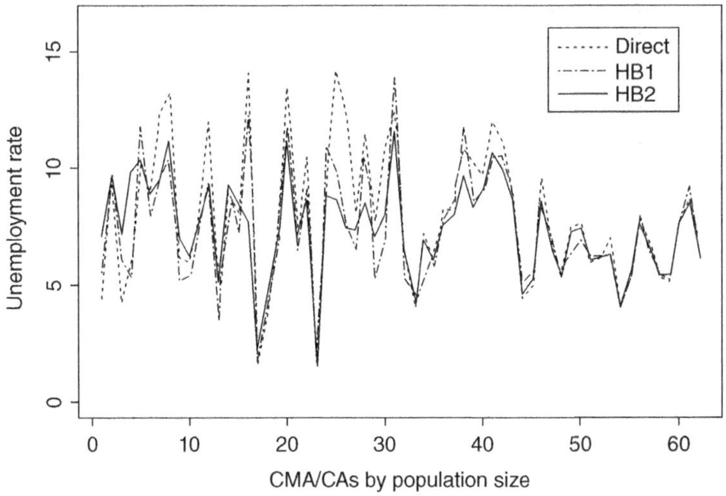

Figure 10.3 Direct, Cross-sectional HB (HB2) and Cross-Sectional and Time Series HB (HB1) Estimates. *Source*: Adapted from Figure 2 in You, Rao, and Gambino (2003).

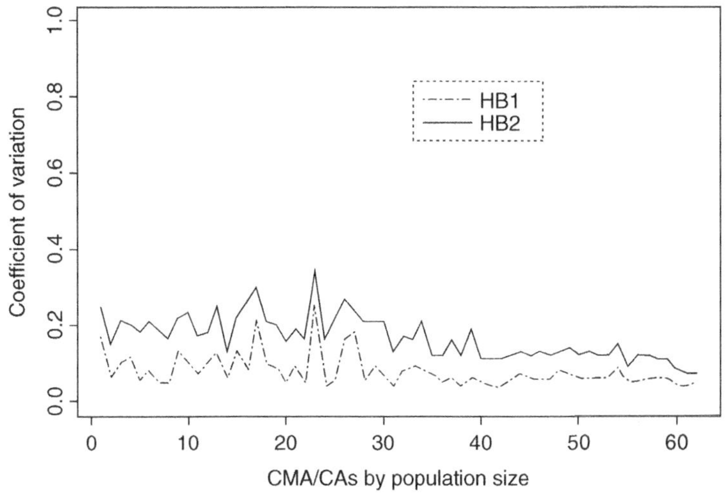

Figure 10.4 Coefficient of Variation of Cross-sectional HB (HB2) and Cross-Sectional and Time Series HB (HB1) Estimates. *Source*: Adapted from Figure 3 in You, Rao, and Gambino (2003).

Figure 10.4 displays the CVs of HB1 and HB2; note that we have already compared the CVs of HB2 and DIRECT in Figure 10.1. It is clear from Figure 10.4 that HB1 leads to significant reduction in CV relative to HB2, especially for the smaller CAs.

Example 10.9.2. U.S. Unemployment Rates. Datta et al. (1999) studied an extension of the random walk model (10.9.1) and applied their model to estimate monthly unemployment rate for 49 U.S. states and the District of Columbia ($m = 50$), using CPS estimates as $\hat{\theta}_{it}$ and Unemployment Insurance (UI) claims rate as z_{it}; New York state was excluded from the study because of very unreliable UI data. They considered the data for the period January 1985–December 1988 ($T = 48$) to calculate HB estimates, $\hat{\theta}_{iT}^{\text{HB}}$, for the current time point T. The HB version of the Datta et al. (1999) model may be written as

$$
\begin{aligned}
&\text{(i)} \quad \hat{\boldsymbol{\theta}}_i | \boldsymbol{\theta}_i \overset{\text{ind}}{\sim} N_T(\boldsymbol{\theta}_i, \boldsymbol{\Psi}_i),\\
&\text{(ii)} \quad \theta_{it} = \beta_{0i} + \beta_{1i} z_{it} + u_{it} + \sum_{u=1}^{12} a_{itu}\gamma_{iu} + \sum_{k=1}^{4} b_{itk}\lambda_{ik},\\
&\qquad \text{where } \gamma_{i,12} = -\sum_{u=1}^{11}\gamma_{iu} \text{ and } \lambda_{i4} = -\sum_{k=1}^{3}\lambda_{ik},\\
&\text{(iii)} \quad \beta_{0i} \overset{\text{iid}}{\sim} N(\beta_0, \sigma_0^2), \beta_{1i} \overset{\text{iid}}{\sim} N(\beta_1, \sigma_1^2), \boldsymbol{\gamma}_i \sim N_{11}(\boldsymbol{\gamma}, \boldsymbol{\Sigma}_\gamma),\\
&\qquad \boldsymbol{\lambda}_i \sim N_3(\boldsymbol{\lambda}, \boldsymbol{\Sigma}_\lambda), u_{it}|u_{i,t-1}, \sigma^2 \sim N(u_{i,t-1}, \sigma^2),\\
&\text{(iv)} \quad \text{Independent gamma priors on } \sigma_0^{-2}, \sigma_1^{-2} \text{ and } \sigma^{-2}, \text{ independent Wishart}\\
&\qquad \text{priors on } \boldsymbol{\Sigma}_\gamma^{-1} \text{and } \boldsymbol{\Sigma}_\lambda^{-1}, \text{ and flat priors on } \beta_0 \text{ and } \beta_1. \qquad (10.9.3)
\end{aligned}
$$

The last two terms in (ii) of model (10.9.3) account for seasonal variation in monthly unemployment rates, where $a_{itu} = 1$ if $t = u$, $a_{itu} = 0$ otherwise; $a_{it,12} = 1$ if $t = 12, 24, 36, 48$, $a_{it,12} = 0$ otherwise, and $\boldsymbol{\gamma}_i = (\gamma_{i1}, \ldots, \gamma_{i,11})^T$ represents random month effects. Similarly, $b_{itk} = 1$ if $12(k-1) < t \le 12k$, $b_{itk} = 0$ otherwise, and $\boldsymbol{\lambda}_i = (\lambda_{i1}, \lambda_{i2}, \lambda_{i3})^T$ represent random year effects. The random effects $\beta_{0i}, \beta_{1i}, \boldsymbol{\gamma}_i, \boldsymbol{\lambda}_i$, and u_{it} are assumed to be mutually independent.

Datta et al. (1999) used $L = 10$ parallel runs to draw MCMC samples $\{\boldsymbol{\eta}^{(k)}\}$ from the joint posterior distribution of the random effects $\{\beta_{0i}, \beta_{1i}, u_{it}, \boldsymbol{\gamma}_i, \boldsymbol{\lambda}_i\}$, and the model parameters. To check the overall fit of the model (10.9.3), they estimated the posterior predictive p-value using the discrepancy measure $T(\hat{\boldsymbol{\theta}}, \boldsymbol{\eta})$ of Example 10.3.2. The estimated posterior predictive p-value, $\widehat{pp} = 0.614$, indicates a good fit of the model to the CPS time series and cross-sectional data. If the seasonal effects are deleted from the model (10.9.3), then the posterior predictive p-value becomes $\widehat{pp} = 0.758$. This increase suggests that the inclusion of seasonal effects is important, probably because of the long time series used in the study ($T = 48$). Note that in Example 10.3.2, the time series is short ($T = 6$).

Datta et al. (1999) also calculated the divergence measure proposed by Laud and Ibrahim (1995) and given by

$$d(\hat{\boldsymbol{\theta}}_*, \hat{\boldsymbol{\theta}}_{\text{obs}}) = \frac{1}{nD_0}\sum_{k=1}^{D_0} \|\hat{\boldsymbol{\theta}}_*^{(k)} - \hat{\boldsymbol{\theta}}_{\text{obs}}\|^2, \qquad (10.9.4)$$

where $n = mT$, D_0 is the total number of MCMC samples, $\hat{\boldsymbol{\theta}}_{\text{obs}}$ is the vector of observed direct estimates with blocks $\hat{\boldsymbol{\theta}}_{i,\text{obs}}$, $\hat{\boldsymbol{\theta}}_{*}^{(k)}$ is the vector of hypothetical replications with blocks $\hat{\boldsymbol{\theta}}_{i*}^{(k)}$, with $\hat{\boldsymbol{\theta}}_{i*}^{(k)}$ drawn from $N_T(\boldsymbol{\theta}_i^{(k)}, \boldsymbol{\Psi}_i)$. For the CPS data, $d(\hat{\boldsymbol{\theta}}_*, \hat{\boldsymbol{\theta}}_{\text{obs}}) = 0.0007$ using model (10.9.3), while it increased to $d_{\text{FH}}(\hat{\boldsymbol{\theta}}_*, \hat{\boldsymbol{\theta}}_{\text{obs}}) = 1.63$ using separate Fay–Herriot models. If the seasonal effects γ_{iu} and λ_{ik} are deleted from the model (10.9.3), then $d(\hat{\boldsymbol{\theta}}_*, \hat{\boldsymbol{\theta}}_{\text{obs}}) = 0.0008$, which is close to the value 0.0007 under the full model.

Datta et al. (1999) calculated the HB estimates, $\hat{\theta}_{iT}^{\text{HB}}$, and the posterior variances, $V(\theta_{iT}|\hat{\boldsymbol{\theta}})$, for the current time point T, under model (10.9.3), and then compared the values to the CPS values, $\hat{\theta}_{iT}$ and ψ_{iT}, and the HB estimates for the Fay–Herriot model fitted only to the current month cross-sectional data. The HB standard errors, $\sqrt{V(\theta_{iT}|\hat{\boldsymbol{\theta}})}$, under the model (10.9.3) were significantly smaller than the corresponding CPS standard errors, $\sqrt{\psi_{iT}}$, uniformly over time and across all states. However, HB was more effective for states with fewer sample households ("indirect-use" states). The reduction in standard error under the Fay–Herriot model for the current period relative to CPS standard error was sizeable for indirect-use states, but it is less than 10% (and as small as 0.1% in some months) for "direct-use" states with larger sample sizes.

10.10 MULTIVARIATE MODELS

10.10.1 Area Level Model

In Section 8.1, we considered EBLUP estimation under the multivariate Fay–Herriot model (4.4.3). An HB version of (4.4.3) is given by

$$\begin{aligned}
&\text{(i)}\quad \hat{\boldsymbol{\theta}}_i|\boldsymbol{\theta}_i \overset{\text{ind}}{\sim} N_r(\boldsymbol{\theta}_i, \Psi_i), \text{ where} \\
&\qquad \hat{\boldsymbol{\theta}}_i = (\hat{\theta}_{i1}, \ldots, \hat{\theta}_{ir})^T \text{is the vector of direct estimators and} \Psi_i \text{is known,} \\
&\text{(ii)}\quad \boldsymbol{\theta}_i|\boldsymbol{\beta}, \Sigma_v \sim N_r(\mathbf{Z}_i\boldsymbol{\beta}, \Sigma_v), \\
&\text{(iii)}\quad f(\boldsymbol{\beta}, \Sigma_v) = f(\boldsymbol{\beta}) f(\Sigma_v) \quad \text{with} \quad f(\boldsymbol{\beta}) \propto 1, f(\Sigma_v) \propto 1.
\end{aligned} \tag{10.10.1}$$

Datta et al. (1996) derived the Gibbs conditionals, $[\boldsymbol{\beta}|\boldsymbol{\theta}, \boldsymbol{\Sigma}_v, \hat{\boldsymbol{\theta}}]$, $[\boldsymbol{\theta}_i|\boldsymbol{\beta}, \boldsymbol{\Sigma}_v, \hat{\boldsymbol{\theta}}]$, and $[\boldsymbol{\Sigma}_v|\boldsymbol{\beta}, \boldsymbol{\theta}, \hat{\boldsymbol{\theta}}]$, under (10.10.1) and showed that all the conditionals have closed forms. They also obtained Rao–Blackwell estimators of the posterior means $E(\boldsymbol{\theta}_i|\hat{\boldsymbol{\theta}})$ and the posterior variances $V(\boldsymbol{\theta}_i|\hat{\boldsymbol{\theta}})$.

Example 10.10.1. Median Income. Datta et al. (1996) used the multivariate Fay–Herriot model (10.10.1) to estimate the median incomes of four-person families in the U.S. states (small areas). Here $\boldsymbol{\theta}_i = (\theta_{i1}, \theta_{i2}, \theta_{i3})^T$ with θ_{i1}, θ_{i2}, and θ_{i3} denoting the true median income of four, three, and five-person families in state i, and θ_{i1}'s are the parameters of interest. The adjusted census median income, z_{ij1}, and base-year census median income, z_{ij2}, for the three groups $j = 1, 2, 3$ were used as explanatory variables. The matrix $\mathbf{Z}_i$ consists of three rows $(\mathbf{z}_{i1}^T, \mathbf{0}_3^T, \mathbf{0}_3^T)$, $(\mathbf{0}_3^T, \mathbf{z}_{i2}^T, \mathbf{0}_3^T)$, and $(\mathbf{0}_3^T, \mathbf{0}_3^T, \mathbf{z}_{i3}^T)$, where $\mathbf{z}_{ij}^T = (1, z_{ij1}, z_{ij2})$ and $\mathbf{0}_3^T = (0, 0, 0)$. Direct survey estimates $\hat{\theta}_i$ of $\boldsymbol{\theta}_i$ and associated sampling covariance matrices $\boldsymbol{\Psi}_i$ were obtained from the 1979 Current Population Survey (CPS). HB estimates of θ_{i1} were obtained using univariate, bivariate, and trivariate Fay–Herriot models, denoted as HB^1, HB^2, and HB^3, respectively. Two cases for the bivariate model were studied: $\boldsymbol{\theta}_i = (\theta_{i1}, \theta_{i2})^T$ with the corresponding $\mathbf{Z}_i$ and $\boldsymbol{\theta}_i = (\theta_{i1}, \theta_{i3})^T$ with the corresponding $\mathbf{Z}_i$. Denote the HB estimates for the two cases as HB^{2a} and HB^{2b}, respectively. In terms of posterior variances, HB^{2b} obtained from the bivariate model that uses only the median incomes of four- and five-person families performed better than the other estimates. HB estimates based on the multivariate models performed better than HB^1 based on the univariate model.

Datta et al. (1996) also conducted an external evaluation by treating the census estimates for 1979, available from the 1980 census data, as true values. Table 10.6 reports the absolute relative error averaged over the states ($\overline{\text{ARE}}$); the direct estimates $\hat{\theta}_{i1}$; and the HB estimates HB^3, HB^{2a}, HB^{2b}, and HB^1. It is clear from Table 10.6 that all HB estimates performed similarly in terms of $\overline{\text{ARE}}$ and outperformed the direct estimates.

10.10.2 Unit Level Model

In Section 8.6, we studied EBLUP estimation under the multivariate nested error regression model (4.5.1). Datta, Day, and Maiti (1998) studied HB estimation for this model. An HB version of the model may be expressed as

$$\text{(i)}\quad \mathbf{y}_{ij}|\mathbf{B}, \boldsymbol{\Sigma}_e \overset{\text{ind}}{\sim} N_r(\mathbf{B}\mathbf{x}_{ij} + \mathbf{v}_i, \boldsymbol{\Sigma}_e), \quad j = 1, \ldots, n_i, \quad i = 1, \ldots, m,$$

$$\text{(ii)}\quad \mathbf{v}_i|\boldsymbol{\Sigma}_v \overset{\text{iid}}{\sim} N_r(\mathbf{0}, \boldsymbol{\Sigma}_v),$$

(iii) Flat prior on $\mathbf{B}$ and independent Wishart priors on $\boldsymbol{\Sigma}_v^{-1}$ and $\boldsymbol{\Sigma}_e^{-1}$.

(10.10.2)

Datta et al. (1998) derived Gibbs conditionals under (10.10.2) and showed that all the conditionals have closed forms. They also considered improper priors of the form $f(\mathbf{B}, \boldsymbol{\Sigma}_v, \boldsymbol{\Sigma}_e) \propto |\boldsymbol{\Sigma}_v|^{-a_v/2}|\boldsymbol{\Sigma}_e|^{-a_e/2}$ under which Gibbs conditionals are proper but the

TABLE 10.6 Average Absolute Relative Error (ARE%): Median Income of Four-Person Families

Direct	HB^1	HB^{2a}	HB^{2b}	HB^3
4.98	2.07	2.04	2.06	2.02

Source: Adapted from Table 11.2 in Datta et al. (1996).

joint posterior may be improper. They obtained necessary and sufficient conditions on a_v and a_e that ensure property of the joint posterior. Datta et al. (1998) also obtained Rao–Blackwell estimators of the posterior means $E(\overline{\mathbf{Y}}_i|\mathbf{y})$ and the posterior variance matrix $V(\overline{\mathbf{Y}}_i|\mathbf{y})$, where $\overline{\mathbf{Y}}_i$ is the vector of finite population means for the ith area.

The multivariate model was applied to the county corn and soybeans data, y_{ij1}, y_{ij2} of Battese et al. (1988). In terms of posterior variances, the HB estimates under the multivariate model performed better than the HB estimates under the univariate model for corn as well as soybeans. We refer the reader to Datta et al. (1998) for details of MCMC implementation and model fit.

10.11 DISEASE MAPPING MODELS

In Section 9.6, we studied EBLUP estimation for three models useful in disease mapping applications: Poisson-gamma, log-normal, and CAR-normal models. In this section, we study HB estimation for these models, assuming priors on the model parameters. We also consider extensions to two-level models.

10.11.1 Poisson-Gamma Model

Using the notation in Section 9.6, let θ_i, y_i and e_i denote, respectively, the relative risk (RR), observed and expected number of cases (deaths) over a given period in the ith area ($i = 1, \ldots, m$). An HB version of the Poisson-gamma model, given in Section 9.6.1, may be written as

$$\begin{aligned}
&\text{(i)} \quad y_i|\theta_i \overset{\text{ind}}{\sim} \text{Poisson}(e_i\theta_i),\\
&\text{(ii)} \quad \theta_i|\alpha, \nu \overset{\text{iid}}{\sim} G(\nu, \alpha),\\
&\text{(iii)} \quad f(\alpha, \nu) \propto f(\alpha)f(\nu),\\
&\qquad \text{with } f(\alpha) \propto 1/\alpha; \quad \nu \sim G(a = 1/2, b), \quad b > 0; \qquad (10.11.1)
\end{aligned}$$

see Datta, Ghosh, and Waller (2000). The joint posterior $f(\boldsymbol{\theta}, \alpha, \nu|\mathbf{y})$ is proper if at least one y_i is greater than zero. It is easy to verify that the Gibbs conditionals are given by

$$\begin{aligned}
&\text{(i)} \quad [\theta_i|\alpha, \nu, \mathbf{y}] \overset{\text{ind}}{\sim} G(y_i + \nu, e_i + \alpha),\\
&\text{(ii)} \quad [\alpha|\boldsymbol{\theta}, \nu, \mathbf{y}] \sim G\left(m\nu, \sum_{i=1}^{m} \theta_i\right),\\
&\text{(iii)} \quad f(\nu|\boldsymbol{\theta}, \alpha, \mathbf{y}) \propto \left(\prod_{i=1}^{m} \theta_i^{\nu-1}\right) \exp(-b\nu)\alpha^{\nu m}/\Gamma^m(\nu). \qquad (10.11.2)
\end{aligned}$$

MCMC samples can be generated directly from (i) and (ii) of (10.11.2), but we need to use the M–H algorithm to generate samples from (iii) of (10.11.2). Using the MCMC samples $\{(\theta_1^{(k)}, \ldots, \theta_m^{(k)}, \nu^{(k)}, \alpha^{(k)}); k = d+1, \ldots, d+D\}$, posterior quantities of interest may be computed; in particular, the posterior mean $E(\theta_i|\mathbf{y})$ and posterior variance $V(\theta_i|\mathbf{y})$ for each area $i = 1, \ldots, m$.

10.11.2 Log-Normal Model

An HB version of the basic log-normal model, given in Section 9.6.2, may be written as

$$\text{(i)}\quad y_i|\theta_i \overset{\text{ind}}{\sim} \text{Poisson}(e_i\theta_i),$$

$$\text{(ii)}\quad \xi_i = \log(\theta_i)|\mu, \sigma^2 \overset{\text{iid}}{\sim} N(\mu, \sigma^2),$$

$$\text{(iii)}\quad f(\mu, \sigma^2) \propto f(\mu)f(\sigma^2) \text{ with}$$

$$f(\mu) \propto 1; \quad \sigma^{-2} \sim G(a, b), \quad a \geq 0, \quad b > 0. \qquad (10.11.3)$$

The joint posterior $f(\boldsymbol{\theta}, \mu, \sigma^2|\mathbf{y})$ is proper if at least one y_i is greater than zero. It is easy to verify that the Gibbs conditionals are given by

$$\text{(i)}\quad f(\theta_i|\mu, \sigma^2, \mathbf{y}] \propto \theta_i^{y_i - 1} \exp\left[-e_i\theta_i - \frac{1}{2\sigma^2}(\xi_i - \mu)^2\right],$$

$$\text{(ii)}\quad [\mu|\boldsymbol{\theta}, \sigma^2, \mathbf{y}] \sim N\left(\frac{1}{m}\sum_{i=1}^m \xi_i, \frac{\sigma^2}{m}\right),$$

$$\text{(iii)}\quad [\sigma^2|\boldsymbol{\theta}, \mu, \mathbf{y}] \sim G\left(\frac{m}{2} + a, \frac{1}{2}\sum_{i=1}^m (\xi_i - \mu)^2 + b\right); \qquad (10.11.4)$$

see Maiti (1998). MCMC samples can be generated directly from (ii) and (iii) of (10.11.4), but we need to use the M–H algorithm to generate samples from (i) of (10.11.4). We can express (i) as $f(\theta_i|\mu, \sigma^2, \mathbf{y}) \propto k(\theta_i)h(\theta_i|\mu, \sigma^2)$, where $k(\theta_i) = \exp(-e_i\theta_i)\theta_i^{y_i}$ and $h(\theta_i|\mu, \sigma^2) \propto g'(\theta_i)\exp\{-(\xi_i - \mu)^2/2\sigma^2\}$ with $g'(\theta_i) = \partial g(\theta_i)/\partial\theta_i$ and $g(\theta_i) = \log(\theta_i)$. We can use $h(\theta_i|\mu, \sigma^2)$ to draw the candidate, θ_i^*, noting that $\theta_i = g^{-1}(\xi_i)$ and $\xi_i|\mu, \sigma^2 \sim N(\mu, \sigma^2)$. The acceptance probability used in the M–H algorithm is then given by $a(\theta^{(k)}, \theta_i^*) = \min\{k(\theta_i^*)/k(\theta_i^{(k)}), 1\}$.

As noted in Section 9.6.2, the basic log-normal model with Poisson counts, y_i, readily extends to the case of covariates $\mathbf{z}_i$ by changing (ii) of (10.11.3) to $\xi_i|\boldsymbol{\beta}, \sigma^2 \sim N(\mathbf{z}_i^T\boldsymbol{\beta}, \sigma^2)$. Furthermore, we change (iii) of (10.11.3) to $f(\boldsymbol{\beta}, \sigma^2) \propto f(\boldsymbol{\beta})f(\sigma^2)$ with

$f(\boldsymbol{\beta}) \propto 1$ and $\sigma^{-2} \sim G(a, b)$. Also, the basic model can be extended to allow spatial covariates. An HB version of the spatial CAR-normal model is given by

(i) $y_i|e_i \sim \text{Poisson}(e_i\theta_i)$,

(ii) $\xi_i|\xi_{j(j\neq i)}, \rho, \sigma^2 \sim N\left[\mu + \rho\sum_{\ell=1}^{m} q_{i\ell}(\xi_\ell - \mu), \sigma^2\right]$,

(iii) $f(\mu, \sigma^2, \rho) \propto f(\mu)f(\sigma^2)f(\rho)$ with

$$f(\mu) \propto 1; \qquad \sigma^{-2} \sim G(a, b),\ a \geq 0, b > 0; \qquad \rho \sim U(0, \rho_0), \tag{10.11.5}$$

where ρ_0 denotes the maximum value of ρ in the CAR model, and $Q = (q_{i\ell})$ is the "adjacency" matrix of the map with $q_{i\ell} = q_{\ell i}$, $q_{i\ell} = 1$ if i and ℓ are adjacent areas and $q_{i\ell} = 0$ otherwise.

Maiti (1998) obtained the Gibbs conditionals. In particular, $[\mu|\boldsymbol{\theta}, \sigma^2, \rho, \mathbf{y}]$ is distributed as normal, $[\sigma^{-2}|\boldsymbol{\theta}, \mu, \rho, \mathbf{y}]$ gamma, $[\rho|\boldsymbol{\theta}, \mu, \sigma^2, \mathbf{y}]$ truncated normal, and $[\theta_i|\theta_{j(j\neq i)}, \mu, \sigma^2, \rho, \mathbf{y}]$ does not admit a closed form in the sense that the conditional is known only up to a multiplicative constant. MCMC samples can be generated directly from the first three conditionals, but we need to use the M–H algorithm to generate samples from the conditionals $[\theta_i|\theta_{j(j\neq i)}, \mu, \sigma^2, \rho, \mathbf{y}]$, $i = 1, \ldots, m$.

Example 10.11.1. Lip Cancer. In Example 9.6.1, EB estimation was applied to lip cancer counts, y_i, in each of 56 counties of Scotland. Maiti (1998) applied HB estimator to the same data using the log-normal and the CAR-normal models. The HB estimates $E(\theta_i|\mathbf{y})$ of lip cancer incidence are very similar for the two models, but the standard errors, $\sqrt{V(\theta_i|\mathbf{y})}$, are smaller for the CAR-normal model as it exploits the spatial structure of the data.

Ghosh et al. (1999) proposed a spatial log-normal model that allows covariates $\mathbf{z}_i$, given by

(i) $y_i|e_i \overset{\text{ind}}{\sim} \text{Poisson}(e_i\theta_i)$

(ii) $\xi_i = \mathbf{z}_i^T\boldsymbol{\beta} + u_i + v_i$ where $\mathbf{z}_i^T\boldsymbol{\beta}$ does not include an intercept term,

$v_i \overset{\text{iid}}{\sim} N(0, \sigma_v^2)$ and the u_i's have joint density

$$f(\mathbf{u}) \propto (\sigma_u^2)^{-m/2}\exp\left[-\sum_{i=1}\sum_{\ell\neq i}(u_i - u_\ell)^2 q_{i\ell}/(2\sigma^2)\right],\ \text{where } q_{i\ell} \geq 0$$

for all $1 \leq i \neq \ell \leq m$,

(iii) $\boldsymbol{\beta}, \sigma_u^2$ and σ_v^2 are mutually independent with $f(\boldsymbol{\beta}) \propto 1$,

$$\sigma_u^{-2} \sim G(a_u, b_u) \text{ and } \sigma_v^{-2} \sim G(a_v, b_v). \tag{10.11.6}$$

Ghosh et al. (1999) showed that all the Gibbs conditionals admit closed forms except for $[\theta_i | \theta_{\ell(\ell \neq i)}, \boldsymbol{\beta}, \boldsymbol{\mu}, \sigma_u^2, \sigma_v^2, \mathbf{y}]$. They also established conditions for the propriety of the joint posterior; in particular, we need $b_u > 0, b_v > 0$.

Example 10.11.2. Leukemia Incidence. Ghosh et al. (1999) applied the HB method, based on the model (10.11.6), to leukemia incidence estimation for $m = 281$ census tracts (small areas) in an eight-county region of upstate New York. Here $q_{i\ell} = 1$ if i and ℓ are neighbors and $q_{i\ell} = 0$ otherwise, and $\mathbf{z}_i$ is a scalar ($p = 1$) variable z_i denoting the inverse distance of the centroid of the ith census tract from the nearest hazardous waste site containing trichloroethylene (TCE), a common contaminant of ground water. We refer the reader to Ghosh et al. (1999) for further details.

10.11.3 Two-Level Models

Let y_{ij} and n_{ij} denote, respectively, the number of cases (deaths) and the population at risk in the jth age class in the ith area ($j = 1, \ldots, J$, $i = 1, \ldots, m$). Using the data $\{y_{ij}, n_{ij}\}$, it is of interest to estimate the age-specific mortality rates τ_{ij} and the age-adjusted rates $\sum_{j=1}^{n_i} a_j \tau_{ij}$, where the a_j's are specified constants. The basic assumption is

$$y_{ij} | \tau_{ij} \overset{\text{ind}}{\sim} \text{Poisson}(n_{ij}\tau_{ij}). \tag{10.11.7}$$

Nandram, Sedransk, and Pickle (1999) studied HB estimation under different linking models:

$$\log(\tau_{ij}) = \mathbf{z}_j^T \boldsymbol{\beta} + v_i, \quad v_i | \sigma_v^2 \overset{\text{iid}}{\sim} N(0, \sigma_v^2), \tag{10.11.8}$$

$$\log(\tau_{ij}) = \mathbf{z}_j^T \boldsymbol{\beta}_i, \quad \boldsymbol{\beta}_i | \boldsymbol{\beta}, \boldsymbol{\Delta} \overset{\text{iid}}{\sim} N_p(\boldsymbol{\beta}, \boldsymbol{\Delta}), \tag{10.11.9}$$

$$\log(\tau_{ij}) = \mathbf{z}_j^T \boldsymbol{\beta}_i + \delta_j, \quad \boldsymbol{\beta}_i | \boldsymbol{\beta}, \boldsymbol{\Delta} \overset{\text{iid}}{\sim} N_p(\boldsymbol{\beta}, \boldsymbol{\Delta}), \delta_j \overset{\text{iid}}{\sim} N(0, \sigma^2), \tag{10.11.10}$$

where $\mathbf{z}_j$ is a $p \times 1$ vector of covariates and δ_j is an "offset" corresponding to age class j. Nandram et al. (1999) assumed the flat prior $f(\boldsymbol{\beta}) \propto 1$ and proper diffuse (i.e., proper with very large variance) priors for σ_v^2, $\boldsymbol{\Delta}$, and σ^2. For model selection, they used the posterior EPD, the posterior predictive value, and measures based on the cross-validation predictive densities (see Section 10).

Example 10.11.3. Mortality Rates. Nandram et al. (1999) applied the HB method to estimate age-specific and age-adjusted mortality rates for U.S. Health Service Areas (HSAs). They studied one of the disease categories, all cancer for white males, presented in the 1996 Atlas of United States Mortality. The number of HSAs (small areas), m, is 798 and the number of age categories, J, is 10: $0-4, 5-14, \ldots, 75-84, 85$ and higher, coded as $0.25, 1, \ldots, 9$. The vector of auxiliary variables is given by $\mathbf{z}_j = [1, j-1, (j-1)^2, (j-1)^3, \max\{0, ((d-1) - \text{knot})^3\}]^T$ for $j \geq 2$ and $\mathbf{z}_1 = [1, 0.25, (0.25)^2, (0.25)^3, \max\{0, (0.25 - \text{knot})^3\}]^T$, where the value of the "knot" was obtained by maximizing the likelihood based on marginal deaths, $y_j = \sum_{i=1}^m y_{ij}$, and population at risk, $n_j = \sum_{i=1}^m n_{ij}$, where $y_j|n_j, \tau_j \stackrel{\text{ind}}{\sim} \text{Poisson}(n_j\tau_j)$ with $\log(\tau_j) = \mathbf{z}_j^T\boldsymbol{\beta}$. The auxiliary vector $\mathbf{z}_j$ was used in the Atlas model based on a normal approximation to $\log(r_{ij})$ with mean $\log(\tau_{ij})$ and matching linking model given by (10.11.9), where $r_{ij} = y_{ij}/n_{ij}$ is the crude rate. Nandram et al. (1999) used unmatched sampling and linking models based on the Poisson sampling model (10.11.7) and the linking models (10.11.8)–(10.11.10). We denote these models as Models 1, 2, and 3, respectively.

Nandram et al. (1999) used the MCMC samples generated from the three models to calculate the values of the posterior expected predictive deviance $E[\Delta(\mathbf{y}; \mathbf{y}_{\text{obs}})|\mathbf{y}_{\text{obs}}]$ using the chi-square measure $\Delta(\mathbf{y}, \mathbf{y}_{\text{obs}}) = \sum_{i=1}^m \sum_{j=1}^{n_i} (y_{ij} - y_{ij,\text{obs}})^2/(y_{ij} + 0.5)$. They also calculated the posterior predictive p-values, using $T(\mathbf{y}, \boldsymbol{\tau}) = \sum_{i=1}^m \sum_{j=1}^{n_i} (y_{ij} - n_{ij}\tau_{ij})^2/(n_{ij}\tau_{ij})$, the standardized cross-validation residuals

$$d_{2,ij}^* = \frac{r_{ij,\text{obs}} - E(r_{ij}|\mathbf{y}_{(ij),\text{obs}})}{\sqrt{V(r_{ij}|\mathbf{y}_{(ij),\text{obs}})}}, \tag{10.11.11}$$

where $\mathbf{y}_{(ij),\text{obs}}$ denotes all elements of $\mathbf{y}_{\text{obs}}$ except for $y_{ij,\text{obs}}$; see (10.2.28). The residuals $d_{2,ij}^*$ were summarized by counting (a) the number of elements (i,j) such that $|d_{2,ij}^*| \geq 3$, called "outliers", and (b) the number of HSAs, i, such that $|d_{2,ij}^*| \geq 3$ for at least one j, called "# of HSAs".

Table 10.7 reports the posterior EPD, the posterior predictive p-value, the number of outliers according to the above definition, and the number of HSAs for Models 1–3. It is clear from Table 10.7 that Model 1 performs poorly with respect to all the four measures. Overall, Model 3 with the random age coefficient, δ_j, provides the best fit to the data, although Models 2 and 3 are similar with respect to EPD.

Based on Model 3, Nandram et al. (1999) produced cancer maps of HB estimates of age-specific mortality rates for each age group j. Note that the HB estimate of the mortality rate τ_{ij} is given by the posterior mean $E(\tau_{ij}|\mathbf{y})$. The maps revealed that the mortality rates, for all age classes, are highest among the Appalachian mountain range (Mississippi to West Virginia) and the Ohio River Valley (Illinois to Ohio). Also, the highest rates formed more concentrated clusters in the middle and older age groups (e.g., 45–54), whereas the youngest and oldest age groups exhibited more scattered patterns.

TABLE 10.7 Comparison of Models 1–3: Mortality Rates

Model	EPD	p-value	Outliers	# of HSAs
1	22,307	0.00	284	190
2	16,920	0.00	136	93
3	16,270	0.32	59	54

Source: Adapted from Table 1 in Nandram et al. (1999).

Nandram, Sedransk, and Pickle (2000) used models and methods similar to those in Nandram et al. (1999) to estimate age-specific and age-adjusted mortality rates for chronic obstructive pulmonary disease for white males in HSAs.

10.12 *TWO-PART NESTED ERROR MODEL

Pfeffermann, Terryn, and Moura (2008) studied the case of a response variable, y, taking either the value 0 or a value drawn from a continuous distribution. For example, in the assessment of adult literacy, y is either zero indicating illiteracy or a positive score measuring the literacy level. Pfeffermann et al. (2008) used a twofold, two-part model to estimate the mean of the literacy scores and the proportion of positive scores in each village (subarea) belonging to a district (area) in Cambodia. For simplicity, we consider only a onefold, two-part model here.

Let y_{ij} denote the response (e.g., literacy score) associated with the jth unit in ith area, $\mathbf{z}_{ij}$ a vector of covariates for that unit, and $p_{ij} = \Pr(y_{ij} > 0|\mathbf{z}_{ij}, u_i)$, ($j = 1, \ldots, N_i$, $i = 1, \ldots, m$). The probabilities p_{ij} are then modeled as

$$\text{logit}(p_{ij}) = \mathbf{z}_{ij}^T \boldsymbol{\gamma} + u_i, \quad u_i \stackrel{\text{iid}}{\sim} N(0, \sigma_u^2). \tag{10.12.1}$$

Furthermore, consider another vector of covariates $\mathbf{x}_{ij}$, possibly different from $\mathbf{z}_{ij}$, such that

$$y_{ij}|\mathbf{x}_{ij}^T, v_i, y_{ij} > 0 \sim N(\mathbf{x}_{ij}^T \boldsymbol{\beta} + v_i, \sigma_e^2) \tag{10.12.2}$$

with $v_i \stackrel{\text{iid}}{\sim} N(0, \sigma_v^2)$. Note that (10.12.2) implies

$$E(y_{ij}|\mathbf{x}_{ij}, \mathbf{z}_{ij}, v_i, u_i) = (\mathbf{x}_{ij}^T \boldsymbol{\beta} + v_i)p_{ij}. \tag{10.12.3}$$

The conditional likelihood for the two-part model (10.12.1) and (10.12.2) may be written as

$$L = \prod_{i=1}^{m} \prod_{j \in s_i} p_{ij}^{a_{ij}} [f(y_{ij}|\mathbf{x}_{ij}, v_i, y_{ij} > 0)]^{a_{ij}} (1 - p_{ij})^{1-a_{ij}}, \tag{10.12.4}$$

where $a_{ij} = 1$ if $y_{ij} > 0$, $a_{ij} = 0$ otherwise, and $p_{ij} = E(a_{ij})$. Pfeffermann et al. (2008) introduced a correlation between u_i and v_i by assuming that

$$u_i|v_i \sim N(K_v\, v_i, \sigma_{u|v}^2), \quad i = 1, \ldots, m. \tag{10.12.5}$$

Note that the unknown model parameters are the regression coefficients $\boldsymbol{\beta}$, $\boldsymbol{\gamma}$, and K_v and the variances σ_u^2, σ_e^2, and $\sigma_{u|v}^2$.

By specifying diffuse priors on all model parameters, MCMC samples $\{(\mathbf{v}^{(b)}, \mathbf{u}^{(b)}, \boldsymbol{\beta}^{(b)}, \boldsymbol{\gamma}^{(b)}, K_v^{(b)}, \sigma_v^{2(b)}, \sigma_e^{2(b)}, \sigma_{u|v}^{2(b)}); b = 1, \ldots, B\}$ are generated from the joint posterior distribution using WinBUGS to implement M–H within Gibbs algorithm, where $\mathbf{u} = (u_1, \ldots, u_m)^T$ and $\mathbf{v} = (v_1, \ldots, v_m)^T$. It follows from (10.12.1) and (10.12.3) that the HB estimator of the area mean $\overline{Y}_i$ is given by

$$\hat{\overline{Y}}_i^{\text{HB}} = N_i^{-1}\left\{\sum_{j\in s_i} y_{ij} + \sum_{j\in r_i}\left[\frac{1}{B}\sum_{b=1}^{B} y_{ij}^{(b)}\right]\right\} =: \frac{1}{B}\sum_{b=1}^{B}\hat{\overline{Y}}_i^{\text{HB}}(b), \tag{10.12.6}$$

where $y_{ij}^{(b)} = (\mathbf{x}_{ij}^T\boldsymbol{\beta}^{(b)} + v_i^{(b)})p_{ij}^{(b)}$ and $\text{logit}(p_{ij}^{(b)}) = \mathbf{z}_{ij}^T\boldsymbol{\gamma}^{(b)} + u_i^{(b)}$. Similarly, the HB estimator of the area proportion $P_i = N_i^{-1}\sum_{j=1}^{N_i} a_{ij}$ is given by

$$\hat{P}_i^{\text{HB}} = N_i^{-1}\left\{\sum_{j\in s_i} a_{ij} + \sum_{j\in r_i}\left[\frac{1}{B}\sum_{b=1}^{B} p_{ij}^{(b)}\right]\right\} =: \frac{1}{B}\sum_{b=1}^{B}\hat{P}_i^{\text{HB}}(b). \tag{10.12.7}$$

Posterior variances of $\overline{Y}_i$ and P_i and credible intervals on $\overline{Y}_i$ and P_i are similarly obtained from the generated values $\hat{\overline{Y}}_i^{\text{HB}}(b)$ and $\hat{P}_i^{\text{HB}}(b)$, $b = 1, \ldots, B$. The HB estimators (10.12.6) and (10.12.7) are more efficient than those given in Pfeffermann et al. (2008) because they make use of (10.12.1) and (10.12.3).

10.13 BINARY DATA

In Section 9.5, we studied EB models for binary responses y_{ij}. In this section, we study HB estimation for these models, assuming priors on the model parameters. We also consider extensions to logistic linear mixed models.

10.13.1 Beta-Binomial Model

An HB version of the beta-binomial model studied in Section 9.5.1 is given by

(i) $y_i|p_i \overset{\text{ind}}{\sim} \text{binomial}(n_i, p_i)$,

(ii) $p_i|\alpha, \beta \overset{\text{iid}}{\sim} \text{beta}(\alpha, \beta),\ \alpha > 0, \beta > 0$,

(iii) α and β mutually independent of

$$\alpha \sim G(a_1, b_1),\ a_1 > 0, b_1 > 0,$$

$$\beta \sim G(a_2, b_2),\ a_2 > 0, b_2 > 0. \tag{10.13.1}$$

The Gibbs conditional $[p_i|p_{\ell(\ell\neq i)}, \alpha, \beta, \mathbf{y}]$ is $\text{beta}(y_i + \alpha, n_i - y_i + \beta)$ under model (10.13.1), but the remaining conditionals $[\alpha|\mathbf{p}, \beta, \mathbf{y}]$ and $[\beta|\mathbf{p}, \alpha, \mathbf{y}]$ do not admit closed forms. He and Sun (1998) showed that the latter two conditionals are log-concave if $a_1 \geq 1$ and $a_2 \geq 1$. Using this result, they used adaptive rejection sampling to generate samples from these conditionals. BUGS version 0.5 handles log-concave conditionals using adaptive rejection sampling (Section 10.2.4).

Example 10.13.1. Turkey Hunting. He and Sun (1998) applied the beta-binomial model (10.13.1) to data collected from a mail survey on turkey hunting conducted by Missouri Department of Conservation. The purpose of this survey was to estimate hunter success rates, p_i, and other parameters of interest for $m = 114$ counties in Missouri. Questionnaires were mailed to a random sample of 5151 permit buyers after the 1994 spring turkey hunting season; the response rate after these mailings was 69%. A hunter was allowed at most 14 one-day trips, and each respondent reported the number of trips, whether a trip was success or not, and the county where hunted for each trip. From this data, a direct estimate of p_i is computed as $\hat{p}_i = y_i/n_i$, where n_i is the total number of trips in county i and y_i is the number of successful trips. The direct estimate is not reliable for counties with small n_i; for example, $n_i = 11$ and $y_i = 1$ for county 11. The direct estimate of success rate for the state, $\hat{p} = 10.1\%$, is reliable because of the large overall sample size; the average of HB estimates, $\hat{p}_i^H$, equals 10.2%. For the counties with small n_i, the HB estimates shrink toward $\hat{p}$. For example, $\hat{p}_i = 9.1\%$ and $\hat{p}_i^{\text{HB}} = 10.8\%$ for county 11.

You and Reiss (1999) extended the beta-binomial HB model (10.13.1) to a twofold beta-binomial HB model. They applied the twofold model to estimate the proportions, p_{ij}, within groups, j, in each area i. They also derived Rao–Blackwell estimators of the posterior means and posterior variances. Using data from Statistics Canada's 1997 Homeowner Repair and Renovation Survey, they obtained HB estimates of the response rates, p_{ij}, within response homogeneity groups (RHGs), j, in each province i. The number of RHGs varied widely across provinces; for example, 153 RHGs in Ontario compared to only 8 RHGs in Prince Edward Island. Sample sizes, n_{ij}, varied from 1 to 201. As a result, direct estimates of response rates p_{ij} are unreliable for the RHGs with small sample sizes. HB standard errors were substantially smaller than the direct standard errors, especially for the RHGs with small n_{ij}. For example, the sample size is only 4 for the RHG 17 in Newfoundland, and the direct standard error is 0.22 compared to HB standard error equal to 0.03.

10.13.2 Logit-Normal Model

As noted in Section 9.5.2, the logit-normal model readily allows covariates, unlike the beta-binomial model. We first consider the case of area-specific covariates, $\mathbf{z}_i$, and then study the case of unit level covariates, $\mathbf{x}_{ij}$.

Area Level Covariates An HB version of the logit-normal model with area level covariates may be expressed as

$$
\begin{aligned}
&\text{(i)}\quad y_i|p_i \overset{\text{ind}}{\sim} \text{binomial}(n_i, p_i),\\
&\text{(ii)}\quad \xi_i = \text{logit}(p_i) = \mathbf{z}_i^T\boldsymbol{\beta} + v_i,\ \text{ with } v_i \overset{\text{iid}}{\sim} N(0, \sigma_v^2)\\
&\text{(iii)}\quad \boldsymbol{\beta} \text{ and } \sigma_v^2 \text{ are mutually independent of } f(\boldsymbol{\beta}) \propto 1\\
&\qquad \text{and } \sigma_v^{-2} \sim G(a, b), a \geq 0, b > 0.
\end{aligned}
\tag{10.13.2}
$$

It is easy to verify that the Gibbs conditionals corresponding to the HB model (10.13.2) are given by

$$\text{(i)}\quad [\boldsymbol{\beta}|\mathbf{p},\sigma_v^2,\mathbf{y}] \sim N_p\left[\boldsymbol{\beta}^*,\sigma_v^2\left(\sum_{i=1}^m \mathbf{z}_i\mathbf{z}_i^T\right)^{-1}\right]$$

$$\text{(ii)}\quad [\sigma_v^2|\boldsymbol{\beta},\mathbf{p},\mathbf{y}] \sim G\left[\frac{m}{2}+a,\frac{1}{2}\sum_{i=1}^m(\xi_i-\mathbf{z}_i^T\boldsymbol{\beta})^2+b\right].$$

$$\text{(iii)}\quad f(p_i|\boldsymbol{\beta},\sigma_v^2,\mathbf{y}) \propto h(p_i|\boldsymbol{\beta},\sigma_v^2)k(p_i), \tag{10.13.3}$$

where $\boldsymbol{\beta}^* = (\sum_{i=1}^m \mathbf{z}_i\mathbf{z}_i^T)^{-1}\sum_{i=1}^m \mathbf{z}_i\xi_i$, $k(p_i) = p_i^{y_i}(1-p_i)^{n_i-y_i}$ and

$$h(p_i|\boldsymbol{\beta},\sigma_v^2) \propto g'(p_i)\exp\left\{-\frac{(\xi_i-\mathbf{z}_i^T\boldsymbol{\beta})^2}{2\sigma_v^2}\right\}, \tag{10.13.4}$$

where $g'(p_i) = \partial g(p_i)/\partial p_i$ with $g(p_i) = \text{logit}(p_i)$. It is clear from (10.13.3) that the conditionals (i) and (ii) admit closed forms while (iii) has form similar to (10.4.3). Therefore, we can use $h(p_i|\boldsymbol{\beta},\sigma_v^2)$ to draw the candidate p_i^*, noting that $p_i = g^{-1}(\xi_i)$ and $\xi_i|\boldsymbol{\beta},\sigma_v^2 \sim N(\mathbf{z}_i^T\boldsymbol{\beta},\sigma_v^2)$. In this case, the acceptance probability used in the M–H algorithm is given by $a(p_i^{(k)},p_i^*) = \min\{k(p_i^*)/k(p_i^{(k)}),1\}$.

The HB estimate of p_i and the posterior variance of p_i are obtained directly from the MCMC samples $\{(p_1^{(k)},\ldots,p_m^{(k)},\boldsymbol{\beta}^{(k)},\sigma_v^{2(k)}); k=d+1,\ldots,d+D\}$ generated from the joint posterior $f(p_1,\ldots,p_m,\boldsymbol{\beta},\sigma_v^2|\mathbf{y})$, as

$$\hat{p}_i^{\text{HB}} \approx \frac{1}{D}\sum_{k=d+1}^{d+D} p_i^{(k)} = p_i^{(\cdot)} \tag{10.13.5}$$

and

$$V(p_i|\hat{p}) \approx \frac{1}{D-1}\sum_{k=d+1}^{d+D}(p_i^{(k)}-p_i^{(\cdot)})^2. \tag{10.13.6}$$

Unit Level Covariates In Section 9.5.2, we studied EB estimation for a logistic linear mixed model with unit level covariates $\mathbf{x}_{ij}$, assuming that the model holds for the sample $\{(y_{ij},\mathbf{x}_{ij}); j\in s_i, i=1,\ldots,m\}$. An HB version of this model may be expressed as

(i) $y_{ij}|p_{ij} \overset{\text{ind}}{\sim} \text{Bernoulli}(p_{ij})$,

(ii) $\xi_{ij} = \text{logit}(p_{ij}) = \mathbf{x}_{ij}^T\boldsymbol{\beta} + v_i$, with $v_i \overset{\text{iid}}{\sim} N(0,\sigma_v^2)$

(iii) $\boldsymbol{\beta}$ and σ_v^2 are mutually independent with

$$f(\boldsymbol{\beta}) \propto 1 \text{ and } \sigma_v^{-2} \sim G(a,b), a \geq 0, b > 0. \tag{10.13.7}$$

An alternative prior on $\boldsymbol{\beta}$ and σ_v^2 is the flat prior $f(\boldsymbol{\beta}, \sigma_v^2) \propto 1$.

Let $\{(v_1^{(k)}, \ldots, v_m^{(k)}, \boldsymbol{\beta}^{(k)}, \sigma_v^{2(k)}); k = d+1, \ldots, d+D\}$ denote the MCMC samples generated from the joint posterior $f(v_1, \ldots, v_m, \boldsymbol{\beta}, \sigma_v^2|\mathbf{y})$. Then the HB estimate of the finite population proportion P_i is obtained as

$$\hat{P}_i^{\text{HB}} \approx \frac{1}{N_i}\left[\sum_{j\in s_i} y_{ij} + \frac{1}{D}\sum_{k=d+1}^{d+D}\sum_{\ell\in r_i} p_{i\ell}^{(k)}\right], \tag{10.13.8}$$

where $\text{logit}(p_{ij}^{(k)}) = \mathbf{x}_{ij}^T\boldsymbol{\beta}^{(k)} + v_i^{(k)}$ and r_i is the set of nonsampled units in the ith area. Similarly, the posterior variance $V(P_i|\mathbf{y})$ is obtained as

$$V(P_i|\mathbf{y}) \approx N_i^{-2}\frac{1}{D}\sum_{k=d+1}^{d+D}\left[\sum_{\ell\in r_i} p_{i\ell}^{(k)}(1-p_{i\ell}^{(k)}) + \left(\sum_{\ell\in r_i} p_{i\ell}^{(k)}\right)^2\right]$$
$$-N_i^{-2}\left[\frac{1}{D}\sum_{k=d+1}^{d+D} p_{i\ell}^{(k)}\right]^2; \tag{10.13.9}$$

see (9.5.21). The total $Y_i = N_i P_i$ is estimated by $N_i\hat{P}_i^{\text{HB}}$ and its posterior variance is given by $N_i^2 V(P_i|\mathbf{y})$. Note that the $\mathbf{x}_{i\ell}$-values for $\ell \in r_i$ are needed to implement (10.13.8) and (10.13.9).

The Gibbs conditionals corresponding to the HB model (10.13.7) are given by

(i) $f(\beta_1|\beta_2, \ldots, \beta_p, \mathbf{v}, \mathbf{y}) \propto \prod_{i=1}^{m}\prod_{j\in s_i} p_{ij}^{y_{ij}}(1-p_{ij})^{1-y_{ij}}$,

(ii) $f(\beta_u|\beta_1, \ldots, \beta_{u-1}, \beta_{u+1}, \ldots, \beta_p, \mathbf{v}, \mathbf{y}) \propto \prod_{i=1}^{m}\prod_{j\in s_i} p_{ij}^{y_{ij}}(1-p_{ij})^{1-y_{ij}}$,

(iii) $f(v_i|\boldsymbol{\beta}, \sigma_v^2, \mathbf{y}, v_1, \ldots, v_{i-1}, v_{i+1}, \ldots, v_m)$

$$\propto \prod_{i=1}^{m}\prod_{j\in s_i} p_{ij}^{y_{ij}}(1-p_{ij})^{1-y_{ij}}\exp\left[-v_i^2/(2\sigma_v^2)\right],$$

(iv) $$[\sigma_v^{-2}|\boldsymbol{\beta}, \mathbf{v}, \mathbf{y}] \sim G\left(\frac{m}{2} + a, \frac{1}{2}\sum_{i=1}^{m} v_i^2 + b\right). \tag{10.13.10}$$

It is clear from (10.13.10) that (i)–(iii) do not admit closed forms. Farrell (2000) used the griddy-Gibbs sampler (Ritter and Tanner 1992) to generate MCMC samples from (i) to (iv). We refer the reader to Farrell (2000) for details of the griddy-Gibbs sampler with regard to the HB model (10.13.7). Alternatively, M–H within Gibbs may be used to generate MCMC samples from (i) to (iv) of (10.13.10).

Example 10.13.2. Simulation Study. Farrell (2000) conducted a simulation study on the frequentist performance of the HB method and the approximate EB method of MacGibbon and Tomberlin (1989); see Section 9.5.2 for the approximate EB method. He treated a public use microdata sample of the 1950 United States Census as the population and calculated the true local area female participation rates, P_i, for $m = 20$ local areas selected with probability proportional to size and without replacement from $M = 52$ local areas in the population. Auxiliary variables, $\mathbf{x}_{ij}$, related to p_{ij}, were selected by a stepwise logistic regression procedure. Both unit (or individual) level and area level covariates were selected. Unit level variables included age, marital status, and whether the individual had children or not. Area level variables included average age, proportions of individuals in various marital status categories, and proportion of individuals having children.

Treating the $m = 20$ sampled areas as strata, $R = 500$ independent stratified random samples with $n_i = 50$ individuals in each stratum were drawn. The HB estimates $\hat{P}_i^{\text{HB}}(r)$ and the approximate EB estimates $\hat{P}_i^{\text{EB}}(r)$ were then calculated from each simulation run $r = 1, \ldots, 500$. Using these estimates, the absolute differences

$$\text{AD}_i^{\text{HB}} = \frac{1}{R}\sum_{r=1}^{R} |\hat{P}_i^{\text{HB}}(r) - P_i|, \quad \text{AD}_i^{\text{EB}} = \frac{1}{R}\sum_{r=1}^{R} |\hat{P}_i^{\text{EB}}(r) - P_i|$$

and the mean absolute differences $\overline{\text{AD}}^{\text{HB}} = m^{-1}\sum_{i=1}^{m} \text{AD}_i^{\text{HB}}$ and $\overline{\text{AD}}^{\text{EB}} = m^{-1}\sum_{i=1}^{m} \text{AD}_i^{\text{EB}}$ were obtained. In terms of mean absolute difference, HB performed better than EB, obtaining $\overline{\text{AD}}^{\text{HB}} = 0.0031$ compared to $\overline{\text{AD}}^{\text{EB}} = 0.0056$. Moreover, the absolute difference AD_i^{HB} was smaller than AD_i^{EB} for 16 of the 20 local areas. However, these results should be interpreted with caution because Farrell (2000) used the gamma prior on σ_v^{-2} with both $a = 0$ and $b = 0$, which results in an improper joint posterior.

10.13.3 Logistic Linear Mixed Models

A logistic linear mixed HB model is given by

$$\text{(i)}\quad y_{ij}|p_{ij} \stackrel{\text{ind}}{\sim} \text{Bernoulli}(p_{ij}),$$

$$\text{(ii)}\quad \xi_{ij} = \text{logit}(p_{ij}) = \mathbf{x}_{ij}^T\boldsymbol{\beta} + \mathbf{z}_{ij}^T\mathbf{v}_i \text{ with } \mathbf{v}_i \stackrel{\text{iid}}{\sim} N_q(\mathbf{0}, \boldsymbol{\Sigma}_v),$$

$$\text{(iii)}\quad f(\boldsymbol{\beta}, \boldsymbol{\Sigma}_v) \propto f(\boldsymbol{\Sigma}_v). \tag{10.13.11}$$

Zeger and Karim (1991) proposed the Jeffrey's prior

$$f(\boldsymbol{\Sigma}_v) \propto |\boldsymbol{\Sigma}_v|^{-(q+1)/2}. \tag{10.13.12}$$

Unfortunately, the choice (10.13.12) leads to an improper joint posterior distribution even though all the Gibbs conditionals are proper (Natarajan and Kass 2000). A simple choice that avoids this difficulty is the flat prior $f(\boldsymbol{\Sigma}_v) \propto 1$. An alternative prior is obtained by assuming a Wishart distribution on $\boldsymbol{\Sigma}_v^{-1}$ with parameters reflecting lack of information on $\boldsymbol{\Sigma}_v$. Natarajan and Kass (2000) proposed different priors on $\boldsymbol{\Sigma}_v$ that ensure propriety of the joint posterior.

In the context of small area estimation, Malec et al. (1997) studied a two-level HB model with class-specific covariates, $\mathbf{x}_j$, when the population is divided into classes j. Suppose there are M areas (say, counties) in the population and $y_{ij\ell}$ denotes the binary response variable associated with the ℓth individual in class j and area i ($\ell = 1, \ldots, N_{ij}$). The two-level population HB model of Malec et al. (1997) may be expressed as

$$\begin{aligned}
&\text{(i)} \quad y_{ij\ell}|p_{ij} \stackrel{\text{ind}}{\sim} \text{Bernoulli}(p_{ij})\\
&\text{(ii)} \quad \xi_{ij} = \text{logit}(p_{ij}) = \mathbf{x}_j^T \boldsymbol{\beta}_i\\
&\qquad \boldsymbol{\beta}_i = \mathbf{Z}_i\boldsymbol{\alpha} + \mathbf{v}_i; \quad \mathbf{v}_i \stackrel{\text{iid}}{\sim} N_q(\mathbf{0}, \boldsymbol{\Sigma}_v)\\
&\text{(iii)} \quad f(\boldsymbol{\alpha}, \boldsymbol{\Sigma}_v) \propto 1,
\end{aligned} \tag{10.13.13}$$

where $\mathbf{Z}_i$ is a $p \times q$ matrix of area level covariates. We assume that the model (10.13.13) holds for the sample $\{(y_{ij\ell}, \mathbf{x}_j, \mathbf{Z}_i); \ell \in s_{ij}, i = 1, \ldots, m\}$, where $m \leq M$ is the number of sampled areas and s_{ij} is the sample of n_{ij} individuals in class j and sampled area i.

Let $\{(\boldsymbol{\beta}_1^{(k)}, \ldots, \boldsymbol{\beta}_m^{(k)}, \boldsymbol{\alpha}^{(k)}, \boldsymbol{\Sigma}_v^{(k)}); k = d+1, \ldots, d+D\}$ denote the MCMC samples generated from the joint posterior $f(\boldsymbol{\beta}_1^{(k)}, \ldots, \boldsymbol{\beta}_m^{(k)}, \boldsymbol{\alpha}, \boldsymbol{\Sigma}_v|\mathbf{y})$. For a nonsampled area $i = m+1, \ldots, M$, given $\boldsymbol{\alpha}^{(k)}, \boldsymbol{\Sigma}_v^{(k)}$, we generate $\boldsymbol{\beta}_i^{(k)}$ from $N(\mathbf{Z}_i\boldsymbol{\alpha}^{(k)}, \boldsymbol{\Sigma}_v^{(k)})$, assuming that $\mathbf{Z}_i$ for $i = m+1, \ldots, M$ is observed. Using this value of $\boldsymbol{\beta}_i$, we calculate $p_{ij}^{(k)} = \exp\,(\mathbf{x}_j^{(k)}\boldsymbol{\beta}_i^{(k)})/[1 + \exp\,(\mathbf{x}_j^{(k)}\boldsymbol{\beta}_i^{(k)})]$ for all areas $i = 1, \ldots, M$.

Suppose that we are interested in a finite population proportion corresponding to a collection of areas, say $I \subset \{1, \ldots, M\}$, and a collection of classes, say J; that is,

$$\begin{aligned}
P(I,J) &= \sum_{i\in I}\sum_{j\in J}\sum_{\ell=1}^{N_{ij}} y_{ij} \Big/ \sum_{i\in I}\sum_{j\in J} N_{ij},\\
&=: Y(I,J)/N(I,J).
\end{aligned} \tag{10.13.14}$$

Then the HB estimate of the proportion $P(I,J)$ is given by $\hat{P}^{\text{HB}}(I,J) = \hat{Y}^{\text{HB}}(I,J)/N(I,J)$, where $\hat{Y}^{\text{HB}}(I,J)$ is obtained as

$$\hat{Y}^{\text{HB}}(I,J) = \sum_{i\in I}\sum_{j\in J}\sum_{\ell\in s_{ij}} y_{ij\ell} + \frac{1}{D}\sum_{k=d+1}^{d+D}\left[\sum_{i\in I}\sum_{j\in J}(N_{ij} - n_{ij})p_{ij}^{(k)}\right]. \tag{10.13.15}$$

Similarly, the posterior variance is $V[P(I,J)|\mathbf{y}] = V[Y(I,J)|\mathbf{y}]/[N(I,J)]^2$, where $V[Y(I,J)|\mathbf{y}]$ is obtained as

$$V[Y(I,J)|\mathbf{y}] \approx \frac{1}{D}\sum_{k=d+1}^{d+D}\left\{\sum_{i\in I}\sum_{j\in J}(N_{ij}-n_{ij})p_{ij}^{(k)}(1-p_{ij}^{(k)}) + \left[\sum_{i\in I}\sum_{j\in J}(N_{ij}-n_{ij})p_{ij}^{(k)}\right]^2\right\}$$

$$-\left[\frac{1}{D}\sum_{k=d+1}^{d+D}\sum_{i\in I}\sum_{j\in J}(N_{ij}-n_{ij})p_{ij}^{(k)}\right]^2. \qquad (10.13.16)$$

Note that $n_{ij} = 0$ if an area $i \in I$ is not sampled.

Example 10.13.3. Visits to Physicians. Malec et al. (1997) applied the two-level model (10.13.13) to estimate health-related proportions from the National Health Interview Survey (NHIS) data for the 50 states and the District of Columbia and for specified subpopulations within these 51 areas. The NHIS sample consisted of 200 primary sampling units (psu's) and households sampled within each selected psu. Each psu is either a county or a group of contiguous counties, and the total sample consisted of approximately 50,000 households and 120,000 individuals.

Selection of Variables Individual-level auxiliary variables included demographic variables such as race, age, and sex and socioeconomic variables such as highest education level attained. County-level covariates, such as mortality rates, counts of hospitals and hospital beds, and number of physicians by field of specialization, were also available.

The population was partitioned into classes defined by the cross-classification of race, sex, and age (in 5-year groups). Reliable estimates of the counts, N_{ij}, were available for each county i and class j. The vector of covariates, $\mathbf{x}_j$, used in (ii) of the HB model (10.13.13), is assumed to be the same for all individuals in class j. A particular binary variable y, denoting the presence/absence of at least one doctor visit within the past year, was used to illustrate the model fitting and HB estimation of proportions $P(I,J)$.

The auxiliary variables, $\mathbf{x}_j$, and the area level covariates in the matrix $\mathbf{Z}_i$, used in the linking model (ii), were selected in two steps using PROC LOGISTIC in SAS. In the first step, the variation in the $\boldsymbol{\beta}_i$'s was ignored by setting $\boldsymbol{\beta}_i = \boldsymbol{\beta}$ and the elements of $\mathbf{x}$ were selected from a list of candidate variables and their interactions, using the model $\text{logit}(p_j) = \mathbf{x}_j^T\boldsymbol{\beta}$ together with $y_{ij\ell} \overset{\text{ind}}{\sim} \text{Bernoulli}(p_j)$. The following variables and interactions were selected by the SAS procedures: $\mathbf{x}_j = (1, x_{0,j}, x_{15,j}, x_{25,j}, x_{55,j}, a_j x_{15,j}, a_j x_{25,j}, b_j)^T$, where a_j and b_j are (0,1) variables with $a_j = 1$ if class j corresponds to males, $b_j = 1$ if class j corresponds to whites, and $x_{t,j} = \max(0, c_j - t)$, $t = 0, 15, 20, 25$ with c_j denoting the center point of the age group in class j; for example, age $c_j = 42.5$ if class j corresponds to age group $[40, 45]$, and then $x_{15,j} = \max(0, 42.5 - 15) = 27.5$.

The variables $\mathbf{x}_j$ selected in the first step were then used in the combined model $\text{logit}(p_{ij}) = \mathbf{x}_j^T \mathbf{Z}_i \boldsymbol{\alpha}$ and $y_{ij\ell} \overset{\text{ind}}{\sim} \text{Bernoulli}(p_{ij})$ to select the covariates $\mathbf{Z}_i$, using the SAS procedure. Note that the combined model is obtained by setting $\boldsymbol{\Sigma}_v = \mathbf{0}$. For the particular binary variable y, the choice $\mathbf{Z}_i = \mathbf{I}$ captured between county variation quite well, that is, county level covariates are not needed. But for other binary variables, county-level variables may be needed. Based on the above two-step selection of variables, the final linking model is given by

$$\begin{aligned}\text{logit}(p_{ij}) = \beta_{1i} + \beta_{2i}x_{0,j} + \beta_{3i}x_{15,j} + \beta_{4i}x_{25,j} + \beta_{5i}x_{55,j}\\ + \beta_{6i}a_j x_{15,j} + \beta_{7i}a_j x_{25,j} + \beta_{8i}b_j,\end{aligned} \quad (10.13.17)$$

where $\boldsymbol{\beta}_i = (\beta_{1i}, \ldots, \beta_{8i})^T = \boldsymbol{\alpha} + \mathbf{v}_i$.

Model Fit To determine the model fit, two kinds of cross-validation were conducted. In the first kind, the sample individuals were randomly divided into five groups, while, in the second kind, the sample counties were randomly divided into groups. Let s_{ih} be the set of individuals in the hth group in county i, and $\bar{y}_{ih}$ be the corresponding sample mean. Denote the expectation and variance of $\bar{y}_{ih}$ with respect to the full predictive density by $E_2(\bar{y}_{ih})$ and $V_2(\bar{y}_{ih}) = E_2[E_2(\bar{y}_{ih}) - \bar{y}_{ih}]^2$, respectively. Malec et al. (1997) compared $D_{ih}^2 = [E_2(\bar{y}_{ih}) - \bar{y}_{ih}]^2$ to $V_2(\bar{y}_{ih})$ using the following summary measure:

$$C = \left[\sum_{i=1}^{m}\sum_{h=1}^{5} V_2(\bar{y}_{ih}) \Big/ \sum_{i=1}^{m}\sum_{h=1}^{5} D_{ih}^2\right]^{1/2}. \quad (10.13.18)$$

If the assumed model provides an adequate fit, then $|C - 1|$ should be reasonably small. The values of C were calculated for both types of cross-validation and for each of several subpopulations. These values clearly indicated adequacy of the model in the sense of C values close to 1.

The measure C^2 may also be regarded as the ratio of the average posterior variance to the average MSE of HB estimator, treating $\bar{y}_{ih}$ as the "true" value. Under this interpretation, the posterior variance is consistent with the MSE because the C values are close to 1.

Comparison of Estimators The HB method was compared to the EB method and also to synthetic estimation. In the EB method, the posterior mean and posterior variance were obtained from (10.13.15) and (10.13.16) by setting $\boldsymbol{\alpha}$ and $\boldsymbol{\Sigma}_v$ equal to the corresponding ML estimates and then generating the MCMC samples. Similarly, for synthetic estimation, posterior mean and associated posterior variance were obtained from (10.13.15) and (10.13.16) by setting $\boldsymbol{\beta}_i = \mathbf{Z}_i\boldsymbol{\alpha}$ and $\boldsymbol{\Sigma}_v = \mathbf{0}$ and then generating the MCMC samples. As expected, the EB estimates were close to the corresponding HB estimates, and the EB standard errors were considerably smaller than the corresponding HB standard errors. Synthetic standard errors were also much smaller than the corresponding HB standard errors, but the relative differences in the point estimates were small.

Malec et al. (1997) also conducted an external evaluation in a separate study. For this purpose, they used a binary variable, health-related partial work limitation, which was included in the 1990 U.S. Census of Population and Housing Long Form. Treating the small area census proportions for this variable as "true" values, they compared the estimates corresponding to alternative methods and models to the true values.

Folsom, Shah, and Vaish (1999) used a different logistic linear mixed HB model to produce HB estimates of small area prevalence rates for states and age groups for up to 20 drug use related binary outcomes, using data from pooled National Household Survey on Drug Abuse (NHSDA) surveys. Let $y_{aij\ell}$ denote the value of a binary variable, y, for the ℓth individual in age group $a = 1, \ldots, 4$ belonging to the jth cluster of the ith state. Assume that $y_{aij\ell}|p_{aij\ell} \overset{\text{ind}}{\sim} \text{Bernoulli}(p_{aij\ell})$. A linking model of the logistic linear mixed model type was assumed:

$$\text{logit}(p_{aij\ell}) = \mathbf{x}_{aij\ell}^T \boldsymbol{\beta}_a + v_{ai} + u_{aij}, \tag{10.13.19}$$

where $\mathbf{x}_{aij\ell}$ denotes a $p_a \times 1$ vector of auxiliary variables associated with age group a and $\boldsymbol{\beta}_a$ is the associated regression parameters. Furthermore, the vectors $\mathbf{v}_i = (v_{1i}, \ldots, v_{4i})^T$ and $\mathbf{u}_{ij} = (u_{1ij}, \ldots, u_{4ij})^T$ are assumed to be mutually independent with $\mathbf{v}_i \sim N_4(\mathbf{0}, \boldsymbol{\Sigma}_v)$ and $\mathbf{u}_{ij} \sim N_4(\mathbf{0}, \boldsymbol{\Sigma}_u)$. The model parameters $\boldsymbol{\beta}, \boldsymbol{\Sigma}_u$, and $\boldsymbol{\Sigma}_v$ are assumed to obey the following prior:

$$f(\boldsymbol{\beta}, \boldsymbol{\Sigma}_u^{-1}, \boldsymbol{\Sigma}_v^{-1}) \propto f(\boldsymbol{\Sigma}_u^{-1}) f(\boldsymbol{\Sigma}_v^{-1}), \tag{10.13.20}$$

where $f(\boldsymbol{\Sigma}_u^{-1})$ and $f(\boldsymbol{\Sigma}_v^{-1})$ are proper Wishart densities.

The population model is assumed to hold for the sample (i.e., absence of sample selection bias), but survey weights, $w_{aij\ell}$, were introduced to obtain pseudo-HB estimates and pseudo-HB standard errors, similarly as in Section 10.5.4. We refer the reader to Folsom et al. (1999) for details on MCMC sampling, selection of covariates and validation studies.

10.14 *MISSING BINARY DATA

Nandram and Choi (2002) extended the results of Section 10.13.1 on beta-binomial model to account for missing binary values y_{ij}. Let R_{ij} denote the response indicator associated with jth individual in ith area, where $R_{ij} = 1$ if the individual responds and $R_{ij} = 0$ otherwise. The extended model assumes that

$$y_{ij}|p_i \overset{\text{iid}}{\sim} \text{Bernoilli}(p_i), \tag{10.14.1}$$

$$R_{ij}|\{p_{Ri}, y_{ij} = 0\} \overset{\text{iid}}{\sim} \text{Bernoilli}(p_{Ri}), \tag{10.14.2}$$

$$R_{ij}|\{p_{Ri}, \gamma_i, y_{ij} = 1\} \overset{\text{iid}}{\sim} \text{Bernoilli}(\gamma_i p_{Ri}). \tag{10.14.3}$$

This model allows the response indicator R_{ij} depend on y_{ij}. When $\gamma_i = 1$, we have ignorable nonresponse in the sense that the probability of response does not depend on y_{ij}. The parameter γ_i in (10.14.3) reflects uncertainty about ignorability of the response mechanism for the ith area.

Nandram and Choi (2002) used a beta linking model, $p_i|\alpha, \beta \sim \text{beta}(\alpha, \beta)$, $\alpha > 0$, $\beta > 0$ on p_i, as in Section 10.13.1. Furthermore, p_{Ri} is assumed to obey a beta distribution, $\text{beta}(\alpha_R, \beta_R)$, and γ_i obeys a truncated $G(\nu, \nu)$ distribution in $0 < \gamma_i < p_{Ri}^{-1}$. Suitable proper priors are specified for the model parameters. Let $R_i = \sum_{j=1}^{n_i} R_{ij}$ the number of respondents in area i, $\mathbf{r}$ be the vector of response indicators R_{ij}, $\mathbf{p} = (p_1, \ldots, p_m)^T$, $\mathbf{p}_R = (p_{R1}, \ldots, p_{Rm})^T$, $\boldsymbol{\pi}_R = (\pi_{R1}, \ldots, \pi_{Rm})^T$, where $\pi_{Ri} = \gamma_i p_{Ri}$, and $\mathbf{z} = (z_1, \ldots, z_m)^T$, where $z_i = \sum_{j=R_i+1}^{n_i} y_{ij}$ is the unknown y-total among nonrespondents in area i. MCMC samples are then generated from the joint posterior of the model parameters $\boldsymbol{\theta} = (\alpha, \beta, \alpha_R, \beta_R, \nu)^T$ and the random effects $\mathbf{p}$, $\mathbf{p}_R$, $\boldsymbol{\pi}_R$, and $\mathbf{z}$.

Generation of a sample is done in three steps: (i) Integrate out $\mathbf{p}$, $\mathbf{p}_R$, and $\boldsymbol{\pi}_R$ from the joint posterior density $f(\boldsymbol{\theta}, \mathbf{p}, \mathbf{p}_R, \boldsymbol{\pi}_R, \mathbf{z}|\mathbf{y}, \mathbf{r})$ to obtain the marginal posterior density $f(\boldsymbol{\theta}, \mathbf{z}|\mathbf{y}, \mathbf{r})$. (ii) Generate $(\boldsymbol{\theta}, \mathbf{z})$ from $f(\boldsymbol{\theta}, \mathbf{z}|\mathbf{y}, \mathbf{r})$ using a M–H algorithm and a sample importance resampling (SIR) algorithm. (iii) Obtain (p_i, p_{Ri}, π_{Ri}), $i = 1, \ldots, m$, from the posterior conditional density of (p_i, p_{Ri}, π_{Ri}) given the value of $(\boldsymbol{\theta}, \mathbf{z})$ generated from step (ii) and the data $(\mathbf{y}, \mathbf{r})$. Repeat steps (i)–(iii) a large number of times L to generate $\{p_i^{(\ell)}; \ell = 1, \ldots, L\}$ and take the mean $L^{-1}\sum_{\ell=1}^{L} p_i^{(\ell)}$ as the HB estimate of p_i. The HB estimates of p_{Ri} and γ_i are obtained similarly. We refer the reader to Nandram and Choi (2002) for further details.

Nandram and Choi (2002) applied their HB method to binary data from the U.S. National Crime Survey (NCS) consisting of $m = 10$ domains determined by urbanization, type of place, and poverty level (areas). Domain sample sizes n_i ranged from 10 to 162. Nonresponse in these domains ranged from 9.4% to 16.9%. The parameter p_i here refers to the proportion of households reporting at least one crime in domain i.

Nandram and Choi (2010) proposed a two-part model to estimate mean body mass index (BMI) for small domains from the U.S. National Health and Nutrition Examination Survey (NHANES III) in the presence of nonignorable nonresponse. Part 1 of the model specifies a Bernouilli(p_{Rij}) distribution for the response indicator R_{ij} with $\text{logit}(p_{Rij}) = \beta_{0i} + \beta_{1i} y_{ij}$ and $(\beta_{0i}, \beta_{1i})^T$ are iid and obeying a bivariate normal distribution, where y_{ij} is the BMI for the jth individual in area i. Part 2 of the model assumes a two-level linear mixed model based on covariates related to BMI. The model is extended to include sample selection probabilities into the nonignorable nonresponse model to reflect the higher selection probabilities for black non-Hispanics and Hispanic Americans in NHANES III. HB inferences on the mean BMI for each area are implemented using sophisticated M–H algorithms to draw MCMC samples.

10.15 NATURAL EXPONENTIAL FAMILY MODELS

In Section 4.6.4 of Chapter 4, we assumed that the sample statistics y_{ij} ($j = 1, \ldots, n_i$, $i = 1, \ldots, m$), given the θ_{ij}'s, are independently distributed with probability density function belonging to the natural exponential family with canonical parameters θ_{ij} and known scale parameters $\phi_{ij}(> 0)$:

$$f(y_{ij}|\theta_{ij}) = \exp\left[\frac{1}{\phi_{ij}}[\theta_{ij}y_{ij} - a(\theta_{ij})] + b(y_{ij}, \phi_{ij})\right], \tag{10.15.1}$$

where $a(\cdot)$ and $b(\cdot)$ are known functions. For example, $\theta_{ij} = \text{logit}(p_{ij}), \phi_{ij} = 1$ if $y_{ij} \sim \text{binomial}(n_{ij}, p_{ij})$. The linking model on the θ_{ij}'s is given by

$$\theta_{ij} = \mathbf{x}_{ij}^T\boldsymbol{\beta} + v_i + u_{ij}, \tag{10.15.2}$$

where v_i, and u_{ij} are mutually independent of $v_i \overset{\text{iid}}{\sim} N(0, \sigma_v^2)$ and $u_{ij} \overset{\text{iid}}{\sim} N(0, \sigma_u^2)$, and $\boldsymbol{\beta}$ is a $p \times 1$ vector of covariates without the intercept term. For the HB version of the model, we make an additional assumption that the model parameters $\boldsymbol{\beta}, \sigma_v^2$, and σ_u^2 are mutually independent of $f(\boldsymbol{\beta}) \propto 1$, $\sigma_v^{-2} \sim G(a_v, b_v)$, and $\sigma_u^{-2} \sim G(a_u, b_u)$. The objective here is to make inferences on the small area parameters; in particular, evaluate the posterior quantities $E(\theta_{ij}|\mathbf{y})$, $V(\theta_{ij}|\mathbf{y})$ and $\text{Cov}(\theta_{ij}, \theta_{\ell k}|\mathbf{y})$ for $j \neq k$ or $i \neq \ell$. For example, $\theta_{ij} = \text{logit}(p_{ij})$, where p_{ij} denotes the proportion associated with a binary variable in the jth age-sex group in the ith region.

Ghosh et al. (1998) gave sufficient conditions for the propriety of the joint posterior $f(\boldsymbol{\theta}|\mathbf{y})$. In particular, if y_{ij} is either binomial or Poisson, the conditions are: $b_v > 0$, $b_u > 0$, $\sum_{i=1}^m n_i - p + a_u > 0$, $m + a_v > 0$ and $\sum_{j=1}^{n_i} y_{ij} > 0$ for each i. The posterior is not identifiable (that is, improper) if an intercept term is included in the linking model (10.15.2).

The Gibbs conditionals are easy to derive. In particular, $[\boldsymbol{\beta}|\boldsymbol{\theta}, \mathbf{v}, \sigma_v^2, \sigma_u^2, \mathbf{y}]$ is p-variate normal, $[v_i|\boldsymbol{\theta}, \boldsymbol{\beta}, \sigma_v^2, \sigma_u^2, \mathbf{y}] \overset{\text{ind}}{\sim}$ normal, $[\sigma_v^{-2}|\boldsymbol{\theta}, \boldsymbol{\beta}, \mathbf{v}, \sigma_u^2, \mathbf{y}] \sim$ gamma, and $[\sigma_u^{-2}|\boldsymbol{\theta}, \boldsymbol{\beta}, \mathbf{v}, \sigma_v^2, \mathbf{y}] \sim$ gamma, but $[\theta_{ij}|\boldsymbol{\beta}, \mathbf{v}, \sigma_v^2, \sigma_u^2, \mathbf{y}]$ does not admit a closed-form density function. However, $\log[f(\theta_{ij}|\boldsymbol{\beta}, \mathbf{v}, \sigma_v^2, \sigma_u^2, \mathbf{y})]$ is a concave function of θ_{ij}, and therefore one can use the adaptive rejection sampling scheme of Gilks and Wild (1992) to generate samples.

Ghosh et al. (1998) generalized the model given by (10.15.1) and (10.15.2) to handle multicategory data sets. They applied this model to a data set from Canada on exposures to health hazards. Sample respondents in 15 regions of Canada were asked whether they experienced any negative impact of health hazards in the work place. Responses were classified into four categories: 1 = yes, 2 = no, 3 = not exposed, and 4 = not applicable or not stated. Here it was desired to make inferences on the category proportions within each age–sex class j in each region i.

10.16 CONSTRAINED HB

In Section 9.8.1 of Chapter 9, we obtained the constrained Bayes (CB) estimator $\hat{\theta}_i^{\text{CB}}$, given by (9.8.6), that matches the variability of the small area parameters θ_i. This estimator, $\hat{\theta}_i^{\text{CB}}(\boldsymbol{\lambda})$, depends on the unknown model parameters $\boldsymbol{\lambda}$ of the two-stage model $\hat{\theta}_i|\theta_i \stackrel{\text{ind}}{\sim} f(\hat{\theta}_i|\theta_i, \lambda_1)$ and $\theta_i \stackrel{\text{iid}}{\sim} f(\theta_i|\lambda_2)$. Replacing λ by a suitable estimator $\hat{\lambda}$, we obtained an empirical CB (ECB) estimator $\hat{\theta}_i^{\text{ECB}}$.

We can use a constrained HB (CHB) estimator instead of the ECB estimator. The CHB estimator, $\hat{\theta}_i^{\text{CHB}}$, is obtained by minimizing the posterior squared error $E[\sum_{i=1}^m (\theta_i - t_i)^2|\hat{\theta}]$ subject to $t_{\cdot} = \hat{\theta}_{\cdot}^{\text{HB}}$ and $(m-1)^{-1}\sum_{i=1}^m (t_i - t_{\cdot})^2 = E[(m-1)^{-1}\sum_{i=1}^m (\theta_i - \theta_{\cdot})^2|\hat{\theta}]$. Solving this problem, we obtain

$$t_{i\,\text{opt}} = \hat{\theta}_i^{\text{CHB}} = \hat{\theta}_{\cdot}^{\text{HB}} + a(\hat{\theta})(\hat{\theta}_i^{\text{HB}} - \hat{\theta}_{\cdot}^{\text{HB}}), \tag{10.16.1}$$

where $\hat{\theta}_i^{\text{HB}}$ is the HB estimator for area i, $\hat{\theta}_{\cdot}^{\text{HB}} = \sum_{i=1}^m \hat{\theta}_i^{\text{HB}}/m$ and

$$a(\hat{\theta}) = \left[1 + \frac{\sum_{i=1}^m V(\theta_i - \theta_{\cdot}|\hat{\theta})}{\sum_{i=1}^m (\hat{\theta}_i^{\text{HB}} - \hat{\theta}_{\cdot}^{\text{HB}})^2}\right]^{1/2}, \tag{10.16.2}$$

where $\theta_{\cdot} = \sum_{i=1}^m \theta_i/m$. The proof of (10.16.1) follows along the lines of Section 9.12.3.

The estimator $\hat{\theta}_i^{\text{CHB}}$ depends on the HB estimators $\hat{\theta}_i^{\text{HB}}$ and the posterior variances $V(\theta_i - \theta_{\cdot}|\hat{\theta})$, which can be evaluated using MCMC methods. The estimator $\hat{\theta}_i^{\text{CHB}}$ usually employs less shrinking toward the overall average compared to $\hat{\theta}_i^{\text{ECB}}$ (Ghosh and Maiti 1999).

An advantage of the CHB approach is that it readily provides a measure of uncertainty associated with $\hat{\theta}_i^{\text{CHB}}$. Similar to the posterior variance $V(\theta_i|\hat{\theta})$ associated with $\hat{\theta}_i^{\text{HB}}$, we use the posterior MSE, $E[(\theta_i - \hat{\theta}_i^{\text{CHB}})^2|\hat{\theta}]$, as the measure of uncertainty associated with $\hat{\theta}_i^{\text{CHB}}$. This posterior MSE can be decomposed as

$$\begin{aligned} E[(\theta_i - \hat{\theta}_i^{\text{CHB}})^2|\hat{\theta}] &= E[(\theta_i - \hat{\theta}_i^{\text{HB}})^2|\hat{\theta}] + (\hat{\theta}_i^{\text{HB}} - \hat{\theta}_i^{\text{CHB}})^2 \\ &= V(\theta_i|\hat{\theta}) + (\hat{\theta}_i^{\text{HB}} - \hat{\theta}_i^{\text{CHB}})^2. \end{aligned} \tag{10.16.3}$$

It is clear from (10.16.3) that the posterior MSE is readily obtained from the posterior variance $V(\theta_i|\hat{\theta})$ and the estimators $\hat{\theta}_i^{\text{HB}}$ and $\hat{\theta}_i^{\text{CHB}}$. On the other hand, it appears difficult to obtain a nearly unbiased estimator of MSE of the ECB estimator $\hat{\theta}_i^{\text{ECB}}$. The jackknife method used in Chapter 9 is not readily applicable to $\hat{\theta}_i^{\text{ECB}}$.

10.17 *APPROXIMATE HB INFERENCE AND DATA CLONING

We return to the general two-stage modeling set up of Section 10.1, namely the conditional density $f(\mathbf{y}|\boldsymbol{\mu}, \lambda_1)$ of $\mathbf{y}$ given $\boldsymbol{\mu}$ and λ_1, the conditional density $f(\boldsymbol{\mu}|\lambda_2)$ of $\boldsymbol{\mu}$ given λ_2, and the prior density $f(\boldsymbol{\lambda})$ of $\boldsymbol{\lambda} = (\boldsymbol{\lambda}_1^T, \boldsymbol{\lambda}_2^T)^T$. The joint posterior density of $\boldsymbol{\mu}$ and $\boldsymbol{\lambda}$ may be expressed as

$$\begin{aligned} f(\boldsymbol{\mu}, \boldsymbol{\lambda}|\mathbf{y}) &= f(\boldsymbol{\mu}|\boldsymbol{\lambda}, \mathbf{y})f(\boldsymbol{\lambda}|\mathbf{y}) \\ &= f(\mathbf{y}|\boldsymbol{\mu}, \boldsymbol{\lambda}_1)f(\boldsymbol{\mu}|\boldsymbol{\lambda}_2)f(\boldsymbol{\lambda})/f_1(\mathbf{y}) \end{aligned} \tag{10.17.1}$$

where $f_1(\mathbf{y})$ is the marginal density of $\mathbf{y}$. The posterior density of λ is then given by

$$f(\lambda|\mathbf{y}) = L(\lambda;\mathbf{y})f(\lambda)/f_1(\mathbf{y}), \tag{10.17.2}$$

where $L(\lambda;\mathbf{y})$ is the likelihood function of the data $\mathbf{y}$, given by

$$L(\lambda;\mathbf{y}) = \int f(\mathbf{y}|\boldsymbol{\mu},\lambda_1)f(\boldsymbol{\mu}|\lambda_2)d\boldsymbol{\mu}. \tag{10.17.3}$$

A first-order approximation to the posterior density $f(\lambda|\mathbf{y})$ is given by the Normal density with mean equal to the ML estimate $\hat{\lambda}$ of λ and covariance matrix $I(\hat{\lambda})$, the Fisher information matrix evaluated at $\hat{\lambda}$. Note this approximation does not depend on the prior density of λ. Lele, Nadeem, and Schmuland (2010) suggested the use of this approximation in (10.1.1) to generate MCMC samples $\{(\boldsymbol{\mu}^{(r)},\lambda^{(r)}); r = 1, \ldots, R\}$ by employing WinBUGS and use the $\boldsymbol{\mu}^{(r)}$-values to make posterior inferences on $\boldsymbol{\mu}$. In particular, credible intervals, posterior means and posterior variances for $\boldsymbol{\mu}$ may be computed. Kass and Steffey (1989) in fact used this approximation to get an analytical approximation to the posterior variance of θ_i in the basic area level model (10.3.1), leading to $V_{KS}(\theta_i|\hat{\boldsymbol{\theta}})$ given by (9.2.26).

For complex two-stage models, the computation of ML estimates of model parameters λ can be difficult. Lele et al. (2010) proposed a "data cloning" method that exploits MCMC generation to compute ML estimate $\hat{\lambda}$ and its asymptotic covariance matrix $I^{-1}(\hat{\lambda})$ routinely. This method essentially creates a large number, K, of clones of the data, $\mathbf{y}$, and takes the likelihood function of the cloned data $\mathbf{y}^{(K)} = (\mathbf{y}^T, \ldots, \mathbf{y}^T)^T$ as $[L(\lambda;\mathbf{y})]^K$.

The posterior density of λ given the cloned data $\mathbf{y}^{(K)}$ is equal to

$$f(\lambda|\mathbf{y}^{(K)}) = [L(\lambda;\mathbf{y})]^K f(\lambda)/f_1(\mathbf{y}^{(K)}), \tag{10.17.4}$$

where $f_1(\mathbf{y}^{(K)})$ is the marginal density of $\mathbf{y}^{(K)}$. Under regularity conditions, $f(\lambda|\mathbf{y}^{(K)})$ is approximately the normal density with mean $\hat{\lambda}$ and covariance matrix $K^{-1}I^{-1}(\hat{\lambda})$. Hence, this distribution is nearly degenerate at $\hat{\lambda}$ if K is large. The mean of the posterior distribution given in (10.17.4) is taken as the ML estimate $\hat{\lambda}$ and K times the posterior covariance matrix as the asymptotic covariance matrix of $\hat{\lambda}$. We simply generate MCMC samples $\{\lambda^{(r)}; r = 1, \ldots, R\}$ from $f(\lambda|\mathbf{y}^{(K)})$ and take $\lambda^{(\cdot)} = R^{-1}\sum_{r=1}^{R}\lambda^{(r)}$ as $\hat{\lambda}$ and $R^{-1}\sum_{r=1}^{R}(\lambda^{(r)} - \lambda^{(\cdot)})(\lambda^{(r)} - \lambda^{(\cdot)})^T$ as the asymptotic covariance matrix of $\hat{\lambda}$, provided the parameter space is continuous (Lele et al. 2010). Torabi and Shokoohi (2015) applied the above approximate HB and data cloning approach to complex two-stage models, involving semi-parametric linear, logistic and Poisson mixed models based on spline approximations to the mean function. Their simulation results indicate good performance of the method in terms of coverage probability and absolute relative bias of the MSE estimator obtained from MCMC samples.

10.18 PROOFS

10.18.1 Proof of (10.2.26)

Noting that $f(\mathbf{y}) = f(\mathbf{y}, \boldsymbol{\eta})/f(\boldsymbol{\eta}|\mathbf{y})$, we express $f(y_r|\mathbf{y}_{(r)})$ as

$$f(y_r|\mathbf{y}_{(r)}) = \frac{f(\mathbf{y})}{f(\mathbf{y}_{(r)})} = \frac{1}{\displaystyle\int \frac{f(\mathbf{y}_{(r)}, \boldsymbol{\eta})}{f(\mathbf{y}, \boldsymbol{\eta})} f(\boldsymbol{\eta}|\mathbf{y}) d\boldsymbol{\eta}}$$

$$= \frac{1}{\displaystyle\int \frac{1}{f(y_r|\mathbf{y}_{(r)}, \boldsymbol{\eta})} f(\boldsymbol{\eta}|\mathbf{y}) d\boldsymbol{\eta}}. \tag{10.18.1}$$

The denominator of (10.18.1) is the expectation of $1/f(y_r|\mathbf{y}_{(r)}, \boldsymbol{\eta})$ with respect to $f(\boldsymbol{\eta}|\mathbf{y})$. Hence, we can estimate (10.18.1) by (10.2.26) using the MCMC output $\{\boldsymbol{\eta}^{(k)}\}$.

10.18.2 Proof of (10.2.32)

We write $E[a(y_r)|\mathbf{y}_{(r),\text{obs}}]$ as

$$E[a(y_r)|\mathbf{y}_{(r),\text{obs}}] = E_1 E_2[a(y_r)] = E_1[b_r(\boldsymbol{\eta})],$$

where E_2 is the expectation over y_r given $\boldsymbol{\eta}$ and E_1 is the expectation over $\boldsymbol{\eta}$ given $\mathbf{y}_{(r),\text{obs}}$. Note that we have assumed conditional independence of y_r and $\mathbf{y}_{(r)}$ given $\boldsymbol{\eta}$. Therefore,

$$E_1[b_r(\boldsymbol{\eta})] = \int b_r(\boldsymbol{\eta}) f(\boldsymbol{\eta}|\mathbf{y}_{(r),\text{obs}}) d\boldsymbol{\eta}$$

$$= f(y_{r,\text{obs}}|\mathbf{y}_{(r),\text{obs}}) \int \frac{b_r(\boldsymbol{\eta})}{f(y_{r,\text{obs}}|\boldsymbol{\eta})} f(\boldsymbol{\eta}|\mathbf{y}_{\text{obs}}) d\boldsymbol{\eta}, \tag{10.18.2}$$

noting that (i) $f(\boldsymbol{\eta}|\mathbf{y}_{(r)}) = f(\mathbf{y}_{(r)}|\boldsymbol{\eta}) f(\boldsymbol{\eta})/f(\mathbf{y}_{(r)})$, (ii) $f(\boldsymbol{\eta}) = f(\mathbf{y}) f(\boldsymbol{\eta}|\mathbf{y})/f(\mathbf{y}|\boldsymbol{\eta})$, (iii) $f(\mathbf{y})/f(\mathbf{y}_{(r)}) = f(\mathbf{y}|\mathbf{y}_{(r)})$, and (iv) $f(\mathbf{y}|\boldsymbol{\eta}) = f(\mathbf{y}_{(r)}, y_r|\boldsymbol{\eta}) = f(\mathbf{y}_{(r)}|\boldsymbol{\eta}) f(y_r|\boldsymbol{\eta})$. The integral in (10.18.2) is the expectation of $b_r(\boldsymbol{\eta})/f(y_{r,\text{obs}}|\boldsymbol{\eta})$ with respect to $f(\boldsymbol{\eta}|\mathbf{y}_{\text{obs}})$. Hence, (10.18.2) may be estimated by (10.2.32) using the MCMC output $\{\boldsymbol{\eta}^{(k)}\}$.

10.18.3 Proof of (10.3.13)–(10.3.15)

The Gibbs conditional $[\boldsymbol{\beta}|\boldsymbol{\theta}, \sigma_v^2, \hat{\boldsymbol{\theta}}]$ may be expressed as

$$f(\boldsymbol{\beta}|\boldsymbol{\theta}, \sigma_v^2, \hat{\boldsymbol{\theta}}) = \frac{f(\boldsymbol{\beta}, \boldsymbol{\theta}, \sigma_v^2|\hat{\boldsymbol{\theta}})}{\int f(\boldsymbol{\beta}, \boldsymbol{\theta}, \sigma_v^2|\hat{\boldsymbol{\theta}}) d\boldsymbol{\beta}}$$

$$\propto f(\boldsymbol{\beta}, \boldsymbol{\theta}, \sigma_v^2|\hat{\boldsymbol{\theta}}) \tag{10.18.3}$$

because the denominator of (10.18.3) is constant with respect to $\boldsymbol{\beta}$. Retaining terms involving only $\boldsymbol{\beta}$ in $f(\boldsymbol{\beta}, \boldsymbol{\theta}, \sigma_v^2|\hat{\boldsymbol{\theta}})$ and letting $\tilde{\mathbf{z}}_i = \mathbf{z}_i/b_i$ and $\tilde{\theta}_i = \theta_i/b_i$, we get

$$\log[f(\boldsymbol{\beta}|\boldsymbol{\theta}, \sigma_v^2, \hat{\boldsymbol{\theta}})] = \text{const} - \frac{1}{2\sigma_v^2}\boldsymbol{\beta}^T\left(\sum_{i=1}^m \tilde{\mathbf{z}}_i\tilde{\mathbf{z}}_i^T\right)\boldsymbol{\beta} - 2\sum_{i=1}^m \tilde{\theta}_i\, \tilde{\mathbf{z}}_i^T\boldsymbol{\beta}$$

$$= \text{const} - \frac{1}{2\sigma_v^2}\left[(\boldsymbol{\beta} - \boldsymbol{\beta}^*)^T\left(\sum_{i=1}^m \tilde{\mathbf{z}}_i\tilde{\mathbf{z}}_i^T\right)^{-1}(\boldsymbol{\beta} - \boldsymbol{\beta}^*)\right] \quad (10.18.4)$$

where $\boldsymbol{\beta}^* = (\sum_{i=1}^m \tilde{\mathbf{z}}_i\tilde{\mathbf{z}}_i^T)^{-1}\sum_{i=1}^m \tilde{\mathbf{z}}_i\tilde{\theta}_i$. It follows from (10.18.4) that $[\boldsymbol{\beta}|\boldsymbol{\theta}, \sigma_v^2, \hat{\boldsymbol{\theta}}]$ is $N_p[\boldsymbol{\beta}^*, \sigma_v^2(\sum_{i=1}^m \tilde{\mathbf{z}}_i\tilde{\mathbf{z}}_i^T)^{-1}]$. Similarly,

$$\log[f(\theta_i|\boldsymbol{\beta}, \sigma_v^2, \hat{\boldsymbol{\theta}})] = \text{const} - \frac{1}{2}\left[\frac{\theta_i^2}{\gamma_i\psi_i} - 2\frac{\theta_i\hat{\theta}_i}{\psi_i} - 2\frac{\theta_i\, \mathbf{z}_i^T\boldsymbol{\beta}}{\sigma_v^2 b_i^2}\right]$$

$$= \text{const} - \frac{1}{2\gamma_i\psi_i}[\theta_i - \hat{\theta}_i^B(\boldsymbol{\beta}, \sigma_v^2)]^2. \quad (10.18.5)$$

It now follows from (10.18.5) that $[\theta_i|\boldsymbol{\beta}, \sigma_v^2, \hat{\boldsymbol{\theta}}]$ is $N[\hat{\theta}_i^B(\boldsymbol{\beta}, \sigma_v^2), \gamma_i\psi_i]$. Similarly,

$$\log[f(\sigma_v^2|\boldsymbol{\beta}, \boldsymbol{\theta}, \hat{\boldsymbol{\theta}})] = \text{const} - \frac{1}{2\sigma_v^2}\sum_{i=1}^m (\tilde{\theta}_i - \tilde{\mathbf{z}}_i^T\boldsymbol{\beta})^2$$

$$- \frac{b}{\sigma_v^2} + \log\left[\frac{1}{(\sigma_v^2)^{m/2+a-1}}\right]. \quad (10.18.6)$$

It now follows from (10.18.6) that the Gibbs conditional $[\sigma_v^{-2}|\boldsymbol{\beta}, \boldsymbol{\theta}, \hat{\boldsymbol{\theta}}]$ is $G[m/2 + a, \sum_{i=1}^m (\tilde{\theta}_i - \tilde{\mathbf{z}}_i^T\boldsymbol{\beta})^2/2 + b]$.

REFERENCES

Ambrosio Flores, L. and Iglesias Martínez, L. (2000), Land Cover Estimation in Small Areas Using Ground Survey and Remote Sensing, *Remote Sensing of Environment*, **74**, 240–248.

Anderson, T.W. (1973), Asymptotically Efficient Estimation of Covariance Matrices with Linear Covariance Structure, *Annals of Statistics*, **1**, 135–141.

Anderson, T.W. and Hsiao, C. (1981), Formulation and Estimation of Dynamic Models Using Panel Data, *Journal of Econometrics*, **18**, 67–82.

Ansley, C.F. and Kohn, R. (1986), Prediction Mean Squared Error for State Space Models with Estimated Parameters, *Biometrika*, **73**, 467–473.

Arima, S., Datta, G.S., and Liseo, B. (2015), Accounting for Measurement Error in Covariates in SAE: An Overview, in M. Pratesi (Ed.), *Analysis of Poverty Data by Small Area Methods*, Hoboken, NJ: John Wiley & Sons, Inc., in print.

Arora, V. and Lahiri, P. (1997), On the Superiority of the Bayes Method over the BLUP in Small Area Estimation Problems, *Statistica Sinica*, **7**, 1053–1063.

Baíllo, A. and Molina, I. (2009), Mean-Squared Errors of Small-Area Estimators Under a Unit-Level Multivariate Model, *Statistics*, **43**, 553–569.

Banerjee, M. and Frees, E.W. (1997), Influence Diagnostics for Linear Longitudinal Models, *Journal of the American Statistical Association*, **92**, 999–1005.

Banerjee, S., Carlin, B.P., and Gelfand, A.E. (2004), *Hierarchical Modeling and Analysis for Spatial Data*, New York: Chapman and Hall.

Bankier, M. (1988), Power Allocation: Determining Sample Sizes for Sub-National Areas, *American Statistician*, **42**, 174–177.

Small Area Estimation, Second Edition. J. N. K. Rao and Isabel Molina.

Bartoloni, E. (2008), Small Area Estimation and the Labour Market in Lombardy's Industrial Districts: a Mathematical Approach, *Scienze Regionali*, **7**, 27–54.

Bates, D., Maechler, M., Bolker, B., Walker, S., Bojesen Christensen, R.H., Singmann, H., and Dai, B. (2014), *lme4: Linear mixed-effects models using Eigen and S4*, R package version 1.1-7.

Battese, G.E., Harter, R.M., and Fuller, W.A. (1988), An Error Component Model for Prediction of County Crop Areas Using Survey and Satellite Data, *Journal of the American Statistical Association*, **83**, 28–36.

Bayarri, M.J. and Berger, J.O. (2000), P Values for Composite Null Models, *Journal of the American Statistical Association*, **95**, 1127–1142.

Beaumont, J.-F., Haziza, D., and Ruiz-Gazen, A. (2013), A Unified Approach to Robust Estimation in Finite Population Sampling, *Biometrika*, **100**, 555–569.

Beckman, R.J., Nachtsheim, C.J., and Cook, R.D. (1987), Diagnostics for Mixed-Model Analysis of Variance, *Technometrics*, **1987**, 413–426.

Béland, Y., Bailie, L., Catlin, G., and Singh, M.P. (2000), An Improved Health Survey Program at Statistics Canada, *Proceedings of the Section on Survey Research Methods*, Washington, DC: American Statistical Association, pp. 671–676.

Bell, W.R. (1999), *Accounting for Uncertainty About Variances in Small Area Estimation*, Bulletin of the International Statistical Institute.

Bell, W.R. (2008), Examining Sensitivity of Small Area Inferences to Uncertainty About Sampling Error Variances, *Proceedings of the Survey Research Section*, American Statistical Association, pp. 327–334.

Bell, W.R., Datta, G.S., and Ghosh, M. (2013), Benchmarking Small Area Estimators, *Biometrika*, **100**, 189–202.

Bell, W.R. and Huang, E.T. (2006), Using the *t*-distribution to Deal with Outliers in Small Area Estimation, in *Proceedings of Statistics*, Canada Symposium on Methodological Issues in Measuring Population Health, Statistics Canada, Ottawa, Canada.

Berg, E.J. and Fuller, W.A. (2014), Small Area Estimation of Proportions with Applications to the Canadian Labour Force Survey, *Journal of Survey Statistics and Methodology*, **2**, 227–256.

Berger, J.O. and Pericchi, L.R. (2001), Objective Bayesian Methods for Model Selection: Introduction and Comparison, in P. Lahiri (ed.), *Model Selection*, Notes - Monograph Series, Volume **38**, Beachwood, OH: Institute of Mathematical Statistics.

Besag, J.E. (1974), Spatial Interaction and the Statistical Analysis of Lattice Systems (with Discussion), *Journal of the Royal Statistical Society, Series B*, **35**, 192–236.

Bilodeau, M. and Srivastava, M.S. (1988), Estimation of the MSE Matrix of the Stein Estimator, *Canadian Journal of Statistics*, **16**, 153–159.

Brackstone, G.J. (1987), Small Area Data: Policy Issues and Technical Challenges, in R. Platek, J.N.K. Rao, C.-E. Särndal, and M.P. Singh (Eds.), *Small Area Statistics*, New York: John Wiley & Sons, Inc., pp. 3–20.

Brandwein, A.C. and Strawderman, W.E. (1990), Stein-Estimation: The Spherically Symmetric Case, *Statistical Science*, **5**, 356–369.

Breckling, J. and Chambers, R. (1988), M-quantiles, *Biometrika*, **75**, 761–771.

Breidt, F.J., Claeskens, G., and Opsomer, J.D. (2005), Model-Assisted Estimation for Complex Surveys Using Penalized Splines, *Biometrika*, **92**, 831–846.

Breslow, N. and Clayton, D. (1993), Approximate Inference in Generalized Linear Mixed Models, *Journal of the American Statistical Association*, **88**, 9–25.

Brewer, K.R.W. (1963), Ratio Estimation and Finite Populations: Some Results Deducible from the Assumption of an Underlying Stochastic Process, *Australian Journal of Statistics*, **5**, 5–13.

Brooks, S.P. (1998), Markov Chain Monte Carlo Method and Its Applications, *Statistician*, **47**, 69–100.

Brooks, S.P., Catchpole, E.A., and Morgan, B.J. (2000), Bayesian Animal Survival Estimation, *Statistical Science*, **15**, 357–376.

Browne, W.J. and Draper, D. (2006), A Comparison of Bayesian and Likelihood-Based Methods for Fitting Multilevel Models, *Bayesian Analysis*, **1**, 473–514.

Butar, F.B. and Lahiri, P. (2003), On Measures of Uncertainty of Empirical Bayes Small-Area Estimators, *Journal of Statistical Planning and Inference*, **112**, 63–76.

Calvin, J.A. and Sedransk, J. (1991), Bayesian and Frequentist Predictive Inference for the Patterns of Care Studies, *Journal of the American Statistical Association*, **86**, 36–48.

Carlin, B.P. and Louis, T.A. (2008), *Bayes and Empirical Bayes Methods for Data Analysis* (3rd ed.), Boca Raton, FL: Chapman and Hall/CRC.

Carroll, R.J., Ruppert, D., Stefanski, L.A., and Crainiceanu, C.M. (2006), *Measurement Error in Nonlinear Models: A Modern Perspective.* (2nd ed.), Boca Raton, FL: Chapman & Hall/CRC.

Casady, R.J. and Valliant, R. (1993), Conditional Properties of Post-stratified Estimators Under Normal Theory, *Survey Methodology*, **19**, 183–192.

Casella, G. and Berger, R.L. (1990), *Statistical Inference*, Belmonte, CA: Wadsworth & Brooks/Cole.

Casella, G. and Hwang, J.T. (1982), Limit Expressions for the Risk of James-Stein Estimators, *Canadian Journal of Statistics*, **10**, 305–309.

Casella, G., Lavine, M., and Robert, C.P. (2001), Explaining the Perfect Sampler, *American Statistician*, **55**, 299–305.

Chambers, R.L. (1986), Outlier Robust Finite Population Estimation. *Journal of the American Statistical Association*, **81**, 1063–1069.

Chambers, R.L. (2005), What If… ? Robust Prediction Intervals for Unbalanced Samples. S3RI Methodology Working Papers M05/05, Southampton Statistical Sciences Research Institute, Southampton, UK.

Chambers, R., Chandra, H., Salvati, N., and Tzavidis, N. (2014), Outlier Robust Small Area Estimation, *Journal of the Royal Statistical Society, Series B*, **76**, 47–69.

Chambers, R., Chandra, H., and Tzavidis, N. (2011), On Bias-Robust Mean Squared Error Estimation for Pseudo-Linear Small Area Estimators, *Survey Methodology*, **37**, 153–170.

Chambers, R.L. and Clark, R.G. (2012), *An Introduction to Model-Based Survey Sampling with Applications*, Oxford: Oxford University Press.

Chambers, R.L. and Feeney, G.A. (1977), Log Linear Models for Small Area Estimation, Unpublished paper, Australian Bureau of Statistics.

Chambers, R. and Tzavidis, N. (2006), M-quantile Models for Small Area Estimation, *Biometrika*, **93**, 255–268.

Chandra, H. and Chambers, R. (2009), Multipurpose Weighting for Small Area Estimation, *Journal of Official Statistics*, **25**, 379–395.

Chandra, H., Salvati, N., and Chambers, R. (2007), Small Area Estimation for Spatially Correlated Populations - A Comparison of Direct and Indirect Model-Based Estimators, *Statistics in Transition*, **8**, 331-350.

Chatrchi, G. (2012), *Robust Estimation of Variance Components in Small Area Estimation*, Masters Thesis, School of Mathematics and Statistics, Carleton University, Ottawa, Canada.

Chatterjee, S., Lahiri, P., and Li, H. (2008), Parametric Bootstrap Approximation to the Distribution of EBLUP and Related Prediction Intervals in Linear Mixed Models, *Annals of Statistics*, **36**, 1221–1245.

Chattopadhyay, M., Lahiri, P., Larsen, M., and Reimnitz, J. (1999), Composite Estimation of Drug Prevalences for Sub-State Areas, *Survey Methodology*, **25**, 81–86.

Chaudhuri, A. (1994), Small Domain Statistics: A Review, *Statistica Neerlandica*, **48**, 215–236.

Chaudhuri, A. (2012), *Developing Small Domain Statistics: Modelling in Survey Sampling*, Saarbrücken: LAP LAMBERT Academic Publishing GMbH & Co. KG.

Chen, M-H., Shao, Q-M., and Ibrahim, J.G. (2000), *Monte Carlo Methods in Bayesian Computation*, New York: Springer-Verlag.

Chen, S. and Lahiri, P. (2003), A Comparison of Different MSPE Estimators of EBLUP for the Fay-Herriot Model, *Proceedings of the Section on Survey Research Methods*, Washington, DC: American Statistical Association, pp. 903–911.

Chen, S., Lahiri, P. and Rao, J.N.K. (2008). Robust mean squared prediction error estimators of EBLUP of a small area total under the Fay-Herriot model, Unpublished manuscript.

Cheng, Y.S., Lee, K-O., and Kim, B.C. (2001), Adjustment of Unemployment Estimates Based on Small Area Estimation in Korea, Technical Report, Department of Mathematics, KAIST, Taejun, Korea.

Chib, S. and Greenberg, E. (1995), Understanding the Metropolis-Hastings Algorithm, *American Statistician*, **49**, 327–335.

Choudhry, G.H., Rao, J.N.K., and Hidiroglou, M.A. (2012), On Sample Allocation for Efficient Domain Estimation, *Survey Methodology*, **38**, 23–29.

Christiansen, C.L. and Morris, C.N. (1997), Hierarchical Poisson Regression Modeling, *Journal of the American Statistical Association*, **92**, 618–632.

Christiansen, C.L., Pearson, L.M., and Johnson, W. (1992), Case-Deletion Diagnostics for Mixed Models, *Technometrics*, **34**, 38–45.

Chung, Y.S., Lee, K-O., and Kim, B.C. (2003), Adjustment of Unemployment Estimates Based on Small Area Estimation in Korea, *Survey Methodology*, **29**, 45–52.

Claeskens, G. and Hart, J.D. (2009), Goodness-of-fit tests in mixed models, *Test*, **18**, 213–239.

Clayton, D. and Bernardinelli, L. (1992), Bayesian Methods for Mapping Disease Risk, in P. Elliot, J. Cuzick, D. English, and R. Stern (Eds.), *Geographical and Environmental Epidemiology: Methods for Small-Area Studies*, London: Oxford University Press.

Clayton, D. and Kaldor, J. (1987), Empirical Bayes Estimates of Age-Standardized Relative Risks for Use in Disease Mapping, *Biometrics*, **43**, 671–681.

Cochran, W.G. (1977), *Sampling Techniques* (3rd ed.), New York: John Wiley & Sons, Inc..

Cook, R.D. (1977), Detection of Influential Observations in Linear Regression, *Technometrics*, **19**, 15–18.

Cook, R.D. (1986), Assessment of Local Influence, *Journal of the Royal Statistical Society, Series B*, **48**, 133–155.

Costa, A., Satorra, A., and Ventura, E. (2004), Using Composite Estimators to Improve both Domain and Total Area Estimation, *SORT*, **28**, 69–86.

Cowles, M.K. and Carlin, B.P. (1996), Markov Chain Monte Carlo Convergence Diagnostics: A Comparative Review, *Journal of the American Statistical Association*, **91**, 883–904.

Cressie, N. (1989), Empirical Bayes Estimation of Undercount in the Decennial Census, *Journal of the American Statistical Association*, **84**, 1033–1044.

Cressie, N. (1991), Small-Area Prediction of Undercount Using the General Linear Model, *Proceedings of Statistics Symposium 90: Measurement and Improvement of Data Quality*, Ottawa, Canada: Statistics Canada, pp. 93–105.

Cressie, N. (1992), REML Estimation in Empirical Bayes Smoothing of Census Undercount, *Survey Methodology*, **18**, 75–94.

Cressie, N. and Chan, N.H. (1989), Spatial Modelling of Regional Variables, *Journal of the American Statistical Association*, **84**, 393–401.

Daniels, M.J. and Gatsonis, C. (1999), Hierarchical Generalized Linear Models in the Analysis of Variations in Health Care Utilization, *Journal of the American Statistical Association*, **94**, 29–42.

Daniels, M.J. and Kass, R.E. (1999), Nonconjugate Bayesian Estimation of Covariance Matrices and its Use in Hierarchical Models, *Journal of the American Statistical Association*, **94**, 1254–1263.

Das, K., Jiang, J., and Rao, J.N.K. (2004), Mean Squared Error of Empirical Predictor, *Annals of Statistics*, **32**, 818–840.

Dass, S.C., Maiti, T., Ren, H., and Sinha, S. (2012), Confidence Interval Estimation of Small Area Parameters Shrinking Both Means and Variances, *Survey Methodology*, **38**, 173–187.

Datta, G.S. (2009), Model-Based Approach to Small Area Estimation, in D. Pfeffermann and C.R. Rao, (Eds.), *Sample Surveys: Inference and Analysis, Handbook of Statistics*, Volume 29B, Amsterdam: North-Holland, pp. 251–288.

Datta, G.S., Day, B., and Basawa, I. (1999), Empirical Best Linear Unbiased and Empirical Bayes Prediction in Multivariate Small Area Estimation, *Journal of Statistical Planning and Inference*, **75**, 169–179.

Datta, G.S., Day, B., and Maiti, T. (1998), Multivariate Bayesian Small Area Estimation: An Application to Survey and Satellite Data, *Sankhyā, Series A*, **60**, 1–19.

Datta, G.S., Fay, R.E., and Ghosh, M. (1991), Hierarchical and Empirical Bayes Multivariate Analysis in Small Area Estimation, in *Proceedings of Bureau of the Census 1991 Annual Research Conference*, U.S. Bureau of the Census, Washington, DC, pp. 63–79.

Datta, G.S. and Ghosh, M. (1991), Bayesian Prediction in Linear Models: Applications to Small Area Estimation, *Annals of Statistics*, **19**, 1748–1770.

Datta, G.S., Ghosh, M., Huang, E.T., Isaki, C.T., Schultz, L.K., and Tsay, J.H. (1992), Hierarchical and Empirical Bayes Methods for Adjustment of Census Undercount: The 1988 Missouri Dress Rehearsal Data, *Survey Methodology*, **18**, 95–108.

Datta, G.S., Ghosh, M., and Kim, Y-H. (2001), Probability Matching Priors for One-Way Unbalanced Random Effect Models, Technical Report, Department of Statistics, University of Georgia, Athens.

Datta, G.S., Ghosh, M., Nangia, N., and Natarajan, K. (1996), Estimation of Median Income of Four-Person Families: A Bayesian Approach, in W.A. Berry, K.M. Chaloner, and J.K. Geweke (Eds.), *Bayesian Analysis in Statistics and Econometrics*, New York: John Wiley & Sons, Inc., pp. 129–140.

Datta, G.S., Ghosh, M., Smith, D.D., and Lahiri, P. (2002), On the Asymptotic Theory of Conditional and Unconditional Coverage Probabilities of Empirical Bayes Confidence Intervals, *Scandinavian Journal of Statistics*, **29**, 139–152.

Datta, G.S., Ghosh, M., Steorts, R., and Maples, J.J. (2011), Bayesian Benchmarking with Applications to Small Area Estimation, *Test*, **20**, 574–588.

Datta, G.S., Ghosh, M., and Waller, L.A. (2000), Hierarchical and Empirical Bayes Methods for Environmental Risk Assessment, in P.K. Sen and C.R. Rao (Eds.), *Handbook of Statistics*, Volume 18, Amsterdam: Elsevier Science B.V., pp. 223–245.

Datta, G.S., Hall, P., and Mandal, A. (2011), Model Selection and Testing for the Presence of Small-Area Effects, and Application to Area-Level Data, *Journal of the American Statistical Association*, **106**, 362–374.

Datta, G.S., Kubokawa, T., Molina, I., and Rao, J.N.K. (2011), Estimation of Mean Squared Error of Model-Based Small Area Estimators, *Test*, **20**, 367–388.

Datta, G.S. and Lahiri, P. (1995), Robust Hierarchical Bayes Estimation of Small Area Characteristics in the Presence of Covariates, *Journal of Multivariate Analysis*, **54**, 310–328.

Datta, G.S. and Lahiri, P. (2000), A Unified Measure of Uncertainty of Estimated Best Linear Unbiased Predictors in Small Area Estimation Problems, *Statistica Sinica*, **10**, 613–627.

Datta, G.S., Lahiri, P., and Maiti, T. (2002), Empirical Bayes Estimation of Median Income of Four-Person Families by State Using Time Series and Cross-Sectional Data, *Journal of Statistical Planning and Inference*, **102**, 83–97.

Datta, G.S., Lahiri, P., Maiti, T., and Lu, K.L. (1999), Hierarchical Bayes Estimation of Unemployment Rates for the U.S. States, *Journal of the American Statistical Association*, **94**, 1074–1082.

Datta, G.S. and Mandal, A. (2014), Small Area Estimation with Uncertain Random Effects, in Paper presented at *SAE 2014 Conference*, Poznan, Poland, September 2014.

Datta, G.S., Rao, J.N.K., and Smith, D.D. (2005), On Measuring the Variability of Small Area Estimators under a Basic Area Level Model, *Biometrika*, **92**, 183–196.

Datta, G.S., Rao, J.N.K., and Torabi, M. (2010), Pseudo-Empirical Bayes Estimation of Small Area Means Under a Nested Error Linear Regression Model with Functional Measurement Errors, *Journal of Statistical Planning and Inference*, **140**, 2952–2962.

Dawid, A.P. (1985), Calibration-Based Empirical Probability, *Annals of Statistics*, **13**, 1251–1274.

Dean, C.B. and MacNab, Y.C. (2001), Modeling of Rates Over a Hierarchical Health Administrative Structure, *Canadian Journal of Statistics*, **29**, 405–419.

Deming, W.E. and Stephan, F.F. (1940), On a Least Squares Adjustment of a Sample Frequency Table When the Expected Marginal Totals are Known, *Annals of Mathematical Statistics*, **11**, 427–444.

Dempster, A.P., Laird, N.M., and Rubin, D.B. (1977), Maximum Likelihood from Incomplete Data Via the EM Algorithm, *Journal of the Royal Statistical Society, Series B*, **39**, 1–38.

Dempster, A.P. and Ryan, L.M. (1985), Weighted Normal Plots, *Journal of the American Statistical Association*, **80**, 845–850.

DeSouza, C.M. (1992), An Appropriate Bivariate Bayesian Method for Analysing Small Frequencies, *Biometrics*, **48**, 1113–1130.

Deville, J.C. and Särndal, C.-E. (1992), Calibration Estimation in Survey Sampling, *Journal of the American Statistical Association*, **87**, 376–382.

Diallo, M. (2014), *Small Area Estimation: Skew-Normal Distributions and Time Series*, Unpublished Ph.D. Dissertation, Carleton University, Ottawa, Canada.

Diallo, M. and Rao, J.N.K. (2014), Small Area Estimation of Complex Parameters Under Unit-level Models with Skew-Normal Errors, *Proceedings of the Survey Research Section*, Washington, DC: American Statistical Association.

Diao, L., Smith, D.D., Datta, G.S., Maiti, T., and Opsomer, J.D. (2014), Accurate Confidence Interval Estimation of Small Area Parameters Under the Fay-Herriot Model. *Scandinavian Journal of Statistics*, **41**, 497–515.

Dick, P. (1995), Modelling Net Undercoverage in the 1991 Canadian Census, *Survey Methodology*, **21**, 45–54.

Draper, N.R. and Smith, H. (1981), *Applied Regression Analysis* (2nd ed.), New York: John Wiley & Sons, Inc..

Drew, D., Singh, M.P., and Choudhry, G.H. (1982), Evaluation of Small Area Estimation Techniques for the Canadian Labour Force Survey, *Survey Methodology*, **8**, 17–47.

Efron, B. (1975), Biased Versus Unbiased Estimation, *Advances in Mathematics*, **16**, 259–277.

Efron, B. and Morris, C. (1972a), Limiting the Risk of Bayes and Empirical Bayes Estimators, Part II: The Empirical Bayes Case, *Journal of the American Statistical Association*, **67**, 130–139.

Efron, B. and Morris, C. (1972b), Empirical Bayes on Vector Observations: An Extension of Stein's Method, *Biometrika*, **59**, 335–347.

Efron, B. and Morris, C. (1973), Stein's Estimation Rule and its Competitors - An Empirical Bayes Approach, *Journal of the American Statistical Association*, **68**, 117–130.

Efron, B. and Morris, C. (1975), Data Analysis Using Stein's Estimate and Its Generalizations, *Journal of the American Statistical Association*, **70**, 311–319.

Elbers, C., Lanjouw, J.O., and Lanjouw, P. (2003), Micro-Level Estimation of Poverty and Inequality. *Econometrica*, **71**, 355–364.

Erciulescu, A.L. and Fuller, W.A. (2014), Parametric Bootstrap Procedures for Small Area Prediction Variance, *Proceedings of the Survey Research Methods Section*, Washington, DC: American Statistical Association.

Ericksen, E.P. and Kadane, J.B. (1985), Estimating the Population in Census Year: 1980 and Beyond (with discussion), *Journal of the American Statistical Association*, **80**, 98–131.

Ericksen, E.P., Kadane, J.B., and Tukey, J.W. (1989), Adjusting the 1981 Census of Population and Housing, *Journal of the American Statistical Association*, **84**, 927–944.

Esteban, M.D., Morales, D., Pérez, A., and Santamaría, L. (2012), Small Area Estimation of Poverty Proportions Under Area-Level Time Models, *Computational Statistics and Data Analysis*, **56**, 2840–2855.

Estevao, V., Hidiroglou, M.A., and Särndal, C.-E. (1995), Methodological Principles for a Generalized Estimation Systems at Statistics Canada, *Journal of Official Statistics*, **11**, 181–204.

Estevao, V., Hidiroglou, M.A., and You Y. (2014), *Area Level Model, Unit Level, and Hierarchical Bayes Methodology Specifications*, Statistical Research and Innovation Division, Internal Statistics Canada Document, 240 pages.

Fabrizi, E., Salvati, N., and Pratesi, M. (2012), Constrained Small Area Estimators Based on M-quantile Methods, *Journal of Official Statistics*, **28**, 89–106.

Falorsi, P.D., Falorsi, S., and Russo, A. (1994), Empirical Comparison of Small Area Estimation Methods for the Italian Labour Force Survey, *Survey Methodology*, **20**, 171–176.

Falorsi, P.D. and Righi, P. (2008), A Balanced Sampling Approach for Multi-Way Stratification Designs for Small Area Estimation, *Survey Methodology*, **34**, 223–234.

Farrell, P.J. (2000), Bayesian Inference for Small Area Proportions, *Sankhyā, Series B*, **62**, 402–416.

Farrell, P.J., MacGibbon, B., and Tomberlin, T.J. (1997a), Empirical Bayes Estimators of Small Area Proportions in Multistage Designs, *Statistica Sinica*, **7**, 1065–1083.

Farrell, P.J., MacGibbon, B., and Tomberlin, T.J. (1997b), Bootstrap Adjustments for Empirical Bayes Interval Estimates of Small Area Proportions, *Canadian Journal of Statistics*, **25**, 75–89.

Farrell, P.J., MacGibbon, B., and Tomberlin, T.J. (1997c), Empirical Bayes Small Area Estimation Using Logistic Regression Models and Summary Statistics, *Journal of Business & Economic Statistics*, **15**, 101–108.

Fay, R.E. (1987), Application of Multivariate Regression to Small Domain Estimation, in R. Platek, J.N.K. Rao, C.-E. Särndal, and M.P. Singh (Eds.), *Small Area Statistics*, New York: John Wiley & Sons, Inc., pp. 91–102.

Fay, R.E. (1992), Inferences for Small Domain Estimates from the 1990 Post Enumeration Survey, Unpublished Manuscript, U.S. Bureau of the Census.

Fay, R.E. and Herriot, R.A. (1979), Estimation of Income from Small Places: An Application of James–Stein Procedures to Census Data, *Journal of the American Statistical Association*, **74**, 269–277.

Fellner, W.H. (1986), Robust Estimation of Variance Components, *Technometrics*, **28**, 51–60.

Ferretti, C. and Molina, I. (2012), Fast EB Method for Estimating Complex Poverty Indicators in Large Populations. *Journal of the Indian Society of Agricultural Statistics*, **66**, 105–120.

Folsom, R., Shah, B.V., and Vaish, A. (1999), Substance Abuse in States: A Methodological Report on Model Based Estimates from the 1994–1996 National Household Surveys on Drug Abuse, *Proceedings of the Section on Survey Research Methods*, Washington, DC: American Statistical Association, pp. 371–375.

Foster, J., Greer, J., and Thorbecke, E. (1984), A Class of Decomposable Poverty Measures. *Econometrica*, **52**, 761–766.

Freedman, D.A. and Navidi, W.C. (1986), Regression Methods for Adjusting the 1980 Census (with discussion), *Statistical Science*, **18**, 75–94.

Fuller, W.A. (1975), Regression Analysis for Sample Surveys, *Sankhyā, Series C*, **37**, 117–132.

Fuller, W.A. (1987), *Measurement Error Models*, New York: John Wiley & Sons, Inc..

Fuller, W.A. (1989), Prediction of True Values for the Measurement Error Model, in *Conference on Statistical Analysis of Measurement Error Models and Applications*, Humboldt State University.

Fuller, W.A. (1999), Environmental Surveys Over Time, *Journal of Agricultural, Biological and Environmental Statistics*, **4**, 331–345.

Fuller, W.A. (2009), *Sampling Statistics*, New York: John Wiley & Sons, Inc..

Fuller, W.A. and Battese, G.E. (1973), Transformations for Estimation of Linear Models with Nested-Error Structure, *Journal of the American Statistical Association*, **68**, 626–632.

Fuller, W.A. and Goyeneche, J.J. (1998), *Estimation of the State Variance Component*, Unpublished manuscript.

Fuller, W.A. and Harter, R.M. (1987), The Multivariate Components of Variance Model for Small Area Estimation, in R. Platek, J.N.K. Rao, C.-E. Särndal, and M.P. Singh (Eds.), *Small Area Statistics*, New York: John Wiley & Sons, Inc., pp. 103–123.

Ganesh, N. and Lahiri, P. (2008), A New Class of Average Moment Matching Priors, *Biometrika*, **95**, 514–520.

Gelfand, A.E. (1996), Model Determination Using Sample-Based Methods, in W.R. Gilks, S. Richardson, and D.J. Spiegelhalter (Eds.), *Markov Chain Monte Carlo in Practice*, London: Chapman and Hall, pp. 145–161.

Gelfand, A.E. and Smith, A.F.M. (1990), Sample-Based Approaches to Calculating Marginal Densities, *Journal of the American Statistical Association*, **85**, 972–985.

Gelfand, A.E. and Smith, A.F.M. (1991), Gibbs Sampling for Marginal Posterior Expectations, *Communications in Statistics - Theory and Methods*, **20**, 1747–1766.

Gelman, A. (2006), Prior Distributions for Variance Parameters in Hierarchical Models, *Bayesian Analysis*, **1**, 515–533.

Gelman, A. and Meng, S-L. (1996), Model Checking and Model Improvement, in W.R. Gilks, S. Richardson, and D.J. Spiegelhalter (Eds.), *Markov Chain Monte Carlo in Practice*, London: Chapman and Hall, pp. 189–201.

Gelman, A. and Rubin, D.B. (1992), Inference from Iterative Simulation Using Multiple Sequences, *Statistical Science*, **7**, 457–472.

Ghangurde, P.D. and Singh, M.P. (1977), Synthetic Estimators in Periodic Household Surveys, *Survey Methodology*, **3**, 152–181

Ghosh, M. (1992a), Hierarchical and Empirical Bayes Multivariate Estimation, in M. Ghosh and P.K. Pathak (Eds.), *Current Issues in Statistical Inference: Essays in Honor of D. Basu, IMS Lecture Notes - Monograph Series*, Volume 17, Beachwood, OH: Institute of Mathematical Statistics.

Ghosh, M. (1992b), Constrained Bayes Estimation with Applications, *Journal of the American Statistical Association*, **87**, 533–540.

Ghosh, M. and Auer, R. (1983), Simultaneous Estimation of Parameters, *Annals of the Institute of Statistical Mathematics, Part A*, **35**, 379–387.

Ghosh, M. and Lahiri, P. (1987), Robust Empirical Bayes Estimation of Means from Stratified Samples, *Journal of the American Statistical Association*, **82**, 1153–1162.

Ghosh, M. and Lahiri, P. (1998), Bayes and Empirical Bayes Analysis in Multistage Sampling, in S.S. Gupta and J.O. Berger (Eds.), *Statistical Decision Theory and Related Topics IV*, Volume 1, New York: Springer-Verlag, pp. 195–212.

Ghosh, M. and Maiti, T. (1999), Adjusted Bayes Estimators with Applications to Small Area Estimation, *Sankhyā, Series B*, **61**, 71–90.

Ghosh, M. and Maiti, T. (2004), Small Area Estimation Based on Natural Exponential Family Quadratic Variance Function Models and Survey Weights, *Biometrika*, **91**, 95–112.

Ghosh, M., Maiti, T., and Roy, A. (2008), Influence Functions and Robust Bayes and Empirical Bayes Small Area Estimation, *Biometrika*, **95**, 573–585.

Ghosh, M., Nangia, N., and Kim, D. (1996), Estimation of Median Income of Four-Person Families: A Bayesian Time Series Approach, *Journal of the American Statistical Association*, **91**, 1423–1431.

Ghosh, M., Natarajan, K., Stroud, T.W.F., and Carlin, B.P. (1998), Generalized Linear Models for Small Area Estimation, *Journal of American Statistical Association*, **93**, 273–282.

Ghosh, M., Natarajan, K., Waller, L.A., and Kim, D. (1999), Hierarchical Bayes GLMs for the Analysis of Spatial Data: An Application to Disease Mapping, *Journal of Statistical Planing and Inference*, **75**, 305–318.

Ghosh, M. and Rao, J.N.K. (1994), Small Area Estimation: an Appraisal (with Discussion), *Statistical Science*, **9**, 55–93.

Ghosh, M. and Sinha, K. (2007), Empirical Bayes Estimation in Finite Population Sampling Under Functional Measurement Error Models, *Scandinavian Journal of Statistics*, **33**, 591–608.

Ghosh, M., Sinha, K., and Kim, D. (2006), Empirical and Hierarchical Bayesian Estimation in Finite Population Sampling Under Structural Measurement Error Models, *Journal of Statistical Planning and Inference*, **137**, 2759–2773.

Ghosh, M. and Steorts, R. (2013), Two-Stage Bayesian Benchmarking as Applied to Small Area Estimation, *Test*, **22**, 670–687.

Gilks, W.R., Richardson, S., and Spiegelhalter, D.J. (Eds.) (1996), *Markov Chain Monte Carlo in Practice*, London: Chapman and Hall.

Gilks, W.R. and Wild, P. (1992), Adaptive Rejection Sampling for Gibbs Sampling, *Applied Statistics*, **41**, 337–348.

Godambe, V.P. and Thompson, M.E. (1989), An Extension of Quasi-likelihood Estimation (with Discussion), *Journal of Statistical Planning and Inference*, **22**, 137–152.

Goldstein, H. (1975), A Note on Some Bayesian Nonparametric Estimates, *Annals of Statistics*, **3**, 736–740.

Goldstein, H. (1989), Restricted Unbiased Iterative Generalized Least Squares Estimation, *Biometrika*, **76**, 622–623.

Gonzalez, M.E. (1973), Use and Evaluation of Synthetic Estimates, *Proceedings of the Social Statistics Section*, American Statistical Association, pp. 33–36.

Gonzalez, J.F., Placek, P.J., and Scott, C. (1996), Synthetic Estimation of Followback Surveys at the National Center for Health Statistics, in W.L. Schaible (ed.), *Indirect Estimators in U.S. Federal Programs*, New York: Springer–Verlag, pp. 16–27.

Gonzalez, M.E. and Waksberg, J. (1973), Estimation of the Error of Synthetic Estimates, in Paper presented at the *First Meeting of the International Association of Survey Statisticians*, Vienna, Austria.

González-Manteiga, W., Lombardía, M.J., Molina, I., Morales, D., and Santamaría, L. (2008a), Bootstrap Mean Squared Error of a Small-Area EBLUP, *Journal of Statistical Computation and Simulation*, **78**, 443–462.

González-Manteiga, W., Lombardía, M.J., Molina, I., Morales, D., and Santamaría, L. (2008b), Analytic and Bootstrap Approximations of Prediction Errors Under a Multivariate FayHerriot Model, *Computational Statistics and Data Analysis*, **52**, 5242–5252.

González-Manteiga, W., Lombardía, M.J., Molina, I., Morales, D., and Santamaría, L. (2010), Small Area Estimation Under Fay-Herriot Models with Nonparametric Estimation of Heteroscedasticity, *Statistical Modelling*, **10**, 215–239.

Griffin, B. and Krutchkoff, R. (1971), Optimal Linear Estimation: An Empirical Bayes Version with Application to the Binomial Distribution, *Biometrika*, **58**, 195–203.

Griffiths, R. (1996), Current Population Survey Small Area Estimations for Congressional Districts, *Proceedings of the Section on Survey Research Methods*, Washington, DC: American Statistical Association, pp. 314–319.

Groves, R.M. (1989), *Survey Errors and Survey Costs*, New York: John Wiley & Sons, Inc..

Groves, R.M. (2011), Three Eras of Survey Research, *Public Opinion Quarterly*, **75**, 861–871.

Hall, P. and Maiti, T. (2006a), Nonparametric Estimation of Mean-Squared Prediction Error in Nested-Error Regression Models, *Annals of Statistics*, **34**, 1733–1750.

Hall, P. and Maiti, T. (2006b), On Parametric Bootstrap Methods for Small Area Prediction, *Journal of the Royal Statistical Society, Series B*, **68**, 221–238.

Han, B. (2013), Conditional Akaike Information Criterion in the Fay-Herriot Model, *Survey Methodology*, **11**, 53–67.

Han, C.-P. and Bancroft, T.A. (1968), On Pooling Means When Variance is Unknown, *Journal of the American Statistical Association*, **63**, 1333–1342.

Hansen, M.H., Hurwitz, W.N., and Madow, W.G. (1953), *Sample Survey Methods and Theory I*, New York: John Wiley & Sons, Inc..

Hansen, M.H., Madow, W.G., and Tepping, B.J. (1983), An Evaluation of Model-Dependent and Probability Sampling Inferences in Sample Surveys, *Journal of the American Statistical Association*, **78**, 776–793.

Hartigan, J. (1969), Linear Bayes Methods, *Journal of the Royal Statistical Society, Series B*, **31**, 446–454.

Hartless, G., Booth, J.G., and Littell, R.C. (2000), Local Influence Diagnostics for Prediction of Linear Mixed Models, Unpublished manuscript.

Hartley, H.O. (1959), *Analytic Studies of Survey Data, in a Volume in Honor of Corrado Gini*, Rome, Italy: Instituto di Statistica.

Hartley, H.O. (1974), Multiple Frame Methodology and Selected Applications, *Sankhyā, Series C*, **36**, 99–118.

Hartley, H.O. and Rao, J.N.K. (1967), Maximum Likelihood Estimation for the Mixed Analysis of Variance Model, *Biometrika*, **54**, 93–108.

Harvey, A.C. (1990), *Forecasting, Structural Time Series Models and the Kalman Filter*, Cambridge: Cambridge University Press.

Harville, D.A. (1977), Maximum Likelihood Approaches to Variance Component Estimation and to Related Problems, *Journal of the American Statistical Association*, **72**, 322–340.

Harville, D.A. (1991), Comment, *Statistical Science*, **6**, 35–39.

Harville, D.A. and Jeske, D.R. (1992), Mean Squared Error of Estimation or Prediction Under General Linear Model, *Journal of the American Statistical Association*, **87**, 724–731.

Hastings, W.K. (1970), Monte Carlo Sampling Methods Using Markov Chains and Their Applications, *Biometrika*, **57**, 97–109.

He, X. (1997), Quantile Curves Without Crossing, *American Statistician*, **51**, 186–192.

He, Z. and Sun, D. (1998), Hierarchical Bayes Estimation of Hunting Success Rates, *Environmental and Ecological Statistics*, **5**, 223–236.

Hedayat, A.S. and Sinha, B.K. (1991), *Design and Inference in Finite Population Sampling*, New York: John Wiley & Sons, Inc..

Henderson, C.R. (1950), Estimation of Genetic Parameters (Abstract), *Annals of Mathematical Statistics*, **21**, 309–310.

Henderson, C.R. (1953), Estimation of Variance and Covariance Components, *Biometrics*, **9**, 226–252.

Henderson, C.R. (1963), Selection Index and Expected Genetic Advance, in *Statistical Genetics and Plant Breeding*, Publication 982, Washington, DC: National Academy of Science, National Research Council, pp. 141–163.

Henderson, C.R. (1973), Maximum Likelihood Estimation of Variance Components, Unpublished manuscript.

Henderson, C.R. (1975), Best Linear Unbiased Estimation and Prediction Under a Selection Model, *Biometrics*, **31**, 423–447.

Henderson, C.R., Kempthorne, O., Searle, S.R., and von Krosigk, C.N. (1959), Estimation of Environmental and Genetic Trends from Records Subject to Culling, *Biometrics*, **13**, 192–218.

Hidiroglou, M.A. and Patak, Z. (2009), An Application of Small Area Estimation Techniques to the Canadian Labour Force Survey, *Proceedings of the Survey Research Methods Section*, Statistical Society of Canada.

Hill, B.M. (1965), Inference About Variance Components in the One-Way Model, *Journal of the American Statistical Association*, **60**, 806–825.

Hobert, J.P. and Casella, G. (1996), The Effect of Improper Priors on Gibbs Sampling in Hierarchical Linear Mixed Models, *Journal of the American Statistical Association*, **9l**, 1461–1473.

Hocking, R.R. and Kutner, M.H. (1975), Some Analytical and Numerical Comparisons of Estimators for the Mixed A.O.V. Model, *Biometrics*, **31**, 19–28.

Holt, D. and Smith, T.M.F. (1979), Post-Stratification, *Journal of the Royal Statistical Society, Series A*, **142**, 33–46.

Holt, D., Smith, T.M.F., and Tomberlin, T.J. (1979), A Model-based Approach to Estimation for Small Subgroups of a Population, *Journal of the American Statistical Association*, **74**, 405–410.

Huang, E.T. and Fuller, W.A. (1978), Non-negative Regression Estimation for Sample Survey Data, *Proceedings of the Social Statistics Section*, Washington, DC: American Statistical Association, pp. 300–3005.

Huber, P.J. (1972), The 1972 Wald Lecture Robust Statistics: A Review. *Annals of Mathematical Statistics*, **43**, 1041–821.

Huggins, R.M. (1993), A Robust Approach to the Analysis of Repeated Measures. *Biometrics*, **49**, 715–720.

Hulting, F.L. and Harville, D.A. (1991), Some Bayesian and Non-Bayesian Procedures for the Analysis of Comparative Experiments and for Small Area Estimation: Computational Aspects, Frequentist Properties and Relationships, *Journal of the American Statistical Association*, **86**, 557–568.

IASS Satellite Conference (1999), *Small Area Estimation*, Latvia: Central Statistical Office.

Ireland, C.T. and Kullback, S. (1968), Contingency Tables with Given Marginals, *Biometrika*, **55**, 179–188.

Isaki, C.T., Huang, E.T., and Tsay, J.H. (1991), Smoothing Adjustment Factors from the 1990 Post Enumeration Survey, *Proceedings of the Social Statistics Section*, Washington, DC: American Statistical Association, pp. 338–343.

Isaki, C.T., Tsay, J.H., and Fuller, W.A. (2000), Estimation of Census Adjustment Factors, *Survey Methodology*, **26**, 31–42.

Jacqmin-Gadda, H., Sibillot, S., Proust, C., Molina, J.-M., and Thiébaut, R. (2007), Robustness of the Linear Mixed Model to Misspecified Error Distribution, *Computational Statistics and Data Analysis*, **51**, 5142–5154.

James, W. and Stein, C. (1961), Estimation with Quadratic Loss, *Proceedings of the 4th Berkeley Symposium on Mathematical Statistics and Probability*, University of California Press, Berkeley, pp. 361–379.

Jiang, J. (1996), REML Estimation: Asymptotic Behavior and Related Topics, *Annals of Statistics*, **24**, 255–286.

Jiang, J. (1997), A Derivation of BLUP – Best Linear Unbiased Predictor, *Statistics & Probability Letters*, **32**, 321–324.

Jiang, J. (1998), Consistent Estimators in Generalized Linear Mixed Models, *Journal of the American Statistical Association*, **93**, 720–729.

Jiang, J. (2001), Private Communication to J.N.K. Rao.

Jiang, J. and Lahiri, P. (2001), Empirical Best Prediction for Small Area Inference with Binary data, *Annals of the Institute of Statistical Mathematics*, **53**, 217–243.

Jiang, J. and Lahiri, P. (2006), Mixed Model Prediction and Small Area Estimation, *Test*, **15**, 1–96.

Jiang, J., Lahiri, P., and Wan, S-M. (2002), A Unified Jackknife Theory, *Annals of Statistics*, **30**, 1782–1810.

Jiang, J., Lahiri, P., Wan, S-M., and Wu, C-H. (2001), Jackknifing the Fay-Herriot Model with an Example, Unpublished manuscript.

Jiang, J., Lahiri, P., and Wu, C-H. (2001), A Generalization of Pearson's Chi-Square Goodness-of-Fit Test with Estimated Cell Frequencies, *Sankhyā, Series A*, **63**, 260–276.

Jiang, J. and Nguyen, T. (2012), Small Area Estimation via Heteroscedastic Nested-Error Regression, *Canadian Journal of Statistics*, **40**, 588–603.

Jiang, J., Nguyen, T., and Rao, J.S. (2009), A Simplified Adaptive Fence Procedure, *Statistics and Probability Letters*, **79**, 625–629.

Jiang, J., Nguyen, T., and Rao, J.S. (2011), Best Predictive Small Area Estimation, *Journal of the American Statistical Association*, **106**, 732–745.

Jiang, J., Nguyen, T., and Rao, J.S. (2014), Observed Best Prediction Via Nested-error Regression with Potentially Misspecified Mean and Variance, *Survey Methodology*, in print.

Jiang, J. and Rao, J.S. (2003), Consistent Procedures for Mixed Linear Model Selection, *Sankhyā, Series A.*, **65**, 23–42.

Jiang, J., Rao, J.S., Gu, I., and Nguyen, T. (2008), Fence Methods for Mixed Model Selection, *Annals of Statistics*, **36**, 1669–1692.

Jiang, J. and Zhang, W. (2001), Robust Estimation in Generalized Linear Mixed Models, *Biometrika*, **88**, 753–765.

Jiongo, V.D., Haziza, D., and Duchesne, P. (2013), Controlling the Bias of Robust Small-area Estimation, *Biometrika*, **100**, 843–858.

Kackar, R.N. and Harville, D.A. (1981), Unbiasedness of Two-stage Estimation and Prediction Procedures for Mixed Linear Models, *Communications in Statistics, Series A*, **10**, 1249–1261.

Kackar, R.N. and Harville, D.A. (1984), Approximations for Standard Errors of Estimators of Fixed and Random Effects in Mixed Linear Models, *Journal of the American Statistical Association*, **79**, 853–862.

Kalton, G., Kordos, J., and Platek, R. (1993), *Small Area Statistics and Survey Designs Volume I: Invited Papers: Volume II: Contributed Papers and Panel Discussion*, Warsaw, Poland: Central Statistical Office.

Karunamuni, R.J. and Zhang, S. (2003), Optimal Linear Bayes and Empirical Bayes Estimation and Prediction of the Finite Population Mean, *Journal of Statistical Planning and Inference*, **113**, 505–525.

Kass, R.E. and Raftery, A. (1995), Bayes Factors, *Journal of the American Statistical Association*, **90**, 773–795.

Kass, R.E. and Steffey, D. (1989), Approximate Bayesian Inference in Conditionally Independent Hierarchical Models (Parametric Empirical Bayes Models), *Journal of the American Statistical Association*, **84**, 717–726.

Kim, H., Sun, D., and Tsutakawa, R.K. (2001), A Bivariate Bayes Method for Improving the Estimates of Mortality Rates with a Twofold Conditional Autoregressive Model, *Journal of the American Statistical Association*, **96**, 1506–1521.

Kish, L. (1999), Cumulating/Combining Population Surveys, *Survey Methodology*, **25**, 129–138.

Kleffe, J. and Rao, J.N.K. (1992), Estimation of Mean Square Error of Empirical Best Linear Unbiased Predictors Under a Random Error Variance Linear Model, *Journal of Multivariate Analysis*, **43**, 1–15.

Kleinman, J.C. (1973), Proportions with Extraneous Variance: Single and Independent Samples, *Journal of the American Statistical Association*, **68**, 46–54.

Koenker, R. (1984), A Note on L-Estimators for Linear Models, *Statistics and Probability Letters*, **2**, 323–325.

Koenker, R. and Bassett, G. (1978), Regression Quantiles, *Econometrica*, **46**, 33–50.

Kokic, P., Chambers, R., Breckling, J., and Beare, S. (1997), A Measure of Production Performance, *Journal of Business and Economic Statistics*, **15**, 445–451.

Kott, P.S. (1990), Robust Small Domain Estimation Using Random Effects Modelling, *Survey Methodology*, **15**, 3–12.

Lahiri, P. (1990), "Adjusted" Bayes and Empirical Bayes Estimation in Finite Population Sampling, *Sankhyā, Series B*, **52**, 50–66.

Lahiri, P. and Maiti, T. (2002), Empirical Bayes Estimation of Relative Risks in Disease Mapping. *Calcutta Statistical Association Bulletin*, **1**, 53–211.

Lahiri, P. and Pramanik, S. (2013), Estimation of Average Design-based Mean Squared Error of Synthetic Small Area Estimators, *Unpublished paper*.

Lahiri, P. and Rao, J.N.K. (1995), Robust Estimation of Mean Squared Error of Small Area Estimators, *Journal of the American Statistical Association*, **82**, 758–766.

Lahiri, S.N., Maiti, T., Katzoff, M., and Parson, V. (2007), Resampling-Based Empirical Prediction: An Application to Small Area Estimation. *Biometrika*, **94**, 469–485.

Laird, N.M. (1978), Nonparametric Maximum Likelihood Estimation of a Mixing Distribution, *Journal of the American Statistical Association*, **73**, 805–811.

Laird, N.M. and Louis, T.A. (1987), Empirical Bayes Confidence Intervals Based on Bootstrap Samples, *Journal of the American Statistical Association*, **82**, 739–750.

Laird, N.M. and Ware, J.H. (1982), Random-Effects Models for Longitudinal Data, *Biometrics*, **38**, 963–974.

Lange, N. and Ryan, L. (1989), Assessing Normality in Random Effects Models, *Annals of Statistics*, **17**, 624–642.

Langford, I.H., Leyland, A.H., Rasbash, J., and Goldstein, H. (1999), Multilevel Modelling of the Geographical Distribution of Diseases, *Applied Statistics*, **48**, 253–268.

Laud, P. and Ibrahim, J.G. (1995), Predictive Model Selection, *Journal of the Royal Statistical Society, Series B*, **57**, 247–262.

Lehtonen, R., Särndal, C.-E., and Veijanen, A. (2003), The Effect of Model Choice in Estimation for Domains Including Small Domains, *Survey Methodology*, **29**, 33–44.

Lehtonen, R. and Veijanen, A. (1999), Domain Estimation with Logistic Generalized Regression and Related Estimators, *Proceedings of the IASS Satellite Conference on Small Area Estimation*, Riga: Latvian Council of Science, pp. 121–128.

Lehtonen, R. and Veijanen, A. (2009), Design-Based Methods of Estimation for Domains and Small Areas, in D. Pfeffermann and C.R. Rao, (Eds.), *Sample Surveys: Inference and Analysis, Handbook of Statistics*, Volume 29B, Amsterdam: North-Holland, pp. 219–249.

Lele, S.R., Nadeem, K., and Schmuland, B. (2010), Estimability and Likelihood Inference for Generalized Linear Mixed Models Using Data Cloning, *Journal of the American Statistical Association*, **105**, 1617–1625.

Levy, P.S. (1971), The Use of Mortality Data in Evaluating Synthetic Estimates, *Proceedings of the Social Statistics Section*, American Statistical Association, pp. 328–331.

Li, H. and Lahiri, P. (2010), An Adjusted Maximum Likelihood Method for Solving Small Area Estimation Problems, *Journal of Multivariate Analysis*, **101**, 882–892.

Liu, B., Lahiri, P., and Kalton, G. (2014), Hierarchical Bayes Modeling of Survey-Weighted Small Area Proportions, *Survey Methodology*, **40**, 1–13.

Lohr, S.L. (2010), *Sampling: Design and Analysis*, Pacific Grove, CA: Duxbury Press.

Lohr, S.L. and Prasad, N.G.N. (2003), Small Area Estimation with Auxiliary Survey Data, *Canadian Journal of Statistics*, **31**, 383–396.

Lohr, S.L. and Rao, J.N.K. (2000), Inference for Dual Frame Surveys, *Journal of the American Statistical Association*, **95**, 271–280.

Lohr, S.L. and Rao, J.N.K.. (2009), Jackknife Estimation of Mean Squared Error of Small Area Predictors in Non-Linear Mixed Models, *Biometrika*, **96**, 457–468.

Longford, N.T. (2005), *Missing Data and Small-Area Estimation*, New York: Springer-Verlag.

Longford, N.T. (2006), Sample Size Calculation for Small-area Estimation, *Survey Methodology*, **32**, 87–96.

López-Vizcaíno, E., Lombardía, M.J., and Morales, D. (2013), Multinomial-based Small Area Estimation of Labour Force Indicators, *Statistical Modelling*, **13**, 153–178.

López-Vizcaíno, E., Lombardía, M.J., and Morales, D. (2014), Small Area Estimation of Labour Force Indicators Under a Multinomial Model with Correlated Time and Area Effects, *Journal of the Royal Statistical Society, Series A*, doi: 10.1111/rssa.12085.

Louis, T.A. (1984), Estimating a Population of Parameter Values Using Bayes and Empirical Bayes Methods, *Journal of the American Statistical Association*, **79**, 393–398.

Luery, D.M. (2011), Small Area Income and Poverty Estimates Program, in *Proceedings of 27th SCORUS Conference*, Jurmala, Latvia, pp. 93–107.

Lunn, D.J., Thomas, A., Best, N., and Spiegelhalter, D. (2000), WinBUGS - A Bayesian Modeling Framework: Concepts, Structures and Extensibility, *Statistics and Computing*, **10**, 325–337.

MacGibbon, B. and Tomberlin, T.J. (1989), Small Area Estimation of Proportions Via Empirical Bayes Techniques, *Survey Methodology*, **15**, 237–252.

Maiti, T. (1998), Hierarchical Bayes Estimation of Mortality Rates for Disease Mapping, *Journal of Statistical Planning and Inference*, **69**, 339–348.

Malec, D., Davis, W.W., and Cao, X. (1999), Model-Based Small Area Estimates of Overweight Prevalence Using Sample Selection Adjustment, *Statistics in Medicine*, **18**, 3189-3200.

Malec, D., Sedransk, J., Moriarity, C.L., and LeClere, F.B. (1997), Small Area Inference for Binary Variables in National Health Interview Survey, *Journal of the American Statistical Association*, **92**, 815–826.

Mantel, H.J., Singh, A.C., and Bureau, M. (1993), Benchmarking of Small Area Estimators, in *Proceedings of International Conference on Establishment Surveys*, Washington, DC: American Statistical Association, pp. 920–925.

Marchetti, S., Giusti, C., Pratesi, M., Salvati, N., Giannotti, F., Pedreschi, D., Rinzivillo, S., Pappalardo, L., and Gabrielli, L. (2015), Small Area Model-based Estimators Using Big Data Sources, *Journal of Official Statistics*, in print.

Marhuenda, Y., Molina, I., and Morales, D. (2013), Small Area Estimation with Spatio-Temporal Fay-Herriot Models, *Computational Statistics and Data Analysis*, **58**, 308–325.

Maritz, J.S. and Lwin, T. (1989), *Empirical Bayes Methods* (2nd ed.), London: Chapman and Hall.

Marker, D.A. (1995), *Small Area Estimation: A Bayesian Perspective*, Unpublished Ph.D. Dissertation, University of Michigan, Ann Arbor, MI.

Marker, D.A. (1999), Organization of Small Area Estimators Using a Generalized Linear Regression Framework, *Journal of Official Statistics*, **15**, 1–24.

Marker, D.A. (2001), Producing Small Area Estimates From National Surveys: Methods for Minimizing Use of Indirect Estimators, *Survey Methodology*, **27**, 183–188.

Marshall, R.J. (1991), Mapping Disease and Mortality Rates Using Empirical Bayes Estimators, *Applied Statistics*, **40**, 283–294.

McCulloch, C.E. and Searle, S.R. (2001), *Generalized, Linear, and Mixed Models*, New York: John Wiley & Sons, Inc..

Metropolis, N., Rosenbluth, A.W., Rosenbluth, M.N., Teller, A.H., and Teller, E. (1953), Equations of State Calculations by Fast Computing Machine, *Journal of Chemical Physics*, **27**, 1087–1097.

Meza, J.L. and Lahiri, P. (2005), A Note on the C_p Statistic Under the Nested Error Regression Model, *Survey Methodology*, **31**, 105–109.

Militino, A.F., Ugarte, M.D., and Goicoa, T. (2007), A BLUP Synthetic Versus and EBLUP Estimator: an Empirical Study of a Small Area Estimation Problem, *Journal of Applied Statistics*, **34**, 153–165.

Militino, A.F., Ugarte, M.D., Goicoa, T., and González-Audícana, M. (2006), Using Small Area Models to Estimate the Total Area Occupied by Olive Trees, *Journal of Agricultural, Biological, and Environmental Statistics*, **11**, 450–461.

Mohadjer, L., Rao, J.N.K., Lin, B., Krenzke, T., and Van de Kerckove, W. (2012), Hierarchical Bayes Small Area Estimates of Adult Literacy Using Unmatched Sampling and Linking Models, *Journal of the Indian Agricultural Statistics*, **66**, 55–63.

Molina, I. and Marhuenda, Y. (2013), *sae: Small Area Estimation*, R package version 1.0-2.

Molina, I. and Marhuenda, Y. (2015), sae: An R Package for Small Area Estimation, *R Journal*, in print.

Molina, I. and Morales, D. (2009), Small Area Estimation of Poverty Indicators, *Boletín de Estadística e Investigación Operativa*, **25**, 218–225.

Molina, I., Nandram, B., and Rao, J.N.K. (2014), Small Area Estimation of General Parameters with Application to Poverty Indicators: A Hierarchical Bayes Approach. *Annals of Applied Statistics*, **8**, 852–885.

Molina, I. and Rao, J.N.K. (2010), Small Area Estimation of Poverty Indicators. *Canadian Journal of Statistics*, **38**, 369–385.

Molina, I., Rao, J.N.K., and Datta, G.S. (2015), Small Area Estimation Under a Fay-Herriot Model with Preliminary Testing for the Presence of Random Effects, *Survey Methodology*, in print.

Molina, I., Rao, J.N.K., and Hidiroglou, M.A. (2008), SPREE Techniques for Small Area Estimation of Cross-classifications, Unpublished manuscript.

Molina, I., Saei, A., and Lombardía, M.J. (2007), Small Area Estimates of Labour Force Participation Under a Multinomial Logit Mixed Model, *Journal of the Royal Statistical Society, Series A*, **170**, 975–1000.

Molina, I., Salvati, N., and Pratesi, M. (2008), Bootstrap for Estimating the MSE of the Spatial EBLUP, *Computational Statistics*, **24**, 441–458.

Morris, C.N. (1983a), Parametric Empirical Bayes Intervals, in G.E.P. Box, T. Leonard, and C.F.J. Wu, (Eds.), *Scientific Inference, Data Analysis, and Robustness*, New York: Academic Press, pp. 25–50.

Morris, C.N. (1983b), Parametric Empirical Bayes Inference: Theory and Applications, *Journal of the American Statistical Association*, **78**, 47–54.

Morris, C.N. (1983c), Natural Exponential Families with Quadratic Variance Function: Statistical Theory, *Annals of Mathematical Statistics*, **11**, 515–529.

Morris, C. and Tang, R. (2011), Estimating Random Effects via Adjustment for Density Maximization, *Statistical Science*, **26**, 271–287.

Moura, F.A.S. and Holt, D. (1999), Small Area Estimation Using Multilevel Models, *Survey Methodology*, **25**, 73–80.

Mukhopadhyay, P. (1998), *Small Area Estimation in Survey Sampling*, New Delhi: Narosa Publishing House.

Mukhopadhyay, P. and McDowell, A. (2011), *Small Area Estimation for Survey Data Analysis Using SAS Software*, Paper 336-2011, Cary, NC: SAS Institute.

Müller, S., Scealy, J.L., and Welsh, A.H. (2013), Model Selection in Linear Mixed Models, *Statistical Science*, **28**, 135–167.

Nandram, B. and Choi, J.W. (2002), Hierarchical Bayes Nonresponse Models for Binary Data from Small Areas with Uncertainty About Ignorability, *Journal of the American Statistical Association*, **97**, 381–388.

Nandram, B. and Choi, J.W. (2010), A Bayesian Analysis of Body Mass Index Data from Small Domains Under Nonignorable Nonresponse and Selection, *Journal of the American Statistical Association*, **105**, 120–135.

Nandram, B., Sedransk, J., and Pickle, L. (1999), Bayesian Analysis of Mortality Rates for U.S. Health Service Areas, *Sankhyā, Series B*, **61**, 145–165.

Nandram, B., Sedransk, J., and Pickle, L. (2000), Bayesian Analysis and Mapping of Mortality Rates for Chronic Obstructive Pulmonary Disease, *Journal of the American Statistical Association*, **95**, 1110–1118.

Natarajan, R. and Kass, R.E. (2000), Reference Bayesian Methods for Generalized Linear Mixed Models, *Journal of the American Statistical Association*, **95**, 227–237.

Natarajan, R. and McCulloch, C.E. (1995), A Note on the Existence of the Posterior Distribution for a Class of Mixed Models for Binomial Responses, *Biometrika*, **82**, 639–643.

National Center for Health Statistics (1968), *Synthetic State Estimates of disability*, P.H.S. Publications 1759, Washington, DC: U.S. Government Printing Office.

National Institute on Drug Abuse (1979), *Synthetic Estimates for Small Areas, Research Monograph 24*, Washington, DC: U.S. Government Printing Office.

National Research Council (2000), Small-Area Estimates of School-Age Children in Poverty: Evaluation of Current Methodology, in C.F. Citro and G. Kalton (Eds.), *Committee on National Statistics*, Washington, DC: National Academy Press.

Opsomer, J.D., Claeskens, G., Ranalli, M.G., Kauermann, G., and Breidt, F.J. (2008), Nonparametric Small Area Estimation Using Penalized Spline Regression, *Journal of the Royal Statistical Society, Series B*, **70**, 265–286.

Patterson, H.D. and Thompson, R. (1971), Recovery of Inter-Block Information When Block Sizes are Unequal, *Biometrika*, **58**, 545–554.

Peixoto, J.L. and Harville, D.A. (1986), Comparisons of Alternative Predictors Under the Balanced One-Way Random Model, *Journal of the American Statistical Association*, **81**, 431–436.

Petrucci, A. and Salvati, N. (2006), Small Area Estimation for Spatial Correlation in Watershed Erosion Assessment. *Journal of Agricultural, Biological and Environmental Statistics*, **11**, 169–182.

Pfeffermann, D. (2002), Small Area Estimation – New Developments and Directions, *International Statistical Review*, **70**, 125–143.

Pfeffermann, D. (2013), New Important Developments in Small Area Estimation, *Statistical Science*, **28**, 40–68.

Pfeffermann, D. and Barnard, C. (1991), Some New Estimators for Small Area Means with Applications to the Assessment of Farmland Values, *Journal of Business and Economic Statistics*, **9**, 73–84.

Pfeffermann, D. and Burck, L. (1990), Robust Small Area Estimation Combining Time Series and Cross-Sectional Data, *Survey Methodology*, **16**, 217–237.

Pfeffermann, D. and Correa, S. (2012), Empirical Bootstrap Bias Correction and Estimation of Prediction Mean Square Error in Small Area Estimation, *Biometrika*, **99**, 457–472.

Pfeffermann, D., Feder, M., and Signorelli, D. (1998), Estimation of Autocorrelations of Survey Errors with Application to Trend Estimation in Small Areas, *Journal of Business and Economic Statistics*, **16**, 339–348.

Pfeffermann, D., Sikov, A. and Tiller, R. (2014), Single- and Two-Stage Cross-Sectional and Time Series Benchmarking, *Test*, **23**, 631–666.

Pfeffermann, D. and Sverchkov, M. (2007), Small-Area Estimation Under Informative Probability Sampling of Areas and Within Selected Areas, *Journal of the American Statistical Association*, **102**, 1427–1439.

Pfeffermann, D., Terryn, B., and Moura, F.A.S. (2008), Small Area Estimation Under a Two-Part Random Effects Model with Application to Estimation of Literacy in Developing Countries, *Survey Methodology*, **34**, 235–249.

Pfeffermann, D. and Tiller, R.B. (2005), Bootstrap Approximation to Prediction MSE for State-Space Models with Estimated Parameters, *Journal of Time Series Analysis*, **26**, 893–916.

Pfeffermann, D. and Tiller, R.B. (2006), Small Area Estimation with State-Space Models Subject to Benchmark Constraints, *Journal of the American Statistical Association*, **101**, 1387–1397.

Pinheiro, J., Bates, D., DebRoy, S., Sarkar, D., and R Core Team (2014), *nlme: Linear and Nonlinear Mixed Effects Models*, R package version 3.1-118.

Platek, R., Rao, J.N.K., Särndal, C.-E., and Singh, M.P. (Eds.) (1987), *Small Area Statistics*, New York: John Wiley & Sons, Inc..

Platek, R. and Singh, M.P. (Eds.) (1986), *Small Area Statistics: Contributed Papers*, Ottawa, Canada: Laboratory for Research in Statistics and Probability, Carleton University.

Porter, A.T., Holan, S.H., Wikle, C.K., and Cressie, N. (2014), Spatial Fay-Herriot Model for Small Area Estimation with Functional Covariates, *Spatial Statistics*, **10**, 27–42.

Portnoy, S.L. (1982), Maximizing the Probability of Correctly Ordering Random Variables Using Linear Predictors, *Journal of Multivariate Analysis*, **12**, 256–269.

Prasad, N.G.N. and Rao, J.N.K. (1990), The Estimation of the Mean Squared Error of Small-Area Estimators, *Journal of the American Statistical Association*, **85**, 163–171.

Prasad, N.G.N. and Rao, J.N.K. (1999), On Robust Small Area Estimation Using a Simple Random Effects Model, *Survey Methodology*, **25**, 67–72.

Pratesi, M. and Salvati, N. (2008), Small Area Estimation: The EBLUP Estimator Based on Spatially Correlated Random Area Effects, *Statistical Methods and Applications*, **17**, 113–141.

Purcell, N.J. and Kish, L. (1979), Estimates for Small Domain, *Biometrics*, **35**, 365–384.

Purcell, N.J. and Kish, L. (1980), Postcensal Estimates for Local Areas (or Domains), *International Statistical Review*, **48**, 3–18.

Purcell, N.J. and Linacre, S. (1976), Techniques for the Estimation of Small Area Characteristics, Unpublished Paper, Canberra: Australian Bureau of Statistics.

Raghunathan, T.E. (1993), A Quasi-Empirical Bayes Method for Small Area Estimation, *Journal of the American Statistical Association*, **88**, 1444–1448.

Randrianasolo, T. and Tillé, Y. (2013), Small Area Estimation by Splitting the Sampling Weights, *Electronic Journal of Statistics*, **7**, 1835-–1855.

Rao, C.R. (1971), Estimation of Variance and Covariance Components – MINQUE Theory, *Journal of Multivariate Analysis*, **1**, 257–275.

Rao, C.R. (1976), Simultaneous Estimation of Parameters – A Compound Decision Problem, in S.S. Gupta and D.S. Moore (Eds.), *Statistical Decision Theory and Related Topics II*, New York: Academic Press, pp. 327–350.

Rao, J.N.K. (1965), On Two Simple Schemes of Unequal Probability Sampling Without Replacement, *Journal of the Indian Statistical Association*, **3**, 173–180.

Rao, J.N.K. (1974), Private Communication to T.W. Anderson.

Rao, J.N.K. (1979), On Deriving Mean Square Errors and Their Non-Negative Unbiased Estimators in Finite Population Sampling, *Journal of the Indian Statistical Association*, **17**, 125–136.

Rao, J.N.K. (1985), Conditional Inference in Survey Sampling, *Survey Methodology*, **11**, 15–31.

Rao, J.N.K. (1986), Synthetic Estimators, SPREE and Best Model Based Predictors, in *Proceedings of the Conference on Survey Research Methods in Agriculture*, U.S. Department of Agriculture, Washington, DC, pp. 1–16.

Rao, J.N.K. (1992), Estimating Totals and Distribution Functions Using Auxiliary Information at the Estimation Stage, in *Proceedings of the Workshop on Uses of Auxiliary Information in Surveys*, Statistics Sweden.

Rao, J.N.K. (1994), Estimating Totals and Distribution Functions Using Auxiliary Information at the Estimation Stage, *Journal of Official Statistics*, **10**, 153–165.

Rao, J.N.K. (1999), Some Recent Advances in Model-based Small Area Estimation, *Survey Methodology*, **25**, 175–186.

Rao, J.N.K. (2001a), EB and EBLUP in Small Area Estimation, in S.E. Ahmed and N. Reid (Eds.), *Empirical Bayes and Likelihood Inference, Lecture Notes in Statistics*, Volume 148, New York: Springer-Verlag, pp. 33–43.

Rao, J.N.K. (2001b), Small Area Estimation with Applications to Agriculture, in *Proceedings of the Second Conference on Agricultural and Environmental Statistical Applications*, Rome, Italy: ISTAT.

Rao, J.N.K. (2003a), *Small Area Estimation*, Hoboken, NJ: John Wiley & Sons, Inc..

Rao, J.N.K. (2003b), Some New Developments in Small Area Estimation, *Journal of the Iranian Statistical Society*, **2**, 145–169.

Rao, J.N.K. (2005), Inferential Issues in Small Area Estimation: Some New Developments, *Statistics in Transition*, **7**, 523–526.

Rao, J.N.K. (2008), Some Methods for Small Area Estimation, *Rivista Internazionale di Scienze Sociali*, **4**, 387–406.

Rao, J.N.K. and Choudhry, G.H. (1995), Small Area Estimation: Overview and Empirical Study, in B.G. Cox, D.A. Binder, B.N. Chinnappa, A. Christianson, M.J. Colledge, and P.S. Kott (Eds.), *Business Survey Methods*, New York: John Wiley & Sons, Inc., pp. 527–542.

Rao, J.N.K. and Molina, I. (2015), Empirical Bayes and Hierarchical Bayes Estimation of Poverty Measures for Small Areas, in M. Pratesi (Ed.) *Analysis of Poverty Data by Small Area Methods*, Hoboken, NJ: John Wiley & Sons, Inc., in print.

Rao, J.N.K. and Scott, A.J. (1981), The Analysis of Categorical Data from Complex Sample Surveys: Chi-squared Tests for Goodness of Fit and Independence in Two-Way Tables, *Journal of the American Statistical Association*, **76**, 221–230.

Rao, J.N.K. and Singh, A.C. (2009), Range Restricted Weight Calibration for Survey Data Using Ridge Regression, *Pakistani Journal of Statistics*, **25**, 371–384.

Rao, J.N.K., Sinha, S.K., and Dumitrescu, L. (2014), Robust Small Area Estimation Under Semi-Parametric Mixed Models, *Canadian Journal of Statistics*, **42**, 126–141.

Rao, J.N.K., Wu, C.F.J., and Yue, K. (1992), Some Recent Work on Resampling Methods for Complex Surveys, *Survey Methodology*, **18**, 209–217.

Rao, J.N.K. and Yu, M. (1992), Small Area Estimation by Combining Time Series and Cross-Sectional Data, in *Proceedings of the Section on Survey Research Method*, American Statistical Association, pp. 1–9.

Rao, J.N.K. and Yu, M. (1994), Small Area Estimation by Combining Time Series and Cross-Sectional Data, *Canadian Journal of Statistics*, **22**, 511–528.

Rao, P.S.R.S. (1997), *Variance Component Estimation*, London: Chapman and Hall.

R Core Team (2013), *R: A Language and Environment for Statistical Computing*, Vienna, Austria: R Foundation for Statistical Computing, http://www.R-project.org/.

Richardson, A.M. and Welsh, A.H. (1995), Robust Estimation in the Mixed Linear Model. *Biometrics*, **51**, 1429–1439.

Ritter, C. and Tanner, T.A. (1992), The Gibbs Stopper and the Griddy-Gibbs Sampler, *Journal of the American Statistical Association*, **87**, 861–868.

Rivest, L.-P. (1995), A Composite Estimator for Provincial Undercoverage in the Canadian Census, in *Proceedings of the Survey Methods Section, Statistical Society of Canada*, pp. 33–38.

Rivest, L.-P. and Belmonte, E. (2000), A Conditional Mean Squared Error of Small Area Estimators, *Survey Methodology*, **26**, 67–78.

Rivest, L.-P. and Vandal, N. (2003), Mean Squared Error Estimation for Small Areas when the Small Area Variances are Estimated, in *Proceedings of the International Conference on Recent Advances in Survey Sampling*, Technical Report No. 386, Ottawa, Canada: Laboratory for Research in Statistics and Probability, Carleton University.

Roberts, G.O. and Rosenthal, J.S. (1998), Markov-Chain Monte Carlo: Some Practical Implications of Theoretical Results, *Canadian Journal of Statistics*, **26**, 5–31.

Robinson, G.K. (1991), That BLUP Is a Good Thing: The Estimation of Random Effects, *Statistical Science*, **6**, 15–31.

Robinson, J. (1987), Conditioning Ratio Estimates Under Simple Random Sampling, *Journal of the American Statistical Association*, **82**, 826–831.

Royall, R.M. (1970), On Finite Population Sampling Theory Under Certain Linear Regression, *Biometrika*, **57**, 377–387.

Royall, R.M. (1976), The Linear Least-Squares Prediction Approach to Two-stage Sampling, *Journal of the American Statistical Association*, **71**, 657–664.

Rubin-Bleuer, S. and You, Y. (2013), A Positive Variance Estimator for the Fay-Herriot Small Area Model, SRID2-12-OOIE, Statistics Canada.

Rueda, C., Menéndez, J.A., and Gómez, F. (2010), Small Area Estimators Based on Restricted Mixed Models, *Test*, **19**, 558–579.

Ruppert, D., Wand, M.P., and Carroll, R.J. (2003), *Semiparametric Regression*, New York: Cambridge University Press.

Rust, K.F. and Rao, J.N.K. (1996), Variance Estimation for Complex Surveys Using Replication Techniques, *Statistical Methods in Medical Research*, **5**, 283–310.

Saei, A. and Chambers, R. (2003), Small area estimation under linear and generalized linear mixed models with time and area effects, Working Paper M03/15. Southampton: Southampton Statistical Sciences Research Institute, University of Southampton.

Ehsanes Saleh, A.K.Md.E. (2006), *Theory of Preliminary Test and Stein-Type Estimation with Applications*, New York: John Wiley & Sons, Inc..

Salibian-Barrera, M. and Van Aelst, S. (2008), Robust Model Selection Using Fast and Robust Bootstrap, *Computational Statistics and Data Analysis*, **52**, 5121–5135.

Sampford, M.R. (1967), On Sampling Without Replacement with Unequal Probabilities of Selection, *Biometrika*, **54**, 499–513.

Särndal, C.E. and Hidiroglou, M.A. (1989), Small Domain Estimation: A Conditional Analysis, *Journal of the American Statistical Association*, **84**, 266–275.

Särndal, C.E., Swensson, B., and Wretman, J.H. (1989), The Weighted Regression Technique for Estimating the Variance of Generalized Regression Estimator, *Biometrika*, **76**, 527–537.

Särndal, C.E., Swensson, B., and Wretman, J.H. (1992), *Model Assisted Survey Sampling*, New York: Springer-Verlag.

SAS Institute Inc. (2013), *SAS/STAT 13.1 User's Guide*, Cary, NC: SAS Institute.

Schaible, W.L. (1978), Choosing Weights for Composite Estimators for Small Area Statistics, *Proceedings of the Section on Survey Research Methods*, Washington, DC: American Statistical Association, pp. 741–746.

Schaible, W.L. (Ed.) (1996), *Indirect Estimation in U.S. Federal Programs,* New York: Springer-Verlag.

Schall, R. (1991), Estimation in Generalized Linear Models with Random Effects, *Biometrika*, **78**, 719—727.

Schirm, A.L. and Zaslavsky, A.M. (1997), Reweighting Households to Develop Microsimulation Estimates for States, *Proceedings of the 1997 Section on Survey Research Methods*, American Statistical Association, pp. 306–311.

Schmid, T. and Münnich, R.T. (2014), Spatial Robust Small Area Estimation, *Statistical Papers*, **55**, 653–670.

Schoch, T. (2012), Robust Unit Level Small Area Estimation: A Fast Algorithm for Large Data Sets, *Austrian Journal of Statistics*, **41**, 243–265.

Searle, S.R., Casella, G., and McCulloch, C.E. (1992), *Variance Components*, New York: John Wiley & Sons. Inc..

Shao, J. and Tu, D. (1995), *The Jackknife and Bootstrap*, New York: Springer-Verlag.

Shapiro, S.S. and Wilk, M.B. (1965), An Analysis of Variance Test for Normality (Complete Samples), *Biometrika*, **52**, 591–611.

Shen, W. and Louis, T.A. (1998), Triple-Goal Estimates in Two-Stage Hierarchical Models, *Journal of the Royal Statistical Society, Series B*, **60**, 455–471.

Singh, A.C. and Folsom, R.E. (2001), Benchmarking of Small Area Estimators in a Bayesian Framework, Unpublished manuscript.

Singh, A.C., Mantel, H.J., and Thomas, B.W. (1994), Time Series EBLUPs for Small Areas Using Survey Data, *Survey Methodology*, **20**, 33–43.

Singh, A.C. and Mian, I.U.H. (1995), Generalized Sample Size Dependent Estimators for Small Areas, *Proceedings of the 1995 Annual Research Conference*, U.S. Bureau of the Census, Washington, DC, pp. 687–701.

Singh, A.C., Stukel, D.M., and Pfeffermann, D. (1998), Bayesian Versus Frequentist Measures of Error in Small Area Estimation, *Journal of the Royal Statistical Society, Series B*, **60**, 377–396.

Singh, B., Shukla, G., and Kundu, D. (2005), Spatio-Temporal Models in Small Area Estimation, *Survey Methodology*, **31**, 183–195.

Singh, M.P., Gambino, J., and Mantel, H.J. (1994), Issues and Strategies for Small Area Data, *Survey Methodology*, **20**, 3–22.

Singh, M.P. and Tessier, R. (1976), Some Estimators for Domain Totals, *Journal of the American Statistical Association*, **71**, 322–325.

Singh, R. and Goel, R.C. (2000), Use of Remote Sensing Satellite Data in Crop Surveys, Technical Report, Indian Agricultural Statistics Research Institute, New Delhi, India.

Singh, R., Semwal, D.P., Rai, A., and Chhikara, R.S. (2002), Small Area Estimation of Crop Yield Using Remote Sensing Satellite Data, *International Journal of Remote Sensing*, **23**, 49–56.

Singh, T., Wang, S., and Carroll, R.J. (2015a), Efficient Small Area Estimation When Covariates are Measured with Error Using Simulation Extrapolation, Unpublished Technical Report.

Singh, T., Wang, S., and Carroll, R.J. (2015b), Efficient Corrected Score Estimators for the Fay-Herriot Model with Measurement Error in Covariates, Unpublished Technical Report.

Sinha, S.K. and Rao, J.N.K. (2009), Robust Small Area Estimation, *Canadian Journal of Statistics*, **37**, 381–399.

Sinharay, S. and Stern, H. (2002), On the Sensitivity of Bayes Factors to the Prior Distributions, *American Statistician*, **56**, 196–201.

Skinner, C.J. (1994), Sampling Models and Weights, *Proceedings of the Section on Survey Research Methods*, Washington, DC: American Statistical Association, pp. 133–142.

Skinner, C.J. and Rao, J.N.K. (1996), Estimation in Dual Frame Surveys with Complex Designs, *Journal of the American Statistical Association*, **91**, 349–356.

Slud, E.V. and Maiti, T. (2006), Mean-squared Error Estimation in Transformed Fay-Herriot Models, *Journal of the Royal Statistical Society, Series B*, **68**, 239–257.

Smith, T.M.F. (1983), On the Validity of Inferences from Non-Random Samples, *Journal of the Royal Statistical Society, Series A*, **146**, 394–403.

Spiegelhalter, D.J., Best, N.G., Gilks, W.R., and Inskip, H (1996), Hepatitis B: A Case Study in MCMC Methods, in W.R. Gilks, S. Richardson, and D.J. Spiegelhalter (Eds.), *Markov Chain Monte Carlo in Practice*, London: Chapman and Hall, pp. 21–43.

Spiegelhalter, D.J., Thomas, A., Best, N.G., and Gilks, W.R. (1997), *BUGS: Bayesian Inference Using Gibbs Sampling*, Version 6.0, Cambridge: Medical Research Council Biostatistics Unit.

Spjøtvoll, E. and Thomsen, I. (1987), Application of Some Empirical Bayes Methods to Small Area Statistics, *Bulletin of the International Statistical Institute,* **2**, 435–449.

Srivastava, M.S. and Bilodeau, M. (1989), Stein Estimation Under Elliptical Distributions, *Journal of Multivariate Analysis*, **28**, 247–259.

Stasny, E., Goel, P.K., and Rumsey, D.J. (1991), County Estimates of Wheat Production, *Survey Methodology*, **17**, 211–225.

Stefan, M. (2005), *Contributions à l'Estimation pour Petits Domains*, Ph.D. Thesis, Université Libre de Bruxelles.

Stein, C. (1981), Estimation of the Mean of a Multivariate Normal Distribution, *Annals of Statistics*, **9**, 1135–1151.

Steorts, R. and Ghosh, M. (2013), On Estimation of Mean Squared Error of Benchmarked Empirical Bayes Estimators, *Statistica Sinica*, **23**, 749–767.

Stukel, D.M. (1991), *Small Area Estimation Under One and Two-fold Nested Error Regression Models*, Unpublished Ph.D. Thesis, Carleton University, Ottawa, Canada.

Stukel, D.M. and Rao, J.N.K. (1997), Estimation of Regression Models with Nested Error Structure and Unequal Error Variances Under Two and Three Stage Cluster Sampling, *Statistics & Probability Letters*, **35**, 401–407.

Stukel, D.M. and Rao, J.N.K. (1999), Small-Area Estimation Under Two-Fold Nested Errors Regression Models, *Journal of Statistical Planning and Inference*, **78**, 131–147.

Sutradhar, B.C. (2004), On Exact Quasi-Likelihood Inference in Generalized Linear Mixed Models, *Shankya*, **66**, 263–291.

Swamy, P.A.V.B. (1971), *Statistical Inference in Random Coefficient Regression Models*, Berlin: Springer-Verlag.

Thompson, M.E. (1997), *Theory of Sample Surveys*, London: Chapman and Hall.

Tillé, Y. (2006), *Sampling Algorithms, Springer Series in Statistics*, New York: Springer-Verlag.

Tiller, R.B. (1982), Time Series Modeling of Sample Survey Data from the U.S. Current Population Survey, *Journal of Official Statistics*, **8**, 149–166.

Torabi, M., Datta, G.S., and Rao, J.N.K. (2009), Empirical Bayes Estimation of Small Area Means Under a Nested Error Linear Regression Model with Measurement Errors in the Covariates, *Scandinavian Journal of Statistics*, **36**, 355–368.

Torabi, M. and Rao, J.N.K. (2008), Small Area Estimation under a Two-Level Model, *Survey Methodology*, **34**, 11–17.

Torabi, M. and Rao, J.N.K. (2010), Mean Squared Error Estimators of Small Area Means Using Survey Weights, *Canadian Journal of Statistics*, **38**, 598–608.

Torabi, M. and Rao, J.N.K. (2014), On Small Area Estimation under a Sub-Area Model, *Journal of Multivariate Analysis*, **127**, 36–55.

Torabi, M. and Shokoohi, F. (2015), Non-Parametric Generalized Linear Mixed Models in Small Area Estimation, *Canadian Journal of Statistics*, **43**, 82–96.

Tsutakawa, R.K., Shoop, G.L., and Marienfield, C.J. (1985), Empirical Bayes Estimation of Cancer Mortality Rates, *Statistics in Medicine*, **4**, 201–212.

Tzavidis, N., Marchetti, S., and Chambers, R. (2010), Robust Estimation of Small-Area Means and Quantiles, *Australian and New Zealand Journal of Statistics*, **52**, 167–186.

Ugarte, M.D., Goicoa, T., Militino, A.F., and Durban, M. (2009), Spline Smoothing in Small Area Trend Estimation and Forecasting, *Computational Statistics and Data Analysis*, **53**, 3616–3629.

Vaida, F. and Blanchard, S. (2005), Conditional Akaike Information for Mixed-Effect Models, *Biometrika*, **92**, 351–370.

Valliant, R., Dorfman, A.H., and Royall, R.M. (2001), *Finite Population Sampling and Inference: A Prediction Approach*, New York: John Wiley & Sons, Inc..

Verret, F, Rao, J.N.K., and Hidiroglou, M.A. (2014), Model-Based Small Area Estimation Under Informative Sampling, *Survey Methodology*, in print.

Wang, J. (2000), *Topics in Small Area Estimation with Applications to the National Resources Inventory*, Unpublished Ph.D. Thesis, Iowa State University, Ames.

Wang, J. and Fuller, W.A. (2003), The Mean Squared Error of Small Area Predictors Constructed with Estimated Area Variances, *Journal of the American Statistical Association*, **98**, 718–723.

Wang, J., Fuller, W.A., and Qu, Y. (2008), Small Area Estimation Under Restriction, *Survey Methodology*, **34**, 29–36.

Wolter, K.M. (2007), *Introduction to Variance Estimation* (2nd ed.), New York: Springer-Verlag.

Woodruff, R.S. (1966), Use of a Regression Technique to Produce Area Breakdowns of the Monthly National Estimates of Retail Trade, *Journal of the American Statistical Association*, **61**, 496–504.

Wu, C.F.J. (1986), Jackknife, Bootstrap and Other Resampling Methods in Regression Analysis (with discussion), *Annals of Statistics*, **14**, 1261–1350.

Ybarra, L.M.R. and Lohr, S.L. (2008), Small Area Estimation When Auxiliary Information is Measures with Error, *Biometrika*, **95**, 919–931.

Yoshimori, M. and Lahiri, P. (2014a), A New Adjusted Maximum Likelihood Method for the Fay-Herriot Small Area Model, *Journal of Multivariate Analysis*, **124**, 281–294.

Yoshimori, M. and Lahiri, P. (2014b), Supplementary material to Yoshimori, M. and Lahiri, P. (2014a). Unpublished note.

Yoshimori, M. and Lahiri, P. (2014c), A Second-Order efficient Empirical Bayes Confidence Interval, *Annals of Statistics*, **42**, 1233–1261.

You, Y. (1999), *Hierarchical Bayes and Related Methods for Model-Based Small Area Estimation*, Unpublished Ph.D. Thesis, Carleton University, Ottawa, Canada.

You, Y. (2008), An Integrated Modeling Approach to Unemployment Rate Estimation for Sub-Provincial Areas of Canada, *Survey Methodology*, **34**, 19–27.

You, Y. and Chapman, B. (2006), Small Area Estimation Using Area Level Models and Estimated Sampling Variances, *Survey Methodology*, **32**, 97–103.

You, Y. and Rao, J.N.K. (2000), Hierarchical Bayes Estimation of Small Area Means Using Multi-Level Models, *Survey Methodology*, **26**, 173–181.

You, Y. and Rao, J.N.K. (2002a), A Pseudo-Empirical Best Linear Unbiased Prediction Approach to Small Area Estimation Using Survey Weights, *Canadian Journal of Statistics*, **30**, 431–439.

You, Y. and Rao, J.N.K. (2002b), Small Area Estimation Using Unmatched Sampling and Linking Models, *Canadian Journal of Statistics*, **30**, 3–15.

You, Y. and Rao, J.N.K. (2003), Pseudo Hierarchical Bayes Small Area Estimation Combining Unit Level Models and Survey Weights, *Journal of Statistical Planning and Inference*, **111**, 197–208.

You, Y., Rao, J.N.K., and Hidiroglou, M.A. (2013), On the Performance of Self-Benchmarked Small Area Estimators Under the Fay-Herriot Area Level Model, *Survey Methodology*, **39**, 217–229.

You, Y., Rao, J.N.K., and Gambino, J. (2003), Model-Based Unemployment Rate Estimation for the Canadian Labour Force Survey: A Hierarchical Bayes Approach, *Survey Methodology*, **29**, 25–32.

You, Y. and Reiss, P. (1999), Hierarchical Bayes Estimation of Response Rates for an Expenditure Survey, *Proceedings of the Survey Methods Section*, Statistical Society of Canada, pp. 123–128.

You, Y. and Zhou, Q.M. (2011), Hierarchical Bayes Small Area Estimation Under a Spatial Model with Application to Health Survey Data, *Survey Methodology*, **37**, 25–37.

Zaslavsky, A.M. (1993), Combining Census, Dual-System, and Evaluation Study Data to Estimate Population Shares, *Journal of the American Statistical Association*, **88**, 1092–1105.

Zeger, S.L. and Karim, M.R. (1991), Generalized Linear Models with Random Effects: A Gibbs Sampling Approach, *Journal of the American Statistical Association*, **86**, 79–86.

Zewotir, T. and Galpin, J.S. (2007), A Unified Approach on Residuals, Leverages and Outliers in the Linear Mixed Model, *Test*, **16**, 58–75.

Zhang, D. and Davidian, M. (2001), Linear Mixed Models with Flexible Distributions of Random Effects for Longitudinal Data, *Biometrics*, **57**, 795–802.

Zhang, L.-C. (2007), Finite Population Small Area Estimation, *Journal of Official Statistics*, **23**, 223–237.

Zhang, L.-C. and Chambers, R.L. (2004), Small Area Estimates for Cross-Classifications, *Journal of the Royal Statistical Society, Series B*, **66**, 479–496.

AUTHOR INDEX

SUBJECT INDEX

Small Area Estimation, Second Edition. J. N. K. Rao and Isabel Molina.

WILEY SERIES IN SURVEY METHODOLOGY

Established in Part by WALTER A. SHEWHART AND SAMUEL S. WILKS

The ***Wiley Series in Survey Methodology*** covers topics of current research and practical interests in survey methodology and sampling. While the emphasis is on application, theoretical discussion is encouraged when it supports a broader understanding of the subject matter.

The authors are leading academics and researchers in survey methodology and sampling. The readership includes professionals in, and students of, the fields of applied statistics, biostatistics, public policy, and government and corporate enterprises.

ALWIN · Margins of Error: A Study of Reliability in Survey Measurement
BETHLEHEM · Applied Survey Methods: A Statistical Perspective
*BIEMER, GROVES, LYBERG, MATHIOWETZ, and SUDMAN · Measurement Errors in Surveys
BIEMER and LYBERG · Introduction to Survey Quality
BRADBURN, SUDMAN, and WANSINK ·Asking Questions: The Definitive Guide to Questionnaire Design—For Market Research, Political Polls, and Social Health Questionnaires, *Revised Edition*
BRAVERMAN and SLATER · Advances in Survey Research: New Directions for Evaluation, No. 70
CHAMBERS and SKINNER (editors · Analysis of Survey Data
COCHRAN · Sampling Techniques, *Third Edition*
CONRAD and SCHOBER · Envisioning the Survey Interview of the Future
COUPER, BAKER, BETHLEHEM, CLARK, MARTIN, NICHOLLS, and O'REILLY (editors) · Computer Assisted Survey Information Collection
COX, BINDER, CHINNAPPA, CHRISTIANSON, COLLEDGE, and KOTT (editors) · Business Survey Methods
*DEMING · Sample Design in Business Research
DILLMAN · Mail and Internet Surveys: The Tailored Design Method
FULLER · Sampling Statistics
GROVES and COUPER · Nonresponse in Household Interview Surveys
GROVES · Survey Errors and Survey Costs
GROVES, DILLMAN, ELTINGE, and LITTLE · Survey Nonresponse
GROVES, BIEMER, LYBERG, MASSEY, NICHOLLS, and WAKSBERG · Telephone Survey Methodology
GROVES, FOWLER, COUPER, LEPKOWSKI, SINGER, and TOURANGEAU · Survey Methodology, *Second Edition*
*HANSEN, HURWITZ, and MADOW · Sample Survey Methods and Theory, Volume 1: Methods and Applications
*HANSEN, HURWITZ, and MADOW · Sample Survey Methods and Theory, Volume II: Theory
HARKNESS, VAN DE VIJVER, and MOHLER · Cross-Cultural Survey Methods
KALTON and HEERINGA · Leslie Kish Selected Papers
KISH · Statistical Design for Research
*KISH · Survey Sampling
KORN and GRAUBARD · Analysis of Health Surveys
LEPKOWSKI, TUCKER, BRICK, DE LEEUW, JAPEC, LAVRAKAS, LINK, and SANGSTER (editors) · Advances in Telephone Survey Methodology

*Now available in a lower priced paperback edition in the Wiley Classics Library.

LESSLER and KALSBEEK · Nonsampling Error in Surveys

LEVY and LEMESHOW · Sampling of Populations: Methods and Applications, *Fourth Edition*

LYBERG, BIEMER, COLLINS, de LEEUW, DIPPO, SCHWARZ, TREWIN (editors) · Survey Measurement and Process Quality

MAYNARD, HOUTKOOP-STEENSTRA, SCHAEFFER, VAN DER ZOUWEN · Standardization and Tacit Knowledge: Interaction and Practice in the Survey Interview

PORTER (editor) · Overcoming Survey Research Problems: New Directions for Institutional Research, No. 121

PRESSER, ROTHGEB, COUPER, LESSLER, MARTIN, MARTIN, and SINGER (editors) · Methods for Testing and Evaluating Survey Questionnaires

RAO · Small Area Estimation

REA and PARKER · Designing and Conducting Survey Research: A Comprehensive Guide, *Third Edition*

SARIS and GALLHOFER · Design, Evaluation, and Analysis of Questionnaires for Survey Research

SÄRNDAL and LUNDSTRÖM · Estimation in Surveys with Nonresponse

SCHWARZ and SUDMAN (editors) · Answering Questions: Methodology for Determining Cognitive and Communicative Processes in Survey Research

SIRKEN, HERRMANN, SCHECHTER, SCHWARZ, TANUR, and TOURANGEAU (editors) · Cognition and Survey Research

SUDMAN, BRADBURN, and SCHWARZ · Thinking about Answers: The Application of Cognitive Processes to Survey Methodology

UMBACH (editor) · Survey Research Emerging Issues: New Directions for Institutional Research No. 127

VALLIANT, DORFMAN, and ROYALL · Finite Population Sampling and Inference: A Prediction Approach

Printed and bound by CPI Group (UK) Ltd, Croydon, CR0 4YY
16/04/2025
14658520-0004